Classification Methods for Remotely Sensed Data

The third edition of the bestselling *Classification Methods for Remotely Sensed Data* covers current state-of-the-art machine learning algorithms and developments in the analysis of remotely sensed data. This book is thoroughly updated to meet the needs of readers today and provides six new chapters on deep learning, feature extraction and selection, multisource image fusion, hyperparameter optimization, accuracy assessment with model explainability, and object-based image analysis, which is relatively a new paradigm in image processing and classification. It presents new AI-based analysis tools and metrics together with ongoing debates on accuracy assessment strategies and XAI methods.

New in this edition:

- Provides comprehensive background on the theory of deep learning and its application to remote sensing data.
- Includes a chapter on hyperparameter optimization techniques to guarantee the highest performance in classification applications.
- Outlines the latest strategies and accuracy measures in accuracy assessment and summarizes accuracy metrics and assessment strategies.
- Discusses the methods used for explaining inherent structures and weighing the features of ML and AI algorithms that are critical for explaining the robustness of the models.

This book is intended for industry professionals, researchers, academics, and graduate students who want a thorough and up-to-date guide to the many and varied techniques of image classification applied in the fields of geography, geospatial and earth sciences, electronic and computer science, environmental engineering, etc.

Classification Methods for Remotely Sensed Data

Third Edition

Taskin Kavzoglu, Brandt Tso, and Paul M. Mather

CRC Press
Taylor & Francis Group
Boca Raton London New York

CRC Press is an imprint of the
Taylor & Francis Group, an **Informa** business

Designed cover image: © iStockPhoto, bluebay2014

Third edition published 2025
by CRC Press
2385 NW Executive Center Drive, Suite 320, Boca Raton FL 33431

and by CRC Press
4 Park Square, Milton Park, Abingdon, Oxon, OX14 4RN

CRC Press is an imprint of Taylor & Francis Group, LLC

© 2025 Taskin Kavzoglu, Brandt Tso, and Paul M. Mather

First edition published by CEC Press 2001
Second edition published by CRC Press 2009

Library of Congress Cataloging-in-Publication Data
Names: Kavzoglu, Taskin, author. | Tso, Brandt, author. | Mather, Paul M., author.
Title: Classification methods for remotely sensed data / Taskin Kavzoglu,
Brandt Tso, and Paul M. Mather.
Description: Third edition. | Boca Raton, FL : CRC Press, 2025. |
Revised edition of: Classification methods for remotely sensed data /
Brandt Tso, Paul M. Mather. 2nd ed. 2009. | Includes bibliographical references and index.
Identifiers: LCCN 2024008647 (print) | LCCN 2024008648 (ebook) |
ISBN 9781032573939 (hardback) | ISBN 9781032573953 (paperback) |
ISBN 9781003439172 (ebook)
Subjects: LCSH: Remote sensing. | Pattern recognition systems.
Classification: LCC G70.4 .T784 2025 (print) | LCC G70.4 (ebook) |
DDC 621.36/78–dc23/eng/20240514
LC record available at https://lccn.loc.gov/2024008647
LC ebook record available at https://lccn.loc.gov/2024008648

ISBN: 9781032573939 (hbk)
ISBN: 9781032573953 (pbk)
ISBN: 9781003439172 (ebk)

DOI: 10.1201/9781003439172

Typeset in Times
by codeMantra

*This book is dedicated to the memory of Prof. Paul Mather OBE (1944–2021),
who truly was a pioneer in remote sensing. He has put great effort into publishing
the previous versions of this book. I feel honored to have worked with him and
happy to take the responsibility of preparing the third edition of the book.*

*The book is also dedicated to the loving memory of my parents, especially
my mother, who passed away while I was writing it. Throughout my
career, she stood as my strongest supporter, was aware of the fact
that this book was being written, encouraged me to write it, and
would be the proudest mother on Earth to see it in print today.*

Taskin Kavzoglu

Contents

Preface to the Third Edition

The first edition of this book was published in 2001, and almost a decade has passed since the publication of the second edition. Considering its high reputation in the remote sensing field, the third edition of *Classification Methods for Remotely Sensed Data* provides up-to-date information on a wide range of methods covering the topics ranging from data preparation to accuracy assessment. Within these years, we have witnessed rapid progress and the emergence of new practices in the field. Scientific methodologies have undergone a noticeable transformation, shifting from computational science to a more data-driven paradigm. The transition has led to a substantial transformation in the analysis of remote sensing images, incorporating advanced machine learning techniques and new learning paradigms. Replacing conventional methods, machine learning methods have dominated the field with their superior performance in diverse remote sensing tasks. Consequently, some methodologies, such as modeling context and texture using fuzzy set theory, have lost their importance and relevance in image classification.

During this period, shallow neural networks have evolved into deep learning models, support vector machines (SVMs) into more robust variants, and simple decision trees into ensemble methods. To cover all these advancements, some existing chapters have been removed, and others updated. For the inclusion of state-of-the-art technologies and developments, six new chapters were added. Two critical chapters on feature engineering are provided for dimensionality reduction and multisource image fusion. Additionally, four new chapters cover deep learning, object-based image analysis, hyperparameter optimization, and accuracy assessment with explainable artificial intelligence (XAI). As a result, it is aimed to comprehensively cover all recent concepts of remote sensing image analysis.

With all these comprehensive updates and the addition of new chapters, the current version of this book features up-to-date and compelling content, suitable for students, research scientists, and professionals in the remote sensing industry who have a basic understanding of remote sensing principles, image processing techniques, and the applications of remotely sensed data. For readers seeking comprehensive guidance, this book offers recommendations for best practices. It further presents a sample accuracy assessment exercise, featuring a variety of measures and local/global XAI techniques in a case study. This book outlines methods and approaches that have universal relevance, reaching beyond engineering to encompass a spectrum of scientific fields. The new edition includes color figures to facilitate a clearer understanding of the content. It encompasses a comprehensive historical timeline of developments and offers complete reference lists for students and researchers to delve into their respective projects.

Chapter 1 presents an updated edition, incorporating recent technological advancements in remote sensing technologies within the optical and microwave regions of the electromagnetic spectrum. This chapter provides foundational knowledge on the principles governing atmospheric interactions with electromagnetic energy as well as the acquisition and preprocessing of remotely sensed data.

Chapter 2 introduces the principles of pattern recognition, offering a taxonomical overview of various techniques and the fundamental principles behind traditional unsupervised and supervised methods. It highlights the significance of fuzzy classification, presenting methods such as fuzzy C-means and fuzzy maximum likelihood classifiers. Additionally, this chapter delves into topics like spectral unmixing and ensemble classifiers. The discussion extends to emphasize the importance of incorporating texture, content, and ancillary data for image classification.

Chapter 3 delves into a crucial aspect of feature engineering, focusing on dimensionality reduction for high-dimensional remote sensing data through feature extraction and feature selection techniques. This chapter presents the theories behind principal components analysis, minimum/maximum autocorrelation factors (MAF), maximum noise fraction (MNF) transformation, independent component analysis, and projection pursuit as part of feature extraction. In addition, it explains various feature selection techniques, covering filter-based methods, wrappers, and embedded methods. This chapter further explores the application of greedy search algorithms, simulated annealing, and separability indices.

Chapter 4 addresses an important research topic in the remote sensing field – multisource image fusion. In addition to the mathematical theories of image fusion methods, assessment of fused image quality, dealing with source reliability, and performance evaluation of fusion methods are presented. The detailed discussion also covers the classification of multisource data using the stacked-vector method and extends into Bayesian classification theory.

Chapter 5 introduces revised material on support vector machines (SVMs). Following the exposition of mathematical theories underpinning the technique, recently developed variations such as relevance vector machines, twin SVMs, and deep SVMs are discussed, taking into account contemporary literature and advancements in the analysis of remotely sensed imagery.

Chapter 6 includes updated content on decision trees that have evolved into robust ensemble derivatives. After introducing the fundamental theory behind decision trees in the initial sections, detailed theories of advanced ensemble methods, including canonical correlation forest, extreme gradient boosting, light gradient boosting machines, and gradient boosting machines are presented. In the conclusion section of this chapter, a current literature review on the popularity of ensemble methods based on the Web of Science database is also provided.

Chapter 7 introduces the groundbreaking paradigm in remote sensing – deep learning, which forms the core of this book. Considering the complexity of the underlying mathematical theory behind deep learning models, fundamental theories and information are initially provided. Subsequently, popular neural network architectures, including convolutional neural networks, recurrent neural networks, vision transformers, and generative adversarial networks, are presented along with their theories and specific applications. Moreover, new learning paradigms (transfer learning, semi-supervised learning, reinforcement learning, active learning, and multitask learning) that have evolved in the last two decades for processing high-dimensional and limited data are also described to inform readers about the latest developments. Finally, popular applications of deep learning in the remote sensing field (semantic segmentation, object detection, scene classification, and change detection) are discussed with reference to their recent applications.

Chapter 8 describes a new and important paradigm – object-based image analysis (OBIA). The demand for high-level and accurate information extraction has shifted the focus of research toward this new paradigm, enhancing the foundation for image analysis and introducing a vital connection between remote sensing and GIS. This chapter presents the fundamental theory of the OBIA approach with special emphasis on segmentation methods (clustering-based, thresholding-based, edge-based, region-based, and hybrid methods), which are discussed considering the current taxonomy. Furthermore, the metrics and algorithms used to assess segmentation quality are reviewed.

Chapter 9 outlines the concept and practice of hyperparameter optimization (HPO), a crucial aspect in achieving optimal performance from machine learning algorithms. The effectiveness of a machine learning method, whether it is at the forefront or exhibits weaker performance, is closely linked to the discovery of the optimal hyperparameter configuration. This chapter lays out the fundamental principles of major HPO approaches and explores pivotal hyperparameters inherent in common machine learning models that require tuning. A comprehensive examination then delves into a diverse array of state-of-the-art optimization approaches specifically designed to address HPO challenges.

Chapter 10 discusses the fundamentals and critical issues pertaining to the accuracy assessment and explainability of machine learning methods, often characterized as "black-box". This has perennially been a major concern for researchers in the remote sensing field. This chapter comprehensively presents both traditional and state-of-the-art methods, accompanied by a case study serving as an illustrative example for the readers of this book. This chapter makes a significant contribution by offering guidelines on best practices for accuracy assessment.

Taskin Kavzoglu
Kocaeli, 2024

Preface to the Second Edition

The first edition of this book was written between 1998 and 2000 and was published in 2001. In the 10 years that have elapsed since the start of this project, developments in image classification technology have been considerable. To keep this book relevant and up-to-date, two new chapters have been added, and the existing chapters have been updated. The new chapters cover the topics of support vector machines (SVMs) and decision trees, both of which are now the subject of research articles in the major journals. SVMs represent a recent development in the computational aspects of image classification, and one of the problems they face is in the allocation of data to one of several (rather than one of two) classes. A number of approaches to this problem are presented in Section 4.4, but further experimentation is required. The number of potential approaches to the use of decision trees is considerable, and these are covered in Chapter 6. Developments such as boosting and random forest generation are described in Section 6.8. Lopping branches that do not contribute to the effectiveness of the decision tree is an important aspect of the design of the tree, which is covered in Section 6.7.

Acknowledgment is due to our institutions, the Management College, National Defense University, and to The University of Nottingham, for providing support while this book was being revised. Brandt Tso also recognizes the benefits of his year as a postdoctoral fellow in the Remote Sensing Laboratory, Naval Postgraduate School, Monterey, California.

Preface to the First Edition

We classify objects in order to make sense of our environment, by reducing a multiplicity of phenomena to a relatively small number of general classes. On a country walk, for example, you might point to cows, trees, tractors, or swans. What you are actually doing is identifying an observed object and allocating it to a preexisting class or giving it a name. Before setting out on the walk, you knew that swans existed, and you could specify their characteristics. When you saw a large white bird, possibly swimming in a canal or river, with an orange and black beak, you compared those characteristics to those of a swan and thus identified the bird, giving it the name or label of *swan*. We must be careful, therefore, to distinguish between the definition of the classes to which objects may belong and the identification or labeling of individual objects and to avoid confusion between the two meanings of the word *classification* – the definition of categories of objects and the assignment or allocation of individual objects to these classes.

The example of the swan can also help to define other concepts. First, you must already have a model (or idealized representation) of the key features of a swan before you can recognize one. You learned, presumably in your childhood, the names of categories and subcategories of animals, plants, and other objects. Now you use that knowledge to identify and name the things you see and hear. In the literature of classification, this approach is termed *supervised learning*, meaning that you have divided the phenomena of interest into a number of *a priori* groups. You have observed a number of examples from each group and have characterized them in terms of a number of discriminating features. The sample set is called *training data*, and this approach is known as *supervised classification*. In fact, it is supervised identification because it is assumed that the classification (the definition of the groups and their characteristics) has been defined before any previously unknown objects were identified.

An alternative approach, known as *unsupervised classification* or *clustering*, is also widely used. In this approach, it is assumed that (1) you have little knowledge of the characteristics of the data set, (2) you wish to determine whether any natural groupings exist in those data, and, if so, (3) whether they can be identified in terms of phenomena of interest. In a sense, this procedure is akin to exploring the data (and visualization methods can help considerably in the process), whereas the supervised approach is inductive.

This book is about pattern recognition for remotely sensed data. We prefer the term *pattern recognition* to *classification* because the latter term can be misleading, as noted above. However, the two terms are used in this book, partly for reasons of tradition. A *pattern* is a set of measurements made on an object. It can be described as a mathematical vector of measurements. For example, a person's height and weight can be represented by the vector [192, 50] (in cm and kg, respectively). If a supervised approach is used, then the pattern is compared in some way to members of the sets of patterns that define the categories of interest, and the given pattern is assigned to one of these categories (one of which may be "unknown"). This approach can be described as inductive. Alternatively, a clustering strategy may be used that is based on the similarity between patterns, in order to determine whether any distinct groups of patterns exist in the data.

In Earth observation by remote sensing, the objects to be labeled are normally the individual pixels forming a multispectral or hyperspectral image. Each pixel is represented by a pattern vector consisting of a set of measurements, one per image band plus, possibly, other measurements such as texture. If each spectral band is represented by one axis of a multidimensional space (the *feature space*), then the pixel can be represented as a point in that space. For simplicity, let the number of features be two, and let the x-axis represent the first feature and the y-axis represent the second. A pixel with a feature vector of [1, 5] can therefore be shown on a graph as a point with Cartesian coordinates [1, 5]. Now imagine that all the pixels in the two-band image have been plotted on the graph, and that they fall into clearly defined groups. We can separate these groups by lines or curves. These lines or curves are called *decision boundaries*, for they show the positions of the boundaries of individual categories. If a point lies on one side of the boundary, it is given a label such as "A", whereas if it lies at the other side of the boundary, it is given the label "B". In higher-dimensional problems, the lines and curves become *hyperplanes* and *hypersurfaces*. So the labeling problem can be thought of as one that involves the positioning of hyperplanes or hyper-surfaces, representing decision boundaries, in a multidimensional feature space. The algorithm that determines the position of the pixel with respect to the decision boundaries, and thus allocates a specific label to that pixel, is called a *decision rule*. The word *classifier* is widely used as a synonym for the term *decision rule*.

The use of pattern recognition methods in remote sensing has a long history. Air photo inter-preters were perhaps the first to use intuitive methods to determine the information contained in reconnaissance photographs, and these methods continue to be of great importance, particularly in the gathering of military intelligence, as the human eye–brain combination can make decisions and judgments on complex problems in milliseconds, using experience as a guide. Automatic methods are more suitable for the routine processing of images that show predictable patterns. Such methods have been developed and applied in a number of disciplines, ranging from speech and handwriting recognition, industrial process management, and medical diagnosis, as well as in the collection of military intelligence. The main distinguishing characteristic of Earth observation data is its vol-ume. Hence, methods that can be applied in other applications may not be suited to the analysis of remotely sensed data because of the computational requirements. A further point to note is that there is often a discrepancy between the dimensionality of remotely sensed data sets and the volume of training data that is available. Where training data are sparse relative to the dimensionality of the data, it becomes difficult to estimate the characteristics of each training class, and so errors may become significant. This phenomenon of increasing error with increasing data dimensionality is sometimes known as the *Hughes effect*.

Advances in technology have led to rapid developments in methods of pattern recognition, lead-ing to the formulation of new and more sophisticated decision rules. Some of those new methods have been introduced into the field of remote sensing and have shown encouraging results. A further feature of remote sensing applications in recent years is the use of combinations of data derived from different sensors or from different time periods, plus terrain and other data extracted from geographical information system (GIS) databases. In addition, the spectral information contained in remotely sensed images is often augmented by derived measures such as values of texture and context. In dealing with multidata sources, a significant problem is the considerable increase in the computational cost. Other problems, such as data scale and data reliability, must also be considered. There is an increasing interest in seeking methods for efficiently manipulating multisource data in order to increase classification accuracy. It should always be remembered, though, that sophisticated algorithms cannot compensate for lack of training data or an inadequate definition of the problem (in terms of the number and nature of the classes to be recognized relative to the scale of the study).

Texture is the tonal variation within an area. A simple example that illustrates the concept is the pattern on a carpet. If we treat each pattern as a whole, then the carpet can easily be described. If the carpet is seen as a set of small rectangular units, then the problem of describing its properties is more difficult. In some cases, texture information seems to be more effective than tonal information to describe the objects, and one can develop *texture features* corresponding to different kinds of patterns to improve the performance of a classifier.

Contextual information describes how the object of interest may be affected by its neighbors. For instance, English words starting with the letter "q" are more likely to be immediately followed by the letter "u" than "z" or "c". In the case of classification of an agricultural area, a pixel labeled as "carrot" is more likely to be surrounded by pixels of the same class rather than by other classes such as "water" or "wheat". The ability to model such contextual behavior may reduce confusion in the classification process.

The decision to write this book was triggered by our experience in attempting to use new methods of describing and labeling pixels in a remotely sensed image. While a number of valuable but more general textbooks are available for undergraduate use, we know of no coherent source of advanced information and guidance in the area of pattern recognition for research scientists and postgraduate research students in remote sensing, together with students taking advanced remote sensing courses. We hope that this book will contribute to the increased understanding and adoption of recently developed techniques of pattern recognition, and that it will provide readers with a link between the remote sensing literature and that of statistics, artificial intelligence, and computing. We do suggest, however, that attention be paid to experimental design and definition of the problem, for an advanced pattern recognition procedure is no substitute for thinking about the problem and defining an appropriate set of features.

Chapter 1 introduces the basic concepts of remote sensing in the optical and microwave regions of the electromagnetic spectrum. This chapter is intended to introduce the field of remote sensing to readers with little or no background in this area, and it can be omitted by readers with adequate background knowledge of remote sensing.

Chapter 2 introduces the principles of pattern recognition. Traditional decision rules, including the supervised minimal distance classifier, Gaussian maximum likelihood, and unsupervised clustering techniques are described, together with other methods such as fuzzy-based procedures and decision trees. This chapter also contains brief accounts of dimension reduction methods, including orthogonal transforms, the assessment of classification accuracy, and the principles underlying the choice of training data.

Chapter 3 describes widely used neural network models and architectures including the multilayer perceptron (also called the feedforward neural network), Kohonen's self-organized feature map, counterpropagation, the Hopfield network, and networks based upon adaptive resonance theory (ART).

Chapter 4 deals with pattern recognition techniques based on fuzzy systems. The main topics of this chapter are the construction of fuzzy rules, fuzzy mapping functions, and the corresponding decision processes.

Chapter 5 presents a survey of methods of quantifying image texture, including fractal- and multifractal-based theory, the multiplicative autoregressive random field model, the gray-level co-occurrence matrix, and frequency domain filtering.

Chapter 6 addresses the theory and the application of Markov random fields. The main application of Markov random fields is to model contextual relationships. Other related topics, including function formulation, image restoration, robust estimation in the presence of noise (outliers), and the derivation of Markov-based texture measures, are also presented.

Chapter 7 provides several approaches for dealing with multisource data. The methods described include the extension of Bayesian classification theory, evidential reasoning, and Markov random fields.

No one is more aware of a book's deficiencies and inadequacies than its authors. Even Socrates, after a lifetime of learning, is reported to have been impressed by the extent of his own ignorance. Had more time and space been available, this book would have contained a longer account of the use of wavelets in texture analysis, and of the applications of decision tree classifiers. Publishers, who live in the real world, impose constraints of time and space, while authors naturally attempt to rewrite their manuscripts every month in order to include the latest developments. We hope that we have reached a happy compromise that should satisfy most readers. We have included references to further work in order to guide the more advanced reader toward the relevant literature.

Most of the research underlying the ideas presented in this book was carried out, while Dr. Brandt Tso was a postgraduate student, and later a postdoctoral fellow, in the School of Geography, The University of Nottingham, under the supervision of Professor Paul M. Mather. The second author provided encouragement, support, contributions to the first three chapters, and numerous rewrites of the draft. We realize that any book written by human authors is necessarily flawed, and we accept responsibility for any errors that may be contained in these pages.

The School of Geography, The University of Nottingham, provided computing facilities as well as a stimulating and encouraging environment for research. The second author is grateful to his many postgraduate research students from different parts of the world who have, over the past decade or so, educated and trained him in many areas of remote sensing. In particular, he would like to thank Valdir Veronese, Taskin Kavzoglu, Carlos Vieira, and Mahesh Pal, who have carried out research projects in areas relevant to the subject matter of this book. The contribution of others, while not directly related to the topic of image classification, has helped by broadening the intellectual debate within my research group, as well as helping in many other ways. We also thank Dr. M. Koch of Boston University for help and guidance with the Red Sea Hills data set, which is used in a number of examples in this book. The help, good humor, and patience of Tony Moore of Taylor & Francis are greatly appreciated. Finally, both authors recognize the contributions of their families and dedicate this book to them.

Brandt Tso
Taipei, 2000

Paul M. Mather
Nottingham, 2000

Acknowledgments

I owe much recognition to people who deserve my heartfelt appreciation for their support during the course of writing this book. These individuals are postgraduate students under my supervision, Alihan Teke, Hasan Tonbul, Elif Ozlem Yilmaz, and Esra Yildirim. This book cannot be concluded and completed without their efforts and contributions. Also, I would like to deeply thank my colleagues Dr. Ismail Colkesen, Dr. Oguz Gungor, Dr. Mahesh Pal, Dr. Umut G. Sefercik, and Dr. Elif Sertel for their valuable and constructive comments on the drafts of the chapters.

I appreciate the continuous collaboration and editorial support provided by Irma Britton and Chelsea Reeves and production personnel at CRC Press throughout this project.

I want to express my special appreciation to my wife, Şebnem, and my sons for standing by me with constant support and encouragement throughout a challenging year where my busy schedule left me with limited time.

Finally, the authorship of this book would not have been possible without the invaluable guidance and knowledge bestowed upon me by Prof. Paul Mather, to whom I owe a great debt. He was not just an outstanding scientist but also an excellent mentor to his students. His significant contribution to my scientific journey is undeniable, and this book stands as a tribute to my respect for him and my commitment to preserving his legacy.

Taskin Kavzoglu
Gebze Technical University
February 2024

Authors

Professor Taskin Kavzoglu is a senior researcher in remote sensing with more than 25 years of research experience in Earth observation and remote sensing. He received his B.Sc. in Geodesy and Photogrammetry from Karadeniz Technical University (Turkiye), M.Sc. in Geographical Information Systems, and Ph.D. in Remote Sensing from Nottingham University (UK) under the supervision of Prof. Paul Mather. The best Ph.D. thesis award was given to his thesis titled "An Investigation of the Design and Use of Feed-forward Artificial Neural Networks in the Classification of Remotely Sensed Images" by the Remote Sensing & Photogrammetry Society (RSPSoc) in the UK. Currently, he is a full-time professor at Gebze Technical University (Turkiye). He is a member of the Turkish Academy of Sciences since 2015. He has published numerous papers aligned with his research interests, including Earth observation, remote sensing of natural resources, deep learning, machine learning, analysis of the environmental effects of climate change (e.g., forest fires and marine mucilage), change detection, mapping of forested lands, object-based image analysis, and landslide susceptibility mapping. He has conducted scientific projects on urban change detection, monitoring forest species using remote sensing technologies and establishment of smart urban information systems.

Dr. Brandt Tso has served as a scientific officer in the Taiwan military service, studying modeling and pattern recognition techniques since 1988. Dr. Tso completed his Master's degree in Information Science at the Management College, National Defense University, Taiwan. Between 1994 and 1998, Dr. Tso was awarded a scholarship by the Taiwan government to pursue a Ph.D. in the School of Geography, The University of Nottingham, UK, under the supervision of Professor Paul M. Mather, concentrating on the field of remotely sensed data classification. In 2003, Dr. Tso was an invited postdoctoral fellow in the Remote Sensing Laboratory, Physics Department, Naval Postgraduate School, Monterey, California, USA, to study more complex remotely sensed data classification skills. Currently, Dr. Tso is an associate professor in the Information Science department, Management College, National Defense University, Taiwan. His main areas of research interests include remotely sensed data recognition, real-scene image retrieval, and machine learning algorithms. He has published numerous research papers relating to these fields.

Professor Paul M. Mather graduated with a degree in Geography from the University of Cambridge in 1966. He then moved to The University of Nottingham to conduct research for his Ph.D. in geomorphology, which was awarded in 1969. As a lecturer, senior lecturer, and full professor (1988), his attention was focused on the use of multivariate analysis in physical geography, a subject on which he published a detailed and well-received monograph in 1976. By the 1980s, his interest in multivariate analysis had branched out to include remote sensing. In 1987, the first edition of his book Computer Processing of Remotely Sensed Data was published. It is now in its third edition (2004) with a fourth edition in preparation. He has always been fascinated by the applications of computers in physical geography and the environmental sciences, and is a proficient Fortran programmer. He retired in 2006 and was made Emeritus Professor. He received the Back Award from the Royal Geographical Society for his work in remote sensing in 1992, and in 2002 was awarded the Order of the British Empire (OBE) by Her Majesty Queen Elizabeth II for services to remote sensing. He has lectured in a number of countries around the world.

1 Fundamentals of Remote Sensing

Remote sensing is a scientific and technological discipline that involves obtaining information about the physical and chemical attributes of an object or event by measuring its reflected and emitted radiation at a distance, typically from aircraft or satellites. Remote sensing offers cost-effective, rapid, and repetitive spatial and temporal data coverage through imagery, providing a synoptic view that enables continuous monitoring of the Earth's surface. Remotely sensed image data are widely used in a wide range of applications at local (e.g., land cover and land use mapping, disaster monitoring, urban planning, and precision farming) to global (e.g., monitoring the effects of global warming on land, oceans, and glaciers) scales. Over the past few decades, significant advancements have taken place in the field of remote sensing, involving both the technological revolution in sensor capabilities and the evolution of methods and approaches utilized for analyzing the data generated by these sensors. At present, we have the capability to obtain optical satellite images with a spatial resolution of 30 cm, daily image acquisition by several systems, coverage images at and with the recent launch of the HotSat-1 thermal sensor, we will soon have access to thermal imagery with a resolution of 3.5 m, clearly demonstrating the rapid advancements in remote sensing sensor technology. Increased spatial, spectral, and radiometric resolutions have expanded the horizons of research possibilities, but they have concurrently brought about the challenge of managing large volumes of data (big data analytics). In parallel to these developments, scientific literature has witnessed the emergence of state-of-the-art methodologies that incorporate diverse machine learning and deep learning algorithms. In particular, deep learning has been a driving force for the advancements in the processing of remotely sensed data. This book provides a complete reference and insight into the theories and applications of these methodologies.

The pixel values in an image, once quantized, can be transformed into physical radiance values and correlated with a specific characteristic of the sensed surface. An example of this approach is the calibration of thermal infrared imagery to produce maps of temperature fields, such as the sea surface temperature. Thus, thematic information is required for other applications. A thematic map shows the spatial variation of a particular element, such as land surface elevation, soil type, geology, or vegetation. The term pattern recognition is used to describe the procedures involved in relating vectors of measurements that are spatially referenced to individual pixel locations to the types or categories into which the phenomenon of interest is subdivided. If, for example, the phenomenon of interest is agricultural crops, then the categories are the individual crop types. Each crop type is represented in the thematic image by a numerical label. Vectors of measurements that are spatially referenced to individual pixel locations include image pixel values plus derived values such as texture, coherence, or context, as well as other geographical data (i.e., ancillary data) that can be related to the pixel location, such as terrain elevation and slope, geology, and soil type.

Digital thematic maps derived from remotely sensed imagery can be represented in two ways, using either the raster or the vector models. The vector model uses the classical cartographic representation of map objects in terms of points, lines, and areas. Using this model, a continuously varying spatial attribute such as terrain elevation is represented by contour lines and spot heights, while an attribute such as soil type or underlying geology is represented in terms of boundary lines enclosing areas that are homogeneous with respect to the property of interest and at the chosen scale of observation. The raster model represents spatial attributes in terms of their values over a contiguous set of small individual areas (and usually square). Thus, variations in land surface elevation over a region of interest are represented by numerical values stored in a rectangular grid

DOI: 10.1201/9781003439172-1

or raster, each element of which is the average elevation of the ground area represented by that element or cell of the grid. This representation of terrain height variation is known as a digital elevation model (DEM). Similarly, variations in geology in a study region are stored in raster format as a set of labels. Each cell in the data array is given a numeric label that is linked to a description. For example, pixels with Label 1 may be described as Carboniferous Limestone, while pixels given the Label 2 might be described as Silurian Grit. In this book, we are concerned with spatial data (specifically, remotely sensed images) that are represented in terms of rectangular grids or rasters.

In the examples given in the preceding paragraphs, each cell of the raster stores either a physical value (such as elevation in meters above a given datum) or a label (such as 1 or 2, indicating rock types such as granite and basalt). In practice, rasters (and particular raster data sets containing images for display) are generally stored in the form of integer rather than floating-point numbers in order to conserve storage space. Thus, a DEM may be represented in terms of an array of integers in the range 0–255 (8-bit), with each integer value representing a range of land surface elevations. The idea is similar to the use of a key in a printed map in which elevation is generally shown in shades of green and brown, with the map key showing the relationship between these hues and specific elevation ranges. In the same way, the values 0–255 contained in the DEM are connected to a range of real elevation values by the use of a table rather than a key. The table, known as a lookup table (LUT), allows the user to determine the actual range of elevations denoted by a particular label, such as 215. In some applications, the physical elevation value is required (for instance, if slope angle is to be calculated). Other questions can be answered by using the counts or labels directly. If label 18 is used to represent land with a surface elevation between 200 and 205 m, then the locations of such areas can be achieved by searching the raster for all values of 18 rather than converting the raster labels back to physical values and searching for cells holding numbers in the range of 200–205.

Remotely sensed images are stored and processed in raster form. Each element of the raster is known as a pixel, and the value contained in any pixel location is simply a quantized count or a label rather than a physical value. For some applications, such as pattern recognition, these counts can be used directly as we are interested in interpixel similarities and differences. In other applications, such as sea surface temperature determination, the pixel counts must be converted to physical values of radiance or reflectance, with corrections applied for factors such as sensor calibration changes and atmospheric influences. The range of quantized counts used in a raster representation of an image ranges from 0 to 255 (8-bit representation, for images derived from sensors such as Landsat Enhanced Thematic Mapper+(ETM+ and SPOT High Resolution Visible [HRV])), to 0–4,095 (12-bit representation, as used for New AstroSat Optical Modular Instruments [NAOMI] and Pléiades Neo). For example, raw synthetic aperture radar data are commonly represented in terms of two 16-bit integers per pixel, with the first integer representing the real part and the second integer representing the imaginary part of a complex number. Whatever the precision (8-, 10-, or 16-bit), these pixel values are stored as rectangular rasters and can be held in a computer in the form of a two-dimensional array. The data set used in pattern recognition consists of a number of co-registered raster images representing, for example, the measurements in the individual bands of a multispectral or hyperspectral image, the ground elevation (DEM), or some other spatial property of interest. The number of features used to represent terrain conditions or characteristics is known as the dimensionality of the data. Multispectral and radar data have low dimensionality; for instance, SPOT NAOMI produces four bands of data, and Landsat 8–9 OLI generates eight bands in wavelengths ranging from optical to thermal infrared, plus a panchromatic band, while the radar satellite TanDEM-X operates in a single band with two polarization modes. Hyperspectral sensors such as PRISMA and CHRIS Proba have the ability to collect data in tens or hundreds of narrow spectral bands. In the classification of high-dimensional remotely sensed data, a notable issue is the scarcity of samples (referred to as "training data") in comparison with the dimensionality of the feature space. In the case of conventional statistical classifiers, this limitation poses challenges in estimating statistical parameters, such as the mean and covariance matrix for statistical classifiers.

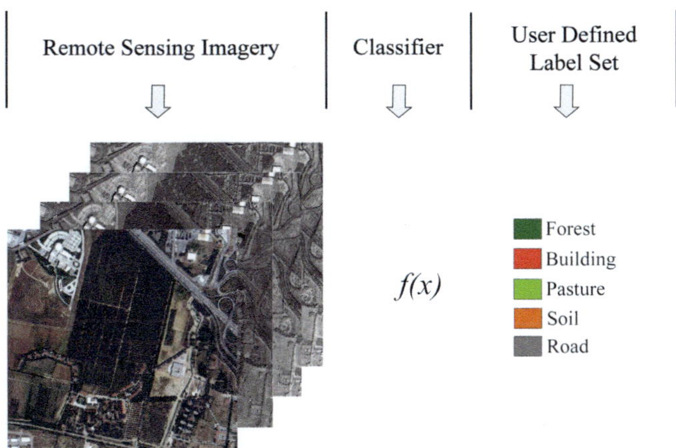

FIGURE 1.1   Concept of the classifier as a link between an image (left) and a set of labels.

Pattern recognition in the domain of remote sensing aspires to associate each object or pixel in the study area with one or more elements from a user-defined label set. This transformation of radiometric information in the image into thematic information, such as vegetation type, can be conceptualized as a mapping function that establishes a linkage between the raw data and the user-defined label set. A simple example is shown in Figure 1.1. Normally, each object or pixel is linked to a single label. However, it is also possible to perform a one-to-many mapping so that a given pixel can be associated with more than one label, with the differing degrees of association between the pixel and each label being expressed as probabilities of membership. Alternatively, a many-to-one scheme will link groups of pixels to a single label. This approach can be used, for example, to give the same label to all of the pixels in a single agricultural field.

Each application generally requires a different methodology, and each methodology is likely to generate different results. If reliable results are to be obtained, the analyst should understand the behavior of the method being used in order to achieve a satisfactory performance. For instance, the performance of a statistical procedure is strongly affected by the accuracy of the estimates of parameters such as the mean vector and the variance-covariance matrix for each class, which are obtained from samples of pixels called training data sets. The design of the architecture of a deep learning model has an equally important impact on the performance of the network.

This book aims to provide a survey of pattern recognition methodology for use with remotely sensed imagery. Besides describing traditional approaches, more advanced techniques using artificial neural networks, deep learning, support vector machines, decision trees, and object-based image analysis are introduced. In the main part of this introductory chapter, the principles of remote sensing in the optical and microwave regions of the spectrum are described. Atmospheric interactions with Earth surface features and fundamental optical and microwave systems are discussed. Some important preprocessing techniques, such as corrections for atmospheric and topographic effects and noise filtering models, are also presented. These techniques help to improve thematic accuracy in some kinds of applications, for example, in change detection and crop monitoring.

1.1   INTRODUCTION TO REMOTE SENSING

Remote sensing employs sensors on drones, aircraft, or satellites to capture electromagnetic energy emitted or scattered from the Earth's surface. The electromagnetic spectrum encompasses a broad range of wavelengths, measured in micrometers. Identified as wavebands, discrete sets of continuous wavelengths, including near-infrared, mid-infrared, thermal infrared, or microwave regions, demonstrate unique characteristics. Figure 1.2 illustrates a portion of the electromagnetic spectrum,

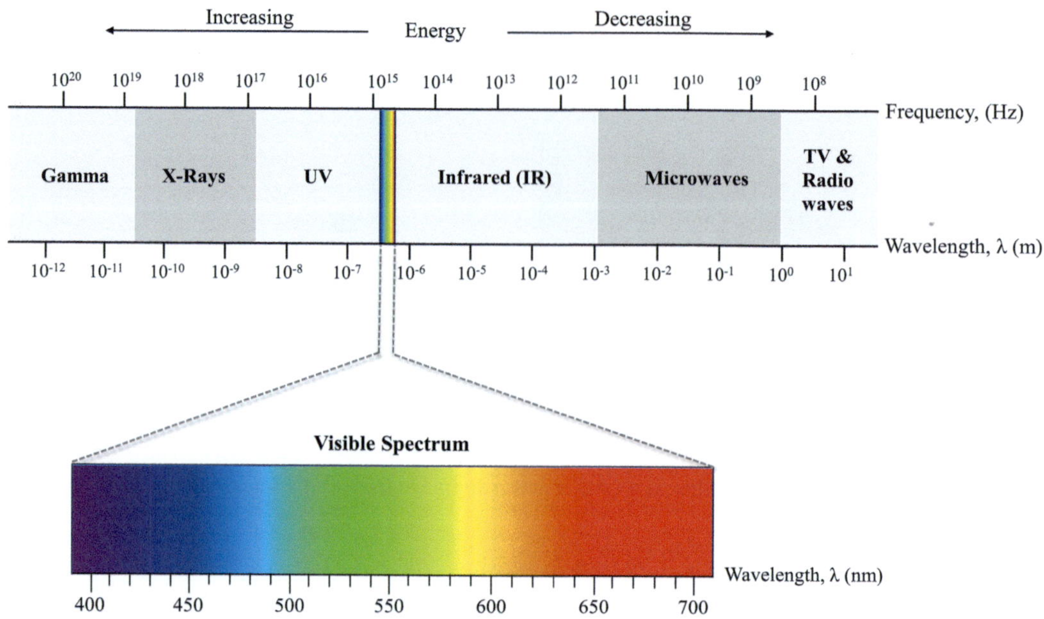

FIGURE 1.2 Regions of the electromagnetic spectrum that are of interest in remote sensing applications.

highlighting that the visible waveband (0.4–0.7 μm), perceivable by human eyes, represents only a fraction of the entire spectrum. Remote sensing instruments are specifically engineered to operate within defined wavebands, selected based on the distinctive traits of the intended target. Thus, a sensor designed to detect electromagnetic radiation in the visible spectrum that has been reflected by chlorophyll in ocean waters will use different wavebands than a sensor designed to detect the characteristics of soils. Apart from the reflectance characteristics of the target, an important factor governing the choice of waveband is the effect of interactions between the atmosphere and electromagnetic radiation. Such radiation passes downward through the atmosphere on its way from the Sun to the Earth and reflected radiation passes upward through the atmosphere on its way from Earth to the sensor. Absorption and scattering are the main mechanisms that alter the intensity and direction of electromagnetic radiation within the atmosphere. In some regions of the optical spectrum, these mechanisms (principally absorption) ensure that remote sensing is impossible. Spectral regions of lower absorption are known as atmospheric windows (Figure 1.3), though it should be remembered that scattering and absorption affect all wavebands in the optical spectrum to a greater or lesser degree, and these effects are variable both in space and in time.

1.1.1 Atmospheric Interactions

Electromagnetic radiation interacts with the Earth's atmosphere, the degree of interaction depending on the wavelength of the radiation, and the local characteristics of the atmosphere. The basic interactions are known as scattering and absorption. Scattering that is more likely to appear at shorter wavelengths occurs when the incoming electromagnetic radiation interacts with particles and gas molecules and then redirected from its natural path. Three types of scattering exist, which are Rayleigh, Mie, and nonselective scattering. The most common scattering behavior is Rayleigh scattering, which is the main cause of haze in remotely sensed imagery. It results when the sizes of particles and gases are smaller than the wavelength of the incident radiation. Due to the Rayleigh scattering, the sky appears blue; otherwise, it would appear black. The effect of scatter is the addition of upwelling atmospheric radiance, or path radiance to the radiance of the ground object. The amount of upwelling atmospheric radiance is dependent upon several factors,

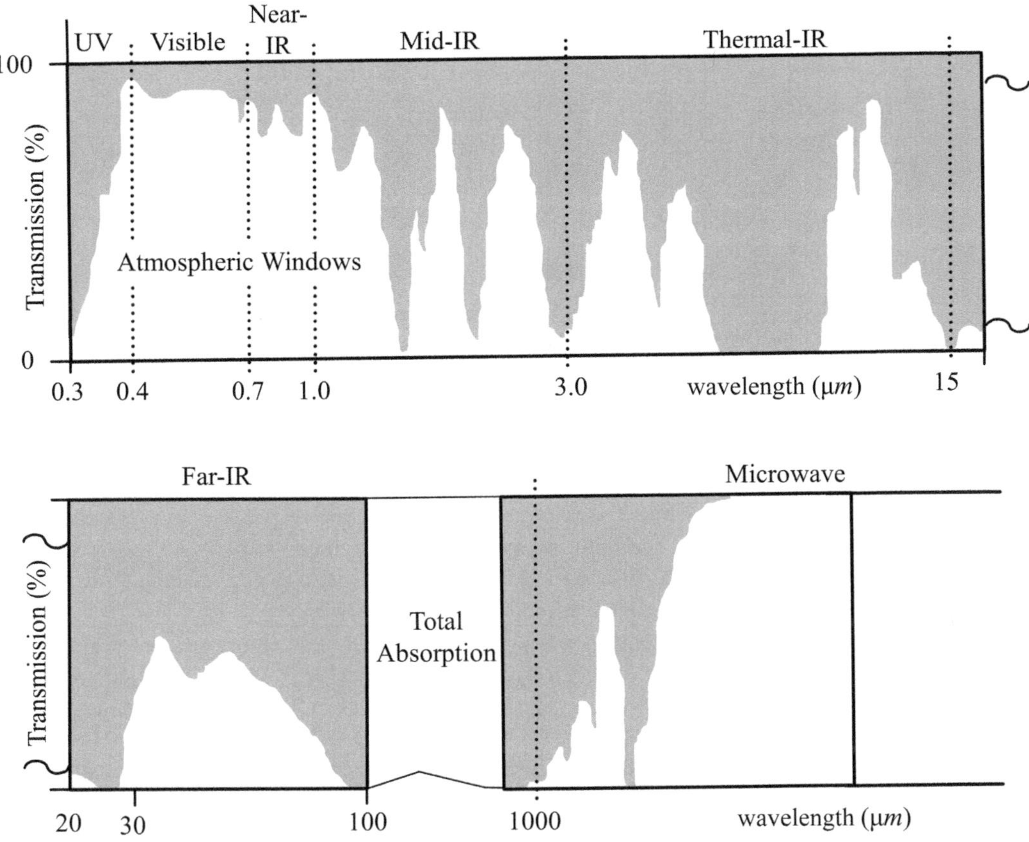

FIGURE 1.3 Atmospheric windows (unshaded). The vertical axis is atmospheric transmission (%). The horizontal axis shows the wavelengths in micrometers on a logarithmic scale.

including the altitude of the sensor, atmospheric conditions, solar zenith angle, angle of view from the nadir and azimuth with respect to the Sun.

The atmosphere absorbing about 23% of the incoming solar radiation has different levels of absorption at different wavelengths. The most effective atmospheric constituents absorbing radiation are ozone, water vapor, and carbon dioxide. Ozone in the stratosphere absorbs the harmful ultraviolet radiation coming from the Sun, thus preventing this radiation from reaching to Earth's surface. Absorption bands of water vapor are in wavelength ranges centered at 1.38, 1.87, and 2.7 μm. Regions of the spectrum that have a relatively high transmission are called atmospheric (or transmission) windows (Figure 1.3). The energy in some regions (e.g., from 15 to 10^3 μm) is almost completely absorbed by the atmosphere. These wavelengths cannot, therefore, be used for remote sensing as there will not be any or enough amount of energy to be recorded. Wavebands with a high transmission could be potential candidates for remote sensing missions.

1.1.2 REFLECTANCE PROPERTIES OF SURFACE MATERIALS

The choice of wavebands for remote sensing in the optical region is also affected by the characteristics of the surface material. The energy that is incident upon a target can be separated into three components, namely, energy that is transmitted, absorbed, and reflected. Surface material reflectance characteristics may be quantified by spectral reflectance, which is a percentage measure obtained simply by dividing reflected energy in a given waveband by incident energy. The quantity of the reflected energy depends mainly on three factors: the magnitude of the incident energy,

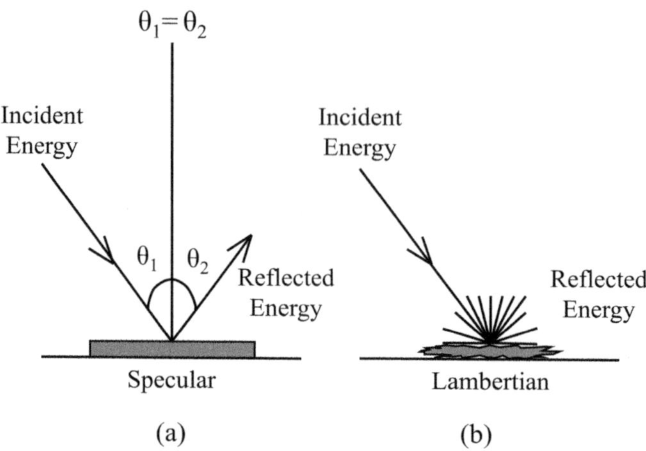

FIGURE 1.4 (a) Specular and (b) Lambertian reflectance.

the roughness of the material, and the material type. Normally, the first two factors are regarded as constants. Therefore, only the third factor (i.e., the material type) is considered. However, it is worthwhile to first describe how roughness affects the reflected energy.

Surface roughness is a wavelength-dependent phenomenon. Given the same material, the longer the wavelength, the smoother the material appears. For a perfectly smooth (specular) surface, reflected energy travels only in one direction such that the reflection angle is the same as the incidence angle. In the case of a perfectly rough (Lambertian) surface, the incident energy is reflected equally in all directions (Figure 1.4). However, in practical applications, most surface materials act neither as specular nor as Lambertian reflectors; their roughness lies somewhere between these two extremes.

Figure 1.5 shows the average reflectance over the optical region of the spectrum for five ideal surface materials observed in urban and suburban areas: bare soil, clear water, concrete, asphalt, and green vegetation. This graph shows how these surface materials can be separated in terms of their reflectance spectra. It is apparent that vegetation reflectance varies considerably across these wavebands. The lowest reflectance values occur at 0.4 μm (i.e., in the blue waveband), while the highest reflectance values occur around the near-infrared and part of the mid-infrared bands. The high near-infrared reflectance of vegetation, combined with its low reflectance in the red waveband, is used in the construction of vegetation indices. One such index, the NDVI (Normalized Difference Vegetation Index), is described in the following text. The reflectance spectrum of bare soil, in contrast, shows reflectance increasing smoothly with wavelength. Its reflectance in the visible and mid-infrared wavebands is greater than that of vegetation. Starting from about 10% reflectance at 0.4 μm, the spectral reflectance of asphalt increases steadily, whereas that of concrete is much higher particularly in near- and mid-infrared regions.

Recent interest in global environmental change has led to the development of global climate models, which require inputs describing land cover types. The AVHRR sensor carried by the National Oceanographic and Atmospheric Administration (NOAA) satellite series (from NOAA-6 to NOAA-18) and the MODIS sensor carried by Terra provide data that are useful for large-scale studies of terrestrial vegetation (Eidenshink and Faundeen, 1994; Lim and Kafatos, 2002; Neupane et al., 2022). A variety of mathematical combinations (e.g., subtraction, ratioing) have been developed to characterize the spatial distribution of vegetation and its condition. The best known of these vegetation indices is the NDVI, defined as

$$\text{NDVI} = \frac{\left(\text{Near-infrared band}\right) - \left(\text{Red band}\right)}{\left(\text{Near-infrared band}\right) + \left(\text{Red band}\right)} \tag{1.1}$$

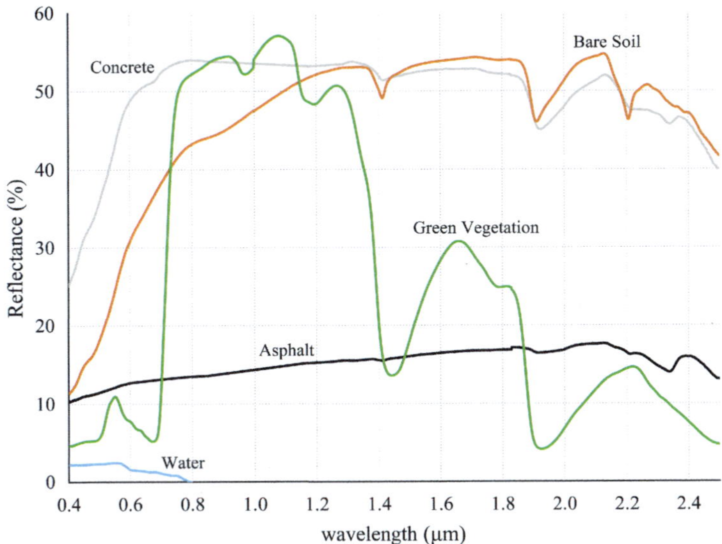

FIGURE 1.5 Typical spectral reflectance (%) of five major surface materials, collected by in-situ measurements using an ASD FieldSpec 3 FR spectroradiometer.

This index is based on the observation that vegetation has a high reflectance in the near-infrared band, while reflectance is lower in the red band (Figure 1.5). The NDVI is less sensitive to the changes in atmospheric conditions than are other indices and is effective in estimating leaf area index (LAI), chlorophyll concentration in leaves, plant productivity, and plant stress. Therefore, it has been widely applied for vegetation monitoring (Holben and Kimes, 1986; Vicente-Serrano et al., 2016; Huang et al., 2021). Data from other sensors that provide red and near-infrared (near-IR) images can also be used to generate NDVI images. For instance, in the case of Landsat 4–7 images, the NDVI is estimated based on bands 4 and 3, for Landsat 8–9 based on bands 5 and 4, and for Sentinel-2 based on bands 8 and 4. Researchers have made efforts to develop new spectral indices to quantitatively assess the vegetation cover. Bannari et al. (1995) gave a good review of these indices. The use of vegetation indices under the conditions of different sensors, satellite product versions, atmospheric and Sun-target-sensor geometry corrections, compositing algorithms, and the use of quality assurance and control flags is also discussed in Zeng et al. (2022).

Since the 1960s, remote sensing has demonstrated its promise for describing the maritime environment with particular emphasis on sea surface features, mapping of shorelines, wetlands, and coral reefs (Brewington et al., 2014; Cracknell et al., 2001). Water surfaces studied by remote sensing range from wetlands to lakes and oceans that are vulnerable to rapid changes particularly due to anthropogenic activities and climate change effects. Periodic monitoring of coastal water quality is of critical importance for the effective management of water resources and the sustainability of marine ecosystems. It should be highlighted that for many wetlands, remote sensing appears to be the only solution for obtaining data and monitoring changes. On the other hand, a fundamental problem in aquatic environments is the appearance of harmful algal blooms destroying the ecological habitat, threatening human health, and causing massive fish die-offs, which is also a major study in remote sensing. In recent years, abrupt changes in the physical and biogeochemical conditions in seas and oceans have caused marine mucilage, also called sea snot, which destroy the balance in the aquatic ecosystem, adversely affecting all species in the sea, particularly the ones on the seabed (Kavzoglu et al., 2021; Kavzoglu and Goral, 2022; Colkesen et al., 2023). Inspired by the idea of the NDVI, several indices were developed for monitoring the water surfaces, including the Floating Algae Index (FAI), Normalized Difference Water Index (NDWI), Automated Water Extraction Index (AWEI), Automated Mucilage Extraction Index (AMEI). Knowledge of surface material reflectance characteristics provides us with a principle on the basis of which suitable wavebands to scan the Earth's surface for a particular mission can be selected (e.g., for vegetation monitoring, sea surface observation, or lithological identification). Such knowledge also provides an important basis to make the objects more distinguishable in terms of multiband image manipulations such as overlay, subtraction, or ratioing.

1.1.3 Spatial, Spectral, and Radiometric Resolution

The resolution of a remote sensing instrument is usually expressed in terms of its spatial, spectral, and radiometric resolution. Temporal resolution is also considered as an additional resolution for the remote sensing data. The higher the spatial resolution, the smaller the ground objects that can be distinguished. The spatial resolution is related to the instantaneous field of view (IFOV) of the sensor, which denotes the size of the area from which the sensor receives the energy at the given instant in time. In Figure 1.6, the energy transmission path to the sensor takes a conelike shape, and the ground resolution is roughly equivalent to the diameter of the circle formed by the intersection of this cone and the ground surface. As the sensor scanning area moves away from the nadir, the larger the IFOV will be (Figure 1.6), and thus results in the distortion of the resulting image as the spatial scale decreases from the left and right edges of the image toward the center. The correction for this scale distortion effect (and other effects due principally to the platform's orbit characteristics and to the eastward rotation of the Earth during scanning) is called geometric correction. This correction is usually performed by constructing a transform (using either an empirical procedure based on least squares methods or an analytical procedure using orbital information) in order to map ground coordinates to their corresponding image coordinates, and vice versa. A review of local and global methods of geometric correction and image registration is provided by Brown (1992), Zitová and Flusser (2003), and Le Moigne et al. (2011).

The images obtained from satellite or aircraft platforms are affected by three impulse response functions: atmosphere, optics, and sampling aperture. The outcome is an image characterized by

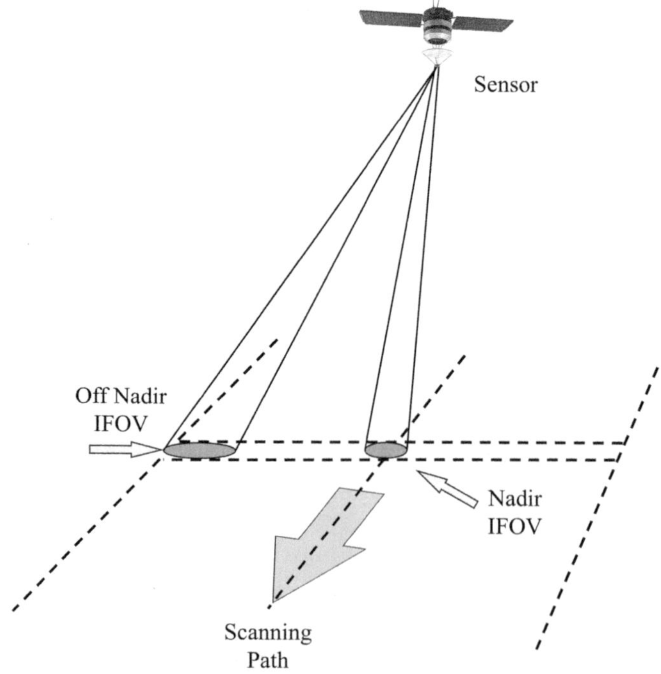

FIGURE 1.6　Variation in the size of the instantaneous field of view (IFOV) of a sensor across a scan line.

diminished contrast, primarily influenced by atmospheric effects, and a loss of detail primarily linked to the scanning aperture. In the application of remotely sensed images, the typical assumption is that the spectral content of a pixel is exclusively derived from the instrument's IFOV. Nevertheless, a fraction of the spectral signal attributed to a particular pixel is influenced by the surrounding pixels, a consequence of the instrument's point spread function (PSF). Various factors, such as the instrument's optics, electronics, and sample spacing, impact the attributes of the PSF. Additionally, image-motion smear and the varying influence of the atmosphere can contribute to these characteristics (Kavzoglu, 2004). It is a common practice to apply a modular transfer function compensation process to deal with the disruptive effects of the PSF.

One's instinctive feeling might be that the finer the spatial resolution, the better the results of the analysis. However, in practice, this may not always be the case. The choice of spatial resolution should depend on what we want to see. For instance, in everyday life, we recognize a human face from the combination of features such as eyes, nose, and lips. Therefore, the appropriate spatial resolution should be set at a level to allow us to recognize each feature in the context of the whole face. It is arguable whether an increase in the spatial resolution of our eyes is likely to improve our identification of faces, once a certain limit has been reached. Thus, the choice of spatial resolution is case-dependent, which is discussed in the following subsection.

Reducing the spatial resolution in terms of narrowing the instrument's IFOV also affects a related issue. The use of a smaller IFOV implies that the quantity of energy received by a detector is less (because the area from which energy is collected is smaller, and the time available for the sensor to detect the upwelling energy is also shorter). It follows that the instrument's sensitivity to changes in the levels of energy will decrease, and thus, the sensor may not be able to distinguish slight energy differences along a scan line. In other words, the radiometric resolution is degraded. A smaller IFOV may thus result in a worse signal-to-noise ratio, which means that the signal is contaminated by more noise as IFOV decreases (assuming that the same radiometric resolution applies). Although such a decrease in signal-to-noise ratio can be compensated for by enlarging the scanning bandwidth (i.e., the region of the electromagnetic spectrum from which the sensor receives energy), this will cause a reduction in the spectral resolution (i.e., the sensor's ability to quantify spectral differences). Overall, it can be concluded that enhancement of both spatial and radiometric resolution cannot be achieved together without a level of compromise.

Spectral resolution is an important characteristic of imaging instruments that record the electromagnetic energy from different parts of the spectrum as discrete spectral bands. It refers to the number and width of the spectral bands or channels used to identify and discriminate Earth's surface features. While multispectral remote sensing systems record energy in multiple portions of the spectrum, hyperspectral sensors collect data with more spectral bands than multispectral sensors, usually more than a hundred spectral bands (e.g., AVIRIS and Hyperion sensors record data at 224 and 220 bands, respectively). A spectral band that collects signals across a wide spectrum effectively averages all signals within that range. A panchromatic image, as a good example, records the upwelling radiance over a broader spectrum range compared with multispectral bands, enabling it to capture signals from a smaller area on the ground. For instance, the spectral range of the red band in the Landsat 8 OLI sensor is between 0.636 and 0.673 µm, while that of the panchromatic band is between 0.503 and 0.676 µm. Therefore, it can be deduced that when there are equal numbers of spectral bands from different sensors, the one with narrower bands should be preferred to delineate specific information. Depending on the characteristics of the sensor, the recorded signals contain some degree of noise resulting from various sources, including atmospheric interactions and the electronic noise of the sensor. The purity or quality of an image is directly related to the signal-to-noise ratio (SNR) of the recording sensor. Increasing the spectral resolution leads to a decrease in the SNR of the sensor output because the radiance magnitude (the signal strength) within narrower spectral bands is less than that of wider bands. A careful selection of spectral bands substantially increases the probability of extracting the intended information. In many cases, identification and extraction of critical information can only be achieved when narrow wavebands exist. With the use of appropriate spectral bands of hyperspectral data providing narrow wavebands, many studies have become possible, including detecting plant diseases, physicochemical properties of water, and soil content analysis.

Radiometric resolution of an imaging system refers to its ability to discriminate minute energy differences using a single wavelength at varying intensities. The radiometric resolution of a sensor, which determines the digital numbers (DN values) it records, depends on its sensitivity to the magnitude of electromagnetic radiation. Data are recorded in the form of positive DN within the range from zero to the system's designed resolution. DN values denote the number of discrete levels, quantified in powers of 2. For instance, pixels in an 8-bit image would encompass DN values ranging from 0 to 255, offering a total of 256 variations. A higher radiometric resolution permits the identification of the most subtle energy variations, while images with lower resolution can only detect substantial differences. In other words, a sensor with finer radiometric resolution is more attuned to minute changes in reflected or emitted energy. Remote sensing images with high radiometric resolution enhance the accuracy of Earth surface feature analysis by delivering comprehensive spectral information. In the 1990s and 2000s, many satellites had an 8-bit resolution, but contemporary sensors can now provide 12-bit radiometric resolution images, offering 16 times better range. This improvement widens the distance between similar landscape features; thus, discrimination between spectrally similar classes can be achieved, which definitely helps to improve the classification accuracy.

In addition to spatial, spectral, and radiometric resolution, the temporal resolution of a satellite sensor, the revisit period for the same location, can be critical, particularly for monitoring natural and anthropogenic events occurring during a period. In many real-world classification scenarios, particularly including vegetation types, the information content of an image from a single acquisition is insufficient; hence, data acquired at different dates are required as complementary information to discriminate land surface features and, at the same time, to capture ecosystem dynamics. Due to the incorporation of time-varying spectral information, multitemporal imagery is mainly required to conduct change detection analysis. Before the classification process, it is advised to conduct a preliminary analysis on the spectral signatures of the land use and land cover objects considered in the classification and then decide on the requirement of images from different dates, even from different sensors. In general, remote sensing analysts are intuitively aware of the requirement of additional information when they consider the problems that have spectrally similar objects or species of the same type (e.g., different species and clones of poplar trees). To increase the temporal resolution, several satellites can share exactly the same orbit. For instance, Sentinel-2A and Sentinel-2B occupy the same orbit separated by 180° to supply images at 5-day intervals.

1.1.4 SCALE ISSUES IN REMOTE SENSING

Advancements in sensor design have enabled the acquisition of data with a high level of precision and quality similar to aerial photography using modern high-resolution sensors. This precision and quality empower the examination of thematic information even in the most complex natural landscapes among various land cover classes (Köhl et al., 2006). The intricate spatial and temporal dynamics within landscapes, as captured through a sequence of satellite images, exhibit a high degree of complexity due to the combination of natural and man-made elements, along with challenges such as shadows and occlusions. Pixels within these images often encompass a mixture of man-made and natural objects, contingent on the sensor's spatial resolution and the characteristics of the area being observed. Therefore, the problem of determining the most appropriate scale (pixel size) for a particular study has been a major concern in remote sensing. According to Quattrochi and Goodchild (1997), "of all words that some degree of specialized scientific meaning, scale is one of the most ambiguous and overworked". Theoretically, the scale represents the window of perception, the filter, or the measuring tool through which a landscape may be perceived (Levin, 1992). It is also associated with the structure, extent, and strength of spatial dependence within the remotely sensed data (Zhang et al., 2014). In June 1996, the University Consortium for Geographic Information Systems (UCGIS) endorsed scale as a research priority in GIS science. The scale issue is of considerable importance in remote sensing and image processing studies because it refers to the spatial resolution of the imaging instrument, which represents the size of the area corresponding to the IFOV of the sensor. It is important for researchers to recognize and differentiate among specific scale-related terms, such as geographical scale, operational scale, measurement scale, and cartographic scale, to fully grasp the methodologies and ongoing discussions related to this subject.

Since each scene contains different-sized objects, the use of a single scale may not be relevant to identifying the objects (Kavzoglu et al., 2017). The choice of an appropriate scale or spatial resolution, for a particular application, depends on several factors, namely, the information desired about the ground scene, the analysis methods to be used to extract the information, and the spatial structure of the scene itself (Woodcock and Strahler, 1987). For instance, it is impractical and irrelevant to use 1-m or higher resolution imagery to conduct a region-based or continental-based study, considering the scale of the objects under consideration, generalized land cover classes (e.g., vegetation, soil, and water), and the extreme computational burden. For a particular remotely sensed imagery, a graph (called variogram) showing how the local variance of a digital image for a scene changes as the pixel size changes can be used to select the appropriate scale. The appropriate scale can be estimated using fractals, geographical variance, local variance, wavelets, and spatial statistics (Woodcock et al., 1988; Curran and Atkinson, 2002; Zhang et al., 2014). It should be highlighted that conclusions drawn at a specific scale are valid only for that scale and should not be used to reach conclusions at other scales. In summary, the choice of spatial resolution is either case-dependent or problem-specific.

Any study using spatial information may face the integration of data from different sources with different scales and resolutions. This is certainly a critical issue for both GIS and remote sensing studies because they require high accurate results that are often used in decision-making processes. Most GIS and remote sensing software will not allow overlaying of data at different resolutions and projection systems. Researchers should be aware of the possible outputs of incorrect actions of map overlay and integration of data at different map scales. Combining images with significantly different resolutions should be avoided since it is not possible to integrate a finer detailed image with a coarser detailed one due to inconsistencies in geometry, attributes, and semantics (Zhang et al., 2014). As underlined by Goodchild (1999), further research is required to delve into the relationship between the level of geographical detail and related concepts along with their manipulation and management in GIS. It is crucial to gain a deeper understanding of the consequences when utilizing data with an inappropriate scale and to develop automated methods for scale adjustments.

With the introduction of high-resolution images in the late 1990s, object-based image analysis (OBIA) has become an effective solution for dealing with intra-object heterogeneity existing in these images. Thus, the main task in classification has become the determination of objects through segmentation, instead of concentrating on the labeling of each pixel in the image. Segmentations are usually applied with a single-scale value. However, remote sensing images include complex landscape structures, that is, a mosaic of LULC classes, and each scene is unique in terms of its spectral and spatial patterns and thus requires different scales for each landscape feature (Johnson, 2013). In other words, since each scene contains different-sized objects, the use of a single scale may not be relevant for identifying the land cover classes. When a single scale produced by a global-scale parameter is selected and applied for segmentation, the segments will be either too coarse or too fine for the geographical objects. Therefore, each landscape class in the scene requires different scales to define landscape objects, which can be applied through region-based or hierarchical multiscale segmentations to increase recognition accuracy (i.e., classification accuracy) (Kavzoglu et al., 2017; Zhang et al., 2020). When we consider a remotely sensed image that includes large water and forested areas together with urbanized land, each class category covers fields of different sizes within class homogeneity. While water and forest areas are better represented by large-scale values, buildings require small-scale values. In the multiresolution segmentation algorithm, the scale parameter or homogeneity criterion is the most influential factor in regulating the average size of image objects. It defines the maximum allowable diversity in the resulting image objects, and selecting a higher value for the scale parameter yields larger objects. Scale parameters that generate segments larger than the land cover features within an image lead to under-segmentation. In this case, segments encompass multiple types of land use and land cover, and the resulting classification results in poor generalization capabilities and low classification accuracies. In an optimal scale search, the fundamental criteria are to maximize intra-segment homogeneity and inter-segment heterogeneity. Quality metrics measuring segment quality can be employed to evaluate the suitability of the scale or other techniques, including the estimation of the scale parameter (ESP) tool, which is based on the rate of change of local variance estimations. This issue and other parametrization considerations are comprehensively discussed in Chapter 8. More discussion on the scale issue can be found in Quattrochi and Goodchild (1997), Weng (2014), Zhang et al. (2014), and Quattrochi et al. (2017).

1.2 OPTICAL REMOTE SENSING SYSTEMS

Over the last two decades, the most widely used optical remote sensing systems have been the Landsat TM, ETM+ and OLI, the SPOT HRV and HRG, IKONOS, OrbView-3 (later renamed as GeoEye), Terra ASTER, WorldView-2 and WorldView-3 for high-resolution data, and MODIS, and the NOAA AVHRR instruments for low-resolution data, mainly employed in regional to global studies. However, special emphasis should be given to Sentinel missions operated by the European Commission as part of the Copernicus program that provides free data and other products for scientific studies. Sentinel-1 carries a synthetic aperture radar (SAR) that operates on the C-band, while Sentinel-2 carries a multispectral sensor providing 13 reflective bands at different spatial resolutions over the visible, NIR, and SWIR regions: 10-m spatial resolution (VIS and NIR), 20-m spatial resolution (red-edge, NIR, and SWIR), and 60-m spatial resolution (coastal aerosol, water vapor, SWIR-Cirrus). The Sentinel-2 mission launched in 2015 is based on a constellation of two satellites, providing images at 5-day intervals. Sentinel-3 mission provides information on sea surface topography, sea and land surface temperature, and ocean and land surface color with high accuracy. The Sentinel Application Platform (SNAP) not only provides access to data of Sentinel missions (Sentinel-1, Sentinel-2, and Sentinel-3) but also serves toolboxes for data processing.

Other remote sensing satellites carrying optical sensors have been launched by a number of national space agencies and private companies, and these have also gradually become the main sources for serving scientific study and environmental monitoring purposes. Examples are the Chinese-Brazilian remote sensing system (CBERS), the Japanese ALOS program, the Indian IRS and CartoSat series, the Turkish RASAT and Göktürk series, the Korean KOMPSAT series, Chinese GaoFen, and Jilin programs. While high-resolution sensing devices (4 m in a multispectral mode, 1 m or less in a panchromatic mode) were the main data sources in the 2000s (e.g., the QuickBird, OrbView, and IKONOS), we now have very high-resolution sensors with spatial resolution reaching to 30 cm, provided by Pleiades Neo, WorldView-3 and WorldView-4 MS sensors. Additional eight SWIR bands of WorldView-3 make the imagery from this instrument suitable for many studies. Since 2021, another instrument providing images at six spectral bands with a 30-cm spatial resolution is the Pléiades Neo constellation. Radar systems are also becoming more numerous (ESA's C-band ERS-1, ERS-2, Sentinel-1, JAXA's L-band ALOS PALSAR-1 and PALSAR-2, Canada's C-band Radarsat-1 and Radarsat-2, the German X-band TerraSAR-X and TanDEM-X, and the Italian X-band COSMO-SkyMed).

Considerable interest is also being shown in the application of hyperspectral imagery. Whereas multispectral sensors such as the Landsat ETM+ collect upwelling radiation in a small number of broad wavebands (seven in the case of the ETM+ instrument), a hyperspectral sensor collects data in a large number of very narrow wavebands. Hyperspectral images offer more detail on both spectral and spatial domains; therefore, they are invaluable for accurate segmentation and classification tasks. An example is the DAIS instrument, which collects data in 79 bands in the wavelength range of 0.4–12.6 µm. The width of each spectral band varies from 15 to 20 nm in the visible to 2 µm in the middle infrared. The Compact Airborne Spectrographic Imager (CASI) instrument allows data to be collected for any 545-nm segment of the 0.4- to 1.0-µm region in 288 bands, with the bands spaced at intervals of ~1.9 nm. NASA's AVIRIS acquires data in 224 bands in the range of 0.38–1.5 µm with a bandwidth of 10 nm. Up until the present time, hyperspectral data have been usually collected by aircraft-mounted sensors. NASA's Earth Observer-1 (EO-1), launched in 2000 and terminated in 2017, is the first orbiting spacecraft to carry a hyperspectral imager, Hyperion. The instrument collected data over a narrow swath in 220 bands of 10 µm width. The large number of spectral bands produced by hyperspectral sensors (i.e., the high dimensionality of the data) posed significant problems in the pattern recognition process. The high-dimensionality problems and their solutions are discussed in Chapter 3. Currently, the hyperspectral camera (HYC) of the PRISMA satellite, launched in 2019, has 237 bands, the HysIS launched by the Indian Space Research Organization (ISRO) in 2018 has 316 bands, and the CHRIS Proba-1 sensor, launched in 2001, has 62 bands. According to the survey study by Qian (2020), 26 spaceborne hyperspectral imaging systems have been launched to their orbits of Earth, Moon, Venus, and Mars. The number of launched hyperspectral satellites has recently shown a pace. Two initiatives that should be mentioned here aim to provide daily satellite images. With PlanetScope, SkySat, and the to-be-launched Pelican High Resolution constellations, Planet Labs has about 200 satellites providing daily images with a spatial resolution of 3 m. On the other hand, Capella Space provides radar data with a constellation of satellites and delivers one of the most frequent, timely, and high-quality commercial SAR imagery products.

An exhaustive list of sensors and satellites is not supplied here; such lists tend to become outdated very quickly. The advent of the World Wide Web (WWW) means that researchers, teachers, and students can readily obtain up-to-date information via the Internet. Professional and learned societies, such as the Remote Sensing and Photogrammetry Society, maintain WWW pages that provide links to international and national space agencies and projects. The Committee for Earth Observation Satellites (CEOS) also maintains an information site. Readers can access data through major data providers via USGS Earth Explorer, Copernicus Open Access Hub, NASA Earthdata Search, MAXAR Open Data Program, Geo-Airbus Defense, Data ARchives and Transmission System (DARTS) for JAXA archive.

Details of the operation of the main types of sensors carried by remote sensing satellites can be found in textbooks (e.g., Lillesand et al., 2015; Mather and Koch, 2011; Jensen, 2015), and readers who may be unfamiliar with these details are referred to one of these sources.

1.3 ATMOSPHERIC CORRECTION

Electromagnetic energy detected by remote sensing instruments (especially those that operate in the optical region of the spectrum) consists of a mixture of energy reflected from or emitted by the ground surface and energy that has been scattered within or emitted by the atmosphere. The magnitude of the electromagnetic energy in the visible and near-infrared region of the spectrum that is detected by a sensor above the atmosphere is dependent on the magnitude of incoming solar energy (irradiance), which is attenuated by the process of atmospheric absorption, and by the reflectance characteristics of the ground surface. Hence, energy received by the sensor is a function of incident energy (irradiance), target reflectance, atmospherically scattered energy (path radiance), and atmospheric absorption. Interpretation and analysis of remotely sensed images in the optical region of the spectrum are based on the assumption that the values associated with the image pixels accurately represent the spatial distribution of ground surface reflectance and that the magnitude of such reflectance is related to the physical, chemical, or biological properties of the ground surface. Clearly, this is not the case unless corrections are applied to take account of variations in solar irradiance and in the magnitude of atmospheric absorption and scattering, as well as in the sensitivity of the detectors used in the remote sensing instrument. The response of these detectors to a uniform input tends to change over time. Correction for these effects is vital if thematic images of a given area are to be compared over time, for example, over a crop-growing season. In atmospheric corrections, a common practice, when it is possible, is to collect in-situ data with hand-held spectrometers, which are free from atmospheric effects, and calibrate the images with these data.

The necessity for atmospheric correction depends on the objectives of the analysis. In general, land cover identification exercises that are based on single-date images do not require atmospheric correction, as pixels are being compared with other pixels within the image in terms of similarity, for example. The validity of this statement depends also on the quality of the image (e.g., an image displaying severe haze effects, spatially varying haze phenomena, or cloud-cloud shadow effects may be unsuitable for classification). Atmospheric correction and sensor calibration are necessary when multisensor or multidate images are being classified, or where the aim of pattern recognition is to identify land cover change over time, to ensure that pixel values are comparable from one image to the next in a temporal sequence. A more radical view is taken by Smith and Milton (1999), who emphasize that "to collect remotely sensed data of lasting quantitative value then data must be calibrated to physical units such as reflectance".

The value recorded for each pixel in a remotely sensed image is a function of the sensor-detected radiance. Owing to the atmospheric interaction, this apparent radiance is the combination of the contribution of the target object and the atmospheric effect. Their relationship can be approximated as

$$L_{\text{app}} = \frac{\rho TE}{\pi} + L_p. \tag{1.2}$$

Here, L_{app} denotes the apparent radiance received by the sensor, L_p is the path radiance, ρ is the target reflectance (%), T is the atmospheric transmittance (%), and E is the solar irradiance on the target. Radiance is expressed in units of W/m² s/r µ/m, and irradiance is expressed in the units of W/m² µ/m. As these two terms are not expressed in equivalent units, solar irradiance is converted into equivalent solar radiance by introducing the term π into the denominator. This conversion is based on the assumption that the target behaves as a Lambertian reflector (as described in Figure 1.4) (Mackay et al., 1994).

In Equation (1.2), only the first term ρ contains information about the target. The atmosphere contributes the second term, the path radiance L_p, which varies inversely in magnitude with wavelength. In the case of multispectral images, the magnitude of the L_p term in the visible bands will be higher than that in the near- or mid-IR bands.

In addition to land use and land cover features, atmospheric correction of water bodies (ocean, coastal, and inland waters) has been a challenging research task due to additional disturbing effects (e.g., whitecaps and sun glint); hence, some specific correction methods have been developed for this purpose. Comparative studies by Martins et al. (2017), Pereira-Sandoval et al. (2019), Warren et al. (2019), and Bui et al. (2022) tested the performances of various atmospheric correction methods, including the Second Simulation of a Satellite Signal in the Solar Spectrum (6SV), Atmospheric Correction for OLI "lite" (ACOLITE) and Sentinel-2 Correction (Sen2Cor), image correction for atmospheric effects (iCOR), and Case 2 Regional Coast Color processor (C2RCC). Evaluating the suitability and restrictions of the current atmospheric correction approaches is of considerable importance across diverse water categories and atmospheric circumstances.

1.3.1 Dark Object Subtraction

Two kinds of methods are most commonly used for atmospheric effect correction. The first kind consists of the dark object subtraction techniques (Chavez, 1988; Ouaidrari and Vermote, 2001), which involve the subtraction of a constant value (offset) from all pixels in a given spectral band.

These methods are based on the assumption that some pixels in the image should have a reflectance of zero and that the values recorded for these zero pixels result from atmospheric scattering. Thus, these pixel values represent the effects of atmospheric scattering. For example, the reflectance of deep clear water in the near-IR waveband is near zero. Dark object subtraction methods assume that nonzero pixel values over deep clear water areas in the near-IR band are contributed by the path radiance L_p and that the path radiance is spatially constant (meaning that a single value of path radiance is subtracted from all pixel values in the image). In case of the visible bands, one may use shadow areas due to topography as dark objects. In effect, one is using the histogram offset as a measure of atmospheric path radiance, as shown in Figure 1.7.

The empirical line method is described and evaluated by Smith and Milton (1999). The method requires that ground measurements of surface reflectance of dark and bright areas in the image are taken simultaneously with the overflight, though if the areas are spectrally stable, the measurements need not be simultaneous (though they must be taken under similar atmospheric and illumination conditions). The image data are converted to radiance using the standard calibration, and surface reflectance R is regressed against sensor radiance L to give a relationship of the form $R=(L-a)\times s$. The term a represents atmospheric radiance, which is subtracted from apparent radiance L before conversion to reflectance. Other descriptions of the empirical line methods are found in Moran et al. (2001) and Karpouzli and Malthus (2003).

1.3.2 Modeling Techniques

The methods described in Section 1.3.1 above involve the subtraction of the same pixel value from the whole image. They are easy to apply but only provide an approximate correction. If the magnitude of the ground-leaving reflectance for each pixel is required, a more sophisticated method based on modeling techniques is necessary (e.g., Tanré et al., 1986, 1990; Vermote et al., 1997). These methods attempt to model atmospheric interactions, through which one can retrieve estimates of true target reflectance.

1.3.2.1 Modeling the Atmospheric Effect

The 5S model (*S*imulation of the *S*ensor *S*ignal in the *S*olar *S*pectrum) developed by Tanré et al. (1986, 1990) is the best-known atmospheric calibration model. An improved version is the 6S

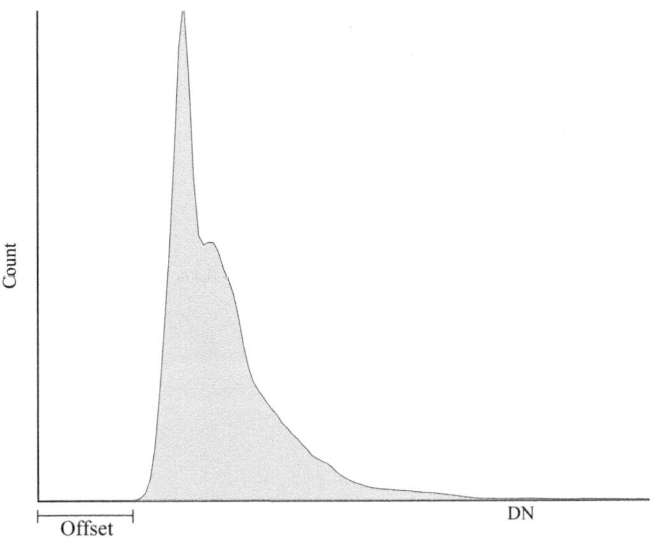

FIGURE 1.7 Estimate of path radiance is the image histogram offset from zero.

(*Second Simulation of the Sensor Signal in the Solar Spectrum*) model (Vermote et al., 1997), which is capable of simulating a non-Lambertian surface to model the signal measured by the sensor. The 6S model also includes data for calculating atmospheric absorption using an increased number of atmospheric gases. In what follows, the core of the model is described in order to provide an introduction to atmospheric modeling.

Assume unit solar irradiance incident on top of the atmosphere. A fraction of the incident solar irradiance is scattered from the path between the Sun and the ground target into the atmospheric volume, with the remainder of the radiation being incident on the ground target as direct solar radiation (Figure 1.8a). This transmitted fraction, denoted by $T(\theta_s)$, is defined as

$$T(\theta_s) = \exp\left(\frac{-\tau}{\cos(\theta_s)}\right). \tag{1.3}$$

The term τ is known as the optical depth, and θ_s is the solar zenith angle, which is illustrated in Figure 1.9.

A fraction of the solar radiation that is scattered into the atmosphere will also appear to contribute to the illumination of the ground target (Figure 1.8c, marked "diffuse") and will compensate

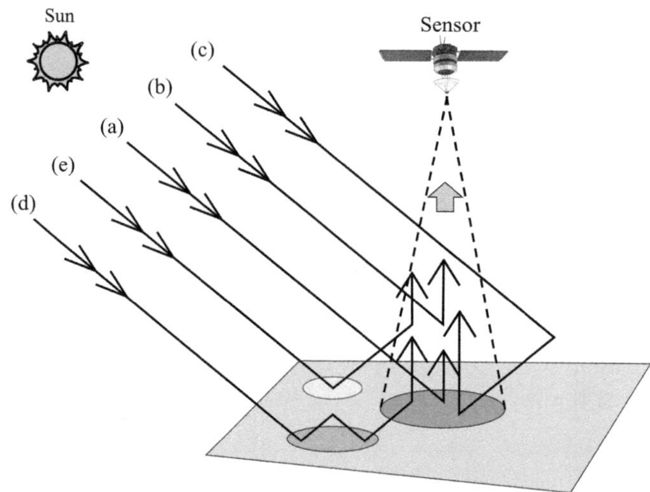

FIGURE 1.8 Five kinds of radiative interaction with the atmosphere. (a) Direct, (b) path, (c) diffuse, (d) multiple, and (e) neighbor.

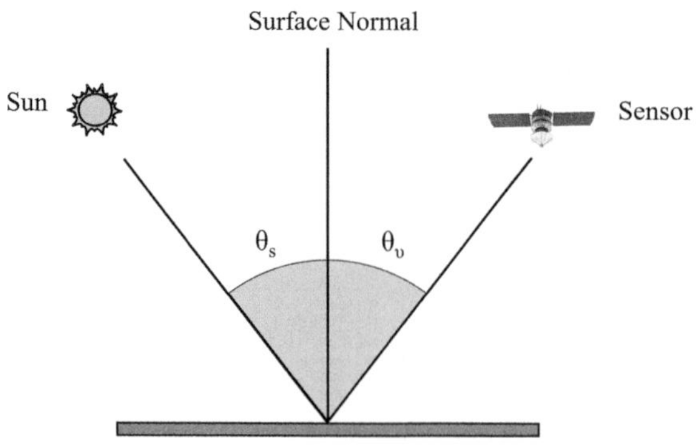

FIGURE 1.9 Solar and sensor view angles on a horizontal surface.

for some of the attenuation of the direct beam. If we denote this diffuse skylight as $t_d(\theta_s)$, then the fraction of solar irradiance incident at the ground target becomes

$$T(\theta_s) = \exp\left(\frac{-\tau}{\cos(\theta_s)}\right) + t_d(\theta_s). \tag{1.4}$$

A second scattering flux that should also be taken into consideration is the trapping mechanism. The effect of this mechanism corresponds to the successive reflection and scattering of solar radiation between the ground target neighborhood and the atmosphere so that the radiation then becomes incident upon the ground target, as shown in Figure 1.8d and e. The magnitude of this effect depends on the spherical albedo of the atmosphere, S, and the surface reflectance, ρ_s. The illumination at the ground target now becomes

$$\frac{1}{1-\rho_s S} \times T(\theta_s). \tag{1.5}$$

The proportion of the solar radiation reflected from the ground target is expressed as

$$\frac{\rho_s}{1-\rho_s S} \times T(\theta_s) \tag{1.6}$$

Consider a sensor that is receiving the reflectance from the ground target. The reflectance is generated by two main sources: one is the contribution of the total solar radiation reflected by the ground target and directly transmitted from the surface to the sensor, while the other is the contribution from the target neighborhood, which is scattered into the field of view of the sensor. The reflectance received by the sensor can thus be expressed by

$$\frac{\rho_s}{1-\rho_s S} \times T(\theta_s) \times T(\theta_v). \tag{1.7}$$

In this equation, θ_v is the sensor view zenith angle, and Equation (1.4) can be again applied to express $T(\theta_v)$ as

$$T(\theta_v) = \exp\left(\frac{-\tau}{\cos(\theta_v)}\right) + t_d(\theta_v). \tag{1.8}$$

The sensor also receives a fraction of the solar radiation that has been scattered out of the downward solar beam into the sensor's field of view without interaction with the ground target, as shown in Figure 1.8b. This component is the atmospheric reflectance, denoted by the function $\rho_a(\theta_s, \theta_v, \phi)$, where ϕ is the relative azimuth between the Sun and the sensor. Therefore, the apparent reflectance ρ^* at the sensor is

$$\rho^* = \frac{\rho_s}{1-\rho_s S} \times T(\theta_s) \times T(\theta_v) + \rho_a(\theta_s, \theta_v, \phi). \tag{1.9}$$

Equation (1.9) is a linear equation that specifies the relationship between the apparent reflectance ρ^* and the surface reflectance ρ_s. The term $T(\theta_s) \times T(\theta_v)$ is the total atmospheric transmittance along the sun-target-sensor path.

A second atmospheric interaction should be considered, namely, the process of absorption. In the solar (optical) spectrum, atmospheric gaseous absorption is principally due to the presence of ozone (O_3), oxygen (O_2), water vapor (H_2O), and carbon dioxide (CO_2). Both O_2 and CO_2 are uniformly mixed in the atmosphere and are constant in terms of their concentrations, whereas H_2O and

O_3 concentrations vary with time and geographical location. If $T_g(\theta_s, \theta_v)$ denotes atmospheric gas transmittance after absorption, then Equation (1.9) can be modified to give

$$\rho^* = T_g(\theta_s, \theta_v) \times \left[\frac{\rho_s}{1 - \rho_s S} \times T(\theta_s) \times T(\theta_v) + \rho_a(\theta_s, \theta_v, \phi) \right]. \tag{1.10}$$

To retrieve the surface target reflectance, ρ_s, Equation (1.10) is further expanded as

$$\frac{\rho^*}{T_g(\theta_s, \theta_v)} - \rho_a(\theta_s, \theta_v, \phi) = \frac{\rho_s}{1 - \rho_s S} \times T(\theta_s) \times T(\theta_v)$$

$$\Rightarrow \frac{\rho^*}{T_g(\theta_s, \theta_v) \times T(\theta_s) \times T(\theta_v)} - \frac{\rho_a(\theta_s, \theta_v, \phi)}{T(\theta_s) \times T(\theta_v)} = \frac{\rho_s}{1 - \rho_s S} \tag{1.11}$$

$$\Rightarrow \frac{\rho^*}{T_g(\theta_s, \theta_v) \times T(\theta_s) \times T(\theta_v)} - \frac{\rho_a(\theta_s, \theta_v, \phi)}{T(\theta_s) \times T(\theta_v)} = \frac{\rho_s}{1 - \rho_s S}.$$

If we define

$$A = \frac{1}{T_g(\theta_s, \theta_v) \times T(\theta_s) \times T(\theta_v)}, \quad \text{and} \quad B = -\frac{\rho_a(\theta_s, \theta_v, \phi)}{T(\theta_s) \times T(\theta_v)},$$

then Equation (1.11) eventually gives

$$\rho_s = \frac{A \times \rho^* + B}{\left[1 + S \times \left(A \times \rho^* + B \right) \right]}. \tag{1.12}$$

Thus, knowing the apparent reflectance ρ^*, the spherical albedo S, and the coefficients A and B, the surface target reflectance ρ_s can be obtained. Methods for obtaining these parameters are illustrated in Section 1.3.2.2.

1.3.2.2 Steps in Atmospheric Correction

The procedure to obtain estimates of the ground target reflectance involves three steps, as illustrated in Figure 1.10. The first step, converting the pixel value to radiance, is sensor-dependent. For instance, in the case of Landsat TM images, the relationship between the pixel values and apparent radiance L_{app} is expressed as

$$L_{\text{app}} = A_i \times DN + B_i. \tag{1.13}$$

In the case of SPOT HRV data, the relationship is

$$L_{\text{app}} = DN/A_i, \tag{1.14}$$

where A_i and B_i are the calibration gain and offset for band i, respectively. These calibration coefficients can be found either from the literature (e.g., Gellman et al., 1993; Thome et al., 1997; Teillet et al., 2001; Meygret, 2005) or from image header files. Note that header files may contain prelaunch calibrations that may differ significantly from the actual calibrations due to sensor degradation over time.

   The second step, conversion from apparent radiance L_{app} to apparent reflectance ρ^*, is based on the observation that in the case of 100% reflectance, the radiance measured by the sensor is the result of the multiplication of equivalent solar radiance (E/π), the cosine of solar zenith angle

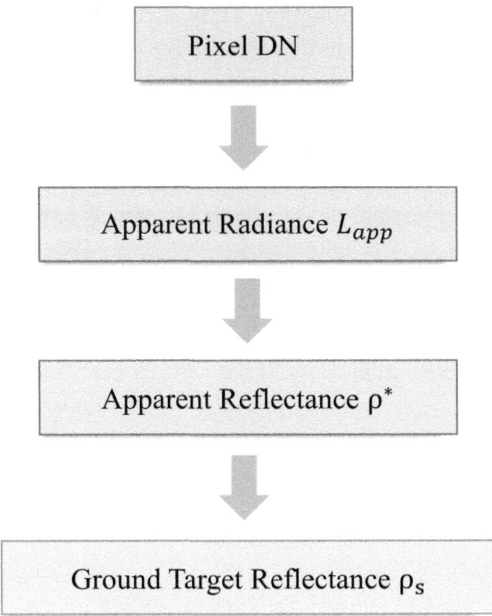

FIGURE 1.10 Procedure for retrieving ground radiance.

$(\cos(\theta_s))$, and the Earth-to-Sun distance multiplicative factor d (Kowalik and Marsh, 1982). The factor d is measured in astronomical units (au) and is described further below. One au is equal to the average Earth-to-Sun distance. On January 3, at perihelion, the Earth-to-Sun distance is ~0.983 au, and on July 5, at aphelion, the Earth-to-Sun distance is about 1.0167 au.

If the required coefficients are known, then the variation in sensor-detected apparent radiance L_{app} is caused by the difference in reflectance. One then obtains the following relation:

$$L_{app} = \frac{\rho^* \times \cos(\theta_s) \times E}{d\pi}$$

or, equivalently,

$$\rho^* = \frac{d \times L_{app}}{\cos(\theta_s) \times E / \pi}. \tag{1.15}$$

The value of the solar zenith angle can be retrieved from the image header file. The distance multiplicative factor d is used to compensate for the variation in solar irradiance E caused by the change in distance between the Sun and the Earth. It is obtained from

$$d = au^2. \tag{1.16}$$

An alternative way to approximate Equation (1.16) is given by

$$d = \left(1 - 0.01673 \times \cos\left(0.9856 \times (JD - 4)\right)\right)^2 \tag{1.17}$$

where the term JD denotes the Julian day (e.g., in the case of February 5, $JD = 36$).

The introduction of the factor d into Equation (1.15) is justified by the following observations. Since at perihelion, the magnitude of solar irradiance is greater than that at aphelion, if a sensor obtains the same radiance L_{app} at perihelion and aphelion, respectively, the apparent reflectance ρ^*

at perihelion should be less than that at aphelion. Thus, if the Sun-to-Earth distance is smaller than the average Sun-to-Earth distance, one should use a lower (<1) weighting factor d, and as the Earth is approaching aphelion, a higher value of d should be used.

The final step – that of converting apparent reflectance to ground target reflectance – uses Equation (1.12). We already know the apparent reflectance ρ^* from Equation (1.15), while other parameters such as the spherical albedo, S, and the coefficients A and B can be obtained by running either the 5S or the 6S model. An example is given in Table 1.1, which shows part of the output generated by the 5S model. The spherical albedo S and relevant parameters for deriving A and B are displayed in bold type. For a detailed description of 5S and 6S usage, readers are referred to Tanré et al. (1986, 1990) and Vermote et al. (1997). Once ground reflectance is retrieved, one may use Equation (1.15) to retrieve radiance followed by using either Equation (1.13) or (1.14) to convert the radiance back to a corrected pixel value.

1.4 CORRECTION FOR TOPOGRAPHIC EFFECTS

Normally, the surface being measured by the remote sensor is assumed to be flat with a Lambertian reflectance behavior. Under this assumption, the magnitude of the radiance detected by the sensor is affected only by variations in the solar zenith angle, the wavelength, and the atmospheric interaction. The atmospheric correction model introduced in Section 1.3 is also based on such ideal assumptions, which may be invalid in the case of rugged terrain because the solar incidence angle will vary with topographic properties and will further contribute to differences in the level of radiance detected by the sensor. This is known as topographic effect. More specifically, the topographic effect can be defined as the variation in radiance exhibited by inclined surfaces compared with radiance from a horizontal surface as a function of the orientation of the surface relative to the radiation source. Moreover, if we assume non-Lambertian reflectance for the surface being measured, the sensor position is another important variable that should be considered.

TABLE 1.1

Results from the 5S Model Used for Atmospheric Correction (see the terms in Equations 1.10 and 1.11 for bold values)

	Downward	Upward	Total
Global gas transmission	0.961	0.965	$\mathbf{0.932} = T_g(\theta s, \theta_v)$
Water transmission	0.988	0.989	0.980
Ozone transmission	0.975	0.978	0.954
CO_2 transmission	1.000	1.000	1.000
O_2 transmission	0.999	0.999	0.998
NO_2 transmission	1.000	1.000	1.000
CH_4 transmission	1.000	1.000	1.000
CO transmission	1.000	1.000	1.000
Rayleigh scattering transmission	0.967	0.971	0.939
Aerosol scattering transmission	0.953	0.962	0.917
Total scattering transmission	0.922	0.934	$\mathbf{0.861} = T(\theta_s) T(\theta_v)$

	Rayleigh	Aerosols	Total
Spherical albedo	0.048	0.087	$\mathbf{0.123} = S$
Optical depth in total	0.054	0.362	0.416
Optical depth plane	0.054	0.362	0.416
Atmospheric reflectance	0.019	0.017	$\mathbf{0.037} = \rho_a(\theta s, \theta v, \varphi)$
Phase function	0.993	0.099	0.215
Single-scattering albedo	1.000	0.990	0.992

Calibration for topographic effects is intended to normalize the sensor-detected signal difference caused by the topographic variation. Various techniques (e.g., Smith et al., 1980; Colby, 1991; Dymond and Shepherd, 1999; Riano et al., 2003; Law and Nichol, 2004) have been published. Here, we present two approaches, using band ratioing and the Minnaert model, due to their better performance (Smith et al., 1980; Colby, 1991; Law and Nichol, 2004).

Band ratioing (Colby, 1991; Mather and Koch, 2011) is the most commonly used method for reducing the topographic effect, as well as extracting specific information about Earth's surface features, including vegetation, hydrological, and lithological structures. Colby (1991) uses a Landsat TM band 5/4 ratio image to study the effectiveness of topographic effect calibration by comparing several sample sites having the same vegetation cover but with differences in topography. In other words, the same vegetation cover should have a similar spectral response, irrespective of location; thus, the differences between sample sites are assumed to be caused by topographic effects. Results showed that the variance of spectral response between sample sites in ratio image was lower than that obtained from original TM band 4 and 5 images. Thus, it was deduced that band ratioing partially compensated for the topographic effects. Furthermore, Holben and Justice (1981) conclude that spectral band ratioing reduced the topographic effect by more than a factor of 6 (i.e., 83%) on their radiance data sets. On the other hand, minimizing the topographic effect permits hydrothermally altered mineral collections to be mapped and classified. For instance, Odunuga et al. (2023) used different band ratios of Landsat 8 to effectively extract sites including ferric iron oxides, ferrous iron oxides, and hydroxyl mineral deposits.

Smith et al. (1980) presented two empirical photometric functions for studying the effect of topography on the radiance field. The first such function, also known as the cosine model, is based on a Lambertian reflectance assumption, while the second function assumes that reflectance is non-Lambertian. Although this is a relatively old model, it generates quite robust calibration results (Colby, 1991; Law and Nichol, 2004).

Figure 1.11 shows the geometric relationships among the Sun, the sensor, and an arbitrary surface element. The Lambertian model assumes that the surface reflects the incident radiation uniformly in all directions. If we treat wavelength as a constant and ignore atmospheric interactions, the variation in radiance detected by the sensor is mainly caused by the local incidence angle

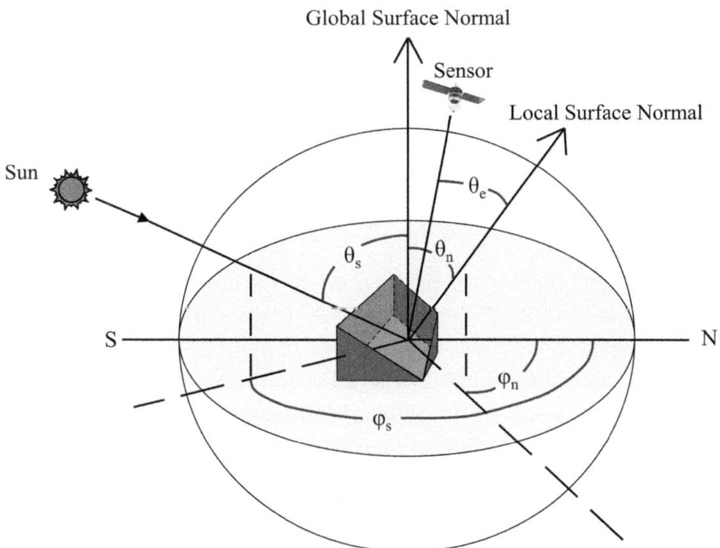

FIGURE 1.11 Geometrical relationships among the Sun, the sensor, and the target position (Smith et al., 1980).

θ_i (i.e., the angle formed between the solar radiation path and local surface normal). In this case, sensor-detected radiance L can be normalized in terms of

$$L_n = L / \cos(\theta_i), \tag{1.18}$$

where L_n denotes the normalized radiance. The $\cos(\theta_i)$ term can be derived from the spherical law as follows:

$$\cos(\theta_i) = \cos(\theta_s)\cos(\theta_n) + \sin(\theta_s)\sin(\theta_n)\cos(\phi_s - \phi_n) \tag{1.19}$$

where θ_s is the solar zenith angle, θ_n is the angle of slope of the terrain surface, ϕ_s is the solar azimuth angle, and ϕ_n is the surface aspect of the slope angle (see Figure 1.11). A slope is defined as a plane tangent to the surface containing two components: one is gradient, which specifies the rate of change in elevation, and the other is aspect, which measures the direction of the gradient. The values of several parameters are required to solve these equations, namely, solar zenith angle θ_s, solar azimuth angle ϕ_s, the surface slope θ_n, and aspect ϕ_n. Both θ_s and ϕ_s can be obtained from the image header file, while slope and aspect can be derived by co-registering the image with a DEM. A variety of approaches can be employed to calculate the slope and aspect from a DEM (Skidmore, 1989; Jones, 1998).

The correction may not be effective in steep terrain where incidence angles are near 90° (Teillet et al., 1982). In the case of the non-Lambertian reflectance assumption, Smith et al. (1980) suggest the following function including the Minnaert correction to correct for the topographic effect:

$$L \times \cos(\theta_e) = L_n \times \left[\cos(\theta_i)\cos(\theta_e)\right]^k \tag{1.20}$$

where θ_e is the effective view angle (Figure 1.11), $\cos(\theta_i)$ is defined in Equation (1.19), and k is known as the Minnaert constant (Minnaert, 1941) describing the bidirectional reflection distribution function of the surface, the type of scattering dependence, and surface roughness, respectively. A Lambertian surface is defined by a k value of 1.0, and Equation (1.20) then reduces to Equation (1.18).

To solve Equation (1.20), one needs to relate the effective view angle θ_e and k (the method for deriving $\cos(\theta_i)$ is described in Equation (1.19)). If Landsat TM or MSS imagery is used, θ_e can be regarded as the same as θ_n (i.e., the slope of the terrain surface) since Landsat has a narrow view angle. In the case of SPOT HRV data, which can acquire imagery through an angle of ±27°, θ_e should be set to $\theta_e = \theta_n$, the satellite view angle. To estimate the Minnaert constant k, Equation (1.20) is converted into logarithmic form as

$$\log(L \times \cos(\theta_e)) = k \times \log(\cos(\theta_i)\cos(\theta_e)) + \log(L_n) \tag{1.21}$$

The term k is then equal to the slope of the regression line of the plot made by the samples $\log(\cos(\theta_i)\cos(\theta_e))$ plotted on the x-axis and $\log(L\cos(\theta_e))$ plotted on the y-axis.

To calibrate for the topographic effect using a non-Lambertian assumption is more complicated than that based on a Lambertian reflectance assumption. As far as the computational cost and calibration accuracy are concerned, Smith et al. (1980) suggest that when surface slopes are <25° and effective illumination angles are <45°, then the Lambertian assumption is more valid. Under such circumstances, one can use Equation (1.18) to carry out topographic effect correction, and the calibration accuracy should be preserved. If either of the above conditions is not satisfied, the use of Equation (1.20) is recommended. It should be mentioned that the Minnaert method does not consider diffuse illumination. It is reported that correction of the image data for terrain illumination effects produces superior classification performance (Parlow, 1996; Vanonckelen et al., 2013; Dong et al., 2020).

It should also be appreciated that surface topographic variations will also cause distortions in the geometry of images. The map to which the image is referenced represents the relationship between features reduced to some datum such as sea level, while the image shows the actual terrain surface. If the terrain surface is significantly above sea level, then the image pixel position will be displaced by an amount proportional to the pixel's elevation above sea level (or whatever datum is used). In an extensive comparative study considering 20 topographic correction and 11 atmospheric correction algorithms, Vanonckelen et al. (2013) concluded that the influence of topographic correction was higher than atmospheric correction on the classification accuracy.

1.5   REMOTE SENSING IN THE MICROWAVE REGION

The word radar is an acronym derived from the phrase "Radio Detection and Ranging". Imaging microwave sensors are known as imaging radars. These instruments transmit a signal in the wavelength range of ~3 cm to 1.33 m and receive reflection (backscatter) from the target. The level of backscatter for each pixel over the imaged area is recorded, and the set of pixels forms the radar image. Remote sensing in the microwave region differs from optical remote sensing in a number of ways, the most important of which are as follows:

1. Radar backscatter is related to the roughness and electrical conductivity of the target. This information is complementary to that which is acquired by optical and thermal sensors.
2. Energy in the microwave region can penetrate clouds, mist, and water vapor.
3. Microwave imaging radars are active, not passive, instruments, and thus can operate independently of solar illumination, enabling acquisition day and night.

An increasing number of spaceborne radar systems are now in orbit, and a revolutionary development took place in 2007 in resolution side with the launch of the German satellite TerraSAR-X and the Italian COSMO-SkyMed. These satellites made 1-m spatial resolution possible in radar imaging. By 2018, the Capella X-band radar mission was launched, and 0.25-m azimuth resolution was achieved in radar imaging. Considering these developments, it is probable that radar imagery will play an increasingly important role in supporting our understanding and monitoring of our environment. The main disadvantage of active microwave systems vis-à-vis optical systems is their power requirements, for the sensor transmits and receives energy. Optical sensors passively detect reflected solar radiation. Passive microwave sensors work on a similar principle to passive optical systems operating in the thermal region and detect microwave radiation produced by the target. They are used to determine properties such as surface temperature and humidity by recording the microwave energy emitted by objects. In passive microwave, radiometers and scanners are used, not antennas.

In the following sections, active radar sensing will be discussed. In Section 1.6, basic ideas underlying the use of radar images, including geometric effects and the main factors affecting surface reflection or backscatter at radar wavelengths, are introduced. Section 1.7 considers the extraction of surface information from backscattered signals. One of the main problems associated with the interpretation of radar imagery is the presence of noise or radar speckle. The use of filters to reduce the noise effect is described in Section 1.8.

1.6   RADAR FUNDAMENTALS

An active radar system repetitively transmits short microwave energy pulses (normally of the order of microseconds, i.e., 10^{-6} second, denoted by μs) toward the area to be imaged. The energy pulse is likely to scatter in all directions when it reaches the surface. Part of the backscattered energy is reflected back to the radar antenna and recorded for later processing. Normally, each energy pulse has a duration of between 10 and 50 μs and utilizes a small range of microwave wavelengths. A waveform can be characterized in terms of its wavelength and amplitude. Wavelength is defined as the distance between two adjacent crests or troughs of the waves (Figure 1.12). Amplitude measures the strength of an electromagnetic wave in terms of the maximum distance achieved by the waveform relative to the mean position (shown by the horizontal line in Figure 1.12). The amplitude may be a function of a complex signal including both magnitude and the phase; this point is discussed further below.

Frequency, rather than wavelength, can also be used to describe wavebands. Frequency is the number of oscillations per unit time or number of wavelengths that pass a point per unit time. One can obtain the frequency f (normally of the order of gigahertz, GHz) corresponding to wavelength λ in terms of

$$f = \frac{c}{\lambda} \tag{1.22}$$

where c is the velocity of light (3×10^8 m/s).

The description of radar operation in the following pages is based on the most widely used radar system installed on aircraft platforms, the side-looking airborne radar (SLAR). Spaceborne imaging radars operate on a similar basis, but use a synthetic rather than a real antenna. They are known as synthetic aperture radars (SAR).

1.6.1 SLAR Image Resolution

SLAR transmits and receives microwave energy using an antenna located on the side of the platform. The area imaged by the sensor is thus a strip of ground parallel to the flight track (known as the azimuth direction). SLAR image resolution is mainly dependent on pulse duration and antenna beam width, which is the ground area "illuminated" by the radar pulse at a given instant in time (Figure 1.13). Pulse duration affects the resolution in the range (cross-track) direction, while antenna

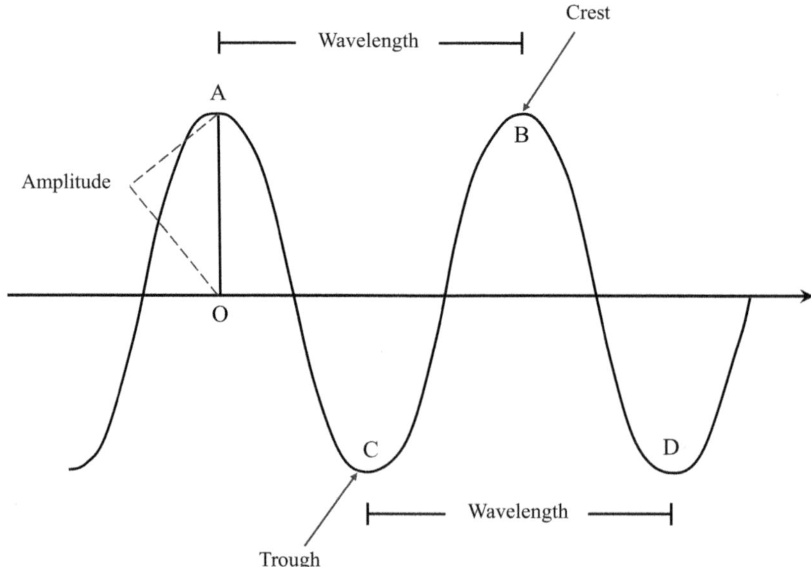

FIGURE 1.12 Wavelength is the distance between two adjacent crests (e.g., A and B) or troughs (e.g., C and D). The amplitude of a waveform (distance AO) measures the "power" or information carried by the wave.

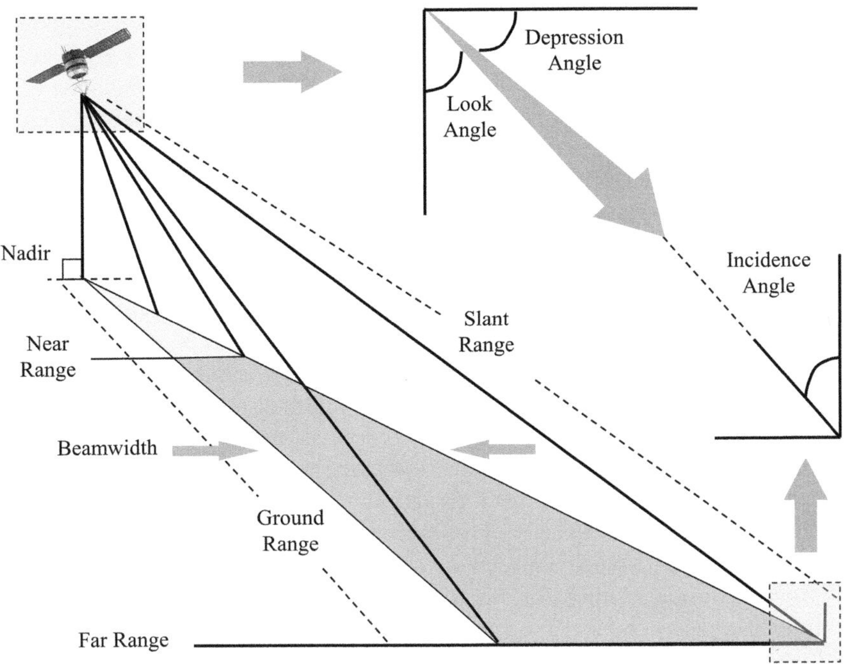

FIGURE 1.13   Some basic parameters of a SLAR system.

beam width controls the azimuth (along-track) resolution. The ground range resolution and azimuth resolution are computed by

$$\text{Range resolution} = \frac{c\tau}{2\cos\theta} \tag{1.23}$$

$$\text{Azimuth resolution} = \beta \times d \tag{1.24}$$

The term c is defined in Equation (1.22), while τ is the pulse duration, θ is the depression angle defined as the angle between the horizontal plane and the direction of emitted microwave energy (Figure 1.13), d is the ground range, and β is the antenna beam width. Range resolution can also be analyzed in terms of incidence angle or look angle (Figure 1.13). Incidence angle is defined as the angle between the radar beam and a line perpendicular to the illuminated surface. The look angle is complementary to the depression angle. If the illuminated surface is assumed to be flat, then one can also regard the incidence angle as the complement of the depression angle. The antenna beam width β, physical antenna length L, and wavelength λ are related as follows:

$$\beta = \frac{\lambda}{L}. \tag{1.25}$$

The combination of azimuth and range resolution determines the ground resolution of each pixel on a radar image.

It can be inferred from Equation (1.23) that the shorter the pulse duration τ, or the smaller the value of θ, the finer the range resolution. The depression angle θ varies across an image. The value of θ in the near range is relatively larger than that in the far range (Figure 1.13). Thus, the ground range resolution will also vary with respect to θ.

Equation (1.24) shows that the smaller the values of β and d, the finer the azimuth resolution will be. Thus, the near-ground range has a higher resolution than the far-ground range because both are smaller in the near range than in the far range (Figure 1.13). According to Equation (1.25), one can use a long antenna length and short wavelength to obtain finer azimuth resolution. However, shorter wavelengths are more likely to be affected by the atmosphere, and furthermore, antenna length is constrained by physical limitations. For instance, to obtain an antenna beam width of 1×10^3 m at a wavelength of 20 cm (L-band), the required physical antenna length will be 200 m (Equation 1.25). Clearly, if finer azimuth resolution is sought in terms of increasing antenna length, then serious practical difficulties will be encountered. An alternative strategy is to use a SAR, where the term aperture means the opening used to collect the reflected energy that is used to generate an image. In the case of radar, this opening is the antenna, while in the case of a camera, the opening is the shutter opening.

SAR increases the antenna length not in physical terms but by synthesizing a long antenna using the forward motion of a short antenna, a process that requires more complicated and expensive technology. SAR uses the Doppler principle in order to synthesize a longer antenna. The Doppler effect is the change in wave frequency as a function of the relative velocities of the transmitter and reflector. A radar sensor can image a given target repeatedly from successive locations, as illustrated in Figure 1.14. Here, the frequency of the waveform reflected by the target will increase from location a to b because the distance between the sensor and the object is reducing. As the platform moves away from the target, from b to c, the frequency of the returned signal decreases. SAR uses the Doppler information to compute frequency shifts and thus determine the location and scattering properties of the target.

1.6.2   Geometric Effects on Radar Images

A radar image is generated from the timing data for transmitted energy to be returned to the radar antenna. This timing delay is dependent on the distance between the radar antenna and the target. This distance is the slant range (Figure 1.13), which is the path along which the microwave energy

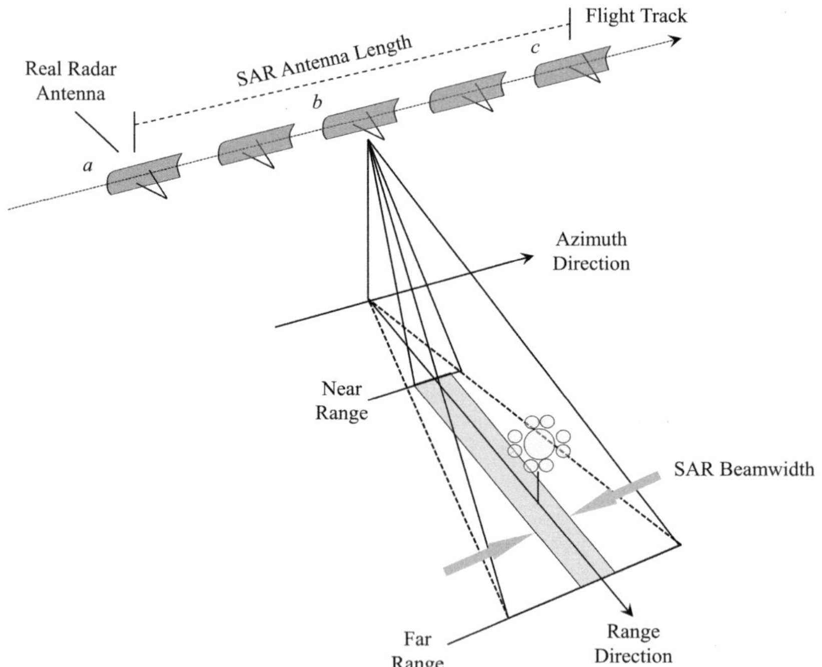

FIGURE 1.14 Concept of the synthetic aperture.

travels. Therefore, every target located on the terrain being observed by the radar will be mapped onto the slant range domain. Because of this slant range mapping, radar imagery is likely to be affected by geometric distortions. The most common distortions are those of layover, foreshortening, and shadow.

The layover effect results when the top of an illuminated target is seen by the radar as the bottom, and the bottom of the target is recorded by the radar as the top. This phenomenon occurs when the time for the microwave energy to travel from the antenna to the top of an object is less than the time needed to travel to the bottom of the same object. Figure 1.15 shows two targets (a building and a mountain) that are illuminated by a radar sensor. The microwave energy transmitted by the radar will reach the tops of both objects (points a and b in Figure 1.15) before the bottoms (points c and d). The antenna will first receive the reflected energy from a and b, then, sometime later, the energy reflected from c and d. After projection onto the slant range domain, the result is called the layover effect.

It might be inferred from the preceding discussion that the higher the isolated target, the greater the layover effect. However, layover is also controlled by another important factor: the angle, θ_f, between the front of the target and the energy path (Figure 1.16). Layover will occur only if θ_f exceeds 90° (see object A in Figure 1.16). If θ_f is smaller than 90°, as in the case of object B in Figure 1.16, then microwave energy will first illuminate the bottom, then the top of the object, and there will be no layover effect.

Foreshortening, like layover, results from the fact that radar is a side-looking sensor. The object labeled B in Figure 1.16 is symmetrical in cross-section, but the angle between its front slope ab and the microwave radiation emitted by the instrument is <90°. Hence, the front slope distance ab appears to be less than the back slope distance bc when projected onto the slant range. Since the front slope also tends to reflect microwave energy more strongly than does the back slope, it will appear to be brighter, steeper, and shorter, while the back slope is shallower and darker. The darker back slope demonstrates another radar image geometry effect, that of shadow, as illustrated in the following text.

Radar shadow is due to the returned energy from targets being affected by the nature of the terrain. A radar image is effectively a representation of returned energy levels plotted against the time

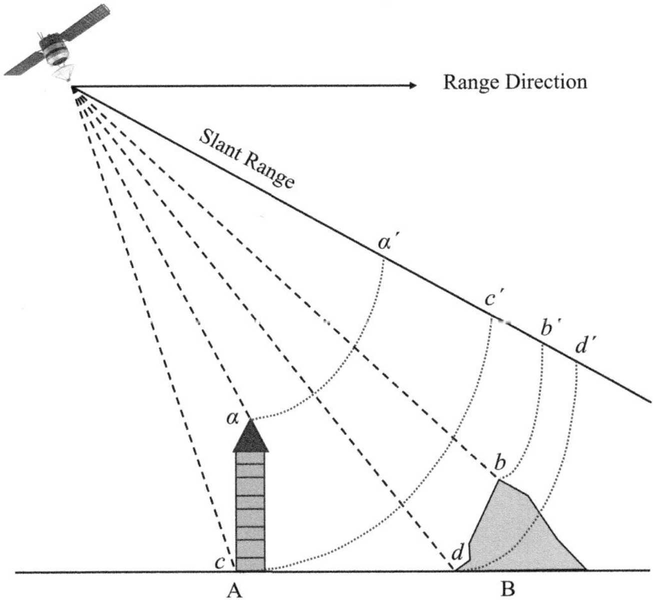

FIGURE 1.15 Layover effect in radar remote sensing.

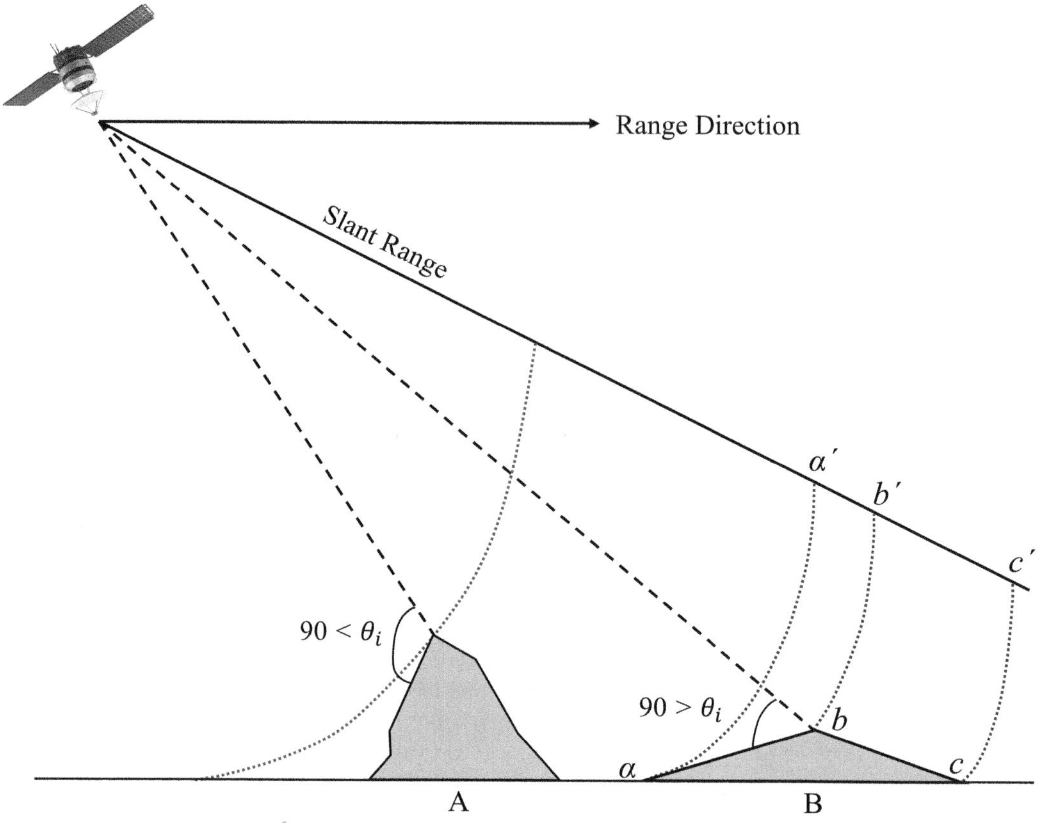

FIGURE 1.16 The angle θ_f controls layover effects (see the text for discussion).

taken for the energy to travel to and from the target. It follows that if, during a certain period, the antenna receives no reflection, then the image area corresponding to this time period will contain zero (dark) values.

The effect of radar shadow is controlled by the target height and angle θ_b (Figure 1.17), which is defined as the angle between the back slope of the target and the horizontal line parallel to the range direction. In Figure 1.17, the angle θ_b of object A is smaller than the corresponding depression angle θ_1. Thus, the back slope of object A is illuminated by the microwave energy. However, since the angle θ_b of object B is larger than the corresponding depression angle θ_2, the radar antenna will not receive any reflection from the back slope of object B, and this period of zero reflection is likely to continue until point a is reached. The resulting radar shadow after projection onto the slant range is also illustrated in Figure 1.17.

The description of radar image distortions given above is a simplification. In real-world applications where terrains are often continuously sloping and rapidly varying, the resulting image will be a combination of different geometric effects. Hence, in order to compensate for these effects, one has to make careful case-by-case analyses (Kropatsch and Strobl, 1990; Goyal et al., 1999). That is, if one knows what effects are occurring at a given pixel, then one can use suitable algorithms to carry out calibration. As in the case of topographic calibration of optical imagery described above, geometric and radiometric correction of radar images requires a co-registration to a DEM. However, the calibration procedures are generally more complicated (Loew and Mauser, 2007; Small et al., 2007; Shimada et al., 2009; Yang et al., 2018; Li et al., 2022).

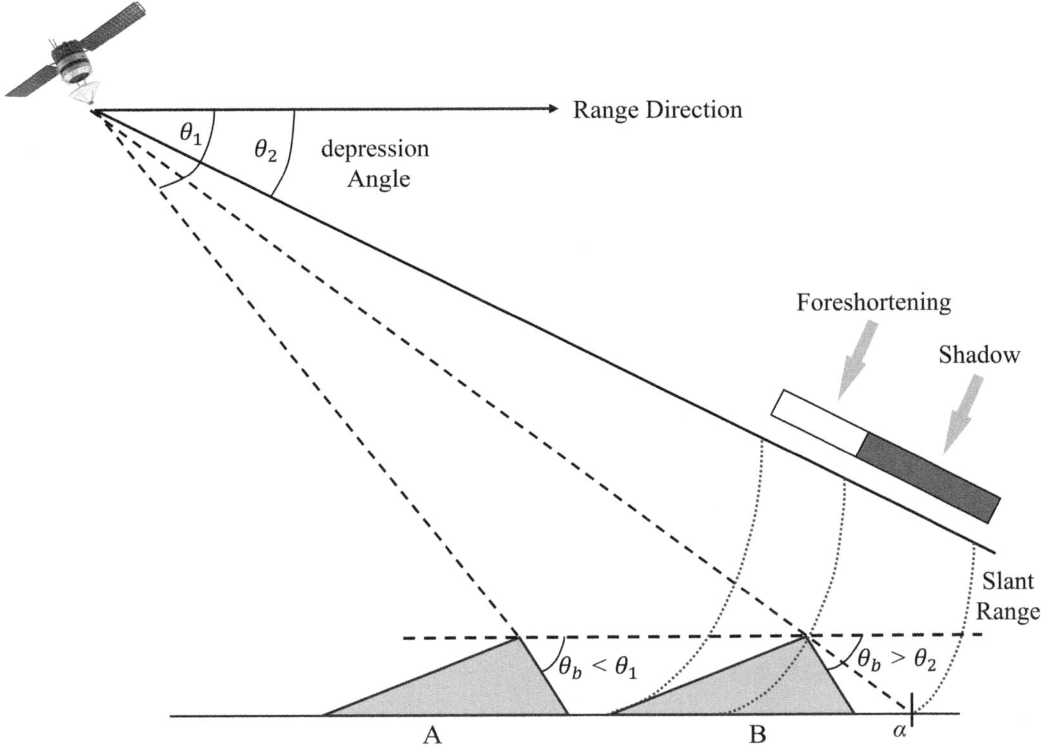

FIGURE 1.17 Showing the relationship between angle θ_j and radar shadow.

1.6.3 Factors Affecting Radar Backscatter

It has already been noted that a radar image is a record of the strength of the backscatter from the targets making up the imaged area. The stronger the backscatter, the brighter the corresponding image element. The level of backscatter is determined by terrain conditions (such as roughness and electrical characteristics), and also by the parameters of the radar system. Understanding the factors affecting radar backscatter can help analyze landscape properties more knowledgeably.

1.6.3.1 Surface Roughness

Radar backscatter is stronger where the ground surface is rough relative to the radar wavelength. The roughness of a surface is dependent on both the wavelength of the incident energy and the angle of incidence. Rough surfaces act as Lambertian reflectors (Section 1.4) so that incident microwave energy is scattered in all directions and a portion is reflected back to the radar antenna. Smooth surfaces are specular, in that they act like a mirror and reflect the incident energy away from the sensor, resulting in extremely weak backscatter (Figure 1.18, point *a*). Specular reflection is most often seen in stagnant water formations such as harbors, dams, and lakes and causes considerable vertical accuracy loss in SAR DEMs (Sefercik and Alkan, 2009). Normally, as the wavelength decreases, the surface appears rougher because smaller facets of the surface contribute to the scattering process, and thus, stronger backscatter results. Likewise, as the wavelength increases, the surface tends to appear smoother. The strength of the backscatter is also affected by the incidence angle. For a given wavelength, as the incidence angle increases, backscatter becomes weaker, and so the illuminated surface appears smoother. Some ground objects can behave like corner reflectors, which can reflect high energy back to

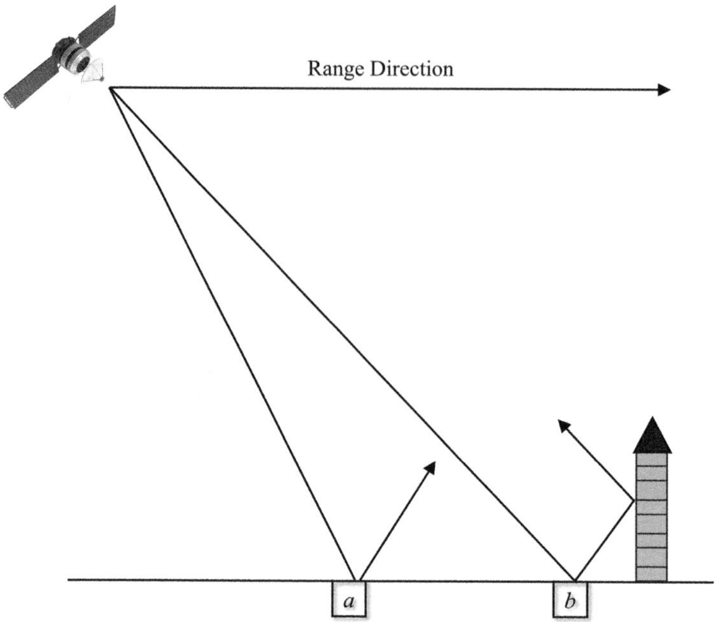

FIGURE 1.18   Microwave energy impinging on a smooth surface (point *a*) at which energy is reflected away from the sensor, and exhibiting a double bounce at point *b* when the reflected energy is reflected by a vertical wall.

the antenna and cause bright spots on the radar image. Such phenomena often occur in urban areas because energy can be returned by means of a double bounce from the corners of buildings (Figure 1.18, point *b*).

1.6.3.2   Surface Conductivity

Highly conductive ground surfaces tend to have higher reflectivity than surfaces with lower conductivities. Water and metal are good conductors. As a result, radar backscatter will be sensitive to metal objects and to the presence of moisture in the illuminated target area, even though the amount of moisture may be small. In a radar image, metal objects such as railway tracks and metal bridges generally result in bright spots. Moisture also affects the depth of microwave energy penetration of the soil surface. If soil contains a large amount of moisture, the signal does not penetrate the soil surface and is reflected back to the radar antenna. If the soil is dry, then the radar signal can penetrate more deeply into the soil surface layer. Wavelength is also another control on the depth of penetration. Lakes and other water bodies might be expected to exhibit high backscatter, but, in fact, the surfaces of rivers and lakes are generally smooth relative to radar wavelengths and act as specular reflectors. The ocean surface is generally rougher, and therefore, the magnitude of backscatter depends on the sea state and on wavelength and depression angle.

1.6.3.3   Parameters of the Radar Equation

The parameters of the radar equation (Van Zyl et al., 1993) are fundamental factors that influence the level of the returned signal. The radar equation is expressed as

$$P_r = \frac{P_t \lambda G_t(\gamma) G_r(\gamma)}{(4\pi)^3 R^4} \sigma^0 A.$$

(1.26)

P_t is the transmitted power from the antenna, λ is the transmitted wavelength, R is the distance to the imaging area, γ is the radar look angle, A is the area on the ground responsible for scattering, G_t and G_r are the transmitted and received antenna gains (describing the system's ability to focus on the transmitted microwave energy) at look angle γ, and σ^0 is the radar backscatter coefficient measured in decibels (dB). All of the parameters in Equation (1.26) affect the received power P_r. However, only σ^0 is related to the properties of the illuminated surface. Thus, the quantized pixel values (0–255) in a radar image are sometimes converted to σ^0 before being interpreted. The received power P_r in Equation (1.26) can also be characterized in terms of other parameters such as the scattering matrix (Equation 1.31) and the coefficient of variation (Equation 1.35), which also relate to the properties of surface objects, and can be used for classification purposes. A discussion of the scattering matrix is presented in the next section.

1.7 IMAGING RADAR POLARIMETRY

The polarimetry theory presented in this section is mainly derived from Evans et al. (1988), Zebker and Van Zyl (1991), Zebker et al. (1987, 1991), Van Zyl et al. (1987, 1993), Kim and Van Zyl (2000), and Hellmann (2002). Knowledge of radar polarimetry enables us to use a variety of features (such as complex-format data, the elements of the scattering matrix, and the coefficient of variation of polarization signature) to perform image interpretation. Some basic concepts are described first.

An electromagnetic wave, besides being described in terms of wavelength and amplitude, can also be characterized using a complex number format (a complex number consists of two components, termed the real and imaginary parts). When the coordinate system is translated into both the real and imaginary axes (Figure 1.19), the wave is described by two parameters, namely, the in-phase I ($I = m \times \cos \varphi$) and quadrature Q ($Q = m \times \sin \varphi$) components. Both the I and Q parameters provide the wave's overall phase φ ($\varphi = \tan^{-1} (Q/I)$) and magnitude m ($m = (I^2 + Q^2)^{0.5}$) as illustrated in Figure 1.19. The radar phase represents the degree of coincidence in time between a repetitive radar signal and a reference signal having the same frequency. Over the complex domain, the amplitude a is expressed as $a = I + iQ$, where $I = (-1)^{0.5}$. The relationship between the magnitude m and a complex

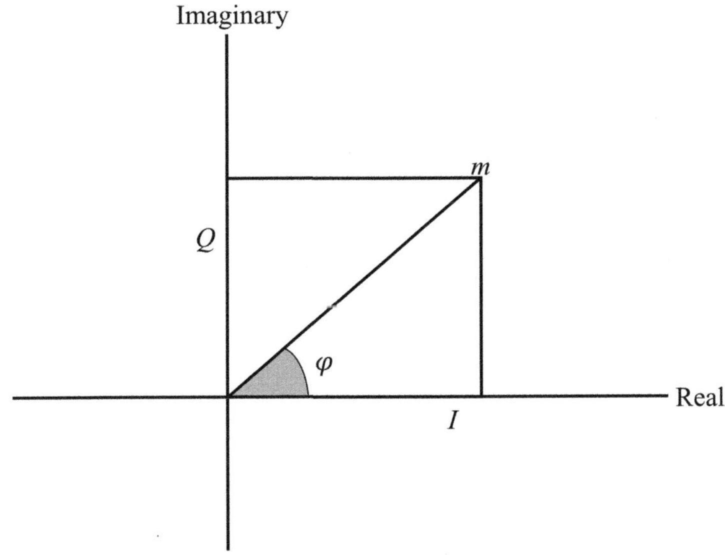

FIGURE 1.19 Wave described by complex (real and imaginary) coordinates.

amplitude a, by definition, can be expressed as $m = |a|$, that is, the absolute value of amplitude. The radar image can thus be formed by using any of the m, I, or Q components. As a result, this kind of radar image is said to be represented in a complex format.

Complex-format SAR images can be used to generate digital surface models (DSMs), classify land cover, and monitor disasters and long-term surface deformations. While interferometric SAR (InSAR) and radargrammetry techniques are used for DSM generation, coherence maps of master-and-slave SAR images are used for land cover classification (Crosetto and Pasquali, 2008; Waske and Braun, 2009; Capaldo et al., 2012; Sefercik, 2013). For disaster and long-term surface deformation monitoring, differential InSAR (DInSAR) and multitemporal DInSAR (MT-DInSAR) techniques, which utilize the phase difference between multiple SAR image acquisitions along the satellite line of sight (LOS), are applied (Stramondo et al., 2010; Lavecchia et al., 2016; Corsa et al., 2022). While the traditional DInSAR technique is preferred to monitor the effects of a sudden disaster such as an earthquake or volcanic eruption by generating interferograms with pre- and post-disaster SAR image processing, MT-DInSAR techniques are more sensitive to monitor millimetric surface deformations that cause disaster in the long term, such as a landslide or urban/mining subsidence. The most preferred MT-DInSAR techniques are Persistent Scatterer Interferometry (PSI) and Small BAseline Subset (SBAS) (Ferretti et al., 2000; Berardino et al., 2002; Casu et al., 2006; Crosetto et al. 2016).

1.7.1 Radar Polarization State

Normally, microwave energy transmitted and received by a radar antenna can travel in all directions perpendicular to the direction of wave propagation. However, most radar systems polarize microwaves in such a way that the transmitted and received waves are restricted to a single plane perpendicular to the direction of wave propagation (Figure 1.20). The polarized wave is therefore transmitted and received in either the horizontal (H) or the vertical (V) plane. Consequently, there are four combinations of transmission and reception for the polarized waves. These combinations are HV, HH, VV, and VH, where HV denotes a wave transmitted in the V direction and received in the H direction. The other combinations can be inferred in a similar manner. Radar imagery generated in terms of HH or VV is called co- or like-polarized imagery, while imagery resulting from HV or VH polarization is called cross-polarized imagery. Cross-polarization detects multiple scattering from the target and thus generally results in weaker backscatter than that measured by a co-polarization configuration. In the next-generation high-resolution SAR missions, such as German TerraSAR-X and Italian Cosmo-SkyMed, quad polarization mode that includes all polarimetric channel combinations (HH, VV, HV, and VH) is available for better object detection. In SAR image processing, the performance of object detection differs according to the polarization mode. For example, in sea surface applications, VV polarization is often preferred due to stronger signal backscatter power, higher SNR, and easier discrimination of outliers from clear seawater (Angelliaume et al. 2018; Sefercik et al., 2024).

The coordinate system shown in Figure 1.20 determines the radar polarimetry, in which horizontally and vertically polarized waves lie in the unit vector $\vec{h}$ and $\vec{v}$ directions, respectively, while unit vector $\vec{z}$ denotes the direction of wave propagation. The relationship among unit vectors $\vec{h}$, $\vec{v}$, and $\vec{z}$ can be represented by

$$\vec{z} = \vec{h} \times \vec{v}. \tag{1.27}$$

The overall electric field can be represented by

$$\vec{E}(z,t) = \Re\left[\left(E_h\vec{h} + E_v\vec{v}\right)e^{-i(\omega t - kz)}\right] \tag{1.28}$$

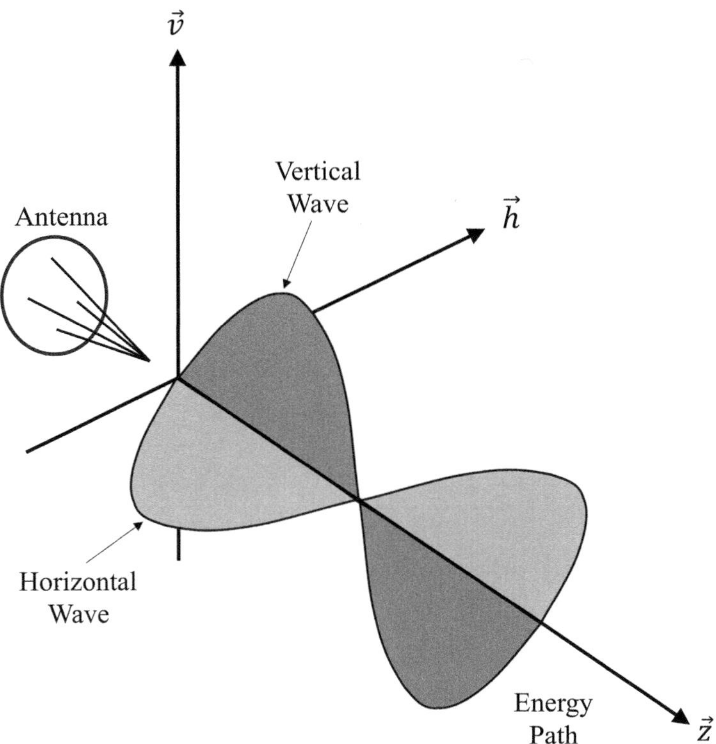

FIGURE 1.20 Polarized microwaves.

where E_h and E_v denote the electric field of the vertical and horizontal plane, respectively, which are expressed by

$$E_h = a_h e^{-i\delta_h}$$

$$E_v = a_v e^{-i\delta_v}.$$

(1.29)

The terms a_h and a_v denote the positive amplitudes in the $\vec{h}$ and $\vec{v}$ directions (Figure 1.21), respectively, and the corresponding phases are δ_h and δ_v relative to the phase factor $\omega t - kz$, where ω is the frequency, t is the time, k is the wave number, and z is the distance traveled in the $\vec{z}$ direction (i.e., the direction of wave propagation).

In the most general case, the electric field vector of a plane monochromatic wave rotates in a plane perpendicular to the direction of microwave energy propagation, and in doing so traces out an ellipse, as shown in Figure 1.21. The wave is said to be elliptically polarized. If one refers to the relative amplitude and phase relationships of the components of a given wave as the elliptic polarization state, Equation (1.28) can be rewritten as

$$\vec{E}(z,t) = a\vec{p}\, \cos\chi\, \sin\left(\omega t - kz + \delta\right) + a\vec{q}\, \sin\chi\, \cos\left(\omega t - kz + \delta\right)$$

(1.30)

where $a^2 = a_h^2 + a_v^2$ is the intensity of the wave, $\delta = \delta_h - \delta_v$ is the phase angle, both $\vec{p}$ and $\vec{q}$ are unit vectors in a coordinate system rotated by angle ψ with respect to $\vec{h}$, and χ is the ellipticity angle. Note that the width of the ellipse is given by the parameter χ so that $\chi = \pm 45°$ results in left- and right-handed circular polarizations, respectively. The orientation parameter ψ determines the orientation of the major axis of the ellipse; if $\chi = 0°$, then the values $\psi = 0°$ or $180°$ represent horizontal polarizations, while $\psi = 90°$ represents vertical polarization.

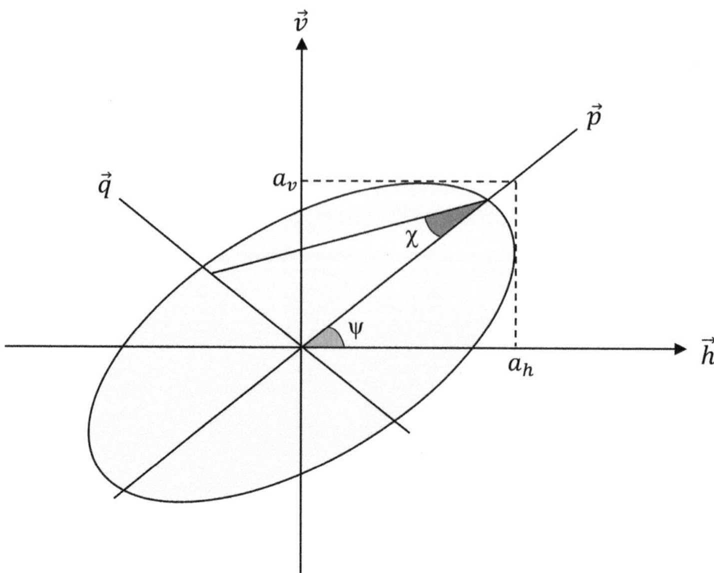

FIGURE 1.21 Elliptical polarization state for polarization synthesis (Evans et al., 1988).

A polarimetric imaging radar measures the magnitude of the backscatter from a target as a vector quantity in such a way that the complex backscattered characteristics of any transmitting and receiving polarization configuration can be determined. Such backscattered characteristics are represented in terms of a scattering matrix [**S**] (Van de Hulst, 1981):

$$[\mathbf{S}] = \begin{pmatrix} \mathbf{S}_{hh} & \mathbf{S}_{hv} \\ \mathbf{S}_{vh} & \mathbf{S}_{vv} \end{pmatrix} \tag{1.31}$$

where element $\mathbf{S}_{hv}$ is determined by measuring both the amplitude and the phase of the electric field where a vertically polarized wave is transmitted and the scattered wave is received in horizontal polarization. The remaining elements are obtained in a similar fashion. The relationship among the electric field E_s of the scattered wave, the electric field E_t of the transmitting wave, and [**S**] is expressed as

$$E_s = \frac{e^{ikr}}{kr}[\mathbf{S}]E_t \Rightarrow \begin{pmatrix} E_h \\ E_v \end{pmatrix}_s = \frac{e^{ikr}}{kr} \begin{pmatrix} \mathbf{S}_{hh} & \mathbf{S}_{hv} \\ \mathbf{S}_{vh} & \mathbf{S}_{vv} \end{pmatrix} \begin{pmatrix} E_h \\ E_v \end{pmatrix}_t \tag{1.32}$$

where r is the distance between the scatterer and the receiving antenna, and k denotes the wave number of the illuminating wave. The scattering matrix thus describes how the ground scatterer transforms the illuminating electric field.

1.7.2 POLARIZATION SYNTHESIS

Knowledge of the scattering matrix [**S**] permits the calculation of the received power P_r for any possible combination of transmitted and received antenna polarizations. This process is called polarization synthesis. The observed power P_r can be derived by evaluating the matrix equation.

$$P_r = K(\lambda, \theta, \phi)\left|E_s[S]E_t\right|^2 \tag{1.33}$$

where

$$K(\lambda,\theta,\phi) = \frac{1}{2}\frac{\lambda^2}{4\pi}\sqrt{\frac{\varepsilon_0}{\mu_0}}\frac{g(\theta,\phi)}{|E_s|^2} \tag{1.34}$$

where $g(\theta,\phi)$ is the antenna gain function, θ and ϕ are the angles between the radar local coordinate system and the scatterer-centered global coordinate system shown in Figure 1.22, $[(\lambda^2/4\pi) \times g(\theta,\phi)]$ is the effective area of the antenna, and ε_0 and μ_0 are the permittivity and permeability of free space (i.e., transmission medium), respectively. Note that both permittivity ε_0 and permeability μ_0 determine the propagation velocity v (or phase velocity) of the wave; the relationship can be expressed as $v = 1/(\varepsilon_0 \times \mu_0)^{0.5}$.

Polarization synthesis can also be expressed in terms of either the Stokes matrix or the covariance matrix. Both of these representations consist of linear combinations of the cross-products of the four basic elements of the scattering matrix. The entries of these matrices can also provide a variety of features for classification purposes.

1.7.3 POLARIZATION SIGNATURES

A particular graphical representation of the variation of received power (or cross-section σ) as a function of polarization is called the polarization signature of an object. A polarization signature is a three-dimensional representation consisting of a plot of synthesized scattering power as a function of the ellipticity and orientation angles (i.e., χ and ψ in Figure 1.21) of the transmitted and received wave. Normally, analyses are based on only two types of polarization signatures, namely, co-polarization and cross-polarization.

Figure 1.23 illustrates the polarization signature based on a theoretical model of a large conducting sphere. Note that at the ellipticity angle $\chi=0°$, the co-polarization signature reaches its highest value of cross-section, while the cross-polarization signature shows the lowest value at the

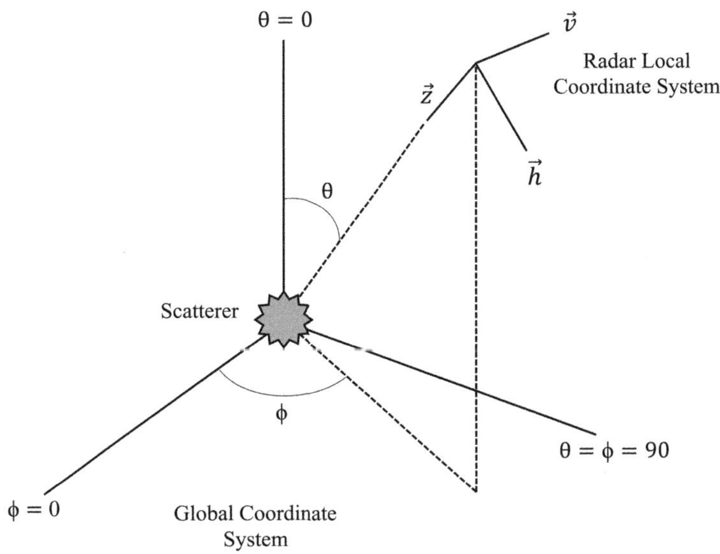

FIGURE 1.22 Local and global coordinate systems determine the angles θ and ϕ.

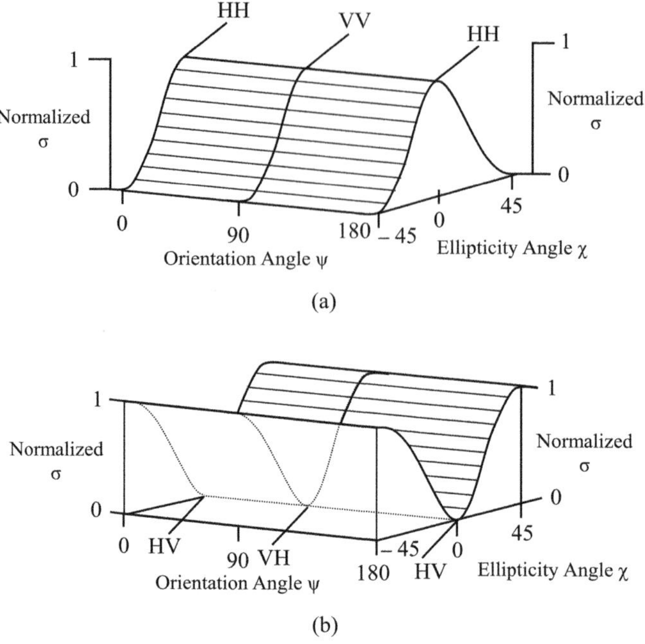

FIGURE 1.23 Polarization signature based on a theoretical model of a large conducting sphere. (a) At an ellipticity angle 0°, the co-polarization signature reaches its highest cross-section value. (b) Cross-polarization value reaches its minimum point at the same ellipticity angle.

same point. There is a measurement of surface roughness in accordance with polarization signature called the coefficient of variation (CoV), which is defined as

$$\text{CoV} = \frac{P_{r\,\text{min}}}{P_{r\,\text{max}}}. \tag{1.35}$$

$P_{r\,\text{min}}$ and $P_{r\,\text{max}}$ each denote minimal power and maximal power occurring within a polarization signature, respectively. Since CoV relates to the surface roughness, it can be used as a discriminating feature in classification. As the value of CoV increases, the measured surface tends to be rougher. The concept of CoV is based on the following observations. The polarization signature for each resolution element represents the sum of the polarization signatures of many individual measurements. If the surface being measured is smooth, the scattering mechanisms from a group of scatterers should be identical. Therefore, the maxima (minima) of a scattering mechanism should coincide with the maxima (minima) of the other scattering mechanisms. When the composite polarization signature is derived, it will produce a composite signature in which there is a large difference in magnitude between maximal and minimal backscatter, and thus, the polarization signature will result in more peak- and valley-like shapes. As a result, the value of the CoV will be small (i.e., closer to 0). Conversely, if the measured ground surface is rough, several different scattering mechanisms may result, the backscatter maxima and minima may occur together from different individual scatterers, and a relatively flat polarization signature shape will be produced (equivalently, CoV will be large, i.e., close to 1).

1.8 RADAR SPECKLE SUPPRESSION

Due to random fluctuations in the signal observed from a spatially extensive target represented by a pixel (or image resolution element), speckle noise is generally present on a radar image. Speckle has the characteristics of a random multiplicative noise (defined below) in the sense that as the average gray level of a local area increases, the noise level increases. In a SAR imaging system, speckle effects are more serious (Lopez-Martinez and Fabregas, 2003; Singh and Shree, 2016). SAR can achieve high resolution in the azimuth direction independent of range, but the presence of speckle decreases the interpretability of the SAR imagery. If such imagery is to be used in classification, then some form of preprocessing to reduce or suppress speckle is necessary. There are two approaches to the suppression of the radar image speckle. The first method is known as the multilook process, while the second method uses filtering techniques to suppress the speckle noise.

1.8.1 Multilook Processing

Radar speckle can be suppressed by averaging several looks (images) to reduce the noise variance. This procedure is called multilook processing. As the radar sensor moves past the target pixel, it obtains multiple looks (i.e., returned samples). If these looks are spaced sufficiently far apart, they can be considered to represent individual observations. The relationships among the radar aperture length L, the resolution R, and the number of independent samples N_s are expressed by (Ulaby et al., 1982, 1986)

$$N_s \approx \frac{R}{0.5L}.$$ (1.36)

For instance, if radar aperture length is 10 m, and the desired spatial resolution is 25 m, then the number of independent samples is 25/5 = 5. Figure 1.24 illustrates the relationship between

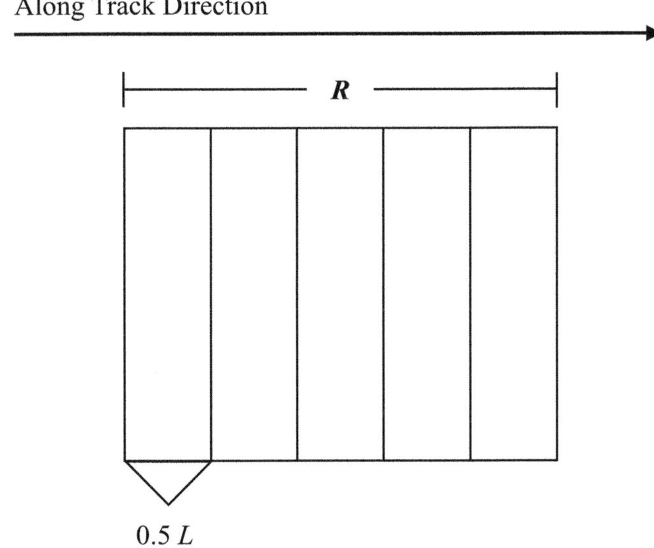

FIGURE 1.24 Relationship between the resolution and the number of looks in a SAR image (see the text for explanations).

the resolution and the number of samples. Although the averaging of independent looks can reduce the noise variance, it also causes degradation in image resolution and loss of significant object details.

1.8.2 FILTERS FOR SPECKLE SUPPRESSION

The second method of speckle suppression uses filtering methods, which fall into two main categories, namely, adaptive and nonadaptive filters. Adaptive filters use weights that are dependent on the degree of speckle in the image, whereas nonadaptive filters use the same set of weights over the entire image. Adaptive filters are more likely to preserve details such as edges or high-texture areas (e.g., forest or urban areas) because the degree of smoothing is dependent on local image statistics.

The best-known nonadaptive filters are those based on the use of the mean or the median. The mean filter uses the same set of smoothing weights for the whole image without regard for differences in image texture, contrast, etc. The median filter does not use a weighting procedure, but is based on the ranking of image pixel values within a specified rectangular window (Mather and Koch, 2011). Both of these filters have a speckle suppression capability, but they also smooth away other high-frequency information. The median is more effective than the mean in eliminating spike noise while retaining sharp edges. Both filters are easily implemented and require less computation than adaptive filters.

In comparison with nonadaptive speckle filters, adaptive speckle filters are more successful in preserving subtle image information. A number of adaptive speckle filters have been proposed, the best known being the Lee filter (Lee, 1980, 1985, 1986), the Kuan filter (Kuan et al., 1987), the Frost filter (Frost et al., 1982), and the Refined Gamma Maximum-A-Posteriori (RGMAP) filter (Touzi et al., 1988; Lopes et al., 1990; Baraldi and Parmiggiani, 1995a). The effectiveness of these adaptive filters is dependent on the following three assumptions (Lee, 1980; Lopes et al., 1990; Baraldi and Parmiggiani, 1995a,b):

1. SAR speckle is modeled as a multiplicative noise (note that the visual effect of multiplicative noise is that the noise level is proportional to the image gray level).
2. The noise and signal are statistically independent.
3. The sample mean and variance of a pixel is equal to its local mean and local variance computed within a window centered on the pixel of interest.

All of the speckle filters described above rely strongly on a good estimate of local statistics (e.g., σ_z and μ_z) from a window. If the window center is located close to the boundary of an image segment (such as a boundary between agricultural fields), the resulting local statistics are likely to be biased and will thus degrade the filtering result. Nezry et al. (1991) noted this point and proposed a refined GMAP filter called the RGMAP filter in which the local statistics extracted from a window do not cross image feature boundaries. Readers are referred to Sheng and Xia (1996) and Liu et al. (2004) for comparisons of the above filters.

In recent years, the wavelet transform has been used for radar imagery denoising (Donoho, 1995; Fukuda and Hirosawa, 1998, 1999; Achim et al., 2003). The wavelet transform can decompose the multiplicative noise, and so simplify the speckle filtering process. Normally, using the wavelet transform for speckle suppression involves three steps as follows. First, the SAR image is translated into a logarithmic domain and is then decomposed by means of the wavelet transform in a multiscale sense. Note that translating the raw SAR image into a logarithmic domain is to convert the multiplicative noise to an additive noise. Second, the empirical wavelet coefficients are shrunk using a thresholding mechanism. Finally, the denoised signal is synthesized from the processed wavelet coefficients through the inverse wavelet transform. It is noted that the quality of the wavelet

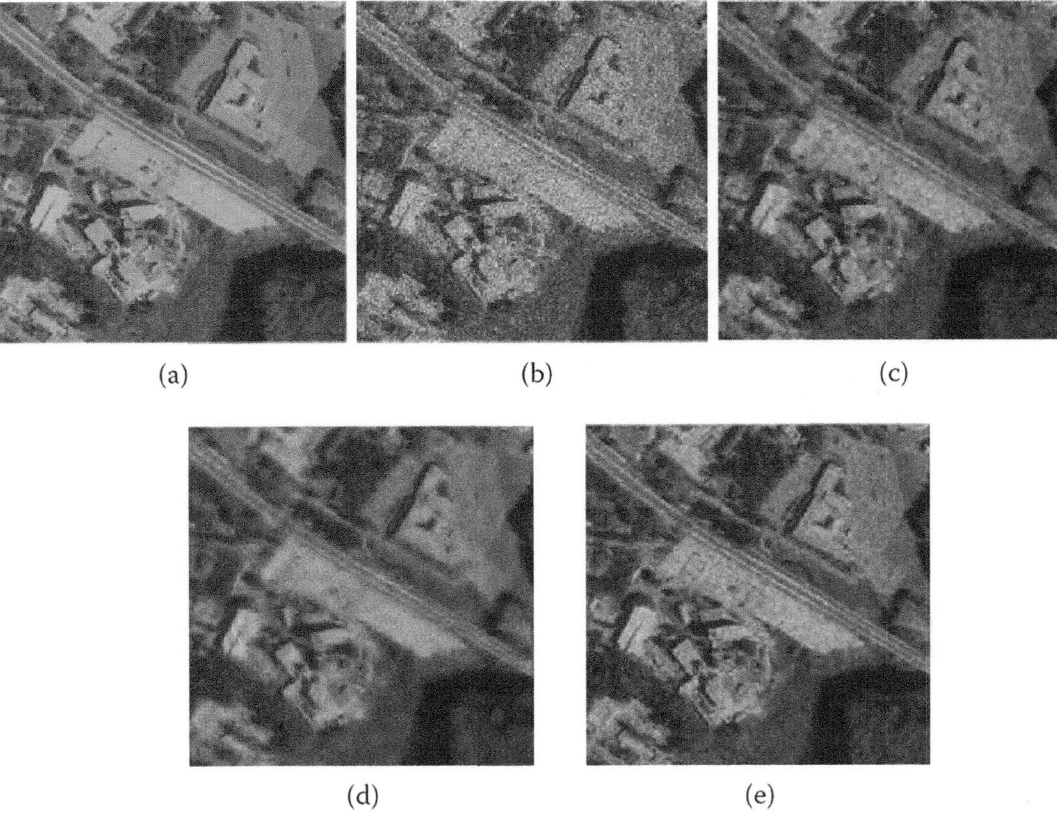

FIGURE 1.25 Effect of various filters on a noise-added SAR image. (a) Original SAR image, (b) noise-added image, (c) GMAP filtering, (d) Wavelet-based filtering developed by Donoho (1995), and (e) Wavelet-based filtering developed by Achim et al. (2003). See text for explanation.

transform for speckle suppression is closely related to the thresholding method used (Donoho, 1995). Some good estimators and modeling techniques proposed by Simoncelli (1999), Pizurica et al. (2001), and Achim et al. (2003) can be applied to solve such thresholding issues and so as to make the speckle suppression quality well controlled. Gagnon and Jouan (1997) and Achim et al. (2003) conducted comparative studies between wavelet-based filters and several statistical adaptive speckle filters described in the previous paragraph and showed that from the perspectives of both qualitative and quantitative measures, the wavelet-based approaches outperform other kinds of filters for speckle removal. Figure 1.25 illustrates the results of speckle suppression in terms of various filters. In Figure 1.25a, the original remotely sensed imagery is shown. The image contaminated by the noise is displayed in Figure 1.25b. The noisy image is then subjected to a GMAP filter (Figure 1.25c), and wavelet-based filters developed by Donoho (1995) (Figure 1.25d) and Achim et al. (2003) (Figure 1.25e), respectively.

Today, many speckle noise suppression filters are used in SAR image processing. The effect of Boxcar, Median, Frost, Gamma MAP, Lee, Lee Sigma, and IDAN speckle suppression filters, which are still widely used today, on a VV polarization SAR image is shown in Figure 1.26 (Parikh et al., 2018; Indriasari et al., 2020). For all, a 3×3 filtering window size was applied.

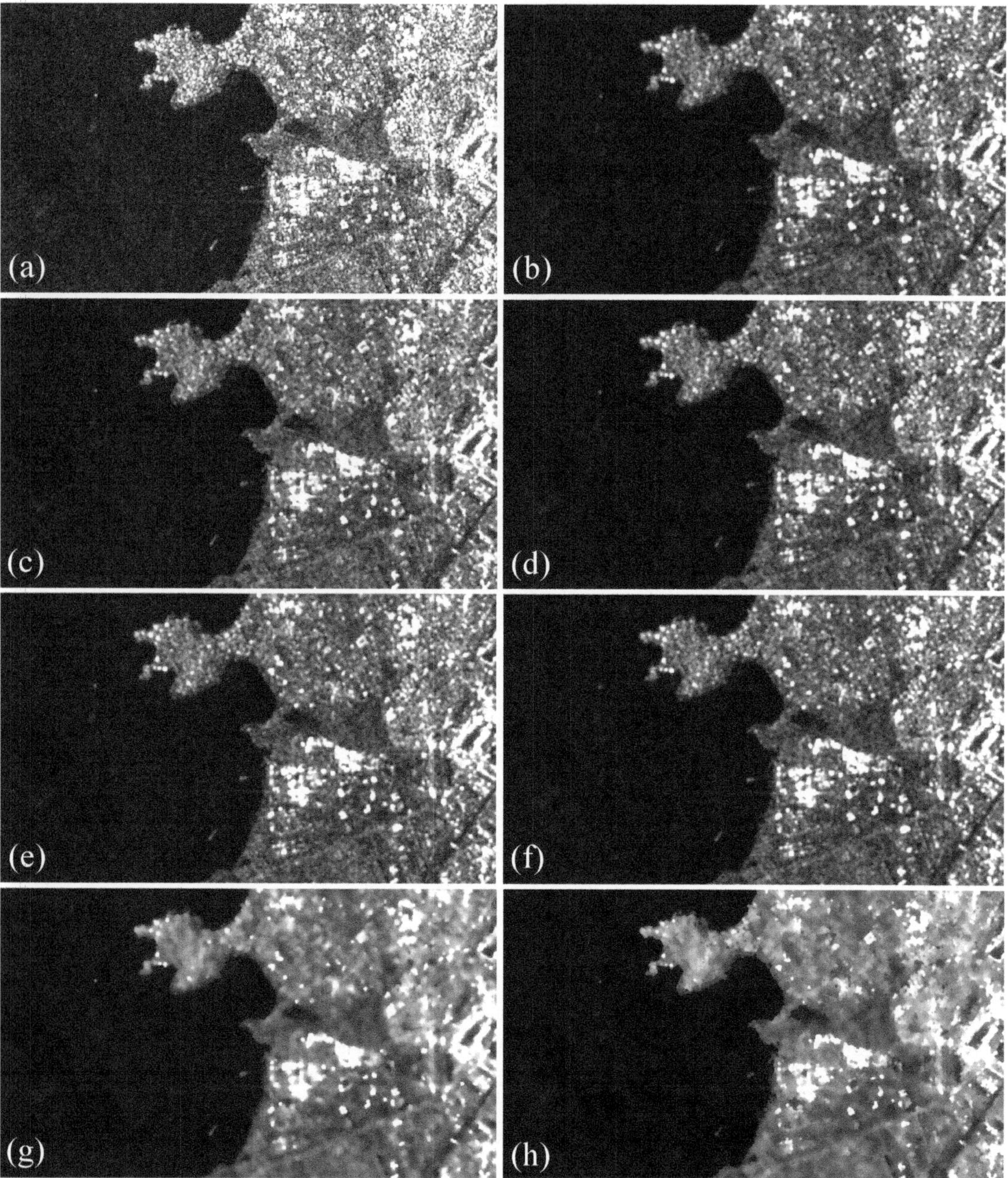

FIGURE 1.26 Effect of widely used speckle noise suppression filters for 3×3 window: (a) original SAR image (VV polarization), (b) Boxcar filtering, (c) Median filtering, (d) Frost filtering, (e) Gamma MAP filtering, (f) Lee filtering, (g) Lee Sigma filtering, and (h) IDAN filtering.

REFERENCES

Achim, A., P. Tsakalides, and A. Bezerianos. 2003. SAR image denoising via Bayesian wavelet shrinkage based on heavy-tailed modeling. IEEE *Transactions on Geoscience and Remote Sensing* 41:1773–1784. https://doi.org/10.1109/TGRS.2003.813488.

Angelliaume, S., P. C. Dubois-Fernandez, C. E. Jones, B. Holt, B. Minchew, E. Amri, and V. Miegebielle. 2018. SAR imagery for detecting sea surface slicks: Performance assessment of polarization-dependent parameters. IEEE *Transactions on Geoscience and Remote Sensing* 56:4237–4257. https://doi.org/10.1109/TGRS.2018.2803216.

Bannari, A., D. Morin, F. Bonn, and A. A. Huete. 1995. A review of vegetation indices. *Remote Sensing Reviews* 13:95–120. https://doi.org/10.1080/02757259509532298.

Baraldi, A., and F. Parmiggiani. 1995a. A refined Gamma MAP SAR speckle filter with improved geometrical adaptivity. IEEE *Transactions on Geoscience and Remote Sensing* 33:1245–1257. https://doi.org/10.1109/36.469489.

Baraldi, A., and F. Parmiggiani. 1995b. An alternative form of the Lee filter for speckle suppression in SAR images. *Computer Vision, Graphics and Image Processing-Graphical Models and Image Processing* 57:75–78. https://doi.org/10.1006/gmip.1995.1008.

Berardino, P., G. Fornaro, R. Lanari, and E. Sansosti. 2002. A new algorithm for surface deformation monitoring based on small baseline differential SAR interferograms. *IEEE Transactions on Geoscience and Remote Sensing* 40:2375–2383. https://doi.org/10.1109/TGRS.2002.803792.

Brewington, L., B. G. Frizzelle, S. J. Walsh, C. F. Mena, and C. Sampedro. 2014. Remote sensing of the marine environment: Challenges and opportunities in the Galapagos Islands of Ecuador. In *The Galapagos Marine Reserve: A Dynamic Social-Ecological System*, (eds.) J. Denkinger, and L. Vinueza, pp. 109–136. Cham, Switzerland: Springer. https://doi.org/10.1007/978-3-319-02769-2_6.

Brown, L. G. 1992. A survey of image registration techniques. *ACM Computing Surveys* 24:325–376. https://doi.org/10.1145/146370.146374.

Bui, Q. T., C. Jamet, V. Vantrepotte, X. Mériaux, A. Cauvin, and M. A. Mograne. 2022. Evaluation of Sentinel-2/MSI atmospheric correction algorithms over two contrasted French coastal waters. *Remote Sensing* 14:1099. https://doi.org/10.3390/rs14051099.

Capaldo, P., M. Crespi, F. Fratarcangeli, A. Nascetti, and F. Pieralice. 2012. DSM generation from high resolution COSMO-SkyMed imagery with radargrammetric model. In *Proceedings of the International Archives of the Photogrammetry, Remote Sensing and Spatial Information Sciences* 38:239–244. https://doi.org/10.5194/isprsarchives-XXXVIII-4-W19-239-2011.

Casu, F., M. Manzo, and R. Lanari. 2006. A quantitative assessment of the SBAS algorithm performance for surface deformation retrieval from DInSAR data. *Remote Sensing of Environment* 102:195–210. https://doi.org/10.1016/j.rse.2006.01.023.

Chavez, P. S., Jr. 1988. An improved dark-object subtraction technique for atmospheric scattering correction of multispectral data. *Remote Sensing of Environment* 24:459–479. https://doi.org/10.1016/0034-4257(88)90019-3.

Colby, J. D. 1991. Topographic normalisation in rugged terrain. *Photogrammetric Engineering and Remote Sensing* 57:531–537.

Colkesen, I., T. Kavzoglu, U. G. Sefercik, and M. Y. Ozturk. 2023. Automated mucilage extraction index (AMEI): A novel spectral water index for identifying marine mucilage formations from Sentinel-2 imagery. *International Journal of Remote Sensing* 44:105–141. https://doi.org/10.1080/01431161.2022.2158049.

Corsa, B., M. Barba-Sevilla, K. Tiampo, and C. Meertens. 2022. Integration of DInSAR time series and GNSS data for continuous volcanic deformation monitoring and eruption early warning applications. *Remote Sensing* 14:784. https://doi.org/10.3390/rs14030784.

Cracknell, A. P., S. K. Newcombe, A. F. Black, and N. E. Kirby. 2001. The ABDMAP (Algal Bloom Detection, Monitoring and Prediction) concerted action. *International Journal of Remote Sensing* 22:205–247. https://doi.org/10.1080/014311601449916.

Crosetto, M., O. Monserrat, M. Cuevas-González, N. Devanthéry, and B. Crippa. 2016. Persistent scatterer interferometry: A review. *ISPRS Journal of Photogrammetry and Remote Sensing* 115:78–89. https://doi.org/10.1016/j.isprsjprs.2015.10.011.

Crosetto, M., and P. Pasquali. 2008. DSM generation and deformation measurement from SAR data. In *Advances in Photogrammetry, Remote Sensing and Spatial Information Sciences*, (eds.) E. Baltsavias, Z. Li, and J. Chen, pp. 157–167. London: Taylor & Francis.

Curran, P. J., and P. M. Atkinson. 2002. Issues of scale and optimal scale size. In *Spatial Statistics for Remote Sensing*, (eds.) A. Stein, F. van der Meer, and B. Gorte, pp. 115–134. Dordrecht: Kluwer Academic. https://doi.org/10.1007/0-306-47647-9_7.

Dong, C., G. Zhao, Y. Meng, B. Li, and B. Peng. 2020. The effect of topographic correction on forest tree species classification accuracy. *Remote Sensing* 12:787. https://doi.org/10.3390/rs12050787.

Donoho, D. L. 1995. Denoising by soft-thresholding. *IEEE Transactions on Information Theory* 41:613–627. https://doi.org/10.1109/18.382009.

Dymond, J. R., and J. D. Shepherd. 1999. Correction of the topographic effect in remote sensing. *IEEE Transactions on Geoscience and Remote Sensing* 37:2618–1619. https://doi.org/10.1109/36.789656.

Eidenshink, J. C., and J. L. Faundeen. 1994. The 1 km AVHRR global land data set: First stages in implementation. *International Journal of Remote Sensing* 15:3443–3462. https://doi.org/10.1080/01431169408954339.

Evans, D. L., T. G. Farr, J. J. Van Zyl, and H. A. Zebker. 1988. Radar polarimetry: Analysis tools and application. IEEE *Transactions on Geoscience and Remote Sensing* 26:774–789. https://doi.org/10.1109/36.7709.

Ferretti, A., C. Prati, and F. Rocca. 2000. Nonlinear subsidence rate estimation using permanent scatterers in differential SAR interferometry. IEEE *Transactions on Geoscience and Remote Sensing* 38:2202–2212. https://doi.org/10.1109/36.86887.

Frost, V. S., J. A. Stiles, K. S. Shanmugan, and J. C. Holtzman. 1982. A model for radar images and its application to adaptive digital filtering of multiplicative noise. IEEE *Transactions on Pattern Analysis and Machine Learning* 4:157–166. https://doi.org/10.1109/TPAMI.1982.4767223.

Fukuda, S., and H. Hirosawa. 1998. Suppression of speckle in synthetic aperture radar images using wavelet. *International Journal of Remote Sensing* 19:507–519. https://doi.org/10.1080/014311698216125.

Fukuda, S., and H. Hirosawa. 1999. Smoothing effect of wavelet-based speckle filtering: The Haar basis case. *IEEE Transactions on Geoscience and Remote Sensing* 37:1168–1172. https://doi.org/10.1109/36.752236.

Gagnon, L., and A. Jouan. 1997. Speckle filtering of SAR images-A comparative study between complex-wavelet-based and standard filters. *Proceedings of the SPIE* 3169:80–91. https://doi.org/10.1117/12.279681.

Gellman, D. I., S. F. Biggar, M. C. Dinguirard, P. J. Henry, M. S. Moran, K. T. Thome, and P. N. Slater. 1993. Review of SPOT-1 and -2 calibration at White Sands from launch to present. *Proceedings of SPIE* 1938:118–125.

Goodchild, M. F. 1999. Future directions in geographic information science. *Geographic Information Sciences* 5:1–8. https://doi.org/10.1080/10824009909480507.

Goyal, S. K., M. S. Seyfried, and P. E. O'Neill. 1999. Correction of surface roughness and topographic effects on airborne SAR in mountainous rangeland areas. *Remote Sensing of Environment* 67:124–136. https://doi.org/10.1016/S0034-4257(98)00077-7.

Hellmann, M. 2002. SAR polarimetry tutorial. https://epsilon.nought.de/. (Accessed August 21, 2023.)

Holben, B. N., and D. Kimes. 1986. Directional reflectance response in AVHRR red and near-infrared bands for three cover types and varying atmospheric conditions. *Remote Sensing of Environment* 19:213–226. https://doi.org/10.1016/0034-4257(86)90054-4.

Holben, B., and C. Justice. 1981. An examination of spectral band ratioing to reduce the topographic effect on remotely sensed data. *International Journal of Remote Sensing* 2:115–133. https://doi.org/10.1080/01431168108948349.

Huang, S., L. Tang, J. P. Hupy, Y. Wang, and G. Shao. 2021. A commentary review on the use of normalized difference vegetation index (NDVI) in the era of popular remote sensing. *Journal of Forestry Research* 32:1–6. https://doi.org/10.1007/s11676-020-01155-1.

Indriasari, N., R. Arief, Kustiyo, M. E. Budiono, H. S. Dyatmika, M. I. Rahayu, A. S. Payani, et al. 2020. Analisa filter spekle single dan multitemporal data Sentinel 1-A. *Proceedings of IOP Conference Series: Earth and Environmental Science* 500:012021.

Jensen, J. R. 2015. *Introductory Digital Image Processing: A Remote Sensing Perspective.* Upper Saddle River, NJ: Pearson Prentice Hall.

Johnson, B. A. 2013. High-resolution urban land-cover classification using a competitive multi-scale object-based approach. *Remote Sensing Letters* 4:131–140. https://doi.org/10.1080/2150704X.2012.705440.

Jones, K. 1998. A comparison of algorithms used to compute hill slope as a property of the DEM. *Computers & Geosciences* 24:315–323. https://doi.org/10.1016/S0098-3004(98)00032-6.

Karpouzli, E., and T. Malthus. 2003. The empirical line method for the atmospheric correction of IKONOS imagery. *International Journal of Remote Sensing* 24:1143–1150. https://doi.org/10.1080/0143116021000026779.

Kavzoglu, T. 2004. Simulating Landsat ETM+ imagery using DAIS 7915 hyperspectral scanner data. *International Journal of Remote Sensing* 25:5049–5067.

Kavzoglu, T., and M. Goral. 2022. Google Earth Engine for monitoring marine mucilage: Izmit bay in spring 2021. *Hydrology* 9:135. https://doi.org/10.3390/hydrology9080135.

Kavzoglu, T., H. Tonbul, I. Colkesen, and U. G. Sefercik. 2021. The use of object-based image analysis for monitoring 2021 marine mucilage bloom in the Sea of Marmara. *International Journal of Environment and Geoinformatics* 8:529–536. https://doi.org/10.30897/ijegeo.990875.

Kavzoglu, T., M. Y. Erdemir, and H. Tonbul. 2017. Classification of semiurban landscapes from very high-resolution satellite images using a regionalized multiscale segmentation approach. *Journal of Applied Remote Sensing* 11:035016. https://doi.org/10.1117/1.JRS.11.035016.

Kim, Y., and J. Van Zyl. 2000. Overview of polarimetric interferometry. In *Proceedings of 2000 IEEE Aerospace Conference*, March 18–25, Big Sky, MT, pp. 231–236. https://doi.org/10.1109/AERO.2000.879850.

Köhl, M., S. Magnussen, and M. Marchetti. 2006. *Sampling Methods, Remote Sensing, and GIS Multiresource Forest Inventory*. Berlin, Heidelberg: Springer. https://doi.org/10.1007/978-3-540-32572-7.

Kowalik, W. S., and S. E. Marsh. 1982. A relation between Landsat Digital Numbers, surface reflectance, and the cosine of the solar zenith angle. *Remote Sensing of Environment* 12:39–55. https://doi.org/10.1016/0034-4257(82)90006-2.

Kropatsch, W. G., and D. Strobl. 1990. The generation of SAR layover and shadow maps from digital elevation models. IEEE *Transactions on Geoscience and Remote Sensing* 28:98–107. https://doi.org/10.1109/36.45752.

Kuan, D. T., A. A. Sawchuk, T. C. Strand, and P. Chavel. 1987. Adaptive restoration of images with speckle. *IEEE Transactions on Acoustics, Speech, and Signal Processing* 35:373–383. https://doi.org/10.1109/TASSP.1987.1165131.

Lavecchia, G., R. Castaldo, R. De Nardis, V. De Novellis, F. Ferrarini, S. Pepe, F. Brozzetti, et al. 2016. Ground deformation and source geometry of the 24 August 2016 Amatrice earthquake (Central Italy) investigated through analytical and numerical modeling of DInSAR measurements and structural-geological data. *Geophysical Research Letters* 43:12–389. https://doi.org/10.1002/2016GL071723.

Law, K. H., and J. Nichol. 2004. Topographic correction for differential illumination effects on IKONOS satellite imagery. In *Proceedings of 20th ISPRS Congress*, July 12–23, Istanbul, Turkey, pp. 641–646.

Le Moigne, J., N. S. Netanyahu, and R. D. Eastman. 2011. *Image Registration for Remote Sensing*. New York: Cambridge University Press.

Lee, J. S. 1980. Digital image enhancement and noise filtering by use of local statistics. IEEE *Transactions on Pattern Analysis and Machine Intelligence* 2:165–168. https://doi.org/10.1109/TPAMI.1980.4766994.

Lee, J. S. 1985. Speckle suppression and analysis for synthetic aperture radar images. *Optical Engineering* 15:380–389.

Lee, J. S. 1986. Speckle suppression and analysis for synthetic aperture radar images. *Optical Engineering* 25:636–643. https://doi.org/10.1117/12.7973877.

Levin, S. A. 1992. The problem of pattern and scale in ecology. *Ecology* 73:1943–1967. https://doi.org/10.2307/1941447.

Li, L., F. Zhang, Y. Shao, Q. Wei, Q. Huang, and Y. Jiao. 2022. Airborne SAR radiometric calibration based on improved sliding window integral method. *Sensors* 22:320. https://doi.org/10.3390/s22010320.

Lillesand, T., R. W. Kiefer, and J. Chipman. 2015. *Remote Sensing and Image Interpretation*, 7th edition. Chichester: John Wiley & Sons.

Lim, C., and M. Kafatos. 2002. Frequency analysis of natural vegetation distribution using NDVI/AVHRR data from 1981 to 2000 for North America: Correlations with SOI. *International Journal of Remote Sensing* 23:3347–3383. https://doi.org/10.1080/01431160110110956.

Liu, Z., A. Liu, C. Wang, and S. Qi. 2004. Statistical ratio rank-ordered differences filter for radar speckle removal. In *Proceedings of the SPIE* 5238:524–531. https://doi.org/10.1117/12.513565.

Loew, A., and W. Mauser. 2007. Generation of geometrically and radiometrically terrain corrected SAR image products. *Remote Sensing of Environment* 106:337–349. https://doi.org/10.1016/j.rse.2006.09.002.

Lopes, A., E. Nezry, and R. Touzi. 1990. Maximum a posteriori speckle filtering and first order texture models in SAR images. In *Proceedings of the International Geoscience and Remote Sensing Symposium (IGARSS'90)*, May 20–24, College Park, MD, pp. 2409–2412. https://doi.org/10.1109/IGARSS.1990.689026.

Lopez-Martinez, C., and X. Fabregas. 2003. Polarimetric SAR speckle model. *IEEE Transactions on Geosciences and Remote Sensing* 41:2232–2242. https://doi.org/10.1109/TGRS.2003.815240.

Mackay, G., M. D. Steven, and J. A. Clarke. 1994. A pseudo-5S code for atmospheric correction of ATSR-2 visible and near-IR land-surface data. In *Proceeding of the 6th International Symposium-Physical Measurements and Signatures in Remote Sensing*, January 17–21, Val d'Isère, France, pp. 103–110.

Martins, V. S., C. C. Barbosa, L. A. D. Carvalho, D. S. Jorge, F. D. L., and E. M. Novo. 2017. Assessment of atmospheric correction methods for Sentinel-2 MSI images applied to Amazon Floodplain Lakes. *Remote Sensing* 9:322. https://doi.org/10.3390/rs9040322.

Mather, P. M., and M. Koch. 2011. *Computer Processing of Remotely-Sensed Images: An Introduction*, 4th edition. Chichester: John Wiley & Sons.

Meygret, A. 2005. Absolute calibration: From SPOT1 to SPOT5. *Proceedings of SPIE* 5882. https://doi.org/10.1117/12.613855.

Minnaert, M. 1941. The reciprocity principle in Lunar photometry. *Astrophysics Journal* 93:403–410. https://doi.org/10.1086/144279.

Moran, M. S., R. Bryanta, K. Thomeb, W. Nia, and Y. Nouvellona. 2001. A refined empirical line approach for reflectance factor retrieval from Landsat-5 TM and Landsat-7 ETM+. *Remote Sensing of Environment* 78:71–82. https://doi.org/10.1016/S0034-4257(01)00250-4.

Neupane, N., M. Peruzzi, A. Arab, S. J. Mayor, J. C. Withey, L. Ries, and A. O. Finley. 2022. A novel model to accurately predict continental-scale timing of forest green-up. *International Journal of Applied Earth Observation and Geoinformation* 108:102747. https://doi.org/10.1016/j.jag.2022.102747.

Nezry, E., A. Lopes, and R. Touzi. 1991. Detection of structure and textural features for SAR images filtering. In *Proceedings of the International Geoscience and Remote Sensing Symposium (IGARSS'91)*, June 03–06, Espoo, Finland, pp. 2169–2172. https://doi.org/10.1109/IGARSS.1991.575470.

Odunuga, S., O. E. G. S. OshoBadru, and S. A. Raji. 2023. Prospecting and mapping iron ore within the Niger-Benue confluence using Landsat 8. *Journal of the Indian Society of Remote Sensing* 51:1523–1539. https://doi.org/10.1007/s12524-023-01717-w.

Ouaidrari, H., and E. F. Vermote. 2001. Operational atmospheric correction of Landsat TM data. *Remote Sensing of Environment* 70:4–15. https://doi.org/10.1016/S0034-4257(99)00054-1.

Parikh, H., S. Patel, and V. Patel. 2018. Analysis of denoising techniques for speckle noise removal in synthetic aperture radar images. In *Proceedings of International Conference on Advances in Computing, Communications and Informatics*, Bangalore, India, pp. 671–677. https://doi.org/10.1109/ICACCI.2018.8554917.

Parlow, E. 1996. Correction of terrain controlled illumination effects in satellite data. In *Progress in Environmental Remote Sensing Research and Applications: Proceedings of the 15th EARSeL Symposium*, Basel, Switzerland, September 4–6, pp.139–145. Boca Raton, FL: CRC Press.

Pereira-Sandoval, M., A. Ruescas, P. Urrego, A. Ruiz-Verdú, J. Delegido, C. Tenjo, X. Soria-Perpinyà, et al. 2019. Evaluation of atmospheric correction algorithms over Spanish inland waters for Sentinel-2 Multi Spectral imagery data. *Remote Sensing* 11:1469. https://doi.org/10.3390/rs11121469.

Pizurica, A., W. Philips, I. Lemahieu, and M. Acheroy. 2001. Despeckling SAR images using wavelets and a new class of adaptive shrinkage estimators. In *Proceedings of IEEE International Conference on Image Processing*, October 7–10, Thessaloniki, Greece, pp. 233–236. https://doi.org/10.1109/ICIP.2001.958467.

Qian, S.-E. 2020. *Hyperspectral Satellites and System Design*. Boca Raton, FL: CRC Press.

Quattrochi, D. A., E. Wentz, N. S. N. Lam, and C. W. Emerson. 2017. *Integrating Scale in Remote Sensing and GIS*. Boca Raton, FL: CRC Press. https://doi.org/10.1201/9781315373720.

Quattrochi, D. A., and M. A. Goodchild. 1997. *Scale in Remote Sensing and GIS*. New York: Lewis Publishers.

Riano, D., E. Chuvieco, J. Salas, and I. Aguado. 2003. Assessment of different topographic corrections in Landsat-TM data for mapping vegetation types. *IEEE Transactions on Geoscience and Remote Sensing* 41:1056–1061. https://doi.org/10.1109/TGRS.2003.811693.

Sefercik, U. G. 2013. Productivity of TerraSAR-X 3D data in urban areas: A case study in Trento. *European Journal of Remote Sensing* 46:597–612.

Sefercik, U. G., and M. Alkan, M. 2009. Advanced analysis of differences between C and X bands using SRTM data for mountainous topography. *Journal of the Indian Society of Remote Sensing* 37:335–349. https://doi.org/10.1007/s12524-009-0044-4.

Sefercik, U. G., I. Colkesen, T. Kavzoglu, N. Ozdogan, and M. Y. Ozturk. 2024. Assessing the physical and chemical characteristics of marine mucilage utilizing in-situ and remote sensing data (Sentinel-1,-2,-3). *PFG-Journal of Photogrammetry, Remote Sensing and Geoinformation Science*. https://doi.org/10.1007/s41064-023-00254-y.

Sheng, Y., and Z. G. Xia. 1996. A comprehensive evaluation of filters for radar speckle suppression. *Proceedings of the International Geoscience and Remote Sensing Symposium (IGARSS'96)*, May 27–31, Lincoln, NE, pp. 1559–1561. https://doi.org/10.1109/IGARSS.1996.516730.

Shimada, M., O. Isoguchi, T. Tadono, and K. Isono. 2009. PALSAR radiometric and geometric calibration. *IEEE Transactions on Geoscience and Remote Sensing* 47:3915–3932.

Simoncelli, E. P. 1999. Bayesian denoising of visual images in the wavelet domain. In *Bayesian Inference in Wavelet-Based Models*, (eds.) P. Muller, and B. Vidakovic, pp. 291–308. New York: Springer. https://doi.org/10.1007/978-1-4612-0567-8_18.

Singh, P., and R. Shree. 2016. Speckle noise: Modelling and implementation. *International Journal of Control Theory and Applications* 9:8717–8727.

Skidmore, A. K. 1989. A comparison of techniques for calculating gradient and aspect from a gridded digital elevation model. *International Journal of Geographical Information Systems* 3:323–334. https://doi.org/10.1080/02693798908941519.

Small, D., A. Schubert, B. Rosich, E. Meier, H. Lacoste, and L. Ouwehand. 2007. Geometric and radiometric correction of ESA SAR products. In *Proceedings of 2007 Envisat Symposium*, April 23–27, Montreux, Switzerland. https://doi.org/10.5167/uzh-77977.

Smith, G. M., and E. J. Milton. 1999. The use of the empirical line method to calibrate remotely sensed data to reflectance. *International Journal of Remote Sensing* 20:2653–2662. https://doi.org/10.1080/014311699211994.

Smith, J. A., T. L. Lin, and K. Ranson. 1980. The Lambertian assumption and Landsat data. *Photogrammetric Engineering and Remote Sensing* 46:1183–1189.

Stramondo, S., M. Chini, C. Bignami, S. Salvi, and S. Atzori. 2010. X-, C-, and L-band DIn SAR investigation of the April 6, 2009, Abruzzi Earthquake. *IEEE Geoscience and Remote Sensing Letters* 8:49–53. https://doi.org/10.1109/LGRS.2010.2051015.

Tanré, D., C. Deroo, P. Duhaut, M. Herman, J. J. Morcette, J. Perbos, and P. Y. Deschamps. 1986. Simulation of the satellite signal in the solar spectrum (5S). In *Technical Report*, Laboratoire d'Optique Atmospherique, Universite des Sciences et Techniques de Lille, 59655 Villeneuve d'ascq Cedex, France.

Tanré, D., C. Deroo, P. Duhaut, M. Herman, J. J. Morcette, J. Perbos, and P. Y. Deschamps. 1990. Description of a computer code to simulate the satellite signal in the solar spectrum: The 5S code. *International Journal of Remote Sensing* 11:659–668. https://doi.org/10.1080/01431169008955048.

Teillet, P. M., B. Guindon, and D. G. Goodenough. 1982. On the slope-aspect correction of multispectral scanner data. *Canadian Journal of Remote Sensing* 8:84–106. https://doi.org/10.1080/07038992.1982.1085 5028.

Teillet, P. M., J. L. Barker, B. L. Markham, R. R. Irish, G. Fedosejevs, and J. C. Storey. 2001. Radiometric cross-calibration of the Landsat-7 ETM+ and Landsat-5 TM sensors based on tandem data sets. *Remote Sensing of Environment* 78:39–54. https://doi.org/10.1016/S0034-4257(01)00248-6.

Thome, K., P. Slater, S. Biggar, B. Markham, and J. Barker. 1997. Radiometric calibration of Landsat. *Photogrammetric Engineering and Remote Sensing* 63:853–858.

Touzi, R., A. Lopes, and P. Bousquet. 1988. A statistical and geometrical edge detector for SAR images. *IEEE Transactions on Geoscience and Remote Sensing* 26:764–773. https://doi.org/10.1109/36.7708.

Ulaby, F. T., A. K. Fung, and R. K. Moore. 1982. *Microwave Remote Sensing: Active and Passive*, Vol. 1–2. Reading, MA: Addison-Wesley.

Ulaby, F. T., A. K. Fung, and R. K. Moore. 1986. *Microwave Remote Sensing: Active and Passive*, Vol. 3. Dedham, MA: Artech House.

Van de Hulst, H. C. 1981. *Light Scattering by Small Particles*. New York: Dover.

Van Zyl, J. J., B. D. Chapman, P. Dubois, and J. C. Shi. 1993. The effect of topography on SAR calibration. IEEE *Transactions on Geoscience and Remote Sensing*. 31:1036–1043. https://doi.org/10.1109/36.263774.

Van Zyl, J. J., H. A. Zebker, and C. Elachi. 1987. Image radar polarization signatures: Theory and observation. *Radio Science* 22:529–543. https://doi.org/10.1029/RS022i004p00529.

Vanonckelen, S., S. Lhermitte, and A. Van Rompaey. 2013. The effect of atmospheric and topographic correction methods on land cover classification accuracy. *International Journal of Applied Earth Observation and Geoinformation* 24:9–21. https://doi.org/10.1016/j.jag.2013.02.003.

Vermote, E. F., D. Tanré, J. L. Deuze, M. Herman, and J. J. Morcette. 1997. Second simulation of the satellite signal in the solar spectrum, 6S: An overview. IEEE *Transactions on Geoscience and Remote Sensing* 35:675–686. https://doi.org/10.1109/36.581987.

Vicente-Serrano, S. M., J. J. Camarero, J. M. Olano, N. Martín-Hernández, M. Peña-Gallardo, M. Tomás-Burguera, A. Gazol, et al. 2016. Diverse relationships between forest growth and the Normalized Difference Vegetation Index at a global scale. *Remote Sensing of Environment* 187:14–29. https://doi.org/10.1016/j.rse.2016.10.001.

Warren, M. A., S. G. H. Simis, V. Martinez-Vicente, K. Poser, M. Bresciani, K. Alikas, E. Spyrakos, et al. 2019. Assessment of atmospheric correction algorithms for the Sentinel-2A MultiSpectral Imager over coastal and inland waters. *Remote Sensing of Environment* 225:267–289. https://doi.org/10.1016/j.rse.2019.03.018.

Waske, B., and M. Braun, M. 2009. Classifier ensembles for land cover mapping using multitemporal SAR imagery. *ISPRS Journal of Photogrammetry and Remote Sensing* 64:450–457. https://doi.org/10.1016/j.isprsjprs.2009.01.003.

Weng, Q. 2014. *Scale Issues in Remote Sensing*. Hoboken, NJ: John Wiley & Sons.

Woodcock, C. E., and A. H. Strahler. 1987. The factor of scale in remote sensing. *Remote Sensing of Environment* 21:311–332. https://doi.org/10.1016/0034-4257(87)90015-0.

Woodcock, C. E., A. H. Strahler, and D. B. Jupp. 1988. The use of variogram in remote sensing: I. Scene models and simulated images. *Remote Sensing of Environment* 25:323–348. https://doi.org/10.1016/0034-4257(88)90108-3.

Yang, J., X. Qiu, C. Ding, and B. Lei. 2018. Identification of stable backscattering features, suitable for maintaining absolute synthetic aperture radar (SAR) radiometric calibration of Sentinel-1. *Remote Sensing* 10:1010. https://doi.org/10.3390/rs10071010.

Zebker, H. A., and J. J. Van Zyl. 1991. Imaging radar polarimetry: A review. *Proceedings of the IEEE* 79:1583–1606. https://doi.org/10.1109/5.118982.

Zebker, H. A., J. J. Van Zyl, and D. N. Held. 1987. Imaging radar polarimetry from wave synthesis. *Journal of Geophysical Research* 92:683–701. https://doi.org/10.1029/JB092iB01p00683.

Zebker, H. A., J. J. Van Zyl, S. L. Durden, and L. Norikane. 1991. Calibrated imaging radar polarimetry: Technique, examples, and applications. *IEEE Transactions on Geoscience and Remote Sensing* 29:942–961. https://doi.org/10.1109/36.101373.

Zeng, Y., D. Hao, A. Huete, B. Dechant, J. Berry, J. M. Chen, J. Joiner, et al. 2022. Optical vegetation indices for monitoring terrestrial ecosystems globally. *Nature Reviews Earth & Environment* 3:477–493. https://doi.org/10.1038/s43017-022-00298-5.

Zhang, J., P. Atkinson, and M. F. Goodchild. 2014. *Scale in Spatial Information and Analysis*. Boca Raton, FL: CRC Press. https://doi.org/10.1201/b16751.

Zhang, X., P. Xiao, and X. Feng. 2020. Object-specific optimization of hierarchical multiscale segmentations for high-spatial resolution remote sensing images. *ISPRS Journal of Photogrammetry and Remote Sensing* 159:308–321. https://doi.org/10.1016/j.isprsjprs.2019.11.009.

Zitová, B., and J. Flusser. 2003. Image registration methods: A survey. *Image and Vision Computing* 21:977–1000. https://doi.org/10.1016/S0262-8856(03)00137-9.

2 Pattern Recognition Principles

A fundamental problem in remote sensing is the characterization and classification of spectral measurements collected at different parts of the electromagnetic spectrum for an object on the Earth's surface. Accurate and up-to-date maps serve as crucial resources for various investigations, spanning from local to global scales. The common practice for acquiring such information from satellite images involves a classification process. Remote sensing classification that deals with high-dimensional data with complex spectral and radiometric characteristics can be achieved with pattern recognition techniques. Remotely sensed image data exhibit a high level of complexity and irregularity so that pixels within these images frequently contain multiple features, a composition influenced by the spatial resolution of the sensor and the configuration of the surveyed area. Therefore, the classification of remotely sensed data into a thematic map continues to be a challenge because many factors may have significant effects on the success of a classification. As the number of categories and the volume of data expand, the classification problem becomes more complex, presenting difficulties in characterizing categories and assigning pixels accurately. Recognizing patterns allows us to make predictions about unseen pixels or objects. In the context of pattern recognition, a pattern is a vector of features describing an object. This pattern is made up of measurements on a set of features, which can be thought of as the axes of a k-dimensional space, called the feature space. Pixels in an image are located in the feature space as a cloud of points. The aim of pattern recognition is to establish a relationship between a pattern and a class label. This relationship label may be one-to-one (producing a hard classification) or one-to-many (producing a fuzzy classification). The features describing the object may be spectral reflectance or emittance values from optical or infrared imagery, radar backscatter values, secondary measurements derived from the image, called ancillary data, such as texture, or geographical features such as terrain elevation, slope, and aspect. The object may be a single pixel or a set of adjacent pixels (i.e., an image object) forming a geographical entity, such as an agricultural field. Finally, the class labels may be known or unknown in the sense that the investigator may, in the case of a known label set, be able to list all of the categories present in the area of study. In other cases, the investigator may wish to determine the number of separable categories and their location and extent.

In the early stage, supervised and unsupervised learning were the two fundamental methods of learning, though some approaches to pattern recognition have employed a combination of both (i.e., hybrid learning). With the introduction of the deep learning paradigm into digital image analysis, new learning strategies have been introduced to train networks with limited training data (semi-supervised learning (SSL) and active learning (AL)) or without any training data (self-supervised learning) (Persello and Bruzzone, 2014). Supervised methods require the user to collect samples to "train" or teach the classifier to determine the decision boundaries in the feature space, and such decision boundaries are significantly affected by the properties and the size of the samples used to train the classifier. For instance, if one decides to use the minimum distance between a pixel and the mean of each class as the classification criterion, one must first collect samples to construct estimates of the class means. The acceptability of the results will depend on how adequately these class means are estimated. The label set selected for supervised classification experiments identifies information classes. The investigator should have sufficient knowledge of the type and the number of information classes that are represented in the study area to allow him or her to collect training samples of pixels from the image that are representative of the information classes. In contrast, unsupervised pattern recognition methods are less dependent on user interaction. Normally, unsupervised classifiers "learn" the characteristics of each class (and possibly even the number of classes) directly from the input data. For instance, if the criterion used to label an object is the minimum distance between the object and the class means, this distance being measured in feature space, the unsupervised procedure will estimate the mean for each class and will refine this mean estimate iteratively (most unsupervised classifiers are iterative in operation). At each iteration, the previous set of estimates of the class means is refined until the process converges, usually when the means remain in the same place in feature space over successive iterations. The result outputs by unsupervised methods are called clusters or, sometimes, data classes. The pattern recognition process is complete when each cluster is identified, that is, linked to a specific information class by the user. In software packages, users are usually asked to define the number of classes or maximum number of classes to be searched in clustering, which can be a difficult task when preliminary knowledge about the study site is unavailable. In recent years, new generation unsupervised algorithms have been proposed based mostly on an evolutionary algorithm (e.g., genetic algorithm, particle swarm optimization) or an improvement in existing conventional methods (e.g., ISODATA) (Banerjee et al., 2015; Senthilnath and Yang, 2016).

Although the unsupervised approach appears to be more elegant and automatic than the supervised procedures, the accuracy of unsupervised methods is generally lower than that achieved by supervised methods. In complex classification experiments, information classes often overlap. In the spectral domain, this implies that the reflectance, emittance, or backscatter characteristics of different classes may be similar. In the spatial domain, the implication is that any one object (a pixel or a field, for example) may contain areas representative of more than one information class. This is the mixed pixel problem. Spectral and spatial overlap of classes is the main barrier to the achievement of high classification accuracy. Even so, some interesting unsupervised algorithms are worthy of investigation as they may reveal useful information concerning the structure of the data set. Such methods can be thought of as exploratory data analyses or even data mining (Witten and Frank, 2005). A further problem with pixel-based classifiers is that radiance (carrying information) that apparently reaches the sensor from a given pixel includes contributions from neighboring pixels, due to atmospheric effects and the properties of the instrument optics (Chapter 1). Townshend et al. (2000) show that by considering this latter effect, improvements in accuracy can be achieved. They note that only where pixel size is small relative to the area of land cover units, will this effect be unimportant.

Hybrid classification approaches taking the advantages of both unsupervised and supervised classification strategies also exist in the literature. In a hybrid classification scheme, unsupervised classification is first performed on the imagery and results are interpreted or analyzed with the ground reference data. A supervised classification is then applied with information gained through unsupervised classification. In recent years, a growing interest has been devoted to SSL and AL approaches that aim to intuitively collect or increase the number of training samples. In these approaches, both labeled (usually limited in size) and unlabeled data are considered together. SSL methods involve jointly using data from both labeled and unlabeled samples during classifier training, while AL methods operate on the principle that the initial training set expands iteratively through an interactive process involving a supervisor, typically a human expert, responsible for assigning the correct label to any queried sample. These state-of-the-art learning strategies have contributed to a significant increase in classification accuracy. Currently, another popular training paradigm applied through deep learning models is self-supervised learning, which does not require any training samples. As an example of this learning strategy, contrastive learning using different augmented views of the same image has been reported to achieve state-of-the-art performance for scene classification (Li et al., 2022; Wang et al., 2023). More detail on learning strategies applied in image classification is given in Section 2.3.

Until the mid-1990s, pattern recognition methods applied to remotely sensed imagery were mostly based on conventional statistical techniques, such as the maximum likelihood or minimum distance procedures, using a pixel-based approach. Although these traditional approaches can perform well in cases where the classes are distinct from each other in the feature space and a large number of training samples are available, their general ability to resolve interclass confusion is limited. As a result, following advances in computer technology, alternative strategies have been proposed, particularly the use of artificial neural networks, decision trees, support vector machines, and methods derived from fuzzy set theory incorporating spatial information. Although support vector machines were as popular as decision trees in earlier applications, robust algorithms based on decision trees including random forest, XGBoost, LightGBM, and NGBoost have lately become more popular machine learning methods for pattern recognition applications.

The neural networks before the introduction of deep learning were simpler; hence, they are today called shallow neural networks having only one or several hidden layers between the input and output layers. Since the 2010s, the deep learning era has been dominating the remote sensing literature. Having more layers and more complex structures than artificial neural networks, deep learning models including deep belief networks, convolutional neural networks (CNNs), and recurrent neural networks have been successfully applied to many fields, such as computer vision, image processing, natural language processing, and autonomous systems. Deep learning networks, particularly requiring high GPU power, are more suitable for solving complex problems, as in the case of hyperspectral imagery containing high-dimensional spectral content with high variance. They also consider contextual and geometric features in their model. As the most popular deep learning model in image classification, CNNs, which can apply both feature extraction and classification thanks to their unique structure, are applied for multidimensional (1D, 2D, and 3D) analyses using kernel functions. Chapter 7 is devoted to the theory and application of neural networks and deep learning with special emphasis on their use in image classification. It should also be mentioned that scene-based classification using deep learning models is currently the state-of-the-art application in remote sensing, which will be also discussed in that chapter.

This chapter introduces the principles of pattern recognition, starting from the terminology used in remote sensing image classification and the taxonomy of classification techniques. Details of the well-known supervised and unsupervised statistical classifiers are then described. Ensemble classifiers and the incorporation of secondary information are also discussed in later chapters.

2.1   A TERMINOLOGICAL INTRODUCTION

In the literature, the terms of taxonomy, identification, and recognition are frequently used to refer to classification. Sneath and Sokal (1973) befittingly differentiate classification from taxonomy in that taxonomy was defined as the theoretical study of classification, including its bases, principles, procedures, and rules. They also define identification as the allocation or assignment of additional unidentified objects to the correct class once a classification has been established. From this definition, identification can be considered as a secondary step following the classification process. In classification exercises, the exact nature of species under investigation should be known *a priori*. According to Pankhurst (1991), "whatever sort of object is in question, it cannot be identified unless there is already a classification of like objects with which the new object can be compared". When the general definition of classification, which is the assignment of unknown elements to known classes, is considered and strictly applied, it is arguable that unsupervised classification is not a proper classification method because there are no known or prespecified classes involved in the process. On the other hand, Cole and King (1968) state that a preliminary ordering of many data before their analysis can be termed "empirical classification", and representation (or conclusions) of the analysis results can be called "genetic classification". In other words, they call the first stage of the classification process "empirical classification" and the second stage of the process "genetic classification". James (1985) notes that, the existence and the structure of the groups to which the object is to be assigned are of secondary importance in the classification analysis. It is the assignment of new cases that concerns us. He also states that classification analysis is sometimes confused with cluster analysis. If one has a mass of currently undifferentiated data and is curious as to whether it has any natural group structure, the method that should be employed is cluster analysis. Basically, cluster analysis attempts to determine any possible groupings in the data. It is not concerned with the problem of classifying new objects into existing groups, as this is the case in classification. Therefore, it should be highlighted that classification analysis itself is not solely concerned with identifying any possible (or inherent) groupings that might be contained within a mass of data.

In some fields such as physical geography, classification problems are concerned with objects that are unique. All objects are unique in some respect and cannot be classified on this basis, for by definition, each unique object would require a separate class. The Earth might be considered to be a unique object in many respects. This is indeed true, as it lies at a unique distance from the Sun and is the home of the human race. On the other hand, the Earth is one of a series of planets that move around the Sun, and thus, it can be classified as a planet. Any unique object is rarely incapable of being subdivided. It is the unique combination of its several elements, which gives it its unique character (Cole and King, 1968). The generality of the classes used in a classification differs depending on the geographical scale and the purpose for which the classification is intended. For example, if a classification is performed to identify a forested region, the classification could be based on discriminating several forest types including deciduous and conifer forests; if a classification is carried out to discriminate general land cover features, one might use the general class of forest with other land cover features, such as grassland, agriculture, and buildings; and if a classification is performed on a global scale, then more general class categories, including vegetation, soil, and water, are considered in the analysis. It should be borne in mind that class subdivisions are totally related to the nature of the data, the classification method, and the purpose of the study (Kavzoglu, 2001).

2.2 TAXONOMY OF CLASSIFICATION TECHNIQUES

In this section, an updated taxonomy for classification techniques used in remote sensing applications is presented. Various survey studies in the literature delve into classification methods used for analyzing remotely sensed data (e.g., Lu and Weng, 2007; Li et al., 2014; Qin and Liu, 2022). Classification techniques may be categorized in terms of four main criteria. Firstly, they can be grouped based on learning strategy considering the involvement of training data. Supervised and unsupervised classifications have been traditional approaches applied in remote sensing. With the introduction of new and advanced methodologies, particularly deep learning–based ones, new learning strategies including SSL, AL, and self-learning were introduced. As discussed above, in supervised classification approaches, the analyst must define the training areas to identify the characteristics of each class. Using the extracted discriminative information, each pixel in the image is assigned to a specific discrete class. Problems related to diagnosis, pattern recognition, and identification are essentially categorized as supervised classification problems, given that the goal is to classify an object into a predefined set of classes (Hand, 1997). Unsupervised classification, conversely, seeks to identify natural groups known as clusters among pixels in the data by analyzing their positions in the feature space. These algorithms, commonly labeled as clustering algorithms, are automated processes that require minimal user input. Unsupervised classification is firstly applied to produce more samples to be used in the training process, and then, supervised classification is performed considering the data set enriched by the previous stage (Bakos and Gamba, 2011; Pati et al., 2020; Pencue-Fierro et al., 2016). SSL is another type of new generation learning strategy combining supervised and unsupervised learning in that compared to the number of labeled data, more samples for which the class label is unknown are included in the supervised learning process. This learning algorithm is particularly useful for scenarios where limited training data are available to train an algorithm with a large number of parameters. By using the newly labeled samples, the model's performance is intended to be maximized while the workload and cost required to collect labeled samples are kept to a minimum. For further information, readers are advised to refer to the studies of Hady and Schwenker (2013), van Engelen and Hoos (2020), and Zhu and Goldberg (2022). Another learning strategy aiming to improve the size and generalization capabilities of the training data is AL. The fundamental idea behind this learning strategy is that a classifier trained with well-defined and representative samples can produce results with similar accuracy compared to the classifier trained with a large training data set. Initial training produces results for unknown pixels, and the user is asked to label the most uncertain ones to maximize generalization capabilities. In recent years, AL has boosted its popularity with its use in deep learning. A survey of AL algorithms and their applications can be found in Tuia et al. (2011), Xu et al. (2020), Lenczner et al. (2022), and Thoreau et al. (2022).

Another distinction among classification methods can be made by considering the underlying philosophy and assumptions. They can be thus classified into statistical and nonstatistical classification methods. As in the cases of the minimum distance and maximum likelihood classifiers, statistical classification procedures employ purely statistical estimations to derive some rules from the data by making some assumptions. The most common assumption of this kind is that the frequency distribution of the data is in the Gaussian form. However, nonstatistical methods, also known as heuristic methods, which consider all training samples in the establishment of decision boundaries, do not make any assumptions about the frequency distribution of the data used and do not use statistical estimates in their estimations. Machine learning methods and deep learning techniques are of this category, which are thoroughly discussed in the following chapters.

Researchers also categorize algorithms as being either "hard" (or crisp) or "soft" (or fuzzy). In a "hard" classification (also called pixel-based classification), each individual pixel is given a single, unambiguous label. This methodology is ideal for cases such as crop classification where agricultural fields are homogeneous in terms of the land cover feature that they contain. In this case, the fields should be large relative to the instantaneous field of view (IFOV) of the sensor; otherwise, it cannot be assumed that a single pixel comprises only a single land cover type. For example, in the case of AVHRR images that have a spatial resolution of 1 km, pixels are likely to be mixed (including more than a single land cover type). The effect of spatial resolution is all dependent upon the nature and scale of variation of the land cover features within the scene. It should be noted that if there are a large number of mixed pixels in the image under analysis, then the scale or the resolution of the image selected is not relevant for the purpose of the study. As stated by Mather (2004), the question of scale is one that bedevils all spatial analysis; also, fuzziness and hardness, heterogeneity, and homogeneity are properties of the landscape at a particular geographical scale of observation that is related to the aims of the investigator. Major drawbacks of "hard" classification are relatively lower classification accuracy levels and poor interpretability of the produced results. In addition, this classification methodology can be an oversimplification of the structure of the data set; in particular, it may be that some objects definitely belong to certain groups, but other objects whose group membership is much less evident (Gordon, 1981). If one is interested in determining the memberships of pixels unambiguously, that is, the relative level of the presence of different classes is in question, "hard" classification methods are clearly irrelevant. For this purpose, the idea of "soft" or "fuzzy" classification should be put into practice. In "soft" classification, instead of assigning a pixel to a certain class, the probability of membership of the pixel for each class is estimated. The decision relating to the labeling of the pixel is left to the analyst. Beyond the application of fuzzy classifiers, an alternative method is to adapt the results of regular "hard" classifiers, creating a fuzzy representation of land cover (Foody, 1996). This methodology provides more detailed membership information, which allows a greater understanding of the nature of each individual pixel. Compared to "hard" classification, "soft" classification methodology is more suitable to the classification problem of land cover features that are generally in a continuous form in nature. A major obstacle in their application lies in the difficulty of accurately assessing the results. The fuzzy C-means algorithm is a prevalent "fuzzy" classification technique in remote sensing, offering versatility for deployment in both unsupervised and supervised classification methodologies. More information on "fuzzy" classification can be found in Mather (1999), Ngo et al. (2015), Nguyen et al. (2015), Feilhauer et al. (2021), and Gao et al. (2022).

Another categorization is made by considering the fundamental unit to be characterized and classified. This brings out three main classification models: pixel-based, object-based, and scene-based classification approaches (Figure 2.1). Pixel-based classification is applied as a basic method of classification in which each pixel is evaluated and classified individually. The result of such a classification can be noisy, with a "salt-and-pepper" appearance, resulting from the fact that some pixels can be atypical, mixed, or unknown, which may not be included in the training data set. Moreover, an increase in spatial resolution increases the internal variation within land parcels. Although it is possible to get detailed classification results using such techniques, the noisiness in the output could be unacceptable in some cases. In object-based classification, each image object or segment defined through a segmentation process is the basic spatial unit to be classified. Object-based image analysis

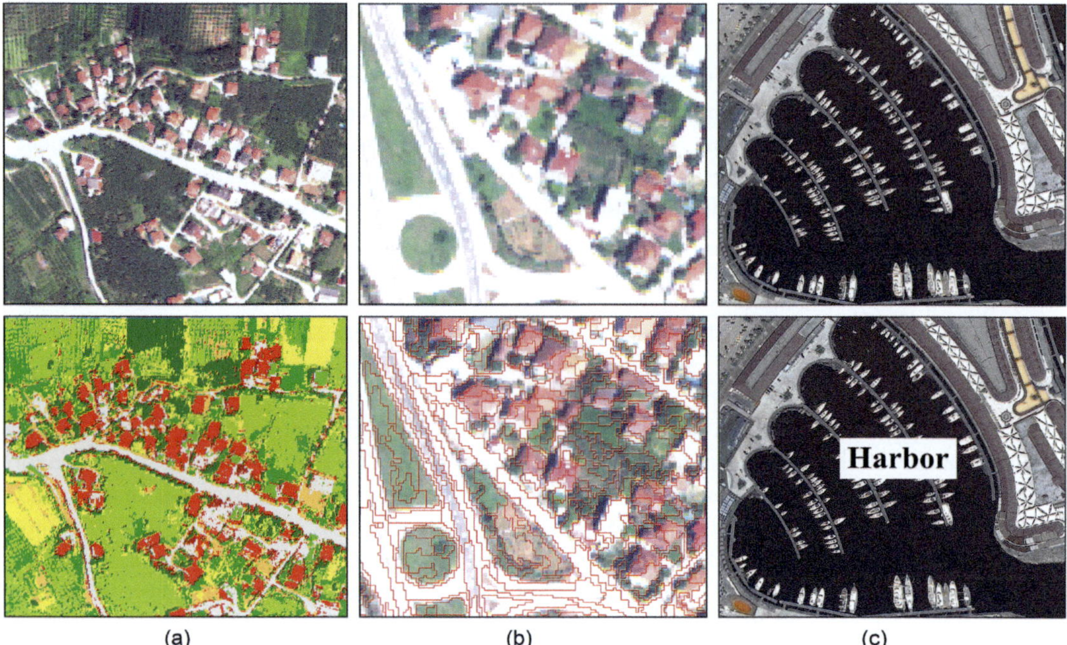

(a)   (b)   (c)

FIGURE 2.1   Major classification approaches. (a) Pixel-based, (b) object-based, (c) scene-based.

may be considered as the extension of the per-field classification that used to be a popular approach in the 1990s. Object-based image analysis (OBIA), or geographical object-based image analysis (GEOBIA), represents a new and evolving paradigm. It is utilized to generate homogeneous image objects and classify them, not just considering spectral features but also incorporating textural, contextual, and geometric attributes. As the number of objects to be classified is much less than the pixels in the image, the computational burden for identification and classification can be lowered. When this approach is applied, it is expected that the noise in the image can be averaged out, which should lead to improvement in the classification accuracy. Due to its importance and widespread use in the remote sensing field, a special chapter (Chapter 8) is devoted to OBIA.

A new paradigm receiving remarkable attention in recent years is remote sensing scene classification. A variety of deep learning techniques and a wide range of data sets have been proposed to apply scene classification, and satisfactory results have been reported. Being a hot research topic for the analysis of aerial and very high-resolution satellite images, scene classification is applied to classify scene images as patches (e.g., 256×256 or 512×512) into semantic classes considering the image contents. The term "scene" is in fact used to describe an image patch cropped from a remotely sensed image and contains distinct semantic information about an object. Various CNN models combined with robust approaches (e.g., vision transformers, feature vectors, convolutional features) have been proposed to boost the performance of a single CNN application (Cheng et al., 2017; Deng et al., 2022). Remote sensing scene classification is considered to be an important research topic, particularly for object detection, vegetation mapping, and natural hazard detection.

2.3   FUNDAMENTAL PATTERN RECOGNITION TECHNIQUES

2.3.1   Unsupervised Methods

As noted earlier, the main difference between the unsupervised and supervised approaches is that unsupervised methods do not require the user to select training data sets to characterize the targets or to train the classifier. Instead, the user specifies only the number of clusters to be generated, and the classifier automatically constructs the clusters by minimizing some predefined error function. Sometimes, the number of clusters can be detected automatically by the classifier (Cheung, 2005). In theory, users do not need to interact with the classifier, which operates independently and automatically. However, in practice, it is more often the case that results are accepted or rejected based on whether or not they meet the user's expectations.

2.3.1.1 The *k*-Means Algorithm

The *k*-means algorithm has been the most popular unsupervised algorithm, being widely used for automatic image segmentation in the field of remote sensing for many years (Ball and Hall, 1967; Mather, 1976, 2004). This algorithm is implemented by recursively migrating a set of cluster means (centers) using the "closest distance to mean" approach until the locations of the cluster means are unchanged, or until the change from one iteration to the next is less than some predefined threshold. Change may also be defined in terms of the number of pixels moving from one cluster to another between iterations, or by the value of a measure of cluster compactness, such as the sum of squares of the deviations of each pixel from the center of its cluster, summed for all classes.

The use of the *k*-means algorithm must estimate the initial number of clusters, say n, present in the data. The algorithm initially determines the location of n cluster means within the feature space, either by generating random feature vectors, or by selecting n pixels at random from the available data, or by using a predefined set of feature vectors. Each pixel is then associated with its nearest cluster center. The nearest distance is defined either by the Euclidean or by the Mahalanobis distance measure, which considers the covariance matrix. At the next stage, the location of each cluster mean is recalculated based on the set of pixels allocated to that center. The process is repeated until the change between iterations becomes less than a user-specified threshold (Figure 2.2).

The calculation of the distance between a pixel and a cluster center normally uses the Euclidean measure D_E represented by

$$D_E^2 = \left(\mathbf{x}_i - \boldsymbol{\mu}_j \right)^2 \tag{2.1}$$

where $\mathbf{x}_i$ is the observed vector of the ith pixel, and $\boldsymbol{\mu}_j$ is the current mean vector of the jth cluster. The dimension of vector $\mathbf{x}_i$ is equal to the number of bands being used as input. For instance, if three-band images are used, then $\mathbf{x}_i$ will be of dimension 3×1. Another choice for the calculation of distance between the pixel and the cluster center is the Mahalanobis distance D_M, given by

$$D_M = \left(\mathbf{x}_i - \boldsymbol{\mu}_j \right)^T \times \mathbf{C}_j^{-1} \times \left(\mathbf{x}_i - \boldsymbol{\mu}_j \right) \tag{2.2}$$

Begin
 Choose the number of clusters and mean change tolerance ε,
 Randomly assign the mean vector μ_j for each cluster j;
 Do while ($|\text{new } \mu_j - \text{old } \mu_j| < \varepsilon$) for all cluster j
  Do for each pixel i
   Calculate the distance D_{ij} between the pixel i and cluster mean μ_j for each cluster j;
   Assign pixel i to the cluster c according to $c = \min\{D_{ij}, \forall \text{ cluster j}\}$.
  EndDo
  Recalculate the cluster mean μ_j for each cluster j based on the pixels allocated to the cluster.
 EndDo
End

FIGURE 2.2 Algorithm for migrating mean clustering process.

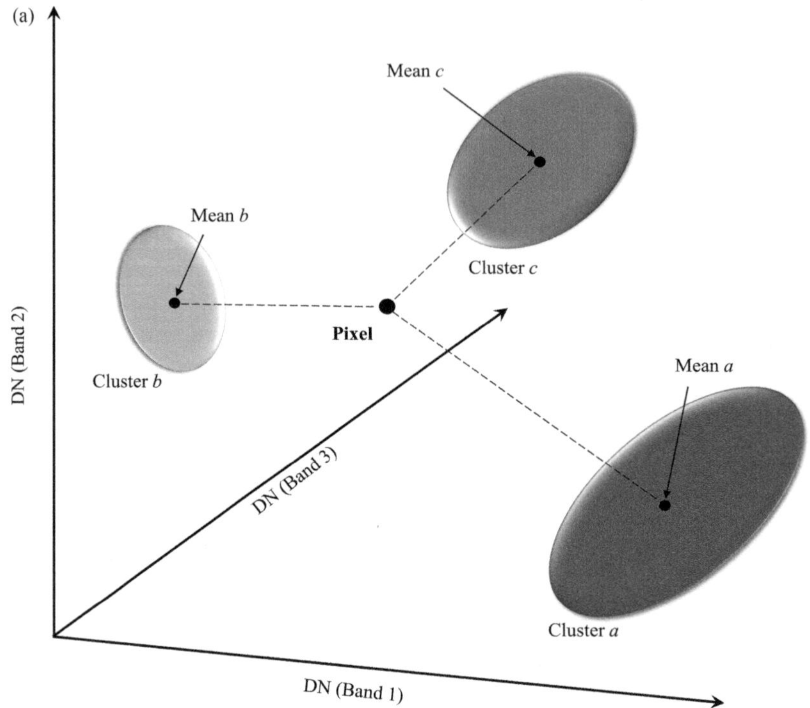

FIGURE 2.3a General cluster shapes (ellipsoidal) using the Mahalanobis distance measure defined in Equation (2.2).

where T denotes the matrix transpose, and $\mathbf{C}_j^{-1}$ is the inverse of the variance-covariance matrix for cluster j, respectively. Note that only pixels belonging to cluster j are used in the calculation of $\mathbf{C}_j$.

The Mahalanobis distance takes into account the shape of the frequency distribution (assumed to be Gaussian) for a given cluster in feature space, resulting in ellipsoidal clusters, whereas the use of the Euclidean distance assumes equal variances and a correlation of 1.0 between the features, giving circular clusters (Figure 2.3). Here, although the distances between the pixel and centers of clusters b and c are the same, the pixel will be assigned to cluster c when the Mahalanobis distance measure is used (Figure 2.3a) because cluster c is of higher value than cluster b in variance-covariance measure (thus the shorter the distance). If the Euclidean distance measure is used, then the decision to place the pixel in cluster b or c will be ambiguous (Figure 2.3b). In addition to the above considerations, the initialization points of the cluster-forming process may also play an important role in generating clustering outcomes. Pena et al. (1999) present a comparative study of different initialization methods for the k-means algorithm.

2.3.1.2 Fuzzy C-Means Clustering

An alternative approach to unsupervised classification using k-means is fuzzy clustering. The main difference between fuzzy clustering and the k-means method is that the resulting clusters generated by fuzzy clustering are no longer "hard" or "crisp" but fuzzy. In other words, each pixel may simultaneously belong to two or more clusters and will have a "membership value" for each cluster. In general, the performance of fuzzy clustering methods is superior to that of the corresponding hard versions, and they are less likely to stick to a local minimum (Bezdek, 1981).

Most analytic fuzzy clustering algorithms are derived from Bezdek's (1981) fuzzy c-means (FCM) algorithm. The FCM algorithm and its derivations have been implemented successfully in many applications, such as pattern classification and image segmentation, especially those in which the final goal is to make a crisp decision. The FCM algorithm uses the probabilistic constraint that the membership probabilities of a data point across classes must sum to one. This constraint

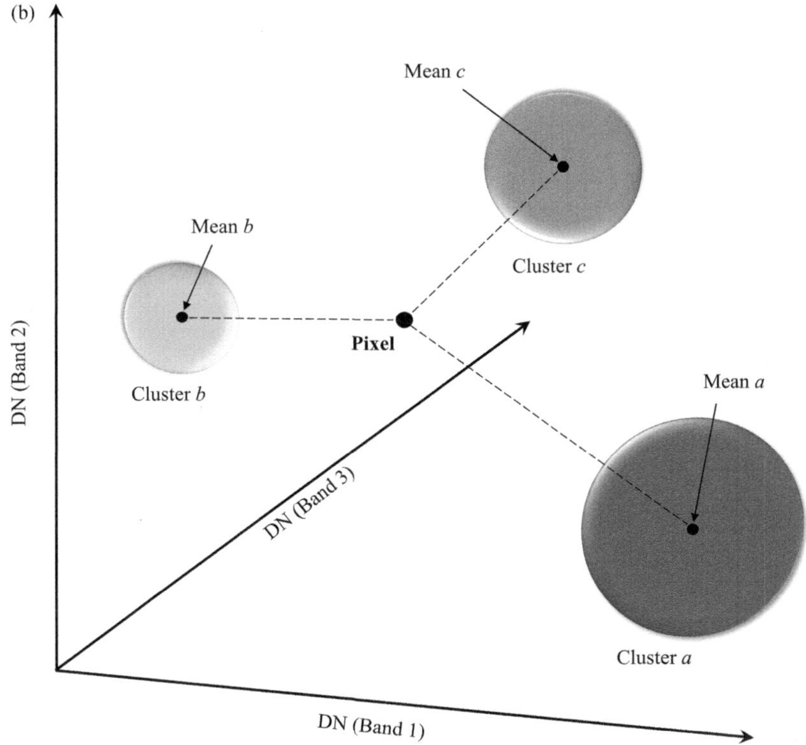

FIGURE 2.3b Circular cluster shape using the Euclidean distance measure defined in Equation (2.1).

comes from generalizing a crisp c-partition of a data set and is used to generate membership-update equations for an iterative algorithm based on the minimization of a least squares type of criterion function. The constraint on membership used in FCM is meant to avoid the trivial solution of all membership probabilities being equal to zero.

Methods such as k-means assume that clusters can be adequately represented as spherical patterns in feature space and that the mean of each cluster can be estimated reasonably. Figure 2.4 illustrates an example of a possible invalid clustering. A more robust clustering method is clearly needed in this and similar situations. The FCM algorithm (Bezdek, 1981; Bezdek et al., 1984) provides a means of overcoming the false clustering problem. This algorithm separates data clusters with fuzzy means and fuzzy boundaries and is less dependent on the initial state of clustering. The algorithm is described as follows.

Let $X = \{x_1, x_2, \ldots, x_n\}$ be a finite subset of R^d, the d-dimensional real number vector space (e.g., where three features are used for clustering, $d=3$). Let the integer c, $n \geq c > 2$, denote the number of fuzzy subsets. Thus, a fuzzy c-partition of X can be represented by a $(c \times n)$ matrix U in which each entry of U, denoted by u_{ik}, satisfies the following two constraints: $u_{ik} \in [0, 1]$ and

$$\sum_{i=1}^{c} u_{ik} = 1, \quad \text{for all } k \tag{2.3}$$

In the case of image classification, n will be the number of pixels, and c is the number of clusters. The resulting matrix of a sample data set comprising 150 samples (30 per class) for 10 clusters in the case of 5 classes is shown in Figure 2.5. The value u_{ik} corresponding to the entry at the location (i, k) stores the kth pixel's membership value for class i. For instance, the second pixel's membership value for cluster 2 ($u_{i=2,k=2}$) is 0.96. Note that all the entries in a given column must sum to 1 as specified in Equation (2.3).

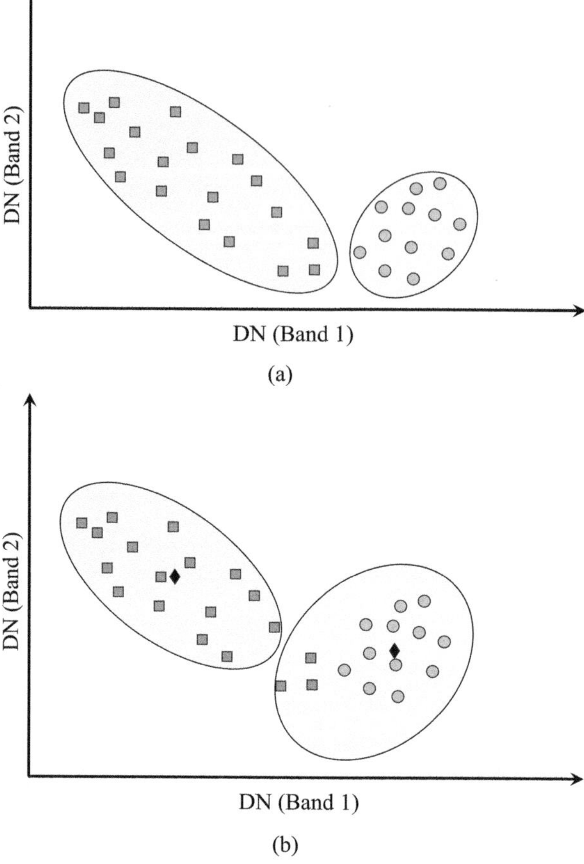

FIGURE 2.4 (a) Two clusters shown in red and blue. (b) Possible invalid clustering result generated by ISODATA. The lozenges (in gray) denote two cluster centers.

The clustering criterion used in the FCM algorithm is based on minimizing the generalized within-groups sum of the square error function J_m:

$$J_m(U,V) = \sum_{k=1}^{n}\sum_{i=1}^{c}(u_{ik}) \times (x_k - v_i)^2 . \tag{2.4}$$

$V = (v_1, v_2, \ldots, v_c)$ is the vector of cluster centers (i.e., the means of the clusters), with $v_i \in R^d$, and m is the membership weighting exponent, $1 \le m < \infty$. The term $(x_k - v_i)^2$ in Equation (2.4) can be replaced by the Mahalanobis distance (Equation 2.2), the calculation of which requires the fuzzy covariance matrix, which is introduced in Equation (2.9). For $m > 1$ and $x_k \neq v_i$, for all i, k, a local minimum of J_m may be achieved under the circumstance

$$u_{ik} = \frac{1}{\sum_{j=1}^{c}\left(\dfrac{|x_k - v_i|}{|x_k - v_j|}\right)^{2/(m-1)}} \quad \text{for all } k \tag{2.5}$$

and the ith cluster mean is calculated from

$$v_i = \frac{\sum_k (u_{ik})^m \times x_k}{\sum_k (u_{ik})^m} \quad \text{for all } i. \tag{2.6}$$

No. of clusters	Number of pixels					
	1	2	3	.	.	$n = 150$
1	0.01	0.00	0.00	.	.	0.01
2	0.32	0.96	0.01	.	.	0.00
3	0.04	0.00	0.00	.	.	0.00
.	.	.	.	.	.	.
.	.	.	.	.	.	.
$c = 10$	0.01	0.00	0.00	.	.	0.75

FIGURE 2.5 Fuzzy c-means clustering membership matrix.

```
Begin
    Choose the number of clusters and mean exponent m, and the false
    tolerance ε,
    Randomly initialize clustering membership matrix U;
Do
    Compute cluster mean vᵢ, for all i, using Equation (2.6);
    Update matrix U using Equation (2.5);
Until ( ‖ Uⁿᵉʷ − Uᵒˡᵈ ‖ < ε)
End
```

FIGURE 2.6 Pseudocode for fuzzy c-means clustering algorithm.

The FCM clustering is thus performed by iteratively applying Equations (2.5) and (2.6). The pseudocode of the algorithm is presented in Figure 2.6.

The effect of the exponent is worthy of further consideration. As $m \rightarrow 1$, the solution of the FCM procedure converges to hardness or crispness (i.e., a pixel's membership grades become closer to 1 or 0, respectively), while the greater the value of m (e.g., two or more), the fuzzier the membership assignments will be (i.e., the membership grades for all clusters are close to each other). Cannon et al. (1986) suggested that for any value of m between 1.3 and 1.8, a high-performance oriented segmentation could be obtained. Nikhil and Bezdek (1995) suggested that the choice of the m value in the interval [1.5, 2.5] gives an acceptable clustering result, whereas Chen and Lee (2001) concluded that an m value of 2.5 gives the best results. Once the clustering process is finished, each pixel can be assigned to the cluster for which the pixel's membership grade is largest, if one wants to perform a crisp interpretation. Alternatively, one can scale membership grades to the interval [0, 255] and produce membership grade images for each class.

The choice of the number of clusters is not always easy, especially when the user does not have any knowledge about the number of information classes. An alternative procedure to solve this issue is to design specific indices to measure clustering performance with a range of cluster numbers. The solution generating the optimal performance indices will usually be adopted. For instance, Pal and Bezdek (1995, 1997) measure clustering validity for different cluster numbers in terms of a within-clusters compactness index and an among-clusters separation index. Kim et al. (2004) propose an approach for determining a suitable number of clusters by minimizing an overlapping index and maximizing a between-clusters separation index. Another interesting algorithm called fuzzy partition – optimum number of clusters (Gath and Geva, 1989) uses a combination of the FCM and fuzzy maximum likelihood algorithms (Section 2.3.2.4) and is illustrated here as an example. The algorithm is similar to the FCM algorithm described, except that the distance measure used to define similarity between pixel x_k and a cluster center v_i is defined as

$$d^2(x_k, v_i) = \frac{\sqrt{|F_i|}}{P_i} \exp\left[\frac{(x_k - v_i)^T F_i^{-1}(x_k - v_i)}{2}\right] \tag{2.7}$$

where P_i is the *a priori* probability of the ith cluster, defined as

$$P_i = \frac{\sum_k u_{ik}}{n} \tag{2.8}$$

n is the number of pixels, and F_i is the fuzzy covariance, expressed by

$$F_i = \sum_k \frac{u_{ik}^m (x_k - v_i)(x_k - v_i)^T}{\sum_k u_{ik}^m} \tag{2.9}$$

and $|F_i|$ denotes the determinant of the ith cluster covariance matrix.

Note that, during the clustering process, the pixel membership grades, the cluster means, and the covariance matrix F_i must be modified at each iteration. Once the algorithm has converged, the process is repeated using a different number of clusters. Finally, the optimal clustering scheme is chosen in terms of the minimization of the overall determinants of the fuzzy covariance matrices (i.e., $|F_i|$) for all i.

The standard FCM method assumes that every information class has been specified so that each pixel can be completely described in terms of its membership in the given set of categories. This is not always the case, as noted by Foody (2000). He suggests that the standard FCM method produces membership grades that are analogous to relative proportions of the specified classes (with the possibility of error if some classes are omitted). The possibilistic c-means and improved probabilistic c-means algorithms (Krishnapuram and Keller, 1993, 1996; Zhang and Leung, 2004) were also proposed as alternatives, as they provide a means of estimating the absolute membership grade of each class independently of all other classes, and less sensitive to noise or outliers.

2.3.2 Supervised Methods

The supervised approach to pixel labeling requires the user to select representative training data for each of a predefined number of classes. Classification performance is highly dependent on how well the user can model the target class distribution. The user's experience can be very helpful in identifying and locating training areas. Ideally, the training areas should be the sites where homogeneous examples of known cover types are found (Townshend, 1981). A supervised statistical classification can be carried out by the following three steps:

1. Define the number and nature of the information classes, and collect sufficient and representative training data for each class,
2. Estimate the required statistical parameters from the training data, and
3. Use an appropriate decision rule.

Although the selection of training data may be tedious, most researchers prefer a supervised approach because it generally provides more accurate class definitions and higher accuracy than unsupervised approaches do. Three statistical classifiers are in widespread use, which are the parallelepiped method, the minimum distance classifier, and the maximum likelihood algorithm.

2.3.2.1 Parallelepiped Method

The parallelepiped method is implemented by defining a parallelepiped-like subspace (i.e., hyper-rectangle) for each class. The boundaries of the parallelepiped, for each feature, can be defined by the minimum and maximum pixel values in the given class, or, alternatively, by a certain number of standard deviations on either side of the mean of the training data for the given class. The decision rule is simply to check whether the point representing a pixel in feature space lies inside any of the parallelepipeds. An example illustrating the specification of the topology of a parallelepiped classifier in the case of a two-dimensional feature space is shown in Figure 2.7.

The parallelepiped method is quick and easy to implement, but errors may arise, particularly when a pixel lies inside more than one parallelepiped or outside all parallelepipeds. These two situations are, in fact, likely to occur, because the distribution of pattern vectors is often quite complex in the feature space. Therefore, it is hard to provide a robust classification performance using this simple method.

2.3.2.2 Minimum Distance Classifier

Like the migrating mean clustering algorithm, the decision rule adopted by the minimum distance classifier to determine a pixel's label is the minimum distance between the pixel and the class centers, measured either by the Euclidean distance or by the Mahalanobis generalized distance (Equations 2.1 and 2.2). Both of these distance measures can be described as dissimilarity coefficients, in that the similarity between objects i and j increases as the distance becomes smaller. The mean spectral vector, or class centroid, for each class (plus the variance-covariance parameters if the Mahalanobis distance measure is used in place of the Euclidean distance) is determined from training data sets. A pixel of unknown identity is labeled by computing the distance between the value of the unknown pixel and each class centroid in turn. The label of the closest centroid is then assigned to the pixel.

An example is shown in Figure 2.8, in which pixel a is allocated to class 2 (i.e., given the label "2") according to the minimum distance between the pixel and the class center. The shape of each class depends on which distance function (Mahalanobis or Euclidean) is used.

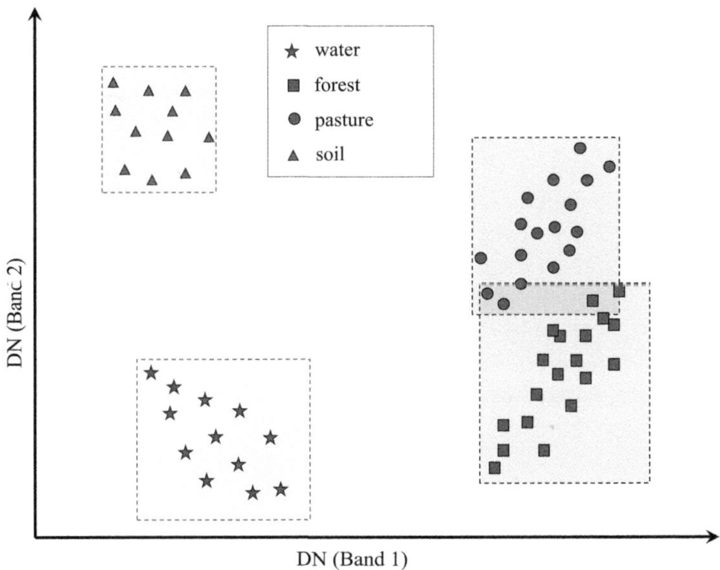

FIGURE 2.7 Each parallelepiped is bounded by minimal and maximal point values within the class.

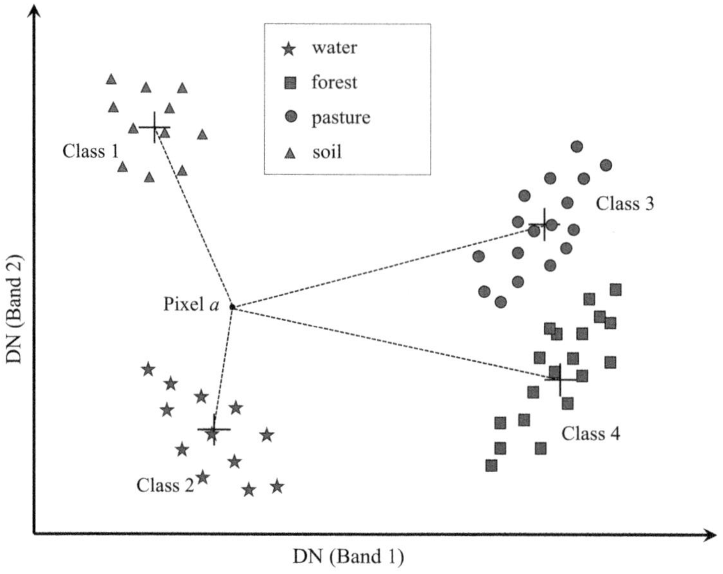

FIGURE 2.8 Example of minimum distance classification of pixel *a*. Crosses represent the mean centers of each class.

This type of classifier is mathematically simple and computationally efficient, but the theoretical basis may not be as robust as that of the maximum likelihood classifier. A comparison reported by Benediktsson et al. (1990) shows that although the minimum distance classification using the Mahalanobis distance measure performs better than the same classifier using the Euclidean distance measure, its accuracy still falls considerably behind that achieved by the maximum likelihood algorithm. Similar to the parallelepiped classifier, this algorithm disregards all the training data and concentrates on the mean spectral value in each band for every class.

2.3.2.3 Maximum Likelihood Classifier

The maximum likelihood (ML) procedure is a supervised statistical approach to pattern recognition. The probability of a pixel belonging to each of a predefined set of classes is calculated, and the pixel is then assigned to the class for which the probability is the highest. ML is based on the Bayesian probability formula:

$$P(\mathbf{x}, w) = P(w|\mathbf{x})\, P(\mathbf{x}) = P(\mathbf{x}|w)\, P(w) \tag{2.10}$$

where $\mathbf{x}$ and w are generally called *events*. $P(\mathbf{x}, w)$ is the probability of coexistence (or intersection) of events $\mathbf{x}$ and w, $P(\mathbf{x})$ and $P(w)$ are the prior probabilities of events x and w, and $P(w|\mathbf{x})$ is the conditional probability of event x given event w. $P(w|\mathbf{x})$ is interpreted in the same manner. If event $\mathbf{x}_i$ is the ith pattern vector, and w_j is information class j, then, according to Equation (2.10), the probability that $\mathbf{x}_i$ belongs to class w_j is given by

$$P(w_j|\mathbf{x}_i) = \frac{P(\mathbf{x}_i|w_j)P(w_j)}{P(\mathbf{x}_i)}. \tag{2.11}$$

Since, in general, $P(\mathbf{x})$ is set to be uniformly distributed (i.e., the probability of occurrence is the same for all pixel features), Equation (2.11) can be rewritten as

$$P(w_j|\mathbf{x}_i) \propto P(\mathbf{x}_i|w_j)P(w_j). \tag{2.12}$$

One can thus allocate pixel i to the class k, which has the largest value of the term $P\left(w_k \mid \mathbf{x}_i\right)$ in Equation (2.12). The classification criterion can be expressed as

$$w_k = \underset{\forall w_j}{\arg\max}\left\{P\left(\mathbf{x}_i \mid w_j\right)P\left(w_j\right)\right\} \tag{2.13}$$

where *arg* denotes argument. The criterion shown in Equation (2.13) is called the Maximum A Posteriori (MAP) solution, which maximizes the product of conditional probability and prior probability. However, on some occasions, the prior probability $P(w)$ is also set to be uniformly distributed (because of a lack of prior knowledge or because one does not know the true distribution). In this case, Equation (2.12) reduces to

$$P\left(w_j \mid \mathbf{x}_i\right) \propto P\left(\mathbf{x}_i \mid w_j\right). \tag{2.14}$$

If we allocate pixel i to that class k that maximizes expression (2.14), the result is called the maximum likelihood solution.

Normally, the conditional probability $P\left(\mathbf{x}_i \mid w_j\right)$ is assumed to follow a Gaussian (or normal) distribution assumption. $P\left(\mathbf{x}_i \mid w_j\right)$ can then be expressed as

$$P\left(\mathbf{x}_i \mid w_j\right) = \frac{1}{\sqrt{2\pi}^\rho \sqrt{\left|\mathbf{C}_j\right|}} \exp\left(-\frac{1}{2} \times \left(\mathbf{x}_i - \boldsymbol{\mu}_j\right)^T \times \mathbf{C}_j^{-1} \times \left(\mathbf{x}_i - \boldsymbol{\mu}_j\right)\right) \tag{2.15}$$

where $\mathbf{C}_j$ is the covariance matrix of class w_j with dimension ρ, $\boldsymbol{\mu}_j$ is the mean vector of class w_j, and $|\mathbf{x}|$ denotes the determinant. In practical applications, Equation (2.15) can be further reduced to the following expression by taking the natural logarithm:

$$\ln\left[P\left(\mathbf{x}_i \mid w_j\right)\right] = \frac{-1}{2}\rho \times \ln(2\pi) \times \frac{-1}{2}\ln\left|\mathbf{C}_j\right| - \frac{1}{2}\left(\mathbf{x}_i - \boldsymbol{\mu}_j\right)^T \times \mathbf{C}_j^{-1} \times \left(\mathbf{x}_i - \boldsymbol{\mu}_j\right). \tag{2.16}$$

Equation (2.16) avoids the computation of the exponential term. Since the term $\rho\ln(2\pi)$ is the same for all classes, it can be regarded as a constant and dropped from the equation without affecting the final ranking of the values of $\ln\left[P\left(\mathbf{x}_i \mid w_j\right)\right]$. Finally, the expression in Equation (2.16) is multiplied by the constant -2 to give

$$-2\ln\left[P\left(\mathbf{x}_i \mid w_j\right)\right] = \ln\left|\mathbf{C}_j\right| + \left(\mathbf{x}_i - \boldsymbol{\mu}_j\right)^T \times \mathbf{C}_j^{-1} \times \left(\mathbf{x}_i - \boldsymbol{\mu}_j\right). \tag{2.17}$$

It is clear that maximizing Equation (2.15) is equivalent to minimizing Equation (2.17). Note that the second term of Equation (2.17) is the Mahalanobis distance, Equation (2.2). Thus, the geometric shape of the cloud formed by a set of pixels belonging to a given class can be described as an ellipsoid. The shape of the ellipsoid depends on the covariance among the features belonging to a feature space. To assess the likelihood probabilities of pixels to be classified, a set of concentric ellipses is employed, with their center positioned at the mean vector of a specific class. In a two-dimensional feature space, the ML function delineates ellipsoidal equiprobability contours, which can be viewed as decision boundaries (see Figure 2.9). The probability of class membership is depicted by these concentric ellipses, with contours indicating a decrease in probability as one moves away from the mean center. Two parameters, the mean vector and the covariance matrix, are used to characterize each class. The importance of choosing a sample size that is adequate to provide an unbiased and efficient estimate of these parameters is considerable (see Section 10.1.2).

The ML classifier assumes that the information class prior probability $P(w)$ is uniformly distributed. However, if one can model $P(w)$ in a suitable way, the classification accuracy could be increased. For example, one may model prior probability as different weights associated with

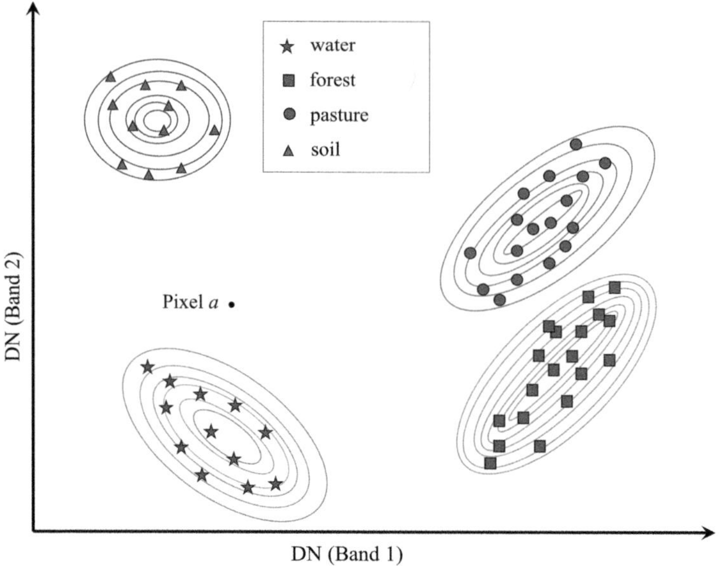

FIGURE 2.9   Example of maximum likelihood classification of pixel a. Note that equiprobability contours show the probability of membership of a class.

each class. A higher weight for a given class implies that there is a higher probability of a pixel receiving the label associated with that class. The effects of modeling $P(w)$ are discussed by Strahler (1980) and Mather (1985, 2004). In recent years, there has been a trend toward modeling the prior probability using a smoothness assumption based on the concept of context. This method is called the smoothness prior. By using the smoothness prior together with the class-conditional probability $P(\mathbf{x}|w)$, one can attempt to perform a MAP classification.

As the performance of conventional classifiers is generally limited by frequency distribution assumptions, in the last two decades, more elegant classifiers such as artificial neural networks and deep learning (Chapter 7), support vector machines (Chapter 5), and decision trees (Chapter 6) have also been introduced into the field of remote sensing imagery classification. These state-of-the-art classifiers draw more attention, because they normally are distribution-free, and are able to show a significant level of improvements over traditional methods introduced previously in this section.

2.3.2.4   Fuzzy Maximum Likelihood Classifier

The crisp ML classification algorithm uses the hard mean and hard covariance matrix of each cluster. In what follows, we show that fuzzy set theory can be extended to the ML algorithm to measure the membership grade of the pixels (Wang, 1990; Maselli et al., 1995; Foody, 1996; Frizzelle and Moody, 2001).

Based on probability theory, if an event A is a precisely (hard) defined set of elements in the universe of discourse Ψ, the probability density function of A denoted by $P(A)$ can be expressed by

$$P(A) = \int_{\Psi} H_A(\mathbf{s}) \tag{2.18}$$

where $\mathbf{s}$ denotes the element in Ψ, and H_A is a hard membership function, that is, $H_A(\mathbf{s}) = 0$ or 1. In the case of image classification, event A is the cluster or class, and $\mathbf{s}$ is a vector of feature values associated with a specific pixel. The membership function H_A thus describes $\mathbf{s}$ either as belonging to A (i.e., membership grade is 1) or not (i.e., membership grade is 0).

If A is regarded as a fuzzy event, which means that set A is a fuzzy subset in Ψ, a probability measure of A becomes

$$P(A) = \int_\Psi \mu_A(\mathbf{s}).\tag{2.19}$$

The term μ_A is the fuzzy membership function as defined in $\mu_G : \mathbf{s} \to [0,1]$. Equation (2.19) is an extension and generalization of Equation (2.18). Even a partial membership value of observation $\mathbf{s}$ in A can provide a contribution to the total probability $P(A)$.

The mean and variance of fuzzy set A relative to a probability measure can similarly be quantified as

$$v_A = \frac{1}{P(A)} \int_\Psi s\mu_A(\mathbf{s})\tag{2.20}$$

and

$$\sigma_A^2 = \int_\Psi (\mathbf{s} - v_A)^2 \mu_A(\mathbf{s}).\tag{2.21}$$

Equations (2.20) and (2.21) determine the fuzzy mean and fuzzy variance, respectively, both of which are derived for the continuous case. In practice, the discrete fuzzy mean v_i and fuzzy covariance matrix $\mathbf{F}_i$ for class i can be expressed as

$$v_i = \frac{\sum_j \mu_i(\mathbf{x}_j)\mathbf{x}_j}{\sum_j \mu_i(\mathbf{x}_j)}\tag{2.22}$$

and

$$\mathbf{F}_i = \frac{\sum_j \mu_i(\mathbf{x}_j) \times (\mathbf{x}_j - v_i) \times (\mathbf{x}_j - v_i)^T}{\sum_j \mu_i(\mathbf{x}_j)}\tag{2.23}$$

where $\mathbf{x}_j$ denotes the feature vector for pixel j. If the exponent m in Equation (2.9) is set to 1, Equation (2.23) becomes equivalent to Equation (2.9), which is used in the optimum clustering algorithm. It can be inferred that when the value of $\mu_i(\mathbf{x}_j)$ becomes either 0 or 1 (i.e., crisp membership function), Equations (2.22) and (2.23) become the conventional crisp mean and covariance matrix formulations.

A fuzzy set is characterized by its membership function. Wang (1990) defines the membership grade for each land cover class based on the ML classification algorithm with a fuzzy mean and fuzzy covariance matrix as shown in Equations (2.22) and (2.23) as follows:

$$\mu_k(\mathbf{x}_j) = \frac{P_k(\mathbf{x}_j)}{\sum_i P_i(\mathbf{x}_j)}\tag{2.24}$$

where k is the land cover class, and probability $P_i(\mathbf{x}_j)$ denotes the class-conditional probability for class i given the observation $\mathbf{x}_j$, except that the crisp mean and covariance matrix in Equation (2.15) are replaced by the fuzzy mean and fuzzy covariance matrix. This is the core of the fuzzy ML algorithm. Calculating membership grades in terms of Equation (2.24) is equivalent to normalizing the probabilities of the pixel to all of the information classes. Although such a method is quite straightforward, its validity requires further investigation.

2.4 SPECTRAL UNMIXING

Pixels in remotely sensed imagery are often formed by the mixture of several land use and land cover classes present at the land surface during sensor acquisition. In addition, low spatial resolution of current hyperspectral sensors on orbit poses problems of mixed pixel content. Such pixels are called mixed pixels, which are abundant throughout the remote sensing images and result in so-called mixed pixel problem in classification practices. Main sources of spectral mixing can be attributed to optical imaging systems, different surface materials in the field of view, variable illumination conditions due to the topographic effects (van der Meer, 1999). It is also worth noting that the effect of the sensor's point spread function (energy transfer from neighboring IFOVs) leads to ambiguity depending on the sensor's characteristics. The abundant spectral detail present in hyperspectral data cubes has the potential to separate mixed pixels in hyperspectral images. Spectral unmixing or spectral mixture analysis, which is a major concern for hyperspectral images, aims to unveil the composition of individual pixels in terms of their constituent proportions of various land cover classes. Unlike the conventional practice of assigning pixels to a single land cover category in hard classification, spectral unmixing overcomes this constraint by offering insight into information at a finer subpixel level. As opposed to representing pixels with a single class label, mixture modeling provides opportunities to represent a number of spectrally pure pixels, commonly termed as endmembers. This enables the extraction of more detailed data, thereby enhancing the decision-making process for users. Due to the representation of pixels with different classes with their contributions, Settle and Drake (1993) use the term of soft classification to designate spectral unmixing models.

Various unmixing models including linear, probabilistic, geometric-optical, stochastic-geometric, spatiotemporal, and fuzzy models have been proposed by researchers (Heinz and Chang, 2001; Keshava and Mustard, 2002; Quintano et al., 2012; Wang et al., 2021). Considering the underlying assumption, Heylen et al. (2014) categorized the mixture models as the linear spectral mixture model (LSMM) and nonlinear model. The LSMM has become popular as it considers each pixel as a linear combination of several endmembers. Therefore, the aim of the LSMM is to establish a set of linear coefficients constructing the mixed pixel spectrum. The observed spectra $y_n \in \mathbb{R}^B$, $n = 1,\ldots,N$ (B is the number of wavelengths, and N is the number of pixels) are a nonnegative linear combination of the endmember spectra $m_j \in \mathbb{R}^B$, $j = 1,\ldots,M$, where M is the number of endmembers (Ma et al., 2019).

$$y_n = \sum_{j=1}^{M} m_j \alpha_{nj} + n_n, \quad \text{s.t.} \quad \alpha_{nj} \geq 0, \quad \sum_{j=1}^{M} \alpha_{nj} = 1, \tag{2.25}$$

where α_{nj} is the proportion (called abundance) for the jth endmember at the nth pixel (with the positivity and sum-to-one constraint), and $n_n \in \mathbb{R}^B$ is additive noise. Also, the endmember set m_j is fixed for all pixels. Various endmember detection algorithms, based on the pixel purity assumption, have been developed to identify pure endmembers. Examples include the pixel purity index, successive projection algorithm, orthogonal subspace projection, and vertex component analysis, and they have found application in the LSMM (e.g., Arngren et al., 2011; Bioucas-Dias et al., 2012; Feng et al., 2016; Ghaffari et al., 2017).

In many cases of hyperspectral unmixing, the LMM may not be valid. Therefore, the Gaussian mixture model (GMM) may be employed for unmixing to represent endmember variability (Zhou et al., 2018). In probability theory, the central limit theorem is based on the fundamental idea that when sample sizes become larger, the distribution of a sample variable takes on the characteristics of a normal distribution. This principle remains valid regardless of the original distribution shape of the population, given that all samples are uniform in size. Thus, for a sufficiently large sample size from a population with a finite level of variance, the mean of all sampled variables from the same population will be approximately equal to the mean of the whole population. From this point of view, the use of Gaussian models for the representation of multivariate data can be justified. Adhering to this philosophy, GMM possesses the ability to represent a wide range of data distributions, making it a suitable tool for multivariate distribution parameterization. The expectation maximization (EM) algorithm or its adaptations are a prevalent method for fitting the GMM to data (McLachlan and Krishnan, 2008). Through the application of an appropriate parametric statistical model and a decision rule optimized by Bayes' theorem, the GMM algorithm presents a systematic methodology for addressing complex elements of unsupervised classification. Particularly, the determination of the number of mixture components can be achieved in a methodically accurate and well-supported manner.

The GMM is built upon the fundamental premise that an observation x represents the realization of a multivariate random variable X with a probability density function (p.d.f.). In the framework of mixture modeling for classification, the distribution of the entire data set is conceptualized as a superpopulation, comprising a finite mixture (weighted sum) of class-conditional densities. In unsupervised statistical learning, the probability density functions (p.d.f.s) of the classes and the number of classes are unknown and need to be estimated. This estimation is performed using a set of training pixels where the class membership is not provided. Once the class densities are estimated, a new pixel can be classified by choosing the probability density function (p.d.f.) of the class that is considered the most probable source, utilizing the MAP probability decision rule. The GMM approach, coupled with Bayesian classification, stands out due to its robust mathematical basis, effective handling of natural noise, and specific applicability to remote sensing.

The GMM can be utilized in both supervised and unsupervised unmixing scenarios. In the supervised approach, a spectral library is used as input to estimate abundance, through either visual interpretation of images or other methods. Conversely, the unsupervised approach involves segmenting the image to identify regions with pure pixels, allowing for unmixing under the assumption that such regions exist. When high spectral mixing is present in the data, geometric approaches can produce unsatisfactory results due to the insufficient presence of spectral vectors within the simplex facets (Ghamisi et al., 2017). Statistical methods offer a potent alternative, often entailing greater computational intricacy compared to geometric methods. Independent component analysis (Section 3.1.4) and Bayesian approaches have been applied for this purpose since they possess the ability to represent statistical fluctuations and enforce prior assumptions.

Spectral unmixing has received great attention from the scientific community in the last three decades and recently evolved into machine learning and neural network methods that have inherent superiority to deal with mixed pixels. Brown et al. (2000) stated that constrained least squares unmixing is equivalent to the linear support vector machines (SVMs) when both models utilize the same thresholding and normalizing operations. When employing a linear SVM to achieve pure pixel classification, it is assumed that there is a two-class problem, and the data points can be divided by a straight line. The primary goal is to discover the linear boundary that provides the most significant separation between the identified classes. The mixed pixels can be used for SVM training after the mixed region is defined (Foody and Mathur, 2006). Wang and Jia (2009) introduced a supervised unmixing approach utilizing an extended SVM and conducted tests to highlight its advantages over the constrained least squares method. They successfully addressed and resolved the spectral variability challenge associated with each endmember, showcasing effective handling, especially in situations with available training samples.

In the late 1990s, researchers started to focus on the evaluation of the potential of artificial neural networks in spectral mixture modeling. One of the earliest studies was conducted by Foody (1996) in that an approach based on relating the activation level of artificial neural network output units was presented to show the strength of class membership. The original activation levels derived were only poorly related to the land cover composition of the pixel. If a class covered more than half of the pixel's area, the activation levels of the output unit associated with that class were high and that for other output units was low. By employing the unit activation function to rescale activation levels, a metric representing the strength of class membership, closely associated with the land cover composition of pixels, was generated. Tatem et al. (2001a,b) suggested a fuzzy classification approach to limit a Hopfield neural network designed as an energy minimization tool. Spectral unmixing for individual pixels is represented through a straightforward spatial clustering function embedded in a Hopfield neural network. Mei et al. (2010) additionally introduced a fully constrained linear unmixing algorithm based on the Hopfield neural network (HNN), ensuring the nonnegative constraint through the activation function of neurons rather than conventional numerical methods. Due to its effectiveness in soft classification of remotely sensed imagery, Hopfield neural networks have also been employed in super-resolution mapping applications (e.g., Tatem et al., 2003; Nguyen et al., 2011; Su, 2019). On the other hand, Licciardi and Del Frate (2011) introduced a neural network–based methodology for extracting pixel abundance from hyperspectral data. The use of multilayer perceptron (MLP) topologies addresses both feature reduction and abundance estimation, offering an efficient solution to handle high-dimensional input vectors in the context of pixel unmixing. The decision to use MLP models is grounded in two primary factors: their demonstrated suitability for classification and inversion problems in terms of topology. Autoassociative networks, tasked with dimensionality reduction, offer a nonlinear principal component analysis of the data. Both the dimensionality reduction process and the final unmixing are carried out by the neural network model. With the advances in computing power and developments in deep learning methodologies, the spectral unmixing problem has been studied using supervised learning and learning the underlying structures by unique architectures, including autoencoders, CNNs, and generative models, details of which will be given in Chapter 7. Theory and application of these deep learning models to spectral unmixing can be found in Zhang et al. (2018), Borsoi et al. (2019), and Deshpande and Bhatt (2021).

2.5 ENSEMBLE CLASSIFIERS

Different approaches to pattern recognition are often viewed as alternative methods, and many researchers have published comparisons among the various procedures to demonstrate that one is "better" than the other in some way. It appears that many of the methods are complementary; some are better in resolving one aspect of the labeling problem, while another method may be superior in another respect. Hence, some interest has been shown in combining the results obtained using different decision rules to improve the overall labeling. This type of system is called an ensemble classifier, committee-based learning, and also multiple classifier system. In these systems, the final decision is not given by considering a single classifier, but a combination of *a posteriori* class probabilities estimated by all classifiers. Disagreement of the classifiers represents the diversity of the ensemble. An ensemble comprises a number of learners called base learners, which can be a decision tree, neural network, or other learning algorithms. Ensemble classifiers are reported to produce better generalization capabilities than base learners (Briem et al., 2002; Pal, 2007; Yan and Shaker, 2011).

If the results from several classifiers are to be amalgamated, then the classifiers should be independent. To be independent, these classifiers must each use an independent feature set or be trained on separate sets of training data. One possibility is to use subsets of the features for each separate classifier, or to use random sampling of the training data to generate p different subsets of training data, where p is the number of classifiers. This fundamental idea constructs the theories of bagging and boosting and random forest. Bagging (Breiman, 1996) is performed by generating bootstraps from the initial data set, building a classifier for each sample, and using voting methods to reach the final prediction. As a sequential ensemble method, boosting (Freund and Schapire, 1999) iteratively adjusts the weights for each model and increases the weight of the ones incorrectly classified. On the other hand, random forest (Breiman, 2001) has been one of the most widely used ensemble classifiers for classification, regression, and feature selection applications. Theories of these algorithms and their variants are discussed in Chapter 6 (Section 6.7).

Cross-validation, which has been widely applied for parameter tuning, is another possibility, in that the test data are subdivided into a number n of approximately equally sized subsets, and $(n-1)$ of these subsets are used to train the classifier. The remaining nth subset is used for estimating the error associated with the labeling. Each of the n subsets is used in turn for testing, with the remaining $(n-1)$ subsets being used for training. Once this cycle of training and testing is completed, the error estimates are combined.

Brief details of combination methods known as voting rules, Bayesian formulation, evidential reasoning, and multiple neural networks are given here and in later chapters. The procedure using voting rules is quite simple. The label outputs by a number of classifiers for a given pixel are collected, and the majority label is selected. This is known as the majority vote rule. A more stringent requirement is that all classifiers agree on a single label, and this approach uses the conservative voting rule. The latter rule is unlikely to produce a fully labeled thematic image if the number of classifiers used is more than two. The former rule may produce conflicts. To avoid the possibility of choosing label A when the evidence supporting A is not much greater than the evidence supporting label B, the method of comparative majority voting can be used. This method requires that the "winning" vote should exceed the runner-up vote by a specified amount, thus ensuring a clear winner. Pixels that are not labeled because no clear winner emerges are labeled as unknown (Vieira, 2000). In an experimental study, Kittler et al. (1998) concluded that the combination rule developed under the most restrictive assumptions – the sum rule – outperforms other classifier combination schemes. The combination of an ensemble of neural networks has also been suggested by researchers to produce high accurate image classifications. The combination of network outputs involves voting rules, belief functions, statistical approaches, Dempster-Shafer evidence theory, and various integration methods (Hansen and Salomon, 1990; Giacinto and Roli, 2001). The question of classifier combination and voting methods in the context of filtering training data is investigated by Brodley and Friedl (1999).

Bayesian formalism can involve a process as simple as averaging. It is used with multiple classifiers that output a probability (or probability-like) estimate of the likelihood of pixel A belonging to class j. Such classifiers include the ML method and artificial neural networks. The probabilities (or pseudoprobabilities) for a pixel for each possible class are accumulated, and the winner is the pixel that has the greatest accumulated probability. As in the case of majority voting, a threshold may be set so that the winning margin is a clear one.

Evidential reasoning, described in Section 4.4, associates a degree of belief with each source of information, and a formal system of rules is used to manipulate the belief function. In the context of pattern recognition, this method is useful in handling multiple sources of data of different kinds and with different levels of accuracy. It can also be used to assess the plausibility of labels assigned to a given pixel by different classifiers.

Further elaboration of multiple classifier approaches can be found in Tumer and Ghosh (1995), Roli et al. (1997), Steele (2000), Smits (2002), Bruzzone et al. (2004), and Dmitriev et al. (2018).

2.6 INCORPORATION OF ANCILLARY INFORMATION

The addition of ancillary (nonspectral) information, such as spatial relationships, may be a more effective way of characterizing the classes of interest, and if added to the spectral information for image classification, the classification accuracy may be improved (Mather et al., 1998; Tso and Olsen, 2005a; Hurskainen et al., 2019). The utilization of spectral information alone in image classification may yield unsatisfactory outcomes, especially in areas characterized by heterogeneous landscapes and scenarios with limited availability of ground reference data. Ancillary data (also termed auxiliary or multisource data) can be extracted directly from an image or from other sources such as digital elevation models, geological and soil maps. Some examples of ancillary features derived from an image are texture, context, and structural relationships. It is worth noting that there is a distinction in the definitions of ancillary and auxiliary data among certain researchers.

Ancillary data, in this context, includes derivatives of remotely sensed data (e.g., spectral indices, textural features), whereas auxiliary data encompass any external geospatial input not derived directly from the imagery (Kavzoglu and Bilucan, 2023).

Incorporating topographic features into image classification in addition to spectral information provides great benefits and increases the produced classification accuracy. For instance, elevation and slope data can help to delineate the forest types in mountainous areas. When spectrally similar land cover classes exist, sometimes due to the number of available spectral bands, elevation data obtained from LiDAR or radar data can be invaluable in identifying the classes with different heights, which can be an indication for different types of vegetation or different phenological stages of vegetation. Elevation data are also utilized to produce a three-dimensional or stereoscopic view by placing the image on top of the three-dimensional surface. Ancillary data are introduced in problem modeling as additional inputs, as in the case of neural networks, they are represented by supplementary neurons in the input layer. In the amalgamation of ancillary data, the main problem is the accurate and reliable geocoding between image data and ancillary data. 1:1.000.000 scale geological maps are sometimes resampled to 10- to 30-meter resolution to match with the Landsat, SPOT, or Sentinel-2 images. Researchers should be cautious about the level of accuracy that an existing map can be resampled.

2.6.1 Use of Texture and Context

Human photo interpreters have long exploited spatial information like texture and context, and attempts have been made to incorporate these attributes of the scene into computer-assisted image classification. While the spectral nature of a pixel is defined by the color (hue) of a pixel in different bands (visible and infrared), textural characteristics are related to the spatial (statistical) distribution of tonal variations within a neighborhood. The context of a pixel or a group of pixels refers to its probability of occurrence based on the nature of the pixels in the remainder of the scene. There is clear evidence that data from future sensors, with their finer spatial resolving power, may not necessarily generate improved classifications when per-pixel classifiers are used (Townshend, 1981; Mather, 2004) because of the higher spectral variability of local areas of the image, which becomes apparent as the resolution becomes finer. In parallel to this statement, the main objective of the OBIA is to maximize the intra-segment homogeneity and inter-segment heterogeneity, which cannot be satisfied with per-pixel classification.

Human beings use both spectral and spatial features to interpret visual information. Spectral features describe variations in tone over a grayscale image, while spatial features reflect the spatial distribution of these tonal variations, which contain two kinds of spatial relationships. One such relationship is tonal variation focused upon the object of interest, representing the structure of the object, while the other measures the broader scale relationship between the object being analyzed and the remainder of the scene. Generally, the words texture and context are used to represent these two forms of spatial relationship. They are frequently used in image interpretation and are generally extracted directly from an image. In visual terms, texture is expressed as the impression of roughness or smoothness created by the variation of tone or repetition of patterns across a surface.

Spectral and textural features are interdependent for, as Haralick et al. (1973) noted, "texture and tone have an inextricable relationship to one another. Tone and texture are always present in an image, although one property can dominate the other at times" depending on the fineness or roughness of the surface of the object and the image resolution relative to the surface roughness of the object. If tonal variation inside a limited range is relatively small, spectral information will dominate. For example, at a given resolution, an image may include areas of relatively constant tone, such as water bodies or expanses of concrete or tarmac. Conversely, if the tonal variation is large and presents meaningful structures, then texture will be dominant, as in images of rocky areas, settlements, and some kinds of clouds.

Texture is an innate property of objects. It contains important information about the structural arrangement of surfaces. The use of texture in addition to spectral features for image classification might be expected to result in some level of accuracy improvement, depending on the spatial resolution of the sensor and the size of the homogeneous area being classified. Where the spatial resolution of the image is fine relative to the scale of tonal variation, texture can be a valuable source of discriminating information. Conversely, where homogeneous regions in the image are small it may prove difficult to estimate texture, for texture is a property of an area rather than of a point. If the area over which texture is measured contains two or more different regions or categories, then texture measures may not prove to be useful.

The operational definition of texture features is difficult. The main texture recognition approaches can be categorized into four groups in terms of their different theoretical backgrounds. The first approach defines texture features that are derived from the Fourier power spectrum of the image under study via frequency domain filtering. Since different textures demonstrate different frequency patterns, it is reasonable to postulate that texture features are related to the distribution of spatial frequency components. The second approach is based on statistics that measure local properties that are thought to be related to texture, for example, the local mean or standard deviation. The third approach is the use of the joint gray-level probability density (Haralick et al., 1973). The final approach is based on modeling the image using assumptions, such as (1) the image being processed possesses fractal properties (Mandelbrot, 1977, 1982), or (2) it can be modeled using a random field model such as the multiplicative autoregressive random (MAR) field (Frankot and Chellappa, 1987). Additionally, Carr and Miranda (1998) reported that the use of the variogram produced classifications with greater accuracy than those based on the gray-level co-occurrence matrix (GLCM). The use of the semi-variogram for the estimation of image texture is discussed in a review article by Atkinson and Lewis (2000). Kupidura et al. (2019) also conducted a comparative study on the contributions of GLCM features, Laplace filters, and granulometric analysis on land use classification.

The principal concept of the GLCM is that the texture information contained in an image is defined by the adjacency relationships that the gray tones in an image have to one another. In other words, it is assumed that the texture information is specified by values f_{ij} within the GLCM, where f_{ij} denotes the frequency of occurrence of two cells of gray tones i and j, respectively, separated by distance d with a specific direction on the image. The joint neighboring pair distribution within the GLCM shows higher values concentrated around its principal diagonal (i.e., in the NW-SE direction, cells (i, j), $i=j$). Values of f_{ij} can be calculated for any feasible direction and distance d. Generally, only four directions corresponding to angles of $0°$, $45°$, $90°$, and $135°$ are used. Haralick et al. (1973) proposed a variety of texture measures based on the GLCM. These texture measures, called textural features, have been employed by a number of researchers (Franklin and Peddle, 1990; Sali and Wolfson, 1992), and the results have generally been successful. Four texture features (angular second moment (ASM), contrast (Con), inverse difference moment (IDM), and entropy (Ent)), which are the most frequently used features by researchers, are described here (Puissant et al., 2005). In what follows, $p(i, j)$ denotes the $(i, j)^{\text{th}}$ entry in a normalized GLCM, and N_G denotes the number of gray levels in the quantized image.

$$\text{ASM} = \sum_i \sum_j \left[p(i,j) \right]^2 \tag{2.26}$$

$$\text{Con} = \sum_{n=0}^{N_G-1} n^2 \times \left[\sum_i \sum_{\substack{j \\ |i-j|=n}} p(i,j) \right] \tag{2.27}$$

$$\text{IDM} = \sum_i \sum_j \frac{1}{1+(i-j)^2} p(i,j) \tag{2.28}$$

$$\text{Ent} = -\sum_i \sum_j p(i,j) \log p(i,j) \tag{2.29}$$

The measure of the ASM produces a higher value when co-occurrence frequency $p(i,j)$ is concentrated in few places in the GLCM, for example, the main diagonal direction. If the $p(i,j)$ are close in value, then ASM will generate a small value. Con measure assigns a higher weight to the $p(i,j)$ that are distant from the main diagonal of the GLCM. When the difference between neighboring pairs becomes large, the contrast increases. In this situation, co-occurrence frequencies $p(i,j)$ will have higher values away from the main diagonal of the GLCM, and consequently, this equation will generate a higher value of Con. On the other hand, IDM will generate higher values for an image containing large homogeneous patches because such an image generates high values on the main diagonal of the GLCM. It gives lower weight to those $p(i,j)$ that are located away from the main diagonal. The entropy measure of D_{ij} outputs a higher value for a homogeneous distribution of $p(i,j)$, and lower otherwise.

There are many studies proving that texture information provides important contributions in remote sensing studies, especially to ones in image classification. For instance, Chen et al. (2004) conducted research to assess the impact of texture window size on classification accuracy. Their findings indicated that texture played a more substantial role in improving accuracy, particularly at finer resolution levels. Lu et al. (2010) conclude that texture features not only reduce the spectral variations within the same land cover classes but also increase the distinguishability between different classes. Thus, significant improvement in classification accuracy was achieved. In a study conducted to estimate forest structure variables, GLCM texture features extracted from an IKONOS-2 image with a spatial resolution of 1 m show promise in estimating coniferous forest variables (Kayitakire et al., 2006). The accurate estimation of top height was achieved using the correlation measure calculated parallel to the tree shadows with a 1-pixel displacement within a 15×15 pixel moving window. Optimal estimates for circumference and age were obtained using the correlation measure, while stem density and basal area were more accurately predicted using the contrast measure. The GLCM orientation parameter was specifically set to the direction perpendicular to the shadow direction. In an effort to streamline the parameter selection process for texture feature extraction, Pathak and Dikshit (2010) introduced a novel approach known as conjoint analysis. The study revealed that in the classification process, the relative importance of texture features and window size outweighed that of quantization level or the choice of image band for extracting texture features.

In the interpretation of visual information, context is very important. It may be derived from spectral, spatial, or even temporal attributes. Effectively employing context enables the removal of potential ambiguities, the retrieval of missing information, and the rectification of errors, as highlighted by Magnussen et al. (2004). The use of context to model the prior probability density functions in order to help in the interpretation of remotely sensed imagery is considered a reasonable procedure since a pixel classified as ocean is likely to be surrounded by pixels of the same class and is unlikely to have neighbors from categories such as pasture or forest. In other words, using the concept of context, pixels are not treated in isolation but are considered to have a relationship with their neighbors. Thus, the relationship between the pixel of interest and its neighbors is treated as being statistically dependent.

Contextual information can be included either in a statistical classifier or in procedures that amend some preliminary classifier output. A simple way to use context is in the form of a majority filter window (Gurney, 1980, 1981). For instance, let a window be centered on a pixel labeled i. If the majority of the pixels within the window belong to class j, the label of the central pixel is altered from i to j. In most cases, the majority filter is used for classification refinement. Although there is some improvement, the increase in classification accuracy is not impressive.

The Markov random field (MRF) is a useful tool for characterizing contextual information and has been widely used in image segmentation and image restoration (Derin and Elliott, 1987; Dubes and Jain, 1989; Tso and Olsen, 2005a). The use of MRF models for linear feature detection has also achieved satisfactory results (Tupin et al., 1998). The practical use of the MRF relies on its relationship to the Gibbs random field (GRF), which provides a tractable way to apply the MRF to deal with context. Moreover, owing to the MRF's local property, the algorithm can be implemented in a highly parallel manner, which makes the MRF more attractive.

Let a set of random variables $d = \{d_1, d_2, \ldots, d_m\}$ be defined on the set S containing m number of sites in which each random variable yi ($1 \le i \le m$) takes a label from label set L. The family d is called a random field. The set S is equivalent to an image containing m pixels; d is a set of pixel DN values, and the label set L depends upon the application. The label set L is equivalent to a set of the user-defined information classes, for example, $L = \{$water, forest, pasture, or residential areas$\}$, while in the case of boundary detection, the label set $L = \{$boundary, nonboundary$\}$. There are many kinds of random field models describing ways of labeling the random variables.

Given the definition of a random field specified earlier, we define the configuration w for the set S as $w = \{d_1 = w_1, d_2 = w_2, \ldots, d_m = w_m\}$, where $w_r \in L$ ($1 \le r \le m$). For convenience, we simplify the notation of w to $w = \{w_1, w_2, \ldots, w_m\}$. A random field, with respect to a neighborhood system, is a MRF if its probability density function satisfies the following three properties:

1. Positivity: $P(w) > 0$ for all possible configurations of w,
2. Markovianity: $P(w_r | w_{S-r}) = P(w_r | w_{Nr})$, and
3. Homogeneity: $P(w_r | w_{Nr})$ is the same for all sites r.

$S - r$ is the set difference (i.e., all pixels in the set S excluding r), w_{S-r} denotes the set of labels at the sites in $S - r$, and Nr denotes the neighbors of site r.

The first property, that of positivity, can usually be satisfied in practice, and the joint probability $P(w)$ can be uniquely determined by local conditional properties as long as the positivity property is sustained (Besag, 1974). Markovianity indicates that labeling of a site r is only dependent on its neighboring sites. The property of homogeneity specifies the conditional probability for the label of a site r, given the labels of the neighboring pixels, regardless of the relative position of site r in S. An MRF may also incorporate other properties such as isotropy. Isotropy is the property of direction independence. That is, the neighboring sites surrounding a site r have the same contributing effect to the labeling of site r.

The conventional neighborhood system employed in image analysis identifies the first-order neighbors of a pixel as the four pixels that share a side with the specified pixel, as depicted in Figure 2.10a. The four pixels situated at the corner boundaries with the pixel of interest, as displayed

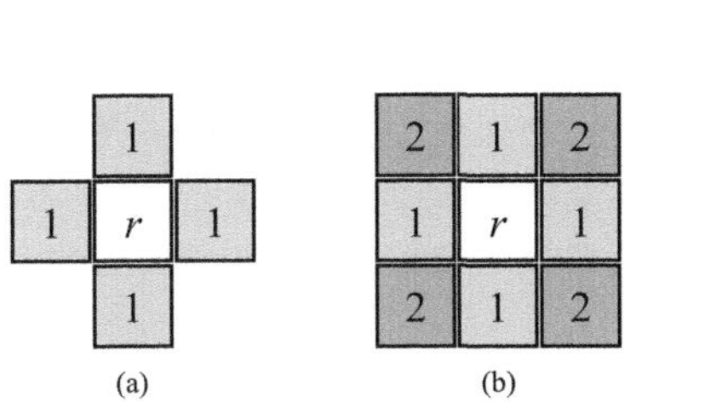

FIGURE 2.10   (a) First-order neighbors of a pixel are the four pixels sharing a side with the given pixel. (b) Second-order neighbors are the four pixels having corner boundaries with the pixel of interest. (c) Higher-order (up to five) neighbors are extended in a similar manner.

in Figure 2.10b, constitute the second-order neighbors. The extension of higher-order neighbors can be carried out in a similar fashion. The neighborhood order up to five is shown in Figure 2.10c. More specifically, when the sites form a regular rectangular lattice, as do pixels in a two-dimensional image, the site $r=(i, j)$ has four nearest (first-order) neighbors, denoted by $N(i, j)=\{(I-1, j), (I+1, j), (i, j-1), (i, j+1)\}$.

The increase in classification accuracy compared with traditional methods demonstrates the practical value of the MRF approach (Tso and Mather, 1999; Tso and Olsen, 2005b; Zheng and Wang, 2014). Moser and Serpico (2013) present a pioneering framework that incorporates both SVMs and MRF models in a unified formulation, specifically designed for spatial contextual classification. The contextual adaptation of SVMs is achieved by connecting the Markovian minimum-energy standard with the implementation of an SVM in a properly transformed dimension. An automated method utilizing hidden MRF and SVM is presented by Ghamisi et al. (2014) for the spectral and spatial classification of hyperspectral images. The use of a Sobel edge detector–based gradient step is implemented to protect the edges in the produced classification map. To enhance classification performance in hyperspectral images, Xia et al. (2015) proposed a contextual classification strategy that integrates rotation forests and Markov random fields. The first step involves applying a decision tree–based rotation forest algorithm to calculate class probabilities based on spectral information. The spatial contextual information, characterized by MRF prior, is then incorporated to enhance the classification results obtained from the rotation forest. This improvement is accomplished by addressing a MAP problem through the α-expansion graph cut optimization method.

The utilization of CNN models, which incorporate local connections and shared weights to address spatial dependencies, has become prevalent. This approach significantly reduces the number of network parameters while leveraging predefined crisp neighborhood systems to extract spatial and contextual information. They have been increasingly employed to incorporate spatial or contextual information, especially with the implementation of 2-D and 3-D CNNs. The MRF model has lately gained prominence for its efficient description of spatial context. Particularly, strategies to enhance accuracy in semantic segmentation incorporate these methods, which can extract additional image information by constructing probability graphs at either the pixel or the object granularity level (Zheng et al., 2021; Yao et al., 2023).

2.6.2 Using Ancillary Multisource Data

As each additional source of information contributes to the characterization of the objects under analysis, the use of ancillary multisource data is of significant importance as they can have a positive contribution to classification accuracy (Zhu et al., 2016). A significant problem arises when one attempts to combine images and ancillary data, namely, the different nature of these data types. Data may be categorical and, even when continuous in nature, may violate the normality assumption required by most statistical classification techniques. Ancillary information can be continuous or categorical (Schepers, 2004). Examples of continuous ancillary data are elevation, slope, and aspect derived from a digital elevation model. Examples of categorical data are land use, soil, and geological types. The different approaches proposed for the combination of spectral and ancillary information include the following: the logical channel approach, stratification, the use of prior probabilities, and file- or object-based classification. The logical channel approach consists of adding each ancillary data set as an additional feature so that the pixel vector is extended by the addition of this external information. This technique was called the logical channel approach (Ricchetti, 2000; Gercek, 2004).

A major issue in the integration of ancillary data into remotely sensed data is the scale compatibility between data of different scales and spatial resolutions. A quick guide for inexperienced users is to combine two sources having similar geometric accuracy, which is in fact a cartographic measure. In geometric rectification of remotely sensed images, a rule of thumb is that a reasonable RMS error estimated from the ground control points should be around half of the spatial resolution of the image (pixel size). Considering this general principle, the geometric accuracy of a 4-m resolution imagery will have about horizontal accuracy of 2 m. On the other hand, the spatial accuracy of cartographic maps is estimated by multiplying the scale with the minimum visible distance on the map (0.2 mm). For instance, if have a map with a scale of 1:10,000, the corresponding field distance will be 10.000×0.2 mm=2 m. Therefore, integration of this map can be performed with the image (4-m spatial resolution). The considerations here can be also applied to estimate the approximate map scale to be produced using remotely sensed imagery.

Stratification involves the subdivision or segmentation of the study area into smaller areas or strata, based on rules derived from external knowledge so that each stratum can be processed independently (Gorte and Stein, 1998; De Bruin and Gorte, 2000). This process is performed before classification with the purpose of increasing the homogeneity of the individual data set to be classified, or to separate different objects that are spectrally similar. Some advantages relating to this kind of approach might be considered: one advantage is the convenience of dealing with smaller data sets; another benefit is the reduction of variability within individual strata. Stratification is effective and easy to implement but is deterministic in the sense that it does not handle uncertainty about the occurrence of certain classes or the gradation between strata. An incorrect stratification can invalidate the entire classification process. Ricchetti (2000) compares the logical channel and stratification methods and is in favor of the logical channel approach.

In most classification processes, the probability of class membership of a given pixel is assumed equal for all classes. However, should data indicate that particular classes have preferences for specific locations within a given area, this inclination can be quantified through prior probabilities of occurrence assigned to each class. This information can then be integrated into the classification procedure. The major difficulty involved in this approach is to define a suitable function of prior probability relating to each class in terms of achieving optimal result. Most experiments rely on the analyst's *ad hoc* decision.

A further application of spatial information is the use of boundary information to define objects prior to classification. Such a classification strategy is suitable for patchy areas (e.g., areas made up of agricultural fields), and might be less affected by boundary pixels, which generally results in classification error. In agricultural crop applications, the objects are fields, the boundaries of which can be derived either by digitizing the maps/images, by applying an edge detection algorithm, or by applying a segmentation approach to the image. Each object (field) is characterized by global statistical parameters and is represented by a unique vector in feature space (Mason et al., 1988; Jansen et al., 1990; Tso and Olsen, 2005b).

In contrast to the mainly *ad hoc* procedures described above, two specific multisource data fusion mechanisms known as an extension of Bayesian theory and evidential reasoning have their particularly practical value because both approaches have a robust theory basis for dealing with multisource data sets. Details of these two methods for managing multisource data are provided in Chapter 4.

2.7 EPILOGUE

In this chapter, an attempt was made to provide an overview of the methods of pattern recognition, concentrating on conventional statistical classifiers for image classification. In the context of image classification, pattern recognition and remote sensing research fields coincide with each other. The range of material covered here may appear to the reader to be daunting; however, the following chapters provide considerably more detail and descriptions of the principal methods of pattern recognition that are summarized in this chapter. The consensus view, in recent years, has emphasized the superiority of machine learning and deep learning methods over statistical methods, largely because of their nonparametric nature (i.e., they do not assume any particular statistical frequency distribution of the data), of the requirement of less training samples, and solving more complex problems includes a high volume of data. More recently, the use of decision trees and other forest-type classifiers (e.g., random forest, rotation forest) has assumed a higher profile, they are nonparametric in nature, but they do not need extensive design and training. Hence, they are known as highly interpretable methods as the output of the tree structure is clear for interpreting the decisions on the class definitions. However, their use of hyperplane decision boundaries parallel to the feature axes may restrict their use to cases in which classes are clearly distinguishable. On the other hand, deep learning models require extensive knowledge and experience in network design and parameter setting as there exist many types of models usually designed for particular problems although considerable improvement in image classification and other image analysis applications has been reported in recent studies. Remote sensing analysts should always consider that the most suitable classification approach for a particular problem is dependent on the nature of the problem, the characteristics of the data set, and the robustness of the algorithm selected for image analysis.

Hyperspectral data present some problems due to their high dimensionality. It is impracticable to consider the collection of large volumes of test and training data when the intrinsic dimensionality of the hyperspectral data set is considerably less than the number of spectral bands. A proportion of the variance exhibited by hyperspectral data is noise, either random or coherent. Moreover, the "curse of dimensionality" has become a major problem with the availability of large volumes of data acquired by new sensors. As underlined by Belgiu and Drăguţ (2016), an efficient classification method needs to address some important challenges, more importantly, the *Hughes phenomenon*, which arises when the number of variables is significantly higher than the number of training samples, the nonlinearity among the variables, imbalanced training samples, excessive noise in the data sets, and computation burden. Hence, it makes sense to consider the use of orthogonal transforms (e.g., principal component analysis), particularly those that explicitly discriminate between signal and noise. Minimum/maximum autocorrelation factors (MAF) and maximum noise fraction (MNF) transformation, described in Chapter 3, also offer the possibility of reducing the size of the data set without compromising accuracy, while avoiding the problems associated with high dimensionality. Other filter-, wrapper-, and embedded-based methods are also discussed in the chapter.

The use of multiple classifiers through ensembling strategies may present a way out of the spiral of increasing complexity. Superior performances were reported with the utilization of XGBoost, NGBoost, and LightGBM in the classification of remotely sensed data. Rather than developing increasingly refined decision rules, it may be sensible to make the best use of what is available. Users of these methods should be aware of the need to ensure that the component classifications are independent. Advanced machine learning algorithms and deep learning models will be discussed in Chapters 5–7.

REFERENCES

Arngren, M., M. N. Schmidt, and J. Larsen. 2011. Unmixing of hyperspectral images using Bayesian non-negative matrix factorization with volume prior. *Journal of Signal Processing Systems* 65:479–496. https://doi.org/10.1007/s11265-010-0533-2.

Atkinson, P. M., and P. Lewis. 2000. Geostatistical classification for remote sensing: An introduction. *Computers and Geosciences* 26:361–371. https://doi.org/10.1016/S0098-3004(99)00117-X.

Bakos, K. L., and P. Gamba. 2011. Hierarchical hybrid decision tree fusion of multiple hyperspectral data processing chains. IEEE *Transactions on Geoscience and Remote Sensing* 49:388–394.

Ball, G. H., and D. J. Hall. 1967. A clustering technique for summarizing multivariate data. *Behavioral Science* 12:153–155. https://doi.org/10.1002/bs.3830120210.

Banerjee, B., F. Bovolo, A. Bhattacharya, L. Bruzzone, S. Chaudhuri, and B. Krishna Mohan. 2015. A new self-training-based unsupervised satellite image classification technique using cluster ensemble strategy. *IEEE Geoscience and Remote Sensing Letters* 12:741–745. https://doi.org/10.1109/LGRS.2014.2360833.

Belgiu, M., and L. Drăguţ. 2016. Random forest in remote sensing: A review of applications and future directions. *ISPRS Journal of Photogrammetry and Remote Sensing* 114:24–31. https://doi.org/10.1016/j.isprsjprs.2016.01.011.

Benediktsson, J. A., P. H. Swain, and O. K. Ersoy. 1990. Neural network approaches versus statistical methods in classification of multisource remote sensing data. *IEEE Transactions on Geoscience and Remote Sensing* 28:540–552. https://doi.org/10.1109/TGRS.1990.572944.

Besag, J. 1974. Spatial interaction and the statistical analysis of lattice systems. *Journal of the Royal Statistical Society* 36:192–236. https://doi.org/10.1111/j.2517-6161.1974.tb00999.x.

Bezdek, J. C. 1981. *Pattern Recognition with Fuzzy Objective Function Algorithms*. New York: Plenum Press.

Bezdek, J. C., R. Ehrlich, and W. Full. 1984. FCM: The fuzzy c-means clustering algorithm. *Computers and Geosciences* 10:191–203.

Bioucas-Dias, J. M., A. Plaza, N. Dobigeon, M. Parente, Q. Du, P. Gader, and J. Chanussot. 2012. Hyperspectral unmixing overview: Geometrical, statistical, and sparse regression-based approaches. *IEEE Journal of Selected Topics in Applied Earth Observations and Remote Sensing* 5:354–379. https://doi.org/10.1109/JSTARS.2012.2194696.

Borsoi, R. A., T. Imbiriba, and J. C. M. Bermudez. 2019. Deep generative endmember modeling: An application to unsupervised spectral unmixing. *IEEE Transactions on Computational Imaging* 6:374–384. https://doi.org/10.1109/TCI.2019.2948726.

Breiman, L. 1996. Bagging predictors. *Machine Learning* 26:123–140. https://doi.org/10.1007/BF00058655.

Breiman, L. 2001. Random forests. *Machine Learning* 45:5–32. https://doi.org/10.1023/A:1010933404324.

Briem, G. J., J. A. Benediktsson, and J. R. Sveinsson. 2002. Multiple classifiers applied to multisource remote sensing data. *IEEE Transactions on Geoscience and Remote Sensing* 40:2291–2299. https://doi.org/10.1109/TGRS.2002.802476.

Brodley, C., and M. A. Friedl. 1999. Identifying mislabeled training data. *Journal of Artificial Intelligence Research* 11:131–167. https://doi.org/10.48550/arXiv.1106.0219.

Brown, M., H. G. Lewis, and S. R. Gunn, 2000. Linear spectral mixture models and support vector machines for remote sensing. *IEEE Transactions on Geoscience and Remote Sensing* 38:2346–2360. https://doi.org/10.1109/36.868891.

Bruzzone, J., R. Gossu, and G. Vernazza. 2004. Detection of land-cover transitions by combining multidate classifiers. *Pattern Recognition Letters* 25:1491–1500. https://doi.org/10.1016/j.patrec.2004.06.002.

Cannon, R. L., J. V. Dave, J. C. Bezdek, and M. M. Trivedi. 1986. Segmentation of a Thematic Mapper image using the fuzzy c-means clustering algorithm. *IEEE Transactions on Geoscience and Remote Sensing* GE-24:400–408.

Carr, J. R., and F. P. Miranda. 1998. The semivariogram in comparison to the co-occurrence matrix for classification of image texture. *IEEE Transactions on Geoscience and Remote Sensing* 36:1945–1952. https://doi.org/10.1109/36.729366.

Chen, C. F., and J. M. Lee. 2001. The validity measurement of fuzzy c-means classification for remotely sensed images. In *Proceedings of the 22nd Asian Conference on Remote Sensing*, November 5–9, Singapore, pp. 208–210.

Chen, D., D. A. Stow, and P. Gong. 2004. Examining the effect of spatial resolution and texture window size on classification accuracy: An urban environment case. *International Journal of Remote Sensing* 25:2177–2192. https://doi.org/10.1080/01431160310001618464.

Cheng, G., J. Han, and X. Lu. 2017. Remote sensing image scene classification: Benchmark and state of the art. In *Proceedings of the IEEE*, 105:1865–1883. https://doi.org/10.1109/JPROC.2017.2675998.

Cheung, Y. M. 2005. On rival penalization controlled competitive learning for clustering with automatic cluster number selection. *IEEE Transactions on Knowledge and Data Engineering* 17:1583–1588. https://doi.org/10.1109/TKDE.2005.184.

Cole, J. P., and C. A. M. King. 1968. *Quantitative Geography: Techniques and Theories in Geography*. London: John Wiley & Sons.

De Bruin, S., and B. G. H. Gorte. 2000. Probabilistic image classification using geological map units applied to land-cover change detection. *International Journal of Remote Sensing* 21:2389–2402. https://doi.org/10.1080/01431160050030529.

Deng, P., K. Xu, and H. Huang. 2022. When CNNs meet vision transformer: A joint framework for remote sensing scene classification. *IEEE Geoscience and Remote Sensing Letters* 19:1–5. https://doi.org/10.1109/LGRS.2021.3109061.

Derin, H., and H. Elliott. 1987. Modeling and segmentation of noisy and textured images using Gibbs random fields. *IEEE Transactions on Pattern Analysis and Machine Intelligence* 9:39–55. https://doi.org/10.1109/TPAMI.1987.4767871.

Deshpande, V. S., and J. S. Bhatt. 2021. A practical approach for hyperspectral unmixing using deep learning. *IEEE Geoscience and Remote Sensing Letters* 19:1–5. https://doi.org/10.1109/LGRS.2021.3127075.

Dmitriev, E. V., V. V. Kozoderov, A. O. Dementyev, and A. N. Safonova. 2018. Combining classifiers in the problem of thematic processing of hyperspectral aerospace images. *Optoelectronics, Instrumentation and Data Processing* 54:213–221. https://doi.org/10.3103/S8756699018030019.

Dubes, R. C., and A. K. Jain. 1989. Random field models in image analysis. *Journal of Applied Statistics* 16:131–164. https://doi.org/10.1080/02664768900000014.

Feilhauer, H., A. Zlinszky, A. Kania, G. M. Foody, D. Doktor, A. Lausch, and S. Schmidtlein. 2021. Let your maps be fuzzy! Class probabilities and floristic gradients as alternatives to crisp mapping for remote sensing of vegetation. *Remote Sensing in Ecology and Conservation* 7:292–305. https://doi.org/10.1002/rse2.188.

Feng, R., Y. Zhong, and L. Zhang. 2016. Adaptive spatial regularization sparse unmixing strategy based on joint MAP for hyperspectral remote sensing imagery. *IEEE Journal of Selected Topics in Applied Earth Observations and Remote Sensing* 9:5791–5805. https://doi.org/10.1109/JSTARS.2016.2570947.

Foody, G. M. 1996. Approaches for the production and evaluation of fuzzy land cover classifications from remotely-sensed data. *International Journal of Remote Sensing* 17:1317–1340. https://doi.org/10.1080/01431169608948706.

Foody, G. M. 2000. Estimation of sub-pixel land cover composition in the presence of untrained classes. *Computers and Geosciences* 26:469–478. https://doi.org/10.1016/S0098-3004(99)00125-9.

Foody, G. M., and A. Mathur. 2006. The use of small training sets containing mixed pixels for accurate hard image classification: Training on mixed spectral responses for classification by a SVM. *Remote Sensing of Environment* 103:179–189.

Franklin, S. E., and D. R. Peddle. 1990. Classification of SPOT HRV imagery and texture features. *International Journal of Remote Sensing* 11:551–556. https://doi.org/10.1080/01431169008955039.

Frankot, R. T., and R. Chellappa. 1987. Lognormal random-field models and their applications to radar image synthesis. *IEEE Transactions on Geoscience and Remote Sensing* 25:195–207. https://doi.org/10.1109/TGRS.1987.289818.

Freund, Y., and R. E. Schapire. 1999. A short introduction to boosting. *Journal of Japanese Society for Artificial Intelligence* 14:771–780.

Frizzelle, B. G., and A. Moody. 2001. Mapping continuous distributions of land cover: A comparison of maximum likelihood estimation and artificial neural networks. *Photogrammetric Engineering and Remote Sensing* 67:693–705.

Gao, Y., Z. Wang, J. Xie, and J. Pan. 2022. A new robust fuzzy c-means clustering method based on adaptive elastic distance. *Knowledge-Based Systems* 237:107769. https://doi.org/10.1016/j.knosys.2021.107769.

Gath, I., and A. B. Geva. 1989. Unsupervised optimal fuzzy clustering. *IEEE Transactions on Pattern Analysis and Machine Intelligence* 11:773–781. https://doi.org/10.1109/34.192473.

Gercek, D. 2004. Improvement of image classification with the integration of topographical data. In *Proceedings of 20th ISPRS Congress*, Istanbul, Turkey, July 12–23, 2004, 35-B8, 53–58. Available: https://www.isprs.org/proceedings/XXXV/congress/yf/papers/929.pdf.

Ghaffari, O., M. J. V. Zoej, and M. Mokhtarzade. 2017. Reducing the effect of the endmembers' spectral variability by selecting the optimal spectral bands. *Remote Sensing* 9:884. https://doi.org/10.3390/rs9090884.

Ghamisi, P., J. A. Benediktsson, and M. O. Ulfarsson. 2014. Spectral-spatial classification of hyperspectral images based on hidden Markov random fields. *IEEE Transactions on Geoscience and Remote Sensing* 52:2565–2574. https://doi.org/10.1109/TGRS.2013.2263282.

Ghamisi, P., N. Yokoya, J. Li, W. Liao, S. Liu, J. Plaza, B. Rasti, and A. Plaza. 2017. Advances in hyperspectral image and signal processing: A comprehensive overview of the state of the art. *IEEE Geoscience and Remote Sensing Magazine* 5:37–78. https://doi.org/10.1109/MGRS.2017.2762087.

Giacinto, G., and F. Roli. 2001. Design of effective neural network ensembles for image classification purposes. *Image and Vision Computing* 19:699–707. https://doi.org/10.1016/S0262-8856(01)00045-2.

Gordon, A. D. 1981. *Classification*. London: Chapman & Hall.

Gorte, B., and A. Stein. 1998. Bayesian classification and class area estimation of satellite images using stratification. *IEEE Transactions on Geoscience and Remote Sensing* 36:803–812. https://doi.org/10.1109/36.673673.

Gurney, C. M. 1980. Threshold selection for line detection algorithms. *IEEE Transactions on Geoscience and Remote Sensing* 18:204–211. https://doi.org/10.1109/TGRS.1980.350274.

Gurney, C. M. 1981. The use of contextual information to improve land cover classification of digital remotely sensed data. *International Journal of Remote Sensing* 2:379–388. https://doi.org/10.1080/01431168108948372.

Hady, M. F. A., and F. Schwenker. 2013. Semi-supervised learning. In *Handbook on Neural Information Processing*, (eds.) M. Bianchini, M. Maggini, and L. Jain, pp. 215–239. Berlin, Heidelberg: Springer. https://doi.org/10.1007/978-3-642-36657-4_7.

Hand, D. J. 1997. *Construction and Assessment of Classification Rules*. New York: John Wiley & Sons.

Hansen, L. K., and P. Salomon. 1990. Neural network ensembles. *IEEE Transactions on Pattern Recognition and Machine Intelligence* 12:993–1001. https://doi.org/10.1109/34.58871.

Haralick, R. M., M. Shanmugam, and I. Dinstein. 1973. Texture features for image classification. *IEEE Transactions on Systems, Man, and Cybernetics* 3:610–621. https://doi.org/10.1109/TSMC.1973.4309314.

Heinz, D. C., and C.-I. Chang. 2001. Fully constrained least squares linear spectral mixture analysis method for material quantification in hyperspectral imagery. *IEEE Transactions on Geoscience and Remote Sensing* 39:529–545. https://doi.org/10.1109/36.911111.

Heylen, R., M. Parente, and P. Gader. 2014. A review of nonlinear hyperspectral unmixing methods. *IEEE Journal of Selected Topics in Applied Earth Observations and Remote Sensing* 7:1844–1868. https://doi.org/10.1109/JSTARS.2014.2320576.

Hurskainen, P., H. Adhikari, M. Siljander, P. K. E. Pellikka, and A. Hemp. 2019. Auxiliary datasets improve accuracy of object-based land use/land cover classification in heterogeneous savanna landscapes. *Remote Sensing of Environment* 233:111354. https://doi.org/10.1016/j.rse.2019.111354.

James, M. 1985. *Classification Algorithms*. London: Collins.

Jansen, L. L. F., M. N. Jarsma, and E. T. M. Linden. 1990. Integrating topographic data with remote sensing for land-cover classification. *Photogrammetric Engineering and Remote Sensing* 56:1503–1506.

Kavzoglu, T. 2001. An investigation of the design and use of feed-forward artificial neural networks in the classification of remotely sensed images. Ph.D. Thesis, School of Geography, The University of Nottingham, Nottingham, UK.

Kavzoglu, T., and F. Bilucan. 2023. Effects of auxiliary and ancillary data on LULC classification in a heterogeneous environment using optimized random forest algorithm. *Earth Science Informatics* 16:415–435. https://doi.org/10.1007/s12145-022-00874-9.

Kayitakire, F., C. Hamel, and P. Defourny. 2006. Retrieving forest structure variables based on image texture analysis and IKONOS-2 imagery. *Remote Sensing of Environment* 102:390–401. https://doi.org/10.1016/j.rse.2006.02.022.

Keshava, N., and J. F. Mustard. 2002. Spectral unmixing. *IEEE Signal Processing Magazine* 19:44–57. https://doi.org/10.1109/79.974727.

Kim, D. W., K. H. Lee, and D. Lee. 2004. On cluster validity index for estimation of the optimal number of Fuzzy clusters. *Pattern Recognition* 37:2009–2025.

Kittler, J., M. Hatef, R. P. W. Duin, and J. Matas. 1998. On combining classifiers. IEEE *Transactions on Pattern Analysis and Machine Intelligence* 20:226–239. https://doi.org/10.1109/34.667881.

Krishnapuram, R., and J. M. Keller. 1993. A possibilistic approach to clustering. *IEEE Transactions on Fuzzy Systems* 1:98–110. https://doi.org/10.1109/91.227387.

Krishnapuram, R., and J. M. Keller. 1996. The possibilistic c-means algorithm: Insights and recommendations. *IEEE Transactions on Fuzzy Systems* 4:385–393. https://doi.org/10.1109/91.531779.

Kupidura, P., K. Osińska-Skotak, K. Lesisz, and A. Podkowa. 2019. The efficacy analysis of determining the wooded and shrubbed area based on archival aerial imagery using texture analysis. *ISPRS International Journal of Geo-Information* 8:450. https://doi.org/10.3390/ijgi8100450.

Lenczner, G., A. Chan-Hon-Tong, B. Le Saux, N. Luminari, and G. Le Besnerais. 2022. DIAL: Deep interactive and active learning for semantic segmentation in remote sensing. *IEEE Journal of Selected Topics in Applied Earth Observations and Remote Sensing* 15:3376–3389. https://doi.org/10.1109/JSTARS.2022.3166551.

Li, M., S. Zang, B. Zhang, S. Li, and C. Wu. 2014. A review of remote sensing image classification techniques: The role of spatio-contextual information. *European Journal of Remote Sensing* 47:389–411. https://doi.org/10.5721/EuJRS20144723.

Li, X., D. Shi, X. Diao, and H. Xu. 2022. SCL-MLNet: Boosting few-shot remote sensing scene classification via self-supervised contrastive learning. *IEEE Transactions on Geoscience and Remote Sensing* 60:1–12. https://doi.org/10.1109/TGRS.2021.3109268.

Licciardi, G. A., and F. Del Frate. 2011. Pixel unmixing in hyperspectral data by means of neural networks. *IEEE Transactions on Geoscience and Remote Sensing* 49:4163–4172. https://doi.org/10.1109/TGRS.2011.2160950.

Lu, D., and Q. Weng. 2007. A survey of image classification methods and techniques for improving classification performance. *International Journal of Remote Sensing* 28:823–870. https://doi.org/10.1080/01431160600746456.

Lu, D., S. Hetrick, and E. Moran. 2010. Land cover classification in a complex urban-rural landscape with QuickBird imagery. *Photogrammetric Engineering & Remote Sensing* 76:1159–1168. https://doi.org/10.14358/PERS.76.10.1159.

Ma, Y., Q. Jin, X. Mei, X. Dai, F. Fan, H. Li, and J. Huang. 2019. Hyperspectral unmixing with gaussian mixture model and low-rank representation. *Remote Sensing* 11:911. https://doi.org/10.3390/rs11080911.

Magnussen, S., P. Boudewyn, and M. Wulder. 2004. Contextual classification of Landsat TM images to forest inventory cover types. *International Journal of Remote Sensing* 25:2421–2440. https://doi.org/10.1080/0143116031000164296.

Mandelbrot, B. B. 1977. *Fractals: Form, Chance, and Dimension*. San Francisco, CA: Freeman.

Mandelbrot, B. B. 1982. *The Fractal Geometry of Nature*. San Francisco, CA: Freeman.

Maselli, F., C. Conese, T. De Filippis, and S. Norcini. 1995. Estimation of forest parameters through fuzzy classification of TM data. *IEEE Transactions on Geoscience and Remote Sensing* 33:77–84. https://doi.org/10.1109/36.368220.

Mason, D. C., D. G. Corr, A. Cross, D. C. Hogg, D. H. Lawrence, M. Petrou, and A. M. Taylor. 1988. The use of digital map data in the segmentation and classification of remotely-sensed images. *International Journal of Geographical Information Systems* 2:195–215. https://doi.org/10.1080/02693798808927896.

Mather, P. M. 1976. *Computational Methods of Multivariate Analysis in Physical Geography*. Chichester: John Wiley & Sons.

Mather, P. M. 1985. A computationally-efficient maximum-likelihood classifier employing prior probabilities for remotely-sensed data. *International Journal of Remote Sensing* 6:369–376. https://doi.org/10.1080/01431168508948456.

Mather, P. M. 1999. Land cover classification revisited. In *Advances in Remote Sensing and GIS Analysis*, (eds.) P. M. Atkinson, and N. J. Tate, pp. 7–16. Chichester: John Wiley and Sons.

Mather, P. M. 2004. *Computer Processing of Remotely-Sensed Images: An Introduction*, 3rd edition. Chichester: John Wiley & Sons.

Mather, P. M., B. Tso, and M. Koch. 1998. An evaluation of combining spectral and textural information for lithological classification. *International Journal of Remote Sensing* 19:587–604.

McLachlan, G. J., and T. Krishnan. 2008. *The EM Algorithm and Extensions*. New York: Wiley. https://doi.org/10.1002/9780470191613.

Mei, S., M. He, Z. Wang, and D. Feng, 2010. Mixture analysis by multichannel Hopfield neural network, *IEEE Geoscience and Remote Sensing Letters* 7:455–459. https://doi.org/10.1109/LGRS.2009.2039114.

Moser, G., and S. B. Serpico. 2013. Combining support vector machines and Markov random fields in an integrated framework for contextual image classification. *IEEE Transactions on Geoscience and Remote Sensing* 51:2734–2752. https://doi.org/10.1109/TGRS.2012.2211882.

Ngo, L. T., D. S. Mai, and W. Pedrycz. 2015. Semi-supervising interval type-2 fuzzy c-means clustering with spatial information for multi-spectral satellite image classification and change detection. *Computers & Geosciences* 83:1–16. https://doi.org/10.1016/j.cageo.2015.06.011.

Nguyen, D. D., L. T. Ngo, L. T. Pham, and W. Pedrycz. 2015. Towards hybrid clustering approach to data classification: Multiple kernels based interval-valued Fuzzy C-Means algorithms. *Fuzzy Sets and Systems* 279:17–39. https://doi.org/10.1016/j.fss.2015.01.020.

Nguyen, Q. M., P. M. Atkinson, and H. G. Lewis. 2011. Super-resolution mapping using Hopfield neural network with panchromatic imagery. *International Journal of Remote Sensing* 32:6149–6176. https://doi.org/10.1080/01431161.2010.507797.

Nikhil, R. P., and J. C. Bezdek. 1995. On cluster validity for the fuzzy c-means model. *IEEE Transactions on Fuzzy Systems* 3(3):370–379.

Pal, M. 2007. Ensemble learning with decision tree for remote sensing classification. *World Academy of Science, Engineering, and Technology* 36:258–260.

Pal, N. R., and J. C. Bezdek. 1995. On cluster validity for fuzzy c-means model. *IEEE Transactions on Fuzzy Systems* 3:370–379.

Pal, N. R., and J. C. Bezdek. 1997. Correction to "On cluster validity for fuzzy c-means model." *IEEE Transactions on Fuzzy Systems* 5:152–153.

Pankhurst, R. J. 1991. *Practical Taxonomic Computing*. Cambridge: Cambridge University Press.

Pathak, V., and O. Dikshit. 2010. A new approach for finding an appropriate combination of texture parameters for classification. *Geocarto International* 25:295–313. https://doi.org/10.1080/10106040903576195.

Pati, C., A. K. Panda, A. K. Tripathy, S. K. Pradhan, and S. Patnaik. 2020. A novel hybrid machine learning approach for change detection in remote sensing images. *Engineering Science and Technology, an International Journal* 2:973–981. https://doi.org/10.1016/j.jestch.2020.01.002.

Pena, J. M., J. A. Lozano, and P. Larranaga. 1999. An empirical comparison of four initialization methods for the k-means algorithm. *Pattern Recognition Letters* 20:1027–1040. https://doi.org/10.1016/S0167-8655(99)00069-0.

Pencue-Fierro, E., Y. Leonairo, T. Solano-Correa, J. C. Corrales-Muñoz, and A. Figueroa-Casas. 2016. A semi-supervised hybrid approach for multitemporal multi-region multisensor Landsat data classification. *IEEE Journal of Selected Topics in Applied Earth Observations and Remote Sensing* 9:5424–5435. https://doi.org/10.1109/JSTARS.2016.2623567.

Persello, C., and L. Bruzzone, 2014. Active and semisupervised learning for the classification of remote sensing images. *IEEE Transactions on Geoscience and Remote Sensing* 52:6937–6956. https://doi.org/10.1109/TGRS.2014.2305805.

Puissant, A., J. Hirsch, and C. Weber. 2005. The utility of texture analysis to improve per-pixel classification for high to very high spatial resolution imagery. *International Journal of Remote Sensing* 26:733–745. https://doi.org/10.1080/01431160512331316838.

Qin, R., and T. Liu. 2022. A review of landcover classification with very-high resolution remotely sensed optical images–Analysis unit, model scalability and transferability. *Remote Sensing* 14:646. https://doi.org/10.3390/rs14030646.

Quintano, C., A. Fernandez-Manso, and Y. E. Shimabukuro. 2012. Spectral unmixing. *International Journal of Remote Sensing* 33:5307–5340. https://doi.org/10.1080/01431161.2012.661095.

Ricchetti, E. 2000. Multispectral satellite image and ancillary data integration for geological classification. *Photogrammetric Engineering and Remote Sensing* 66:429–436.

Roli, F., G. Giacinto, and G. Vernazza. 1997. Comparison and combination of statistical and neural net algorithms for remote sensing image classification. In *Neurocomputation in Remote Sensing Data Analysis*, (eds.) I. Kanellopoulos, G. G. Wilkinson, G. G. Roli, and J. Austin, pp. 117–124. Berlin: Springer-Verlag. https://doi.org/10.1007/978-3-642-59041-2_13.

Sali, E., and H. Wolfson. 1992. Texture classification in aerial photographs and satellite data. *International Journal of Remote Sensing* 13:3395–3408. https://doi.org/10.1080/01431169208904130.

Schepers, J. 2004. Integrating remote sensing and ancillary information into management systems. In *Remote Sensing for Agriculture and the Environment*, (eds.) S. Stamadiadis, J. M. Lynch, and J. S. Schepers, pp. 254–259. Larissa Greece: Peripheral Editions.

Senthilnath, J., and X.-S. Yang. 2016. Multitemporal remote sensing image classification by nature-inspired techniques. In *Bio-Inspired Computation and Applications in Image Processing*, (eds.) X.-S. Yang, and J. P. Papa, pp. 187–219. Elsevier. https://doi.org/10.1016/B978-0-12-804536-7.00009-0.

Settle, J. J., and N. A. Drake. 1993. Linear mixing and the estimation of ground cover proportions. *International Journal of Remote Sensing* 14:1159–1177. https://doi.org/10.1080/01431169308904402.

Smits, P. C. 2002. Multiple classifier systems for supervised remote sensing image classification based on dynamic classifier selection. *IEEE Transactions on Geoscience and Remote Sensing* 40:801–813.

Sneath, P. H. A., and R. R. Sokal. 1973. *Numerical Taxonomy*. San Francisco: W.H. Freeman.

Steele, B. M. 2000. Combining multiple classifiers: An application using spatial and remotely sensed information for land cover type mapping. *Remote Sensing of Environment* 74:545–556. https://doi.org/10.1016/S0034-4257(00)00145-0.

Strahler, A. H. 1980. The use of prior probabilities in maximum likelihood classification of remotely sensed data. *Remote Sensing of Environment* 10:135–163. https://doi.org/10.1016/0034-4257(80)90011-5.

Su, Y. F. 2019. Integrating a scale-invariant feature of fractal geometry into the Hopfield neural network for super-resolution mapping. *International Journal of Remote Sensing* 40:8933–8954. https://doi.org/10.1080/01431161.2019.1624865.

Tatem, A. J., H. G. Lewis, P. M. Atkinson, and M. S. Nixon. 2001a. Multiple-class land-cover mapping at the sub-pixel scale using a Hopfield neural network. *International Journal of Applied Earth Observation and Geoinformation* 3:184–190. https://doi.org/10.1016/S0303-2434(01)85010-8.

Tatem, A. J., H. G. Lewis, P. M. Atkinson, and M. S. Nixon. 2001b. Super-resolution target identification from remotely sensed images using a Hopfield neural network. *IEEE Transactions on Geoscience and Remote Sensing* 39:781–796. https://doi.org/10.1109/36.917895.

Tatem, A. J., H. G. Lewis, P. M. Atkinson, and M. S. Nixon. 2003. Increasing the spatial resolution of agricultural land cover maps using a Hopfield neural network. *International Journal of Geographical Information Science* 17:647–672. https://doi.org/10.1080/1365881031000135519.

Thoreau, R., V. Achard, L. Risser, B. Berthelot, and X. Briottet. 2022. Active learning for hyperspectral image classification: A comparative review. *IEEE Geoscience and Remote Sensing Magazine* 10:256–278. https://doi.org/10.1109/MGRS.2022.3169947.

Townshend, J. R. G. 1981. *Terrain Analysis and Remote Sensing*. London: George Allen & Unwin.

Townshend, J. R. G., C. Huang, S. N. V. Kalluri, R. S. Defries, S. Liang, and K. Yang. 2000. Beware of per-pixel characteristics of land cover. *International Journal of Remote Sensing* 21:839–843. https://doi.org/10.1080/014311600210641.

Tso, B., and P. M. Mather. 1999. Classification of multisource remote sensing imagery using a Genetic Algorithm and Markov Random Fields. *IEEE Transactions on Geoscience and Remote Sensing* 37:1255–1260. https://doi.org/10.1109/36.763284.

Tso, B., and R. C. Olsen. 2005a. Combining spectral and spatial information into hidden Markov models for unsupervised image classification. *International Journal of Remote Sensing* 26:2113–2133. https://doi.org/10.1080/01431160512331337844.

Tso, B., and R. C. Olsen. 2005b. A contextual classification scheme based on MRF model with improved parameter estimation and multiscale fuzzy line process. *Remote Sensing of Environment* 97:127–136. https://doi.org/10.1016/j.rse.2005.04.021.

Tuia, D., M. Volpi, L. Copa, M. Kanevski, and J. Munoz-Mari. 2011. A survey of active learning algorithms for supervised remote sensing image classification. *IEEE Journal of Selected Topics in Signal Processing* 5(3):606–617. https://doi.org/10.1109/JSTSP.2011.2139193.

Tumer, K., and J. Ghosh. 1995. Theoretical foundations of linear and order statistics combiners for neural pattern classifiers. Technical Report TR-95-02-98, The Computer and Vision Research Center, The University of Texas at Austin.

Tupin, F., H. Maitre, J. F. Mangin, J. M. Nicolas, and E. Pechersky. 1998. Detection of linear features in SAR images: Application to road network extraction. *IEEE Transactions on Geoscience and Remote Sensing* 36:434–454. https://doi.org/10.1109/36.662728.

van der Meer, F. 1999. Image classification through spectral unmixing. In *Spatial Statistics for Remote Sensing*, (eds.) A. Stein, F. van der Meer, and B. Gorte, pp. 185–193. Dordrecht, Netherlands: Springer. https://doi.org/10.1007/0-306-47647-9_11.

van Engelen, J. E., and H. H. Hoos. 2020. A survey on semi-supervised learning. *Machine Learning* 109:373–440. https://doi.org/10.1007/s10994-019-05855-6.

Vieira, C. A. O. 2000. Accuracy of remote sensing classification of agricultural crops: A comparative study. Ph.D. Thesis, School of Geography, The University of Nottingham, Nottingham, UK.

Wang, F. 1990. Fuzzy supervised classification of remote sensing images. *IEEE Transactions on Geoscience and Remote Sensing* 28:194–201. https://doi.org/10.1109/36.46698.

Wang, J., Y. Zhong, and L. Zhang. 2023. Change detection based on supervised contrastive learning for high-resolution remote sensing imagery. *IEEE Transactions on Geoscience and Remote Sensing* 61:1–16. https://doi.org/10.1109/TGRS.2023.3236664.

Wang, L., and X. Jia. 2009. Integration of soft and hard classifications using extended support vector machines. *IEEE Geoscience and Remote Sensing Letters* 6:543–547. doi: 10.1109/LGRS.2009.2020924.

Wang, Q., X. Ding, X. Tong, and P. M. Atkinson. 2021. Spatio-temporal spectral unmixing of time-series images. *Remote Sensing of Environment* 259:112407. https://doi.org/10.1016/j.rse.2021.112407.

Witten, I. H., and E. Frank. 2005. *Data Mining: Practical Machine Learning Tools and Techniques*. San Francisco, CA: Morgan Kaufmann.

Xia, J., J. Chanussot, P. Du, and X. He. 2015. Spectral-spatial classification for hyperspectral data using rotation forests with local feature extraction and Markov random fields. *IEEE Transactions on Geoscience and Remote Sensing* 53:2532–2546. https://doi.org/10.1109/TGRS.2014.2361618.

Xu, G., X. Zhu, and N. Tapper. 2020. Using convolutional neural networks incorporating hierarchical active learning for target-searching in large-scale remote sensing images. *International Journal of Remote Sensing* 41:4057–4079. https://doi.org/10.1080/01431161.2020.1714774.

Yan, W. Y., and A. Shaker. 2011. The effects of combining classifiers with the same training statistics using Bayesian decision rules. *International Journal of Remote Sensing* 32:3729–3745. https://doi.org/10.1080/01431161003777197.

Yao, H., L. Zhao, M. Tian, Y. Jin, Z. Hu, Q. Peng, and Q. Qiu. 2023. Semantic segmentation for remote sensing image using the multigranularity object-based Markov random field with blinking coefficient. *IEEE Transactions on Geoscience and Remote Sensing* 61:1–22. https://doi.org/10.1109/TGRS.2023.3301494.

Zhang, J. S., and Y. W. Leung. 2004. Improved possibilistic C-means clustering algorithms. *IEEE Transactions on Fuzzy Systems* 12:209–217. https://doi.org/10.1109/TFUZZ.2004.825079.

Zhang, X., Y. Sun, J. Zhang, P. Wu, and L. Jiao. 2018. Hyperspectral unmixing via deep convolutional neural networks. *IEEE Geoscience and Remote Sensing Letters* 15:1755–1759. https://doi.org/10.1109/LGRS.2018.2857804.

Zheng, C., and L. Wang. 2014. Semantic segmentation of remote sensing imagery using object-based Markov random field model with regional penalties. *IEEE Journal of Selected Topics in Applied Earth Observations and Remote Sensing* 8:1924–1935. https://doi.org/10.1109/JSTARS.2014.2361756.

Zheng, C., Y. Zhang, and L. Wang. 2021. Multigranularity multiclass-layer Markov random field model for semantic segmentation of remote sensing images. *IEEE Transactions on Geoscience and Remote Sensing* 59:10555–10574. https://doi.org/10.1109/TGRS.2020.3033293.

Zhou, Y., A. Rangarajan, and P. D. Gader. 2018. A Gaussian mixture model representation of endmember variability in hyperspectral unmixing. *IEEE Transactions on Image Processing* 27:2242–2256. https://doi.org/10.1109/TIP.2018.2795744.

Zhu, X., and A. B. Goldberg. 2022. *Introduction to Semi-Supervised Learning*. Switzerland: Springer Nature.

Zhu, Z., A. L. Gallant, C. E. Woodcock, B. Pengra, P. Olofsson, T. R. Loveland, S. Jin, et al. 2016. Optimizing selection of training and auxiliary data for operational land cover classification for the LCMAP initiative. *ISPRS Journal of Photogrammetry and Remote Sensing* 122:206–221. https://doi.org/10.1016/j.isprsjprs.2016.11.004.

3 Dimensionality Reduction
Feature Extraction and Selection

Many factors have varying degrees of effects on the success of image classification and the accuracy achieved by the applied classifier. However, the representativeness and the quality of the samples employed in the training, validation, and test stages are of utmost importance. Effective machine learning heavily relies on the choice of data representation for its application. Consequently, a considerable effort in implementing machine learning algorithms involves devising preprocessing pipelines and data transformations, often referred to as feature engineering, to create a data representation that supports efficient machine learning (Bengio et al., 2013). There is a widespread yet unfounded assumption that increasing the features during classification results in higher accuracy. In most cases, this may not be a valid assumption because the addition of features increases the probability of high correlation among features and the inclusion of irrelevant information, which definitely hinders the construction of correct decision boundaries for the identification and classification of landscape features. In other words, the "garbage in, garbage out" principle holds true when irrelevant features are incorporated into a remote sensing process. Moreover, it should always be borne in mind that the addition of each feature will increase the dimension of the data set, and more samples will be needed to estimate the hyperparameters of the selected algorithm.

Collecting samples on the ground through field campaigns is costly and typically requires a substantial amount of time in remote sensing projects. Therefore, two solutions, namely, feature extraction and feature selection, have been applied in the literature to decrease the dimension of multivariate data sets. While the original feature space is projected or transformed into a lower dimension in the feature extraction approach, the most relevant features are sought and selected, or conversely, the least effective features are detected and excluded from the data set through intelligent algorithms in the feature selection approach. Both approaches share the same underlying philosophy that with the elimination of correlated and irrelevant features, a data set at lower dimensions can be obtained, which can produce similar or better classification accuracies. This philosophy is analogous to the principle of Occam's Razor, suggesting that a model with the smallest possible features or parameters that adequately represent the data should be preferred. The mutual concern of reducing the dimensions in both machine learning and pattern recognition is expected, considering their common objective of classification. In pattern recognition, it can affect the efficiency of data acquisition and the accuracy and complexity of the classifiers. This also holds true in machine learning, which faces the additional challenge of extracting useful information from the introduced data.

Image coordinates give the relative location of a pixel in the spatial domain, and given the origin of the coordinate system and the pixel spacing (Δx and Δy), geometrical calculations, such as interpixel distance, can be performed. When we take into account the values associated with a pixel, which form a vector of measurements on a set of selected features, we can think of a space defined not by the x and y or row and column spatial coordinates, but by the features on which the pixel values are measured. These features may be image pixel values in separate wavebands, context or texture measurements, or geographical attributes of the area represented by the pixel, such as mean elevation, slope angle, or slope azimuth (i.e., ancillary data). Feature space is multidimensional and as such cannot be visualized. Nevertheless, standard geometrical measures such as the Euclidean distance as the shortest distance between the two points are still valid.

Figure 3.1a shows that the spatial domain coordinates of the shaded pixel (in row-column representation) are (5, 4). Figure 3.1b shows three co-registered images, perhaps representing reflectance

DOI: 10.1201/9781003439172-3

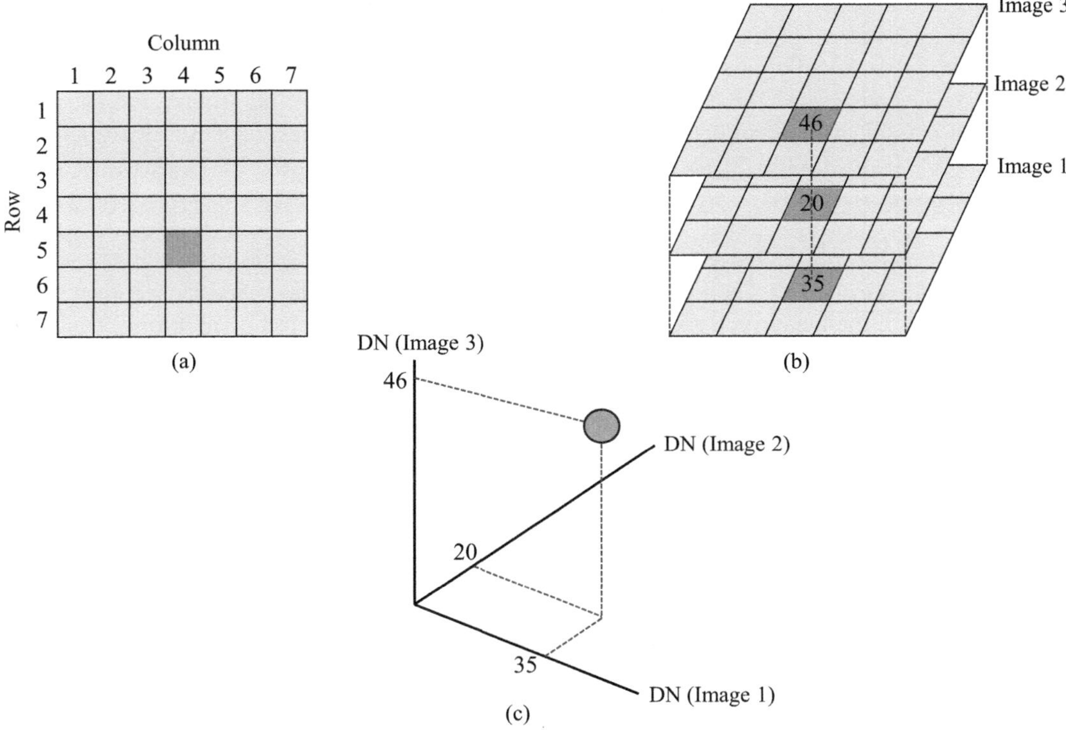

FIGURE 3.1   (a) Image spatial space. (b) Co-registration of three images. (c) Image feature space.

in the green, red, and near-infrared (near-IR) wavebands. The quantized pixel values in these wavebands are {35, 20, 46}, respectively. Figure 3.1c shows a plot of the position of the pixel in a feature space that has three axes defined by these three bands. In the case of more than three wavebands, visualization of data in three-dimensional feature space is only possible with three-waveband combinations, which is cumbersome and insignificant in many cases, considering the fact that an abundance of data is available from recent sensors including the hyperspectral ones. Therefore, computer processing of remotely sensed data is compulsory. At this point, it should be pointed out that when there are a limited number of features, orthogonal transforms can be applied and the first three components representing the majority of the information inherent in the data set can be visualized for better understanding and interpreting the data. This can be particularly important to analyze the quality and representativeness of training samples in supervised image classification.

The foregoing discussion suggests that there might be an optimal number of input features for a given data set to accurately depict the attributes of the training data. The pattern recognition process involves the subdivision of feature space into homogeneous regions separated by decision boundaries. The statistical, machine learning, and knowledge-based methods discussed in this book use different decision rules to define or specify these boundaries. In the fuzzy classification procedure, decision boundaries can overlap. A number of techniques can be used to manipulate or transform the axes of the feature space in order to facilitate classification, for example, by determining a subspace that contains most of the information present in the original feature space. Orthogonal and nonorthogonal transforms, which accomplish this aim to a greater or lesser extent, are described in the following sections.

In a survey of pattern recognition methodology in remote sensing, Landgrebe (1998) noted the need for accurate class statistics if supervised classifiers are to perform in a satisfactory way. The minimum size of a training data set depends to a considerable extent on the number of features used to characterize the objects to be classified. However, the ground reference data set is difficult and also expensive to procure. Landgrebe (1998) points out that the number of possible locations in a feature space depends on the number of features and on the number of quantization levels used for each feature. For example, the Landsat 7 ETM+ produces seven bands of data with 256 levels per band. It is theoretically possible, therefore, for a pixel to occupy any of $(28)7 = 7 \times 10^{16}$ discrete locations in feature space. The benefits of reducing the dimensions of the ground reference data in terms of classification accuracy have been discussed in many studies, including Kavzoglu and Mather (2002), Foody (2009), Millard and Richardson (2015), Colkesen and Kavzoglu (2016, 2019), and Georganos et al. (2018).

Given a fixed training set size, an increase in the dimensionality of the feature space (e.g., by adding additional spectral bands) means that the number of parameters to be estimated in a statistical classifier also increases. The maximum likelihood decision rule (described in Section 2.3.2.3) uses the values of the mean vector $\bar{x}$ and the variance-covariance matrix $\mathbf{C}$. The former has k elements to be estimated, while the latter has $k(k-1)/2 + k$ elements, where k is the number of features, giving a total of $k(k-1)/2 + 2k$ elements. As the value of k increases, the number of parameters to be estimated increases disproportionately. For $k = 6$, the number of parameters is 27. For $k = 12$, this value rises to 90. The fact that efficient estimation of statistical parameters requires a representative sample of a sufficient size is well known; consequently, as the number of parameters increases, then, for a fixed sample size, the efficiency of the estimation decreases, which implies that the confidence limits for each estimate become wider. The effectiveness of the classifier will therefore begin to decrease once a certain number of dimensions are reached. This is known as the *Hughes phenomenon* (Hughes, 1968). It follows that if satisfactory results are to be obtained from the classification of remotely sensed data, the relationship between dimensionality and training sample size must be born in mind. The topic of sampling adequacy is considered further in Section 10.1.

In the literature, two main approaches exist to reduce the number of input features without sacrificing accuracy. The first one is to project the original feature space onto a subspace (i.e., a space of smaller dimensionality). This can be done using different orthogonal and nonorthogonal methodologies. In addition to feature extraction methods, as described in Section 3.1, feature selection algorithms (Section 3.2) that are based on the idea of selecting the "best" possible combination of features considering a fitness or quality metric have been conventional ways for reducing the dimensions of the multivariate data. With the introduction of neural networks, we have more options for selecting the most important or relevant features. Autoencoders, specific types of deep learning architecture used for learning representation of data, specific neural network models, such as a self-organized feature map (SOM), and deep learning models, such as the deep Boltzmann machine (Taherkhani et al., 2018), and deep belief network (Zou et al., 2015) have recently become popular with their superior performances. The use of autoencoders in the context of deep learning is given in Section 7.2.7. On the other hand, major architectures of deep learning will be discussed in Chapter 7.

The second method is to use separability measurements in the input feature space and then select the subfeature dimension in which separability is maximum. The aim is to reduce the feature space dimension without influencing classification accuracy. One way of mitigating the effects of high dimensionality is to determine a subset of the k-dimensional feature space that contains most of the pixels. This is the aim of orthogonal transforms. If the measurements made by two instruments are correlated, then there is a degree of redundancy present in the data. For instance, this can be the case for hyperspectral data collected at narrow spectral ranges adjacent to each other, with a high probability of carrying similar information. When the correlation between two sets of measurements reaches ± 1.0, then one set of measurements is completely redundant. The features used in remote sensing image classification are generally correlated, and a high proportion of the information content of the data can be represented in terms of m dimensions, where $m < k$. Landgrebe (1998) also claims that the data distribution in the reduced-space representation is more likely to be normally distributed than the original data. This observation is an important one if statistical classifiers are used. Cortijo and Pérez de la Blanca (1999) suggest that where the sample size is small, it may be advisable to compute a common covariance matrix so that all training data are used, rather than compute separate covariance matrices for each class.

An alternative approach involves the use of orthogonal transforms, which provide a means of reducing feature space dimensionality. Nielsen (1994), Park et al. (2004), and Tian et al. (2005) provide details and comparisons of these methods. In essence, an orthogonal transform aims to replace the feature set with a derived or synthetic feature set. The covariances of the synthetic features are zero; hence, there is no redundancy. The synthetic features are defined by linear combinations of the observed features, and much of the information content of the observed features is reproduced by m synthetic features, where $m < n$.

In the following sections, orthogonalizing procedures including five conventional feature extraction algorithms are described. For instance, principal component analysis (PCA) is a well-known method of orthogonalizing a data set and of ordering the synthetic features (in this case, the principal components) in terms of their contribution to total variance. However, it is sensitive to the scale of measurement used for each feature, and there is no reason to believe that lower-order principal components represent noise or unwanted information. The minimum/maximum autocorrelation factor (MAF) transform and the maximum noise fraction (MNF) procedures aim specifically to separate information and noise and, unlike PCA, are independent of the measurement scale. Nonorthogonal techniques of independent component analysis (ICA) and projection pursuit are also used to decompose the data set into its components. Wavelet transform, the theory of which is given in Chapter 4 (see Section 4.1.1.5), has been also widely used in feature extraction applications. A wide range of feature selection methods including filter, wrapper, and embedded methods are discussed and reviewed in the following sections. Local (sequential feature selection strategies, recursive feature selection, simulated annealing) and global (genetic algorithm, particle swarm optimization, gray wolf, and dragonfly algorithms) search algorithms are also discussed with their theories and applications.

3.1 FEATURE EXTRACTION

3.1.1 PRINCIPAL COMPONENT ANALYSIS

Principal component analysis (PCA, also known as Karhunen-Loeve analysis) is a general tool for transformation and data reduction in remote sensing image processing. It is, in fact, one of the most popular techniques used in numerous studies, requiring the analysis of high-dimensional data. The new axes formed by PCA are not specified by the user's prior definition of the transformation matrix, but they are derived from the variance-covariance or correlation matrix computed from the data under analysis. The process of PCA can be divided into three steps:

1. Calculation of the variance-covariance (or correlation) matrix of multiband images (e.g., in the case of a four-band image, the covariance matrix has dimension 4×4),
2. Extraction of the eigenvalues and eigenvectors of the matrix, and
3. Transformation of the feature space coordinates using these eigenvectors.

Let M and x denote the multiband image mean and individual pixel value vectors, respectively. The first step, derivation of the covariance matrix $\mathbf{C}$, is expressed as

$$\mathbf{C} = \frac{\sum_{j=1}^{n} (x_j - M)(x_j - M)^T}{n - 1} \tag{3.1}$$

where n is the number of pixels. If the correlation matrix is used, each entry in $\mathbf{C}$ should be further divided by the product of the standard deviations of the features represented by the corresponding row and column. For instance, let c_{12} denote the entry of the first row and second column (i.e., the covariance between image bands 1 and 2) in matrix $\mathbf{C}$; then, the corresponding correlation r_{12} is obtained by

$$r_{12} = \frac{c_{12}}{\sigma_1\sigma_2} \tag{3.2}$$

where σ_1 and σ_2 are the standard deviations of image bands 1 and 2, respectively. The correlations of other entries can be computed in a similar manner.

The second step, the calculation of the eigenvectors of $\mathbf{C}$, is achieved by solving the following equation:

$$(\mathbf{C} - \lambda_i\mathbf{I})\mathbf{A}_i = 0 \tag{3.3}$$

where $\mathbf{A}_i = (a_1, a_2,\ldots, a_k)^T$ is the eigenvector corresponding to the eigenvalue λ_j, k is the total number of feature space dimension, and $\mathbf{I}$ is the identity matrix (i.e., a matrix with diagonal entries set to 1 and off-diagonal entries set to 0). All the eigenvalues λ can be determined by solving $(\mathbf{C} - \lambda_i\mathbf{I}) = 0$. The new coordinate system is formed by the normalized eigenvectors of the variance-covariance (or correlation) matrix. The mapping location f_i of each pixel $\mathbf{x} = (x_1, x_2, \ldots, x_k)$ on the ith principal component is given by

$$f_i = \mathbf{x}\mathbf{A}_i = x_1a_1 + x_2a_2 + \cdots + x_ka_k. \tag{3.4}$$

This is effectively a rotation of the axes of the feature space.

PCA has the property that the first principal component image (PC1, derived from the first eigenvector) represents the maximum amount of the total variance of the data set, and the variances of the remaining principal component images decrease in order, as denoted by the magnitudes of the corresponding eigenvalues. Normally, the variance contained in the last few principal components is small. Therefore, PCA is often used to condense the information in a multiband image set into fewer new bands (represented by the higher-order components), and input them, rather than the raw data, into a classifier, thus reducing the computational demands and possibly improving performance. However, as noted already, there is no reason to assume that lower-order principal components do not contain any discriminating information. Estimated principal components will not always produce components concentrating on the recorded signal. It may be possible that noise in the data set can be represented with a higher variance than certain types of signal components. It may also be of interest to note that the locations of the principal component axes in the feature space are fixed by reference to two constraints. First, the axes are orthogonal. Second, each axis accounts for the maximum variance in the data set given that the influence of any higher-order components has been removed. Relaxation of the second constraint allows the use of orthogonal rotations that might make the resulting components more interpretable. It is also possible to relax the first constraint, but that does not seem to be helpful in the present context.

Figure 3.2 shows an example using a two-band image. Here, the original pixel values in the two images fall in the range [A, B] and [C, D], respectively. After the covariance (or correlation) matrix is formed, the resulting eigenvectors are orientated in the directions [E, F] and [G, H], respectively. The data range [E, F] in the first principal component (PC1) is larger than either [A, B] or [C, D], and thus, a higher contrast image will result. The data range [G, H] of PC2 is much smaller, and the corresponding principal component image will show less variation. However, the information content of the two axes will differ in that PC1 will represent the information that is shared between all or most of the original spectral bands (and may therefore be thought to represent average brightness), whereas the second and subsequent components will contain information that is statistically uncorrelated with brightness. For example, the second principal component of a multispectral image set of a vegetated area may be related to variations in the nature, vigor, and spatial cover of the vegetation.

In the next example, an eight-band WorldView-2 image of the Gebze Technical University campus in Turkey is used (Figure 3.3). The resulting principal component images are shown in Figure 3.4a, and a clear decrease in data variance (information) in terms of histograms is illustrated in Figure 3.4b.

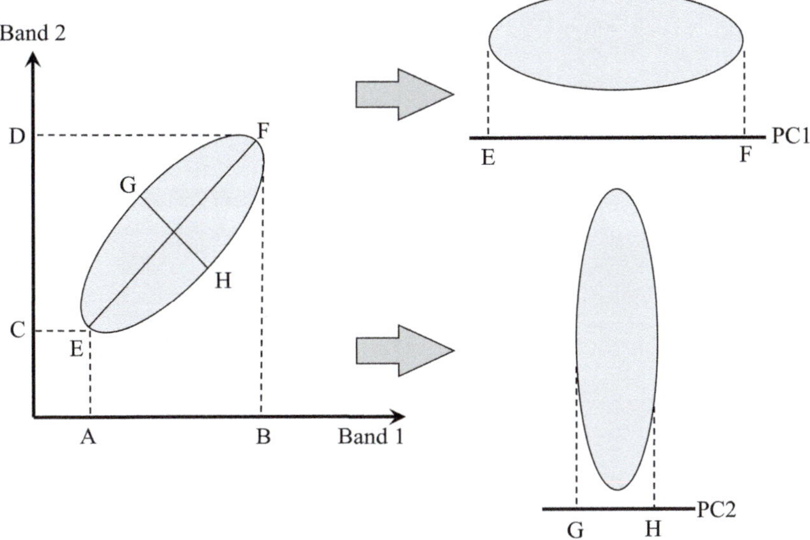

FIGURE 3.2 Two-band example for principal component analysis.

FIGURE 3.3 Eight-band WorldView-2 image of Gebze Technical University campus in Turkey.

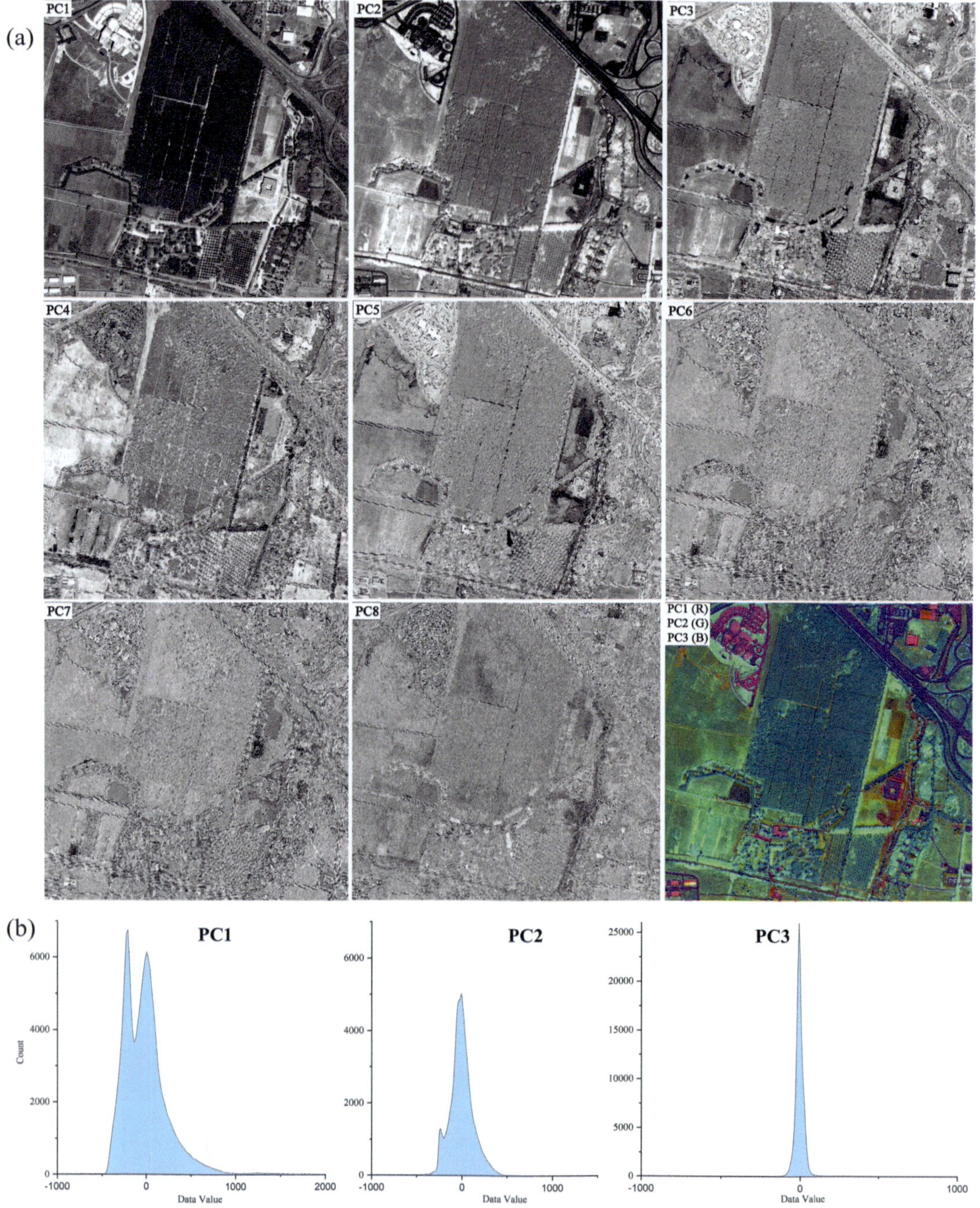

FIGURE 3.4 (a) Principal component images and a false-color composite of the first three components derived from the WV-2 image shown in Figure 3.3. (b) Histograms of the first three principal component images. The histograms show that the variance of the first principal component exceeds that of the second, which in turn is greater than the variance of the third component.

The first three principal components explained 79.80%, 17.99%, and 0.90% of the total cumulative variation in the data, respectively. Based on these statistics, the first three components represent 98.69% of the data set, which provides a good opportunity to visualize and interpret the data.

As noted earlier, PCA can be based either on the matrix of variances and covariances or on the matrix of correlations among the spectral bands of an image set, which can be derived either from sample data or from all pixels of the image set. PCA based on the covariance matrix is sometimes also called unstandardized PCA, while standardized PCA uses the correlation matrix. These matrices have different eigenvectors; thus, the resulting principal component images will also be different. If the images subjected to PCA are not measured in the same scale, the standardized method is also a better choice than the unstandardized method because the correlation matrix normalizes the data onto the same scale (i.e., each spectral band is given unit variance, and hence, inter-band variation is ignored). Users should be aware that the use of the correlation matrix implies the equalization of within-band variance and should also note that the technique does not differentiate between information and noise. In some image sets (e.g., hyperspectral data sets), considerable noise may be present and have a variance that is greater than some sources. As proposed by Roger (1996), the PCA method could be modified to eliminate the noise variance within the data set. Thus, the "noise-adjusted" approach, a theory of which can be found in Mather and Koch (2011), is developed. It should be also noted that PCA is a pixel-wise method of feature extraction and does not consider the spatial nature of the image data. Moreover, the spectral information represented by the original wavebands is lost after the transformation process. Therefore, interpretations of the components should be made considering this important issue.

PCA transformation has played a crucial role in enhancing the performance of image classification, particularly for hyperspectral image (HSI) data that encompass hundreds of adjacent narrow wavebands. Attaining cost-effective and satisfactory classification accuracy is often difficult when using the original HSI bands, which often include irrelevant bands for classification tasks. Moreover, the classification of these images can be challenging when limited training samples are available and several parameters have to be set for the classifier. On the other hand, conventional PCA approaches have some challenges when dealing with HSI classification (Cheriyadat and Bruce, 2003; Fauvel et al., 2009; Uddin et al., 2019; Hashjin and Khazai, 2022). Key challenges include the following: (1) their inability to capture local statistics within specific distributions of HSI data, as they predominantly focus on global variance; (2) their inability to guarantee the revelation of distinct information within the top principal components (PCs); and (3) their susceptibility to the influence of bands within the visible and near-infrared spectral regions (Uddin et al., 2021). A nonlinear variant of PCA has demonstrated the ability to capture a portion of higher-order statistical features, thereby improving the representation of information derived from the initial data set (Schölkopf et al., 1998). In many HSI classification practices, PCA and its derivatives (e.g., kernel PCA, segmented PCA, spectrally segmented PCA, folded PCA) have been adopted to eliminate the irrelevant data and, most importantly, the data with high correlation. It has been reported that classification with selected PCs can produce significantly higher classification accuracies compared with the use of original spectral bands (Fauvel et al., 2009; Xia et al., 2014; Kavzoglu et al., 2018a).

3.1.2 Minimum/Maximum Autocorrelation Factors

A somewhat different approach to the orthogonalization of specifically spatial data is presented by Switzer and Green (1984). They argued that PCA does not differentiate between signal and noise, that it is not explicitly spatial (in the sense that the image data could be rearranged randomly without affecting the results), and that it is scale-dependent. They proposed a method of distinguishing between signal and noise and suggested that their approach should be preferred to filtering because it does not blur the data. The method is based on the principle that information, or signal, is spatially autocorrelated (in the sense that pixels in a neighborhood will tend to have similar values because they represent some geographical object such as a lake or a forest). Conversely, the noise will have a low spatial autocorrelation if systematic noise such as banding is excluded. The Minimum/Maximum Autocorrelation Factors (MAF) procedure is based on the ordering of a set of orthogonal functions in such a way that autocorrelation decreases from the low-order to the high-order functions; that is, autocorrelation is minimized rather than variance being maximized, as is the case with PCA. The lower-order functions may therefore be expected to contain mainly nonautocorrelated noise, while the higher-order components would represent information. The procedure is scale-free in the sense that the same result is achieved irrespective of the scale of measurement used for each feature. This is because the procedure maximizes a ratio – that of signal to noise.

A detailed account of the computational procedures involved in the MAF transform is provided by Nielsen (1994) and Nielsen et al. (1998). The variance-covariance of the total data set, $\mathbf{C}$, is derived, together with the corresponding eigenvalues Λ and eigenvectors P. Next, the variance-covariance of the noise component, $\mathbf{C}_\Delta$, is computed. Switzer and Green (1984) suggest that the original data set is converted to two difference images, one being shifted by one pixel horizontally and the other being shifted by the same amount vertically. The variance-covariance matrices of the two difference images are calculated and pooled to give $\mathbf{C}_\Delta$. Finally, the eigenvalues and eigenvectors of $\Lambda^{-\frac{1}{2}}P^{\mathrm{T}}\mathbf{C}_\Delta P\,\Lambda^{\frac{1}{2}}$ are obtained, and MAF images are derived using procedures equivalent to those described above for the calculation of principal component images (Equation 3.4).

3.1.3 Maximum Noise Fraction (MNF) Transformation

Maximum noise fraction transformation (MNF), which is sometimes known as noise-adjusted PCs, requires estimates of the variance-covariance matrices of the signal, $\mathbf{C}_S$, and noise, $\mathbf{C}_N$, components, and it uses the signal-to-noise ratio to determine the ordering of the MNF components. Thus, the MNF produces components that successively maximize the signal-to-noise ratio, just as PCA generates components that successively maximize the variance. The noise variance can be estimated in several ways. For example, the differences between adjacent pixels can be used, as described in Section 3.1.2. In this case, MNF and MAF produce the same eigenvectors, and consequently the same orthogonally transformed images.

The variance-covariance (dispersion) matrix of the full data set is $\mathbf{C}$, and it is axiomatic that $\mathbf{C} = \mathbf{C}_S + \mathbf{C}_N$. Nielsen (1994) shows that the required eigenvalues and eigenvectors are the solution of the generalized eigenproblem equation $\det\left(\mathbf{C}_N - \lambda\mathbf{C}_S\right) = 0$. Martin and Wilkinson (1971) showed how this equation is solved by reduction to the standard form. The codes of their implementation are available in libraries, such as Netlib (http://netlib.org). The computed eigenvectors are used to derive component images, as described in Section 3.1.1.

Clearly, the nub of the issue is the estimation of $\mathbf{C}_N$. Nielsen (1994) lists five methods, partly derived from Olsen (1993). These are as follows:

1. Simple differencing using the method of Section 3.1.2.
2. Use of a simultaneous autoregressive synthetic aperture radar (SAR) model involving the pixel of interest and its neighbors to the W, NW, N, and NE.
3. Determine the difference between the value of the pixel of interest and the local mean, which is computed for a rectangular window.
4. Method (3), but the local median is used in place of the local mean.
5. Compute the residual from a local quadratic surface based on the pixel values neighboring the pixel of interest.

Nielsen (1994) also considers the problem of periodic noise, such as banding due to differences in sensor calibrations. Banding, which is seen as horizontal striping, is obviously autocorrelated, and so will be identified as part of the signal. Nielsen (1994) suggests that frequency domain filtering be used before the calculation of the noise covariance. The use of wavelets to estimate noise is another possible approach (Mather and Koch, 2011). Other pertinent references to the derivation and use of

MNF are Green et al. (1988) and Lee et al. (1990), who discuss their mathematical details, while Van der Meer and De Jong (2000) show how the results of spectral unmixing are enhanced by the orthogonalization of endmembers. Nielsen et al. (1998) discuss the MAF transformation in the context of change detection.

3.1.4 Independent Component Analysis

Independent component analysis (ICA), which can be regarded as an extension of the PCA, is an effective signal processing technique that extracts independent signals from a composite signal. Despite PCA, ICA is a nonorthogonal decomposition method. The fundamental theory behind ICA is the use of higher-order statistics instead of second-order ones, which provides the ability to extract smaller and more detailed features in the image data. The method is based on the non-Gaussian assumption of the independent variables or features. ICA is a well-known unsupervised blind source separation technique, widely used in remote sensing studies, which aims to search statistically independent components (ICs) by only considering the observation of mixture signals (Lennon et al., 2001). The basic equation of the ICA is similar to that of the linear mixture model, as given below:

$$\mathbf{x} = \mathbf{As} + \mathbf{n} \tag{3.5}$$

where $\mathbf{x} = (x_1, x_2, \ldots, x_m)$ is a vector of observations, $\mathbf{s} = (s_1, s_2, \ldots, s_n)$ is the vector of the latent variables called the independent components, and $\mathbf{A}$ is an unknown constant matrix, called the mixing matrix. The vector $\mathbf{n}$ is noise, which is often omitted. It is assumed that the ICs $\mathbf{s}_i$ are mutually statistically independent. Then, the basic problem becomes the estimation of mixing matrix $\mathbf{A}$ and the realization of the ICs $\mathbf{s}_i$ using only observations of the mixtures $\mathbf{x}_j$ (Hyvärinen, 1998; Hyvärinen et al., 2001).

When an ICA is implemented for dimensionality reduction, an important issue is how to rank the ICs according to their importance since the ICs are generated by random initial projection vectors. Therefore, the IC created earlier is not necessarily more meaningful than those created later. Unlike PCA and MNF methods, which prioritize the components based on the magnitude of eigenvalues or signal-to-noise ratio, ICA has no such criteria for prioritizing generated ICs. To overcome this issue, ICs are ordered based on two measures, namely, skewness and kurtosis. If kurtosis is used as the basis for the decomposition of observations (pixel values) into sources of variability (endmembers in mixture modeling parlance), then fourth-order statistic is applied. It should be noted that when the significance of an independent component is measured by variance, the ICA is reduced to PCA/MNF, and the score of a component for significance is the eigenvalue associated with that component (Wang and Chang, 2006). In recent years, many studies used ICA for various research studies, including dimensionality reduction and data mining (Lennon et al., 2001; Jayaprakash et al., 2020), spectral unmixing (Nascimento and Dias, 2005; Wang et al., 2014), change detection (Zhong and Wang, 2006), image classification (Villa et al., 2011; Falco et al., 2014).

3.1.5 Projection Pursuit

Although the projection pursuit method, an explanatory analysis of large multivariate data sets, was first introduced by Kruskal (1969, 1972), the first successful implementation of the method was conducted by Friedman and Tukey (1974). The method finds hidden structures in multivariate data by applying low-dimensional orthogonal projections. In its original form, it is impractical to explore possible projections exhaustively due to the high dimension of space of projection directions. Projection pursuit is the process of making "interesting" projections by the local optimization over projection directions using some index of "interestingness", which is commonly known as the projection index (Jones and Sibson, 1987). The projection index is a function that associates the projection of the data with the actual value that measures its interestingness. The traditional metrics of skewness and kurtosis, the third and fourth moments of a random variable after standardization, have usually been utilized for the projection index estimations. The higher the index, the more interesting the projection. Therefore, the method searches for the data projection that maximizes the projection index. The robustness of the projection pursuit lies in the fact that it can avoid the "curse of dimensionality" by working in low-dimensional linear projections. Another advantage of the method is that it can ignore irrelevant features.

Projection pursuit methods are computationally intensive since the time and memory requirement increase exponentially when the high-dimensional data are not well scaled. The computational difficulties of projection pursuit within the framework of skewness maximization are discussed by Loperfido (2018). A randomly generated vector generated in the initial setup may converge to a different projection vector. Therefore, the results of each experiment may produce different projection indices. To overcome this difficulty, several solutions including projection index–based prioritized projection pursuit have been suggested (Chang, 2013). It can be stated that most conventional multivariate analysis methods are special cases of projection pursuit (Huber, 1985). Given a sample $\mathbf{x}_1,\ldots,\mathbf{x}_n$ of p-dimensional vectors, the methodology of projection pursuit is as follows (Bickel et al., 2018):

i. Generate the empirical distribution of projections of the observed data into one or more dimensions, that is, the empirical distribution of $\mathbf{z}^T\mathbf{x}_i$, $i = 1,\ldots,n$, where the vector $\mathbf{z}$ belongs to the p-dimensional unit sphere $\mathbb{S}^{p-1}$. The distribution can be illustrated with $\hat{G}_{\mathbf{z},n}$ or $\hat{G}_{\mathbf{z}}$ for short:

$$\hat{G}_{\mathbf{z},n}(t) = \frac{1}{n}\sum_{j=1}^{n}1\left(\mathbf{x}_j^T\mathbf{z} \leq t\right) \tag{3.6}$$

ii. Analyze these distributions while varying the vector $\mathbf{z}$ vary on the unit sphere, identifying projections where $\hat{G}_{\mathbf{z}}$ exhibits notable deviations from a Gaussian distribution.

Friedman and Tukey (1974) put forth the first projection pursuit index, and subsequently, additional indices emerged, incorporating cumulants, entropy, and the distance from a Gaussian distribution (Bickel et al., 2018). In recent years, several studies (e.g., Blanchard et al., 2006; Sasaki et al., 2016; Virta et al., 2016; Loperfido, 2018) proposed algorithms to find non-Gaussian components in a high-dimensional distribution and reduce the computational cost. Projection pursuit has been a popular feature extraction approach, also used in image compression, segmentation, classification, and image enhancement analysis.

3.2   FEATURE SELECTION

Recent advancements in remote sensing satellite technology have introduced improved sensors with enhanced spatial resolution and more spectral bands. The quantity of these spectral bands can vary significantly, ranging from a few bands (e.g., Landsat OLI, Sentinel-2, IRS LISS) to an extensive count exceeding two hundred bands (e.g., Hyperion, AVIRIS, HyMap). Although each of these individual wavebands provides crucial information for comprehending the characteristics of remotely detected objects, numerous bands within the data set exhibit substantial correlation, leading to redundancy. Eliminating such redundancy is imperative for developing more efficient classification techniques. When limited training samples per feature (land use/land cover classes in remote sensing image classification) are available, a reduction in classification accuracy is observed with the addition of more features (Pal and Foody, 2010). Moreover, certain bands may be irrelevant to the specific objectives of the investigation. In classification practices with multispectral data, the effective wavebands may be predicted with prior knowledge or can be determined by analyzing the spectral signatures of the objects under investigation. For instance, thermal bands are not employed in land use and land cover (LULC) classifications because of their inferior spatial resolution, which requires different preprocessing and very limited contribution to existing information; hence, these bands are usually removed from the data set. To support this argument, Thenkabail et al. (2004) comprehensively analyzed this objective by determining the best hyperspectral wavebands for vegetation and agricultural studies in the spectral range of 400–2,500 nm. After applying rigorous data mining techniques, they suggested 22 optimal bands that best characterized and classified vegetation and agricultural crops. The computational requirement of a classifier is highly correlated with the number of features used as input to the classification algorithm. For instance, in the case of n input features, the computational requirements of the maximum likelihood classifier (Section 2.3.2.3) are proportional to n^2. As n increases in magnitude, the computational cost will rise nonlinearly.

The commonly observed trend in literature is the pursuit of potential feature subsets from the entire data set, aiming to identify an optimal subset based on a performance criterion. This process is defined as feature selection (also called factor selection), primarily aiming to exclude bands holding duplicative or irrelevant spectral information. Feature selection is a vital approach not only for remote sensing but also for a wide and active field of research, including bioinformatics, facial recognition, text classification, and medical applications. The use of feature selection techniques has three key advantages. First, classification performance can be improved by reducing the number of bands to a new set of relevant and uncorrelated bands. In the case of neural network models, this issue is quite important for improving the generalization capabilities of the network since, for a given number of training samples, a larger network may have poorer generalization capability than one with fewer inputs. Second, working with a reduced number of inputs results in smaller processing times. Notably, the substantial time required for training poses a significant drawback in the practical use of deep learning models. Finally, since the dimensions of the data are directly linked to the size of the training data, lower-dimensional data sets would be more appropriate in cases where a limited amount of training data are available. Even with dedicated efforts in the field of feature selection (e.g., Liu and Motoda, 2007; Zhao and Liu, 2012; Stańczyk and Jain, 2015), it remains a prevailing subject in remote sensing studies. The advent of diverse methods for remotely gathering data and the introduction of new applications have posed enduring questions related to achieving stable feature selection, efficient removal of redundancy, and proficient use of ancillary data (Wu et al., 2013).

Due to the impracticality of assessing each conceivable subset derived from the complete data set, a variety of search techniques including greedy search and recursive search strategies (Section 3.2.1), metaheuristic algorithms based on genetic algorithms, particle swarm optimization, gray wolf algorithm, and dragonfly algorithm, discussed in the Wrappers section (Section 3.2.5), have been developed and applied in remote sensing. Feature selection algorithms, including separability indices, filter-based methods, wrappers, and ensemble feature selection methods, are all discussed in the following sections. Although numerous studies have been reported in the literature, several studies have been conducted to comprehensively evaluate feature selection algorithms in remote sensing (e.g., Wu et al., 2013; Kumar and Minz, 2014; Vijouyeh and Taşkın, 2016; Ma et al., 2017; Colkesen and Kavzoglu, 2018).

3.2.1 Greedy Search Methods

Several search techniques exist to search for the optimal subset from the data set, which is generally the training data set in image classification experiments. The sequential forward selection technique (SFS) starts by finding the best individual feature and then evaluates the remaining features one at a time to find the second-best feature (i.e., the one that gives higher performance than other candidate features). This process continues iteratively until a desired number of features are selected (Figure 3.5). Unlike the SFS method, sequential backward selection (SBS), also called sequential backward elimination, starts with the entire data set and searches for the feature that has the least effect when it is removed (Figure 3.6). In other words, it excludes each feature one at a time and finds the least effective one by looking at the values of a fitness measure, which can be a separability measure (this is a fast solution)

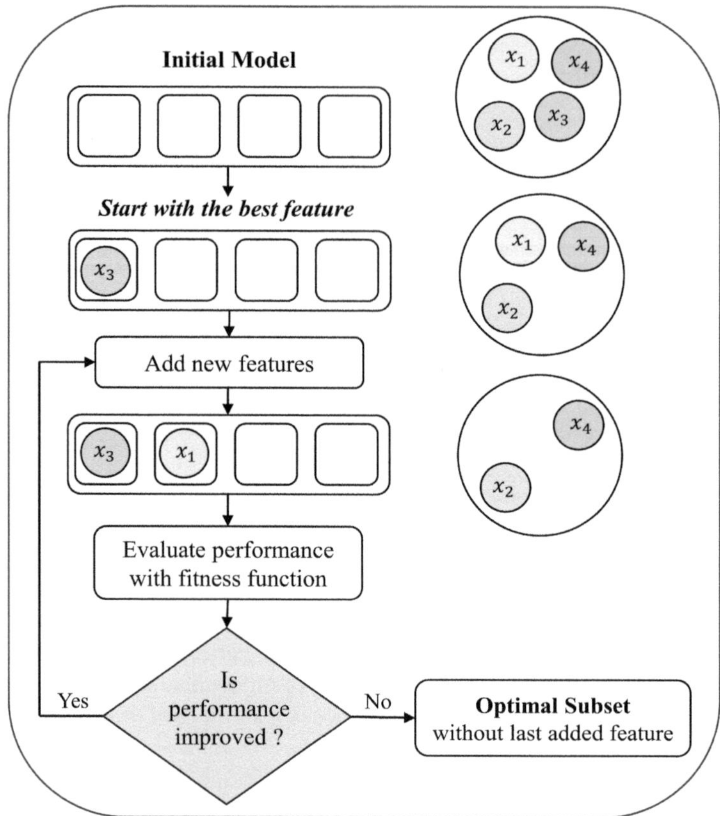

FIGURE 3.5 Working principle of sequential forward selection strategy.

or an overall accuracy metric estimated for a thematic map resulting from a classification process. This process is repeated until the required number of features are selected.

It should be mentioned that it is not guaranteed that this simple method will find the optimal subset. It generally produces a suboptimal solution because of the so-called nesting effect, which results from the fact that in the forward selection process, selected features cannot be discarded later, and in the backward selection process, excluded features cannot be reselected. In fact, both techniques may find different solutions, especially when high-dimensional data sets are considered. Another important point to be emphasized is that using backward searches with classifiers as evaluation functions may cause high complexity problems because more high-dimensional data must be employed in the entire process. This would also result in a greater computing time requirement than SFS. This makes SBS techniques less attractive, especially when a classifier is used for evaluation.

Some extensions of SFS and SBS approaches are described in the literature. One of these extensions introduced by Pudil et al. (1994) is the floating search approach that keeps the feature sets flexible to approximate the optimal solution as much as possible. In other words, the resulting dimensionality in the respective stages of the algorithm is not changing monotonically but is actually "floating" up and down. In a comparative study using variants of the SFS and SBS techniques for the cloud classification problem, Aha and Bankert (1995) report the following conclusions:

- Feature selection improves the accuracy of the classification task,
- SBS does not always outperform SFS,
- Using classifier accuracy as the evaluation function yields better results than using the separability indices.

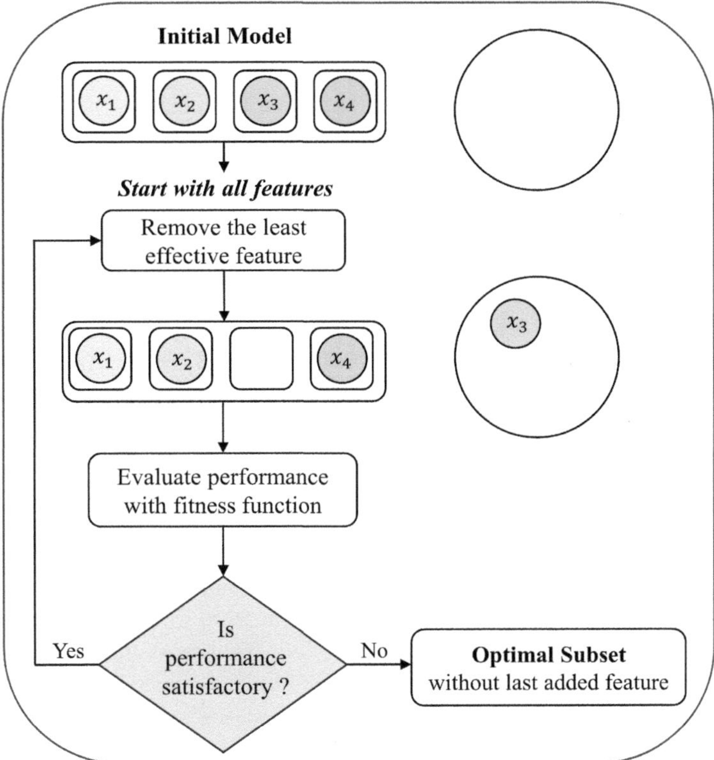

FIGURE 3.6 Procedural framework of the sequential backward selection approach.

SFS is usually preferred when the optimal number of selected features is small, while SBS is preferred otherwise.

In a review study, Siedlecki and Sklansky (1988) underlined the fact that both SFS and SBS can be easily "derailed". For instance, the SFS algorithm can add two features that are subsequently the best ones, but that are bad if used together. To solve this problem, a bidirectional search technique that combines SFS and SBS is proposed.

Another effective method in the search for the optimal subset solution is the branch-and-bound search, which is based on the monotonicity assumption that adding new features is not going to decrease the performance of the new subset. An evaluation (or fitness) function, generally a separability index, must be used to assess the subsets. The fundamental idea behind the algorithm is that if a subset of size greater than m has a performance value lower than an initially set threshold, all sub-subsets of this subset are eliminated as they will have values lower than that of the main subset. Thus, many subsets can be discarded without evaluating their fitness. This is the main advantage of this technique. The threshold value used in the process is either found by one of the simple search techniques or set by the algorithm each time the highest performance is reached. If the solution space is thought of as a tree, branch-and-bound search prunes the tree step by step based on the fitness values.

Due to the superiority of the algorithm, the branch-and-bound algorithm can also be applied to search clusters and nearest neighbors. Details of the algorithm and some of its application areas, such as restricted least squares, maximum likelihood paired comparison ranking, and selection variables in regression, can be found in Hand (1981). It is worth noting that in some practical pattern classification scenarios, the monotonicity assumption may not be satisfied. For example, the addition of irrelevant information may significantly reduce the generalization accuracy of a decision tree classifier (Yang and Honavar, 1998). Moreover, feature subset selection techniques that rely on the monotonicity of the performance criterion can exhibit poor performance with nonlinear classifiers, such as neural networks (Ripley, 1996). This technique can be computationally expensive when handling high-dimensional data sets.

Recursive feature elimination (RFE), which is a type of backward selection of features, has recently become popular for regression and classification problems. RFE begins by building a model on the entire set of features and computing an importance score or fitness function value for each feature (Z-score for linear regression, Gini index for tree-based methods, etc.) with the goal of maximizing some target metric (e.g., AUC, overall accuracy, or F1-score). The least important or "weak" features are then discarded, the model is rebuilt, and importance scores are computed iteratively until the subset size defined by the analyst is reached. Thus, correlations and collinearities that may exist in the data set are removed. When support vector machines (SVMs) are used to evaluate the scores of the subsets, the objective function of the SVM classifier ($(1/2)w^2$) is applied as a feature-ranking criterion to produce a list of features ordered by apparent discriminatory ability. At each step, the coefficients of the weight vector w are used to compute the ranking scores of all remaining features, and the feature with the smallest ranking score $(w_i)^2$ is eliminated, where w_i represents the corresponding ith component of w (Pal and Foody, 2010).

As it is often not known in advance how many features (or which subset of all features) are required for modeling the problem, cross-validation procedures are usually applied to determine the optimal number of features. The root mean square error and standard deviation error of 10-fold cross-validation can be used to determine the feature subset. In most applications, the subset size, which is the tuning parameter for RFE, may not be defined or predicted with confidence. In such cases, it is advisable to visually analyze the importance scores estimated during the feature elimination process. When the score value drops abruptly or decreases to a threshold level defined by the analyst, the process should be terminated. The possibility of overfitting can be mitigated through RFE, which opts for the most vital features. However, the exclusion of important features could result in underfitting. To guarantee that the models are appropriately balanced and adequately fitted, it is essential to evaluate their overall performance using the test data set. Valuable reviews on RFE are provided by Liu and Yu (2005) and Kuhn and Johnson (2019). For further reading, readers are suggested on RFE with random forest (Pullanagari et al., 2018; Yin et al., 2019), SVM-RFE (Pal and Foody, 2010; Colkesen and Kavzoglu, 2016; Ma et al., 2017; Yadav and Pal, 2021), and extreme gradient boosting (XGBoost) with RFE (Xiao et al., 2023).

3.2.2 Simulated Annealing

The concept of simulated annealing (SA), proposed by Metropolis et al. (1953), represents a flexible stochastic search algorithm inspired by the annealing problem in metallurgy. By adopting the gradual heating and cooling approach of annealing, this method permits the initially agitated and disordered atoms within a metal to uncover robust and consistent configurations. In a parallel manner, simulated annealing seeks solutions to various optimization problems by initially subjecting the solution to random modifications (high temperature) and, subsequently, slowly increasing the ratio of greedy improvements (analogous to cooling) until further enhancements become unattainable. The SA algorithm can be used in conjunction with filter and wrapper approaches. Geman and Geman (1984) applied a similar idea to image segmentation in that the Gibbs sampler generates a new label W_r' for each pixel r based on conditional probability $P\left(w_r' \mid w_{S-\{r\}}\right)$ instead of energy change as used by Metropolis et al. (1953), which can be regarded as a "hot bath" version of Metropolis's algorithm (i.e., a temperature parameter is added).

The algorithm starts at a high temperature T. After the Metropolis algorithm converges to equilibrium at the current temperature T, the value of T is decreased according to some carefully defined criterion called a cooling schedule. The process is repeated until the system becomes frozen (i.e., $T \to 0$). It is shown that a high temperature T can increase the probability of w_r being replaced by new class w_r' (because $\Delta < 0$, large T indicates large $\exp(\Delta/T)$), even though the energy $U\left(w_r' \mid d_r\right)$ of the new class w'_r is higher (i.e., the probability is lower) than that of class w_r. As the system cools down and temperature T decreases, only small increases in U are accepted. Near the freezing point $(T \to 0)$, no increase in U can be accepted. The relationship between temperature T and $\exp(\Delta/T)$ is shown in Figure 3.7.

The central idea of the SA algorithm is equivalent to the introduction of noise into the system to divert the search process away from a local minimum. The idea is similar to a process in metallurgy in which a small region of a metal structure is heated until it is pliable enough to be reconstructed into the desired shape. The metal is then cooled very slowly to make sure that it is given enough time to respond. Temperature changes must be very small until the metal has hardened.

Geman and Geman (1984) present proofs of two convergence theorems. The first concerns the convergence of Metropolis's algorithm: if every configuration w is visited infinitely often, the distribution of generated configurations is guaranteed to converge to the Gibbs distribution. The second theorem concerns the temperature T of SA, in which the decreasing sequence of temperatures satisfies the following condition:

$$\lim_{k \to \infty} T_k = 0, \quad \text{and} \quad T_k \geq \frac{m \times \Omega}{\ln(1+k)} \tag{3.7}$$

where

$$\Omega = \max_w U(w) - \min_w U(w) \tag{3.8}$$

and if m is the number of lattices (i.e., pixels), then convergence can be guaranteed. However, decreasing temperature T in terms of Equation (3.7) is too slow for practical applications. One may apply another temperature cooling function f defined as (Dubes and Jain, 1989)

$$T_{k+1} = f(T_k) = \frac{\ln(1+k)}{\ln(2+k)} T_k. \tag{3.9}$$

The initial value of temperature T_0 is usually set to a value of 2 or 3.

The SA algorithm has found relatively limited use in the literature for feature selection (Hegarat-Mascle et al., 1996; Chang et al., 2011; Koukiou, 2022). However, the algorithm's remarkable potential for customization (encompassing annealing schedule, neighbor function, and evaluation function) designates it as an attractive prospect for upcoming research (Stracuzzi, 2008).

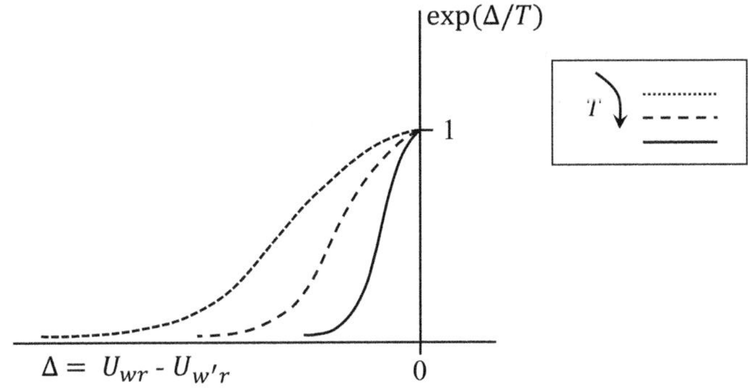

FIGURE 3.7 Probability for class replacement. As T is decreasing, only small increases in energy can be accepted.

3.2.3 SEPARABILITY INDICES

Measurements associated with each class in a multivariate data set exhibit a statistical probability distribution. Varying degrees of intersection occur among these probability distributions, posing the class separability challenge as a function of both mean separation and the statistical distribution of data points. As underlined by Kavzoglu (2001), the fundamental idea behind the separability indices is that the larger the separation between the classes in the feature space, the easier it will be to discriminate between the features because of better decision boundary determination. Thus, a lower error rate (better performance) can be achieved by the classifier following feature selection.

Four separability indices, the divergence index (Singh, 1984), the transformed divergence, the Bhattacharyya distance (B-distance) (Haralick and Fu, 1983), and the Jeffries-Matusita distance, are widely used for feature selection purposes. The divergence index D_{ij} between two classes i and j is derived from the likelihood ratio of any pair of classes. For multivariate Gaussian distributions, D_{ij} for classes i and j based on a subfeature dimension m of the total p dimension is defined by

$$D_{ij} = \frac{\mathrm{tr}\left(\left(\mathbf{C}_i - \mathbf{C}_j\right) \times \left(\mathbf{C}_j^{-1} - \mathbf{C}_i^{-1}\right)\right)}{2} + \frac{\mathrm{tr}\left(\left(\mathbf{C}_i^{-1} + \mathbf{C}_j^{-1}\right) \times \left(\mu_i - \mu_j\right) \times \left(\mu_i - \mu_j\right)^T\right)}{2} \tag{3.10}$$

where tr denotes the trace (sum of the diagonal elements) of the matrices, $\mathbf{C}$ is the class variance-covariance matrix of dimension $m \times m$, μ is the class mean vector, and T denotes the transpose of a matrix. In some cases, a transformation of D_{ij}, called the transformed divergence (TD_{ij}), is used. It is defined by

$$TD_{ij} = 2,000 \times \left[1 - \exp\left(-D_{ij} / 8\right)\right] \tag{3.11}$$

in which the value D_{ij} is projected into the interval 0–2,000. The greater the value of D_{ij} or TD_{ij}, the greater the class separability based on the selected m subfeature dimension. If Equation (3.11) is used, Jensen (2015) points out that above 1,900 indicates good separation, while below 1,700 shows poor separability.

The B-distance, B_{ij}, is defined by (Haralick and Fu, 1983)

$$B_{ij} = \frac{1}{8}\left(\mu_i - \mu_j\right)^T \frac{\left(\mathbf{C}_i + \mathbf{C}_j\right)}{2}\left(\mu_i + \mu_j\right) + \frac{1}{2}\ln\frac{\dfrac{\left(\mu_i - \mu_j\right)}{2}}{\sqrt{\mu_i}\sqrt{\mu_j}}. \tag{3.12}$$

The quantity B_{ij} is computed for every pair of classes given m subfeature dimension. The sum of B_{ij} for every pair of classes (except the class itself) is obtained and is a measure of the overall separability. Although the two separability measures defined by Equations (3.10) and (3.12) are based on different principles, Mausel et al. (1990) find that the divergence index and B-distance generally give similar results. However, Kailath (1967) noted that the B-distance is more appropriate to interclass separability problems than divergence when the class probability distributions are broad. Another metric indexing the B-distance measure is the Jeffries-Matusita (J-M) distance (Equation 3.13), holding a saturating behavior with increasing class separability. However, it tends to suppress high separability values, while overemphasizing the low separability values.

$$(J - M)_{ij} = \sqrt{2\left(1 - e^{-B_{ij}}\right)} \tag{3.13}$$

In addition to the above-mentioned separability indices, several multivariate statistical techniques have been employed to estimate the degree of discrimination between the classes present in a data set. These tests are based on differences in the sample means and variance-covariance matrices of the classes. Hotelling's T^2 statistic and Wilks' Λ criterion are the two statistical measures that can be employed in a feature selection process. When employed as descriptive statistical tests, these statistics assess the discriminating power of a feature (or relative importance of a feature). Alternatively, when they are used as evaluative metrics in the feature selection process, they can control the feature selection process as evaluation or fitness functions (Kavzoglu and Mather, 2002).

3.2.4 Filter-Based Methods

Filter-based methods make decisions about the appropriateness of the features by analyzing solely the general characteristics of the training data, such as the correlations of the features with the class variable. In this type of feature selection algorithm, features are evaluated without involving any learning process. In other words, no classifier is employed in the decision-making process of feature selection. Simplicity, easy coding, and the rapid process make these methods popular; however, their accuracy level is usually higher than that of a wrapper in which a classification system is employed. Conventional state-of-the-art filter methods are discussed in the following subsections along with their theories and applications in the remote sensing field. Filter-based feature selection algorithms can be easily applied using freely available tools, including the scikit-learn library in Python and the Praznik package in R, a collection of tools for information theory–based feature selection and scoring (Kursa, 2021; Kavzoglu and Teke, 2022). The general framework of filter-based feature selection algorithms is shown in Figure 3.8.

3.2.4.1 Correlation-Based Feature Selection

Correlation-based feature selection (CFS), suggested by Hall (1999), is based on the fundamental idea that a successful feature selection can be conducted by selecting features that are highly correlated with the classes and uncorrelated with each other. Elimination of redundant features is

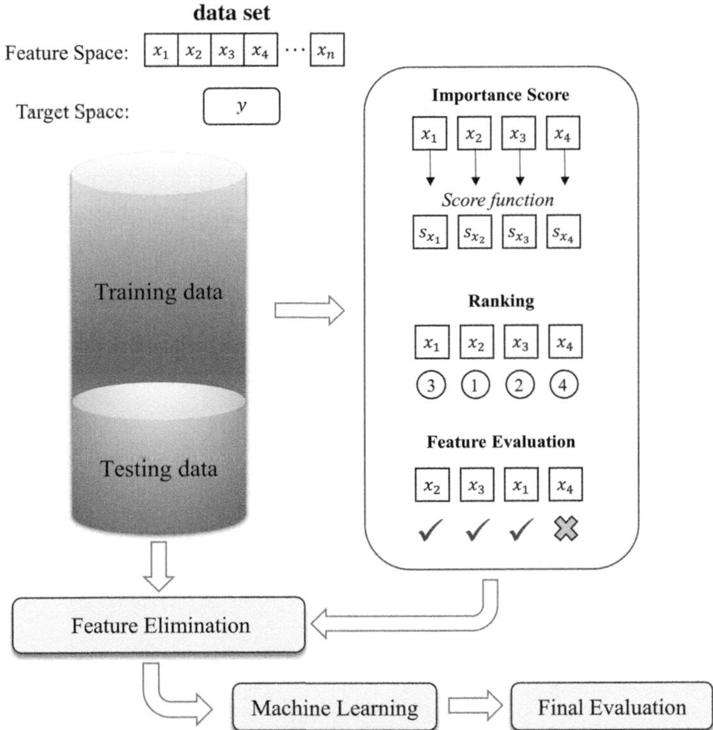

FIGURE 3.8   General working principle of filter-based feature selection.

essential because they often demonstrate significant correlations with one or more of the remaining features (Figure 3.9). Whether a feature is incorporated or not depends on how effectively it predicts classes in parts of the feature space that are not already covered by other features. The feature subset evaluation function of CFS can be given as

$$M_S = \frac{k\overline{r_{cf}}}{\sqrt{k + k(k-1)\overline{r_{ff}}}}$$

(3.14)

where M_S is the heuristic "merit" of a feature subset S consisting of k features, $\overline{r_{cf}}$ is the mean feature-class correlation, and $\overline{r_{ff}}$ is the average feature-feature inter-correlation. While the numerator in the equation indicates the features are in relation to the classes, the denominator shows how much redundancy exists among the features. With estimation of M_S, feature subset can be ranked.

The evaluation criterion of the CFS algorithm exhibits a bias toward feature subsets that display strong predictive capabilities for the class while not showing predictiveness toward each other. This criterion eliminates the irrelevant features by virtue of their weak correlations with the class, whereas redundant features are disregarded because of their expected high correlations with other features. This systematic process ultimately yields a subset of the most optimally chosen features. To reduce the computational complexity, a bidirectional search methodology is suggested, essentially a parallel execution of sequential forward and backward selections. The methodology explores the potential feature subsets through a process of iterative hill climbing, ensuring that features selected through forward selection remain unremoved during backward selection and those discarded through backward selection do not get reintroduced through forward selection (Hall, 1999; Pal and Foody, 2010). At the end of the process, using a correlation matrix to capture feature-class and feature correlations, the CFS algorithm employs a best-first

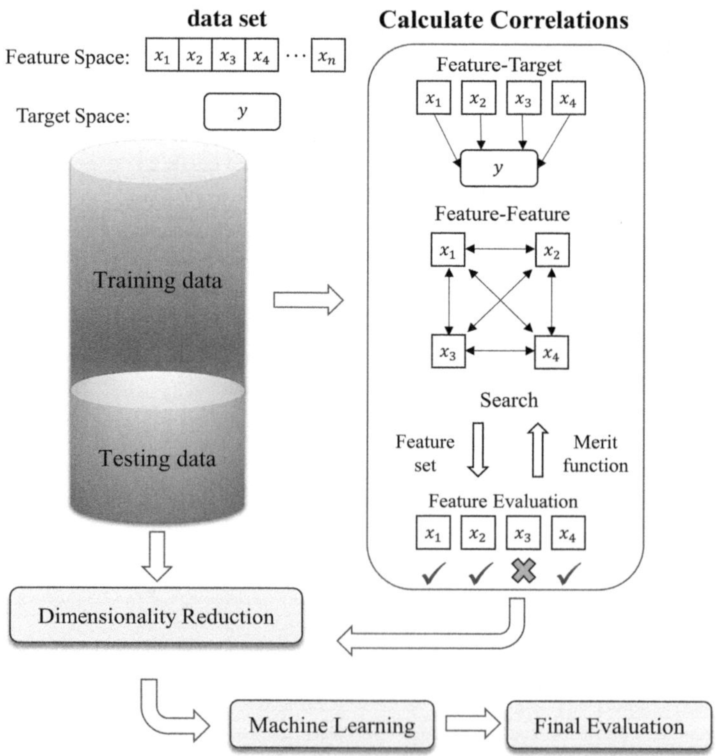

FIGURE 3.9   Operational workflow of the CFS algorithm.

strategy, combining greedy hill climbing with a backtracking facility, to traverse the feature subset space (Witten et al., 2011). In an object-based LULC classification, Colkesen and Kavzoglu (2019) concluded that the CFS algorithm identified the most relevant features of image objects, establishing the optimal feature subset. This led to a considerable reduction in input feature space dimensions by about two-thirds (from 128 to 42 features) and achieved improved classification accuracies across three different machine learning methods. It is important to highlight that the algorithm is applicable not only to its own determinations but also to ordering and identifying optimal features established by other filter-based methods. This is particularly valuable as these methods lack a merit function-like measure to determine the most relevant features (Colkesen and Kavzoglu, 2018).

3.2.4.2 Information Gain

As decision trees are iteratively built by recursively partitioning the data into leaf nodes, it follows that when generating a decision tree, the primary issue encountered by the user is to choose a suitable index to measure the performance of the decision tree (in terms of classification accuracy, for example). The two most frequently used indices in decision tree induction are known as information gain (Quinlan, 1979, 1993) and the Gini impurity index (Breiman et al., 1984), respectively. However, some decision tree induction algorithms may use statistical tests as feature selection criteria. For instance, one may use the chi-square test to measure whether, for a certain subset of the samples, the attribute X is significantly coupled to the information class j. For each attribute X, the chi-square test is performed on the pairs (X, j), and the attribute generating the smallest p-value resulting from the chi-square test is selected as a split basis.

Quinlan (1979, 1993) proposed a measure called information gain for tree induction based on Shannon's entropy measure (Shannon, 1948) used in information theory, which is expressed as ($-\log_2 \times$ probability) bits. Thus, eight equally probable messages each contain $-\log_2(\frac{1}{8})=3$ bits of information. The probability of a set of training data for a particular class is the relative frequency of the observations in that class (e.g., if training data class i contains eight pixels and the total number of training data pixels is 100, then the probability of class i is 0.08). The concept of information theory is used to evaluate all possible tests for the subdivision of the tree, and the test that produces the greatest information gain is selected.

During the tree induction process, for a node t, the information gain for each attribute or feature (spectral band in the case of remote sensing) is calculated based on the split at node t on that attribute. The attribute with the greatest information gain is chosen for splitting at that node. To compute information gain, the entropy $I_E(t)$ for node t must first be calculated, which is expressed by

$$I_E(t) = -\sum_{j=1}^{m} f(t,j)\log_2 f(t,j) \tag{3.15}$$

where $f(t, j)$ is the proportion of training samples belonging to class $j, j \in \{1, 2, ..., m\}$, within node t. Here, m is the number of classes. That is, if node t contains N_t samples, then $f(t, j)$ is calculated by the following expression:

$$f(t,j) = \frac{1}{N_t}\sum_{i=1}^{N_t} \Gamma(y_i, j) \tag{3.16}$$

where

$$\Gamma(y_i, j) = \begin{cases} 1, & \text{if } y_i = j \\ 0, & \text{otherwise} \end{cases}. \tag{3.17}$$

In other words, $I_E(t)$ measures the information conveyed by the distribution of samples within node t. For instance, according to Equation (3.15), if the sample values contained in node t are (0, 1), then $I_E(t)=0$. If the sample values are (0.3, 0.7), then $I_E(t)=0.881$, and if the sample values are (0.5, 0.5), then $I_E(t)=1$. Thus, the more uniform the sample distribution of $I_E(t)$ is, the greater the information content of node t will be (Shannon, 1948). Now, define attribute X (= $\{x_1, x_2, ..., x_r\}$). For a value x_i, suppose that there are n_i samples, and $f\left(t_{X(x_i)}, j\right)$ is the proportion of samples with the value x_i belonging to class j at node t. The entropy $I_E\left(t_{X(x_i)}\right)$ associated with x_i for node t is computed by

$$I_E\left(t_{X(x_i)}\right) = -\sum_{j=1}^{m} f\left(t_{X(x_i)}, j\right) \log_2 f\left(t_{X(x_i)}, j\right). \tag{3.18}$$

Finally, the information gain associated with a split on attribute X is calculated by

$$\text{Gain}(t, X) = I_E(t) - \left(\frac{n_1}{N_t}\right) I_E\left(t_{X(x_1)}\right) - \left(\frac{n_2}{N_t}\right) I_E\left(t_{X(x_2)}\right) \cdots - \left(\frac{n_r}{N_t}\right) I_E\left(t_{X(x_r)}\right). \tag{3.19}$$

Equation (3.19) denotes the difference between the information needed to describe node t both excluding and including attribute X, that is, the gain in information due to attribute X.

Elmaizi et al. (2019) proposed a novel algorithm that evaluates the relevancy of the bands based on the information gain function with the support vector machine classifier. The effectiveness of the information gain algorithm has been investigated in several comparative studies (e.g., Pal, 2013; Colkesen and Kavzoglu, 2018).

3.2.4.3 Gini Impurity Index

The Gini impurity index (named after Italian economist Corrado Gini, and originally used for measuring income inequality) measures the impurity of an input feature with respect to the classes (Breiman et al., 1984). Specifically, the Gini impurity index reaches its minimum (zero) when all attributes in the node fall into a single information class. The Gini index associated with attribute X (= $\{x_1, x_2, ..., x_r\}$) for node t is denoted by $I_G(t_{X(xi)})$ and is expressed as

$$I_G\left(t_{X(x_i)}\right) = 1 - \sum_{j=1}^{m} f\left(t_{X(x_i)}, j\right)^2 \tag{3.20}$$

where $f\left(t_{X(x_i)}, j\right)$ is the proportion of samples with the value x_i belonging to class j at node t as defined in Equation (3.15). The decision tree splitting criterion is based on choosing the attribute with the lowest Gini impurity index of the split. Let a node t be split into r children, n_i be the number of records at child i, and N_t be the total number of samples at node t. The Gini impurity index of the split at node t for attribute X is then computed by

$$\text{Gini}(t, X) = \left(\frac{n_1}{N_t}\right) I_G\left(t_{X(x_1)}\right) + \left(\frac{n_2}{N_t}\right) I_G\left(t_{X(x_2)}\right) + \cdots + \left(\frac{n_r}{N_t}\right) I_G\left(t_{X(x_r)}\right). \tag{3.21}$$

3.2.4.4 Minimum Redundancy-Maximum Relevance

Minimum redundancy-maximum relevance (mRMR), proposed by Ding and Peng (2005) for selecting genes in microarray data, is a filter-based algorithm that involves selecting a group of features that show the highest correlation to a class (relevance), while also displaying the lowest correlation with each other (redundancy). This algorithm ranks attributes based on the criteria of maximum relevance and minimum redundancy. To measure relevance, the F-statistic can be employed for continuous attributes, and mutual information can be used for discrete attributes. Redundancy, on the other hand, can be assessed using the Pearson correlation coefficient for continuous attributes and mutual information for discrete attributes. The mRMR process involves selecting features individually through a greedy search that aims to optimize the objective function, which combines relevance and redundancy factors. There are two common variations of the objective function: MID (Mutual Information Difference criterion) and MIQ (Mutual Information Quotient criterion). These metrics denote the distinction and ratio between relevance and redundancy, respectively. mRMR estimates the redundancy between features and the correlation between features and class based on mutual information $I(x; y)$ that can be defined as

$$I(x; y) = \iint p(x, y) \log \frac{p(x, y)}{p(x)p(y)} \, dx \, dy \tag{3.22}$$

where $p(x, y)$ is the mutual probability density of random variables x and y, and $p(x)$ and $p(y)$ are the marginal probability densities of x and y, respectively.

A sample feature set $S = \{f_1, f_2, \ldots, f_n\}$ and a sample class is c. The relevance between S and c is the mean of all mutual information between each feature f_i and class c and can be expressed as

$$D(S, c) = \frac{1}{|S|} \sum_{f_i \in S} I(f_i, c) \tag{3.23}$$

where $|S|$ is the number of features in S, and $I(f_i, c)$ is the mutual information between the feature f_i and c. The redundancy of all features in S is the mean of all mutual information between feature f_i and feature f_j, and it can be shown as

$$R(S) = \frac{1}{|S|^2} \sum_{f_i, f_j \in S} I(f_i, f_j) \tag{3.24}$$

The mRMR search algorithm seeks the optimal features with maximum relevance $D(S, c)$ and minimal $R(S)$. When dealing with temporal data, the mRMR method for feature selection requires specific preprocessing steps to transform temporal data into a single matrix. Four variants of the algorithm were suggested by Ding and Peng (2005) under the framework of mRMR. Other versions of the algorithm include Fast-mRMR (Ramírez-Gallego et al., 2017), prognostic mRMR (Hamaide and Glineur, 2021), ImRMR (Xie et al., 2023).

3.2.4.5 Chi-Square Test

The chi-square test draws upon the chi-square (χ^2) statistic as its foundation. In this approach, individual features are independently evaluated in relation to their respective class labels (Mingers, 1989; Plackett, 1983). Statistical analysis focuses on the distribution of land class values within spectral bands. A higher computed statistical value signifies that the evaluated spectral band contains more pertinent information about the classes. In the initial phase, χ^2 statistics are computed on the basis of the classes associated with the features. In the subsequent phase, freedom is exercised by employing the chi-merge principle, which estimates the importance level and determines the significance of each feature. This process involves scrutinizing the chi-square values, and iteratively decomposing features until any inconsistencies in the data set are identified.

The Chi-merge algorithm, initially developed by Kerber (1992) and subsequently refined by Liu and Setiono (1995), provides the structural framework. The algorithm arranges the chi-square values calculated for each factor in the data set, offering insights into the dependency of the factor within its corresponding class. This approach facilitates the systematic identification of meaningful relationships between features and classes.

The fundamental principle of the method involves measuring the discrepancy between the actual (observed) and expected values. Equation 3.25 outlines the calculation of this statistical formula. Within the equation, "r" signifies the number of distinct values in a band, "n" represents the number of classes, "$O_{i,j}$" denotes the count of pixels with value "i" in class "j", and "$E_{i,j}$" stands for the anticipated number of samples with value "i" in class "j". The derived χ^2 statistical value offers insights into pixel and class distribution. A higher χ^2 statistical value indicates a stronger correlation between the band in question and the classes.

$$\chi^2 = \sum_{i=1}^{r}\sum_{j=1}^{n} \frac{\left(O_{i,j} - E_{i,j}\right)^2}{E_{i,j}}. \tag{3.25}$$

3.2.4.6 Relief-F

The Relief-F algorithm, a filter-type feature selection method introduced by Kononenko (1994) for binary scenarios, considers the nearest neighbor for the weight estimation. In fact, it is the most popular variant of the Relief algorithm introduced by Kira and Rendell (1992). The algorithm assigns higher weights to the features related to classes, efficiently eliminates less relevant features, and exhibits high efficiency in solving multiclassification problems. In remote sensing applications, the algorithm capitalizes on the ability to differentiate pixel values in a spectral band that exhibits similarity to one another. It operates similarly to the k-nearest neighbor algorithm, where the essence lies in categorizing close distances into the same group. The formulation of the algorithm is illustrated in Equation 3.26. To minimize randomness in feature assessment and obtain the final weight as the determined average value, the entire process needs to be repeated m times.

$$\omega_t^i = \omega_t^{i-1} + \sum_{c \neq \text{class}(x)} \frac{p(x)}{1 - p\left(\text{class}(x)\right)} \sum_{j=1}^{k} \text{diff}_f\left(x, M_j(x)\right)\big/(m \times k)$$
$$- \sum_{j=1}^{k} \text{diff}_f\left(x, H_j(x)\right)\big/(m \times k) \tag{3.26}$$

where t denotes the feature's weight ω_t, diff_i signifies the feature distance between two samples, $p(x)$ represents the class probability of x, $H_j(x)$ stands for a different class example of x, $M_j(x)$ represents the nearest same-class example of x, m is the iteration count, and k signifies the number of nearest neighbors. The algorithm essentially leverages the discerning power of pixel values with resemblances in spectral bands, offering valuable insights for applications in remote sensing. Since the Relief-F method ranks the features and produces an ordered list of feature importances and does not provide information about the number of features to be selected from the list, Kavzoglu and Bilucan (2023) considered the merit score of the CFS algorithm to determine a subset of the features. An improved version of the algorithm called μ-Relief was proposed by Aggarwal et al. (2023). The approach takes into account neighbors that are distant from the mean distance to derive estimates for feature weights, resulting in an enhancement of the algorithm's performance. Thirteen real-world data sets were employed for testing, and validation was conducted on three synthetic data sets. A good review and theoretical background of Relief algorithms can be found in Urbanowicz et al. (2018).

3.2.4.7 Symmetric Uncertainty

The symmetric uncertainty (SU) algorithm, a CFS method, assesses the value of each feature within a subset of features based on the premise that a valuable feature subset exhibits strongly correlated information with classes while maintaining independence among the features (Biesiada and Duch, 2007; Yu and Liu, 2003). SU employs entropy as a metric to quantify the extent of association among distinct features. In this context, SU serves as a gauge of the correlation between a class and a feature, as expressed by Equation (3.27).

$$SU = 2 \times \frac{H(X) + H(Y) - H(X,Y)}{H(X) + H(Y)} \tag{3.27}$$

where $H(X)$ denotes the entropy of a feature, $H(Y)$ represents the entropy of a class, and $H(X,Y)$ denotes the probability value computed for the combination of values X and Y. The resulting statistic varies between 0 and 1. A value of 1 indicates that one variable (either X or Y) can predict the other variable fully, while a value of 0 signifies complete independence between both variables.

3.2.4.8 Fisher's Test

The Fisher's test is one of the simplest filter-based algorithms employed for feature selection (Duda and Hart, 1973; Xie and Wang, 2011). In a data set including two classes, a score is computed using Equation 3.28 for each t feature constituting the data set. The fundamental concept in addressing the feature selection challenge involves determining how effectively any band within the data set can discriminate between the foundational classification classes.

$$F_t = \frac{\left(\mu_1 - \mu_2\right)}{\sigma_1^2 + \sigma_2^2} \tag{3.28}$$

where μ_1 signifies the mean of the t values for the first class; μ_2 denotes the mean of the t values for the second class; σ_1 represents the standard deviation of the first class; and σ_2 signifies the standard deviation of the second class. In remote sensing scenarios, when addressing challenges involving more than two classes, the extended version of Equation (3.28) is utilized, as depicted in Equation (3.29).

$$F_t = \frac{\left(\mu_1 - \sum_{i=2}^{M}\mu_i\right)^2 + \cdots + \left(\mu_1 - \sum_{i=1}^{M}\mu_i\right)^2 + \cdots + \left(\mu_M - \sum_{i=1}^{M-1}\mu_i\right)^2}{\sum_{i=1}^{M}\sigma_i^2} \tag{3.29}$$

where M represents the count of classes. The Fisher's test calculates a ranking for all features within the data set. The algorithm assumes the features are independent from each other. Consequently, each F_t coefficient captures information regarding a specific feature, without accounting for mutual information between others within the data set.

3.2.4.9 OneR

This feature selection algorithm assesses individual features within the data set using the OneR classifier. The OneR classifier ranks features by considering the error rates obtained from the training set. All numerical features are treated as continuous, and a straightforward approach is employed to segment values within specified ranges into multiple value intervals (Holte, 1993). The OneR algorithm generates classification rules grounded in features present within the sample data set. A rule is established for each feature in the training data set to investigate the accuracy of class data prediction using the specific feature.

The core concept behind the OneR algorithm is to identify the feature and its corresponding rule that minimizes misclassifications during predictions. It quantifies the number of misclassifications for each potential rule and ultimately selects the rule with the least misclassification count. The primary goal is to determine the most suitable feature from the feature set F using the OneR algorithm. For the sake of simplicity, let us focus on a binary classification scenario with two class labels: positive and negative:

1. For each feature $f \in F$:
   - Calculate the misclassification count for each unique value x of feature f

$$\text{Misclassified count}(f,x) = \sum_{i=1}^{n} \begin{cases} 1 & \text{if } X_i = x \text{ and } Y_i \neq \text{mode}(Y \text{ for } X = x) \\ 0 & \text{otherwise} \end{cases} \quad (3.30)$$

   - Find the total misclassification count for feature f

$$\text{Total misclassified count}(f) = \sum_{x \text{ unique values of } f} \text{Misclassified count}(f,x) \quad (3.31)$$

2. Choose the feature f^* that minimizes the total misclassification count

$$f^* = \arg\min_{f \in F} \text{Total misclassified count}(f) \quad (3.32)$$

where n is the number of instances in data set D, and $\text{mode}(Y \text{ for } X = x)$ represents the mode (i.e., most frequent) class label of instances where feature f has value x. The feature f^* is the selected feature that the OneR algorithm identifies as the best feature for classification.

3.2.5 WRAPPERS

In contrast to filter-based feature selection, wrapper algorithms involve a learning or a classification process to measure the significance and importance of the training features. The relevancy of the features or variables is evaluated in the classification process. One possibility is to apply the induction algorithm to the entire training set and then assess its performance on the testing set or, alternatively, opt for a cross-validation technique. Wrappers can produce more accurate classification results, but they are relatively slower and require more time to make decisions, particularly for large data sets. Moreover, wrappers usually require repetitive running processes when different learning algorithms are tested and different results are produced with the learning method, which is valid for algorithms starting with randomly defined parameters and having the possibility of sticking to a local minimum. The most widely used wrapper algorithms, including genetic algorithm, particle swarm optimization, and feature selection with SVMs, are discussed in the following subsections. Gray wolf and ant colony optimization techniques have also been employed, but in a limited number of studies (Sameen et al., 2017; Kumar et al., 2021). The working principle of the wrapper algorithms with processing stages is shown in Figure 3.10.

3.2.5.1 Genetic Algorithm

Many practical search and optimization problems require the investigation of multiple maximums or minimums. Genetic algorithms (GAs) have been increasingly used in solving such problems, and their general performance has been empirically shown to be robust (Holland, 1992; Goldberg, 1989). The core principle of GAs is rooted in the principles of natural selection and genetics. GA functions by managing a group of trial structures and creating a population. Each solution is represented by an individual called "chromosome", and each feature in the data set is represented by a "gene" of that chromosome. Each chromosome or trial structure is a finite-length string and generally is represented through special coding techniques, most commonly gray code or binary code in which 0 and 1 are used to represent exclusion and inclusion, respectively.

Once the trial structure and population size have been determined by the user, the GA then starts a series of search processes. At each generation, the survival of the fittest chromosomes in

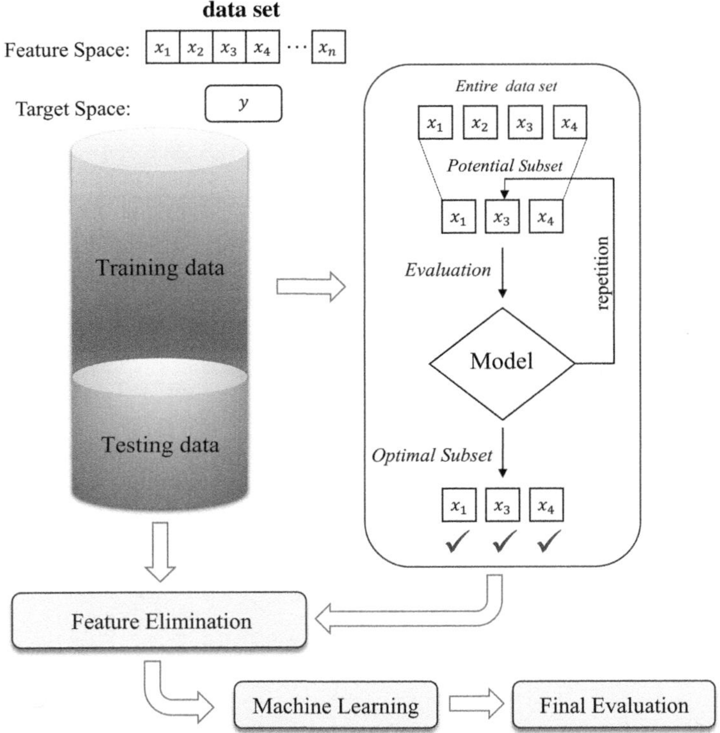

FIGURE 3.10 Operational concept of wrapper algorithms.

the current population is used to create a new population based on both structured and randomized information exchange between the present and the new population. The GA search process can be allowed to run for a prespecified number of iterations, or the search can be continued until the fitness function converges. The whole process of the GA search is shown in Figure 3.11. The GA algorithm is said to have converged if ~95% of the structures forming the population contain the same value.

Why can the performance of GA surpass that of more traditional search methodologies, such as calculus-based, enumerative schemes, and random search algorithms? As Goldberg (1989) notes, there are several significant differences between GA and other algorithms that contribute to GA's promising performance. The first difference is that GA uses multiple starting points for performing the search process and deals with special trial structures, which are codings of the original parameter set. The second reason is that GA uses an objective function (the measure of the fitness of each trial structure) to lead its search process and is not dependent on derivatives or auxiliary knowledge. Finally, the transition from one generation to the succeeding generation is based on probabilistic rather than deterministic rules.

The initial population used in a GA may be chosen at random or composed of heuristically selected initial points. A fitness function is used to evaluate each structure's performance. GA then constructs a new generation using three basic operators: reproduction, crossover, and mutation. During the reproduction process, individuals are randomly selected from this population. Although the selection is random, the expected number of times an individual is selected is associated with the individual's performance. For example, if an individual's performance is three times the average performance of the population, this individual will be present three times in the new generation. As a result, individuals with low performance will be filtered out, and only high-performance individuals will survive. At this point, it should be mentioned that differential evolution, similar method to

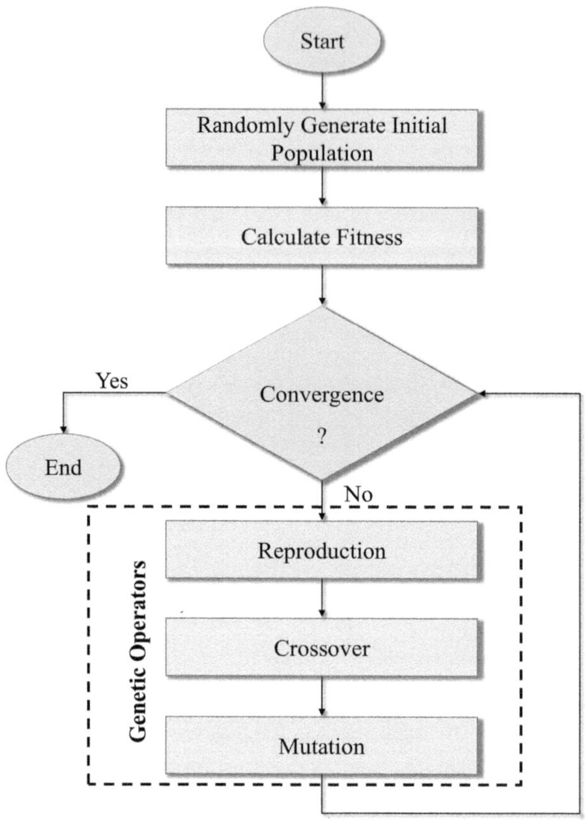

FIGURE 3.11 Flowchart illustrating the standard genetic algorithm procedure.

GA, employs mutation operation that is applied to each individual instead of very limited chromosomes in the population (Storn and Price, 1997). It is a stochastic direct search method that employs three randomly determined indices.

To introduce new individuals into the search process in the new generation, GA uses the procedures of crossover and mutation. Crossover can be regarded as an information exchange between two individuals. For instance, in a simple crossover process for two individuals (both represented by a string of length λ), a cutting point ε within the string is randomly selected. Then, two substrings, addressed from ε to λ, of both individuals are exchanged. This will form two new individuals. Figure 3.12 illustrates the process.

The mutation operation is performed by randomly changing bits within the individual's coding from 1 to 0 or from 0 to 1. The main contribution of the mutation operation is to ensure that the search process by the genetic algorithm will not be limited by local maxima (or minima). Goldberg (1989) suggested that a mutation rate of 0.001 can achieve a relatively good result.

All GA operations discussed earlier have the same purpose, which is to test different yet high-performance similarity templates or schemas. A schema is a subset of strings with similarities at certain string positions (Holland, 1992). For example, suppose there are two individuals; individual A has a binary coded value 1,010, and individual B contains the coding value 1,111. Therefore, one possible schema for these two individuals can be 1*1*, where the symbol * means "don't care". One interesting and important result pointed out by Holland is that if P_i denotes the number of individuals in generation i, the number of schemas being tested will be P_i^3. This result explains why a substantial search improvement can be achieved using a relatively small population size, and the concept is therefore termed implicit parallelism.

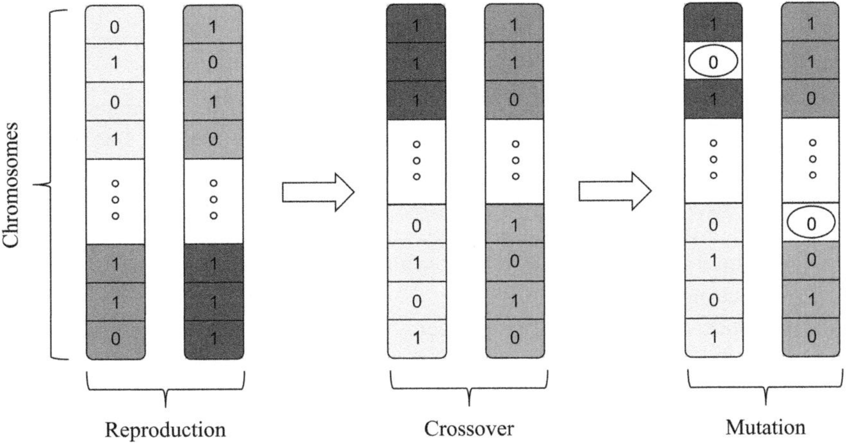

FIGURE 3.12 Reproduction, crossover, and mutation operations in GA algorithms.

During the search process, the progress of GA survival is monitored by a user-defined function called the fitness function. The fitness function can be regarded as controlling GA by providing the ability to evaluate and hence select higher-quality individuals. The overall classification accuracy can be used as a fitness function. On the other hand, for a fast GA process, fitness function can be one of the separability measures. Pal (2013) discussed two hybrid GAs (memetic and steady-state GAs) for feature selection with hyperspectral imagery. With the popularity of the random forest (RF) algorithm in recent years, RF has been employed as a fitness function (i.e., estimating the quality of the subsets) in GA-based feature selection studies (Gasmi et al., 2022; Kavzoglu and Bilucan, 2023). GA has been actively used in the remote sensing field, including image classification (Tso and Mather, 1999; Kavzoglu and Mather, 2002; Lin and Zhang, 2020), parameter optimization (Kavzoglu et al., 2015; Ming et al., 2016), and unsupervised change detection (Celik, 2010; Im et al., 2011; Wu et al., 2015).

3.2.5.2 Particle Swarm Optimization

Particle swarm optimization (PSO), developed on the basis of swarm intelligence and introduced by Kennedy and Eberhart (1995), is an effective evolutionary algorithm that has been successfully applied to obtain the optimal feature subset, consisting of features that exhibit feature complementarity, arising from the challenge posed by feature interaction. Compared with other evolutionary algorithms, such as GAs, PSO is computationally less expensive and hence a popular method for feature selection problems in many fields. It takes inspiration from the cooperative behaviors of creatures such as bird flocking and fish schooling. The core principle underlying PSO is that the optimization of knowledge is facilitated by interactions within a population, where cognitive processes are influenced by both individual and social perspectives.

PSO is based on the principle that each solution can be represented as a particle in the swarm. Each particle has a position in the search space, which is represented by a vector $\mathbf{x}_i\,(x_{i1}, x_{i2}, \ldots, x_{iD})$, where D is the dimensionality of the search space. Particles move in the search space to search for optimal solutions. Therefore, each particle has a velocity, which is represented as $\mathbf{v}_i\,(v_{i1}, v_{i2}, \ldots, v_{iD})$. Throughout the search process, each particle updates its position and velocity based on its own experience and that of its neighbors. The best previous position of the particle is recorded as the personal best *pbest*, and the best position obtained from the population at that point is called *gbest*. Based on *pbest* and *gbest*, the algorithm searches for optimal solutions by updating the velocity and position of each particle according to the following equations (Xue et al., 2013):

$$\mathbf{x}_{id}^{t+1} = \mathbf{x}_{id}^{t} + \mathbf{v}_{id}^{t+1} \tag{3.33}$$

$$\mathbf{v}_{id}^{t+1} = w * \mathbf{v}_{id}^{t} + c_1 * r_1\left(p_{id} - \mathbf{x}_{id}^{t}\right) + c_2 * r_2\left(p_{gd} - \mathbf{x}_{id}^{t}\right) \tag{3.34}$$

where t represents the tth iteration in the evolutionary process. $d \in D$ represents the dth dimension in the search space. w is inertia weight, which is to control the impact of the previous velocities on the current velocity. c_1 and c_2 are acceleration constants. r_1 and r_2 are random values uniformly distributed in [0, 1]. p_{id} and p_{gd} denote the elements of *pbest* and *gbest* in the dth dimension. The velocity is limited by a predefined maximum velocity, v_{max} and $\mathbf{v}_{id}^{t+1} \in [-v_{max}, v_{max}]$. The algorithm stops when a predefined criterion is met, which could be a good fitness value or a predefined maximum number of iterations.

The traditional *pbest* and *gbest* updating mechanism in PSO limits its performance for feature selection. Xue et al. (2013) proposed three novel initialization strategies and three innovative *pbest* and *gbest* updating mechanisms within PSO. The objective was to create unique feature selection approaches that aim to maximize classification performance, minimize the number of features, and reduce computational time. Ye et al. (2022) introduced an improved PSO methodology which is an adaptive PSO with leadership learning (APSOLL), and conducted a comparative analysis of its performance relative to PSO, gray wolf optimizer, Harris Hawks optimization, flower pollination algorithm, salp swarm algorithm, linear PSO, hybrid PSO, and differential evolution. The conclusion drawn from their research was that APSOLL showed superior exploration and exploitation capabilities. PSO has been employed in various classification problems in remote sensing (Ghamisi and Benediktsson, 2015; Akbari et al., 2020; Singh and Singh, 2021; Gupta et al., 2023).

3.2.5.3   Feature Selection with SVMs

The correlation between the number of input features and computational load is evident, but an increase in the number of input bands may not necessarily lead to a substantial improvement in classification accuracy and could, in fact, result in a reduction. It, therefore, becomes attractive to select the optimal subset of features from all those available features to achieve higher accuracy and reduce computational cost. As described in Chapter 5, the decision rule of the SVM algorithm can be given as

$$f(\mathbf{x}) = \text{sign}\left(\sum_{i=1}^{nsv} \alpha_i y_i K(\mathbf{x}, \mathbf{x}_i) + b\right). \tag{3.35}$$

In SVM-based feature selection, if one of the input features does not significantly contribute to the classification problem, the corresponding scaling parameters will be close to zero. Thus, the classification performance is not likely to be affected by removing such insignificant features. Based on such an idea, Chapelle et al. (2002) proposed an approach in combination with the gradient descent algorithm for feature selection by eliminating those features whose scaling parameters are the smallest. Note that both the error estimate functions and the corresponding gradient descent procedures are discussed in Section 4.3. The algorithm terminates when the error estimate cannot be further reduced by the elimination of further spectral bands or features. Rather than base feature selection on the raw features, the algorithm can also be applied to the features transformed by PCA (Section 3.1.1). In this way, the feature space can be condensed further.

Guyon et al. (2002) proposed another idea for feature selection by recursively minimizing the cost function $\frac{1}{2}w^2$ and eventually finding a subset of features that generates the largest margin between classes. The algorithm is similar to backward feature elimination (Kohavi and John, 1997). Note that the features that are eliminated last are not necessarily the ones that are individually most closely controlling the classification accuracy. However, they are likely to be the best-performing subset of features. Also, in order to reduce the computational burden, the algorithm may simultaneously eliminate several features at each iteration. However, this may be at the expense of classification performance degradation.

The feature selection algorithms described above are for solving two-class classification problems. For the multiclass classification problem, the algorithms can be naturally extended using one-against-one or one-against-others (one-against-all) strategies to perform the selection for suitable features. Feature selection with SVM as a wrapper method has lately become popular and several variants (Archibald and Fann, 2007; Maldonado and Weber, 2009). Further readings of the application of SVM-based feature selection are given in Pal (2006), Kuo et al. (2013), and Pal et al. (2021).

3.2.6 Embedded Methods

Several classifiers have feature selection capabilities built into the learning algorithm in a way that they output a ranked list of all features during the classification process itself. This mechanism of feature selection is termed as an embedded method (Figure 3.13). In other words, feature selection is integrated or built into the algorithm in embedded methods. In the training stage, the classifier refines its internal parameters and determines the appropriate weights or importance assigned to each feature to achieve the highest possible classification accuracy. Embedded feature selection methods inherently combine the advantages of both filter and wrapper methods. These algorithms are usually computationally more efficient than wrappers because they simultaneously integrate their models using a two-part objective function with a goodness-of-fit term or a penalty for a larger number of features. Similar to the wrappers, the features selected through embedded methods depend on the specific induction algorithm used. As an example of this category, input pruning for artificial neural networks can be performed on the inputs having a very low influence on the network outputs (Kavzoglu and Mather, 1999). In addition, the weights close to zero can be eliminated without significantly affecting the classification performance (Kavzoglu and Mather, 2002). These

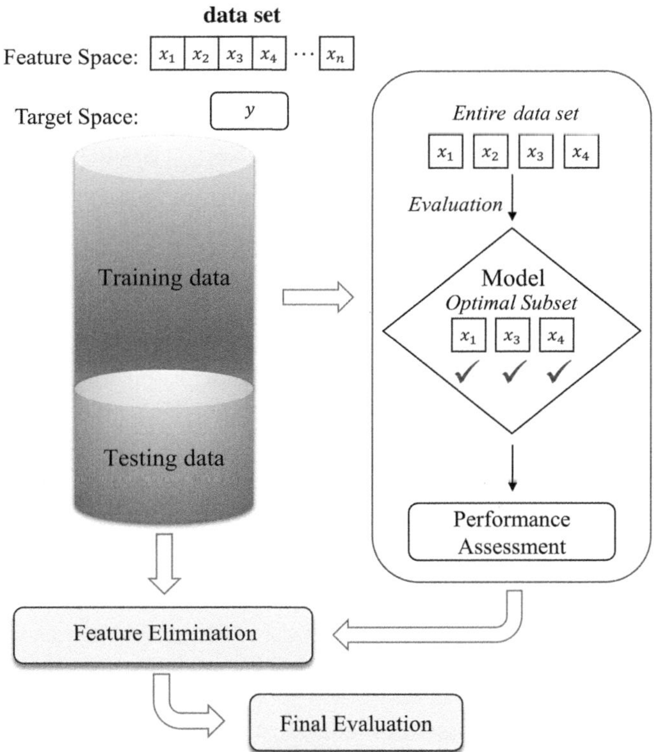

FIGURE 3.13 Working principle of embedded methods.

small weights, even though multiplied by the activities in the preceding layer, may contribute almost nothing to the node in the next layer. In decision trees, some features are selected at each node, and this decision is a constituent element of the algorithm and cannot be simply separated from it. For rough sets, such functions are played by relative reducts, which are subsets of attributes that guarantee the same quality of approximation as the entire set of variables (Stańczyk and Jain, 2015).

3.2.6.1   K-Nearest Neighbor-Based Feature Selection

K-nearest neighbor (k-NN) method is comparatively a simple and nonparametric method applied to classification problems. In addition to classification problems, it has been also employed in feature selection studies. Let $\{F_1, F_2, \ldots, F_m\}$ be the set of m numerical predictor variables, $\{x_1, x_2, \ldots, x_n\}$ be the set of n observations, and $y = (y_1, y_2, \ldots, y_n)$ be the target variable. Then, if we denote the observation in the test set by x_i and the set of indices of k-nearest neighbors of sample x_i by N_i^k, the prediction of target variable p_i is given by

$$p_i = \frac{1}{k} \sum_{j \in N_i^k} y_j. \tag{3.36}$$

Li et al. (2011) proposed a random k-NN algorithm that includes an ensemble of base k-nearest neighbor models, each constructed from a random subset of the input variables. Specifically, a collection of r different k-NN classifiers is generated, and each classifier takes a random subset of the input features. Since k-NN is stable, bootstrapping is not necessary. To rank the importance of the features, a criterion on the Rk-NN framework is specifically defined using the notion of "support". Considering the criterion, a two-stage backward model selection method is then applied. Majority voting is applied for final classifications. It was claimed that without a feature selection process, the random k-NN could produce similar results with RFs in terms of classification accuracy, whereas it is superior when feature selection is put into practice.

Let $F = \{f_1, f_2, \ldots, f_p\}$ be the p input features, and X be the n original input data vectors of length p (i.e., an $n \times p$ matrix). For a given integer $m < p$, denote $F^{(m)} = \{f_{j1}, f_{j2}, \ldots, f_{jm} | f_{jl} \in F, \ 1 \leq l \leq m\}$ a random subset drawn from F with equiprobability. Similarly, let $X^{(m)}$ be the data vectors in the subspace defined by $F^{(m)}$ (i.e., an $n \times m$ matrix). Then, a k-NN$^{(m)}$ classifier is constructed by applying the basic k-NN algorithm to the random collection of features in $X^{(m)}$. A collection of r such base classifiers is then combined to build the final random k-NN classifier (Li et al., 2011).

Another novel wrapper algorithm based on the k-NN algorithm is the sequential random k-NN, which was introduced by Park and Kim (2015). The algorithm resembles the RF method since each base classifier employs a forward sequential selection strategy from a bootstrapped sample and majority voting for predictions. The experimental results showed that higher accuracies were achieved with the proposed algorithm than with the RF algorithm for data sets with high complexity. Another derivative k-NN algorithm was proposed by Bugata and Drotár (2019), in which distance- and attribute-weighted k-nearest neighbors were used, and the gradient descent method was employed to find the function minima. The underlying assumption of this approach is that the target variable is most accurately determined by the most relevant features of the nearest neighbors.

Pal et al. (2021) investigated the feature selection performances of randomized and weighted k-NN methods together with wrapper SVM, RF, and XGBoost using two hyperspectral data. Comparison of the classification results in terms of classification accuracy with the selected features revealed that the performance of the randomized k-NN algorithm was superior to that of the weighted k-NN and other state-of-the-art feature selection methods.

3.2.6.2   Feature Selection with Ensemble Learners

Ensemble learning-based bagging, stacking, and boosting theories refer to the algorithms that combine the predictions from two or more models, which have become popular in classification, regression, and feature selection. The most popular examples of ensemble algorithms include RF, XGBoost, LightGBM, CatBoost, and NGBoost, which are discussed in Section 6.7. Random forests, introduced by Breiman (2001), are popular ensemble learning methods used in classification, regression, and feature selection problems. As a machine learning algorithm, random forests have received increasing attention because of their outstanding classification performances and processing speed. The classifier includes a combination of tree predictors such that each tree makes predictions independently using a random selection of features through replacements (using the bagging approach). The final classification is established on the basis of a majority vote of the trained trees. The algorithm can be easily used to measure the importance of the features by ranking, called out-of-bag estimates. Evaluating the importance of each feature involves permuting the values of the mth feature for the out-of-bag (OOB) samples. The perturbed OOB samples are subsequently reprocessed through each tree. Calculating the average differences in accuracy between the unaltered and perturbed OOB samples across all trees in the random forest produces an importance score for the mth feature. This score acts as a ranking index (Chan and Paelinckx, 2008). Many researchers (Kavzoglu et al., 2018b; Linhui et al., 2021; Kavzoglu and Bilucan, 2023) have tested the performance of random forests in reducing the dimensions of the data sets through feature selection and validated their effectiveness in terms of classification accuracies and computation times.

In gradient boosting methods, the importance score is estimated explicitly for each feature in the data set. The scores for an individual decision tree are determined based on how often each attribute split point enhances the performance measure, with weights assigned according to the number of observations the node is accountable for. The performance measure could be the purity (Gini index) employed for selecting split points or another specialized error function. The feature importance is then averaged across all decision trees within the model. For the XGBoost algorithm, feature importance can be estimated by considering the gain, permutation-based feature importance, and importance with Shapley values (see Section 10.3.1). Some examples of studies employing ensemble learning methods include Zheng et al. (2017), Luo et al. (2021), and Pal et al. (2021).

3.2.6.3 Hilbert-Schmidt Independence Criterion with Lasso

The Hilbert-Schmidt independence criterion Lasso (HSIC-Lasso), proposed by Yamada et al. (2014), is a nonlinear feature selection technique. Unlike the conventional Lasso algorithm, which combines both feature elimination and prediction procedures, HSIC-Lasso is solely designed for feature selection, which aims to choose informative and nonredundant features by utilizing a set of kernel functions and solving a Lasso problem (Afshar and Usefi, 2021).

In scenarios where the relationship between estimators and the target variable is expected to be nonlinear, the linearity assumption of the Lasso algorithm might be inadequate. HSIC-Lasso was developed to address the limitations of the standard Lasso approach. The formulation of HSIC-Lasso is as follows:

$$\min_{\alpha \in \mathbb{R}^p} \frac{1}{2} \left\| \bar{L} - \sum_{k=1}^{p} \alpha_k \bar{K}^{(k)} \right\|_{\text{Frob}}^2 + \lambda \|\alpha\|_1, \quad \alpha_1, \ldots, \alpha_p \geq 0 \tag{3.37}$$

where $\|\cdot\|_{\text{Frob}}$ represents the Frobenius norm, $\|\cdot\|_1$ denotes the ℓ_1 norm, $\bar{K}^{(k)}$ and $\bar{L}$ correspond to the centered Gram matrices, α is the vector of regression coefficients, and λ is the regularization parameter.

Kernel functions play a pivotal role in HSIC-Lasso by transforming the original feature space into a higher-dimensional space that can capture nonlinear relationships more effectively. The centered Gram matrices, $\bar{K}^{(k)}$, are derived from these kernel functions. The regularization term $\lambda \|\alpha\|_1$ introduces a penalty for large coefficient values, promoting sparsity in the selected features and effectively conducting feature selection. The choice of the regularization parameter λ governs the balance between fitting the data and avoiding overfitting. The resulting α solution from the HSIC-Lasso optimization problem signifies the importance of each feature. Features with nonzero coefficients in α are chosen, while those with zero coefficients are essentially excluded. HSIC-Lasso's capacity to handle nonlinearity and its ability to jointly conduct variable selection and independence testing make it a valuable tool for feature engineering in scenarios where traditional linear methods fall short. However, the computational complexity and the challenge of hyperparameter tuning remain considerations when applying HSIC-Lasso in practice. In a comparative study using HSIC-Lasso, Relief-F, and GA-RF for feature selection, Kavzoglu and Bilucan (2023) concluded that HSIC-Lasso provided the highest prediction accuracy at the fastest time, showing its effectiveness and low computational cost in estimations.

3.3 CONCLUDING REMARKS

In theory, the use of high-dimensional data, such as hyperspectral images or multispectral images combined with object features, appears to be a promising approach for multivariate data analysis. However, it introduces several challenges, including the curse of dimensionality. The scientific approach in such cases typically aligns with Occam's Razor principle, favoring simplicity among multiple options. To address the complexities associated with high-dimensional data, researchers predominantly employ two fundamental techniques: feature extraction and feature selection. These approaches aim to eliminate or minimize redundant, irrelevant, and noisy features that can adversely affect the performance of machine learning algorithms. The inclusion of irrelevant or noisy features amplifies the complexity of the problem and makes class separation in the feature space more challenging, which is analogous to the issues of conceptual or logical accuracy in GIS. Using representative feature subsets, which are achieved by selecting critical features and discarding irrelevant ones, enhances the predictive power of machine learning algorithms. Dimensionality reduction substantially simplifies the problem, leading to reduced overfitting risks, faster learning, and more straightforward model interpretation (including the utilization of XAI methods as discussed in Chapter 10). Furthermore, reducing data dimensions also decreases the need for extensive training samples, contributing to models with strong generalization capabilities. Numerous studies have reported improved classification accuracy when employing representative subsets obtained using dimensionality reduction techniques. Feature extraction involves transforming original variables into more informative and concise ones, while feature selection entails identifying and selecting crucial variables based on specific criteria. The foundational principles of major feature extraction and feature selection methods were covered in the previous sections of this chapter.

At this stage, some guidelines will be provided to inexperienced and experienced practitioners of machine learning and deep learning methods. The initial question that often arises is the selection of an appropriate dimensionality reduction approach for a specific remote sensing study. Typically, feature extraction is employed prior to feature selection to generate new variables based on their variance within the data set. Novice users often prefer these techniques due to their simplicity and lack of hyperparameter tuning requirements. However, it is important to recognize that some of these new variables, such as later PCs, may contain vital information about a LULC class represented by only a small number of pixels in an image. Consequently, expertise is crucial when determining the number of new variables or features, because these methods often yield the same number of new features as the original ones. For example, if you have a seven-band image, the PCA method will generate seven new features at the end of the process. On the other hand, when employing deep learning models for high-dimensional data analysis, these models use a feature extractor, which is a specialized type of neural network that learns weights during training. Autoencoders, a fundamental component in constructing deep architectures, and specific types of convolutional neural networks (CNNs) with diverse filters are also employed for this purpose. It is important to acknowledge the substantial progress made in recent years, as feature selection used to involve pruning input nodes in neural networks, especially the multilayer perceptron (Sildir et al., 2020). The integration and advancement of feature selection within deep learning models are currently a prominent and rapidly evolving area, as demonstrated by studies conducted by Zhang et al. (2016),

Singh and Karthikeyan (2022), and Deepa et al. (2023). Consequently, feature extraction and feature selection, as discussed in this chapter, have emerged as critical preprocessing steps for machine learning algorithms, and they continue to hold relevance in the context of deep learning models.

Compared with the feature extraction methods, feature selection techniques are more complex and require more computation burden, but they significantly influence the success of subsequent classification. A recurrent challenge encountered by remote sensing researchers pertains to the choice of the most suitable feature selection method for the specific problem under investigation. Before proceeding, it is pertinent to compare the feature selection approaches, namely, filters, wrappers, and embedded methods (Table 3.1). The table outlines the merits and demerits associated with the three approaches. The advantages and disadvantages of the three approaches are listed in Table 3.1. Filter methods offer straightforward solutions by employing statistical metrics to evaluate the relevance of each feature in relation to the target variables. They proceed to select the most relevant features without requiring any parameter adjustments. These rapid and computationally efficient methods are typically employed in the preprocessing stages and are well suited for quick screening and elimination of irrelevant features from the data sets. Consequently, they may be favored by individuals with limited experience. Nonetheless, their potential drawbacks include the risk of reduced prediction accuracy and susceptibility to issues when dealing with insufficient or imbalanced data sets.

Advanced users are encouraged to explore embedded methods, since they inherently encompass both filter and wrapper approaches, surpassing them in terms of predictive performance. However, the effectiveness of embedded methods tends to diminish when irrelevant features are introduced. In remote sensing, ensemble techniques are commonly employed by combining various feature selection results (Tsymbal et al., 2005; Guan et al., 2014; Kavzoglu and Teke, 2022). Three primary ensemble strategies, namely, data diversity, functional diversity, and hybrid strategies, have been widely used for this purpose. It is important to emphasize that the ideal method for selecting features will differ based on the data set, model, and objectives of the study. There is no universal solution, but the techniques and best practices outlined here can serve as a valuable starting point. It is highly advisable to apply cross-validation to feature selection results to prevent overfitting and identifying the "best" feature subset. Additionally, understanding the strengths and weaknesses of these techniques can assist analysts in selecting the most suitable one for their study, and also by considering the aforementioned considerations.

TABLE 3.1

Comparison of Filter, Wrapper, and Embedded Feature Selection Approaches

Criteria	Filters	Wrappers	Embedded
Underlying theory	Selects the most important features considering each feature separately	Combines related features to create best representative subsets	Classifier adjusts its parameters to determine the weights or importance
Processing speed	Fast	Slowest	Slow
Parameter setting requirement	No	Yes	Yes
Computation burden in training	Low	High	Moderate
Overfitting probability	Low/moderate	High	Low
Generalization capability	Low	High	High
Classifier dependence	No (model agnostic)	Yes (use predictive models)	Yes (in-built feature selection)
Prediction accuracy	Low	High	Highest

REFERENCES

Afshar, M., and H. Usefi. 2021. Dimensionality reduction using singular vectors. *Scientific Reports* 11:3832. https://doi.org/10.1038/s41598-021-83150-y.

Aggarwal, N., U. Shukla, G. J. Saxena, M. Rawat, A. S. Bafila, S. Singh, and A. Pundir. 2023. Mean based relief: An improved feature selection method based on ReliefF. *Applied Intelligence* 53:23004–23028. https://doi.org/10.1007/s10489-023-04662-w.

Aha, D. W., and R. L. Bankert. 1995. A comparative evaluation of sequential feature selection algorithms. In *Proceedings of Fifth International Workshop on Artificial Intelligence and Statistics,* January 4–7, Fort Lauderdale, FL, pp. 1–7.

Akbari, E., A. Darvishi Booloorani, N. Neysani Samany, S. Hamzeh, S. Soufizadeh, and S. Pignatti. 2020. Crop mapping using random forest and particle swarm optimization based on multi-temporal Sentinel-2. *Remote Sensing* 12:1449. https://doi.org/10.3390/rs12091449.

Archibald, R., and G. Fann. 2007. Feature selection and classification of hyperspectral images with support vector machines. *IEEE Geoscience and Remote Sensing Letters* 4:674–677. https://doi.org/10.1109/LGRS.2007.905116.

Bengio, Y., A. Courville, and P. Vincent. 2013. Representation learning: A review and new perspectives. *IEEE Transactions on Pattern Analysis and Machine Intelligence* 35:1798–1828. https://doi.org/10.1109/TPAMI.2013.50.

Bickel, P. J., G. Kur, and B. Nadler. 2018. Projection pursuit in high dimensions. *Proceedings of the National Academy of Sciences* 115:9151–9156. https://doi.org/10.1073/pnas.1801177115.

Biesiada, J., and W. Duch. 2007. Feature selection for high-dimensional data: A Pearson redundancy based filter. In *Computer Recognition Systems 2*, (eds.) M. Kurzynski, E. Puchala, M. Wozniak, and A. Zolnierek, pp. 242–249. Berlin, Heidelberg: Springer. https://doi.org/10.1007/978-3-540-75175-5_30.

Blanchard, G., M. Kawanabe, M. Sugiyama, V. Spokoiny, and K. R. Müller. 2006. In search of non-Gaussian components of a high-dimensional distribution. *Journal of Machine Learning Research* 7:247–282. https://doi.org/10.34657/2555.

Breiman, L. 2001. Random forests. *Machine Learning* 45:5–32. https://doi.org/10.1023/A:1010933404324.

Breiman, L., J. H. Friedman, R. A. Olsen, and C. J. Stone. 1984. *Classification and Regression Trees*. Belmont, CA: Wadsworth.

Bugata, P., and P. Drotár. 2019. Weighted nearest neighbors feature selection. *Knowledge-Based Systems* 163:749–761. https://doi.org/10.1016/j.knosys.2018.10.004.

Celik, T. 2010. Change detection in satellite images using a genetic algorithm approach. *IEEE Geoscience and Remote Sensing Letters* 7:386–390. https://doi.org/10.1109/LGRS.2009.2037024.

Chan, J. C. W., and D. Paelinckx. 2008. Evaluation of random forest and Adaboost tree-based ensemble classification and spectral band selection for ecotope mapping using airborne hyperspectral imagery. *Remote Sensing of Environment* 112:2999–3011. https://doi.org/10.1016/j.rse.2008.02.011.

Chang, C. I. 2013. *Hyperspectral Data Processing: Algorithm Design and Analysis*. Hoboken, NJ: John Wiley & Sons.

Chang, Y.-L., K.-S. Chen, B. Huang, W.-Y. Chang, J. A. Benediktsson, and L. Chang. 2011. A parallel simulated annealing approach to band selection for high-dimensional remote sensing images. *IEEE Journal of Selected Topics in Applied Earth Observations and Remote Sensing* 4:579–590. https://doi.org/10.1109/JSTARS.2011.2160048.

Chapelle, O., V. Vapnik, O. Bousquet, and S. Mukherjee. 2002. Choosing multiple parameters for support vector machines. *Machine Learning* 46:131–159. https://doi.org/10.1023/A:1012450327387.

Cheriyadat, A., and L. M. Bruce. 2003. Why principal component analysis is not an appropriate feature extraction method for hyperspectral data. In *Proceedings of 2003 IEEE International Geoscience and Remote Sensing Symposium,* July 21–25, Toulouse, France, pp. 3420–3422. https://doi.org/10.1109/IGARSS.2003.1294808.

Colkesen, I., and T. Kavzoglu. 2016. Performance evaluation of rotation forest for SVM-based recursive feature elimination using hyperspectral imagery. In *Proceedings of the 8th Workshop on Hyperspectral Image and Signal Processing: Evolution in Remote Sensing (WHISPERS),* August 21–24, Los Angeles, CA, pp. 1–5. https://doi.org/10.1109/WHISPERS.2016.8071792.

Colkesen, I., and T. Kavzoglu. 2018. Selection of optimal object features in object-based image analysis using filter-based algorithms. *Journal of the Indian Society of Remote Sensing* 46:1233–1242. https://doi.org/10.1007/s12524-018-0807-x.

Colkesen, I., and T. Kavzoglu. 2019. Comparative evaluation of decision-forest algorithms in object-based land use and land cover mapping. In *Spatial Modeling in GIS and R for Earth and Environmental Sciences*, (eds.) H. R. Pourghasemi, and C. Gokceoglu, pp. 499–517. Amsterdam: Elsevier. https://doi.org/10.1016/B978-0-12-815226-3.00023-5.

Cortijo, F. J., and N. Pérez de la Blanca. 1999. The performance of regularised discriminant analysis versus non-parametric classifiers applied to high dimensional image classification. *International Journal of Remote Sensing* 20:3345–3365. https://doi.org/10.1080/014311699211372.

Deepa, C., A. Shetty, and A. V. Narasimhadhan. 2023. Knowledge distillation: A novel approach for deep feature selection. *The Egyptian Journal of Remote Sensing and Space Science* 26:63–73. https://doi.org/10.1016/j.ejrs.2022.12.006.

Ding, C., and H. Peng. 2005. Minimum redundancy feature selection from microarray gene expression data. *Journal of Bioinformatics and Computational Biology* 3:185–205. https://doi.org/10.1142/S0219720005001004.

Dubes, R. C., and A. K. Jain. 1989. Random field models in image analysis. *Journal of Applied Statistics* 16:131–164. https://doi.org/10.1080/02664768900000014.

Duda, R. O., and P. E. Hart. 1973. *Pattern Classification and Scene Analysis*. New York: John Wiley & Sons.

Elmaizi, A., H. Nhaila, E. Sarhrouni, A. Hammouch, and C. Nacir. 2019. A novel information gain based approach for classification and dimensionality reduction of hyperspectral images. *Procedia Computer Science* 148:126–134. https://doi.org/10.1016/j.procs.2019.01.016.

Falco, N., J. A. Benediktsson, and L. Bruzzone. 2014. A study on the effectiveness of different independent component analysis algorithms for hyperspectral image classification. *IEEE Journal of Selected Topics in Applied Earth Observations and Remote Sensing* 7:2183–2199. https://doi.org/10.1109/JSTARS.2014.2329792.

Fauvel, M., J. Chanussot, and J. A. Benediktsson. 2009. Kernel principal component analysis for the classification of hyperspectral remote sensing data over urban areas. *EURASIP Journal on Advances in Signal Processing* 1–14. https://doi.org/10.1155/2009/783194.

Foody, G. M. 2009. Sample size determination for image classification accuracy assessment and comparison. *International Journal of Remote Sensing* 30:5273–5291. https://doi.org/10.1080/01431160903130937.

Friedman, J. H., and J. W. Tukey. 1974. A projection pursuit algorithm for exploratory data analysis. *IEEE Transactions on Computers* C-23:881–890. https://doi.org/10.1109/T-C.1974.224051.

Gasmi, A., C. Gomez, A. Chehbouni, D. Dhiba, and M. El Gharous. 2022. Using PRISMA hyperspectral satellite imagery and GIS approaches for soil fertility mapping (FertiMap) in northern Morocco. *Remote Sensing* 14:4080. https://doi.org/10.3390/rs14164080.

Geman, S., and D. Geman. 1984. Stochastic relaxation, Gibbs distributions, and the Bayesian restoration of images. *IEEE Transactions on Pattern Analysis and Machine Intelligence* 6:721–741. https://doi.org/10.1109/TPAMI.1984.4767596.

Georganos, S., T. Grippa, S. Vanhuysse, M. Lennert, M. Shimoni, S. Kalogirou, and E. Wolff. 2018. Less is more: Optimizing classification performance through feature selection in a very-high-resolution remote sensing object-based urban application. *GIScience & Remote Sensing* 55:221–242. https://doi.org/10.1080/15481603.2017.1408892.

Ghamisi, P., and J. A. Benediktsson. 2015. Feature selection based on hybridization of genetic algorithm and particle swarm optimization. *IEEE Geoscience and Remote Sensing Letters* 12:309–313. https://doi.org/10.1109/LGRS.2014.2337320.

Goldberg, D. E. 1989. *Genetic Algorithms in Search, Optimization and Machine Learning*. Reading, MA: Addison-Wesley.

Green, A. A., M. Berman, P. Switzer, and M. D. Craig. 1988. A transformation for ordering multispectral data in terms of image quality with implications for noise removal. *IEEE Transactions on Geoscience and Remote Sensing* 26:65–74. https://doi.org/10.1109/36.3001.

Guan, D., W. Yuan, Y. K. Lee, K. Najeebullah, and M. K. Rasel. 2014. A review of ensemble learning based feature selection. *IETE Technical Review* 31:190–198. https://doi.org/10.1080/02564602.2014.906859.

Gupta, G. R., S. D. Jadhav, and N. M. Pagare. 2023. Binary particle swarm optimization (BPSO) based classification of hyperspectral images. *Advancement in Image Processing and Pattern Recognition* 1:12–18.

Guyon, I., J. Weston, S. Barnhill, and V. Vapnik. 2002. Gene selection for cancer classification using support vector machines. *Machine Learning* 46:389–422. https://doi.org/10.1023/A:1012487302797.

Hall, M. A. 1999. Correlation-based feature selection for machine learning. Ph.D. Thesis, University of Waikato, Hamilton.

Hamaide, V., and F. Glineur. 2021. Unsupervised minimum redundancy maximum relevance feature selection for predictive maintenance: Application to a rotating machine. *International Journal of Prognostics and Health Management* 12:1–14. https://doi.org/10.36001/ijphm.2021.v12i2.2955.

Hand, D. J. 1981. Branch and bound in statistical data analysis. *Journal of the Royal Statistical Society: Series D (The Statistician)* 30:1–13. https://doi.org/10.2307/2987699.

Haralick, R. M., and K. S. Fu. 1983. Pattern recognition and classification. In *Manual of Remote Sensing*, (eds.) R. N. Colwell, D. S. Simonett, and F. T. Ulaby, pp. 881–884. Falls Church, VA: American Society of Photogrammetry.

Hashjin, S. S., and S. Khazai. 2022. A new method to detect targets in hyperspectral images based on principal component analysis. *Geocarto International* 37:2679–2697. https://doi.org/10.1080/10106049.2020.1831625.

Hegarat-Mascle, S. L., D. Vidal-Madjar, and P. Olivier. 1996. Applications of simulated annealing to SAR image clustering and classification problems. *International Journal of Remote Sensing* 17:1761–1776. https://doi.org/10.1080/01431169608948738.

Holland, J. H. 1992. *Adaptation in Natural and Artificial Systems*. Cambridge, MA: The MIT Press.

Holte, R. C. 1993. Very simple classification rules perform well on most commonly used data sets. *Machine Learning* 11:63–91. https://doi.org/10.1023/A:1022631118932.

Huber, P. J. 1985. Projection pursuit. *The Annals of Statistics* 13:435–475. https://doi.org/10.1214/aos/1176349519.

Hughes, G. F. 1968. On the mean accuracy of statistical pattern recognizers. *IEEE Transactions on Information Theory* 14:55–63. https://doi.org/10.1109/TIT.1968.1054102.

Hyvärinen, A. 1998. Independent component analysis in the presence of Gaussian noise by maximizing joint likelihood. *Neurocomputing* 22:49–67. https://doi.org/10.1016/S0925-2312(98)00049-6.

Hyvärinen, A., J. Karhunen, and E. Oja. 2001. *Independent Component Analysis*. New York: John Wiley & Sons.

Im, J., Z. Lu, and J. R. Jensen. 2011. A genetic algorithm approach to moving threshold optimization for binary change detection. *Photogrammetric Engineering and Remote Sensing* 77:167–180. https://doi.org/10.14358/PERS.77.2.167.

Jayaprakash, C., B. B. Damodaran, S. Viswanathan, and K. P. Soman. 2020. Randomized independent component analysis and linear discriminant analysis dimensionality reduction methods for hyperspectral image classification. *Journal of Applied Remote Sensing* 14:036507–036507. https://doi.org/10.48550/arXiv.1804.07347.

Jensen, J. R. 2015. *Introductory Digital Image Processing: A Remote Sensing Perspective*, 4th edition. Glenview, IL: Pearson.

Jones, M. C., and R. Sibson. 1987. What is projection pursuit? *Journal of the Royal Statistical Society: Series A* 150:1–18. https://doi.org/10.2307/2981662.

Kailath, T. 1967. The Divergence and Bhattacharyya distance measures in signal selection. *IEEE Transactions on Communication Technology* 15:52–60. https://doi.org/10.1109/TCOM.1967.1089532.

Kavzoglu, T. 2001. An investigation of the design and use of feed-forward artificial neural networks in the classification of remotely sensed images. Ph.D. Thesis, School of Geography, The University of Nottingham, Nottingham, UK.

Kavzoglu, T., and A. Teke. 2022. Ensemble conditioning factor selection with Markov chain framework for shallow landslide susceptibility mapping in lake Sapanca Basin and its vicinity, Turkey. *Baltic Journal of Modern Computing* 10:224–240. https://doi.org/10.22364/bjmc.2022.10.2.09.

Kavzoglu, T., and F. Bilucan. 2023. Effects of auxiliary and ancillary data on LULC classification in a heterogeneous environment using optimized random forest algorithm. *Earth Science Informatics* 16:415–435. https://doi.org/10.1007/s12145-022-00874-9.

Kavzoglu, T., and P. M. Mather. 1999. Pruning artificial neural networks: an example using land cover classification of multi-sensor images. *International Journal of Remote Sensing* 20:2787–2803. https://doi.org/10.1080/014311699211796.

Kavzoglu, T., and P. M. Mather. 2002. The role of feature selection in artificial neural network applications. *International Journal of Remote Sensing* 23:2919–2937. https://doi.org/10.1080/01431160110107743.

Kavzoglu, T., E. K. Sahin, and I. Colkesen. 2015. Selecting optimal conditioning factors in shallow translational landslide susceptibility mapping using genetic algorithm. *Engineering Geology* 192:101–112. https://doi.org/10.1016/j.enggeo.2015.04.004.

Kavzoglu, T., H. Tonbul, and I. Colkesen. 2018a. Dimensionality reduction for hyperspectral images to improve object-based image classification using feature selection and principal components analysis. In *Proceedings of 39th Asian Conference on Remote Sensing (ACRS'18)*, October 15–19, Lumpur, Malaysia, pp. 2314–2318.

Kavzoglu, T., H. Tonbul, M. Yildiz Erdemir, and I. Colkesen. 2018b. Dimensionality reduction and classification of hyperspectral images using object-based image analysis. *Journal of the Indian Society of Remote Sensing* 46:1297–1306. https://doi.org/10.1007/s12524-018-0803-1.

Kennedy, J., and R. Eberhart. 1995. Particle swarm optimization. In *Proceedings of International Conference on Neural Networks (ICNN'95),* November 27-December 01, Perth, Australia, vol. 4, pp. 1942–1948. https://doi.org/10.1109/ICNN.1995.488968.

Kerber, R. 1992. ChiMerge: Discretization of numeric attributes. In Proceedings of 10th National Conference on Artificial Intelligence, July 12–16, San Jose, CA, pp. 123–128.

Kira, K., and L. A. Rendell. 1992. The feature selection problem: Traditional methods and a new algorithm. In Proceedings of 10th National Conference on Artificial Intelligence, July 12–16, San Jose, CA, vol. 2, pp. 129–134.

Kohavi, R., and G. H. John. 1997. Wrappers for feature subset selection. *Artificial Intelligence Journal* 97:273–324. https://doi.org/10.1016/S0004-3702(97)00043-X.

Kononenko, I. 1994. Estimating attributes: Analysis and extensions of RELIEF. In *Machine Learning: ECML-94,* (eds.) F. Bergadano, and L. De Raedt, pp. 171–182. Berlin, Heidelberg: Springer. https://doi.org/10.1007/3-540-57868-4_57.

Koukiou, G. 2022. Simulated annealing for land cover classification in PolSAR images. *Advances in Remote Sensing* 11:49–61. https://doi.org/10.4236/ars.2022.112004.

Kruskal, J. B. 1969. Toward a practical method which helps uncover the structure of a set of multivariate observations by finding the linear transformation which optimizes a new 'index of condensation'. *Statistical Computation* 1969:427–440. https://doi.org/10.1016/B978-0-12-498150-8.50024-0.

Kruskal, J. B. 1972. Linear transformation of multivariate data to reveal clustering. In *Multidimensional Scaling: Theory and Application in the Behavioral Sciences I: Theory.* New York and London: Seminar Press.

Kuhn, M., and K. Johnson. 2019. *Feature Engineering and Selection: A Practical Approach for Predictive Models.* Boca Raton, FL: CRC Press.

Kumar, A., S. Singh, and A. Kumar. 2021. Grey wolf optimizer and other metaheuristic optimization techniques with image processing as their applications: A review. *IOP Conference Series: Materials Science and Engineering* 1136:012053. https://doi.org/10.1088/1757-899X/1136/1/012053.

Kumar, V., and S. Minz. 2014. Feature selection: A literature review. *Smart Computing Review* 4:211–229. https://doi.org/10.6029/smartcr.2014.03.007.

Kuo, B. C., H. H. Ho, C. H. Li, C. C. Hung, and J. S. Taur. 2013. A kernel-based feature selection method for SVM with RBF kernel for hyperspectral image classification. *IEEE Journal of Selected Topics in Applied Earth Observations and Remote Sensing* 7:317–326. https://doi.org/10.1109/JSTARS.2013.2262926.

Kursa, M. B. 2021. Praznik: High performance information-based feature selection. *SoftwareX* 16:100819. https://doi.org/10.1016/j.softx.2021.100819.

Landgrebe, D. 1998. Information extraction principles and methods for multispectral and hyperspectral image data. In *Information Processing for Remote Sensing,* ed. C. H. Chen, pp. 3–37. River Edge, NJ: World Scientific Publishing. https://doi.org/10.1142/9789812815705_0001.

Lee, J. B., A. S. Woodyatt, and M. Berman. 1990. Enhancement of high spectral resolution remote sensing data by noise-adjusted principal components transform. *IEEE Transactions on Geoscience and Remote Sensing* 28:295–304. https://doi.org/10.1109/36.54356.

Lennon, M., G. Mercier, M. C. Mouchot, and L. Hubert-Moy. 2001. Independent component analysis as a tool for the dimensionality reduction and the representation of hyperspectral images. In *Proceedings of IEEE 2001 International Geoscience and Remote Sensing Symposium (IGARSS'2001),* July 09–13, Sydney, Australia, pp. 2893–2895. https://doi.org/10.1109/IGARSS.2001.978197.

Li, S., E. J. Harner, and D. A. Adjeroh. 2011. Random kNN feature selection-A fast and stable alternative to Random Forests. *BMC Bioinformatics* 12:450. https://doi.org/10.1186/1471-2105-12-450.

Lin, Z., and G. Zhang. 2020. Genetic algorithm-based parameter optimization for EO-1 Hyperion remote sensing image classification. *European Journal of Remote Sensing* 53:124–131. https://doi.org/10.1080/22797254.2020.1747949.

Linhui, L., J. Weipeng, and W. Huihui. 2021. Extracting the forest type from remote sensing images by random forest. *IEEE Sensors Journal* 21:17447–17454. https://doi.org/10.1109/JSEN.2020.3045501.

Liu, H., and H. Motoda. 2007. *Computational Methods of Feature Selection.* Boca Raton, FL: CRC Press. https://doi.org/10.1201/9781584888796.

Liu, H., and L. Yu. 2005. Toward integrating feature selection algorithms for classification and clustering. *IEEE Transactions on Knowledge and Data Engineering* 17:491–502. https://doi.org/10.1109/TKDE.2005.66.

Liu, H., and R. Setiono. 1995. Chi2: Feature selection and discretization of numeric attributes. In *Proceedings of 7th IEEE International Conference on Tools with Artificial Intelligence,* November 05–08, Herndon, VA, pp. 388–391. https://doi.org/10.1109/TAI.1995.479783.

Loperfido, N. 2018. Skewness-based projection pursuit: A computational approach. *Computational Statistics & Data Analysis* 120:42–57. https://doi.org/10.1016/j.csda.2017.11.001.

Luo, M, Y. Wang, Y. Xie, L. Zhou, J. Qiao, S. Qiu, and Y. Sun. 2021. Combination of feature selection and CatBoost for prediction: The first application to the estimation of aboveground biomass. *Forests* 12:216. https://doi.org/10.3390/f12020216.

Ma, L., M. T. Fu, T. Blaschke, M. Li, D. Tiede, Z. Zhou, X. Ma, and D. Chen. 2017. Evaluation of feature selection methods for object-based land cover mapping of unmanned aerial vehicle imagery using random forest and support vector machine classifiers. *ISPRS International Journal of Geo-Information* 6:51. https://doi.org/10.3390/ijgi6020051.

Maldonado, S., and R. Weber. 2009. A wrapper method for feature selection using support vector machines. *Information Sciences* 179:2208–2217. https://doi.org/10.1016/j.ins.2009.02.014.

Martin, R. S., and J. H. Wilkinson. 1971. Reduction of the symmetric eigenproblem $Ax = \lambda Bx$ and related problems to standard form. In *Linear Algebra,* (eds.) J. H. Wilkinson, and C. Reinch, pp. 303–314. Berlin, Heidelberg: Springer. https://doi.org/10.1007/978-3-662-39778-7_21.

Mather, P. M., and M. Koch. 2011. *Computer Processing of Remotely-Sensed Images: An Introduction,* 4th edition. Chichester: John Wiley & Sons. https://doi.org/10.1002/9780470666517.

Mausel, P. W., W. J. Kramber, and J. K. Lee. 1990. Optimum band selection for supervised classification of multispectral data. *Photogrammetric Engineering and Remote Sensing* 56:55–60.

Metropolis, N., A. W. Rosenbluth, M. N. Rosenbluth, A. H. Teller, and E. Teller. 1953. Equations of state calculations by fast computational machine. *Journal of Chemical Physics* 21:1087–1091. https://doi.org/10.1063/1.1699114.

Millard, K., and M. Richardson. 2015. On the importance of training data sample selection in random forest image classification: A case study in peatland ecosystem mapping. *Remote Sensing* 7:8489–8515. https://doi.org/10.3390/rs70708489.

Ming, D., T. Zhou, M. Wang, and T. Tan. 2016. Land cover classification using random forest with genetic algorithm-based parameter optimization. *Journal of Applied Remote Sensing* 10:035021. https://doi.org/10.1117/1.JRS.10.035021.

Mingers, J. 1989. An empirical comparison of pruning methods for decision tree induction. *Machine Learning* 4:227–243. https://doi.org/10.1023/A:1022604100933.

Nascimento, J. M. P., and J. M. B. Dias. 2005. Does independent component analysis play a role in unmixing hyperspectral data? *IEEE Transactions on Geoscience and Remote Sensing* 43:175–187. https://doi.org/10.1109/TGRS.2004.839806.

Nielsen, A. A. 1994. Analysis of regularly and irregularly sampled spatial, multivariate and multi-temporal data. Ph.D. Thesis, Institute of Mathematical Modelling, Technical University of Denmark, Lyngby, Denmark.

Nielsen, A. A., K. Conradsen, and J. J. Simpson. 1998. Multivariate Alteration Detection (MAD) and MAF post-processing in multispectral, bi-temporal image data: New approaches to change detection studies. *Remote Sensing of Environment* 64:1–19. https://doi.org/10.1016/S0034-4257(97)00162-4.

Olsen, S. I. 1993. Estimation of noise in images: An evaluation. *Graphical Models and Image Processing* 55:319–323. https://doi.org/10.1006/cgip.1993.1022.

Pal, M. 2006. Support vector machine-based feature selection for land cover classification: A case study with DAIS hyperspectral data. *International Journal of Remote Sensing* 27:2877–2894. https://doi.org/10.1080/01431160500242515.

Pal, M. 2013. Hybrid genetic algorithm for feature selection with hyperspectral data. *Remote Sensing Letters* 4:619–628. https://doi.org/10.1080/2150704X.2013.777485.

Pal, M., and G. M. Foody. 2010. Feature selection for classification of hyperspectral data by SVM. *IEEE Transactions on Geoscience and Remote Sensing* 48:2297–2307. https://doi.org/10.1109/TGRS.2009.2039484.

Pal, M., T. B. Charan, and A. Poriya. 2021. *K*-nearest neighbour-based feature selection using hyperspectral data. *Remote Sensing Letters* 12:132–141. https://doi.org/10.1080/2150704X.2020.1864051.

Park, C. H., and S. B. Kim. 2015. Sequential random k-nearest neighbor feature selection for high-dimensional data. *Expert Systems with Applications* 42:2336–2342. https://doi.org/10.1016/j.eswa.2014.10.044.

Park, C. H., H. Park, and P. Pardalos. 2004. A comparative study of linear and nonlinear feature extraction methods. In Proceedings of Fourth IEEE International Conference on Data Mining (ICDM'04), November 01–04, Brighton, U.K., pp. 495–498. https://doi.org/10.1109/ICDM.2004.10066.

Plackett, R. L. 1983. Karl Pearson and the chi-squared test. *International Statistical Review/Revue Internationale de Statistique* 51:59–72. https://doi.org/10.2307/1402731.

Pudil, P., J. Novovicovo, and J. Kittler. 1994. Floating search methods in feature selection. *Pattern Recognition Letters* 15:1119–1125. https://doi.org/10.1016/0167-8655(94)90127-9.

Pullanagari, R. R., G. Kereszturi, and L. Yule. 2018. Integrating airborne hyperspectral, topographic, and soil data for estimating pasture quality using recursive feature elimination with random forest regression. *Remote Sensing* 10:1117. https://doi.org/10.3390/rs10071117.

Quinlan, J. R. 1979. Discovering rules by induction from large collections of examples. In *Expert Systems in the Micro-Electronic Age*, ed. D. Michie, pp. 168–201. Edinburgh, Scotland: Edinburgh University Press.

Quinlan, J. R. 1993. *C4.5: Programs for Machine Learning*. San Mateo: Morgan Kaufmann.

Ramírez-Gallego, S., I. Lastra, D. Martínez-Rego, V. Bolón-Canedo, J. M. Benítez, F. Herrera, and A. Alonso-Betanzos. 2017. Fast-mRMR: Fast minimum redundancy maximum relevance algorithm for high-dimensional big data. *International Journal of Intelligent Systems* 32:134–152. https://doi.org/10.1002/int.21833.

Ripley, B. D. 1996. *Pattern Recognition and Neural Networks*. Cambridge: Cambridge University Press. https://doi.org/10.1017/CBO9780511812651.

Roger, R. E. 1996. Principal components transform with simple, automatic noise adjustment. *International Journal of Remote Sensing* 17:2719–2727. https://doi.org/10.1080/01431169608949102.

Sameen, M. I., B. Pradhan, H. Z. M. Shafri, M. R. Mezaal, and H. B. Hamid. 2017. Integration of ant colony optimization and object-based analysis for LiDAR data classification. *IEEE Journal of Selected Topics in Applied Earth Observations and Remote Sensing* 10:2055–2066. https://doi.org/10.1109/JSTARS.2017.2650956.

Sasaki, H., G. Niu, and M. Sugiyama. 2016. Non-Gaussian component analysis with log-density gradient estimation. In *Proceedings of 19th International Conference on Artificial Intelligence and Statistics (AISTATS)*, May 03–11, Cadiz, Spain, pp. 1177–1185.

Schölkopf, B., A. Smola, and K. R. Müller. 1998. Nonlinear component analysis as a kernel eigenvalue problem. *Neural Computation* 10:1299–1319. https://doi.org/10.1162/089976698300017467.

Shannon, C. E. 1948. A mathematical theory of communication. *The Bell System Technical Journal* 27:379–423. https://doi.org/10.1002/j.1538-7305.1948.tb01338.x.

Siedlecki, W., and J. Sklansky. 1988. On automatic feature selection. *International Journal of Pattern Recognition and Artificial Intelligence* 2:197–220. https://doi.org/10.1142/S0218001488000145.

Sildir, H., E. Aydin, and T. Kavzoglu. 2020. Design of feedforward neural networks in the classification of hyperspectral imagery using superstructural optimization. *Remote Sensing* 12:956. https://doi.org/10.3390/rs12060956.

Singh, A. 1984. Some clarifications about the pairwise divergence method in remote sensing. *International Journal of Remote Sensing* 5:623–627. https://doi.org/10.1080/01431168408948845.

Singh, J., and G. Singh, G. 2021. Particle swarm optimization based deep learning model for scene classification of remote sensing images. *International Journal of Research in Engineering, Science and Management* 4:61–65.

Singh, P. S., and S. Karthikeyan. 2022. Enhanced classification of remotely sensed hyperspectral images through efficient band selection using autoencoders and genetic algorithm. *Neural Computing and Applications* 34:21539–21550. https://doi.org/10.1007/s00521-021-06121-4.

Stańczyk, U., and L. C. Jain. 2015. *Feature Selection for Data and Pattern Recognition: An Introduction*. Berlin, Heidelberg: Springer. https://doi.org/10.1007/978-3-662-45620-0.

Storn, R., and K. Price. 1997. Differential evolution: A simple and efficient heuristic for global optimization over continuous spaces. *Journal of Global Optimization* 11:341–359. https://doi.org/10.1023/A:1008202821328.

Stracuzzi, D. J. 2008. Randomized feature selection. In *Computational Methods of Feature Selection*, (eds.) H. Liu, and H. Motoda, pp. 41–62. Boca Raton, FL: Chapman and Hall/CRC.

Switzer, P., and A. A. Green. 1984. Min/Max autocorrelation factors for multivariate spatial imagery. In Technical Report No. 6, Department of Statistics, Stanford University, Stanford, CA.

Taherkhani, A., G. Cosma, and T. M. McGinnity. 2018. Deep-FS: A feature selection algorithm for Deep Boltzmann Machines. *Neurocomputing* 322:22–37. https://doi.org/10.1016/j.neucom.2018.09.040.

Thenkabail, P. S., E. A. Enclona, M. S. Ashton, and B. Van Der Meer. 2004. Accuracy assessments of hyperspectral waveband performance for vegetation analysis applications. *Remote Sensing of Environment* 91:354–376. https://doi.org/10.1016/j.rse.2004.03.013.

Tian, Y., P. Guo, and M. R. Lyu. 2005. Comparative studies on feature extraction methods for multispectral remote sensing image classification. In *Proceedings of IEEE International Conference on Systems, Man and Cybernetics*, October 10–12, Waikoloa, Hawaii, pp. 1275–1279. https://doi.org/10.1109/ICSMC.2005.1571322.

Tso, B., and P. M. Mather. 1999. Classification of multisource remote sensing imagery using a Genetic Algorithm and Markov Random Fields. *IEEE Transactions on Geoscience and Remote Sensing* 37:1255–1260. https://doi.org/10.1109/36.763284.

Tsymbal, A., M. Pechenizkiy, and P. Cunningham. 2005. Diversity in search strategies for ensemble feature selection. *Information Fusion* 6:83–98. https://doi.org/10.1016/j.inffus.2004.04.003.

Uddin, M. P., M. A. Mamun, and M. A. Hossain. 2019. Effective feature extraction through segmentation-based folded-PCA for hyperspectral image classification. *International Journal of Remote Sensing* 40:7190–7220. https://doi.org/10.1080/01431161.2019.1601284.

Uddin, M. P., M. A. Mamun, and M. A. Hossain. 2021. PCA-based feature reduction for hyperspectral remote sensing image classification. *IETE Technical Review* 38:377–396. https://doi.org/10.1080/02564602.2020.1740615.

Urbanowicz, R. J., M. Meeker, W. La Cava, R. S. Olson, and J. H. Moore. 2018. Relief-based feature selection: Introduction and review. *Journal of Biomedical Informatics* 85:189–203. https://doi.org/10.1016/j.jbi.2018.07.014.

Van der Meer, F., and S. M. De Jong. 2000. Improving the results of spectral unmixing of Landsat TM imagery by enhancing the orthogonality of end-members. *International Journal of Remote Sensing* 21:2781–2797. https://doi.org/10.1080/01431160050121249.

Vijouyeh, H. G., and G. Taşkın. 2016. A comprehensive evaluation of feature selection algorithms in hyperspectral image classification. In *Proceedings of IEEE International Geoscience and Remote Sensing Symposium (IGARSS'16)*, July 10–15, Beijing, China, pp. 489–492. https://doi.org/10.1109/IGARSS.2016.7729121.

Villa, A., J. A. Benediktsson, J. Chanussot, and C. Jutten. 2011. Hyperspectral image classification with independent component discriminant analysis. *IEEE Transactions on Geoscience and Remote Sensing* 49:4865–4876. https://doi.org/10.1109/TGRS.2011.2153861.

Virta, J., K. Nordhausen, and H. Oja. 2016. Projection pursuit for non-Gaussian independent components. *arXiv preprint*, arXiv:1612.05445. https://doi.org/10.48550/arXiv.1612.05445.

Wang, J., and C.-I. Chang. 2006. Independent component analysis-based dimensionality reduction with applications in hyperspectral image analysis. *IEEE Transactions on Geoscience and Remote Sensing* 44:1586–1600. https://doi.org/10.1109/TGRS.2005.863297.

Wang, N., B. Du, L. Zhang, and L. Zhang. 2014. An abundance characteristic-based independent component analysis for hyperspectral unmixing. *IEEE Transactions on Geoscience and Remote Sensing* 53:416–428. https://doi.org/10.1109/TGRS.2014.2322862.

Witten, I. H., E. Frank, and M. A. Hall. 2011. *Data Mining: Practical Machine Learning Tools and Techniques*, 3rd edition. Boston, MA: Morgan Kaufmann. https://doi.org/10.1016/C2009-0-19715-5.

Wu, B., C. Chen, T. M. Kechadi, and L. Sun. 2013. A comparative evaluation of filter-based feature selection methods for hyper-spectral band selection. *International Journal of Remote Sensing* 34:7974–7990. https://doi.org/10.1080/01431161.2013.827815.

Wu, L., Y. Wang, J. Long, and Z. Liu. 2015. An unsupervised change detection approach for remote sensing image using principal component analysis and genetic algorithm. In *Image and Graphics,* ed. Y. J. Zhang, pp. 589–602. Cham, Switzerland: Springer. https://doi.org/10.1007/978-3-319-21978-3_52.

Xia, J., J. Chanussot, P. Du and X. He. 2014. (Semi-) supervised probabilistic principal component analysis for hyperspectral remote sensing image classification. *IEEE Journal of Selected Topics in Applied Earth Observations and Remote Sensing* 7:2224–2236. https://doi.org/10.1109/JSTARS.2013.2279693.

Xiao, Y., G. Mo, X. Xiong, J. Pan, C. Wu, and W. Zhai. 2023. DR-XGBoost: An XGBoost model for field-road segmentation based on dual feature extraction and recursive feature elimination. *International Journal of Agricultural and Biological Engineering* 16:169–179. https://doi.org/10.25165/j.ijabe.20231603.8187.

Xie, J., and C. Wang. 2011. Using support vector machines with a novel hybrid feature selection method for diagnosis of erythemato-squamous diseases. *Expert Systems with Applications* 38:5809–5815. https://doi.org/https://doi.org/10.1016/j.eswa.2010.10.050.

Xie, S., Y. Zhang, D. Lv, X. Chen, and J. Liu. 2023. A new improved maximal relevance and minimal redundancy method based on feature subset. *The Journal of Supercomputing* 79:3157–3180. https://doi.org/10.1007/s11227-022-04763-2.

Xue, B., M. Zhang, and W. N. Browne. 2013. Particle swarm optimization for feature selection in classification: A multi-objective approach. *IEEE Transactions on Cybernetics* 43:1656–1671. https://doi.org/10.1109/TSMCB.2012.2227469.

Yadav, Y. and M. Pal. 2021. Effect of search methods on feature selection with hyperspectral data, In *Proceedings of 2021 IEEE International Geoscience and Remote Sensing Symposium,* Brussels, Belgium, pp. 3408–3411. https://doi.org/10.1109/IGARSS47720.2021.9553790.

Yamada, M., W. Jitkrittum, L. Sigal, E. P. Xing, and M. Sugiyama. 2014. High-dimensional feature selection by feature-wise kernelized lasso. *Neural Computation* 26:185–207. https://doi.org/10.1162/NECO_a_00537.

Yang, J., and V. Honavar. 1998. Feature subset selection using a genetic algorithm. In *Feature Extraction, Construction and Subset Selection: A Data Mining Perspective*, (eds.) H. Liu, and H. Motoda, pp. 117–136. Boston, MA: Springer. https://doi.org/10.1007/978-1-4615-5725-8_8.

Ye, Z., Y. Xu, Q. He, M. Wang, W. Bai, and H. Xiao. 2022. Feature selection based on adaptive particle swarm optimization with leadership learning. *Computational Intelligence and Neuroscience* 2022:1825341. https://doi.org/10.1155/2022/1825341.

Yin, J., Q. Feng, T. Liang, B. Meng, S. Yang, J. Gao, J. Ge, M. Hou, J. Liu, W. Wang, H. Yu, and B. Liu. 2019. Estimation of grassland height based on the random forest algorithm and remote sensing in the Tibetan Plateau. *IEEE Journal of Selected Topics in Applied Earth Observations and Remote Sensing* 13:178–186. https://doi.org/10.1109/JSTARS.2019.2954696.

Yu, L., and H. Liu. 2003. Feature selection for high-dimensional data: A fast correlation-based filter solution. In *Proceedings of 20th International Conference on Machine Learning (ICML'03),* August 21–24, Washington, DC, pp. 856–863.

Zhang, P., M. Gong, L. Su, and Z. Li. 2016. Change detection based on deep feature representation and mapping transformation for multi-spatial-resolution remote sensing images. *ISPRS Journal of Photogrammetry and Remote Sensing* 116:24–41. https://doi.org/10.1016/j.isprsjprs.2016.02.013.

Zhao, Z. A., and H. Liu. 2012. *Spectral Feature Selection for Data Mining.* Boca Raton, FL: CRC Press. https://doi.org/10.1201/b11426.

Zheng, H., J. Yuan, and L. Chen. 2017. Short-term load forecasting using EMD-LSTM neural networks with a XGBoost algorithm for feature importance evaluation. *Energies* 10:1168. https://doi.org/10.3390/en10081168.

Zhong, J., and R. Wang. 2006. Multi-temporal remote sensing change detection based on independent component analysis. *International Journal of Remote Sensing* 27:2055–2061. https://doi.org/10.1080/01431160500444756.

Zou, Q., L. Ni, T. Zhang, and Q. Wang. 2015. Deep learning based feature selection for remote sensing scene classification. *IEEE Geoscience and Remote Sensing Letters* 12:2321–2325. https://doi.org/10.1109/LGRS.2015.2475299.

4 Multisource Image Fusion and Classification

Earth-observing satellites currently in operation or planned for the near future carry sensors that operate in the visible, infrared, and microwave regions of the spectrum. They often provide two images for a particular location at a specific instant, called multimodal imaging: a multispectral (MS) image and a panchromatic (PAN) image. Multimodal remotely sensed imagery is now being generated at an unprecedented pace, with sources such as Landsat, MODIS, and Sentinel-2 contributing rich spatial and spectral information that significantly enhances land cover mapping. A crucial element in achieving successful classification is the effective use of abundant multimodal data. Moreover, the widespread availability of GIS means that digital spatial data have become more accessible. Hence, greater attention is now being paid to the use of multisource data in remote sensing image classification by employing image fusion techniques. Such data may consist of images produced by different sensor systems or digital spatial data (known as multimodal data). Image fusion, in this context, can be defined as combining two or more different images from the same or different sensors to construct a new image using a specific algorithm, aiming to retain the highest spectral and spatial information existing in different images. It is also an important step in the super-resolution reconstruction process.

In the field of remote sensing, the integration of lower spatial resolution MS images with higher resolution PAN images, retaining complementary information, is a common practice called pan-sharpening, which is applied to increase the spatial resolution of the MS image by considering the spatial information existing in the PAN image. Fusion of these images becomes feasible on the basis of the major assumption that the images taken on different wavelength bands at the same instance share common geometric information. In theory, fusion methods guarantee the preservation of all critical spatial and spectral information from the original images in the fused image. Furthermore, the fused image should effectively reduce irrelevant features and noise to the greatest extent possible. The choice of image fusion level, whether it is at the pixel, feature, or decision level, depends on the stage at which the fusion process takes place. As pointed out by Yilmaz et al. (2020), it is crucial to understand that there is no such thing as the ultimate "best image fusion method" that simultaneously excels in spectral and spatial fidelity. This recognition serves as a driving force for researchers to advance their methods, striving to achieve the optimal balance between transferring spatial details and maintaining color integrity.

In the historical development of image fusion methods, the first examples were based on component substitution (e.g., principal component analysis (PCA), intensity-hue-saturation (IHS), and the Brovey transforms). More advanced methods based on adaptive, multiresolution frameworks (e.g., wavelets, curvelets, and Laplacian pyramids), unmixing-based methods, and deep learning (DL)-based methods have been proposed to overcome the shortcomings of component substitution methods, such as spectral distortions occurring after the fusion process. State-of-the-art techniques are not without flaws. The resulting images can suffer from aliasing and local dissimilarities. In the last decade, data-driven DL models have dominated the image processing field, offering robust nonlinear representations, and networks processing spatial and spectral information have been suggested.

The assumption is often made that classification accuracy should improve if additional features are incorporated. Such features may be derived from the image data set itself, or from two or more co-registered image sets from different sensors, or from associated geographical information such

DOI: 10.1201/9781003439172-4

as surface elevation, soil type, or drainage pattern. This assumption typically holds true, as a higher inclusion of relevant information in classification tends to decrease the likelihood of interclass confusion. Consequently, it is anticipated that the significance of developing multisource classification methodologies will continue to grow. It should, of course, also be remembered that the use of highly correlated features or features that show no variation between the classes of interest can obscure rather than illuminate the problem (see Chapter 3). One should always bear in mind that weighting is generally applied to input features. If features are standardized to a 0–1 range, for example, variation in feature a is considered equivalent to variation in feature b. Noting that variation can be considered equivalent to information, it is clearly not helpful to include surface elevation as a feature if it varies only slightly over the region of interest and if these small variations are not related to the spatial boundaries of the classes.

The development of remote sensing systems involves careful consideration of several competing constraints, and a significant trade-off exists between IFOV and the signal-to-noise ratio (SNR). In the case of MS and particularly hyperspectral sensors, their narrower spectral ranges compared with those of PAN sensors usually result in lower spatial resolution for a given IFOV. This trade-off allows them to capture more photons and maintain a favorable image SNR. Many sensors, including Landsat ETM+ and OLI, SPOT NAOMI, WorldView-3, and QuickBird, have MS bands with a co-registered higher spatial resolution PAN band. As this type of imagery is the most common one, sharpening the low spatial resolution MS bands with the use of PAN bands has been the major image fusion task, called pan-sharpening, in remote sensing. These techniques increase spatial resolution while simultaneously conserving the spectral data in the MS image, providing the optimal combination of high spectral accuracy and enhanced spatial detail. With the pan-sharpening process, geometric correction can be improved, change detection can be performed with higher accuracy, objects that cannot be identified in coarser resolution can be determined, and, more importantly, the accuracy of image classification can be significantly improved.

A further consideration is the varying reliability and completeness of the different data sources. Therefore, it is necessary to consider the reliability or uncertainty of each data source when the classification of multisource data is attempted. Source reliability factors are important parameters that determine how strongly a given source contributes to the multisource consensus pool. In this sense, the reliability factors are equivalent to the source weighting parameters. If these weighting parameters are not chosen properly, the multisource consensus will give disappointing results though it is based on a theoretically robust mechanism.

In this chapter, the principal approaches and state-of-the-art methods for dealing with multisource classification are addressed. Before introducing classification schemes related to multisource data, techniques for image fusion, particularly current popular DL models, are described. Image fusion is a formal framework for combining images coming from different sources. Several classification methods based on the stacked-layer procedure, the incorporation of topographic data, the extension of Bayesian theory, and evidential reasoning are discussed. We also introduce possible options for source weighting factor assignment, and the experimental results are presented in the final section. In addition to the techniques given in this chapter, numerous extensions of the major methods and some other novel methods exist, including multiscale Kalman filtering (Simone et al., 2002), curvelet transfer and independent component analysis (Nencini et al., 2007; Ghahremani and Ghassemian, 2015), shearlets (Miao et al., 2011), contourlet transform (Li et al., 2017).

4.1 IMAGE FUSION

Image fusion is the process of combining images captured by sensors operating at different wavelengths, simultaneously observing the same scene. The objective is to form a composite image with enriched content, facilitating the user in the detection, recognition, and identification of targets

(Wald, 1999). Concerning spatial and spectral resolution, high spectral resolution is generally advantageous for distinguishing land use and cover types, while high spatial resolution is beneficial for the identification of terrain features and structures. As a result, users may decide to combine images sourced from different origins to capture complementary spectral and spatial features, thereby enhancing the overall content of the fused image. This, in turn, may lead to potential improvements in classification accuracy. Because high-resolution MS images are not available to assess fusion outcomes, the remote sensing literature has established protocols for this purpose. These protocols serve the purpose of appraising the fidelity of the spectral and spatial data within the fused image. Wald's protocol, which is the most popular one, suggests a resolution downgrading process (i.e., degradation to a coarser resolution) to input images for assuming that the performance of the image fusion is not dependent on the spatial resolution of the fused images. Another protocol is Zhou's protocol, which is applied at full resolution of input images without any degradation.

Within the framework of conventional image fusion, the process typically comprises three successive steps. In the spatial domain, this can be accomplished using region- or block-based approaches, whereas in the transform domain, it entails the use of pyramid-, wavelet-, or filtering-based methods. Subsequently, activity levels are quantified, and weight maps are formulated. To measure activity levels, various techniques such as focus measurement, edge detection, and saliency map extraction can be employed. When generating weight maps, the choice of a suitable algorithm is critical and depends on the specific application. Notable examples of weight map calculation methods include normalization, averaging, and operator-based methods, such as max and min. Finally, the fused image is reconstructed by integrating the pertinent information acquired either in the spatial or in the transform domain through the inverse transform. The creation of weight maps and evaluation of activity levels pose significant challenges in image fusion. Neglecting to align the weight map construction process with the requirements of the application can result in a significant deterioration of visual quality in the fused image (Xiao et al., 2020).

Research on multimodal fusion has been actively explored in remote sensing, with a notable focus on the merging of optical and synthetic aperture radar (SAR) imagery, which represents the most commonly explored modalities for image fusion. Image fusion techniques can be classified into two broad categories: transform domain fusion and spatial domain fusion. In fusion methods using transform domains, the initial step is to transform the input images. Subsequently, fusion takes place, and the outcome is reverted to its original form through an inverse transformation. Based on the stage at which the fusion operation takes place, image fusion techniques are divided into three levels: pixel level, feature level, and decision level, which is the most common taxonomy of image fusion (Figure 4.1). In pixel-level fusion, the content of individual images is merged pixel by pixel and followed by an information extraction step. The information obtained from multiple feature sets is combined to form a final decision directly. Various fusion algorithms, ranging from basic weighted averaging to more sophisticated multiscale methods, have been proposed for pixel-level fusion, such as the Brovey-, PCA-, and IHS-based methods. Iterative techniques are applied to compute weights for each input image and every pixel, with the goal of optimizing a cost function. In feature-level fusion, information is independently extracted from each source image and subsequently merged. In decision-level fusion, data are separately extracted from each input, and a determination is made based on specific criteria, such as distinguishing between man-made and natural features to select the content for fusion from each input channel. Each level involves a distinct collection of image processing procedures conducted within various domains and employing diverse mathematical operations. Decision-level fusion, managed through a multiclassifier combination, has become a prominent and highly active research area in the field of pattern recognition. Image fusion methods have also been classified into three categories as component substitution, multiresolution analysis, and variational optimization (Javan et al., 2021). According to the literature, the

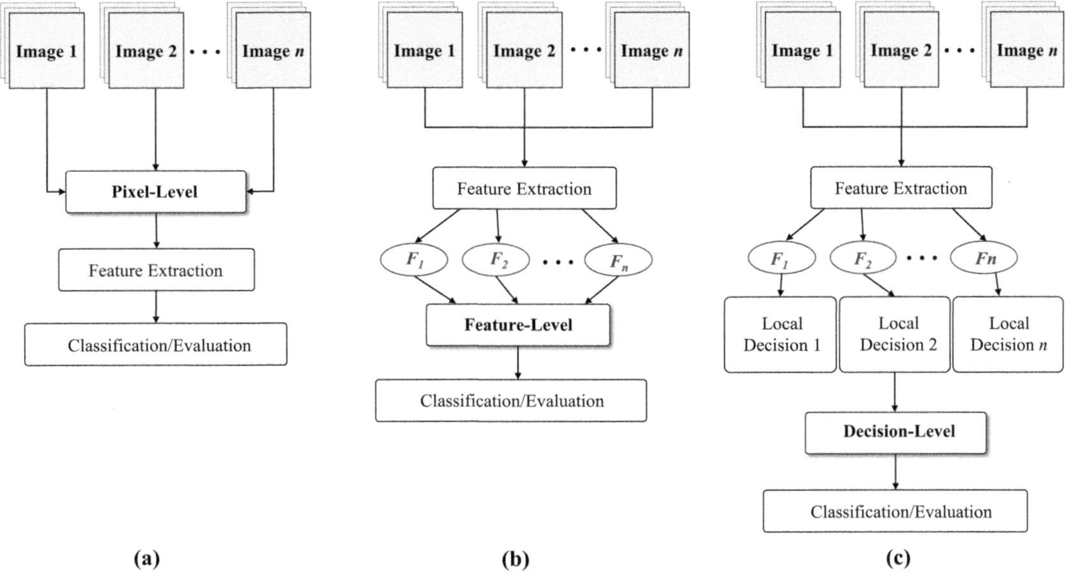

FIGURE 4.1 Taxonomy of image fusion techniques. (a) Pixel level, (b) feature level, and (c) decision level.

most commonly implemented techniques for image fusion are based on wavelets, Gram-Schmidt transformation, PCA, IHS transform, and Brovey's method. However, many methods have been proposed in recent years for image fusion by employing DL strategies. The reader should note that before performing fusion activities, the images have to be preprocessed to obtain consistent image size and geolocation. For instance, in the case of fusing Landsat Operational Land Imager (OLI) with PAN imagery, the OLI imagery is resampled at the PAN image size, and then, the two image sets are co-registered before the fusion algorithm is used.

4.1.1 IMAGE FUSION METHODS

4.1.1.1 PCA-Based Image Fusion

The mathematical theory of PCA is given in Section 3.1.1. Although the PCA method has been chiefly applied for feature extraction practices, it can be also employed in image fusion. In the case of PCA-based fusion, the low-resolution MS imagery is first subjected to PCA to obtain the first principal component (PC1) with maximum variance (the one representing the most information in the data set). PC1 is then replaced by the high-resolution image (a panchromatic image, for instance), which is stretched to give it the same variance and average as PC1 (Figure 4.2). In other words, before the first principal component is replaced by the PAN image, a histogram matching is applied to the PAN image. Band-specific information is attributed to the remaining principal components (eigenvectors), and these components are left in their original state. This combination was then transformed back into the RGB (red, green, blue) domain by inverse PCA transformation to obtain a high-resolution pan-sharpened image. The resulting image is subjected to a rescaling procedure (e.g., 0–255) to obtain the final fused output.

4.1.1.2 IHS-Based Image Fusion

To transfer a color image from an RGB color cube, the IHS transformation, which is a hexcone color model and also called the HSI model, has been a standard method, especially for image

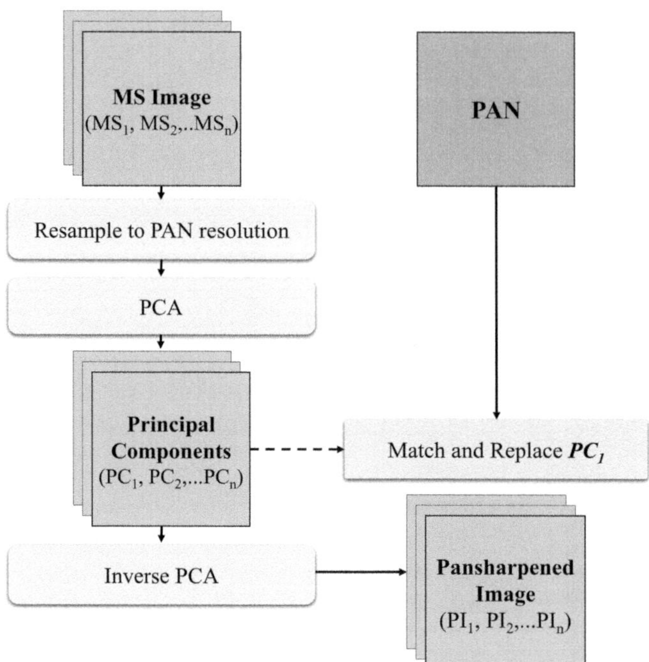

FIGURE 4.2 PCA-based image fusion process.

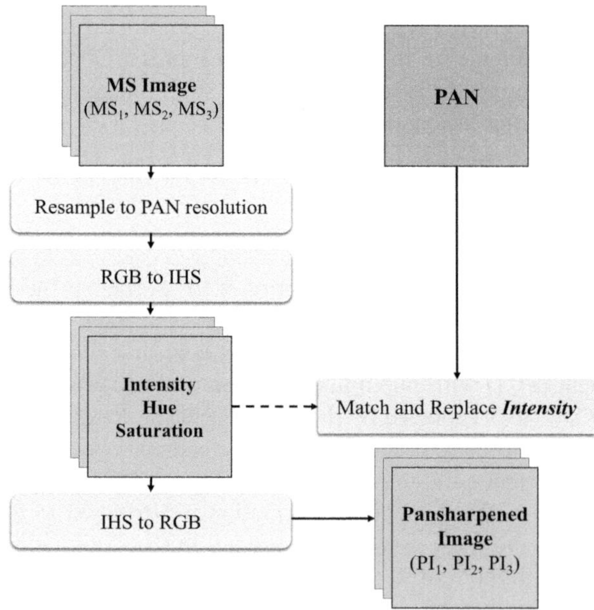

FIGURE 4.3 IHS-based image fusion process.

enhancement and fusion. The workflow of the IHS image fusion process is shown in Figure 4.3. It is relatively simple and computationally efficient when used in sharpening processes. "Intensity" refers to the total brightness of a color, "hue" represents the dominant wavelength of color as we see, and "saturation" is the degree of the purity of color. The values of IHS can be estimated from the RGB values using the below equations (Sabins, 1997):

$$I = R + G + B$$

$$H = \frac{G - B}{I - 3B}$$

$$S = \frac{I - 3B}{I}$$

(4.1)

Alternative representations of the IHS transform are given by Pellemans et al. (1993), Schetselaar (1998), Plataniotis and Venetsanopoulos (2000). It is obvious that the suggestion of different formulations for IHS estimations causes confusion from the user's perspective.

The IHS-based fusion technique initiates by co-registering and resampling low-resolution MS images and PAN images to align with the resolution of the PAN imagery. Subsequently, the low-resolution MS imagery undergoes an IHS transformation, resulting in the extraction of three components from the original image. The next step involves histogram matching the PAN image to the intensity component of the IHS transform, addressing spectral variations resulting from diverse sensors, acquisition dates, or incident angles between the two images. Thus, the global statistics of the two images become similar to maintain the spectral attributes of the original MS data. Afterward, the altered PAN image is substituted for the intensity band, and the information is reverted to its original MS band domain. In other words, the intensity component of the image is then replaced by the high-resolution image channel, which is previously contrast-stretched (in terms of adjusting mean and variance) to match the original intensity component. A reverse IHS transformation is then applied to obtain the final fused image. Note that the standard IHS method is limited to three bands of MS imagery. When more than three bands are available, a feasible approach is to create a generalized IHS transform that incorporates the influence of the near-infrared (NIR) band into the intensity component. In this scenario, the intensity is computed by assigning specific coefficients to each band. The decision regarding these coefficients can be based on the spectral responses of the PAN and MS bands, considering the spectral characteristics of the sensors. On the contrary, Aiazzi et al. (2007) proposed an enhanced strategy for component substitution approaches, including the IHS transformation for image fusion in which a linear regression is performed between PAN and MS bands and "a synthetic intensity" with a minimum mean square error with respect to the reduced PAN is estimated.

4.1.1.3 Brovey Transform

The Brovey transformation is a color normalization method as it involves operations on RGB colors. It represents a simple technique for fusing data from different sensors. This method combines mathematical operations with spectral band normalization, which occurs prior to their multiplication with the PAN image. It retains the spectral properties of each pixel and reconfigures all luminance information into a PAN image of superior resolution. In the case of image fusion using the Brovey transform, the low-resolution imagery is normalized by dividing all individual digital numbers (DN) of the selected band with the sum of the corresponding DN values in all bands and multiplied by the high-resolution band, to incorporate the intensity or spatial component into the output. The mathematical expression of this process for three-band RGB imagery is perhaps clearer:

$$R_{\text{new}} = \frac{R_0 \times P}{(R_0 + G_0 + B_0) / 3}$$

$$G_{\text{new}} = \frac{G_0 \times P}{(R_0 + G_0 + B_0) / 3}$$

(4.2)

$$B_{\text{new}} = \frac{B_0 \times P}{(R_0 + G_0 + B_0) / 3}$$

where R_0, G_0, and B_0, are, respectively, the original DN value of the selected bands in the RGB domain, and P is a higher spatial resolution image for intensity modulation. Despite its simple mathematical theory, it has been a popular image fusion algorithm. For instance, Zhang and Wu (2008) compared the performances of the Brovey transform and wavelet transform in merging SPOT-5 images. Experimental results revealed that the images merged by the Brovey transform showed higher spatial resolution and better spectral features than the original SPOT-5 imagery, and also retained more spectral information compared to wavelet transform. However, there are some studies against this view, particularly in classification tasks (e.g., Hill et al., 1999; Colditz et al., 2006).

4.1.1.4 Gram-Schmidt Transform

The Gram-Schmidt (GS) image fusion approach, which is the most widely used component substitution-based method, relies on the GS orthogonalization algorithm, which was adapted by Laben and Bernard (2000) for the purpose of image fusion. In cases where there is a perfect correlation among the input bands, the GS orthogonalization process would yield a final band containing only zeros. In more realistic scenarios with a high correlation between bands, this process generates a final band with very small values. The GS transform takes nonorthogonal vectors and rotates them to make them orthogonal, similar to PCA. In the context of images, each band (panchromatic, RGB, and infrared) corresponds to a high-dimensional vector with dimensions equal to the number of pixels. The initial step of GS-based image fusion involves creating a low-resolution PAN band by calculating a weighted average of the MS bands, which is then utilized as input for the original MS-to-GS transformation.

$$\text{GS1}(i,j) = \text{SimulatedPan}(i,j) \tag{4.3}$$

where i is the row number and j is the column number, and GS1 is the first GS transform band. In the calculations, the Tth GS band is estimated from the previous $(T-1)$ GS bands as follows (Laben and Bernard, 2000):

$$\text{GST}(i,j) = \left(B_T(i,j) - \mu_T\right) - \sum_{l-1}^{T-1} \Phi(B_T, \text{GS}_l) * \text{GS}_l(i,j) \tag{4.4}$$

where T is the band number being transformed, B is the original band, and μ is the mean of band T given by

$$\mu_T = \frac{\sum_{j=1}^{C}\sum_{i=1}^{R} B_T(i,j)}{C * R} \tag{4.5}$$

where C is the total number of columns in the image, R is the total number of rows in the image, and $\Phi(B_T, \text{GS}_l)$ is given by

$$\Phi(B_T, \text{GS}_l) = \left[\frac{\sigma(B_T, \text{GS}_l)}{\sigma(B_T, \text{GS}_l)^2}\right] \tag{4.6}$$

With this equation, the covariance can be obtained between the given GS band and the original band B, divided by the variance of that GS band. The mean (μ) and standard deviation (σ) of the first GS band are then calculated, and the mean and standard deviation of the high-resolution pan band are also calculated from

$$\sigma_T = \sqrt{\frac{\sum_{j=1}^{C}\sum_{i=1}^{R}\left(B_T(i,j) - \mu_T\right)}{C * R}} \tag{4.7}$$

The PAN image with higher resolution (P) is then subjected to stretching to align its mean digital count (μP) and standard deviation with those of the initial GS band, aiming to preserve the spectral characteristics of the original MS data. As a result of this stretching, the high-resolution PAN band and the first GS band, essentially representing a simulated low-resolution PAN band, exhibit remarkable similarity in both global and local statistical properties. Subsequently, the stretched high-resolution PAN image replaces the first GS band, and the data are retransformed into the original MS band space, yielding $N+1$ MS bands at higher resolution.

In the original form, the GS method uses averaging of the MS components (mode 1) or a low-pass filter (mode 2) to produce the synthetic PAN image at low resolution. It should be noted that the strategy introduced by Aiazzi et al. (2007), mentioned above, has been used as an enhanced GS method. In this strategy, regression optimization is implemented as a preprocessing block. To improve the method further, the intensity component in GS estimations has been estimated using different strategies, including the minimization of the weighted average of the mean square error, using spectral sensitivity curves, linear regression, least squares methods, and genetic algorithm (Maurer, 2013; Pohl and van Genderen, 2016; Yilmaz et al., 2020).

4.1.1.5 Wavelet Transform

It is recognized that most images, especially in the field of remote sensing, contain nonstationary signals (i.e., whose frequency response varies in time or over space). Thus, the Fourier transform may not be suitable for nonstationary signals since the transform is only capable of providing frequency information rather than a time- or space-frequency representation. An alternative way of dealing with such an issue can use the wavelet transform since this is capable of providing the time/space and frequency information simultaneously (i.e., giving a time- or space-frequency representation of the signal). This is the major advantage that the wavelet transform has in comparison with the Fourier transform. Wavelet theory has evolved into a robust and rapidly advancing mathematical basis for a specific group of multiscale representations. Several sophisticated image fusion strategies, employing wavelet transforms, have been introduced and studied in the literature. However, it should be mentioned that the wavelet transform can also be used for a wide range of applications, including denoising, compression, and image classification.

The image is decomposed in the continuous wavelet transform using a set of elementary functions called wavelets, which are created through translations and dilations of a kernel function ψ, identified as the mother wavelet. The wavelet transform displays the multiresolution characteristics of a signal or image by correlating it with different scales and shifts of a mother wavelet function. Formally, we define $a, b \in R$, as the dilation step and translation step of the translation and dilation processes, respectively. The base is then given by

$$\psi_{a,b} = D_a T_b \, \psi \tag{4.8}$$

where $T_b \psi(x) = \psi(x-b)$ and $D_a \psi(x) = |a|^{-1/2} \, \psi(x/a)$ are the translation and dilation operators, respectively (Mallat, 1989; Daubechies, 1992). There are numerous kinds of mother wavelets. However, for $\psi(x)$ to be characterized as a mother wavelet, the following admissibility condition must hold

$$\int_{-\infty}^{\infty} \frac{|\hat{\psi}(\omega)|^2}{|\omega|} d\omega < \infty \tag{4.9}$$

where $\hat{\psi}$ is the Fourier transform of ψ. Once the wavelet has been defined, the continuous wavelet transform of a function $f(x)$ can be generated as

$$W_{\psi_{a,b}} f(x) = \frac{1}{\sqrt{a}} \int_{-\infty}^{\infty} f(x) \bar{\psi}\left(\frac{x-b}{a}\right) dx \tag{4.10}$$

where $\bar{\psi}$ denotes the complex conjugate of ψ (Daubechies, 1992). In practice, when the data being processed are restrained to a discrete image, ψ in a discrete form rather than a continuous wavelet is applied. Discrete wavelet analysis also holds advantages that overcome the wavelet redundancy and computational issues encountered by the continuous wavelet (Burrus et al., 1998). In the case of processing an image, the wavelet transform is generally performed in a multiresolution sense, which requires a scaling function $\varphi(x)$ and the associated mother wavelet $\psi(x)$ in the construction of a complete basis.

The computation is based on the dyadic sequence (2^r), and the so-called two-scale relation (Sheng, 1996) holds for $\varphi(x)$ and $\psi(x)$ as

$$\varphi_{r,k}\left(2^r x\right) = \sum_k h_{r+1}(k)\varphi_{r+1,k}\left(2^{r+1}x - k\right) \tag{4.11}$$

$$\psi_{r,k}\left(2^r x\right) = \sum_k g_{r+1}(k)\varphi_{r+1,k}\left(2^{r+1}x - k\right) \tag{4.12}$$

where $h(k)$ and $g(k)$ are low-pass and high-pass filters, respectively. The above relation shows that the scaling and wavelet functions at a certain scale can be expressed in terms of translated scaling and wavelet functions at the next scale. The two filters are related to each other and are known as a quadrature mirror filter (Mallat and Zhong, 1992). A signal $f(x)$ can be expressed in terms of basis functions

$$f(x) = \sum_k \lambda_r(k)\varphi_{r,k}\left(2^r x - k\right) \tag{4.13}$$

Or, in turn,

$$f(x) = \sum_k \lambda_{r+1}(k)\varphi_{r+1,k}\left(2^{r+1}x - k\right) + \sum_k \delta_{r+1,k}\psi_{r+1,k}\left(2^{r+1}x - k\right) \tag{4.14}$$

where λ and δ are coefficients in the low-pass and high-pass domains, respectively. As the scaling function $\varphi(x)$ and the associated mother wavelet $\psi(x)$ are orthogonal, one can compute the coefficients λ and δ by taking the inner products:

$$\lambda_r(k) = f(x), \varphi_{r,k}(x) \tag{4.15}$$

$$\delta_r(k) = f(x), \psi_{r,k}(x) \tag{4.16}$$

and one can finally obtain the following equations:

$$\lambda_{r+1}(k) = \sum_n \lambda_r(n) h(n - 2k) \tag{4.17}$$

$$\delta_{r+1}(k) = \sum_n \lambda_r(n) g(n - 2k). \tag{4.18}$$

The above formula can be regarded as a downsampling (by 2) convolution. The discrete wavelet transform for two-dimensional imagery is similar in that the transform is first performed along the row direction and then in the column direction. Figure 4.4 shows the process of the wavelet transform for a two-dimensional image. The process is first applied to each row to create an intermediate matrix of row coefficients. The intermediate matrix is then subject to a wavelet transform, this time

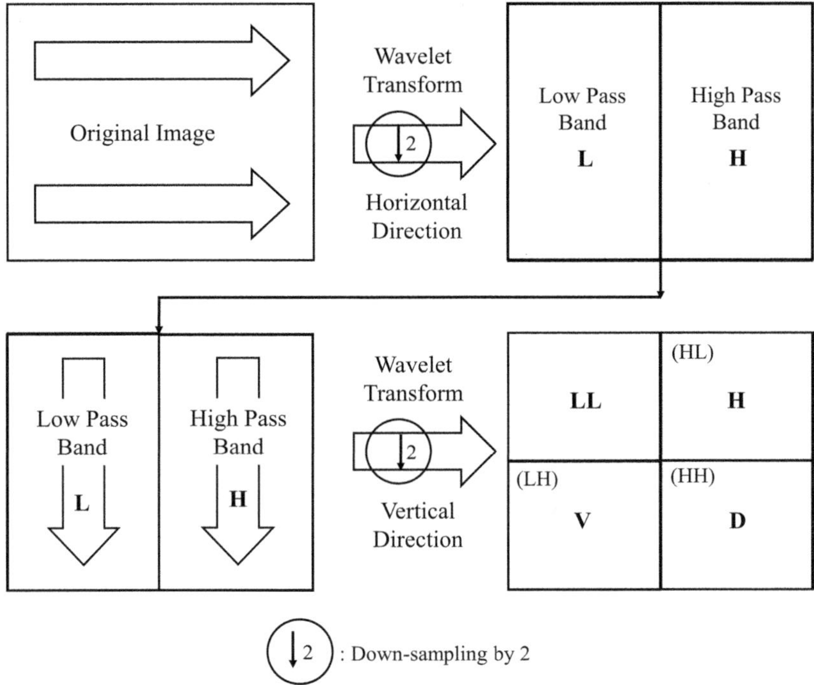

FIGURE 4.4 Example of wavelet transform for two-dimensional imagery.

in a column-wise sense. After the two-dimensional wavelet transform, the resulting image is made of four parts in which HL, LH, and HH bands are all high-frequency wavelet textures, and the LL band is the downsized (1/4) version of the original image. The terms HL, LH, HH, and LL can be interpreted by recalling that the frequency represented by the row wise transform increases from left to right, and for the vertical transform, from top to bottom of the image. The upper left quadrant, after one level of decomposition, is the LL image. The upper right quadrant represents the region in which low-frequency horizontal coefficients are combined with high-frequency vertical coefficients, while the lower left quadrant has high-frequency vertical coefficients and low-frequency horizontal coefficients. The bottom left quadrant combines high-frequency horizontal and vertical coefficients.

Research efforts in remote sensing have extensively explored image fusion methods employing multiresolution analysis based on the discrete wavelet transform (Núñez et al., 1999; Ulfarsson et al., 2003; Gungor and Shan, 2004; Chen et al., 2006; Liu et al., 2020a; Wang et al., 2021). In particular, both Mallat's (Mallat and Zhong, 1992) and the so-called *à trous* algorithms are the most popular approaches utilized for pan-sharpening operations (González-Audícana et al., 2005). Mallat's algorithm is an orthogonal, dyadic, nonsymmetric, decimated, nonredundant algorithm, while the *à trous* algorithm is nonorthogonal, shift-invariant, dyadic, symmetric, undecimated, and redundant in nature. According to Mallat's algorithm (Mallat and Zhong, 1992), after a wavelet transform, the resulting image is made of four parts in which HL, LH, and HH bands are all high-frequency wavelet textures, and the LL band is the downsized (1/4) version of the original image. However, in the case of the *à trous* algorithm, the image decomposition scheme uses a parallelepiped style. In each level of the *à trous* transform, the resulting image has a coarser spatial resolution by the same number of pixels as the original image (i.e., the image size always remains the same with only resolution degraded). Specifically, assume an image is of resolution 2^j, using the *à trous* transform at the nth level, the resulting image will be 2^{j+n} in resolution. For instance, if the original image resolution is of $2^2 (= 4)$ m, then after one pass with the *à trous* transform, the resolution will become

2^{2+1} (= 8) m. The *à trous* image is computed using scaling functions. The spatial detail lost between transformed level 2^{j+n} and 2^{j+n+1} is collected in just one wavelet coefficient image denoted as w^{j+n+1}, which involves horizontal, vertical, and diagonal spatial details at resolution 2^{j+n+1}. For the practical implementation of the *à trous* algorithm for an image, two-dimensional convolution masks associated with the scaling functions are used. The following are the two most popular scaling functions, namely, a 3×3 linear interpolation function

$$\begin{bmatrix} 1/16 & 1/8 & 1/16 \\ 1/8 & 1/4 & 1/8 \\ 1/16 & 1/8 & 1/16 \end{bmatrix} \tag{4.19}$$

and a B3-spline function

$$\begin{bmatrix} 1/256 & 1/64 & 3/128 & 1/64 & 1/256 \\ 1/64 & 1/16 & 3/32 & 1/16 & 1/64 \\ 3/128 & 3/32 & 9/64 & 3/32 & 3/128 \\ 1/64 & 1/16 & 3/32 & 1/16 & 1/64 \\ 1/256 & 1/64 & 3/128 & 1/64 & 1/256 \end{bmatrix} \tag{4.20}$$

In the case of fusion of MS and PAN images, the process of integrating the spatial detail of the PAN image into the MS images can be done in two steps. In the first step, a wavelet transform is applied to both the MS and PAN images to separate the images into a degraded image and wavelet coefficient images (HH, HL, and LH, for instance). In the second step, the wavelet coefficients of the PAN image are substituted for those of the MS image. An inverse wavelet transform is then performed.

It should be noted that either substitutive or additive methods can be applied to perform image fusion. The additive method is implemented by adding the PAN wavelet coefficients to the wavelet coefficients of the MS image bands rather than substituting one image band for another. Both methods may contribute different fusion qualities. González-Audícana et al. (2005) show that if Mallat's (Mallat and Zhong, 1992) algorithm is applied, the quality of the fused images via the substitutive method is similar to that of fused images obtained by the additive approach. However, if the *à trous* algorithm is implemented, then the image quality obtained by the substitutive method is less ideal than that of fused images generated by the additive approach.

Figure 4.5 shows an example of the use of Mallat's wavelet transform to perform image fusion. Once both high-resolution images (e.g., panchromatic, denoted **P** in Figure 4.5) and low-resolution images (e.g., MS, denoted **S** in Figure 4.5) have been subjected to the wavelet transform, the wavelet coefficients (WP1, WP2, and WP3) of the high-resolution image are combined with feature **T**. An inverse wavelet transform is then performed, where feature **T** can be an approximation image **S1** of original low-resolution image **S** (if the substitution method is adopted) or $\mathbf{T} = \mathbf{S1} + \mathbf{P1}$ (if the additive method is used).

The wavelet-based image fusion method can also be incorporated with the PCA and IHS methods introduced above. When an MS image has been transformed into the IHS domain, a wavelet transform is then applied to both the intensity (I) image and PAN image, respectively. The wavelet coefficients of the PAN image are then substituted (in either a substitutive or an additive way) into the intensity image in place of the wavelet coefficients for the intensity image, and an inverse IHS transform is carried out. Where transformation into the PCA domain is used, the fusion step is similar to that just described for the IHS method except that the wavelet coefficients of the PAN image are substituted for the first principal component (i.e., PC1) instead of the intensity image.

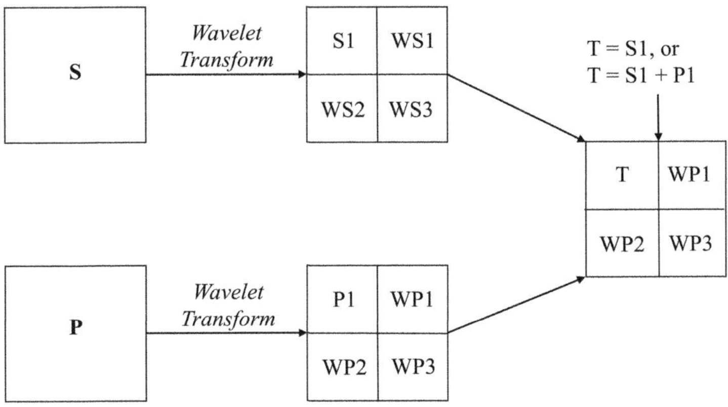

FIGURE 4.5 Image fusion in terms of the wavelet transform using Mallat's algorithm (see text for details).

While the wavelet transform is occasionally seen as a "decorrelator", aiming to render each wavelet coefficient statistically independent, it becomes evident that there are significant interdependencies among these coefficients. These dependencies are characterized by statistical attributes known as clustering and persistence. Persistence implies that wavelet coefficients can maintain strong dependencies across scale, while clustering and locality imply that coefficients can exhibit significant dependencies within scale. As a result, wavelet-domain hidden Markov models modeling the statistical dependencies and non-Gaussian statistics existing in signals have been effectively applied in recent studies (e.g., Smyth et al., 1997; Crouse et al., 1998; Xu et al., 2011; Luo et al., 2017). El-Samie et al. (2012) also describe the integrated IHS for discrete wavelet transform and discrete wavelet frame transform.

4.1.1.6 Deep Learning for Image Fusion

Deep learning (DL), including more layers and sophisticated learning schemes, is an advanced concept of artificial neural networks (ANNs). Recent significant progress in computational capabilities, particularly GPU power, has enabled the training and processing of deep neural networks. In ANN models, essential features are manually extracted from the data, followed by training of a three-layer ANN to determine appropriate weights. Conversely, in the DL-based models, the deep network itself is capable of handling both feature extraction and weight map construction tasks. Consequently, DL-based methods can simplify and automate the intricate process of feature extraction by identifying patterns within the input data. DL-based methods may be divided into four main categories as CNN-based methods, GAN-based methods, Siamese network-based methods, and autoencoder methods (Sun et al., 2020). DL-based models, particularly convolutional neural networks (CNNs), have received great attention from the scientific community, mainly because of their robustness in image processing and image classification tasks in an end-to-end fashion. Several key factors play a central role in this advancement, including the following: (1) the efficient utilization of state-of-the-art GPUs for training; (2) the introduction of fast and efficient training algorithms, batch normalization, residual learning, and the rectified linear unit (ReLU), which expedites convergence while maintaining quality; (3) the ready accessibility of abundant data sets such as ImageNet for training larger models; and (4) the unique nature of the structure enabling high-level patch and feature extraction capabilities. The theories of ANNs and DL models are discussed in Chapter 7. In recent years, many DL models have been employed for supervised and unsupervised image fusion applications. A universal DL-based image fusion framework that encompasses both spatial and transform domains is illustrated in Figure 4.6.

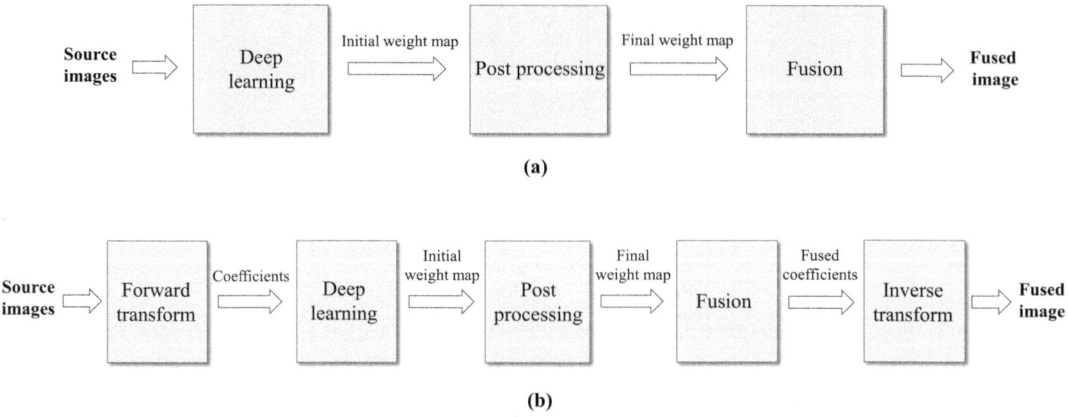

FIGURE 4.6 Deep learning-based image fusion. (a) Spatial domain, (b) transform domain.

In accordance with Figure 4.6, DL-based image fusion in the spatial domain is conducted by the following steps (Xiao et al., 2020):

a. Weight maps are initially derived from the source images through the utilization of either supervised or unsupervised DL models.
b. Utilizing techniques such as thresholding, segmentation, and consistency verification, the final weight maps are calculated.
c. Source images are then combined using final weight maps.

Steps for DL-based image fusion in the transform domain are given below (Xiao et al., 2020):

a. Implement a forward transformation on the source images.
b. Compute initial weight maps by applying either supervised or unsupervised DL techniques.
c. Calculate improved weight maps by applying different methods such as thresholding, segmentation, and consistency verification to eliminate unnecessary artifacts.
d. Integrate decomposed coefficients across multiple scales using weight maps.
e. Reconstruct the fused image using the inverse transform process.

In parallel to other research fields in remote sensing, DL models have been effectively used in image fusion practices, mainly considering the hierarchical features of source image patches. Building on the neural network architecture proposed by Dong et al. (2016) for image super-resolution (SRCNN), a novel study conducted by Masi et al. (2016) applied a three-layer CNN that directly learns an end-to-end mapping between low- and high-resolution images. The inputs by including radiometric indices tailored to features that proved very relevant for applications are augmented. FusionCNN, specifically designed for the fusion of remote sensing images employing CNNs, was proposed by Ye et al. (2019). To attain better fusion quality, low-frequency information of MS is used to improve the PAN image, which is denoted as EPAN. The MS and EPAN images are then input into the trained FusionCNN to produce the fused images.

The fusion of spatiotemporal remote sensing images shows great potential for acquiring remote sensing data with superior spatial and temporal resolutions. Managing the complexities of gradual and abrupt changes in land surface reflectance has been a key challenge in the fusion process. Cutting-edge DL techniques exhibit remarkable proficiency in understanding time-dependent variations associated with land use and land cover activities. Song et al. (2018) suggested a novel spatiotemporal fusion method based on two five-layer CNNs to address the problems of complicated correspondence and large spatial resolution gaps between MODIS and Landsat images. The CNN model is specialized in learning a nonlinear mapping CNN that links MODIS and low-resolution Landsat images, and it further comprehends a super-resolution CNN connecting low-resolution Landsat to original Landsat images. During the prediction phase, a fusion model is employed, incorporating high-pass modulation and a weighting strategy to optimize the utilization of information from previous images, rather than directly adopting the CNN outputs as the fusion result. Chen et al. (2022) introduced a novel spatiotemporal image fusion technique employing multiscale two-stream convolutional neural networks (STFMCNNs). The approach is designed to capture objects of varying sizes through feature learning, utilizing a coarse spatial resolution image and two pairs of coarse and fine spatial resolution (FR) images acquired at different dates. The STFMCNN model may be regarded as a deeper version of SRCNN-based methods. Additionally, a strategy for local fusion was developed to capture local variations by combining two predicted images obtained from each stream. The effectiveness of the proposed model was evaluated on two data sets, revealing promising results in terms of visual and quantitative assessments.

Li et al. (2022) proposed the domain knowledge–guided deep collaborative fusion network (DKDFN), a new DL model that incorporates multimodal data fusion along with prior knowledge, enabling cooperative classification of imbalanced land cover categories. Within this network, a multihead encoder and a multibranch decoder are implemented. The encoder design aims to increase the likelihood of capturing complementary data from multiple modalities. The multibranch decoder operates within a multitask learning framework, conducting land cover classification, semantic segmentation, and reconstruction of remote sensing indices from multiple modes, selected as domain knowledge representatives. Comparisons of different data sets validated the effectiveness of the DKDFN model against six well-known models, namely, U-Net, SegNet, PSPNet, DeepLab, HRNet, and MP-ResNet. Ma et al. (2022) also proposed a dynamic deep network architecture called AMM-FuseNet, which uses multimodal remote sensing images for land cover mapping. In the proposed network, a hybrid strategy incorporates both the channel attention mechanism and densely connected atrous spatial pyramid pooling (DenseASPP). The effectiveness of the AMM-FuseNet model was compared to those of DeepLabV3+, PSPNet, U-Net, SegNet, DenseASPP, and DANet using three data sets. The robustness of the model was particularly tested on cases with limited training samples. Hosseinpour et al. (2022) also introduced a CNN model called the cross-modal gated fusion network (CMGFNet), a comprehensive approach for extracting building footprints from very high-resolution (VHR) remote sensing images and digital surface models (DSMs). CMGFNet extracts features from both RGB and DSM data using dedicated encoders and offers two feature fusion methods: cross-modal fusion and multilevel feature fusion. For cross-modal fusion, a gated fusion module is introduced to efficiently combine information from both modalities. Multilevel feature fusion, on the other hand, integrates deep-layer high-level features with shallower low-level features using a top-down strategy. In addition, we introduce a residual-like depth-wise separable convolution to enhance upsampling performance, reduce parameters, and decrease time complexity in the decoder phase. Extensive experiments on challenging data sets demonstrate the superiority of the CMGFNet over other state-of-the-art models. In a recent publication, Nagarathinam et al. (2023) proposed an innovative approach to remote sensing image fusion that addresses the limitations of existing methods. They devised effective fusion rules as part of their model. The process begins with the use of the low-frequency components from the MS image to enhance the PAN image through an optimized wavelet transform. Subsequently, these improved PAN and MS images are input into the Improved Deep Convolutional Neural Network with Atrous Convolutions (IDCNN-AC) to generate high-quality fused images. To optimize the IDCNN-AC parameters and enhance the fusion performance, the researchers employed Hybrid Harris Hawks Dingo Optimization (HHHDO). The simulation results demonstrate the effectiveness of the proposed image fusion model, as measured by various quantitative metrics.

In recent years, self-attention learning has been introduced within the CNN framework to promote important modality-specific features and, at the same time, to suppress irrelevant information. Nonetheless, CNN-based fusion networks, in conjunction with self-attention, do not effectively interact with cross-modal feature extractors, resulting in the omission of essential shared high-level features across all modalities. The concept of cross-attention is conceived as a solution in which feature extractors from various modalities have the potential to mutually impact one another. Moreover, self-attention learning techniques are based on localized convolution operations. To overcome these issues, Global Attention-based Fusion Network (GAF-Net), equipped with novel self- and cross-attention learning techniques for land cover classification from bimodal remote sensing images, was suggested by Jha et al. (2023). The within-modality feature refinement module was developed using global spectral-spatial attention learning with the query-key-value processing. Thus, the global spatial and channel contexts are considered to generate two-channel attention masks.

Generative adversarial networks (GANs), including two separate neural networks (generative and discriminative networks), have also been extensively employed in image fusion problems. GANs are based on a game theoretic scenario in which both networks compete with each other to optimize a different and apposite loss function. They are usually applied as unsupervised learning algorithms; however, their semi-supervised extension has been recently introduced for training models with labeled and unlabeled data. Liu et al. (2022) provided an extensive review of the use of GANs in remote sensing image fusion. Various specialized GANs have been proposed by researchers, including pan-sharpening GAN (PSGAN) by Liu et al. (2020b), guided colorization of panchromatic images via GANs (PanColorGAN) by Ozcelik et al. (2020), residual encoder-decoder conditional generative adversarial network for pan-sharpening (RED-CGAN) by Shao et al. (2020), perceptual patch generation adversarial network (FDPPGAN) by Pan et al. (2021), and hyperspectral pan-sharpening using 3-D generative adversarial networks (HPGAN) by Xie et al. (2021).

DL, despite its advantages, presents specific challenges that must be addressed within the context of image fusion. One primary issue is that DL-based methods consider original high-resolution MS images as ground truth. Since the relationship between low-resolution and high-resolution MS images tends not to obey simple blur and interpolation operations, it is not reasonable to obtain low-resolution MS images by blurring and downsampling the original high-resolution MS images (Ma et al., 2020). The requirement for a substantial amount of labeled data during the training phase in supervised learning is also a critical issue for DL models since the collection of a representative number of ground reference samples is costly and difficult in most cases (e.g., physical access to the study sites may be restricted or limited to a specific date). Data augmentation techniques, specifically developed for DL, are usually employed to resolve the sample size problem. However, it is important to remember that augmented data sets will retain the biases present in the original data set. Another significant challenge is the extensive computational time required by the current DL-based fusion algorithms, which contrasts sharply with standard fusion algorithms that produce results in a fraction of a second. Consequently, there is a pressing need to develop efficient DL algorithms and fusion strategies to enable the real-time implementation of DL-based image fusion. For further reading on the use of DL models for pan-sharpening, fusion of multispectral and hyperspectral imagery, readers are referred to Liu et al. (2018), Ma et al. (2019), Wang et al. (2020), Zhang et al. (2021), and Meng et al. (2022).

4.1.2 Assessment of Fused Image Quality

Assessment of fused image quality in the spectral domain is initially based on visual inspection. If classification is the major concern, then the classification accuracy would also be one of the indicators used to measure the fusion quality. Alternatively, one may use some or all of the following major quantitative indicators for measuring differences in spectral information between each band of the fused image and of the original image. Singh (2020) reviewed the quantitative and qualitative metrics used in image fusion.

1. **Standard Deviation of the Difference Image (SDD)** (Wald et al., 1997): This metric is estimated between both original and fused images. The lower its value, the better the spectral quality of the fused image.

$$\text{SDD} = \sqrt{\frac{\sum_{i=1}^{n}(X_i - Y_i)^2}{n}}$$

(4.21)

2. **The Correlation Coefficient (CC)**: It estimates the correlation between the fused image and the PAN image. It ranges from −1 to 1, with larger values indicating more similarity between the MS and pan-sharpened images (Wald et al., 1997; Wald, 2002).

$$CC(B_i, B_j) = \frac{\sum_{m=1}^{M}\sum_{n=1}^{N}\left(B_{i(m,n)} - \overline{\overline{B}}_i\right)\left(B_{j(m,n)} - \overline{B}_j\right)}{\sqrt{\sum_{m=1}^{M}\sum_{n=1}^{N}\left(B_{i(m,n)} - \overline{B}_i\right)^2 \sum_{m=1}^{M}\sum_{n=1}^{N}\left(B_{j(m,n)} - \overline{B}_j\right)^2}}$$

(4.22)

where B_i and B_j represent the bands to be fused, $\overline{B}_i$ and $\overline{B}_j$ show the mean values of the pixels, and (i, j) indicates the locations of the pixels.

3. **The ERGAS (Erreur Relative Globale Adimensionnelle de Synthèse) Index** (Wald, 2000): This metric is employed to evaluate the quality of the fused image by determining the normalized average error for each band within the processed image.

$$\text{ERGAS} = 100 \frac{h}{l} \sqrt{\frac{1}{N}\sum_{i=1}^{n}\left(\text{RMSE}^2(B_i) / M_i^2\right)}$$

(4.23)

where h is the resolution of the PAN image, l is the resolution of the MS image, N is the number of spectral bands, B_i is the difference image between the original and the fused images for band i, and M_i is the mean radiance of each spectral band. Lower ERGAS index values are preferred.

4. **The Universal Image Quality Index (Q or UOQI)** (Wang and Bovik, 2002): This is a type of similarity index that characterizes spectral and spatial distortions, and estimated by the following equation:

$$Q = \frac{4 v_{AB} \overline{A}\overline{B}}{\left(\sigma_A^2 + \sigma_B^2\right)\left(\overline{A}^2 + \overline{B}^2\right)}$$

(4.24)

where A and B denote the original and fused images, respectively; $\overline{A}$ and $\overline{B}$ are the means of A and B; σ is the corresponding standard deviation; and v_{AB} is the covariance of A and B. Note that while the CC and Q metrics are monomodal, ERGAS is a multimodal metric.

5. **Structural Similarity Index (SSIM)**: As an improved version of Q index, SSIM measures structural similarity between images through the calculation of directional means and standard deviations within windows of a specified size, systematically moving across the entire image. To obtain the SSIM value for the two monochromatic images, you can compute an average of SSIM values over windows. When determining the SSIM value for the fusion, you would compare a monochromatic representation of the multispectral image (such as the mean value across bands) to the original panchromatic image. Given two local image patches x and y extracted from the original and distorted images,

$$\text{SSIM} = \frac{\left(2\mu_x\mu_y + C_1\right)\left(2\sigma_{xy} + C_2\right)}{\left(\mu_x^2\mu_y^2 + C_1\right)\left(\sigma_x^2 + \sigma_y^2 + C_2\right)} \tag{4.25}$$

where μ_x, σ_x, and σ_{xy} represent the mean, standard deviation, and cross-correlation evaluations, respectively. $C_1 = \left(K_1 L\right)^2$ and $C_2 = \left(K_2 L\right)^2$ are constants used to characterize the saturation effects of the system at low luminance. Note that *SSIM* metric has its enhanced version (Information Content Weighted SSIM – IWSSIM) determining the weights using a logarithmic function of the local variances (Wang and Li, 2011). Thus, higher local variances in specific regions have a greater impact on the overall IWSSIM value.

6. **Spectral Angle Mapper (SAM)**: This metric operates under the assumption that each pixel in remote sensing images corresponds uniquely to a specific land cover, thus belonging to a single class. It quantifies the color disparities between the fused and MS images. The local SAM value is generated as a map by calculating the angular difference between every pixel in the fused image and its corresponding pixel in the MS image. Subsequently, the variance map values are transformed into a linear range spanning from 0 to 255. The following equation is applied to compute the local SAM value:

$$\text{SAM}(x, y) = \frac{(F, M)}{F_2 F M_2} \tag{4.26}$$

where F and M denote the pixels of the fused image and the original MS image, respectively. The global SAM value is estimated by taking the average of all pixels contained in the SAM map. It is measured in degree (°) or radian. A global SAM value of zero signifies the absence of color distortion in the fused image, representing the optimal condition. A drawback of the SAM metric is its incapacity to discern between positive and negative correlations due to its exclusive computation of absolute values (Azarang and Kehtarnavaz, 2022).

7. **Root Average Spectral Error (RASE)**: It is a widely used global distortion metric. Lower RASE values indicate a higher degree of similarity between the fused output and the reference image. This assessment considers the mean values of individual bands as given in the following formula:

$$\text{RASE} = \frac{100}{\sum_{i=1}^{N} \mu(i)} \sqrt{\frac{1}{N} \text{RMSE}(F, M)} \tag{4.27}$$

where $\mu(i)$ represents the mean value of the i-th band. The RASE metric is usually given as a percentage, and the ideal value is expected to be zero.

4.1.3 Performance Evaluation of Fusion Methods

Prasad et al. (2001) use the IHS, PCA, and Brovey methods for fusing IRS LISS-III and PAN data covering the forest area around Hardwar, India, and the results are evaluated in terms of the success in the interpretation of forest features. It is observed that PCA outperforms the IHS and Brovey methods. González-Audícana et al. (2005) perform image fusion based on the wavelet technique for IKONOS MS and PAN images. The images were acquired in October 2000 and cover the irrigated agricultural area of Mendavia (Navarre), in northern Spain. The main crops around the area in the year 2000 were corn, alfalfa, and grapes. Both the *à trous* algorithm and Mallat's algorithm are applied, and the results in terms of several spectral quality measures show that the *à trous* algorithm generally performs better than Mallat's algorithm. González-Audícana et al. (2005) also underlined that due to the orientation emphasis in the horizontal and vertical directions during the decimation process in Mallat's algorithm, the merged images exhibit a visually lower spatial quality than those generated through the *à trous* algorithm. Colditz et al. (2006) use the PCA, IHS, Brovey, and wavelet methods for image fusion, and the fused results are evaluated using classification accuracy. The study imagery is Landsat 7 ETM+ acquired on August 15, 2001, and located around the Würzburg area in central Germany. The results show that the fusion images produced by wavelet and PCA methods achieve better classification results than those generated by using the IHS and Brovey approaches. Teggi et al. (2003) evaluate the performance of the wavelet transform, PCA, and IHS methods for fusing Indian IRS-1C-PAN sensors. The fusion quality is analyzed in terms of correlation, standard deviation (Wald et al., 1997), and classification accuracy. The results suggest that the wavelet transform outperforms both the PCA and IHS methods.

Nikolakopoulos and Oikonomidis (2015) examined the effectiveness of ten widely recognized fusion methods in combining Worldview-2 PAN and MS data. The fused images were assessed for spectral and spatial accuracy through visual examination and various quality indices, including the correlation coefficient (CC), ERGAS, the Universal Image Quality Index (Q), the Expanded Quality Index ($Q4_{expand}$), and entropy values. Findings indicated that the CN, GS, PCA, and wavelet methods produced unsatisfactory results when applied to the downsampled PAN and MS data sets, while yielding acceptable results when utilized on the original data set. The $Q4_{expand}$ quality index was recommended as the principal discriminator for quality assessment due to its ability to provide consistent and reliable results across diverse spectral and spatial input resolutions. Pan-sharpening algorithms are renowned for their effectiveness, consistently yielding stable and high-quality results across various indices. Spectral preservation is particularly notable in the Ehlers and HPF algorithms. Moreover, the LMM, LMVM, and ModIHS fusion techniques, although not emphasizing spectral attributes, retain significance and are applicable in photo-interpretation studies.

Considering 21 case studies, 41 image fusion techniques under the categories of component substitution (CS-based), multiresolution analysis (MRA), variational optimization–based (VO), and hybrid were reviewed and tested by Javan et al. (2021). Results produced in the study revealed that MRA-based methods performed better in terms of spectral quality, whereas most hybrid methods provided the highest spatial quality, and CS-based methods produced the lowest performance both spectrally and spatially. Moreover, the highest spectral quality was achieved with the Additive Wavelet Luminance Proportional Pan-sharpening method, while the highest spatial quality was produced by Generalized IHS with Best Trade-off Parameter with Additive Weights.

Yilmaz et al. (2022) evaluate the performances of 47 conventional and state-of-the-art pan-sharpening methods including DL approaches. The selected methods were grouped into six main categories as MRA-based, component substitution (CS)-based, color-based (CB), DL-based, VO-based, and hybrid techniques. The quantitative evaluation was performed at both reduced scale and full scale. In terms of pan-sharpening performance, the methods falling under the categories of MRA, DL, CB, and VO outperformed the hybrid and CS-based techniques, which displayed comparatively weaker results. To be more specific, the spectral residual of injected details (SR-D), the Generalized Laplacian Pyramid with modulation transfer function (GLP-MTF), Enhanced Context-Based Modular Transfer Function with Generalized Laplacian Pyramid (MTF-GLP-ECB), Target-Adaptive CNN (TA-CNN), Additive Wavelet Luminance Proportional (AWLP), and the À trous Wavelet Transform with the Injection Model 2 (ATWT-M2) methods were ranked as the most successful methods, respectively. On the contrary, the GS Adaptive, PCA, Modified IHS, The Simultaneous Satellite Image Registration and Fusion (SIRF), Surface- and Deep-level Constraint-based Pan-sharpening Network (SDPNet), and IHS-based Discrete Wavelet Transform (IHS-DWT) methods produced the worst performances.

Proposing an adaptive strategy to improve the performances of the Generalized Laplacian Pyramid (GLP) with modulation transfer function (MTF)–matched filter and Context-Based Decision (CBD) injection scheme (MTF-GLP-CBD) and the Sparse Representation of Injected Details (SR-D), Yilmaz (2023) compared the performances of 32 pan-sharpening methods on three data sets. Results of the study revealed that the proposed approach adaptively estimating the gray values in the pan-sharpening output by utilizing variance information obtained from the PAN imagery improved the performances of the MTF-GLP-CBD and SR-D algorithms and outperformed the other pan-sharpening methods after qualitative and quantitative assessment.

4.2 MULTISOURCE CLASSIFICATION USING THE STACKED-VECTOR METHOD

Handling a multisource classification problem can be addressed in a straightforward manner by expanding the dimension of the data vectors to incorporate each source, a method commonly referred to as the stacked-vector or augmented-vector approach. For example, if one has a six-band ETM+ image and a three-band SPOT image, then nine bands can be used together as inputs to the classifier. Alternatively, one may first implement the fusion process to the available imagery to obtain a higher quality of image content in both the spectral and spatial domains, and then implement a stacked-vector method (in such a way that the dimension of data stacked can also be reduced). Although this method is easy to apply, several issues require attention. The initial concern revolves around the measurement scale of each source, as various data sources often exhibit different measurement scales. It is typically advisable to normalize all data on a uniform scale to address this issue, a procedure known as normalization (Figure 4.7). For instance, if data from the optical spectrum are used, one may convert them into ground reflectance units. If radar images are used, one may convert these data into backscatter coefficients (σ^0, Chapter 1). However, where a variety of data sources is used, then the choice of a data normalization scheme may be logically difficult, and data sources demonstrating a larger variation scale could exert a significant impact on the classification process. For instance, if the first data source in a two-source case has a range of [0, 32], and the second has a range of [0, 16,384], then the second data source is more likely to dominate the classification process due to the effects of the measurement scale. Some classification procedures require explicit normalization of the input data (e.g., a feedforward neural network). Other methods, such as maximum likelihood (ML), use a hidden form of normalization. ML uses estimates of the variance-covariance matrices (S_i) of the classes to generate class membership probabilities. The calculation of S_i involves the subtraction of the class mean of each feature from the feature measurements.

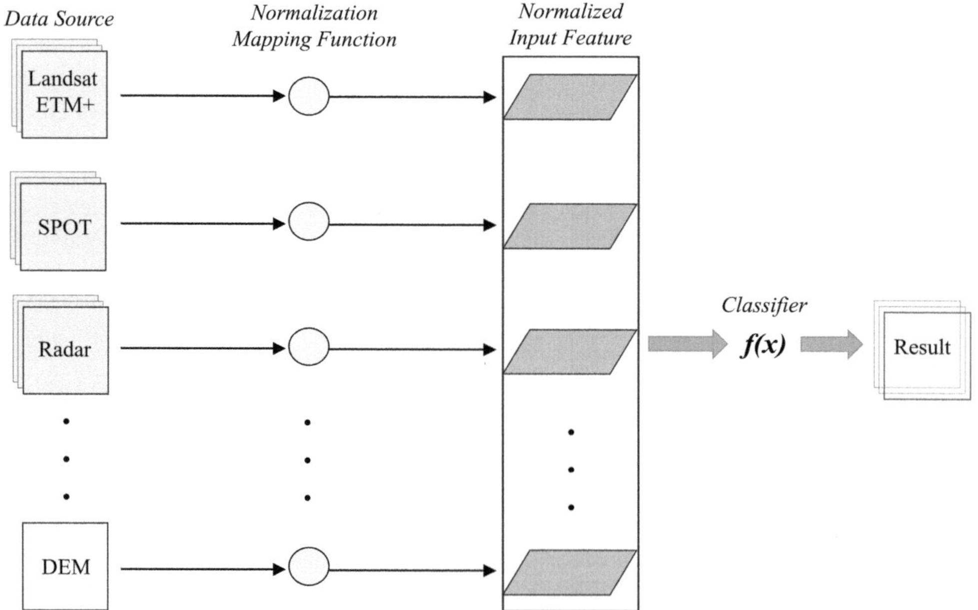

FIGURE 4.7 Stacked-vector classification process.

A second issue is that the stacked-vector method may not be practical in terms of computational cost when the number of data vectors is large. For example, if one uses the statistical Gaussian ML method for classifying N-dimensional variables, the resulting computational cost is proportional to N^2. A way to reduce this problem is to use a feature selection procedure, which involves selecting a subset of data sources. A suitable subset can be selected by using some distance measure, such as the divergence index (Singh, 1984) or the B-distance (Haralick and Fu, 1983), as shown in Chapter 3. Another strategy for feature selection is through the use of data transforms, such as PCA, tasseled cap (Crist and Cicone, 1984), vegetation indices, or the self-organized feature map (SOM). Note that although feature selection can decrease the computational cost of classification, some valuable information may be contained in the discarded data sources. Such a dilemma always results when feature selection techniques are applied.

A third issue is concerned with the reliability (or uncertainty) of the data. The stacked-vector approach treats each data source as being fully reliable (i.e., each source contributes equally to the classifier in the determination of the location of decision boundaries in feature space). This may not always be the case in practice and could hinder achieving higher classification accuracy.

4.3 THE EXTENSION OF BAYESIAN CLASSIFICATION THEORY

Interest in techniques of multisource classification based on the extension of Bayesian theory was triggered by the studies of Lee et al. (1987) and Benediktsson et al. (1990), and later extended by Zhu and Tateishi (2006) to deal with multisource, multitemporal data for land cover mapping. The model proposed by Benediktsson et al. (1990) is a refinement of the method of Lee et al. (1987), in which a more complete statistical mechanism was derived. These two models are introduced next, together with a derivation using the results obtained by Benediktsson et al. (1990) to incorporate the Markov random field (MRF) concept to achieve multisource MRF-MAP classification.

4.3.1 An Overview

The extension of the Bayesian classification theory was compared with the evidential reasoning approach based on the Dempster–Shafer theory by Lee et al. (1987). Both showed satisfactory results for performing multisource data classification, although the method based on the extension of the Bayesian classification theory performed slightly better than the evidential reasoning method (Lee et al., 1987). Benediktsson et al. (1990) adopted extended Bayesian classification theory to classify multisource remotely sensed data. They also compared these results with those achieved by a neural network stacked-vector approach. Both approaches reveal different advantages and disadvantages. Furthermore, Schistad et al. (1994) tested an extended Bayesian classification approach for classifying Landsat TM and SAR images, and their results were better than those achieved by conventional single-source classification. A similar experiment was conducted by Kim and Swain (1995), who used the evidential reasoning approach to manage multisource data and achieve an acceptably good classification result.

Both the extension of Bayesian classification theory and evidential reasoning methods consider each data source as fully independent. Hence, one has to generate probability (or evidence) measures for each information class using each source separately, and then obtain the classification result in terms of probability (or evidence) consensus. The general steps are shown in Figure 4.8 and are explained as follows.

4.3.1.1 Feature Extraction

This is the preparation stage of the classification process, the aim of which is to determine what kind of input features (e.g., pixel gray values, texture measures) should be used in the classification process. The suitable selection of input features can enhance classification accuracy. Certainly, the choice of input features not only depends on the way in which the image was formed (e.g., optical or

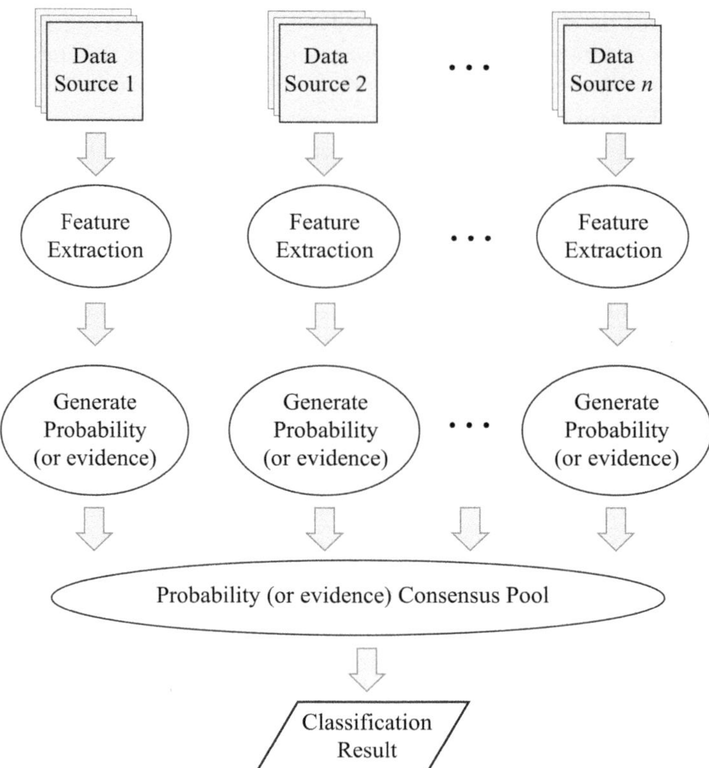

FIGURE 4.8 General steps in multisource classification based on evidential reasoning and extension of Bayesian theory.

microwave sensor), but is also determined by the scale relationship between the ground objects of interest and image resolution. If, for example, image resolution is around 30 m, then in the case of lithological classification, textural information may contribute more significantly to interclass variation than tonal information. In the case of classifying crop types in an agricultural area, textural and tonal information may make no significant difference (Tso, 1997) because of the scale and the size of the object being classified (refer to Section 2.6.1 for a discussion of texture). Where radar images are used, textural features may be more useful than image tonal information (Ulaby et al., 1986).

4.3.1.2 Probability or Evidence Generation

This stage involves the definition of a probability density function or some alternative methodology (e.g., by the use of a feedforward neural network) to generate class-associated probabilities or evidence in which the user has high confidence. If a statistical model is used for generating the probability or statistical evidence, then the choice of probability density function should be source-dependent (Kim and Swain, 1995). For example, optical images can be modeled in terms of a Gaussian probability density function (p.d.f), while radar images may more suitably be modeled as a Gamma distribution (Chapter 1). This step is an alternative way of normalization of the data scale, as each different source is mapped onto the probability domain.

4.3.1.3 Multisource Consensus

In this final stage, all the probability (or evidence) measures are combined to generate a classified image (Figure 4.9). Where the evidential reasoning method is used for evidence consensus, the data sources are considered in a pairwise fashion, while if the consensus mechanism is based on the extension of Bayesian classification theory, all probability measures are combined simultaneously.

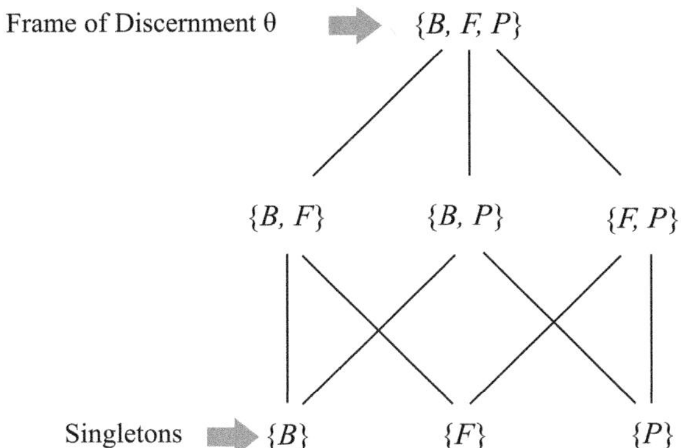

FIGURE 4.9 Representation of subsets of the frame of discernment {B, F, P} made up of bare soil, forest, and pasture.

4.3.2 BAYESIAN MULTISOURCE CLASSIFICATION MECHANISM

The model described in this section is based on Lee et al. (1987). In the general multisource case, a set of observations from n ($n > 1$) different sources is used. Let x_i, $i \in [1, n]$, denote the measure of a specific pixel from measurement source i. The aim is to derive the probability of the pixel belonging to each of c information classes ω_j, $j \in [1, c]$. It is possible to get some information about the prior probability of class ω_j, denoted by $P(\omega_j)$, that is, the probability that an observation will be a member of class ω_j.

According to Bayesian probability theory, the law of conditional probabilities states that

$$P\left(\omega_i \mid X\right) = P\left(\omega_j \mid x_1, x_2, \ldots, x_n\right) = \frac{P\left(x_1, x_2, \ldots, x_n \mid \omega_j\right) \times P\left(\omega_j\right)}{P\left(x_1, x_2, \ldots, x_n\right)} \tag{4.28}$$

where $P\left(\omega_j \mid x_1, x_2, \ldots, x_n\right)$ is known as the conditional or posterior probability that ω_j is the correct class, given the observed data vector $(x_1, x_2, \ldots, x_n)$. $P\left(x_1, x_2, \ldots, x_n \mid \omega_j\right)$ is the p.d.f. associated with the measured data $(x_1, x_2, \ldots, x_n)$ given that $(x_1, x_2, \ldots, x_n)$ are members of class ω_j. $P(x_1, x_2, \ldots, x_n)$ is the p.d.f. of data $(x_1, x_2, \ldots, x_n)$. Assuming class-conditional independence among the sources, one obtains $P\left(x_1, x_2, \ldots, x_n \mid \omega_j\right) = P\left(x_1 \mid \omega_j\right) \times P\left(x_2 \mid \omega_j\right) \times \cdots \times P\left(x_1 \mid \omega_n\right)$, and Equation (4.28) becomes

$$P\left(\omega_j \mid x_1, x_2, \ldots, x_n\right) = \frac{P\left(x_1 \mid \omega_j\right) \times P\left(x_2 \mid \omega_j\right) \times \cdots \times P\left(x_n \mid \omega_j\right) \times P\left(\omega_j\right)}{P\left(x_1, x_2, \ldots, x_n\right)} \tag{4.29}$$

Again, following the law of conditional probabilities,

$$P\left(x_i \mid \omega_j\right) \times P\left(\omega_j\right) = P\left(\omega_j \mid x_i\right) \times P\left(x_i\right)$$

$$\Rightarrow P\left(x_i \mid \omega_j\right) = \frac{P\left(\omega_j \mid x_i\right) \times P\left(x_i\right)}{P\left(\omega_j\right)} \tag{4.30}$$

If Equation (4.30) is substituted into Equation (4.29), one obtains

$$P\left(\omega_j \mid x_1, x_2, \ldots, x_n\right) = P\left(\omega_j \mid x_1\right) \times P\left(\omega_j \mid x_2\right) \times \cdots \times P\left(\omega_j \mid x_n\right) \times P\left(\omega_j\right)^{1-n}$$
$$\times \frac{P(x_1) \times P(x_2) \times \cdots \times P(x_n)}{P\left(x_1, x_2, \ldots, x_n\right)} \tag{4.31}$$

If the inter-source independence assumption has been made such that $P(x_1) \times P(x_2) \times \cdots \times P(x_n) = P(x_1, x_2, \ldots, x_n)$, then Equation (4.31) results in

$$P\left(\omega_j \middle| x_1, x_2, \ldots, x_n\right) \propto P\left(\omega_j \middle| x_1\right) \times P\left(\omega_j \middle| x_2\right) \times \cdots \times P\left(\omega_j \middle| x_n\right) \times P\left(\omega_j\right)^{1-n} \qquad (4.32)$$

Equation (4.32) is only valid under the circumstance that the data sources are mutually independent. The practical value of this assumption is considered further in Section 4.3.5.

When processing multisource information, it is natural to consider the inclusion of a reliability (or uncertainty) measure for each different data source. In other words, one can assign different weights to different sources in order to achieve a more satisfactory multisource consensus. One simple way to adjust the contribution for each data source is to append exponents to the source-specific posterior probabilities (Lee et al., 1987). This modification to Equation (4.32) gives

$$P\left(\omega_j \middle| x_1, x_2, \ldots, x_n\right) \propto P\left(\omega_j \middle| x_1\right)^{\alpha_1} \times P\left(\omega_j \middle| x_2\right)^{\alpha_2} \times \cdots \times P\left(\omega_j \middle| x_n\right)^{\alpha_n} \times P\left(\omega_j\right)^{1-n} \qquad (4.33)$$

where $\alpha_i \in [0,1]$ is the source-specific weighting parameter, which allows one to adjust the contribution for the ith source, and a measure of the sensitivity of $P\left(\omega_j \middle| x_1, x_2, \ldots, x_n\right)$ to changes of the ith posterior probability, which reduces to the following expression:

$$\frac{\delta P\left(\omega_j \middle| x_1, x_2, \ldots, x_n\right)}{P\left(\omega_j \middle| x_1, x_2, \ldots, x_n\right)} \propto \alpha_i \frac{\delta P\left(\omega_j \middle| x_i\right)}{P\left(\omega_j \middle| x_i\right)}. \qquad (4.34)$$

Equation (4.34) shows that the difference in the posterior probability $P\left(\omega_j \middle| x_i\right)$ for source i leads to a difference of α_i in $P\left(\omega_j \middle| x_1, x_2, \ldots, x_n\right)$. The weighting parameter α_i thus represents the level of sensitivity of the data source i.

Given the above definition of α_i, it is clear that if α_i is equal to the value 1, then the data source i is fully reliable. As α_i tends to 0, the source i is considered to be less reliable. Note that in the case $\alpha_i = 0$, one should not include any contribution from source i. Equation (4.33) cannot provide a satisfactory mechanism to cope with this situation. For instance, if there are a total of five data sources, and only the weighting parameters α_1 and α_2 for the first and second data sources are nonzero, the posterior probability derived from Equation (4.33) will use $P(w_j)^{-4}$ (where the exponent -4 is obtained from $1-n$, and the number of data sources $n=5$ in this case) as the prior probability. A refined Bayesian multisource classification mechanism was proposed by Benediktsson et al. (1990) to deal with this difficulty.

4.3.3 A Refined Multisource Bayesian Model

Consider that one more source is being added to the classification analysis. Equation (4.32) then becomes (Benediktsson et al., 1990)

$$P\left(\omega_j \middle| x_1, x_2, \ldots, x_n, x_{n+1}\right) \propto P\left(\omega_j \middle| x_1\right) \times P\left(\omega_j \middle| x_2\right) \times \cdots \times P\left(\omega_j \middle| x_{n+1}\right) \times P\left(\omega_j\right)^{-n} \qquad (4.35)$$

By dividing Equation (4.35) by Equation (4.32), the probability ratio is seen to be

$$P\left(\omega_j \middle| x_{n+1}\right) \middle/ P\left(\omega_j\right) \qquad (4.36)$$

This is the contribution of the $(n+1)$th source, and therefore, Equation (4.32) can be modified as follows:

$$\begin{aligned} P\left(\omega_j \middle| x_1, x_2, \ldots, x_n\right) \propto &\left[P\left(\omega_j \middle| x_1\right) \middle/ P\left(\omega_j\right)\right] \times \left[P\left(\omega_j \middle| x_2\right) \middle/ P\left(\omega_j\right)\right] \times \cdots \\ &\times \left[P\left(\omega_j \middle| x_n\right) \middle/ P\left(\omega_j\right)\right] \times P\left(\omega_j\right) \end{aligned} \qquad (4.37)$$

For reliability measurements, we can again use the idea of appending exponents to each data source, and Equation (4.38) can be rewritten as

$$
P\left(\omega_j \middle| x_1, x_2, \ldots, x_n\right) \propto \left[P\left(\omega_j \middle| x_1\right)\middle/P\left(\omega_j\right)\right]^{\alpha_1} \times \left[P\left(\omega_j \middle| x_2\right)\middle/P\left(\omega_j\right)\right]^{\alpha_2} \times \cdots
$$
$$
\times \left[P\left(\omega_j \middle| x_n\right)\middle/P\left(\omega_j\right)\right]^{\alpha_n} \times P\left(\omega_j\right)
$$

(4.38)

Clearly, Equation (4.38) provides a more satisfactory relationship for dealing with multisource fusion problems in comparison with Equation (4.33). The reason is that when a source is found to be fully unreliable (i.e., $\alpha_i = 0$), both posterior and associated prior probabilities should be rejected simultaneously as shown in Equation (4.38). However, it is worthwhile to note that if prior probability is not used (which is equivalent to regarding $P(\omega_j)$ as being uniformly distributed for all information class j), then Equations (4.33) and (4.38) are equivalent.

The sensitivity measure as given by Equation (4.34) is converted to the form

$$
\frac{\delta P\left(\omega_j \middle| x_1, x_2, \ldots, x_n\right)}{P\left(\omega_j \middle| x_1, x_2, \ldots, x_n\right)} = \alpha_i \frac{\delta P\left(\omega_j \middle| x_i\right)\middle/P\left(\omega_j\right)}{P\left(\omega_j \middle| x_i\right)\middle/P\left(\omega_j\right)}.
$$

(4.39)

Another way to understand the behavior of α_i more clearly is to represent Equation (4.38) in logarithmic form (Benediktsson et al., 1990):

$$
\ln P\left(\omega_j \middle| x_1, x_2, \ldots, x_n\right) \propto \ln P\left(\omega_j\right) + \sum_i \alpha_i \ln\left[P\left(\omega_j \middle| x_i\right)\middle/P\left(\omega_j\right)\right]
$$

(4.40)

It is now apparent that α_i is a coefficient that corresponds to the weighting parameter associated with data source i.

4.3.4 MULTISOURCE CLASSIFICATION USING THE MRF

In most remote sensing image classification experiments, the prior probability $P(w)$ is rarely used because of the problem of properly modeling the prior probability. However, the smooth contextual concept can be used to model prior probability using smooth prior modeling in terms of MRF to achieve an MRF-MAP classification. The concept of MRF can also be applied to multisource classification.

The method for deriving an MRF-MAP estimate, as described in Section 2.6.1, requires the specification of a posterior energy function. The posterior energy shown in Equation (4.38) is valid only for classifying a single data source. In the case of multisource classification, one has to construct a multisource posterior energy function. Equation (4.40) can be used to achieve this aim. According to Bayesian probability theory, it can be shown that $\left[P\left(\omega_j \middle| x_i\right)\middle/P\left(\omega_j\right)\right] \propto P\left(x_i \middle| \omega_j\right)$ so that the right-hand side of Equation (4.40) can be expressed as

$$
\ln P\left(\omega_j \middle| x_1, x_2, \ldots, x_n\right) \propto \ln P\left(\omega_j\right) + \sum_i \alpha_i \ln P\left(x_i \middle| \omega_j\right)
$$

(4.41)

or, equivalently,

$$
U\left(\omega_j \middle| x_1, x_2, \ldots, x_n\right) \propto U\left(\omega_j\right) + \sum_i \alpha_i U\left(x_i \middle| \omega_j\right)
$$

(4.42)

$U\left(\omega_j \middle| x_1, x_2, \ldots, x_n\right)$ denotes the multisource posterior energy, and $U\left(\omega_j\right)$ and $U\left(x_i \middle| \omega_j\right)$ are the prior energy and the class-conditional energy. Using this definition of multisource posterior energy, one can use the classification algorithms to perform multisource MRF-MAP estimation (note that one has to substitute the multisource posterior energy in place of the original single-source posterior energy in the algorithm being applied). Schistad et al. (1996) use a similar equation in multisource classification.

From the foregoing description, it can be seen that not only will the posterior probability of each source influence the result of multisource consensus, but the weighting parameter α_i will also contribute to the final decision. The problem is: how to specify a reliability measure in order to obtain a more reliable multisource fusion result. Note also that in the single-source MRF-MAP estimate, one has to choose a suitable potential parameter for the energy functions in order to obtain a good classification result. Here, in the case of multisource MRF-MAP estimation, both the potential parameter and the source-associated weighting factors have to be determined.

4.3.5 ASSUMPTION OF INTER-SOURCE INDEPENDENCE

The assumption of inter-source independence means that the joint probability distributions for the measurements of sources are mutually independent, and the final joint probability function can be expressed as the product of each class-conditional probability function. Although this might not always be the case, the validity of the assumption is hard to determine.

In practical situations, multiple data sources usually contain complex but unknown interactions. For example, in the case of multisource classification, the data set may consist of optical and radar images and a digital terrain model. If there is no reliable information concerning the relationship between surface spectral reflectance (or backscattering) and terrain parameters (such as ground slope, surface shape), then a lack of knowledge of these interactions will force us to ignore any inter-source relationships, and therefore to treat these data sources as independent variables. However, rather than being a shortcoming in the data consensus analysis, the inter-source independence assumption does provide an easy way to perform classification using multiple data sources.

4.4 EVIDENTIAL REASONING

The mathematical theory of evidence involves the integration of various data sources to derive a cohesive inference about the labeling of pixels. Originating in the 1960s with Dempster, the theory underwent further expansion by Shafer in 1979. This comprehensive framework is now recognized as the Dempster-Shafer (D-S) theory of evidence. Garvey et al. (1981) discuss applications using the D-S theory of aggregating evidential knowledge. Barnett (1981) mentions some of the computational issues involved in reducing the computational requirements of the method. Gordon and Shortliffe (1985) and Shafer and Logan (1987) propose modified approaches that are mainly based on a hierarchical evidence space. The D-S theory has also been related to the field of artificial intelligence systems (Gouvernet et al., 1980; Friedman, 1994; Strat, 1984) with promising results.

The evidential reasoning approach also provides a valuable theoretical basis for dealing with the remotely sensed multisource classification problem. The approach has been tested in several studies (e.g., Lee et al., 1987; Srinivasan and Richards, 1990; Wilkinson and Mégier, 1990; Kim and Swain, 1995; Franklin et al., 2002). Key challenges in the evidential reasoning method revolve around the generation of the mass of evidence and the assessment of uncertainty, influencing the weight assigned to each data source. The basic concepts of D-S theory are described in the following sections.

4.4.1 CONCEPT DEVELOPMENT

An essential characteristic of the D-S theory is its departure from the Bayesian restriction, ensuring that belief in a hypothesis does not automatically entail commitment of the remaining belief to its opponents; that is, belief in event w is equivalent to $P(w)$ so that belief in the concept *Not w* is equal to $1 - P(w)$. In many real applications, evidence partially in favor of one particular hypothesis is not necessarily evidence against alternative hypotheses. In D-S theory, complete knowledge is represented as unity (i.e., the sum of all labeling possibilities is equal to one), which includes the case of uncertainty. In other words, the measure of belief assigned to each hypothesis may be less than one.

If there is some evidence in favor of hypothesis *w*, then the remaining belief (which is equivalent to the value of uncertainty) will be assigned to the whole hypothesis space, not simply to its opponents. A simple example may illustrate this idea.

Suppose that an analyst is trying to classify an image that involves labeling pixels as belonging to one of three classes {*B*, *F*, *P*}, where *B* denotes bare soil, *F* denotes forest, and *P* denotes pasture. In D-S theory, this set {*F*, *B*, *P*} is generally called a frame of discernment, and is denoted by the symbol θ. The number of all possible subsets of the frame of discernment θ is equal to $2^{|\theta|}$, where $|\theta|$ denotes the number of one-element subsets (called singletons). In our case, there are a total of three singletons in $|\theta|$, and therefore, the total number of subsets of θ is $2^3 = 8$, as shown in Figure 4.9. Note that the empty set { } is one of these subsets, but is not displayed in Figure 4.9.

D-S theory uses a number within the range of [0, 1] to indicate the degree of belief in a hypothesis, given a piece of evidence. The number that expresses the degree to which the evidence supports a particular hypothesis is represented by a function called the basic probability assignment (*bpa*); the resulting quantity of evidence is generally called the measure of mass or mass of evidence and is denoted by $m(\psi)$, where ψ is any subset of θ. The sum of $m(\psi)$, for $\forall\psi$, must be equal to 1 (this is a rather important constraint). In our case, ψ can be one of the sets shown in Figure 4.10. If $m(\psi) = q$ and no other belief is assigned to the subsets of θ, then $m(\theta) = 1 - q$; that is, the remaining belief is assigned to the frame of discernment θ, rather than ψ's opponents. This point has already been emphasized in the previous section. An example is given below.

Suppose that a source of evidence indicates that a particular pixel of interest has 30% support for the belief that the pixel belongs to class {*B*}, 30% support for the belief that the pixel belongs to class {*F*}, and 40% support for the belief that the pixel belongs to class {*P*}. Note again that the sum of all $m(\psi)$ must equal 1. If one assumes that the source of evidence is fully reliable (i.e., the uncertainty is zero), then using the theory of evidence symbolism, one obtains

$$m(\{B,F,P\}) = \langle 0.3, 0.3, 0.4 \rangle. \tag{4.43}$$

Suppose one is somewhat uncertain about the reliability of this data evidence and is only willing to commit oneself to label the pixel with, for example, 80% confidence. Then, one should change the previous *bpa* by multiplying each element by 0.8, which leads to:

$$m(\{B,F,P,\theta\}) = \langle 0.24, 0.24, 0.32, 0.2 \rangle. \tag{4.44}$$

Note that the quantity of *bpa* expressed by θ is denoted by $m(\theta)$ and is expressed in this case by

$$1 - (0.24 + 0.24 + 0.32) = 0.2 \tag{4.45}$$

which is the measure of uncertainty or ignorance; in our case, 80% confidence leads to 20% uncertainty, and this is reflected in the labeling process.

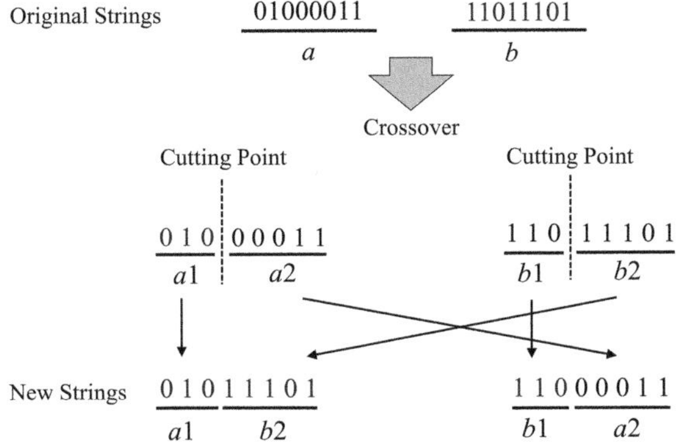

FIGURE 4.10 Crossover operation used by GA to generate new search strings.

4.4.2 Belief Function and Belief Interval

In what follows, both the belief function (or support) and the plausibility of each labeling proposition within evidential reasoning are described. A belief function, denoted by Bel, for a hypothesis ψ is defined as the sum of the mass of evidence that is committed to the ψ and to its subsets (if any), while the plausibility of ψ, denoted by Pl, is defined as one minus the Bel committed to ψ's contradiction, denoted by Bel($\sim\psi$). It is more convenient to explain such definitions in terms of mathematical representation:

$$\text{Bel}(\psi) = \sum_{\omega \in \psi} m(\omega), \quad \text{and} \quad \text{Pl}(\psi) = 1 - \text{Bel}(\sim\psi). \tag{4.46}$$

For instance, using the example discussed earlier, the Bel for hypothesis $\{B, F\}$ can be represented by

$$\text{Bel}(\{B,F\}) = m(\{B\}) + m(\{F\}) + m(\{B,F\}) \tag{4.47}$$

and it can be found that Bel(ψ) is a measure of the total amount of belief in ψ including all ψ's subsets. When ψ is a singleton, Bel(ψ)=$m(\psi)$. For instance, Bel($\{B\}$)=$m(\{B\})$. The plausibility for hypothesis $\{B, F\}$ can be represented by

$$\text{Pl}(\{B,F\}) = 1 - \text{Bel}(\sim\{B,F\}) = 1 - \text{Bel}(\{P\}). \tag{4.48}$$

Bel(ψ) can thus be interpreted as the minimum amount of evidence that a pixel is properly labeled with ψ, while $Pl(\psi)$ can be interpreted as the maximal extent to which the current evidence could allow one to believe ψ (Shafer, 1979). Note that Pl(ψ) provides another choice for us to make a decision when no evidence exactly supports hypothesis ψ.

If a pixel belongs to ψ, then the probability $P(\psi)$ may lie somewhere between the interval

$$\lfloor \text{Bel}(\psi), \text{Pl}(\psi) \rfloor \quad \text{or} \quad \lfloor \text{Bel}(\psi),\ 1 - \text{Bel}(\sim\psi) \rfloor \tag{4.49}$$

with the range Pl(ψ) − Bel(ψ), which in turn means that

$$\text{Pl}(\psi) \geq P(\psi) \geq \text{Bel}(\psi). \tag{4.50}$$

That is, Pl(ψ) and Bel(ψ) provide the upper bound and lower bound of the probability of the subset. Recall from Equation (4.43) that the belief, plausibility, and interval for each class are

$$
\begin{aligned}
&\text{Belief}: &&\text{Bel}(\{B\}) = 0.24 \\
& &&\text{Bel}(\{F\}) = 0.24 \\
& &&\text{Bel}(\{P\}) = 0.32 \\
&\text{Plausibility}: &&\text{Pl}(\{B\}) = 1 - \text{Bel}(\{F\}) - \text{Bel}(\{P\}) = 0.44 \\
& &&\text{Pl}(\{F\}) = 1 - \text{Bel}(\{B\}) - \text{Bel}(\{P\}) = 0.44 \\
& &&\text{Pl}(\{P\}) = 1 - \text{Bel}(\{B\}) - \text{Bel}(\{F\}) = 0.52 \\
&\text{Interval}: &&\text{Pl}(\{B\}) - \text{Bel}(\{B\}) = 0.2 \\
& &&\text{Pl}(\{F\}) - \text{Bel}(\{F\}) = 0.2 \\
& &&\text{Pl}(\{P\}) - \text{Bel}(\{P\}) = 0.2
\end{aligned}
\tag{4.51}
$$

In Equation (4.51), the interval is equal to $m(\theta)$, and one may finally choose class P as the label of the pixel of interest because in Equation (4.51), each interval is the same, and the class pasture (i.e., subset $\{P\}$) has the highest belief and plausibility in comparison with the other two classes.

Now consider a more complicated instance. If one holds the following evidence for each hypothesis:

$$m(\{B\}) = 0.19$$

$$m(\{F\}) = 0.19$$

$$m(\{B,F\}) = 0.1$$

$$m(\{P\}) = 0.32$$

$$m(\{\theta\}) = 0.2$$

then the belief, plausibility, and interval become

$$
\begin{aligned}
\text{Belief}: \quad & \mathrm{Bel}(\{B\}) = 0.19 \\
& \mathrm{Bel}(\{F\}) = 0.19 \\
& \mathrm{Bel}(\{B,F\}) = 0.48 \\
& \mathrm{Bel}(\{P\}) = 0.32 \\
\text{Plausibility}: \quad & \mathrm{Pl}(\{B\}) = 0.49 \\
& \mathrm{Pl}(\{F\}) = 0.49 \\
& \mathrm{Pl}(\{B,F\}) = 0.68 \\
& \mathrm{Pl}(\{P\} = 0.52 \\
\text{Interval}: \quad & \mathrm{Pl}(\{B\}) - \mathrm{Bel}(\{B\}) = 0.3 \\
& \mathrm{Pl}(\{F\}) - \mathrm{Bel}(\{F\}) = 0.3 \\
& \mathrm{Pl}(\{B,F\}) - \mathrm{Bel}(\{B,F\}) = 0.2 \\
& \mathrm{Pl}(\{P\}) - \mathrm{Bel}(\{P\}) = 0.2
\end{aligned}
\tag{4.52}
$$

In such circumstances, if one considers only single-class labeling, one may still choose $\{P\}$, because its belief and plausibility are the highest without considering the set $\{B, F\}$. However, it may be sensible to collect more evidence rather than to generate a label for the pixel using the single-class evidence alone. If other evidence is available, the evidence combination problem will arise. This issue is considered in Section 4.4.3.

To investigate further the relationship between belief and plausibility, one can speculate that plausibility is always equal to or greater than belief, that is

$$\mathrm{Pl}(\psi) \geq \mathrm{Bel}(\psi),$$

or equivalently,

$$\text{Bel}(\sim \psi) \geq \text{Bel}(\psi) \tag{4.53}$$

which leads to

$$1 \geq \text{Bel}(\psi) + \text{Bel}(\sim \psi) \tag{4.54}$$

Equation (4.54) is derived on the basis of the observation that both Bel(ψ) and Bel($\sim\psi$) have no subset in common, and Bel(ψ) and Bel($\sim\psi$) are each made by the sum of the *bpa* of its own subsets. If we let $\psi = P$, it follows that

$$1 - \text{Bel}(\psi) - \text{Bel}(\sim \psi) = 1 - \text{Bel}(\{P\}) - \text{Bel}(\{-P\}) = 1 - \text{Bel}(\{P\}) - \text{Bel}(\{B,F\})$$

$$= 1 - m(P) - m(B) - m(F) - m(\{B,F\}) \tag{4.55}$$

$$= m(\{B,P\}) + m(\{F,P\}) + m(\{B,F,P\}).$$

Equation (4.55) specifies what the interval or uncertainty is, and it is then apparent that $1 - \text{Bel}(\psi) - \text{Bel}(\sim\psi)$ can be greater than 0, which confirms Equation (4.54).

4.4.3 EVIDENCE COMBINATION

In most cases, more than one set of evidence is available, and the decision to label a pixel is made based on the accumulation of all of the evidence. D-S theory acknowledges such a requirement and provides a formal proposal for multievidence management.

The aggregation of multiple belief functions is called Dempster's orthogonal sum, or Dempster's rule of combination (Shafer, 1979). Let Bel_a and Bel_b denote two belief functions, and let m_a and m_b be their corresponding *bpas*. Dempster's orthogonal sum generates a new *bpa*, denoted by $m_a \oplus m_b$, which represents the result of combing m_a and m_b. The result of the accumulation of both belief functions, denoted by $\text{Bel}_a \oplus \text{Bel}_b$, is derived from $m_a \oplus m_b$. If $m(\psi)$ denotes the new aggregated *bpa*, the combination rule can be specified by

$$m(\psi) = K \sum_{\xi \cap \zeta = \psi} m_a(\xi) \times m_b(\zeta) \tag{4.56}$$

where K is a normalizing constant, which is defined by

$$K^{-1} = 1 - \sum_{\xi \cap \zeta = \varnothing} m_a(\xi) \times m_b(\zeta). \tag{4.57}$$

The accumulation of *bpa* $m_a : [m_a(\xi_1), m_a(\xi_2), ..., m_a(\theta)]$ and *bpa* $m_b : [m_b(\zeta_1), m_a(\zeta_2), ..., m_a(\theta)]$ is calculated by considering all products in the form of $m_a(\xi) \times m_b(\zeta)$, where ξ and ζ are individually varied over all subsets of Bel_a and Bel_b. Note that the resulting sum of $m_a(\xi) \times m_b(\zeta)$ is equal to one, as is required by the definition of a *bpa*.

Equations (4.56) and (4.57) illustrate that $m(\psi)$ is formed by the orthogonal summation of all products $m_a(\xi) \times m_b(\zeta)$, which have an intersection ψ. The normalizing constant K is formed by the reciprocal of the sum of all products $m_a(\xi) \times m_b(\zeta)$ for which there is no intersection. This normalizing constant ensures that no contribution is committed to the null set (i.e., $\xi \cap \zeta = \varnothing$). The evidence combination process is illustrated by two examples, which are described below.

TABLE 4.1

Example of Dempster's Rule Combination

$m_a \rightarrow m_b \downarrow$	$m_a(\{B, F\}) = (0.7)$	$m_a(\theta) = (0.3)$
$m_b(\{B, P\}) = (0.6)$	$\{B\} = (0.42)$	$\{B, P\} = (0.18)$
$m_b(\theta) = (0.4)$	$\{B, F\} = (0.28)$	$\{\theta\} = (0.12)$

In the first example, suppose that one observation supports $\{B, F\}$ to the degree of 0.7 (i.e., m_a), whereas another observation disconfirms $\{F\}$ to the degree of 0.6 (i.e., m_b) (note that the situation here is the same to confirm $\{B, P\}$ to the degree of 0.6). For illustrative purposes, it is convenient to use an intersection table with the values assigned by m_a and m_b along the rows and columns, respectively, and only nonzero values are taken into consideration. We define the entry (r, c) in the table as the intersection of the subsets in row r and column c. The result of the rule combination is shown in Table 4.1.

In the example, each subset appears only once in the intersection table and no null intersections occur. Hence, $m_a \oplus m_b$ for each intersection is easily calculated as shown in each entry. Once the calculation of $m_a \oplus m_b$ is completed, the belief and plausibility of the combination $Bel_a \oplus Bel_b$ can be derived as

$$\mathrm{Bel}(\{B\}) = m_a \oplus m_b(\{B\}) = 0.42$$

$$\mathrm{Bel}(\{B, P\}) = m_a \oplus m_b(\{B\}) + m_a \oplus m_b(\{P\}) + m_a \oplus m_b(\{B, P\})$$

$$= 0.42 + 0 + 0.18 = 0.6$$

$$\mathrm{Bel}(\{B, F\}) = m_a \oplus m_b(\{B\}) + m_a \oplus m_b(\{F\}) + m_a \oplus m_b(\{B, F\})$$

$$= 0.42 + 0 + 0.28 = 0.7$$

The belief for other subsets is zero. The plausibilities for subsets $\{B\}$, $\{B, F\}$, and $\{B, P\}$ are all equivalent to one, because

$$\mathrm{Bel}(\sim \{B\}) = \mathrm{Bel}(\sim \{B, F\}) = \mathrm{Bel}(\sim \{B, P\}) = 0.$$

The second example is an extension of the first. Suppose that there is a third observation that confirms $\{P\}$ to the degree of 0.7, denoted by m_c, and let $m_a \oplus m_b = m_d$. The orthogonal sum $m_c \oplus m_d$ is illustrated in Table 4.2. There are two null intersections (i.e., ϕ). One contains the value 0.294, and the other, 0.196. Therefore, the normalization constant K and the final evidence combination results are computed by

$$K^{-1} = 1 - (0.294 + 0.196) = 0.51$$

$$m_c \oplus m_d(\{B\}) = 0.126 / 0.51 = 0.247$$

$$m_c \oplus m_d(\{P\}) = (0.126 + 0.084) / 0.51 = 0.412$$

$$m_c \oplus m_d(\{B, F\}) = 0.084 / 0.51 = 0.165$$

$$m_c \oplus m_d(\{B, P\}) = 0.054 / 0.51 = 0.106$$

$$m_c \oplus m_d(\theta) = 0.036 / 0.51 = 0.07$$

and for all other subsets, $m_c \oplus m_d = 0$

TABLE 4.2

Extended Example of Dempster's Rule Combination

$m_d \rightarrow m_c\downarrow$	$m_d(\{B\}) = (0.42)$	$m_d(\{B, F\}) = (0.28)$	$m_d(\{B, P\}) = (0.18)$	$m_d(\theta) = (0.12)$
$mc(\{P\})(0.7)$	$\phi = 0.294$	$\phi = 0.196$	$\{P\} = 0.126$	$\{P\} = 0.084$
$m_c(\theta)(0.3)$	$\{B\} = 0.126$	$\{B, F\} = 0.084$	$\{B, P\} = 0.054$	$\theta = 0.036$

Both belief and plausibility can be derived in a similar way, as explained in example one, and so will not be duplicated here. Readers may also wish to combine *bpas* using the above examples but in a different sequence. For example, one may compute the orthogonal sum first using evidence 1 and 3 and then combine with evidence 2. This exercise should show that the results of evidence combination are the same, regardless of the ordering.

4.4.4 DECISION RULES FOR EVIDENTIAL REASONING

Where a statistical classifier such as ML is used, the label assigned to a given pixel is the one for which the class membership likelihood function is greatest. This is the most straightforward choice. However, in evidential reasoning, the decision rules are more complicated because the decision space involves two elements (i.e., belief and plausibility). According to Shafer (1979), there are three possible choices:

1. **Belief-Driven**: Label is chosen on the basis of maximal belief.
2. **Plausibility-Driven**: Label is chosen on the basis of maximal plausibility.
3. **Mean of the Interval**: Label is chosen on the basis of the maximal average of belief and plausibility values.

These three choices may generate different results. If we look at the example given above, with three classes (bare soil $\{B\}$, forest $\{F\}$, and pasture $\{P\}$), it is possible to compute the belief and plausibility values for each.

$$\text{Bel}\{B\} = m(\{B\}), \quad \text{Pl}\{B\} = 1 - m(\{F\}) - m(\{P\})$$

$$\text{Bel}\{F\} = m(\{F\}), \quad \text{Pl}\{F\} = 1 - m(\{B\}) - m(\{P\}) \tag{4.58}$$

$$\text{Bel}\{P\} = m(\{P\}), \quad \text{Pl}\{P\} = 1 - m(\{B\}) - m(\{F\})$$

It is clear that if $m(\{B\}) > m(\{F\})$, that is, $\text{Bel}\{B\} > \text{Bel}\{F\}$, then one can also conclude that $\text{Pl}\{B\} > \text{Pl}\{F\}$. In this situation, two of the decision rules (i.e., belief-driven and plausibility-driven) produce the same result. Thus, it shows that one still has just one choice to make the decision if one tends to perform the single-class labeling.

4.5 DEALING WITH SOURCE RELIABILITY

The methods described above show promise for handling the multisource classification problem. The issue concerning the source weighting factors is not fully resolved, however. In practical circumstances, each available data source may not be fully certain and complete. Therefore, it is necessary to weigh each of the sources so that the final classification reflects our knowledge of the reliability of each of the data sources.

The process for the determination of weighting factors is similar to that of computing reliability in consensus theory for managing different opinion sources (Winkler, 1968; McConway, 1980;

French, 1985). The choice of source weighting factors will have a significant effect on the results of a multisource classification because the contribution of each source will be reduced or enhanced in proportion to the weight. It can be seen that the results of the classification process depend both on the information provided by the source and on the source weighting factor.

For example, if two sources, denoted by A and B, respectively, are used for labeling a pixel of interest, source A may suggest that the pixel belongs to class a with probability 0.8 and to class b with probability 0.2. However, source B indicates that the pixel belongs to class b with probability 1.0 and assigns zero probability to the pixel belonging to class a. If both sources have been determined as fully reliable (i.e., their associated weights are unity), the result based on both statistical Bayesian and evidential reasoning approaches for source combination will label the pixel as class b. If we have determined that the weight of source B is 0.5, and that of source A is unity, then the pixel will be labeled as class a. Several possible methods for measuring the source weighting parameters are described below.

4.5.1 USING CLASSIFICATION ACCURACY

The derivation of data source weighting parameters from classification accuracy measurements is an instinctive approach. A data source should be assigned a higher weight if the resulting classification accuracy is high. However, if the classification derived from the data source is unsatisfactory, the data are considered to be relatively unreliable and should be assigned a lower weight. For example, if the classification accuracy using data source A is 80%, and using data source B is only 50%, then the probability measure or evidence derived from data source A should be assigned a higher weight. This method is very easy to apply, but one should note that the 80% and 50% classification accuracies do not necessarily mean that the source weighting parameters are equal to 0.8 and 0.5, respectively, since different methods can be used for analyzing classification accuracy (Chapter 2), and these accuracy analysis tools are likely to provide different measures. The reliability measure in terms of classification accuracy can thus only give us a rough guide.

4.5.2 USE OF CLASS SEPARABILITY

The second approach to the measurement of source weighting uses the concept of the separability of the information classes. One can assign a higher weight to a source that contributes more to the separability of the information classes, while sources that contribute little to interclass separability can be assigned low weights. The estimates of the weighting parameter values are therefore based on the contribution to the statistical separability of each data source. Approaches to the estimation of separability, using distance, divergence, or separability functions, are discussed in Chapter 3. One problem may result from the employment of separability as a weighting parameter measure. The assumption underlying each approach is that the data distribution for each class is Gaussian, and consequently that interclass relationships are fully described in the covariance matrices. However, not all data sources are Gaussian in their distribution; for example, the frequency distribution of land elevation measurements is often skewed. In such cases, the use of separability measures may be misleading.

4.5.3 DATA INFORMATION CLASS CORRESPONDENCE MATRIX

Consider that there are a total of I information classes denoted by $\{\alpha_1, \alpha_2, ..., \alpha_I\}$, and that there are J data classes (e.g., J clusters) denoted by $\{\beta_1, \beta_2, ..., \beta_J\}$. The relationship between information classes α_i and data classes β_j can be represented using conditional probabilities, denoted by $P(\alpha_i/\beta_j)$, which can be used to construct a J by I correspondence matrix M, expressed as

$$M = \begin{bmatrix} P(\alpha_1/\beta_1) & P(\alpha_2/\beta_1) & \cdots & P(\alpha_{I-1}/\beta_1) & P(\alpha_I/\beta_1) \\ P(\alpha_1/\beta_2) & P(\alpha_1/\beta_2) & \cdots & P(\alpha_{I-1}/\beta_2) & P(\alpha_I/\beta_2) \\ \cdot & \cdot & \cdots & \cdot & \cdot \\ \cdot & \cdot & \cdots & \cdot & \cdot \\ \cdot & \cdot & \cdots & \cdot & \cdot \\ P(\alpha_1/\beta_J) & P(\alpha_1/\beta_J) & \cdots & P(\alpha_{I-1}/\beta_J) & P(\alpha_I/\beta_J) \end{bmatrix} \tag{4.59}$$

The objective of the correspondence matrix is to express the strength of the relationship between data classes and information classes. This kind of relationship is sometimes called equivocation. A higher weight is assigned to the data source if a data class strongly supports the information class. A data source should have a lower weight if there is no such strong link between some data classes and information classes. After constructing the correspondence matrix, a meaningful measure is needed to quantify the overall relationships.

In the case in which the linkage between the data classes and the information classes is high, there will be a unique conditional probability in each row of matrix M containing the value of 1, with the remaining entries taking the value 0. Conversely, if a data class is simultaneously linking to several information classes, then in the extreme case, each entry of matrix M will be equal (i.e., there will be no correspondence between the data classes and information classes). A method of expressing these relationships is the entropy measure of Shannon and Weaver (1963). For a particular data class β_j, entropy measure E is given by

$$E(\alpha/\beta_j) = -\sum_i P(\alpha_i/\beta_j) \times \ln P(\alpha_i/\beta_j) \tag{4.60}$$

and the overall measure for all observed data classes and information classes can be expressed as

$$E(\alpha/\beta) = -\sum_i \sum_j P(\alpha_i/\beta_j) \times \ln P(\alpha_i/\beta_j). \tag{4.61}$$

It should be noted that as entropy is being used to quantify the relationship between data classes and information classes, the higher the value, the less weight should be allocated to the data source.

The methodology discussed above only provides a rough estimation of data source weighting parameters. One has to determine a mapping function to convert the measured results into the weighting parameters. It is not always easy to define such a mapping function, and the determination of the weighting parameter still relies on the analyst's *ad hoc* decisions, which cannot guarantee an optimal solution. Moreover, when the MRF-MAP estimate is attempted, the parameter selection issue will become more complicated because both weighting and potential parameters must be determined simultaneously.

A straightforward approach to achieve optimal parameter assignments is to perform an exhaustive search of the total parameter space and then select the best solution. Searching for the best combination of weights can be time-consuming. For example, if there are 10 data sources, and if the weights are defined within the range [0, 1] with steps of 1/256, the total search space will be 256^{10}. A more efficient search tool is needed to reduce the computational cost. As suggested by Yilmaz et al. (2020), the genetic optimal search algorithm appears to be a valuable option for finding optimal solutions.

4.6 CONCLUDING REMARKS AND FUTURE TRENDS

In remote sensing applications that often consider two or more images capturing the same scene, image fusion is an essential prerequisite, constituting the initial phase of most processing operations. By merging images possessing diverse spatial and spectral resolutions, we obtain more informative images that contribute to improving classification accuracy, change detection performance, and human visual perception. The fundamental principle underpinning all fusion algorithms is the preservation of both the spectral and spatial information inherent in the original images. Despite the historical dominance of panchromatic (PAN) and multispectral (MS) image fusion in remote sensing, the current emphasis has shifted toward the fusion of hyperspectral data with images from different sensors, collectively termed as multimodal data. Traditional image fusion methods predominantly employ component substitution strategies, adaptive techniques, and multiresolution frameworks. Continuous efforts have been made to address various significant challenges associated with these algorithms by refining their improved versions. However, every algorithm possesses inherent limitations rooted in its mathematical foundations. Since the mid-2010s, deep learning-based algorithms have emerged as the dominant force in the field of image fusion, aligning with trends observed in other facets of remote sensing. The preceding sections have explored the theories and applications of key conventional and deep learning-based methods. To present a comprehensive overview, this section addresses the deficiencies associated with these techniques and outlines the future trajectories of image fusion, which is also discussed in several research (e.g., Fotso Kamga et al., 2021; Paul and Pati, 2021; Zhang et al., 2021; Karim et al., 2022; Soniminde and Biradar, 2023). Most fusion methods encounter issues such as insufficient preservation of spectral details, inadequate incorporation of spatial details, registration errors, and computational complexities. Consequently, research in this domain remains a vibrant focal point for pattern recognition and remote sensing practitioners. At this point, it should be underlined that many physical phenomena interfere during the acquisition of a remotely sensed image, and they need to be considered in a successful image fusion framework (Thomas et al., 2008). The primary challenges confronting the current methods are outlined below:

- Considering that PAN and MS images originate from distinct sensors, the likelihood of encountering registration errors is an ongoing concern. Image registration represents an indispensable initial phase that must be executed before initiating the image fusion process. The purpose of image registration is to achieve geometric alignment between one image and another, and it serves as a pivotal element in all image fusion techniques. A comprehensive review of image registration in remote sensing is available in Paul and Pati (2021). Within the context of image registration that relies on mutual information, research has been conducted to address the problem of interpolation artifacts. It has been established that basic parametric geometric transformations may not be well suited for effectively modeling deformations. Furthermore, when investigating interpolation artifacts, it is necessary to consider a wider range of similarity measures beyond mutual information. The explanations currently available in the literature for the artifacts observed in registration functions do not appear to provide a comprehensive understanding that is universally applicable to all similarity measures and image varieties. Inglada et al. (2007) delved into the characterization of these artifacts across various similarity measures and interpolation schemes and concluded that the interpolation has a smoothing effect that depends on the applied shift and that the interpolation has a denoising effect. In recent years, DL models have been successfully applied to improve the performance of iterative, intensity-based registration approaches, such as by employing reinforcement learning to registration (Simonovsky et al., 2016; Xu et al., 2022). Hence, the use of DL methods stands out as the most promising approach for tackling the image registration issue, primarily due to their robust learning attributes.

- Recent progress in deep learning-based image fusion demonstrates considerable potential for future advancements in this field. The primary trajectory in this field will involve devising more sophisticated fusion techniques using DL models. Although CNNs, GANs, and autoencoder methods have been predominant in recent research, each approach possesses its own set of strengths and weaknesses. Within the field of remote sensing, diverse challenges are encountered, with the adequacy of a large training data set being the primary concern for DL models. Consequently, the forthcoming directions are expected to focus on implementing self-supervised learning that incorporates both labeled and unlabeled data, as well as employing self-attention networks. These strategies aim to enhance both the quantity and quality of training samples, thus avoiding the issue of overfitting. Self-Attention Generative Adversarial Adaptation Network (SaGAAN) proposed by Zhao et al. (2020) is an example of such an approach. The self-attention framework enhances the potential of DL architectures to emphasize the significant dynamic attributes in time-series remote sensing imagery, accounting for the interdependencies among the data. On the other hand, deep domain adaptation networks are a recent development, yet they hold significant promise in addressing issues related to the need for extensive training data in remote sensing. Their potential lies in transferring valuable knowledge from one domain to another. Due to the unique conditions of each image related to illumination, viewing angle, misalignment, and image quality, the transfer learning concept performs poorly when the target domain is typically subject to different conditions (Fotso Kamga et al., 2021). Hence, these models can be successfully adapted to different conditions of the input images employed in the image fusion process.
- The objective evaluation of any application or method has posed significant challenges. While several recognized fusion evaluation metrics are available for assessing the performance of image fusion techniques, there is no universally accepted metric for evaluating the quality of image fusions resulting from pixel-based, frequency-based, and deep learning-based fusion techniques. Visual assessment of the fusion quality, despite its subjectivity, remains a prevalent method. Image fusion quality metrics are derived either with the use of a reference image or ground reference data, or by applying nonreference measures such as entropy, edge-based, feature-based, and correlation-based statistics. A common practice involves combining subjective and objective evaluations when reporting the outcomes of image fusion procedures. Selecting the most suitable metric can be a challenging endeavor, especially for researchers and practitioners who are less experienced. This concern is expected to become increasingly significant when assessing the outcomes of future state-of-the-art methods.
- Another crucial factor in validating the reliability of a new fusion algorithm is the necessity for real-world case studies and situations. Similar to the necessity for benchmark data sets in assessing classifier performance using hyperspectral data (e.g., University of Pavia, Indian Pines, Salinas), standardized data sets are also indispensable for evaluating pan-sharpening, spatiotemporal fusion, and various multimodal image fusion scenarios. At present, the process of accessing and obtaining suitable data sets, including hyperspectral, multispectral, LiDAR, and SAR data, and ensuring their compatibility for fusion purposes, can be quite demanding. Therefore, one of the future objectives in the field of remote sensing will involve the compilation of extensive data sets for the evaluation of image fusion methods with a focus on reliability.
- When compared to other supervised and unsupervised models, deep learning-based models have been recognized as the most suitable architectures for image fusion. However, it is important to note that in addition to a substantial training data set, they also necessitate considerable computational resources, particularly GPU power, for effective processing. As the array of image types and model complexity increases, researchers are increasingly concerned about the need to modernize the computer infrastructure. The duration of

training can range from hours to days in certain situations, depending on the computing unit's capabilities, model complexity, and the size of the training data set. Cloud computing, as a revolutionary concept, holds the promise of advancing the field of image fusion soon by reducing reliance on individual computational resources. The widespread adoption of distributed computing techniques, once limited by the availability of robust computers, has become more prevalent due to the increasing availability of cloud computing infrastructure services. Within this framework, cloud computing technology can facilitate the handling of data cubes, while the array of algorithms can be personalized to meet the specific requirements of applications.

- The continuous advancement of UAV systems and sensors mounted on these platforms, along with the widespread use of LiDAR data, has opened up new possibilities for fusing 3D geometric information with multispectral, hyperspectral, and radar data. The incorporation of three-dimensional (3D) mapping applications holds significant promise in the realm of remote sensing, including applications for crop classification, creation of 3D canopy models, object detection, and change analysis. At present, dense point clouds can be easily derived from digital images captured by UAV cameras. The fusion of remotely sensed images into 3D scenarios brings forth novel developments in image classification to characterize the dynamic natural environment in terms of morphological and physiological features (Jurado et al., 2022). It is expected that future studies will delve into 3D data fusion with the aim of enhancing the current accuracy standards.

REFERENCES

Aiazzi, B., S. Baroni, and M. Selva. 2007. Improving component substitution pan-sharpening through multivariate regression of MS+Pan data. *IEEE Transactions on Geoscience and Remote Sensing* 45:3230–3239. https://doi.org/10.1109/TGRS.2007.901007.

Azarang, A., and N. Kehtarnavaz. 2022. *Image Fusion in Remote Sensing: Conventional and Deep Learning Approaches*. Morgan & Claypool. https://doi.org/10.1007/978-3-031-02256-2.

Barnett, J. A. 1981. Computational methods for a mathematical theory of evidence. In *Proceedings of 7th International Conference on Artificial Intelligence,* August 24–28, Vancouver, Canada, pp. 868–875.

Benediktsson, J. A., P. H. Swain, and O. K. Ersoy. 1990. Neural network approaches versus statistical methods in classification of multisource remote sensing data. *IEEE Transactions on Geoscience and Remote Sensing* 28:540–552. https://doi.org/10.1109/TGRS.1990.572944.

Burrus, C. S., R. A. Gopinath, and H, Guo. 1998. *Introduction to Wavelets and Wavelet Transforms: A Primer.* Upper Saddle River, NJ: Prentice Hall.

Chen, Y. H., L. Den, J. Li, X. Li, and P. J. Shi. 2006. A new wavelet-based image fusion method for remotely sensed data. *International Journal of Remote Sensing* 27:1465–1476. https://doi.org/10.1080/01431160500474365.

Chen, Y., K. Shi, Y. Ge, and Y. Zhou. 2022. Spatiotemporal remote sensing image fusion using multiscale two-stream convolutional neural networks. *IEEE Transactions on Geoscience and Remote Sensing* 60:1–12. https://doi.org/10.1109/TGRS.2021.3069116.

Colditz, R. R., T. Wehrmann, M. Bachmann, K. Steinnocher, M. Schmidt, G. Strunz, and S. Dech. 2006. Influence of image fusion approaches on classification accuracy: A case study. *International Journal of Remote Sensing* 27:3311–3335. https://doi.org/10.1080/01431160600649254.

Crist, E. P., and R. C. Cicone. 1984. A physically-based transformation of thematic mapper data: The TM Tasselled Cap. *IEEE Transactions on Geoscience and Remote Sensing* 22:256–263. https://doi.org/10.1109/TGRS.1984.350619.

Crouse, M. S., R. D. Nowak, and R. G. Baraniuk. 1998. Wavelet-based statistical signal processing using hidden Markov models. *IEEE Transactions on Signal Processing* 46:886–902. https://doi.org/10.1109/78.668544.

Daubechies, I. 1992. *Ten Lectures on Wavelets,* 2nd edition. Philadelphia, PA: Society for Industrial and Applied Mathematics (SIAM). https://doi.org/10.1137/1.9781611970104.

Dong, C., C. C. Loy, K. He, and X. Tang. 2016. Image super-resolution using deep convolutional networks. *IEEE Transactions on Pattern Analysis and Machine Intelligence* 38:295–307. https://doi.org/10.1109/TPAMI.2015.2439281.

El-Samie, F. E. A., M. M. Hadhoud, and S. E. El-Khamy. 2012. *Image Super-Resolution and Applications.* Boca Raton, FL: CRC Press.

Fotso Kamga, G. A., L. Bitjoka, T. Akram, A. Mengue Mbom, S. Rameez Naqvi, and Y. Bouroubi. 2021. Advancements in satellite image classification: Methodologies, techniques, approaches and applications. *International Journal of Remote Sensing* 42:7662–7722. https://doi.org/10.1080/01431161.2021.1954261.

Franklin, S. E., D. R. Peddle, and J. A. Dechka. 2002. Evidential reasoning with Landsat TM, DEM and GIS data for landcover classification in support of grizzly bear habitat mapping. *International Journal of Remote Sensing* 23:4633–4652. https://doi.org/10.1080/01431160110113971.

French, S. 1985. Group consensus probability distributions: A critical survey. In *Bayesian Statistics 2*, (eds.) J. M. Bernardo, M. H. DeGroot, D. V. Lindley, and A. F. M. Smith, pp. 183–202. New York: North Holland.

Friedman, J. H. 1994. An overview of predictive learning and function approximation. In *From Statistics to Neural Networks*, (eds.) V. Cherkassky, J. Friedman, and H. Wechsler, pp. 1–55. Berlin, Heidelberg: Springer. https://doi.org/10.1007/978-3-642-79119-2_1.

Garvey, T. D., J. D. Lowrance, and M. A. Fischler. 1981. An inference technique for integrating knowledge from disparate sources. In *Proceedings 7th International Conference on Artificial Intelligence*, August 24–28, Vancouver, Canada, pp. 319–325.

Ghahremani, M., and H. Ghassemian. 2015. Remote-sensing image fusion based on curvelets and ICA. *International Journal of Remote Sensing* 36:4131–4143. https://doi.org/10.1080/01431161.2015.1071897.

González-Audícana, M., X. Otazu, O. Fors, and A. Seco. 2005. Comparison between Mallat's and the 'a trous' discrete wavelet transform based algorithms for the fusion of multispectral and panchromatic images. *International Journal of Remote Sensing* 26:595–614. https://doi.org/10.1080/01431160512331314056.

Gordon, J., and E. H. Shortliffe. 1985. Method for managing evidential reasoning in a hierarchical hypothesis space. *Artificial Intelligence* 26:323–357. https://doi.org/10.1016/0004-3702(85)90064-5.

Gouvernet, J., J. Ayme, S. Sanchez, E. Mattei, and F. Giraud. 1980. Diagnosis assistance in medical genetics based on belief functions and a tree structured thesaurus: A conversational mode realization. In *Proceedings of MEDINFO 80*, September 29–October 4, Tokyo, Japan, p. 798.

Gungor, O., and J. Shan. 2004. Evaluation of satellite image fusion using wavelet transform. In *Proceedings of 20th Congress ISPRS*, July 12–23, Istanbul, Turkey, pp. 12–13.

Haralick, R. M., and K. S. Fu. 1983. Pattern recognition and classification. In *Manual of Remote Sensing*, (eds.) R. N. Colwell, D. S. Simonett, and F. T. Ulaby, pp. 881–884. Falls Church, VA: American Society of Photogrammetry.

Hill, J., C. Diemer, O. Stöver, and T. Udelhoven. 1999. A local correlation approach for the fusion of remote sensing data with different spatial resolutions in forestry applications. *International Archives of Photogrammetry and Remote Sensing* 32:781–789.

Hosseinpour, H., F. Samadzadegan, and F. D. Javan. 2022. CMGFNet: A deep cross-modal gated fusion network for building extraction from very high-resolution remote sensing images. *ISPRS Journal of Photogrammetry and Remote Sensing* 184:96–115. https://doi.org/10.1016/j.isprsjprs.2021.12.007.

Inglada, J., V. Muron, D. Pichard, and T. Feuvrier. 2007. Analysis of artifacts in subpixel remote sensing image registration. *IEEE Transactions on Geoscience and Remote Sensing* 45:254–264. https://doi.org/10.1109/TGRS.2006.882262.

Javan, F. D., F. Samadzadegan, S. Mehravar, A. Toosi, R. Khatami, and A. Stein. 2021. A review of image fusion techniques for pan-sharpening of high-resolution satellite imagery. *ISPRS Journal of Photogrammetry and Remote Sensing* 171:101–117. https://doi.org/10.1016/j.isprsjprs.2020.11.001.

Jha, A., S. Bose, and B. Banerjee. 2023. GAF-Net: Improving the performance of remote sensing image fusion using novel global self and cross attention learning. In *Proceedings of IEEE/CVF Winter Conference on Applications of Computer Vision (WACV'2023)*, January 02–07, Waikoloa, HI, pp. 6354–6363. https://doi.org/10.1109/WACV56688.2023.00629.

Jurado, J. M., A. López, L. Pádua, and J. J. Sousa. 2022. Remote sensing image fusion on 3D scenarios: A review of applications for agriculture and forestry. *International Journal of Applied Earth Observation and Geoinformation* 112:102856. https://doi.org/10.1016/j.jag.2022.102856.

Karim, S., G. Tong, J. Li, A. Qadir, U. Farooq, and Y. Yu. 2022. Current advances and future perspectives of image fusion: A comprehensive review. *Information Fusion* 90:185–217. https://doi.org/10.1016/j.inffus.2022.09.019.

Kim, H., and P. H. Swain. 1995. Evidential reasoning approach to multisource-data classification in remote sensing. *IEEE Transactions on Systems, Man, and Cybernetics* 25:1257–1265. https://doi.org/10.1109/21.398687.

Laben, C. A., and V. B. Bernard. 2000. Process for enhancing the spatial resolution of multispectral imagery using pan-sharpening. US Patent 6,011,875 filed April 29, 1998, and issued January 4, 2000.

Lee, T., J. A. Richards, and P. H. Swain. 1987. Probabilistic and evidential approaches for multisource data analysis. *IEEE Transactions on Geoscience and Remote Sensing* 25:283–293. https://doi.org/10.1109/TGRS.1987.289800.

Li, S., Q. Guo, and A. Li. 2022. Pan-sharpening based on CNN+ pyramid transformer by using no-reference loss. *Remote Sensing* 14:624. https://doi.org/10.3390/rs14030624.

Li, S., X. Kang, L. Fang, J. Hu, and H. Yin. 2017. Pixel-level image fusion: A survey of the state of the art. *Information Fusion* 33:100–112. https://doi.org/10.1016/j.inffus.2016.05.004.

Liu, D., F. Yang, H. Wei, and P. Hu. 2020a. Remote sensing image fusion method based on discrete wavelet and multiscale morphological transform in the IHS color space. *Journal of Applied Remote Sensing* 14:016518. https://doi.org/10.1117/1.JRS.14.016518.

Liu, P., J. Li, L. Wang, and G. He. 2022. Remote sensing data fusion with generative adversarial networks: State-of-the-art methods and future research directions. *IEEE Geoscience and Remote Sensing Magazine* 10:295–328. https://doi.org/10.1109/MGRS.2022.3165967.

Liu, Q., H. Zhou, Q. Xu, X. Liu, and Y. Wang. 2020b. PSGAN: A generative adversarial network for remote sensing image pan-sharpening. *IEEE Transactions on Geoscience and Remote Sensing* 59:10227–10242. https://doi.org/10.1109/TGRS.2020.3042974.

Liu, Y., X. Chen, Z. Wang, Z. J. Wang, R. K. Ward, and X. Wang. 2018. Deep learning for pixel-level image fusion: Recent advances and future prospects. *Information Fusion* 42:158–173. https://doi.org/10.1016/j.inffus.2017.10.007.

Luo, X., Z. Zhang, B. Zhang, and X. Wu. 2017. Image fusion with contextual statistical similarity and nonsubsampled Shearlet transform. *IEEE Sensors Journal* 17:1760–1771. https://doi.org/10.1109/JSEN.2016.2646741.

Ma, J., W. Yu, C. Chen, P. Liang, X. Guo, and J. Jiang. 2020. Pan-GAN: An unsupervised pan-sharpening method for remote sensing image fusion. *Information Fusion* 62:110–120. https://doi.org/10.1016/j.inffus.2020.04.006.

Ma, L., Y. Liu, X. Zhang, Y. Ye, G. Yin, and B. A. Johnson. 2019. Deep learning in remote sensing applications: A meta-analysis and review. *ISPRS Journal of Photogrammetry and Remote Sensing* 152:166–177. https://doi.org/10.1016/j.isprsjprs.2019.04.015.

Ma, W., O. Karakuş, and P. L. Rosin. 2022. AMM-FuseNet: Attention-based multi-modal image fusion network for land cover mapping. *Remote Sensing* 14:4458. https://doi.org/10.3390/rs14184458.

Mallat, S. G. 1989. A theory for multi-resolution signal decomposition: The wavelet representation. *IEEE Transactions on Pattern Analysis and Machine Intelligence* 11:674–693. https://doi.org/10.1109/34.192463.

Mallat, S., and S. Zhong. 1992. Characterization of signals from multiscale edges. *IEEE Transactions on Pattern Analysis and Machine Intelligence* 8:679–698. https://doi.org/10.1109/34.142909.

Masi, G., D. Cozzolino, L. Verdoliva, and G. Scarpa. 2016. Pan-sharpening by convolutional neural networks. *Remote Sensing* 8:594. https://doi.org/10.3390/rs8070594.

Maurer, T. 2013. How to pan-sharpen images using the gram-schmidt pan-sharpen method: A recipe. *International Archives of the Photogrammetry, Remote Sensing and Spatial Information Sciences* 40:239–244. https://doi.org/10.5194/isprsarchives-XL-1-W1-239-2013.

McConway, K. J. 1980. The combination of experts' opinions in probability assessment: Some theoretical considerations. Ph.D. Thesis, University College London, London, UK.

Meng, X., N. Wang, F. Shao, and S. Li. 2022. Vision transformer for pan-sharpening. *IEEE Transactions on Geoscience and Remote Sensing* 60:1–11. doi: 10.1109/TGRS.2022.3168465.

Miao, Q. G., C. Shi, P. F. Xu, M. Yang, and Y. B. Shi. 2011. A novel algorithm of image fusion using shearlets. *Optics Communications* 284:1540–1547. https://doi.org/10.1016/j.optcom.2010.11.048.

Nagarathinam, S., A. Vasuki, and K. Paramasivam. 2023. Deep remote fusion: Development of improved deep CNN with atrous convolution-based remote sensing image fusion. *The Imaging Science Journal* 1–21. https://doi.org/10.1080/13682199.2023.2206761.

Nencini, F., A. Garzelli, S. Baronti, and L. Alparone. 2007. Remote sensing image fusion using the curvelet transform. *Information Fusion* 8:143–156. https://doi.org/10.1016/j.inffus.2006.02.001.

Nikolakopoulos, K., and D. Oikonomidis. 2015. Quality assessment of ten fusion techniques applied on Worldview-2. *European Journal of Remote Sensing* 48:141–167. https://doi.org/10.5721/EuJRS20154809.

Núñez, J., X. Otazu, O. Fors, A. Prades, V. Palá, and R. Arbiol. 1999. Multiresolution-based image fusion with additive wavelet decomposition. *IEEE Transactions on Geoscience and Remote Sensing* 37:1204–1211. https://doi.org/10.1109/36.763274.

Ozcelik, F., U. Alganci, E. Sertel and G. Unal. 2020. Rethinking CNN-based pan-sharpening: Guided colorization of panchromatic images via GANs. *IEEE Transactions on Geoscience and Remote Sensing* 59:3486–3501. https://doi.org/10.1109/TGRS.2020.3010441.

Pan, Y., D. Pi, J. Chen, and H. Meng. 2021. FDPPGAN: Remote sensing image fusion based on deep perceptual patchGAN. *Neural Computing and Applications* 33:9589–9605. https://doi.org/10.1007/s00521-021-05724-1.

Paul, S., and U. C. Pati. 2021. A comprehensive review on remote sensing image registration. *International Journal of Remote Sensing* 42:5396–5432. https://doi.org/10.1080/01431161.2021.1906985.

Pellemans, A. H., R. W. Jordans, and R. Allewijn. 1993. Merging multispectral and panchromatic SPOT images with respect to the radiometric properties of the sensor. *Photogrammetric Engineering and Remote Sensing* 59:81–87.

Plataniotis, K. N., and A. N. Venetsanopoulos. 2000. *Color Image Processing and Applications*. Berlin, Heidelberg: Springer. https://doi.org/10.1007/978-3-662-04186-4.

Pohl, C., and J. van Genderen. 2016. *Remote Sensing Image Fusion: A Practical Guide*. Boca Raton, FL: CRC Press. https://doi.org/10.1201/9781315370101.

Prasad, N., S. Saran, S. P. S. Kushwaha, and P. S. Roy. 2001. Evaluation of various image fusion techniques and imaging scales for forest features interpretation. *Research Communication* 81:1218–1224.

Sabins, F. F. 1997. *Remote Sensing: Principles and Interpretation*, 3rd edition. San Francisco: W.H. Freeman.

Schetselaar, E. M. 1998. Fusion by the IHS transform: Should we use cylindrical or spherical coordinates? *International Journal of Remote Sensing* 19:759–765. https://doi.org/10.1080/014311698215982.

Schistad, A. H., A. K. Jain, and T. Taxt. 1994. Multisource classification of remotely sensed data: Fusion of Landsat TM and SAR images. *IEEE Transactions on Geoscience and Remote Sensing* 32:768–778. https://doi.org/10.1109/36.298006.

Schistad, A. H., T. Taxt, and A. K. Jain. 1996. A Markov random field model for classification of multisource satellite imagery. *IEEE Transactions on Geoscience and Remote Sensing* 34:100–113. https://doi.org/10.1109/36.481897.

Shafer, G. 1979. *A Mathematical Theory of Evidence*. Princeton, NJ: Princeton University Press. https://doi.org/10.2307/j.ctv10vm1qb.

Shafer, G., and R. Logan. 1987. Implementing Dempster's rule for hierarchical evidence. *Artificial Intelligence* 33:271–298. https://doi.org/10.1016/0004-3702(87)90040-3.

Shannon, C. E., and W. Weaver. 1963. *The Mathematical Theory of Communication*. Chicago, IL: University of Illinois Press.

Shao, Z., Z. Lu, M. Ran, L. Fang, J. Zhou, and Y. Zhang. 2020. Residual encoder-decoder conditional generative adversarial network for pan-sharpening. *IEEE Geoscience and Remote Sensing Letters* 17:1573–1577. https://doi.org/10.1109/LGRS.2019.2949745.

Sheng, Y. 1996. Wavelet transform. In *The Transforms and Applications Handbook* (ed.) A. D. Poularikas, pp. 747–827. Boca Raton, FL: CRC Press.

Simone, G., A. Farina, F. C. Morabito, S. B. Serpico, and L. Bruzzone. 2002. Image fusion techniques for remote sensing applications. *Information Fusion* 3:3–15. https://doi.org/10.1016/S1566-2535(01)00056-2.

Simonovsky, M., B. Gutiérrez-Becker, D. Mateus, N. Navab, and N. Komodakis. 2016. A deep metric for multimodal registration. In *Medical Image Computing and Computer-Assisted Intervention - MICCAI 2016*, (eds.) S. Ourselin, L. Joskowicz, M. Sabuncu, G. Unal, and W. Wells, pp. 10–18. Cham, Switzerland: Springer. https://doi.org/10.1007/978-3-319-46726-9_2.

Singh, A. 1984. Some clarifications about the pairwise divergence method in remote sensing. *International Journal of Remote Sensing* 5:623–627. https://doi.org/10.1080/01431168408948845.

Singh, V. 2020. Image fusion metrics: A systematic study. In *Proceedings of 2nd International Conference on IoT, Social, Mobile, Analytics & Cloud in Computational Vision & Bio-Engineering (ISMAC-CVB 2020)*, October 29–30, Thoothukudi, India, pp. 239–245. https://dx.doi.org/10.2139/ssrn.3734738.

Smyth, P., D. Heckerman, and M. I. Jordan. 1997. Probabilistic independence networks for hidden Markov probability models. *Neural Computation* 9:227–269. https://doi.org/10.1162/neco.1997.9.2.227.

Song, H., Q. Liu, G. Wang, R. Hang, and B. Huang. 2018. Spatiotemporal satellite image fusion using deep convolutional neural networks. *IEEE Journal of Selected Topics in Applied Earth Observations and Remote Sensing* 11:821–829. https://doi.org/10.1109/JSTARS.2018.2797894.

Soniminde, N., and M. Biradar. 2023. Progress of image fusion in the digital era: A comprehensive survey of techniques, applications, and future directions. Available at SSRN: https://ssrn.com/abstract=4524933 or https://dx.doi.org/10.2139/ssrn.4524933.

Srinivasan, A., and J. A. Richards. 1990. Knowledge-based techniques for multi-source classification. *International Journal of Remote Sensing* 11:505–525. https://doi.org/10.1080/01431169008955036.

Strat, T. M. 1984. Continuous belief functions for evidential reasoning. In *Proceedings of Fourth National Conference on Artificial Intelligence (AAAI-84)*, August 6–10, Austin, TX, pp. 308–313.

Sun, C., C. Zhang, and N. Xiong. 2020. Infrared and visible image fusion techniques based on deep learning: A review. *Electronics* 9:2162. https://doi.org/10.3390/electronics9122162.

Teggi, S., R. Cecchi, and F. Serafina. 2003. TM and IRS-1C-PAN data fusion multiresolution decomposition methods based on the 'a tròus' algorithm. *International Journal of Remote Sensing* 24:1287–1301. https://doi.org/10.1080/01431160210144561.

Thomas, C., T. Ranchin, L. Wald, and J. Chanussot. 2008. Synthesis of multispectral images to high spatial resolution: A critical review of fusion methods based on remote sensing physics. *IEEE Transactions on Geoscience and Remote Sensing* 46:1301–1312. https://doi.org/10.1109/TGRS.2007.912448.

Tso, B. 1997. Investigation of alternative strategies for incorporating spectral, textural and contextual information in remote sensing image classification. Ph.D. Thesis, School of Geography, The University of Nottingham, Nottingham, U.K.

Ulaby, F. T., F. Kouyate, B. Brisco, and L. Williams. 1986. Texture information in SAR images. *IEEE Transactions on Geoscience and Remote Sensing* 24:235–245. https://doi.org/10.1109/TGRS.1986.289643.

Ulfarsson, M. O., J. A. Benediktsson, and J. R. Sveinsson. 2003. Data fusion and feature extraction in the wavelet domain. *International Journal of Remote Sensing* 24:3933–3945. https://doi.org/10.1080/0143116031000103790.

Wald, L. 1999. Some terms of reference in data fusion. *IEEE Transactions on Geosciences and Remote Sensing* 37:1190–1193. https://doi.org/10.1109/36.763269.

Wald, L. 2000. Quality of high resolution synthesized images: Is there a simple criterion. In *Proceedings of the Third Conference "Fusion of Earth Data: Merging Point Measurements, Raster Maps and Remotely Sensed Images"*, France, January 26–28, Sophia Antipolis, France, pp. 99–103.

Wald, L. 2002. *Data Fusion: Definitions and Architectures*. Paris: Ecole des Mines de Paris.

Wald, L., T. Ranchin, and M. Mangolini. 1997. Fusion of satellite images of different spatial resolutions: Assessing the quality of resulting images. *Photogrammetric Engineering and Remote Sensing* 63:691–699.

Wang, C., W. Sun, D. Fan, X. Liu, and Z. Zhang. 2021. Adaptive feature weighted fusion nested U-Net with discrete wavelet transform for change detection of high-resolution remote sensing images. *Remote Sensing* 13:4971. https://doi.org/10.3390/rs13244971.

Wang, Z., and A. C. Bovik. 2002. A universal image quality index. *IEEE Signal Processing Letters* 9:81–84.

Wang, Z., and Q. Li. 2011. Information content weighting for perceptual image quality assessment. *IEEE Transactions on Image Processing* 20:1185–1198. https://doi.org/10.1109/TIP.2010.2092435.

Wang, Z., B. Chen, R. Lu, H. Zhang, H. Liu, and P. K. Varshney. 2020. FusionNet: An unsupervised convolutional variational network for hyperspectral and multispectral image fusion. *IEEE Transactions on Image Processing* 29:7565–7577. https://doi.org/10.1109/TIP.2020.3004261.

Wilkinson, G. G., and J. Mégier. 1990. Evidential reasoning in a pixel classification hierarchy: A potential method for integrating image classifiers and expert system rules based on geographic context. *International Journal of Remote Sensing* 11:1963–1968. https://doi.org/10.1080/01431169008955152.

Winkler, R. L. 1968. The consensus of subjective probability distributions. *Management Science* 15:B61–B75. https://doi.org/10.1287/mnsc.15.2.B61.

Xiao, G., D. P. Bavirisetti, G. Liu, and X. Zhang. 2020. Image fusion based on machine learning and deep learning. In *Image Fusion*, pp. 325–352, Singapore: Springer. https://doi.org/10.1007/978-981-15-4867-3_7.

Xie, W., Y. Cui, Y. Li, J. Lei, Q. Du, and J. Li. 2021. HPGAN: Hyperspectral pan-sharpening using 3-D generative adversarial networks. *IEEE Transactions on Geoscience and Remote Sensing* 59:463–477. https://doi.org/10.1109/TGRS.2020.2994238.

Xu, H., J. Ma, J. Yuan, Z. Le, and W. Liu. 2022. RFNet: Unsupervised network for mutually reinforcing multi-modal image registration and fusion. In *Proceedings of the IEEE/CVF Conference on Computer Vision and Pattern Recognition (CVPR)*, June 18–24, New Orleans, LA, pp. 19679–19688. https://doi.org/10.1109/CVPR52688.2022.01906.

Xu, M., H. Chen, and P. K. Varshney. 2011. An image fusion approach based on Markov random fields. *IEEE Transactions on Geoscience and Remote Sensing* 49:5116–5127. https://doi.org/10.1109/TGRS.2011.2158607.

Ye, F., X. Li, and X. Zhang. 2019. FusionCNN: A remote sensing image fusion algorithm based on deep convolutional neural networks. *Multimedia Tools and Applications* 78:14683–14703. https://doi.org/10.1007/s11042-018-6850-3.

Yilmaz, C. S., V. Yilmaz, and O. Gungor. 2022. A theoretical and practical survey of image fusion methods for multispectral pan-sharpening. *Information Fusion* 79:1–43. https://doi.org/10.1016/j.inffus.2021.10.001.

Yilmaz, V. 2023. Adaptive hybrid pansharpening: A novel approach for combining two methods to achieve superior pansharpening performance. *International Journal of Remote Sensing* 44:4301–4325. https://doi.org/10.1080/01431161.2023.2234095.

Yilmaz, V., C. Serifoglu Yilmaz, O. Güngör, and J. Shan. 2020. A genetic algorithm solution to the Gram-Schmidt image fusion. *International Journal of Remote Sensing* 41:1458–1485. https://doi.org/10.1080/0143116 1.201-9.1667553.

Zhang, H., H. Xu, X. Tian, J. Jiang, and J. Ma. 2021. Image fusion meets deep learning: A survey and perspective. *Information Fusion* 76:323–336. https://doi.org/10.1016/j.inffus.2021.06.008.

Zhang, N., and Q. Wu. 2008. Effects of Brovey transform and wavelet transform on the information capacity of SPOT-5 imagery. *Proceedings of the SPIE* 6623:257–261. https://doi.org/10.1117/12.791423.

Zhao, W., X. Chen, J. Chen, and Y. Qu. 2020. Sample generation with self-attention generative adversarial adaptation network (SaGAAN) for hyperspectral image classification. *Remote Sensing* 12:843. https://doi.org/10.3390/rs12050843.

Zhu, L., and R. Tateishi. 2006. Fusion of multisensor multitemporal satellite data for land cover mapping. *International Journal of Remote Sensing* 27:903–918. https://doi.org/10.1080/0143116031000139818.

5 Support Vector Machines

In the field of pattern recognition, support vector machines (SVMs) hold an important position, as a focal point of considerable interest and a subject of active exploration among researchers since 1998. SVMs represent a group of theoretically superior machine learning algorithms. The development of SVMs was the result of the exploration and formalization of learning machine capacity control and overfitting issues (Vapnik, 1979, 1995, 1998). Although SVMs can be said to have emerged in the late 1970s (Vapnik, 1979), they did not receive substantial attention until the late 1990s. The attractiveness of SVMs is their ability to minimize the so-called structural risk or classification errors when solving classification problems. Unlike conventional statistical maximum likelihood methods that empirically minimize misclassification error in an empirical sense, which is directly determined by the distribution of training samples, the structural risk minimization concept adopted by SVMs minimizes the probability of misclassifying a previously unseen data point drawn randomly from a fixed but unknown probability distribution (Vapnik, 1995, 1998). Such a property is also different from the decision boundary–forming logic of the ANNs, as discussed in Chapter 7. SVMs, which originated from investigations in statistical learning theory, focus on managing the trade-off between structural complexity and empirical risk (Jayadeva and Chadra, 2007). Specifically, SVM training always finds a global minimum. Their simple geometric interpretation, as shown later in this chapter, provides a basis for further investigation (Burges, 1998).

To address the supervised classification and regression problems, SVMs serve as nonparametric statistical methods that avoid assumptions about the underlying data distribution. The purpose of the SVM training algorithm is to find a hyperplane that separates the data set into distinct predefined classes, adhering to the patterns within the training examples. The term "optimal separating hyperplane" describes the boundary that minimizes misclassification during the training phase. Using the kernel trick, as detailed in upcoming sections, the original data can be projected into higher-dimensional space, providing a solution for managing linear nonseparable problems. Within this space, it establishes a linear separation between the two classes. Transformation to a higher-dimensional space generally scatters the data, making it easier to identify a linear division. Consequently, nonlinear separations within the initial data space can become linear within the higher-dimensional space. The SVM learning process involves discovering the best decision boundary to distinguish training patterns and subsequently segregate test data under the same conditions. SVMs exhibit considerable strength because they utilize the structural risk principle, leading to enhanced generalization and decreased errors during training (Tanveer et al., 2022). It is noteworthy that the selection of a suitable kernel function is of critical importance in the design of SVMs (Kavzoglu and Colkesen, 2009). Although numerous approaches have been developed to classify remotely sensed data, SVMs have seen an increased level of interest, particularly for the classification of satellite imagery. The sustained interest in SVMs can be attributed to their advancements through active and semi-supervised learning, as well as their combined usage with other advanced algorithms (e.g., Bayesian learning, genetic algorithms in Section 3.2.5.1, particle swarm optimization in Section 3.2.5.2, and fuzzy clustering). In a comprehensive meta-analysis of supervised pixel-based image classification practices, Khatami et al. (2016) concluded that SVM algorithms, including the original SVMs and their variants, outperformed all other classifiers in pairwise comparisons, establishing the method as the strongest competitor to deep learning methods.

SVMs were originally linear binary classifiers that allocate the labels +1 and −1. The core operation of SVMs is to construct a separating hyperplane (i.e., a decision boundary) based on the properties of the training samples, specifically their distribution in feature space. Such a separating

DOI: 10.1201/9781003439172-5

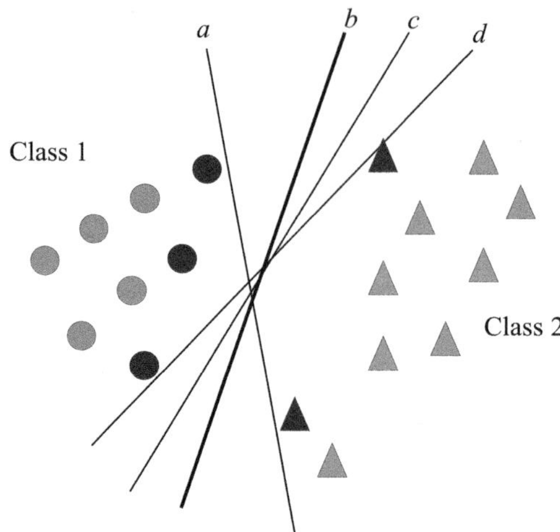

FIGURE 5.1 Hyperplane *b* separates the two classes with the maximal margin.

hyperplane is subject to the condition that the margin of separation between class +1 and class −1 samples is maximized (Vapnik, 1979). Figure 5.1 illustrates this idea in which hyperplane *b* separates the two classes with a maximal margin, whereas hyperplanes *a*, *c*, and *d* do not fulfill the requirements of minimizing structural risk. As the reader will see in the later mathematical derivations, not necessarily all training samples contribute to the building of the hyperplane, but normally only a subset of training samples is chosen as support vectors. This attribute is unique to SVMs. As shown in Figure 5.1, only the black points are support vectors that define the hyperplane *b* separating the two classes with a maximal distance.

The accuracy achieved with SVMs significantly depends on hyperparameter optimization and the selection of the suitable kernel function. Consequently, the tuning method chosen for these issues plays a crucial role in achieving the best possible performance. The widely adopted technique for adjusting various hyperparameters is the grid search algorithm, which is discussed in Section 5.3.3. By incrementally adjusting hyperparameters within a broad spectrum of values with a uniform step size, the performance of each combination is evaluated using a performance metric. This approach can suffer from a severe computational burden as it trains the SVM model for all parameter combinations, especially when more than two classes are to be separated. Gradient descent search algorithms (Section 5.3.4) offer a solution for dealing with an extensive set of parameters. This approach involves navigating the parameter space, starting from an initial point with random parameter values, and making adjustments based on the slope and its direction on the error surface. To avoid searching the entire parameter space and speed up the process, other alternatives, including class separability measures, which are often rooted in scatter matrix–based measures (Fukunaga, 2013), have been proposed. As a novel separability index, Yin and Yin (2016) developed the expected square distance ratio (ESDR) that considers the exact data distribution. The experimental results revealed that the ESDR index can be used to select an SVM model including any kernel function and any form of data distribution. Moreover, evolutionary methods, such as genetic algorithms (Section 3.2.5.1), particle swarm optimization (Section 3.2.5.2), and simulated annealing (Section 3.2.2), have gained considerable popularity within the scientific community (Friedrichs and Igel, 2005). On the other hand, Wang et al. (2014) proposed a method for parameter selection using SVMs with a Gaussian kernel that selects the kernel parameter that retains a sufficiently large number of potential support vectors in the training data set and eliminates outliers by assigning a special value to the penalty factor. For dealing with large data sets, special learning strategies, including ensemble learning, online learning, and incremental learning, have also been suggested.

To improve the input data for supervised learning and achieve enhanced classification accuracy, practitioners have integrated semi-supervised learning (SSL) and active learning (AL) techniques with SVMs. The objective of SSL is to jointly utilize both labeled and unlabeled data in the classifier's training process. Applications of SVM-based SSL approaches have been successful in the classification of multispectral and hyperspectral remote sensing images, particularly when the number of training samples is limited with respect to the available spectral bands. As a first step, a semi-supervised method based on SVMs called transductive SVMs (TSVMs) was suggested by Vapnik (1998), and variants of TSVMs have been suggested by Bruzzone et al. (2006). SVMs applying SSL principles are specifically called semi-supervised SVMs (S^{3}VM). Progressive semi-supervised support vector machine (PS^{3}VM), an improved version of S^{3}VM, was also proposed by Bruzzone et al. (2007). Another SSL-based algorithm that improves PS^{3}VM by integrating concepts from AL was introduced by Persello and Bruzzone (2014). When integrated with AL, SVMs start with a small set of labeled points and then enhance the model by actively seeking labels for pixels from a larger pool of unlabeled data. This process involves querying the label of the most intriguing or uncertain unlabeled point at each step. On the other hand, active SVMs have recently emerged as a research topic, particularly in studies involving limited training samples, and numerous studies (e.g., Schohn and Cohn, 2000; Mitra et al., 2004; Tuia et al., 2008; Pasolli et al., 2013; Kremer et al., 2014) have reported their effectiveness in classification problems. Many other SVMs, such as support cluster machines (Li et al., 2006; Chi et al., 2008), nonconvex online SVMs (Ertekin et al., 2011), density-based SVMs (Han and Davis, 2012), and multiple kernel SVMs (Yao et al., 2019; Gönen and Alpaydın, 2011), have been developed to enhance the performance of SVMs.

Several modifications have been suggested to improve the performance of SVMs and overcome major deficiencies including computational complexity, particularly for complex data sets, the requirement of hyperparameter tuning, problems in the processing of big data, and imbalanced data. As discussed in Section 5.5, Relevance Vector Machines (RVMs), introduced by Tipping (1999), perform probabilistic modeling via a sparse Bayesian learning framework. The key attributes of RVMs include the capability to offer probabilistic predictions, the utilization of non-Mercer kernel functions, and the absence of a requirement for setting the regularization parameter. Gómez-Chova et al. (2008) proposed a Laplacian SVM (LapSVM) in which the SVM is regularized with the unnormalized graph Laplacian and tested its performance on two challenging problems. This approach integrates spectral graph theory, manifold learning, and kernel-based algorithms within a kernel-based regularization framework, naturally integrating geometric structure. One notable advancement in this domain was the introduction of the Generalized Eigenvalue Proximal SVM (GEPSVM) by Mangasarian and Wild (2001), which provided the basis for the formulation of the Twin Support Vector Machine (TWSVM), discussed in Section 5.6. TWSVM establishes two nonparallel planes, each in close proximity to one of the two classes while maximizing the distance from the other. TWSVMs work by solving two smaller-sized quadratic programming problems (QPPs) instead of a single large QPP, which sets them apart from standard SVMs. The research findings revealed that TWSVM displays superior speed and improved generalization capabilities when compared to both standard SVM and GEPSVM. In recent years, research trends have shifted from conventional machine learning methods to their integration with deep learning; thus, performance improvements have been achieved over standard machine learning and neural network models. SVMs are incorporated with different deep learning models (DeepSVM) to obtain the benefits of two robust approaches, the theoretical foundation of which is discussed in Section 5.7.

Recent applications and extensions of SVMs have been introduced into a variety of research domains, such as pattern recognition (Burges, 1998), handwriting identification (Burges and Schölkopf, 1997; Adankon and Cheriet, 2009), document categorization (Lee et al., 2012), and bioinformatics (Van Messem, 2020). In the field of remote sensing, SVMs have been extensively used, and the results of many studies have shown that using SVMs to deal with classification issues may result in higher accuracy than other classifiers and require fewer training samples. For instance, Huang et al. (2002) analyzed four types of classifiers, including SVMs, maximum likelihood (ML),

ANN, and the decision tree classifier, and concluded that SVMs generate more stable overall accuracies. Pal and Mather (2005) compared SVMs with the ML and ANN methods. The results showed that SVMs achieved a higher level of classification accuracy than either the ML or the ANN classifier and that the SVM could be used with small training data sets and high-dimensional data. Foody and Mathur (2004, 2006) investigated both the characteristics and size of training samples in SVMs and illustrated the training data acquisition strategies (such as targeting on decision boundary or spectral mixed pixels) to allow efficient and accurate image classification using small training samples. In a hyperspectral classification study, Gualtieri and Cromp (1999) suggested that the SVM approach did not suffer from the Hughes effect and used the full dimensionality of the hyperspectral data without the requirement of feature selection, which was disagreed by Pal and Foody (2012) for limited training samples. Mantero et al. (2005) proposed an SVM approach for estimating probability density functions using a recursive procedure to generate prior probability density estimates for known and unknown classes. Results on synthetic and two real data sets prove the effectiveness of the proposed algorithm in the partially supervised classification context with limited training data. Other studies such as Keuchel et al. (2003), Liu et al. (2006), Maulik and Chakraborty (2011), and Kavzoglu et al. (2020) also showed that SVMs are effective in classifying remotely sensed imagery. Reviews on SVM-based methods can be found in Melgani and Bruzzone (2004), Mountrakis et al. (2011), Pal (2011), Maulik and Chakraborty (2017), and Cervantes et al. (2020). Specialized reviews comparing SVM with convolutional neural networks are also discussed in Heydari and Mountrakis (2019) and Kaul and Raina (2022).

This chapter provides the level of knowledge of SVMs required to enable readers to intelligently use SVMs to perform remotely sensed imagery classification. We start with the introductory linear case, which is then followed by solving the nonlinear classification issue in terms of well-known kernel function methods. Methods of dealing with multiclass cases and parameter assignments are also considered. Three popular SVM approaches (RVMs, Twin Support Vector Machines (TWSVM), and Deep Support Vector Machines (DeepSVM)) are described in detail, and information about recently introduced variants is provided.

5.1 LINEAR CLASSIFICATION

5.1.1 THE SEPARABLE CASE

Assume that there are two linearly separable classes. The training data set is represented by pairs $\{\mathbf{x}_i, y_i\}$, $i = 1, \ldots, n$, $y_i \in \{1, -1\}$, $\mathbf{x}_i \in \mathbf{R}^d$, where $\mathbf{x}_i$ are the observed multispectral features, and y_i is the label of the information class for training case i. The label is either $+1$ or -1, representing class 1 or class 2. The support vector classifier aims to build an optimal hyperplane that separates the two classes in such a way that the distance from the hyperplane to the closest training data points in each of the classes is as large as possible. This distance is called the margin. The hyperplane can be represented as the following decision function:

$$\mathbf{w}^T\mathbf{x} + b = 0 \tag{5.1}$$

where $\mathbf{x}$ is a point lying on the hyperplane, $\mathbf{w}$ is normal to the hyperplane, T denotes matrix transposition, and the parameter b denotes bias. The perpendicular distance from the hyperplane to the origin is represented by $|b|/\|\mathbf{w}\|$, where $\|\mathbf{w}\|$ is the Euclidean norm of $\mathbf{w}$. Figure 5.2 illustrates these concepts.

Suppose that all training data satisfy the following constraints:

$$\mathbf{w}^T\mathbf{x}_i + b \geq +1, \quad \text{for} \quad y_i = +1 \tag{5.2}$$

$$\mathbf{w}^T\mathbf{x}_i + b \leq -1, \quad \text{for} \quad y_i = -1 \tag{5.3}$$

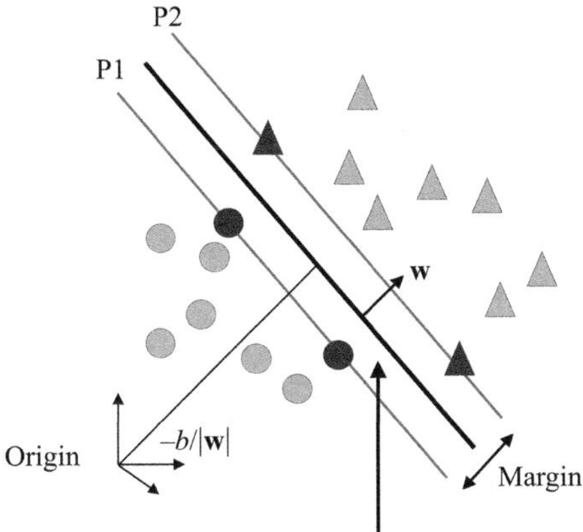

Optimal separating hyperplane: $\mathbf{w}^T\mathbf{x} + b = 0$

FIGURE 5.2   Linearly separable case.

These two equations can be combined to achieve

$$y_i\left(\mathbf{w}^T\times\mathbf{x}_i+b\right)-1\geq 0. \tag{5.4}$$

One can then implicitly define a scale for $(\mathbf{w}\times b)$ to generate two canonical hyperplanes (i.e., $P1$ and $P2$ in Figure 5.2), namely, first, $\mathbf{w}^T\times\mathbf{x}_i+b=1$ for the closest training points lying on the one side of the hyperplane with normal $\mathbf{w}$ and perpendicular distance from the origin $|1-b|/\|\mathbf{w}\|$, and second, $\mathbf{w}^T\times\mathbf{x}_i+b=-1$ for the closest training points on the other side with perpendicular distance from the origin $|-1-b|/\|\mathbf{w}\|$. These training points are referred to as support vectors. In other words, these training samples are central to the establishment of the optimal separating hyperplane. Accordingly, the margin between these two hyperplanes is $2/\|\mathbf{w}\|$. The maximization of this margin in turn leads to solving the following problem:

$$\min\left\{\frac{\|\mathbf{w}\|^2}{2}\right\} \tag{5.5}$$

subject to the inequality constraints shown in Equation (5.4).

Equation (5.4) becomes easier to handle if a primal Lagrangian formulation is used:

$$L_{\text{primal}} = \frac{1}{2}\|\mathbf{w}\|^2 - \sum_{i=1}^{n}\alpha_i\left(y_i\left(\mathbf{w}\times\mathbf{x}_i+b\right)-1\right)+\sum_{i=1}^{n}\alpha_i \tag{5.6}$$

where α_i are positive Lagrangian multipliers. One should then minimize L_{primal} with respect to $\mathbf{w}$ and b. From Wolfe's theorem (Fletcher, 1987), one can take the derivative of L_{primal} with respect to b and $\mathbf{w}$ to obtain

$$\frac{\partial L_{\text{primal}}}{\partial b} = 0 \Rightarrow \sum_{i=1}^{n}\alpha_i y_i = 0 \tag{5.7}$$

$$\frac{\partial L_{\text{primal}}}{\partial \mathbf{w}} = 0 \Rightarrow \mathbf{w} = \sum_{i=1}^{n}\alpha_i y_i \mathbf{x}_i \tag{5.8}$$

and substitute them to the primal formulation (5.6) to achieve the Wolfe dual Lagrangian:

$$L_{\text{dual}} = \sum_{i=1}^{n} \alpha_i - \frac{1}{2} \sum_{i=1}^{n} \alpha_i \alpha_j y_i y_j \mathbf{x}_i \times \mathbf{x}_j. \tag{5.9}$$

The training of the SVMs now involves in maximization of Equation (5.9) with respect to $\alpha_i \geq 0$, subject to the constraint shown in Equation (5.7) with the solution given by Equation (5.8). Notice that there is a Lagrange multiplier α_i for every training point. Those points with $\alpha_i > 0$ are support vectors and lie on one of the parallel hyperplanes $P1$ or $P2$ (Figure 5.2). All other training points have $\alpha_i = 0$ and lie either on (or that side of) $P1$ or $P2$. Specifically, the solutions for $\mathbf{w}$ and b are formulated as

$$\mathbf{w} = \sum_{i=1}^{nsv} \alpha_i y_i \mathbf{x}_i \tag{5.10}$$

$$b = -\frac{1}{2} \mathbf{w} \times (\mathbf{x}_r + \mathbf{x}_s) \tag{5.11}$$

where nsv denotes the number of support vectors, $\mathbf{x}_r$ is the support vector belonging to class $y_r = 1$, and $y_s = -1$ for $\mathbf{x}_s$. Note that bias b may be computed using all the support vectors on the margin for stability concern. Accordingly, in the case of a two-class classification problem, the decision rule shown in Equation (5.1), which separates the two information classes, can be written as

$$f(\mathbf{x}) = \text{sign}\left(\sum_{i=1}^{nsv} \alpha_i y_i \left(\mathbf{x} \times \mathbf{x}_i \right) + b \right). \tag{5.12}$$

This formulation of the SVM optimization problem is called the *hard margin* formulation since no training errors are allowed. All the training samples satisfy the inequality $y_i \times f(\mathbf{x}_i) \geq 1$. In some cases, one may require the separating hyperplane to pass through the origin by choosing a fixed $b = 0$. This variant is called the hard margin SVM without threshold. In this case, the optimization problem remains the same as that in Equation (5.12) except that the constraint $\alpha_i \times y_i = 0$ in Equation (5.7) no longer exists.

5.1.2 THE NONSEPARABLE CASE

Information classes derived from remotely sensed data are not usually completely separated by linear boundaries. Thus, the constraints of Equation (5.4) cannot be satisfied in practice; therefore, the slack variables ξ_i, $i = 1, \ldots, n$, which are proportional to some measure of cost, are introduced to relax the constraints (Cortes and Vapnik, 1995). This is sometimes called the soft margin method. The reader is referred to Figure 5.3 for further illustration of this idea. When slack variables are included, Equation (5.4) becomes

$$y_i \left(\mathbf{w}^T \times \mathbf{x}_i + b \right) \geq 1 - \xi_i; \quad \xi_i \geq 0, \forall i. \tag{5.13}$$

The optimization problem then becomes as

$$\min \left\{ \frac{\|\mathbf{w}\|^2}{2} + C \sum_{i=1}^{n} \xi_i \right\} \tag{5.14}$$

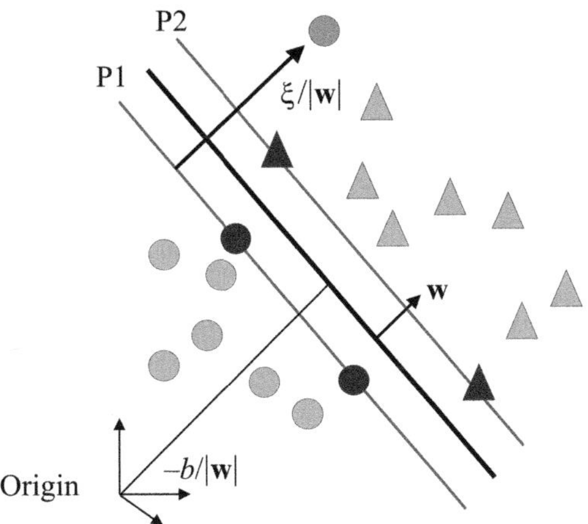

FIGURE 5.3 Linear hyperplanes for the partially separable case.

where C is a penalty parameter to be determined by the user. A larger C means "assign a higher penalty to errors". The first part of Equation (5.14) aims to maximize the margin, while the second part seeks to penalize the training samples located on the "wrong" side of the decision boundary. In other words, a large value of C corresponds to overfitting to the training data and thus reduces the generalization capability. Belousov et al. (2002), however, showed that SVMs can show a high degree of robustness to variations in parameter values.

The primal Lagrangian formulation in Equation (5.6) then becomes

$$L_{\text{primal}} = \frac{1}{2}\|\mathbf{w}\|^2 - \sum_{i=1}^{n}\alpha_i\left(y_i\left(\mathbf{w}^T \times \mathbf{x}_i + b\right) - 1 + \xi_i\right) + C\sum_{i=1}^{n}\xi_i - \sum_{i=1}^{n}\mu_i\xi_i \tag{5.15}$$

where μ_i are the Lagrange multipliers introduced to enforce the positivity of the slack variable ξ_i. The solution to Equation (5.15) is determined by the saddle points of the Lagrangian, by minimizing with respect to $\mathbf{w}$, ξ, and b, and maximizing with respect to $C \geq \alpha_i \geq 0$ and $\mu_i \geq 0$. The Lagrangian multipliers α_i now have an upper bound of C. The upper bound $\alpha_i \leq C$ corresponds to an upper bound on the force that any given support vector is allowed to exert on the hyperplanes $P1$ or $P2$. Note that, in the nonseparable case, the Wolfe dual Lagrangian formulation remains the same as shown in Equation (5.9).

5.2 NONLINEAR CLASSIFICATION AND KERNEL FUNCTIONS

In some cases where a linear hyperplane is unable to appropriately separate the classes, SVMs then adapt a rather old strategy (Aizerman et al., 1964), which maps the raw input data into a higher-dimensional space so as to improve the separability between classes. The SVMs then perform manipulation within this newly mapped feature space. This method is referred to as nonlinear SVMs.

5.2.1 NONLINEAR SVMs

For nonlinear decision surfaces, Boser et al. (1992) proposed that a feature vector, $\mathbf{x}_i \in \mathbf{R}^d$, is mapped into a higher-dimensional Euclidean space (or a generalization of Euclidean space) called Hilbert space (Halmos, 1967; Kolmogorov and Fomin, 1970), to spread the distribution of the training samples in a way that facilitates the fitting of a linear hyperplane. Specifically, the training samples are projected into a higher-dimensional space $\mathcal{H}$, via a nonlinear vector mapping function called $\Phi : \mathbf{R}^d \to \mathcal{H}$. Recall Equation (5.12), which can then be further derived as

$$f(\mathbf{x}) = \operatorname{sign}\left(\sum_{i=1}^{\mathrm{nsv}} \alpha_i y_i \Phi(\mathbf{x}) \cdot \Phi(\mathbf{x}_i) + b \right). \tag{5.16}$$

The computational burden of $(\Phi(\mathbf{x}) \times \Phi(\mathbf{x}_i))$ can be quite expensive in a higher-dimensional space. Vapnik (1995) proposed an alternative to reduce the computational loading via a positive definite kernel function denoted as $K(\mathbf{x}, \mathbf{y})$, such that $K(\mathbf{x}, \mathbf{y}) = \Phi(\mathbf{x}) \times \Phi(\mathbf{y})$. To illustrate such a relationship further, let $K(\mathbf{x}, \mathbf{y}) = (\mathbf{x} \times \mathbf{y})^2$, where $\mathbf{x} = (x_1, x_2)$, and $\mathbf{y} = (y_1, y_2)$. One can then expand the kernel function as

$$\begin{aligned} K(\mathbf{x}, \mathbf{y}) &= (x_1 y_1 + x_2 y_2)^2 \\ &= x_1^2 y_1^2 + 2 x_1 y_1 x_2 y_2 + x_2^2 y_2^2 \\ &= \left(x_1^2, x_2^2, \sqrt{2} x_1 x_2 \right) \times \left(y_1^2, y_2^2, \sqrt{2} y_1 y_2 \right) \\ &= \Phi(\mathbf{x}) \times \Phi(\mathbf{y}) \end{aligned} \tag{5.17}$$

Clearly, according to Equation (5.17), by using the kernel function $K(\mathbf{x}, \mathbf{y}) = (\mathbf{x} \times \mathbf{y})^2$ instead of

$$\Phi(\mathbf{x}) \times \Phi(\mathbf{y}) = \left(x_1^2, x_2^2, \sqrt{2} x_1 x_2 \right) \times \left(y_1^2, y_2^2, \sqrt{2} y_1 y_2 \right),$$

the representation of the data is greatly simplified and the computational burden is reduced. One would only need to use function K in the training algorithm, and there is probably no need to explicitly know what Φ actually is. More examples of kernel functions are discussed in Section 5.2.2. The optimization problem of Equation (5.9) can then be rewritten as

$$L_{\mathrm{dual}} = \sum_{i=1}^{n} \alpha_i - \frac{1}{2} \sum_{i=1}^{n} \sum_{j=1}^{n} \alpha_i \alpha_j y_i y_j K(\mathbf{x}_i \times \mathbf{x}_j) \tag{5.18}$$

and the decision rule expressed in Equation (5.16) now generalizes into

$$f(\mathbf{x}) = \operatorname{sign}\left(\sum_{i=1}^{\mathrm{nsv}} \alpha_i y_i K(\mathbf{x}, \mathbf{x}_i) + b \right). \tag{5.19}$$

In effect, the role of a kernel function is to map the training samples into a higher-dimensional space where the training samples may be spread farther apart. Figure 5.4 illustrates this idea in which data are originally linearly nonseparable in a two-dimensional space, but once mapped into a higher-dimensional feature space $\Phi()$, data may be more easily separated in terms of a hyperplane.

Note that kernel functions being used to fit SVMs should satisfy Mercer's condition (i.e., kernel function $K(x,y)$ should meet the constraint

$$\iint K(x, y) g(x) g(y) d(x) d(y) \geq 0,$$

where $g(x)$ is a square integrable function) to guarantee that a mapping pair $\{\Phi, \mathcal{H}\}$ exists, and consequently, the training process converges (Vapnik, 1995). It is also noted that for the nonseparable

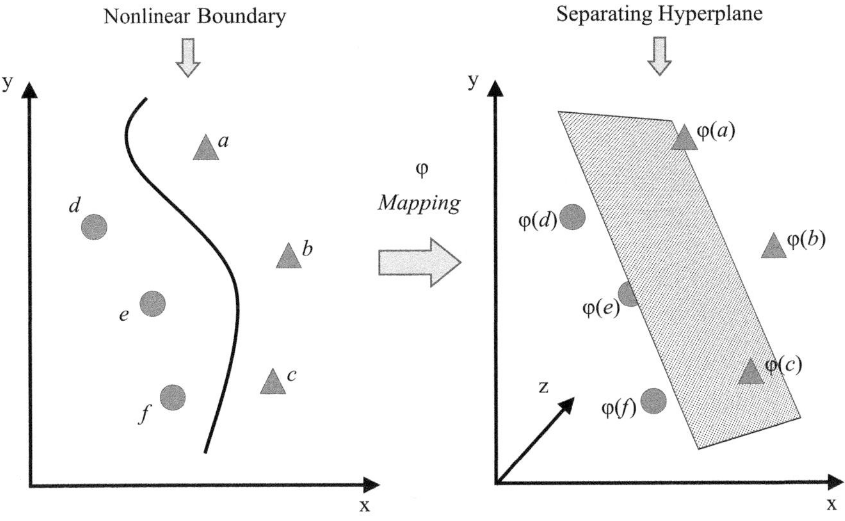

FIGURE 5.4 In the linearly nonseparable case, SVMs map the raw data into a higher dimension to increase the separability between classes.

case, the soft margin SVMs with quadratic penalization of errors can be regarded as a special case of the hard margin version with the modified kernel $K = K + \mathbf{I}/C$, where $\mathbf{I}$ denotes the identity matrix (Cortes and Vapnik, 1995).

5.2.2 Kernel Functions

The discussion in the previous paragraph demonstrates that the purpose of a kernel function is to enable operations to be performed in the current domain (i.e., $K(\mathbf{x},\mathbf{y})$ in Equation (5.19)) rather than the potentially high-dimensional feature space (i.e., $\Phi(\mathbf{x}) \times \Phi(\mathbf{y})$ in Equation (5.16)). This provides a means of resolving the computational issue caused by high dimensionality. In the following discussion, five common kernel functions are introduced:

- Polynomial (homogeneous):

$$K\left(\mathbf{x}_i, \mathbf{x}_j\right) = \gamma\left(\mathbf{x}_i \times \mathbf{x}_j\right)^d \tag{5.20}$$

- Polynomial (inhomogeneous):

$$K\left(\mathbf{x}_i, \mathbf{x}_j\right) = \left(\gamma\left(\mathbf{x}_i \times \mathbf{x}_j\right) + \delta\right)^d, \ \gamma > 0, \delta > 0 \tag{5.21}$$

- Radial basis function:

$$K\left(\mathbf{x}_i, \mathbf{x}_j\right) = \exp\left(-\gamma \mathbf{x}_i - \mathbf{x}_j^2\right), \ \gamma > 0 \tag{5.22}$$

- Gaussian radial basis function:

$$K\left(\mathbf{x}_i, \mathbf{x}_j\right) = \exp\left(-\mathbf{x}_i - \mathbf{x}_j^2 / 2\sigma^2\right) \tag{5.23}$$

- Sigmoid:

$$K\left(\mathbf{x}_i, \mathbf{x}_j\right) = \tanh\left(\gamma\left(\mathbf{x}_i \times \mathbf{x}_j\right) - \delta\right), \ \gamma > 0, \delta > 0 \tag{5.24}$$

Huang et al. (2002) show that both polynomial Equations (5.20) and (5.21) are popular methods for nonlinear mapping with polynomial order $d > 1$, and there is an obvious trend of improved accuracy as d increases. It is also suggested that the value of d has different impacts on kernel performance when different numbers of input variables are used. When dealing with radial basis kernel functions, Schölkopf et al. (1997) show that a Gaussian radial basis function (Equation 5.23) gives excellent results compared to classical radial basis functions (Equation 5.22). The sigmoid kernel function, as shown in Equation (5.24), can be regarded as a particular type of sigmoid neural network. The architecture (number of weights) of the neural network is determined by the number of support vectors used in SVM training. Here, the support vectors correspond to the first layer and the Lagrange multipliers to the weights. Note that the sigmoid kernel function may not satisfy Mercer's condition for certain values of γ and δ (Vapnik, 1995). After a number of experiments on polynomial and radial basis kernels, Kavzoglu and Colkesen (2009) observed that results with polynomial kernel varied with the variation in the user-defined polynomial degree, whereas in comparison with the effect of polynomial degree, the variation in kernel width has little effect on the performance of radial basis functions.

From the above descriptions, it is clear that the performance of SVMs is closely related to the choice of the kernel function. Although certain studies have worked on choosing suitable kernels based on prior knowledge (Burges, 1998), the best choice of kernel for a given problem still requires further investigation. Once the type of kernel is determined, the user must carefully choose the kernel-relating parameters (such as d and γ in Equations (5.21) and (5.22)) and the error penalty C to ensure the most satisfactory classification results.

5.3 PARAMETER DETERMINATION

It is clear from the preceding discussion that the main limitation of the use of SVMs is the need to select suitable parameter values to improve the capability of well-trained SVMs, and eventually to accurately predict unknown data. To select appropriate parameter values, one has to first ensure that the estimate of validation errors is unbiased. Then, using this unbiased error estimate, one can perform parameter adjustments that enhance the generalization capability of the classifier. This section illustrates ways of determining unbiased estimates of errors, followed by a consideration of two kinds of parameter adjustment methodology, known as the grid search and the gradient descent technique, respectively. In recent years, the automated determination of optimal parameters for SVMs, leading to increased classification accuracy and generalization abilities, has been achieved through the utilization of evolutionary algorithms, including genetic algorithms, particle swarm optimization, and their integration with grid search (Friedrichs and Igel, 2005; Guo et al., 2006; Lorena and de Carvalho, 2006; Liu et al., 2006; Yan et al., 2007; Zhang et al., 2009, 2020). Evolutionary algorithms as parameter optimization techniques have been thoroughly discussed in Chapter 9.

5.3.1 T-FOLD CROSS-VALIDATIONS

As mentioned in previous chapters, the available samples are randomly allocated to one of two sets, one set being used for training purposes and the other treated as unknown test data for validation purposes. If the available samples are sufficient, the estimate of the validation error will be less biased. In the case of SVMs, the estimate of the error can be written as

$$P_{\text{error}}^t = \frac{1}{t}\sum_{i=1}^{t}\Gamma\left(-y_i f(\mathbf{x}_i)\right) \tag{5.25}$$

where t is the number of samples for validation, $f(\mathbf{x}_i)$ is the decision function as addressed in Equation (5.19), and Γ is the step function, $\Gamma(\beta) = 1$ when $\beta > 0$ and $\Gamma(\beta) = 0$ otherwise. The parameters that achieve the highest validation accuracy would be the ideal candidates for specifying the final classifier.

Overfitting poses a fundamental challenge to supervised machine learning algorithms, hindering the creation of generalized models as they become overspecific to the training data. The same problem can be encountered more seriously when the error estimation procedure outlined previously in this section is applied. Since the decision boundaries are defined by the support vectors, such one-trial training and validation may not realistically reflect the actual distribution of the unclassified samples, especially in cases where the number of samples is limited. In other words, the resulting SVMs that achieve the highest training accuracy might not successfully generalize to unknown samples. A technique called t-fold cross-validation is generally implemented to overcome this issue.

In the t-fold cross-validation approach, the training samples are separated equally into a user-defined number of subsets, t. During the training phase, the user chooses one of the t subsets for validation and the remaining $(t-1)$ subsets for training. This process is repeated sequentially until all data sets have been tested. The classification errors computed from the validation sets are then averaged over t trials. This method is also called leave-one-out cross-validation when t is equal to the total number of training samples. The t-fold cross-validation accuracy is the percentage of validation samples that are correctly classified, and this measure provides a more objective and accurate estimate of classification accuracy than the one-trial method. According to a study by Luntz and Brailovsky (1969), the leave-one-out procedure can provide an almost unbiased estimate of the expected generalization error regarding validation samples.

5.3.2 Bound on Leave-One-Out Error

Although the use of t-fold and leave-one-out cross-validation to estimate the generalization error is generally recommended, it may nevertheless become computationally expensive since it requires running the training algorithm t times. Thus, the alternative strategy is to "upper bound" or approximate the error by an estimate that is easier to compute. Let f^{all} denote the SVMs trained to form decision boundaries using all the training samples, and f^i denote the decision boundary obtained when sample i is removed; one can derive the expected generalization error for the SVMs as

$$E\left[P_{\text{error}}^{t-1}\right] = \frac{1}{t}\sum_{i=1}^{t}\Gamma\left(-y_i f^i\left(\mathbf{x}_i\right)\right) \tag{5.26}$$

where P_{error}^{t-1} denotes the actual error for SVMs trained on $t-1$ samples, $E\left[P_{\text{error}}^{t-1}\right]$ is the expectation of the actual error over all choices of training set of size $t-1$, and the expectation E is taken over a random choice of training samples. The right-hand side of Equation (5.26) can be further derived as

$$\frac{1}{t}\sum_{i=1}^{t}\Gamma\left(-y_i f^i\left(\mathbf{x}_i\right)\right) = \frac{1}{t}\sum_{i=1}^{t}\Gamma\left(-y_i f^{\text{all}}\left(\mathbf{x}_i\right) + y_i\left(f^{\text{all}}\left(\mathbf{x}_i\right) - f^i\left(\mathbf{x}_i\right)\right)\right). \tag{5.27}$$

Assume the term $y_i\left(f^{\text{all}}\left(\mathbf{x}_i\right) - f^i\left(\mathbf{x}_i\right)\right)$ is bounded by U_i. One can then express the following upper bound on the leave-one-out error as

$$E\left[P_{\text{error}}^{t-1}\right] \leq \frac{1}{t}\sum_{i=1}^{t}\Gamma\left(U_i - 1\right). \tag{5.28}$$

Several studies proposed estimates of the error bound in different ways as addressed in the following section.

Based on the theory of SVMs, it is known that removing a nonsupport vector from the training set does not change the solution calculated by the SVMs (i.e., $U_i = 0$, for nonsupport vector $\mathbf{x}_i$); Vapnik (1995) accordingly derives the bound on the leave-one-out error of SVMs as

$$E\left[P_{\text{error}}^{t-1}\right] \leq \frac{E\left[\text{Number of support vectors}\right]}{t}. \tag{5.29}$$

For SVMs without threshold b (cf. Equation (5.12)), Jaakkola and Haussler (1999) propose another bound on the leave-one-out error:

$$E\left[P_{\text{error}}^{t-1}\right] \le \frac{1}{t}\sum_{i=1}^{t}\Gamma\left(\alpha_i K\left(\mathbf{x}_i,\mathbf{x}_j\right)-1\right) \tag{5.30}$$

where α_i is the Lagrangian multiplier for support vector i, and $K(\mathbf{x}_i, \mathbf{x}_j)$ is the kernel function. Opper and Winther (2000) adopt the concept of linear response theory to propose the following generalization error bound under the assumption that the set of support vectors does not change when removing sample i:

$$E\left[P_{\text{error}}^{t-1}\right] \le \frac{1}{t}\sum_{i=1}^{t}\Gamma\left(\frac{\alpha_i}{\left(\mathbf{K}_{\text{SV}}^{-1}\right)}-1\right) \tag{5.31}$$

where $\mathbf{K}_{\text{SV}}$ is the matrix of dot products between support vectors.

Another expression defining the bound on the leave-one-out error is based on the concept of sphere and radius. Let $\Phi(\mathbf{x}_1)$, $\Phi(\mathbf{x}_2)$, ..., $\Phi(\mathbf{x}_t)$ be the mappings of the training samples in feature space (note that there are a total of t sets, and where $\mathbf{x}_i$ denotes the training set with the ith set excluded), and lying within a sphere of radius R. Here, m is the margin between two hyperplanes. Vapnik (1998) suggests the following bound for the generalization error of the leave-one-out procedure:

$$E\left[P_{\text{error}}^{t-1}\right] \le \frac{1}{t}\frac{R^2}{m^2} \tag{5.32}$$

where the radius R^2 can be derived in terms of maximizing the following the Wolfe duality (Burges, 1998):

$$R^2 = \sum_{i=1}^{t}\beta_i K\left(\mathbf{x}_i\times\mathbf{x}_i\right)-\sum_{i=1}^{t}\sum_{j=1}^{t}\beta_i\beta_j K\left(\mathbf{x}_i\times\mathbf{x}_j\right) \tag{5.33}$$

subject to

$$\sum_{i=1}^{t}\beta_i = 1,$$

$\beta_i \ge 0$ for all i. Note that kernel function $K(\mathbf{x}_i, \mathbf{x}_j)$ is again used to replace $\Phi(\mathbf{x}_i)\times\Phi(\mathbf{x}_j)$.

Under the assumption that the set of support vectors remains the same during the leave-one-out procedure, Vapnik and Chapelle (2000) derive the following error estimate based on the concept of the span of support vectors:

$$E\left[P_{\text{error}}^{t-1}\right] \le \frac{1}{t}\sum_{i=1}^{t}\Gamma\left(\alpha_i S_i^2 -1\right) \tag{5.34}$$

where S_i is the Euclidean distance between the sample $\Phi(\mathbf{x}_j)$ and the set Ω_i, where Ω_i is defined as

$$\Omega_i = \left\{\sum_{i\neq j,\alpha_j>0}\lambda_j\Phi\left(\mathbf{x}_j\right)\right\} \tag{5.35}$$

subject to

$$\sum_{i\neq j}\lambda_j = 1 \quad\text{and}\quad \forall i-j,\ \alpha_j + y_i y_j \alpha_i \lambda_j \ge 0. \tag{5.36}$$

Note that values of λ_i less than 0 are also allowed. The user can then try to minimize the bound on the errors by using one of the error-bounded formulae, that is, Equations (5.29), (5.30), (5.31), (5.32), and (5.34), introduced previously (Pal, 2006).

5.3.3 Grid Search

The question now remaining is how to seek the optimal way of choosing suitable parameter values. The parameters to be determined are the kernel function-related parameters, as shown in Equations (5.20)–(5.24), and the penalty term C in Equations (5.14) and (5.15), respectively. A method called grid search can be implemented to estimate the optimal values of these hyper-parameters (Chang and Lin, 2007). The concept of grid search is similar to that of exhaustive search within a solution space. The user starts by randomly choosing the parameter values that are input to the classifier to evaluate the t-fold cross-validation performance. The parameter values are then increased or decreased, and the performance is reevaluated until all the chosen parameters have been evaluated. If the radial basis kernel function (Equation 5.22) is used to fit SVMs, one has to determine the possible parameter pairs of (γ, C) values and test the corresponding performances. The parameter pair (γ, C) that produces the best cross-validation accuracy is then selected.

Traditionally, a grid search over a bounded space (x, y) begins at one corner of the grid and is evaluated at every grid point separated by a value δ until the opposite corner of the grid is reached. The parameter values that generate the highest classification accuracy are then selected. To conduct a complete grid search using such a step-by-step method may not be practical as the computational demands are considerable. One alternative is initially to evaluate the parameter values over a coarse grid (i.e., bigger δ), and then focus the search on a finer grid (smaller δ) consisting of a small part of the coarse grid. For instance, according to the recommendations made by Chang and Lin (2007), one may choose exponentially growing sequences of pairs (γ, C), $\gamma = 2^{-20}, 2^{-18}, \ldots, 2^6$, and $C = 2^{-10}$, $2^{-8}, \ldots, 2^{10}$ (i.e., $\delta = 2^2$), to roughly identify the best parameter pair. If the pair $(2^{-2}, 2^4)$ is identified to generate the better cross-validation outcome, a finer search with (γ, C), $\gamma = 2^{-0.8}, 2^{-1.0}, \ldots, 2^{3.2}$, and $C = 2^{2.8}, 2^{3.0}, \ldots, 2^{5.2}$ is then used to determine the optimal values.

5.3.4 Gradient Descent Method

If the number of parameters exceeds two, the grid search method becomes unfeasible. It is recognized that as the number of parameters grows sequentially, the solution space will expand exponentially. It is because of the difficulty in efficiently estimating the SVM parameters; there are only limited types of kernels currently available. In the following paragraphs, a technique based on the concept of gradient descent for dealing with large numbers of parameters is introduced (Chapelle et al., 2002). Once the parameter selection issue is resolved, new and possibly more complex kernels may be introduced for classification purposes.

The gradient descent approach searches the parameter space by starting at a random point defined by random values of the parameters and proceeding by changing the parameter values in accordance with the slope (derivative) and the direction of the slope of the error surface. The position on the error surface is defined by parameter values, and the value of the error surface at a given position is the classification error obtained with those parameters. For illustrative purposes, let θ denote the set involving both the SVM kernel parameters and penalty term C. The elements of θ can be best estimated by an iterative procedure (Chapelle et al., 2002). Note that if one determines to use the bound on the leave-one-out error $E\left[P_{error}^{t-1}\right]$, shown in Equations (5.30), (5.31), or (5.34), the step function Γ may cause a problem because Γ is not differentiable. To overcome this problem, Chapelle et al. (2002) recommended the use of a sigmoid function. The readers are referred to Section 7.1.4 for an illustration of the characteristics of the sigmoid function.

To perform the gradient descent procedure, parameter θ_i can be updated in terms of

$$\theta_i^{n+1} = \theta_i^n - \eta \frac{\partial P_{\text{error}}}{\partial \theta_i^n} \tag{5.37}$$

where θ_i^n is the ith parameter at iteration n, η is a small positive constant in the range $0 \leq \eta \leq 1$, and P_{error} denotes the user-defined function for the test error estimate. The readers are recommended to look at Equation (7.5) in Chapter 7 for a more detailed description of the gradient approach. In what follows, according to Chapelle et al. (2002), the different parameter derivative methods regarding various error estimate functions P_{error} are addressed.

If Equation (5.25) is selected as the error estimate function P_{error}, one has to compute the derivative of function $f(\mathbf{x}_i)$ (Equation 5.19), which in turn determines the derivative of α and b with respect to parameter θ_i. Assuming that the samples of nonsupport vectors are removed from the training set, all the remaining samples must lie on the margin, and this situation can be expressed as

$$\underbrace{\begin{pmatrix} K^Y & Y \\ Y^T & 0 \end{pmatrix}}_{H} \begin{pmatrix} \alpha \\ b \end{pmatrix} = \begin{pmatrix} 1 \\ 0 \end{pmatrix} \Rightarrow (\alpha,b)^T = H^{-1}(1,1,\ldots,1,0)^T \tag{5.38}$$

where $K_{ij}^Y = y_i y_j K(\mathbf{x}_i, \mathbf{x}_j)$. The derivatives of those parameters with respect to θ_i can be written as

$$\frac{\partial(\alpha,b)}{\partial \theta_i} = -H^{-1} \frac{\partial H}{\partial \theta_i} H^{-1}(1,1,\ldots,0)^T = -H^{-1} \frac{\partial H}{\partial \theta_i} (\alpha,b)^T. \tag{5.39}$$

Equation (5.39) can also be used for resolving the derivative of the leave-one-out error bound proposed by Jaakkola and Haussler (1999) (Equation 5.30) and Opper and Winther (2000) (Equation 5.31) with $b=0$.

If one uses the bound R^2/m^2 (Vapnik, 1998) in Equation (5.32) for the error estimate, one has to compute the derivative of the square radius R^2 and square margin m^2. According to Equation (5.33), the derivative of the radius with respect to θ_i can be written as follows:

$$\frac{\partial R^2}{\partial \theta_i} = \sum_{i=1}^{n} \beta_i^0 \frac{\partial K(\mathbf{x}_i \times \mathbf{x}_i)}{\partial \theta_i} - \sum_{i=1}^{n} \sum_{j=1}^{n} \beta_i \beta_j \frac{K(\mathbf{x}_i \times \mathbf{x}_j)}{\partial \theta_i} \tag{5.40}$$

where β_i^0 maximizes the Wolfe duality (Equation 5.33). Margin m is defined by

$$m = \frac{1}{\|\mathbf{w}\|} \Rightarrow \frac{R^2}{m^2} = R^2 \|\mathbf{w}\|^2. \tag{5.41}$$

The term $\mathbf{w}$ is the surface normal as defined in Equation (5.10). Vapnik (1998) suggested that $\frac{1}{2}\|\mathbf{w}\|^2$ is equal to the optimization of Equation (5.19) and so the following formulation holds:

$$\frac{1}{2}\|\mathbf{w}\|^2 = \sum_{i=1}^{n} \alpha_i - \frac{1}{2} \sum_{i=1}^{n} \sum_{j=1}^{n} \alpha_i \alpha_j y_i y_j K(\mathbf{x}_i \times \mathbf{x}_j). \tag{5.42}$$

According to Equation (5.41), the derivative of the square margin m^2 with respect to θ_i leads to

$$\frac{\partial \|\mathbf{w}\|^2}{\partial \theta_i} = -\sum_{i=1}^{n} \sum_{j=1}^{n} \alpha_i \alpha_j y_i y_j \frac{\partial K(\mathbf{x}_i \times \mathbf{x}_j)}{\partial \theta_i}. \tag{5.43}$$

Finally, if the error estimate is based on the concept of the span of support vectors (Equation 5.34), the derivative of the square span S_i^2 can be expressed as (Chapelle et al., 2002)

$$\frac{\partial S_i^2}{\partial \theta_i} = S_i^4 \left(\tilde{\mathbf{K}}_{SV}^{-1} \frac{\partial \tilde{\mathbf{K}}_{SV}}{\theta_i} \tilde{\mathbf{K}}_{SV}^{-1} \right) \tag{5.44}$$

where $\tilde{\mathbf{K}}_{SV}$ is the extended matrix of the dot products between the support vectors expressed as

$$\tilde{\mathbf{K}}_{SV} = \begin{pmatrix} \mathbf{K}_{SV} & 1 \\ 1^T & 0 \end{pmatrix} \tag{5.45}$$

where T denotes the transpose of the matrix.

The derivative of each error bound is illustrated above. The reader can then choose the error function of interest and use Equation (5.37) to sequentially tune the SVM parameters so as to achieve improved classification accuracy.

5.4 MULTICLASS CLASSIFICATION

SVMs were originally designed for binary classification; that is, one SVM can only separate two classes, yet most remote sensing applications require dealing with multiple classes. Several approaches, including one-against-one, one-against-others, directed acyclic graph strategies, and multiclass SVMs, have been proposed for applying SVMs to multiclass classifications (Pal, 2008). These methods are described in the following sections.

5.4.1 ONE-AGAINST-ONE, ONE-AGAINST-OTHERS, AND DAG

The one-against-one classification strategy was proposed by Knerr et al. (1990), who used pairwise comparisons between all k classes. Thus, each classifier is trained exclusively on two of the k classes. All possible one-against-one class combinations are evaluated from the training set of k classes. A total of $k(k-1)/2$ classifiers will be needed to perform the task. The decision rule for labeling a pixel is in terms of majority voting. The pixel will be given the label of the class with the most votes. In the case of a tie, a tie-breaking strategy may be adopted. One may select at random one of the classes that are tied, or refer to the nearest neighbor pixel's class to break the tie.

The one-against-others strategy builds a total of k SVMs for the k classes. Each of the SVMs is trained to classify one class against all other classes. Pixel $\mathbf{x}$ is then given the label of the class with the highest value of $\mathbf{w}_i \Phi(\mathbf{x}) + b_i$, $I = 1$ to k (Vapnik, 1995).

Platt et al. (2000) proposed the use of directed acyclic graph (DAG) SVMs for multiclass classification. The training phase of DAG-SVMs is the same as the one-against-one method, that is, by building $k(k-1)/2$ classifiers, where k is the number of classes. These $k(k-1)/2$ classifiers then form the internal nodes of a DAG, with each node being a binary SVM using the ith and jth classes. Starting from the root of the DAG, a classified test sample is compared at the node and then, depending on the output value of the node, directed to either the left or right node at the next layer, until a leaf is achieved. Note that the choice of class order to form DAG-SVMs can be arbitrary without significantly affecting classification accuracy. One advantage of using DAG-SVMs is the substantial improvements in both training and evaluation time in comparison with the one-against-one and one-against-others algorithms (Platt et al., 2000). Figure 5.5 illustrates four-class DAG-SVMs for classifying roads, trees, bare soil, and vegetation; in total, six binary classifiers are built. At each level, the DAG-SVM expels one of the classes and then heads to the node containing the survival class pair until a final decision is achieved.

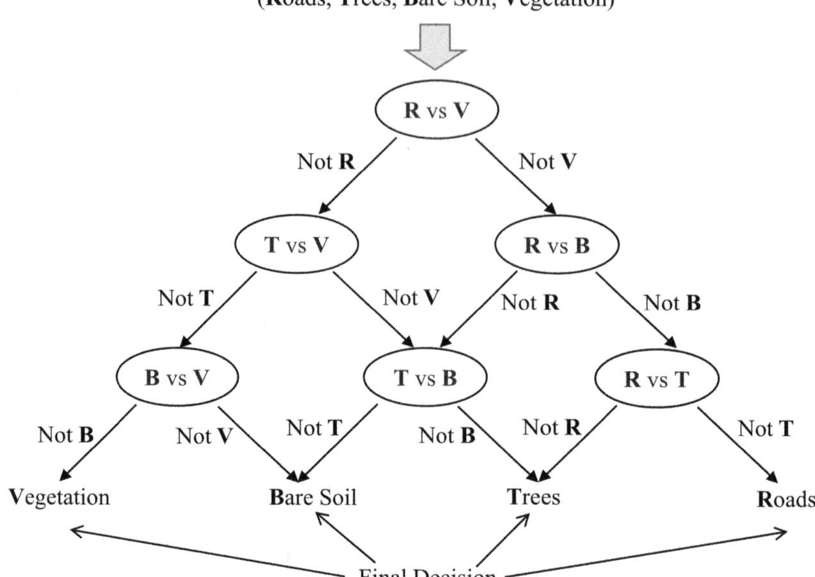

FIGURE 5.5   A DAG-SVM for classifying roads, trees, bare soil, and vegetation.

5.4.2 MULTICLASS SVMs

5.4.2.1 Vapnik's Approach

Vapnik (1998) developed an approach to multiclass SVMs by modifying the optimization problem of binary SVMs. Such a multiclass SVM is similar to the one-against-others approach by using a single optimization approach. Suppose that we have a total of k classes. The multiclass SVMs construct k two-class rules in which the ith hyperplane $w_i^T \Phi(\mathbf{x}) + b$ separates the training samples of class i from other classes. Therefore, although there are k decision functions or hyperplanes, the solution may be obtained by solving one problem. The formulation of multiclass SVMs is expressed as

$$\min \left\{ \frac{1}{2} \sum_{i=1}^{k} \mathbf{w}_i^T \mathbf{w}_i + \sum_{j=1}^{n} \sum_{i \neq y_j} \xi_j^i \right\} \tag{5.46}$$

under the constraints that

$$\mathbf{w}_{y_j}^T \Phi(\mathbf{x}_j) + b_{y_j} \geq \mathbf{w}_i^T \Phi(\mathbf{x}_j) + b_i + 2 - \xi_j^i$$

$$\xi_j^i \geq 0, \ j = 1, \ldots, n; \ i \in \{1, \ldots, k\}/y_j \tag{5.47}$$

where n is the number of training samples. The decision function is then

$$\arg \max_{i=1 \text{ to } k} \left(\mathbf{w}_i^T \Phi(\mathbf{x}) + b_i \right). \tag{5.48}$$

Hsu and Lin (2002) show that the derivation of a solution to Equation (5.48) requires one to select more than k variables at each iteration, which is not an easy condition to fulfill. Hence, to ensure that a valid solution is obtained, the element

$$\sum_{i=1}^{k} b_i^2$$

is introduced into Equation (5.46) as

$$\min\left\{\frac{1}{2}\sum_{i=1}^{k}\left(\mathbf{w}_i^T\mathbf{w}_i + b_i^2\right) + C\sum_{j=1}^{n}\sum_{i\neq y_j}\xi_j^i\right\}$$
$$= \min\left\{\frac{1}{2}\sum_{i=1}^{k}\begin{bmatrix}\mathbf{w}_i^T & b_i\end{bmatrix}\times\begin{bmatrix}\mathbf{w}_i \\ b_i\end{bmatrix} + C\sum_{j=1}^{n}\sum_{i\neq y_j}\xi_j^i\right\}$$

(5.49)

under the constraints

$$\begin{bmatrix}\mathbf{w}_{y_j}^T & b_{y_j}\end{bmatrix}\times\begin{bmatrix}\Phi(\mathbf{x}_j) \\ 1\end{bmatrix} \geq \begin{bmatrix}\mathbf{w}_i^T & b_j\end{bmatrix}\times\begin{bmatrix}\Phi(\mathbf{x}_j) \\ 1\end{bmatrix} + 2 - \xi_j^i$$
$$\xi_j^i \geq 0, \quad j = 1,\ldots,n; \quad i \in \{1,\ldots,k\}/y_j.$$

(5.50)

Equation (5.49) can be modified into the following formulation (Hsu and Lin, 2002):

$$\min\left\{\sum_{r=1}^{n}\sum_{s=1}^{n}\left(\frac{1}{2}c_s^{y_r}A_rA_s - \sum_{i=1}^{k}\alpha_r^i\alpha_s^{y_r} + \frac{1}{2}\sum_{i=1}^{k}\alpha_r^i\alpha_s^i\right)\left(K_{r,s}+1\right) - 2\sum_{r=1}^{n}\sum_{i=1}^{k}\alpha_r^i\right\}$$

(5.51)

with $0 \leq \alpha_r^i \leq C$, $\alpha_r^{y_r} = 0$, where

$$A_r = \sum_{i=1}^{k}\alpha_r^i, \quad c_s^{y_r} = \begin{cases} 1 & \text{if } y_r = y_s \\ 0 & \text{if } y_r \neq y_s \end{cases}.$$

(5.52)

The decision function then becomes

$$f(\mathbf{x}) = \underset{i=1,\ldots,k}{\arg\max}\left(\sum_{r=1}^{n}\left(c_r^i A_r - \alpha_r^i\right)\times\left(K(\mathbf{x}_r,\mathbf{x})+1\right)\right).$$

(5.53)

5.4.2.2 Methodology of Crammer and Singer

Crammer and Singer (2002) propose an approach to the solution of SVM multiclass classification issues based on the following optimization function:

$$\min\left\{\frac{1}{2}\sum_{i=1}^{k}\mathbf{x}_i^T\mathbf{x}_i + C\sum_{j=1}^{n}\xi_i\right\}$$

(5.54)

under the constraints that

$$\mathbf{w}_{y_j}^T\Phi(\mathbf{x}_j) - \mathbf{w}_i^T\Phi(\mathbf{x}_j) \geq e_j^i - \xi_j \quad j = 1,\ldots,n$$

(5.55)

where $e_j^i = 1 - \delta_{y_j,i}$, $\xi_j \geq 0$, and $\delta_{yj,i}$ is defined as

$$\delta_{y_j,i} = \begin{cases} 1 & \text{if } y_j = i \\ 0 & \text{if } y_j \neq i \end{cases}.$$

(5.56)

The decision function is

$$\arg\max_{i=1,\ldots,k} = \mathbf{w}_i^T \Phi(\mathbf{x}) \tag{5.57}$$

The dual formulation of Equation (5.54) is

$$\min f(\alpha) = \frac{1}{2} \sum_{r=1}^{n} \sum_{s=1}^{n} K_{r,s} \bar{\alpha}_r^T \bar{\alpha}_s + \sum_{r=1}^{n} \bar{\alpha}_r^T \bar{e}_i \tag{5.58}$$

under the constraints that

$$\sum_{i=1}^{k} \alpha_r^i = 0, \quad r = 1,\ldots,n, \quad \alpha_r^i \leq C, \quad \text{if} \quad y_i \neq m, \quad \text{and} \quad \alpha_r^i \leq 0, \quad \text{otherwise} \tag{5.59}$$

where

$$\bar{\alpha}_r = \left[\alpha_r^1,\ldots,\alpha_r^k\right]^T, \quad \text{and} \quad \bar{e}_r = \left[e_r^1,\ldots,e_r^k\right]^T. \tag{5.60}$$

Then one obtains

$$\mathbf{w}_i = \sum_{r=1}^{n} \alpha_r^i \Phi(\mathbf{x}_r) \tag{5.61}$$

and the decision function in Equation (5.57) can be written as

$$\arg\max_{i=1,\ldots,k} \sum_{r=1}^{n} \alpha_r^i K(\mathbf{x}_r, \mathbf{x}). \tag{5.62}$$

The multiclass solution proposed by Crammer and Singer (2002) is much simpler than that of Vapnik (1998) (Equation 5.53). Even so, according to the comparative study presented by Hsu and Lin (2002), both multiclass SVMs (Crammer and Singer, 2002; Vapnik, 1998) suffer from slow convergence, and classification accuracy is inferior to that achieved by the one-against-one and DAG-SVM approaches.

Several other multiclass SVM algorithms have been proposed in the literature. For instance, Platt (1999) introduced an iterative algorithm, called sequential minimal optimization (SMO), for solving the dual SVM optimization problem. Dias and Neto (2016) suggest genetic algorithms for training easily (GATE) support vector machines, which use a single-objective genetic algorithm. The initial stage involves creating a population of solutions in the GA, where each individual represents the Lagrange multiplier vector. MC2ESVM, proposed by Rosales-Pérez et al. (2018), is a cooperative evolutionary method for multiclass classification that trains a multiclass SVM in one step. This classification method forms its multiclass classifier from the support vectors of each class. The method adopts a cooperative coevolutionary algorithm, with each subpopulation optimizing support vectors for each class while considering other subpopulations to address the multiclass problem.

5.5 RELEVANCE VECTOR MACHINES

The RVMs, proposed by Tipping (1999), are a significant development in kernel-based machine learning algorithms that extends the principles of SVMs to perform probabilistic modeling using a sparse Bayesian learning framework for regression and classification. Unlike traditional SVMs, which search a subset of data points called support vectors, RVMs assess the relevance of every data point within a regression or classification problem. With the assignment of relevance weights to each

                                                                                      183

data point, RVMs effectively perform feature selection during model training, filtering out unessential features and emphasizing the crucial ones. The algorithm automatically identifies relevant data points (with nonzero weights) and irrelevant data points (with zero weights), indicating the importance of each data point in prediction. Applying a probabilistic framework, RVMs not only forecast target values but also offer an indication of their confidence level in these predictions, providing a measure of certainty. This results in a model that is both efficient and compact, proficient in capturing intricate data relationships. The advantages of RVMs over SVMs can be given as the availability of probabilistic predictions, the use of arbitrary kernel functions (non-Mercer kernels), and not requiring the setting of the regularization parameter (Pal, 2009, 2011). In summary, the RVM framework combines sparse modeling, feature selection, and probabilistic reasoning, resulting in a versatile machine learning algorithm suitable for application across a range of domains.

RVMs, similar to SVMs, were developed specifically for binary classification tasks. In a two-class classification utilizing RVMs, the objective is to predict the posterior probability of belonging to one of the classes (0 or 1) for a given input $\mathbf{x}$ with class labels $\mathbf{c}$. The assignment of a case is then determined by allocating it to the class with the highest likelihood of membership. The linear model is generalized by applying the logistic sigmoid function $\sigma(y) = 1/(1+e^{-y})$ to $y(\mathbf{x})$. By adopting Bernoulli distribution $(p(\mathbf{c} \mid \mathbf{x}))$, the likelihood can be written as

$$p(\mathbf{c} \mid \mathbf{w}) = \prod_{i=1}^{n} \sigma\{(y(\mathbf{x}_i; \mathbf{w}))\}^{c_i} \left[1 - \sigma\{(y(\mathbf{x}_i; \mathbf{w}))\}\right]^{1-c_i}. \tag{5.63}$$

The weights cannot be integrated analytically. Therefore, the approximation method of MacKay (1992), which is based on Laplace's theory, is employed in three steps (Tipping, 2001):

1. When α is held constant, the "most probable" weights $\mathbf{w}_{\mathrm{MP}}$, indicating the mode of the posterior distribution, are identified. Since $p(\mathbf{w} \mid \mathbf{c}, \alpha) \propto P(\mathbf{c} \mid \mathbf{w}) p(\mathbf{w} \mid \alpha)$, this corresponds to finding the maximum, over $\mathbf{w}$, of

$$\log\{P(\mathbf{c} \mid \mathbf{w}) p(\mathbf{w} \mid \alpha)\} - \sum_{n=1}^{N} \left[c_n \log y_n + (1-c_n)\log(1-y_n)\right] - \frac{1}{2}\mathbf{w}^T \mathbf{A}\mathbf{w}, \tag{5.64}$$

1. where $y_n = \sigma\{y(x_n; \mathbf{w})\}$ and $\mathbf{A} = \mathrm{diag}(\alpha_1, \alpha_2, \ldots, \alpha_N)$. The equation indicates a penalized (regularized) logistic log-likelihood function and requires iterative maximization. Then, the reweighted least squares method is applied to estimate $\mathbf{w}_{\mathrm{MP}}$.
2. Laplace's method is simply a quadratic approximation of the log-posterior around its mode. The estimation is differentiated twice to give

$$\nabla_{\mathbf{w}} \nabla_{\mathbf{w}} \log p(\mathbf{w} \mid \mathbf{c}, \alpha)\big|_{\mathbf{w}_{\mathrm{MP}}} = -(\Phi^T \mathbf{B}\Phi + \mathbf{A}), \tag{5.65}$$

2. where $\mathbf{B} = \mathrm{diag}(\beta_1, \beta_2, \ldots, \beta_N)$ is a diagonal matrix with $\beta_N = \sigma\{y(\mathbf{x}_n)\}[1 - \sigma\{y(\mathbf{x}_n)\}]$. The next step involves negating and inverting this value to derive the covariance Σ for a Gaussian approximation to the posterior over weights centered at $\mathbf{w}_{\mathrm{MP}}$.
3. Using the statistics Σ and $\mathbf{w}_{\mathrm{MP}}$ of the Gaussian approximation, the hyperparameters α are updated. At the mode of $p(\mathbf{w} \mid \mathbf{c}, \alpha)$ using Equation (5.65) and the fact that $\nabla_{\mathbf{w}} \log p(\mathbf{w} \mid \mathbf{c}, \alpha)\big|_{\mathbf{w}_{\mathrm{MP}}} = 0$, covariance and weights can be written as

$$\Sigma = (\Phi^T \mathbf{B}\Phi + \mathbf{A})^{-1}, \tag{5.66}$$

$$\mathbf{w}_{\mathrm{MP}} = \Sigma\Phi^T \mathbf{B} \tag{5.67}$$

The expressions presented here are effectively solving a generalized least squares problem. For a more in-depth understanding, readers should refer to Tipping (2001).

The variational solution for RVMs was developed by Bishop and Tipping (2000) to improve the performance of the original RVMs with the Bayesian framework but at a higher computational cost. To overcome the two major shortcomings of RVMs, namely, slow training process and possible poor performance when they are applied to large-scale data sets, Liu et al. (2017) suggested the incorporation of the discrete AdaBoost method into RVMs (DAB-RVMs) and all features boosting RVMs (AFB-RVMs) to speed up the learning process. In addition, Wang et al. (2013) developed another version called smooth RVMs, which uses a wavelet kernel for analysis of ground surface settlement, and Zhang et al. (2014) proposed adaptive RVMs for slope stability inference. Because of their simplicity in parameterization, RVMs have attracted much interest in remote sensing (Pal and Foody, 2012). It is mainly applied to classification problems, and its performance has been reported to be comparable to that of SVMs (Demir and Erturk, 2007; Dong and Zhao, 2010; Mianji and Zhang, 2010; Samui et al., 2012). On the other hand, Braun et al. (2011) conducted a comparative evaluation, contrasting the effectiveness of SVMs, RVMs, and import vector machines, another alternative form of SVMs introduced by Zhu and Hastie (2005). Their findings identified RVMs as the most reliable in terms of both producers' and users' accuracies.

5.6 TWIN SUPPORT VECTOR MACHINES

The traditional SVM strictly requires searching for parallel hyperplanes, which will meet the challenges of solving many classification problems. To overcome this shortcoming, Mangasarian and Wild (2001) proposed the Generalized Eigenvalue Support Vector Machine (GEPSVM), which can construct nonparallel hyperplanes by solving generalized eigenvalue problems. Eigenvectors corresponding to the smallest eigenvalues of two generalized eigenvalue problems define the nonparallel planes. This approach ensures that data points from each class are in proximity to one of the two nonparallel planes. Jayadeva and Chadra (2007) extended the idea of using nonparallel planes for binary data and proposed a new classifier called the Twin Support Vector Machine (TWSVM), which strives to build two nonparallel hyperplanes by solving two concise QPPs. This strategy of solving two smaller-sized QPPs rather than one large QPP makes TWSVMs work faster than standard SVMs.

The TWSVM classifier is derived by addressing the following set of QPPs, which can be given as

$$\text{(TWSVM1)} \quad \min_{w^{(1)},b^{(1)},q} \frac{1}{2}\left(Aw^{(1)}+e_1b^{(1)}\right)^T\left(Aw^{(1)}+e_1b^{(1)}\right)+c_1e_2^Tq$$

$$\text{subject to} -\left(Bw^{(1)}+e_2b^{(1)}\right)+q \geq e_2, q \geq 0, \text{ and} \tag{5.68}$$

$$\text{(TWSVM2)} \quad \min_{w^{(2)},b^{(2)},q} \frac{1}{2}\left(Bw^{(2)}+e_2b^{(2)}\right)^T\left(Bw^{(2)}+e_2b^{(2)}\right)+c_2e_1^Tq$$

$$\text{subject to} \quad \left(Aw^{(2)}+e_1b^{(2)}\right)+q \geq e_1, q \geq 0, \tag{5.69}$$

where $c_1, c_2 > 0$ are parameters, and e_1 and e_2 are vectors of ones of appropriate dimensions (Jayadeva and Chadra, 2007).

Depending on its distance from the two nonparallel hyperplanes, a new data point is assigned to either class +1 or −1. The first term in the objective function of Equation (5.68) or (5.69) represents the sum of the squared distances between the hyperplane and the points of a particular class. Therefore, reducing this term ensures that the hyperplane remains near the points of that class (class +1). The constraints require the hyperplane to maintain a minimum distance of 1 from points of the other class (class -1). Error variables are employed to calculate errors in instances where the hyperplane is closer than the specified minimum distance of 1. The second term of the objective function acts to reduce the sum of error variables, thus minimizing misclassification arising from points belonging to class -1.

The Lagrangian corresponding to the problem TWSVM1 (Equation 5.68) is given by

$$L\left(w^{(1)}, b^{(1)}, q, \alpha, \beta\right) = \frac{1}{2}\left(Aw^{(1)} + e_1 b^{(1)}\right)^T \left(Aw^{(1)} + e_1 b^{(1)}\right)$$
$$+ c_1 e_2^T q - \alpha^T\left(-\left(Bw^{(1)} + e_2 b^{(1)}\right) + q - e_2\right) - \beta^T q, \tag{5.70}$$

where $\alpha = \left(\alpha_1, \alpha_2, \ldots, \alpha_{m_2}\right)^T$ and $\beta = \left(\beta_1, \beta_2, \ldots, \beta_{m_2}\right)^T$ are the vectors of the Lagrange multipliers. The Karush-Kuhn-Tucker (KKT) conditions essential for the optimality of TWSVM1 are presented as follows:

$$A^T\left(Aw^{(1)} + e_1 b^{(1)}\right) + B^T \alpha = 0 \tag{5.71}$$

$$e_1^T\left(Aw^{(1)} + e_1 b^{(1)}\right) + e_2^T \alpha = 0 \tag{5.72}$$

$$c_1 e_2 - \alpha - \beta = 0 \tag{5.73}$$

$$-\left(Bw^{(1)} + e_2 b^{(1)}\right) + q \geq e_2, q \geq 0 \tag{5.74}$$

$$\alpha^T\left(-\left(Bw^{(1)} + e_2 b^{(1)}\right) + q - e_2\right) = 0, \quad \beta^T q = 0 \tag{5.75}$$

$$\alpha \geq 0, \ \beta \geq 0 \tag{5.76}$$

Since $\beta \geq 0$, considering Equation (5.73), α can be written as

$$0 \leq \alpha \leq c_1. \tag{5.77}$$

Combining Equations (5.71) and (5.72) gives

$$\left[A^T e_1^T\right]\left[Ae_1\right]\left[w^{(1)}, b^{(1)}\right]^T + \left[B^T e_2^T\right]\alpha = 0. \tag{5.78}$$

$$H = \left[Ae_1\right], \ G = \left[Be_2\right], \tag{5.79}$$

and the augmented vector $u = \left[w^{(1)}, b^{(1)}\right]^T$. With these notations, Equation (5.78) may be rewritten as

$$H^T Hu + G^T \alpha = 0, \quad i.e., \quad u = -\left(H^T H\right)^{-1} G^T \alpha. \tag{5.80}$$

A regularization term εI, $\varepsilon > 0$, is introduced to deal with problems for possible ill-conditioning of $H^T H$. Note that I denotes an identity matrix of appropriate dimensions. Then, Equation (5.80) can be modified to

$$u = -\left(H^T H + \varepsilon I\right)^{-1} G^T \alpha. \tag{5.81}$$

Using Equation (5.70) and the KKT conditions, the Wolfe duality of TWSVM1 and TWSVM2 can be given as follows:

$$\text{(DTWSVM1)} \quad \max_{\alpha} e_2^T \alpha - \frac{1}{2} \alpha^T G \left(H^T H \right)^{-1} G^T \alpha$$

$$\text{subject to} \qquad 0 \le \alpha \le c_1. \tag{5.82}$$

$$\text{(DTWSVM2)} \quad \max_{\gamma} e_1^T \gamma - \frac{1}{2} \gamma^T P \left(Q^T Q \right)^{-1} P^T \gamma$$

$$\text{subject to} \qquad 0 \le \gamma \le c_2. \tag{5.83}$$

where $P = [Ae_1]$, $Q = [Be_2]$, and the augmented vector $v = \left[w^{(2)}, b^{(2)} \right]^T$, which is given by

$$v = \left(Q^T Q \right)^{-1} P^T \gamma. \tag{5.84}$$

The matrices $H^T H$ and $Q^T Q$ are in size $(n + 1) \times (n + 1)$, where n mostly has a much smaller value than the number of patterns of classes 1 and −1. Since the vectors u and v are available from Equations (5.80) and (5.84), the separating planes are obtained as

$$x^T w^{(1)} + b^{(1)} = 0 \quad \text{and} \quad x^T w^{(2)} + b^{(2)} = 0 \tag{5.85}$$

In recent years, robust adaptations for TWSVM have emerged. One such approach is the least squares TWSVM (LS-TWSVM) algorithm, formulated by Kumar and Gopal (2009) to accelerate the TWSVM training process. LS-TWSVM offers a considerable advantage over TWSVM by working solely with linear equations instead of the QPPs present in TWSVM, ultimately lowering the computational complexity of the model. Gao et al. (2011) introduced the L1 Norm Least Squares Twin Support Vector Machine (NLS-TWSVM), which automatically selects significant features to reinforce the algorithm for handling high-dimensional data sets. To address the issue of overfitting, an advancement called improved LS-TWSVM (ILS-TWSVM) was introduced by Xu et al. (2012). This enhancement aimed to increase classification accuracy by adhering to the principle of structural risk minimization. It has been noted that ILS-TWSVM exhibits increased speed and delivers comparable generalization and computational time as LS-TWSVM. Qi et al. (2013a) proposed the Robust Twin Support Vector Machine (R-TWSVM) using second-order cone programming formulations for classification, which can deal with data with measurement noise efficiently and provide time efficiency. Qi et al. (2013b) also introduced the Structural Twin Support Vector Machine (S-TWSVM) for classification, which leverages prior structural data to improve classification accuracy. Xie et al. (2013) expanded the application of TWSVM to multiclass problems and suggested the One-Versus-All TWSVM (OVA-TWSVM), employing a one-versus-all (OVA) approach to solve k-class problems by creating k-TWSVM. Khemchandani et al. (2018) introduced the Angle-based Twin Support Vector Machine (ATWSVM), where the second hyperplane is positioned close to its respective class, and the angle between the normals of the two hyperplanes is maximized to ensure maximum separation between the hyperplanes. Comprehensive literature surveys on TWSVMs can be found in Huang et al. (2018) and Tanveer et al. (2022).

5.7　DEEP SUPPORT VECTOR MACHINES

Kernel methods and deep learning (DL) are recognized as two of the most successful and remarkable machine learning techniques because of their wide-ranging reported success in various applications. SVMs that utilize kernel methods are renowned for their excellence in achieving superior results in two-class classification and regression tasks by capturing nonlinear patterns within the data set. Using the kernel trick, DL models acquire intricate nonlinear features in high-dimensional spaces. The selection of an appropriate kernel function plays a pivotal role in SVM applications. Standard SVMs, which feature only one adjustable layer, are perceived as a shallow model, whereas multiple kernel learning involves combining and selecting kernels for different learning tasks. DL architectures consist of multiple intermediate (hidden) layers, introducing nonlinearity into the model, rather than a single hidden layer, to learn the characteristics of complex problems, mapping the input to the output directly from data. The training cost of SVMs significantly escalates with sample size, rendering them less viable for large-scale problems. Conversely, DL models maintain a linear cost dependency on sample size, making them a more efficient solution for managing big data. Incorporating the advantages of both paradigms, deep kernel learning, a technique bridging kernel methods and DL, has proven to be a robust alternative learning technique that uses intricate feature representations by combining the nonparametric flexibility of kernel methods with the structural properties of DL (Wang et al., 2021). In other words, deep kernel learning combines the structural properties of DL models with the nonparametric adaptability of kernel methods.

Currently, hybridization of both techniques in various forms is an emerging research area, and there has been a growing interest in integrating DL and SVMs. As a result, SVMs and TWSVM have been extended to the field of DL. As one of the earliest attempts to combine SVMs with DL, Kim et al. (2013) aimed to train a deep structure with two-class SVMs, but no reasonable criteria exist to determine outliers or network depth. Kim et al. (2015) extended their earlier work by introducing another DL structure with a support vector data description (SVDD), as suggested by Tax and Duin (2004), to obtain both the representational power of the DL and the generalization performance of the SVDD. On the other hand, Li et al. (2014) presented the Deep Twin Support Vector Machine (DTWSVM), which uses two TWSVMs in the hidden layer, recording four-dimensional location information for the data set, to extract features based on the projection principle, which is derived from multilayer perceptron (MLP). It should be noted that the proposed network excludes a DL network to extract deep features. In early studies, hidden representations of deep networks were first obtained through supervised or unsupervised approaches and subsequently used as inputs for SVMs. Tang (2013), on the other hand, proposed to train all layers of deep networks using the backpropagation algorithm through top-level SVMs. The Softmax layer was replaced with a linear support vector machine, allowing the learning process to minimize a margin-based loss instead of the cross-entropy loss. Erfani et al. (2016) presented a scalable and computationally efficient hybrid model in which an unsupervised deep belief network is trained to extract generic underlying features and one-class SVMs are trained from the features learned by the network. Experimental results revealed that the proposed model achieved a level of anomaly detection performance similar to that of a deep autoencoder. For hyperspectral image classification, Okwuashi and Ndehedehe (2020) applied the DeepSVM using four kernel functions and conducted a comparative study using Indian Pines and University of Pavia data sets. The research findings indicate the efficiency of the proposed DeepSVM in contrast to other classification algorithms. Similarly, Kalaiarasi and Maheswari (2021) investigated the effectiveness of deep linear and nonlinear proximal support vector machines (PSVMs) for hyperspectral image classification. The results revealed that the developed deep PSVM (linear and nonlinear) was superior in terms of classification accuracy when compared to other techniques using the same hyperspectral data sets. Sun and Xie (2023) proposed a DL version of the Non-Parallel Hyperplane Support Vector Machine (NPHSVM) for classification. The efficacy of the proposed approach was demonstrated through experiments on UCI data sets, surpassing other state-of-the-art algorithms used for comparison. For further readings on DL integration with SVMs, readers are referred to Díaz-Vico et al. (2020) and Wang et al. (2021).

The formulation of the DeepSVM algorithm involves a multilayer architecture that includes several hidden layers (Figure 5.6). In the DeepSVM network structure, $X_1, X_2, \ldots, X_n$ represent the input layer data points. The multiple hidden layers consist of $SVM_{11}, SVM_{12}, \ldots, SVM_{1k}$, $SVM_{21}, SVM_{22}, \ldots, SVM_{2k}$, and $SVM_{n1}, SVM_{n2}, \ldots, SVM_{nk}$, while $F_1(X), F_2(X), \ldots, F_n(X)$ represent

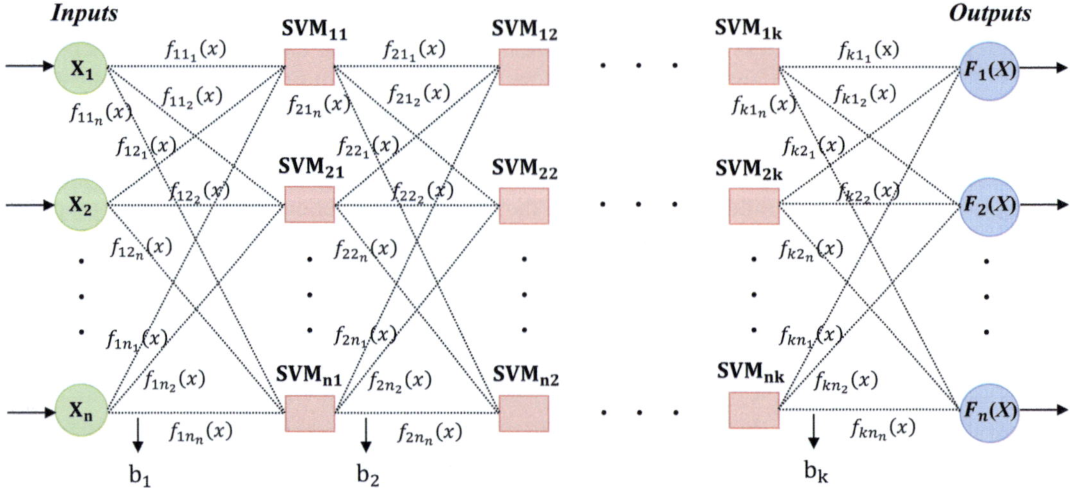

FIGURE 5.6 General architecture of the Deep Support Vector Machine (DeepSVM).

the output layer points. For X_1, the output for training $\text{SVM}_{11}, \text{SVM}_{12}, \ldots, \text{SVM}_{1k}$ is $F_1(X)$, and for X_n, the output for training $\text{SVM}_{n1}, \text{SVM}_{n2}, \ldots, \text{SVM}_{nk}$ is $F_n(X)$. The network weights are represented as $f(x)$. All the various $f(x)$ are evaluated in the hidden layers, which are multiple layers that connect all inputs with outputs. The total net input to each hidden layer neuron can be expressed as (Okwuashi and Ndehedehe, 2020)

$$\text{net}_{h1} = f_{11_1}(x) \cdot X_1 + f_{11_2}(x) \cdot X_2 + \cdots + f_{11_n}(x) \cdot X_n + b_1$$

$$\text{net}_{h2} = f_{12_1}(x) \cdot X_1 + f_{12_2}(x) \cdot X_2 + \cdots + f_{12_n}(x) \cdot X_n + b_1$$

$$\vdots$$

$$\text{net}_{hn} = f_{1n_1}(x) \cdot X_1 + f_{1n_2}(x) \cdot X_2 + \cdots + f_{1n_n}(x) \cdot X_n + b_1$$

$$(5.86)$$

The output for each input neuron is computed using the logistic activation function as follows:

$$\text{out}_{h1} = \frac{1}{1 + e^{-\text{net}_{h1}}}$$

$$\text{out}_{h2} = \frac{1}{1 + e^{-\text{net}_{h2}}}$$

$$\vdots$$

$$\text{out}_{hn} = \frac{1}{1 + e^{-\text{net}_{hn}}}$$

$$(5.87)$$

The outputs of the neurons in the hidden layer are utilized as inputs to calculate the neurons in the output layer $\text{net}_{o1_1}, \ldots, \text{net}_{o1_n}, \text{net}_{o2_1}, \ldots, \text{net}_{o2_n}$, and $\text{net}_{on_1}, \ldots, \text{net}_{on_n}$, which can be written as

$$\text{net}_{o1_1} = f_{21_1}(x) \cdot \text{out}_{h1} + f_{21_2}(x) \cdot \text{out}_{h2} + \cdots + f_{21_n}(x) \cdot \text{out}_{hn} + b_2$$

$$\vdots$$

$$\text{net}_{o1_n} = f_{k1_1}(x) \cdot \text{out}_{h1} + f_{k12}(x) \cdot \text{out}_{h2} + \cdots + f_{k1_n}(x) \cdot \text{out}_{hn} + b_2$$

$$\text{net}_{o2_1} = f_{22_1}(x) \cdot \text{out}_{h1} + f_{22_2}(x) \cdot \text{out}_{h2} + \cdots + f_{22_n}(x) \cdot \text{out}_{hn} + b_2$$

$$\vdots \tag{5.88}$$

$$\text{net}_{o2_n} = f_{k2_1}(x) \cdot \text{out}_{h1} + f_{k2_2}(x) \cdot \text{out}_{h2} + \cdots + f_{k2_n}(x) \cdot \text{out}_{hn} + b_2$$

and

$$\text{net}_{on_1} = f_{2n_1}(x) \cdot \text{out}_{h1} + f_{2n_2}(x) \cdot \text{out}_{h2} + \cdots + f_{2n_n}(x) \cdot \text{out}_{hn} + b_k$$

$$\vdots \tag{5.89}$$

$$\text{net}_{on_n} = f_{kn_1}(x) \cdot \text{out}_{h1} + f_{kn_2}(x) \cdot \text{out}_{h2} + \cdots + f_{kn_n}(x) \cdot \text{out}_{hn} + b_k$$

To keep it simple, let us delve into the case of $\text{net}_{o1_1}, \ldots, \text{net}_{o1_n}$, where the output is calculated using the logistic activation function, as shown below:

$$\text{net}_{on_1} = f_{2n_1}(x) \cdot \text{out}_{h1} + f_{2n_2}(x) \cdot \text{out}_{h2} + \cdots + f_{2n_n}(x) \cdot \text{out}_{hn} + b_k$$

$$\vdots$$

$$\text{net}_{on_n} = f_{kn_1}(x) \cdot \text{out}_{h1} + f_{kn_2}(x) \cdot \text{out}_{h2} + \cdots + f_{kn_n}(x) \cdot \text{out}_{hn} + b_k$$

$$\text{out}_{o1_1} = \frac{1}{1 + e^{-\text{net}_{o1_1}}}$$

$$\vdots \tag{5.90}$$

$$\text{out}_{o1_n} = \frac{1}{1 + e^{-\text{net}_{o1_n}}}$$

The error in computing the output out_{o1_1} specifically for X_1 can be determined by subtracting the calculated output out_{o1_1} from the known value of $F_1(X)$ as

$$E_{o1} = \sum_{i=1}^{n} \frac{1}{2}\left(F_1(X) - \text{output}_{o1_i}\right) \tag{5.91}$$

Similarly, the total error is derived from the summation of all the computed errors $E_{o1}, E_{o2}, \ldots, E_{on}$ as

$$E_{\text{total}} = E_{o1} + E_{o2} + \cdots + E_{on} \tag{5.92}$$

Through the application of the backpropagation algorithm, each $f(x)$ within the network undergoes updates to bring the actual output closer to the target output $F(X)$, leading to the reduction of errors for both individual output neurons and the entire network. For example, $f_{11_1}(x)$ can be computed as the gradient of $\partial E_{\text{total}}$ as

$$\frac{\partial E_{\text{total}}}{f_{11_1}(x)} = \frac{\partial \text{net}_{o1_1}}{\partial f_{11_1}(x)} \cdot \frac{\partial \text{out}_{o1_1}}{\partial \text{net}_{o1_1}} \cdot \frac{\partial E_{\text{total}}}{\partial \text{out}_{o1_1}} \tag{5.93}$$

The updated function $f_{11}^{(\text{new})}(x)$ can be computed as

$$f_{11_1}^{(\text{new})}(x) = f_{11_1}(x) - \lambda \cdot \frac{\partial E_{\text{total}}}{\partial f_{11_1}(x)} \tag{5.94}$$

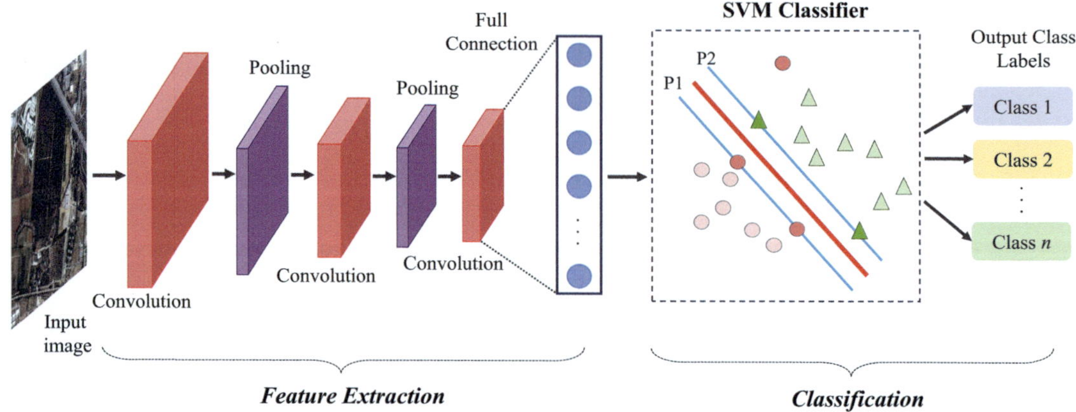

FIGURE 5.7 General framework for the integration of SVM with convolutional neural networks.

where λ is the learning rate used to adjust the network weights. In the same way, adjustments will be made to all the weights $f(x)$ within the network, and this process will be repeated iteratively using Equation (5.86) until E_{total} becomes zero or infinitesimal.

The combination of SVMs with neural networks, particularly convolutional neural networks (CNNs) as outlined in Chapter 7, has attracted increased interest because of the advantages of neural networks in processing image features. Although classification of extracted image features is an essential aspect of DL, SVMs can effectively classify these data and take the scores of data as a basis for evaluation. In classification scenarios, the Softmax activation function is commonly employed by most DL models to make predictions, with an objective of minimizing cross-entropy loss. SVMs have replaced the Softmax activation function in most DL models and are used as the final classification unit of a CNN, achieving better classification performance. SVMs are also employed to replace the fully connected network used for classification tasks at the final stage of a DL algorithm. A generic framework for this type of CNN and SVM integration is shown in Figure 5.7.

5.8 CONCLUDING REMARKS

Because of its robust theoretical foundation and superior generalization capabilities, among other strengths, SVMs have been extensively integrated into pattern recognition applications. Numerous studies have emphasized that SVMs are superior to conventional classification techniques. In the field of remote sensing, SVMs have been extensively used for classification and their capabilities are expanded by using its other robust variants, particularly those that include DL models. Despite limitations in parameter selection, algorithmic complexity, and managing multiclass and imbalanced data sets, SVMs have seen significant adoption in real-world classification tasks because of their strong theoretical framework and exceptional generalization abilities. It is worth noting that SVMs might not be preferred in contexts with extensive data sets, since specific implementations require extensive training time. Similarly, in instances with imbalanced data sets, SVMs might yield inferior outcomes compared with DL models. The performance of spectral SVMs is surpassed by deep networks, especially with the use of CNNs, which demonstrate superior capabilities compared to other DL models. However, this advantage diminishes when providing SVM with enhanced features extracted from the data, such as spatial filters (Heydari and Mountrakis, 2019). In the literature, most research has employed the grid search method as a strategy for hyperparameter tuning to achieve the best classification accuracy. Furthermore, the use of handcrafted features alongside spectral features offers benefits; however, there is a lack of consensus regarding the best practices for their application. On the other hand, the existence of numerous approaches leaves practitioners with an open issue about which variant suits specific applications. It is important to note that there is no standardized approach to ensure the optimal solution for SVM applications in classifying remotely sensed data.

This chapter contains the knowledge necessary for readers to implement SVMs as classifiers. It is clear that the major disadvantage of SVMs like that of ANN classifiers is the user-defined parameters have a strong effect on their performance. This may restrict the capabilities of the SVMs to resolve more complex classification problems. However, based on the experiments and studies mentioned in this chapter, the high-level performance of SVMs has attracted attention and interest. Strategies to handle the multiclass concern, techniques for parameter selection, and identification of appropriate input features (especially beneficial in handling hyperspectral data) are thoroughly explained. A comprehensive outline of the theoretical and mathematical foundations of the three primary SVM variations used in classification is provided. Through in-depth evaluations and surveys, it is suggested that incorporating the SVM method into the framework of DL will likely become a major focus for the analysis of remotely sensed data in the future.

REFERENCES

Adankon, M. M., and M. Cheriet. 2009. Model selection for the LS-SVM. Application to handwriting recognition. *Pattern Recognition* 42:3264–3270. https://doi.org/10.1016/j.patcog.2008.10.023.

Aizerman, M. A., E. M. Braverman, and L. I. Rozoner. 1964. Theoretical foundations of the potential function method in pattern recognition learning. *Automation and Remote Control* 25:821–837.

Belousov, A. I., S. A. Verzakov, and J. von Frese. 2002. A flexible classification approach with optimal generalisation performance: Support vector machines. *Chemometrics and Intelligent Laboratory Systems* 64:15–25. https://doi.org/10.1016/S0169-7439(02)00046-1.

Bishop, C. M., and M. Tipping. 2000. Variational relevance vector machines. In *Proceedings of 16th Conference on Uncertainty in Artificial Intelligence (UAI'2000)*, June 30–July, Stanford, CA, pp. 346–53.

Boser, B. E., I. M. Guyon, and V. N. Vapnik. 1992. A training algorithm for optimal margin classifiers. In *Proceedings of the 5th Annual Workshop on Computational Learning Theory*, July 27–29, Pittsburgh, PA, pp. 144–152. https://doi.org/10.1145/130385.130401.

Braun, A. C., U. Weidner, and S. Hinz. 2011. Support vector machines, import vector machines and relevance vector machines for hyperspectral classification - A comparison. In *Proceedings of 3rd Workshop on Hyperspectral Image and Signal Processing: Evolution in Remote Sensing (WHISPERS)*, July 6–9, Lisbon, Portugal, pp. 1–4. https://doi.org/10.1109/WHISPERS.2011.6080861.

Bruzzone, L., M. Chi, and M. Marconcini. 2006. A novel transductive SVM for semisupervised classification of remote-sensing images. *IEEE Transactions on Geoscience and Remote Sensing* 44:3363–3373. https://doi.org/10.1109/TGRS.2006.877950.

Bruzzone, L., M. Chi, and M. Marconcini. 2007. Semisupervised support vector machines for classification of hyperspectral remote sensing images. In *Hyperspectral Data Exploitation: Theory and Applications*, (ed.) C.-I. Chang, pp. 275–311. Hoboken, NJ: John Wiley & Sons. https://doi.org/10.1002/9780470124628.ch11.

Burges, C. J. C. 1998. A tutorial on support vector machines for pattern recognition. *Data Mining and Knowledge Discovery* 2:121–167. https://doi.org/10.1023/A:1009715923555.

Burges, C. J. C., and B. Schölkopf. 1997. Improving the accuracy and speed of support vector machines. In *Advances in Neural Information Processing Systems*, (eds.) M. C. Mozer, M. I. Jordan, and T. Petsche, pp. 375–381. Cambridge, MA: MIT Press.

Cervantes, J., F. Garcia-Lamont, L. Rodríguez-Mazahua, and A. Lopez. 2020. A comprehensive survey on support vector machine classification: Applications, challenges and trends. *Neurocomputing* 408:189–215. https://doi.org/10.1016/j.neucom.2019.10.118.

Chang, C. C., and C. J. Lin. 2007. LIBSVM: A library for support vector machines. *ACM Transactions on Intelligent Systems and Technology* 2:1–27. https://doi.org/10.1145/1961189.1961199.

Chapelle, O., V. Vapnik, O. Bousquet, and S. Mukherjee. 2002. Choosing multiple parameters for support vector machines. *Machine Learning* 46:131–159. https://doi.org/10.1023/A:1012450327387.

Chi, M., Q. Qian, and J. A. Benediktsson. 2008. Cluster-based ensemble classification for hyperspectral remote sensing images. In *Proceedings of IEEE International Geoscience and Remote Sensing Symposium (IGARSS)*, Jully 7–11, Boston, MA, pp. 209–212. https://doi.org/10.1109/IGARSS.2008.4778830.

Cortes, C., and V. Vapnik. 1995. Support vector networks. *Machine Learning* 20:273–297. https://doi.org/10.1007/BF00994018.

Crammer, K., and Y. Singer. 2002. On the learnability and design of output codes for multiclass problems. *Machine Learning* 47:201–233. https://doi.org/10.1023/A:1013637720281.

Demir, B., and S. Erturk. 2007. Hyperspectral image classification using relevance vector machines. *IEEE Geoscience and Remote Sensing Letters* 4:586–590. https://doi.org/10.1109/LGRS.2007.903069.

Dias, M. L. D., and A. R. R. Neto. 2016. Evolutionary support vector machines: A dual approach. In *Proceedings of IEEE Congress on Evolutionary Computation (CEC)*, July 24–29, Vancouver, Canada, pp. 2185–2192. https://doi.org/10.1109/CEC.2016.7744058.

Díaz-Vico, D., J. Prada, A. Omari, and J. Dorronsoro. 2020. Deep support vector neural networks. *Integrated Computer-Aided Engineering* 27:389–402. https://doi.org/10.3233/ICA-200635.

Dong, C., and H. Zhao. 2010. Hyperspectral image classification and application based on relevance vector machine. *Journal of Remote Sensing* 14:1273–1278.

Erfani, S. M., S. Rajasegarar, S. Karunasekera, and C. Leckie. 2016. High-dimensional and large-scale anomaly detection using a linear one-class SVM with deep learning. *Pattern Recognition* 58:121–134. https://doi.org/10.1016/j.patcog.2016.03.028.

Ertekin, S., L. Bottou, and C. L. Giles. 2011. Nonconvex online support vector machines. *IEEE Transactions on Pattern Analysis and Machine Intelligence* 33:368–381. https://doi.org/10.1109/tpami.2010.109.

Fletcher, R. 1987. *Practical Methods of Optimization*, 2nd edition, Chichester, UK: John Wiley & Sons.

Foody, G. M., and A. Mathur. 2004. Toward intelligent training of supervised image classifications: Directing training data acquisition for SVM classification. *Remote Sensing of Environment* 93:107–117. https://doi.org/10.1016/j.rse.2004.06.017.

Foody, G. M., and A. Mathur. 2006. The use of small training sets containing mixed pixels for accurate hard image classification: Training on mixed spectral responses for classification by a SVM. *Remote Sensing of Environment* 103:179–189. https://doi.org/10.1016/j.rse.2006.04.001.

Friedrichs, F., and C. Igel. 2005. Evolutionary tuning of multiple SVM parameters. *Neurocomputing* 64:107–117. https://doi.org/10.1016/j.neucom.2004.11.022.

Fukunaga, K. 2013. *Introduction to Statistical Pattern Recognition*. Cambridge, MA: Academic Press. https://doi.org/10.1016/C2009-0-27872-X.

Gao, S., Q. Ye, and N. Ye. 2011. 1-Norm least squares twin support vector machines. *Neurocomputing* 74:3590–3597. https://doi.org/10.1016/j.neucom.2011.06.015.

Gómez-Chova, L., G. Camps-Valls, J. Muñoz-Marí, and J. Calpe. 2008. Semisupervised image classification with Laplacian support vector machines. *IEEE Geoscience and Remote Sensing Letters* 5:336–340. https://doi.org/10.1109/LGRS.2008.916070.

Gönen, M., and E. Alpaydın. 2011. Multiple kernel learning algorithms. *The Journal of Machine Learning Research* 12:2211–2268.

Gualtieri, J. A. and R. F. Cromp. 1999. Support vector machines for hyperspectral remote sensing classification. In *Proceedings 27th AIPR Workshop: Advances in Computer-Assisted Recognition,* October 14–16, Washington, DC, pp. 221–232. https://doi.org/10.1117/12.339824.

Guo, X. C., Y. C. Liang, C. G. Wu, and C. Y. Wang. 2006. PSO-based hyper-parameters selection for LS-SVM classifiers. In *Neural Information Processing,* (eds.) I. King, J. Wang, L. W. Chan, and D. Wang, pp. 1138–1147. Berlin, Heidelberg: Springer. https://doi.org/10.1007/11893257_124.

Halmos, P. R. 1967. *A Hilbert Space Problem Book*. Princeton, NJ: D. Van Nostrand Company, Inc.

Han, B., and L. S. Davis. 2012. Density-based multifeature background subtraction with support vector machine. *IEEE Transactions on Pattern Analysis and Machine Intelligence* 34:1017–1023. https://doi.org/10.1109/TPAMI.2011.243.

Heydari, S. S., and G. Mountrakis. 2019. Meta-analysis of deep neural networks in remote sensing: A comparative study of mono-temporal classification to support vector machines. *ISPRS Journal of Photogrammetry and Remote Sensing* 152:192–210. https://doi.org/10.1016/j.isprsjprs.2019.04.016.

Hsu, C. W., and C. J. Lin. 2002. A comparison of methods for multiclass support vector machines. *IEEE Transactions on Neural Networks* 13:415–425. https://doi.org/10.1109/72.991427.

Huang, C., B. Wylie, L. Yang, C. Homer, and G. Zylstra. 2002. Derivation of a tasselled cap transformation based on Landsat 7 at-satellite reflectance. *International Journal of Remote Sensing* 23:1741–1748. https://doi.org/10.1080/01431160110106113.

Huang, H., X. Wei, and Y. Zhou. 2018. Twin support vector machines: A survey. *Neurocomputing* 300:34–43. https://doi.org/10.1016/j.neucom.2018.01.093.

Jaakkola, T. S., and D. Haussler. 1999. Probabilistic kernel regression models. In *Proceedings of the Seventh International Workshop on Artificial Intelligence and Statistics, January 3–6,* Fort Lauderdale, FL.

Jayadeva, R. K., and S. Chadra. 2007. Twin support vector machines for pattern classification. *IEEE Transactions on Pattern Analysis and Machine Intelligence* 29:905–910. https://doi.org/10.1109/TPAMI.2007.1068.

Kalaiarasi, G., and S. Maheswari. 2021. Deep proximal support vector machine classifiers for hyperspectral images classification. *Neural Computing and Applications* 33:13391–13415. https://doi.org/10.1007/s00521-021-05965-0.

Kaul, A., and S. Raina. 2022. Support vector machine versus convolutional neural network for hyperspectral image classification: A systematic review. *Concurrency and Computation: Practice and Experience* 34:e6945. https://doi.org/10.1002/cpe.6945.

Kavzoglu, T., and I. Colkesen. 2009. A kernel functions analysis for support vector machines for land cover classification. *International Journal of Applied Earth Observation and Geoinformation* 11:352–359. https://doi.org/10.1016/j.jag.2009.06.002.

Kavzoglu, T., F. Bilucan, and A. Teke. 2020. Comparison of support vector machines, random forest and decision tree methods for classification of Sentinel-2A image using different band combinations. In *Proceedings of 41st Asian Conference on Remote Sensing (ACRS),* November 9–11, Deqing, China, pp. 1–8.

Keuchel, J., S. Naumann, M. Heiler, and A. Siegmun. 2003. Automatic land cover analysis for Tenerife by supervised classification using remotely sensed data. *Remote Sensing of Environment* 86:530–541. https://doi.org/10.1016/S0034-4257(03)00130-5.

Khatami, R., G. Mountrakis, and S. V. Stehman. 2016. A meta-analysis of remote sensing research on supervised pixel-based land-cover image classification processes: General guidelines for practitioners and future research. *Remote Sensing of Environment* 177:89–100. https://doi.org/10.1016/j.rse.2016.02.028.

Khemchandani, R., P. Saigal, and S. Chandra 2018. Angle-based twin support vector machines. *Annals of Operations Research* 269:387–417. https://doi.org/10.1007/s10479-017-2604-2.

Kim, S., S. Kavuri, and M. Lee. 2013. Deep network with support vector machines. In *Neural Information Processing,* (eds.) M. Lee, A. Hirose, Z.-G. Hou, and R. Kil, pp. 458–465. Berlin, Heidelberg: Springer. https://doi.org/10.1007/978-3-642-42054-2_57.

Kim, S., Y. Choi, and M. Lee. 2015. Deep learning with support vector data description. *Neurocomputing* 165:111–117. https://doi.org/10.1016/j.neucom.2014.09.086.

Knerr, S., L. Personnaz, and G. Dreyfus. 1990. Single-layer learning revisited: A stepwise procedure for building and training neural network. In *Neurocomputing: Algorithms, Architectures and Applications,* (eds.) F. F. Soulié, and J. Hérault, pp. 41–50. Berlin, Heidelberg: Springer. https://doi.org/10.1007/978-3-642-76153-9_5.

Kolmogorov, A. N., and S. V. Fomin. 1970. *Introductory Real Analysis.* Englewood Cliffs, NJ: Prentice-Hall, Inc.

Kremer, J., K. S. Pedersen, and C. Igel. 2014. Active learning with support vector machines. *Wiley Interdisciplinary Reviews: Data Mining and Knowledge Discovery* 4:313–326. https://doi.org/10.1002/widm.1132.

Kumar, M. A., and M. Gopal. 2009. Least squares twin support vector machines for pattern classification. *Expert Systems with Applications* 36:7535–7543. https://doi.org/10.1016/j.eswa.2008.09.066.

Lee, L. H., C. H. Wan, R. Rajkumar, and D. Isa. 2012. An enhanced support vector machine classification framework by using Euclidean distance function for text document categorization. *Applied Intelligence* 37:80–99. https://doi.org/10.1007/s10489-011-0314-z.

Li, B., M. Chi, J. Fan, and X. Xue. 2006. Support cluster machine. In *Proceedings of 24th International Conference on Machine Learning,* June 20–24 Corvallis, OR, pp. 505–512. https://doi.org/10.1145/1273496.1273560.

Li, D., Y. Tian, and H. Xu. 2014. Deep twin support vector machine. In *Proceedings of IEEE International Conference on Data Mining Workshop,* December 14, Shenzhen, China, pp. 65–73. https://doi.org/10.1109/ICDMW.2014.18.

Liu, D., M. Kelly, and P. Gong. 2006. A spatial-temporal approach to monitoring forest disease spread using multi-temporal high spatial resolution imagery. *Remote Sensing of Environment* 101:167–180. https://doi.org/10.1016/j.rse.2005.12.012.

Liu, F., H. Zhong, and S. H. Li. 2017. Accelerating relevance vector machine for large-scale data on Spark. *ITM Web of Conferences* 12:01024. https://doi.org/10.1051/itmconf/20171201024.

Lorena, A. C., and A. C. P. L. F. de Carvalho. 2006. Multiclass SVM design and parameter selection with genetic algorithms. In *Proceedings of Ninth Brazilian Symposium on Neural Networks (SBRN'06),* October 23–27, Ribeirao Preto, Brazil, pp. 23–28. https://doi.org/10.1109/SBRN.2006.28.

Luntz, A., and V. Brailovsky. 1969. On estimation of characters obtained in statistical procedure of recognition (in Russian). *Technicheskaya Kibernetica* 3, pp. 563–575.

MacKay, D. J. C. 1992. The evidence framework applied to classification networks. *Neural Computation* 4:720–736. https://doi.org/10.1162/neco.1992.4.5.720.

Mangasarian, O. L., and E. W. Wild. 2001. Proximal support vector machine classifiers. In *Proceedings of Seventh ACM SIGKDD International Conference on Knowledge Discovery and Data Mining (KDD'01),* August 26–29, San Francisco, CA, pp. 77–86. https://doi.org/10.1145/502512.502527.

Mantero, P., G. Moser, and S. B. Serpico. 2005. Partially supervised classification of remote sensing images through SVM-based probability density estimation. *IEEE Transactions on Geoscience and Remote Sensing* 43:559–570. https://doi.org/10.1109/TGRS.2004.842022.

Maulik, U., and D. Chakraborty. 2011. A self-trained ensemble with semisupervised SVM: An application to pixel classification of remote sensing imagery. *Pattern Recognition* 44:615–623. https://doi.org/10.1016/j.patcog.2010.09.021.

Maulik, U., and D. Chakraborty. 2017. Remote sensing image classification: A survey of support-vector-machine-based advanced techniques. *IEEE Geoscience and Remote Sensing Magazine* 5:33–52. https://doi.org/10.1109/MGRS.2016.2641240.

Melgani, F., and L. Bruzzone. 2004. Classification of hyperspectral remote sensing images with support vector machines. *IEEE Transactions on Geoscience and Remote Sensing* 42:1778–1790. https://doi.org/10.1109/TGRS.2004.831865.

Mianji, F. A. and Y. Zhang. 2010. Improved hyperspectral land-cover analysis using relevance vector machine. In *Proceedings of IEEE International Conference on Image Processing,* September 26–29, Hong Kong, China, pp. 2281–2284. https://doi.org/10.1109/ICIP.2010.5654306.

Mitra, P., B. U. Shankar, and S. K. Pal. 2004. Segmentation of multispectral remote sensing images using active support vector machines. *Pattern Recognition Letters* 25:1067–1074. https://doi.org/10.1016/j.patrec.2004.03.004.

Mountrakis, G., J. Im, and C. Ogole. 2011. Support vector machines in remote sensing: A review. *ISPRS Journal of Photogrammetry and Remote Sensing* 66:247–259. https://doi.org/10.1016/j.isprsjprs.2010.11.001.

Okwuashi, O., and C. E. Ndehedehe. 2020. Deep support vector machine for hyperspectral image classification. *Pattern Recognition* 103:107298. https://doi.org/10.1016/j.patcog.2020.107298.

Opper, M., and O. Winther. 2000. Gaussian processes and SVM: Mean field and leave-one-out. In *Advances in Large Margin Classifiers,* (eds.) A. J. Smola, P. L. Bartlett, B. Schölkopf, and D. Schuurmans, pp. 311–326. Cambridge, MA: MIT Press.

Pal, M. 2006. Support vector machine-based feature selection for land cover classification: A case study with DAIS hyperspectral data. *International Journal of Remote Sensing* 27:2877–2894. https://doi.org/10.1080/01431160500242515.

Pal, M. 2008. Multiclass approaches for support vector machine based land cover classification. arXiv preprint arXiv:0802.2411.

Pal, M. 2009. Kernel methods in remote sensing: A review. *ISH Journal of Hydraulic Engineering* 15:194–215. https://doi.org/10.1080/09715010.2009.10514975.

Pal, M. 2011. Support vector machines/relevance vector machine for remote sensing classification: A review. arXiv preprint, arXiv:1101.2987. https://doi.org/10.48550/arXiv.1101.2987.

Pal, M., and G. M. Foody. 2012. Evaluation of SVM, RVM and SMLR for accurate image classification with limited ground data. *IEEE Journal of Selected Topics in Applied Earth Observations and Remote Sensing* 5:1344–1355. https://doi.org/10.1109/JSTARS.2012.2215310.

Pal, M., and P. M. Mather. 2005. Support vector machines for classification in remote sensing. *International Journal of Remote Sensing* 26:1007–1011. https://doi.org/10.1080/01431160512331314083.

Pasolli, E., F. Melgani, D. Tuia, F. Pacifici, F., and W. J. Emery. 2013. SVM active learning approach for image classification using spatial information. *IEEE Transactions on Geoscience and Remote Sensing* 52:2217–2233. https://doi.org/10.1109/TGRS.2013.2258676.

Persello, C., and L. Bruzzone. 2014. Active and semisupervised learning for the classification of remote sensing images. *IEEE Transactions on Geoscience and Remote Sensing* 52:6937–6956. https://doi.org/10.1109/TGRS.2014.2305805.

Platt, J. C. 1999. Fast training of support vector machines using sequential minimal optimization. In *Advances in Kernel Methods: Support Vector Learning,* (eds.) B. Schölkopf, C. J. C. Burges, and A. J. Smola, pp. 185–208. Cambridge, MA: MIT Press.

Platt, J. C., N. Cristianini, and J. Shawe-Taylor. 2000. Large margin DAGs for multiclass classification. In *Advances in Neural Information Processing Systems,* (eds.) S. A. Solla, T. K. Leen, and K.-R. Müller, pp. 547–553. Cambridge, MA: MIT Press.

Qi, Z., Y. Tian, and Y. Shi. 2013a. Robust twin support vector machine for pattern classification. *Pattern Recognition* 46:305–316. https://doi.org/10.1016/j.patcog.2012.06.019.

Qi, Z., Y. Tian, and Y. Shi. 2013b. Structural twin support vector machine for classification. *Knowledge-Based Systems* 43:74–81. https://doi.org/10.1016/j.knosys.2013.01.008.

Rosales-Pérez, A., S. Garcia, H. Terashima-Marin, C. A. Coello Coello, and F. Herrera. 2018. MC2ESVM: Multiclass classification based on cooperative evolution of support vector machines. *IEEE Computational Intelligence Magazine* 13:18–29. https://doi.org/10.1109/MCI.2018.2806997.

Samui, P., P. H. Gowda, T. Oommen, T. A. Howell, T. H. Marek, and D. O. Porter. 2012. Statistical learning algorithms for identifying contrasting tillage practices with Landsat Thematic Mapper data. *International Journal of Remote Sensing* 33:5732–5745. https://doi.org/10.1080/01431161.2012.671555.

Schohn, G., and D. Cohn. 2000. Less is more: Active learning with support vectors machines. In *Proceedings of 17th International Conference on Machine Learning (ICML'00)*, June 29–July 2, Stanford, CA, pp. 839–846.

Schölkopf, B., K. K. Sung, C. J. C. Burges, F. Girosi, P. Niyogi, T. Poggio, and V. Vapnik. 1997. Comparing support vector machines with Gaussian kernels to radial basis function classifiers. *IEEE Transactions on Signal Processing* 45:2758–2765. https://doi.org/10.1109/78.650102.

Sun, F., and X. Xie. 2023. Deep non-parallel hyperplane support vector machine for classification. *IEEE Access* 11:7759–7767. https://doi.org/10.1109/ACCESS.2023.3237641.

Tang, Y. 2013. Deep learning using linear support vector machines. In *Proceedings of International Conference on Machine Learning 2013: Challenges in Representation Learning Workshop*. Atlanta, GA.

Tanveer, M., T. Rajani, R. Rastogi, Y. H. Shao, and M. A. Ganaie. 2022. Comprehensive review on twin support vector machines. *Annals of Operations Research*. https://doi.org/10.1007/s10479-022-04575-w.

Tax, D. M. J., and R. P. W. Duin. 2004. Support vector data description. *Machine Learning* 54:45–66. https://doi.org/10.1023/B:MACH.0000008084.60811.49.

Tipping, M. 1999. The relevance vector machine. In *Advances in Neural Information Processing Systems*, (eds.) S. A. Solla, T. K. Leen, and K.-R. Müller, pp. 547–553. Cambridge, MA: MIT Press.

Tipping, M. 2001. Sparse Bayesian learning and the relevance vector machine. *Journal of Machine Learning Research* 1:211–244. https://doi.org/10.1162/15324430152748236.

Tuia, D., F. Ratle, F. Pacifici, A. Pozdnoukhov, M. Kanevski, F. Del Frate, D. Solimini, and W. J. Emery. 2008. Active learning of very-high resolution optical imagery with SVM: Entropy versus margin sampling. In *Proceedings of IEEE International Geoscience and Remote Sensing Symposium*, Jully 7–11, Boston, MA, pp. IV-73–IV-76. https://doi.org/10.1109/IGARSS.2008.4779659.

Van Messem, A. 2020. Support vector machines: A robust prediction method with applications in bioinformatics. In *Principles and Methods for Data Science*, (eds.) A. S. R. S. Rao, and C. R. Rao, pp. 391–466. Amsterdam, The Netherlands: Elsevier. https://doi.org/10.1016/bs.host.2019.08.003.

Vapnik, V. 1979. Estimation of Dependences Based on Empirical Data (*in Russian*). Nauka, Moscow. (English translation: New York: Springer-Verlag, 1982).

Vapnik, V. 1995. *The Nature of Statistical Learning Theory*. New York: Springer-Verlag. https://doi.org/10.1007/978 1 4757 3264 1.

Vapnik, V. 1998. *Statistical Learning Theory*. New York: John Wiley & Sons.

Vapnik, V., and O. Chapelle. 2000. Bounds on error expectation for support vector machines. *Neural Computation* 12:2013–2036. https://doi.org/10.1162/089976600300015042.

Wang, F., B. Gou, and Y. Qin. 2013. Modeling tunneling-induced ground surface settlement development using a wavelet smooth relevance vector machine. *Computers and Geotechnics* 54:125–132. https://doi.org/10.1016/j.compgeo.2013.07.004.

Wang, T., L. Zhang, and W. Hu. 2021. Bridging deep and multiple kernel learning: A review. *Information Fusion* 67:3–13. https://doi.org/10.1016/j.inffus.2020.10.002.

Wang, X., F. Huang, and Y. Cheng. 2014. Super-parameter selection for Gaussian-Kernel SVM based on outlier-resisting. *Measurement* 58:147–153. https://doi.org/10.1016/j.measurement.2014.08.019.

Xie, J., K. Hone, W. Xie, X. Gao, Y., Shi, and X. Liu. 2013. Extending twin support vector machine classifier for multi-category classification problems. *Intelligent Data Analysis* 17:649–664. https://doi.org/10.3233/IDA-130598.

Xu, Y., W. Xi, X. Lv, and R. Guo. 2012. An improved least squares twin support vector machine. *Journal of Information & Computational Science* 9:1063–1071.

Yan, G., K. Lishan, L. Fujiang, and M. Linlu. 2007. Evolutionary support vector machine and its application in remote sensing imagery classification. *Proceedings of SPIE 6752, Geoinformatics 2007: Remotely Sensed Data and Information*, 67523D. https://doi.org/10.1117/12.760804.

Yao, J., H. Wang, L. Wang, and G. Fu. 2019. High precision classification of hyperspectral image based on a hierarchical localized multiple kernel learning method. In *Proceedings of 12th International Congress on Image and Signal Processing, BioMedical Engineering and Informatics (CISP-BMEI)*, October 19–21, Suzhou, China, pp. 1–5 https://doi.org/10.1109/CISP-BMEI48845.2019.8965835.

Yin, S., and J. Yin. 2016. Tuning kernel parameters for SVM based on expected square distance ratio. *Information Sciences* 370:92–102. https://doi.org/10.1016/j.ins.2016.07.047.

Zhang, J., H. Tian, D. Wang, H. Li, and A. M. Mouazen. 2020. A novel approach for estimation of above-ground biomass of sugar beet based on wavelength selection and optimized support vector machine. *Remote Sensing* 12:620. https://doi.org/10.3390/rs12040620.

Zhang, Q., G. Shan, X. Duan, and Z. Zhang. 2009. Parameters optimization of support vector machine based on simulated annealing and genetic algorithm. In *Proceedings of IEEE International Conference on Robotics and Biomimetics (ROBIO)*, December 19–23, Guilin, China, pp. 1302–1306. https://doi.org/10.1109/ROBIO.2009.5420717.

Zhang, Z., Z. Liu, L. Zheng, and Y. Zhang. 2014. Development of an adaptive relevance vector machine approach for slope stability inference. *Neural Computing & Applications* 25:2025–2035. https://doi.org/10.1007/s00521-014-1690-1.

Zhu, J., and T. Hastie. 2005. Kernel logistic regression and the import vector machine. *Journal of Computational and Graphical Statistics* 14:185–205. https://doi.org/10.1198/106186005X25619.

6 Decision Trees

Categorization of data using a hierarchical splitting (or top-down) mechanism has been widely used in the environmental and life sciences. The purpose of using a hierarchical structure for labeling objects is to gain a more comprehensive understanding of relationships between objects at different scales of observation or different levels of detail. Its simplest representation takes the form of an inverted tree in which the different levels of classification are represented by the separate levels of a hierarchy. When applied to multispectral image data, the design of a decision tree is based on knowledge of the spectral properties of each class and the relationships among the classes. The main benefit of using a hierarchical tree structure to perform classification decisions is that the tree structure can be viewed as a white box, which makes it easier to interpret and understand the relation between the inputs and outputs in comparison with artificial neural networks (Chapter 7). Moreover, decision tree algorithms require a small computational cost and can handle a large number of features with a high number of training samples with predictable response times and handle both symbolic and numerical data. Because of their interpretability, as thoroughly discussed in Chapter 10, decision tree-based algorithms have become popular, particularly when investigating problems related to human life and critical decision-making processes.

A decision tree is structured with a root node, interior nodes, and terminal nodes, known as leaf nodes. The nonterminal nodes, including the root and interior nodes, create decision stages that lead to terminal nodes representing the final classification. The process involves following a set of rules from the root node to a terminal node, which assigns a label to the object being classified. A decision about the path to the next node must be made at each nonterminal node. Figure 6.1 illustrates a simple decision tree model constructed with pixel values representing the maize, forest, and hazelnut classes in the blue, green, red, and NIR wavebands. In a basic tree structure, there are three fundamental components known as nodes, branches, and leaves. Nodes represent each attribute, such as a band. The structure also includes branches and leaves, with the final segments termed as leaves and the uppermost segment designated as the root node. The segments connecting the root to the leaves are referred to as branches. The fundamental approach to constructing a tree structure with the pixel values of training data is centered on asking a sequence of questions about the data, adapting actions based on the responses received, and quickly reaching the desired outcome. In this manner, the decision tree establishes decision rules by gathering the answers to these questions. Starting from the root node, which is the primary node in the tree, inquiries are posed to categorize the data and shape the tree structure. This iterative process persists until nodes or leaves lacking branches are identified (Pal and Mather, 2003). Considering the example tree structure in the figure, if the reflectance value in the red band of a pixel with an unknown class label is less than or equal to 0.0253 and the reflectance value in the blue band is less than or equal to 0.0228, it is concluded that the class label of the unknown pixel is forest. The figure also shows the predictions of the tree structure for the test data set consisting of 90 pixels (i.e., 30 pixels for each class). The number of pixels reaching each node and the probabilities of these pixels belonging to the three land covers are shown in Figure 6.1. Only 28 pixels fulfilling the conditions obtained during tree generation could reach the leaf node (i.e., forest class) located at the leftmost part of the tree structure. The probability of these pixels being in the forest class was calculated as 1. On the other hand, the remaining four pixels were assigned to the forest class at different leaf node points as they fulfill different decision rules in the tree structure. It is obvious that the nature of the decisions being made and the sequence of attributes occurring within a tree will affect classification results. Thus, the efficiency and performance of the decision-making approach are strongly affected by the algorithm used for inducting a decision tree.

DOI: 10.1201/9781003439172-6

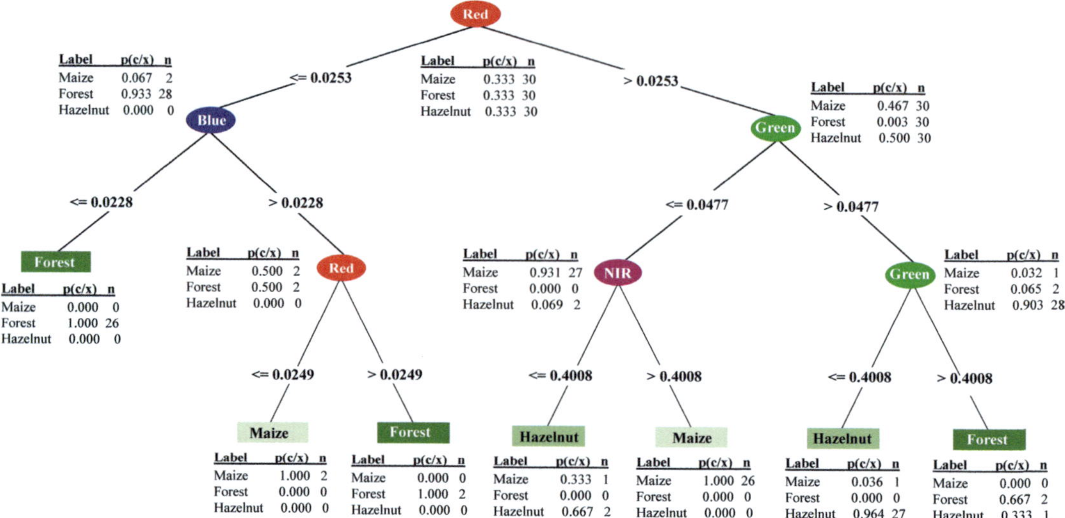

FIGURE 6.1 Simple hierarchical decision tree classifier using blue, green, red, and NIR bands of Sentinel-2 imagery.

Two approaches are used for the design of a decision tree. One is based on the user's knowledge and relies solely on user interaction. This is called the manual design approach. The second approach uses an automatic procedure. In the manual design procedure, statistics for all classes are first computed, and a graph of the spectral range in each band is constructed. From the graph and the statistical parameters, estimates of the decision boundaries are derived, and the tree is designed manually to separate classes in a hierarchical method. The construction of a tree by manual design method is time-consuming and may not provide satisfactory results, particularly when the number of classes is large and there is a spectral overlap between classes.

Swain and Hauska (1977) proposed a heuristic search technique based on a mathematical evaluation function for solving more complex problems. Other strategies for optimal hierarchical tree design have been described by Kulkarni and Laveen (1976) and Kurzynski (1983). The design of an optimum tree classifier, which will be reflected in both performance and computational efficiency, depends on the choice of the tree structure, the choice of features used in each terminal node, and the decision rules for performing the classification at each nonterminal node. Lee and Richards (1985) described a classification strategy in which classes were hierarchically separated in a piecewise linear fashion. In this approach, computational demands increase only linearly with the number of features used, and classification accuracy is claimed to be comparable to that of the maximum likelihood (ML) classifier. Kim and Landgrebe (1991) proposed a hybrid approach for the design of decision tree classifiers. The rationale underlying their technique is that a class must simultaneously be of informational value and separable from other classes. As we know, supervised procedures, based on training samples, can guarantee the former but not the latter; unsupervised procedures can guarantee the latter but not the former. Furthermore, only terminal nodes must have both separable and informational values. Nonterminal nodes are not required to be classes of informational value, but they must represent separable classes. To gain a comprehensive understanding of the manifold methods employed in devising decision tree classifiers, readers should refer to Safavian and Landgrebe (1991), McIver and Friedl (2002), and Costa and Pedreira (2023).

In recent years, ensemble methods have emerged as an approach leveraging multiple learning algorithms to enhance predictive performance. This approach closely mirrors human decision-making, involving the combination and weighting of various individual classifiers to create a classifier that surpasses each of them (Rokach, 2010). Ensemble learning involves the fusion of outcomes from independently trained classifiers known as base classifiers to classify new data. These individual classifiers are trained on distinct subsets, creating a collection of diverse classifiers (Kuncheva and Whitaker, 2003). Ensemble classifiers can produce higher prediction accuracy and outperform individual classifiers (e.g., Gislason et al., 2006; Kavzoglu and Colkesen, 2013). Given their considerable promise, various ensemble methods have been suggested in scholarly works, including widely recognized ones like random forest methods, extreme gradient boosting (XGBoost), light gradient boosting machines (LightGBM), gradient boosting machines (GBM), categorical boosting (CatBoost), and natural gradient boosting (NGBoost). The mathematical foundations of these methods are thoroughly discussed in this chapter.

6.1 ID3, C4.5, AND SEE5.0 DECISION TREES

6.1.1 ID3

Quinlan (1979) proposed a decision induction method called ID3. He noted that the building of a decision tree requires a "divide-and-conquer" strategy that uses a recursive testing procedure with the aim of generating a small tree. ID3 uses information gain (Section 3.2.4.2) as a basis for tree induction (Figure 6.2). Note that the algorithm in Figure 6.2, step 7, uses a technique to recursively self-call program ID3 to build the decision tree. Based on the example shown in Table 6.1, the procedure for using the ID3 algorithm to generate a decision tree is shown in Figure 6.3.

The information gain for *Wetness* is then calculated by

$$\text{Gain}\left(t, \text{wetness}\right) = I_E\left(t\right) - \left(\frac{6}{15}\right) I_E\left(t_{\text{wetness(high)}}\right) - \left(\frac{9}{15}\right) I_E\left(t_{\text{wetness(low)}}\right) \tag{6.1}$$

$$= 0.837 - 0.4 - 0.302 = 0.135.$$

Notations: 1. A: a set of attribute types used as predictors, originally, $A = \{A_1, A_2, ..., A_n\}$, where $A_1, A_2, ..., A_n$ each may represent a spectral band in the case of remotely sensed imagery.

   2. t: the node with corresponding attribute samples A contained inside.

ID3 (A, t)

Begin

1. If t consists of samples all belonging to the same class, label the node with that class then exit the program;
2. If A is empty, label the node t as the most frequent of the class contained in the node t then exit the program;
3. Compute information gain for each attribute type and let A_k be the attribute with largest gain among A;
4. Let $\{A_{k(i)} | i = 1, 2, ..., r\}$ be the values of attribute A_k;
5. Let $\{t_{k(i)} | i = 1, 2, ..., r\}$ be the child-nodes of t consisting respectively of attribute samples according to $A_{k(i)}$;
6. Label node t as A_k and branches labeled $A_{k(1)}, A_{k(2)}, ..., A_{k(r)}$ going respectively to the child-node $t_{k(1)}, t_{k(2)}, ..., t_{k(r)}$;
7. For $i = 1$ to r

  ID3$(A\text{-}\{A_k\}, t_{k(i)})$

 End

End

FIGURE 6.2 Decision tree induction using the ID3 algorithm.

TABLE 6.1

Total of 15 Samples with Attributes A = Brightness, B = Greenness, and C = Wetness

Sample No.	A: *Brightness*	B: *Greenness*	C: *Wetness*	Class: *Target*
1	High	High	Low	Yes
2	High	Low	High	Yes
3	Medium	High	Low	No
4	Low	Low	Low	No
5	Low	High	Low	No
6	Low	Low	High	No
7	Medium	High	High	No
8	High	Low	Low	No
9	High	High	Low	No
10	Low	Low	Low	No
11	High	High	High	No
12	Low	Low	High	Yes
13	Medium	High	Low	No
14	Medium	Low	High	Yes
15	Low	Low	Low	No

See the text for further description.

The gain for other attributes, *Brightness* and *Greenness*, is calculated in a similar way as

$$\text{Gain}\left(t, \text{brightness}\right) = I_E\left(t\right) - \left(\frac{5}{15}\right) I_E\left(t_{\text{brightness(high)}}\right)$$

$$- \left(\frac{4}{15}\right) I_E\left(t_{\text{brightness(medium)}}\right) - \left(\frac{6}{15}\right) I_E\left(t_{\text{brightness(low)}}\right) \tag{6.2}$$

$$= 0.837 - 0.324 - 0.216 - 0.26 = 0.037$$

$$\text{Gain}\left(t, \text{greenness}\right) = I_E\left(t\right) - \left(\frac{7}{15}\right) I_E\left(t_{\text{greenness(high)}}\right) - \left(\frac{8}{15}\right) I_E\left(t_{\text{greenness(low)}}\right) \tag{6.3}$$

$$= 0.837 - 0.276 - 0.509 = 0.052.$$

In Figure 6.3a, attribute *Wetness* is first chosen as the candidate for splitting as it has the largest gain (see Equation 6.1 = 0.135) in comparison with the gains computed for attributes *Brightness* (Equation 6.2 = 0.037) and *Greenness* (Equation 6.3 = 0.052), respectively. Figure 6.3b shows the further tree induction under the circumstance that *Wetness* = high. The corresponding gain for *Brightness* and *Greenness* can be calculated by

$$\text{Gain}\left(t_{\text{wetness(high)}}, \text{brightness}\right) = I_E\left(t_{\text{wetness(high)}}\right) - \left(\frac{2}{6}\right) I_E\left(t_{\text{wethess(high)\&brightness(high)}}\right)$$

$$- \left(\frac{2}{6}\right) I_E\left(t_{\text{wetness(high)\&brightness(medium)}}\right) - \left(\frac{2}{6}\right) I_E\left(t_{\text{wetness(high)\&brightness(low)}}\right)$$

$$= 1 - 0.333 - 0.333 - 0.333 = 0.001 \tag{6.4}$$

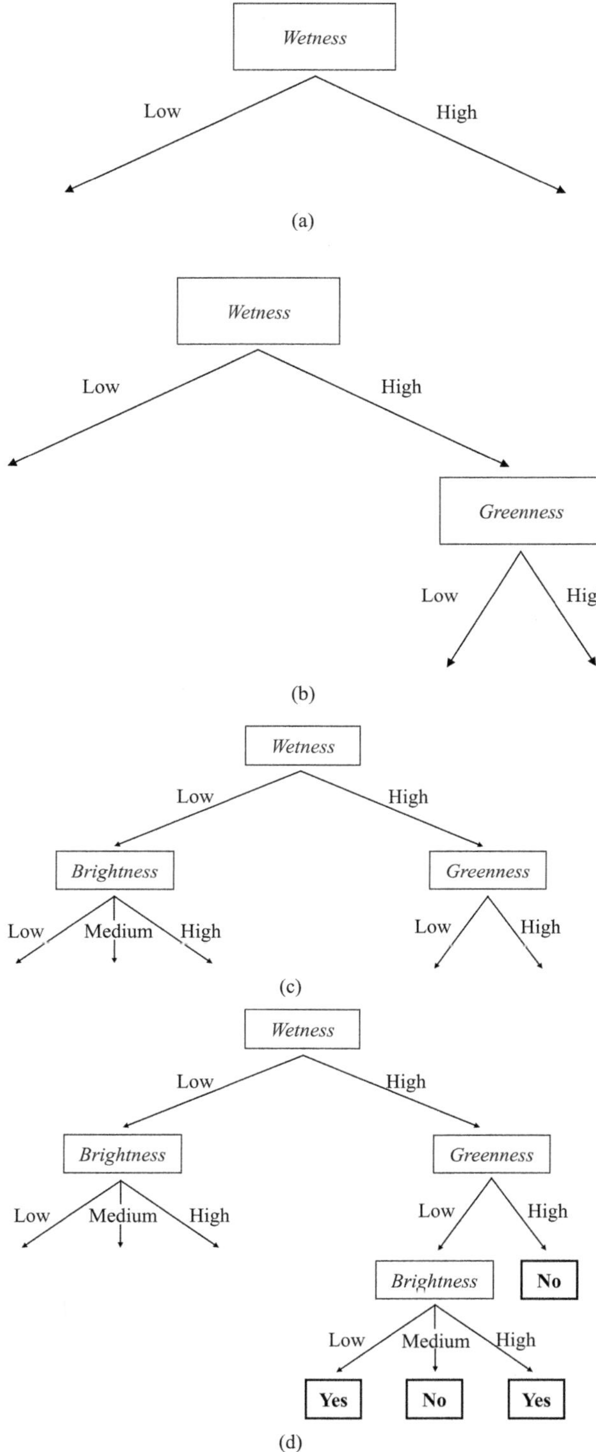

FIGURE 6.3 Tree induction process based on the samples in Table 6.1. (a) Attribute *Wetness* is firstly chosen as a splitting candidate. (b) Tree induction under the circumstance that *Wetness*=high and *Greenness* is selected for further splitting according to gain. When the attribute *Wetness*=low, the attribute *Brightness* is chosen. (c) When attribute *Wetness*=low, the attribute *Brightness* is chosen. (d) The left-hand side of the tree induction is complete. (e) Decision tree induction is finally finished according to the ID3 algorithm.

(Continued)

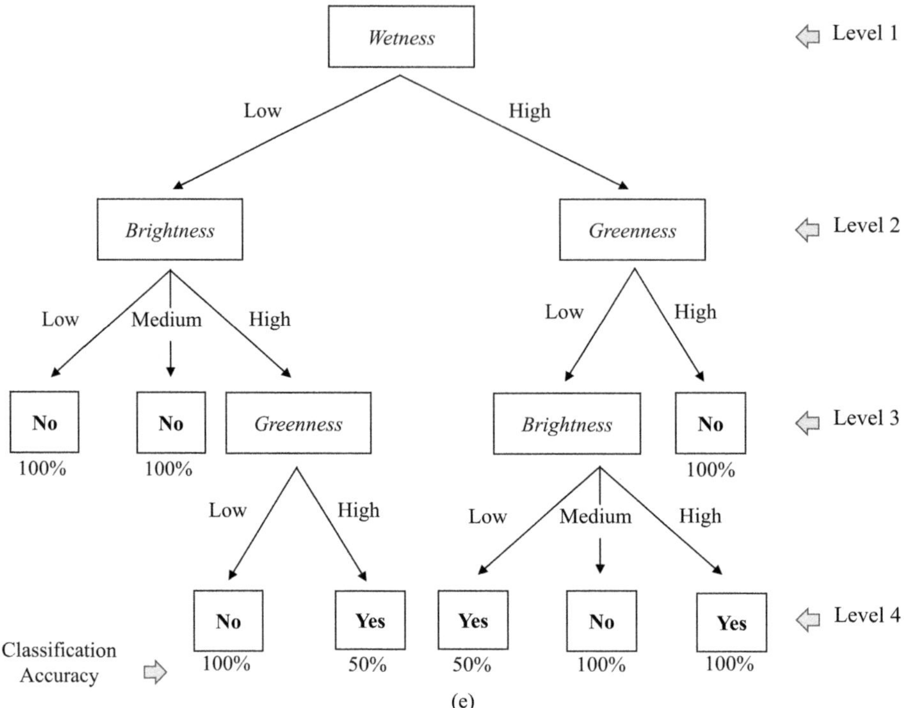

FIGURE 6.3 (Continued)

$$\text{Gain}\left(t_{\text{wetness(high)}},\ \text{greenness}\right) = I_E\left(t_{\text{wetness(high)}}\right) - \left(\frac{2}{6}\right) I_E\left(t_{\text{wetness(high)\&greenness(high)}}\right)$$

$$-\left(\frac{4}{6}\right) I_E\left(t_{\text{wetness(high)\&greenness(low)}}\right)$$

$$= 1 - 0 - 0.54 = 0.46$$

Accordingly, the attribute *Greenness*, with a gain value of 0.46, is the attribute selected for further splitting from the parent node *Wetness*=high. For the circumstance *Wetness*=low, the corresponding gains are computed as

$$\text{Gain}\left(t_{\text{wetness(low)}},\ \text{brightness}\right) = I_E\left(t_{\text{wetness(low)}}\right) - \left(\frac{3}{9}\right) I_E\left(t_{\text{wetness(low)\&brighthess(high)}}\right)$$

$$-\left(\frac{2}{9}\right) I_E\left(t_{\text{wetness(low)\&brightness(medium)}}\right) - \left(\frac{4}{9}\right) I_E\left(t_{\text{wetness(low)\&brightness(low)}}\right)$$

$$= 0.503 - 0.306 - 0 - 0 = 0.197 \tag{6.5}$$

$$\text{Gain}\left(t_{\text{wetness(low)}}, \text{greenness}\right) = I_E\left(t_{\text{wetness(low)}}\right) - \left(\frac{5}{9}\right) I_E\left(t_{\text{wetness(low)\&greenness(high)}}\right)$$

$$- \left(\frac{4}{9}\right) I_E\left(t_{\text{wetness(low)\&greenness(low)}}\right)$$

$$= 0.503 - 0.401 - 0 = 0.102$$

Based on these results, the attribute *Brightness* is selected for splitting in the case *Wetness* = low. The result of further decision tree induction for *Wetness* = low is shown in Figure 6.3c. Similarly, according to the ID3 algorithm, the final decision tree inductions for the attributes *Greenness* and *Brightness* are shown in Figures 6.3d and 6.3e, respectively. The resulting tree is a four-level decision tree. Note that nodes numbered 11 and 12 will generate the classification ambiguity; that is, both nodes choosing either *yes* or *no* will always have 50% classification errors. In such a situation, the user may use more reference data to help make an unequivocal decision. Alternatively, a probability can be used to achieve fuzzy result instead of the originally crisp decision procedure. One should note that ID3 cannot handle numeric and continuous attribute values, is unable to deal with missing (or unknown) attributes, and does not have a pruning facility. Quinlan (1993, 1996) therefore made several improvements to ID3, resulting in the form of an improved decision tree called C4.5.

6.1.2   C4.5

C4.5 is an extension of the ID3 decision tree induction algorithm to account for the several issues not appropriately dealt with by ID3 (Quinlan, 1986, 1993). Significant advancements involve the careful selection of an appropriate attribute selection measure, effective management of data containing missing attributes, addressing the challenges associated with numeric and continuous attribute handling, and the implementation of decision tree pruning. This section will discuss these characteristics, except that the tree pruning issue is left for discussion in Section 6.6.

C4.5 uses either information gain or a normalized version called the gain ratio to choose attributes as candidates for splitting. From Equation (6.5), it is seen that the information gain measure tends to favor attributes that have multiple values. For instance, in an extreme case, if attribute X holds a distinct value for each class, Equation (6.4) will always be 0, which in turn results in maximal gain in Equation (6.5), and thus, this attribute becomes the first candidate for splitting. This feature makes the tree induction process more easily affected by noise. To take care of this issue, Quinlan (1986) proposes an alternative index called the gain ratio, which can be considered as a normalized version of the information gain measure. The gain ratio is defined as

$$\text{Gain\_ratio}(t, X) = \frac{\text{Gain}(t, X)}{\text{Split\_}I(t, X)}, \tag{6.6}$$

where Split_*I*(*t*, *X*) is obtained from

$$\text{Split\_}I(t, X) = -\sum_{i=1}^{r} f(t, x_i) \log_2 f(t, x_i), \tag{6.7}$$

which represents the information encoded in the split of attribute X at node t. In the case of the example shown in Table 6.1, the Split_*I*(*t*, *Brightness*), Split_*I*(*t*, *Greenness*), and Split_*I*(*t*, *Wetness*) are respectively calculated by

$$\text{Split}\_I(t,\text{brightness}) = -\frac{5}{15}\log_2\frac{5}{15} - \frac{4}{15}\log_2\frac{4}{15} - \frac{5}{15}\log_2\frac{5}{15} = 1.565$$

$$\text{Split}\_I(t,\text{greenness}) = -\frac{7}{15}\log_2\frac{7}{15} - \frac{8}{15}\log_2\frac{8}{15} = 0.997 \qquad (6.8)$$

$$\text{Split}\_I(t,\text{wetness}) = -\frac{6}{15}\log_2\frac{6}{15} - \frac{9}{15}\log_2\frac{9}{15} = 0.971.$$

Based on the information gain computed in Equations (6.1), (6.2), and (6.3), the corresponding gain ratio for each attribute is then computed by

$$\text{Gain\_ratio}(t,\text{brightness}) = \frac{\text{Gain}(t,\text{brightness})}{\text{Split}\_I(t,\text{brightness})} = \frac{0.037}{1.565} = 0.024$$

$$\text{Gain}\_\text{ratio}(t,\text{greenness}) = \frac{\text{Gain}(t,\text{greenness})}{\text{Split}\_I(t,\text{greenness})} = \frac{0.052}{0.997} = 0.052 \qquad (6.9)$$

$$\text{Gain\_ratio}(t,\text{wetness}) = \frac{\text{Gain}(t,\text{wetness})}{\text{Split}\_I(t,\text{wetness})} = \frac{0.135}{0.971} = 0.139$$

From these results, the attribute *Wetness* is the first candidate chosen for splitting as it has the highest gain ratio.

To deal with missing attributes, C4.5 uses a probability concept. In the example shown in Table 6.1, suppose that a new sample is inserted, for which *Wetness*=low, *Brightness*=high, and the value for *Greenness* is missing. One then traces the root node *Wetness* in the decision tree (see Figure 6.3e) first down left to node 3 *Brightness* and then to the right path to node 6 *Greenness*. At node 6, since the value of *Greenness* is missing, one has no idea whether the newly added sample should be classified as target (*yes*) or not-target (*no*). To solve this issue, one can find that if one set of unknown *Greenness* is high, the sample belongs to the target (*yes*) or not-target (*no*) will be

TABLE 6.2

15 Samples Are the Same as Shown in Table 6.1 Except that Attribute B = *Greenness* Is Converted to Numeric Type

Sample No.	A: *Brightness*	B: *Greenness*	C: *Wetness*	Class: *Target*
1	High	12	Low	Yes
2	High	10	High	Yes
3	Medium	16	Low	No
4	Low	12	Low	No
5	Low	13	Low	No
6	Low	16	High	No
7	Medium	16	High	No
8	High	19	Low	No
9	High	12	Low	No
10	Low	12	Low	No
11	High	15	High	No
12	Low	12	High	Yes
13	Medium	12	Low	No
14	Medium	12	High	Yes
15	Low	18	Low	No

50% each (because node 12 in Figure 6.3e is 50% in accuracy). If one sets *Greenness* to be low, the sample clearly belongs 100% to not-target (*no*) (node 13 in Figure 6.3e). Thus, it can be concluded for node 6 that under the condition where the value for *Greenness* is missing, the probability of the sample belonging to target (*yes*) is 0.333, and 0.666 for not-target (*no*), respectively. In a case where a crisp rather than a fuzzy decision is required, then based on the calculated probability values, the *no* alternative is selected.

To cope with attributes containing numeric data types (integers or real numbers, for instance) in addition to categorical data types, C4.5 uses the following procedures. The numeric data are first sorted in increasing or decreasing order. The algorithm proceeds by selecting a value, x_i, from the data set and utilizing it to partition the data into two segments: one with values equal to or greater than x_i, and the other with values smaller than x_i. Subsequently, gain ratios are computed for each partition.

The process, choosing the partitioning values and then computing gain ratios, is repeated for each numeric or continuous data item. The partitioning value that produces partitions having the highest gain ratio is then selected. Table 6.2 presents an example in which the attributes are mostly the same as in Table 6.1 except that the data for attribute *Greenness* are turned into numeric values. One can use the same procedure introduced previously to calculate the gain or gain ratio. For *Greenness*, one now has to choose partitioning values one by one and to calculate the corresponding gain ratio. After exhaustive trials for *Greenness* values in Table 6.2, the best partitioning value chosen is $x_i = 13$, which partitions the feature *Greenness* into two categories, one with values <13 (a total of 8 records) and the second with values greater than or equal to 13 (7 records), giving a total of 15 samples. The resulting gain and gain ratio are computed as

$$\text{Gain}\left(t, \text{greenness}\right) = I_E\left(t\right) - \left(\frac{8}{15}\right)I_E\left(t_{\text{greenness }(<13)}\right) - \left(\frac{7}{15}\right)I_E\left(t_{\text{greenness }(\geq 13)}\right)$$

$$= 0.837 - 0.533 - 0 = 0.304 \tag{6.10}$$

$$\text{Gain\_ratio}\left(t, \text{greenness}\right) = \frac{\text{Gain}\left(t, \text{greenness }\right)}{\text{Split\_}I\left(t, \text{greenness }\right)} = \frac{0.304}{0.997} = 0.302$$

Either the gain or gain ratio computed from the attribute *Greenness* now is the largest in comparison with that computed for other attributes. Therefore, the attribute *Greenness* is chosen as the first candidate for splitting. For further tree induction processes, one can iteratively implement the equation for computing either information gain (6.5) or gain ratio (6.18) to perform the tree induction. The final decision tree is shown in Figure 6.4.

6.1.3   SEE5.0 (C5.0)

The successor of C4.5 called SEE5.0 (or C5.0) can operate on several additional data types, as well as those available in C4.5, including dates, times, time stamps, ordered discrete attributes, and case labels. In addressing missing values, SEE5.0 provides the capability to flag values as not applicable and includes tools for defining new attributes as functions derived from other attributes. Additionally, SEE5.0 introduces the important concept of boosting, a technique that combines multiple learning processes with appropriately weighted training class pixels to enhance predictive accuracy. In the boosting process, weights given to incorrectly labeled or unlabeled pixels are increased compared to those assigned to correctly labeled pixels, and the generation of trees is repeated. This iterative cycle is conducted multiple times. Friedl et al. (1999) suggest a limit of 10 iterations, and their results show an accuracy improvement of ~25% with boosting. SEE5.0 comes with the additional capability of allowing the definition of distinct costs for each pair of predicted or actual classes. This proves valuable in practical applications where certain

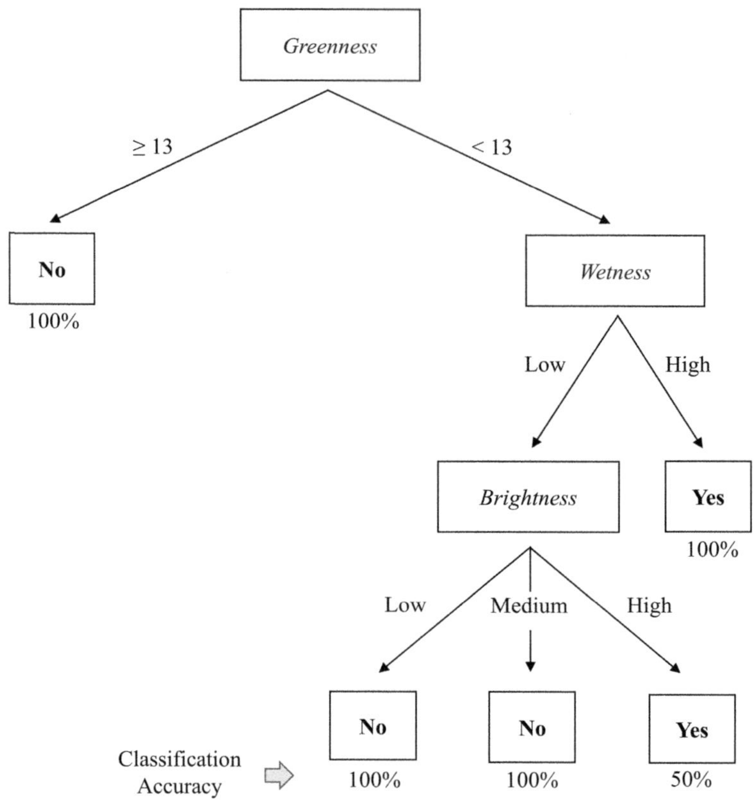

FIGURE 6.4 Tree induction using the values in Table 6.2 as inputs.

classification errors carry more weight than others (C4.5 treats all errors the same). SEE5.0, as a result, constructs classifiers with the objective of minimizing expected misclassification costs rather than error rates. In contemporary data mining scenarios, dealing with data sets of extremely high dimensionality, often containing hundreds or even thousands of attributes, is common. SEE5.0 introduces an automated attribute selection process that entails eliminating attributes with minimal relevance before building a classifier. When handling high-dimensional data, this attribute selection process can yield smaller classifiers, enhance predictive accuracy, and often expedite rule set generation.

6.2 CHAID

CHAID (Kass, 1980; Chi-square Automatic Interaction Detector) is used to study the relationship between dependent variables (i.e., classes) and a series of predictor variables (e.g., spectral bands). The method involves choosing a set of predictors and their interactions to optimally predict the dependent measure. Before using CHAID, one should know that the main purpose of the chi-square test is to look at the relationship between two variables to determine the level of dependency between them. This is the primary characteristic of CHAID using the chi-square test to determine which attribute is the most relevant to the classes being predicted. This characteristic is quite different from the algorithms used in decision trees like the ID3 families and CART (Section 6.3), which use either information gain or the Gini index to choose the optimal splitting attribute. CHAID also uses chi-square to judge if a decision tree should stop growing so as to avoid the issue of overfitting. Therefore, for a CHAID decision tree, there is no need to consider the issue of tree pruning (Section 6.6).

Throughout the tree induction process, CHAID conducts tests on each attribute to determine whether dividing the training pixel sample based on the attribute being tested results in a statistically significant distinction in the dependent measure (i.e., the class). CHAID then builds a contingency table to store the results of these chi-square tests for each attribute-class pair. The most relevant (i.e., the most significant) attribute is first selected, and the samples are split in terms of a rule determined for that attribute (e.g., "if $x > b$ then node I"). After the first split, for each of the split sample groups formed, CHAID then tests if the subgroup could be further significantly split using another attribute. The process continues until the end of the tree induction process. One thus has a series of groups that are maximally different from one another in terms of the attributes. In the general case, CHAID can only deal with nominal (categorical) or ordinal (ordered categories ranked from small to large) attributes. If an attribute is measured on a continuous scale, CHAID has to first turn the continuous values into categorical values and then perform an analysis-of-variance method for optimal tree induction. For instance, if the attribute *Temperature* is measured on a continuous scale from 0 to 100, CHAID may separate these continuous values into category A: 0–9, B: 10–19, …, etc. Such a technique for transforming continuous values into categorical types is called binning. Since binning expresses the original continuous attribute on a coarser scale, the accuracy of the CHAID tree may start declining to a certain extent. It is worth noting that CHAID shares similarities with a forward stepwise regression analysis, bringing along potential challenges like overlooking outliers and being constrained by the t-ratio criterion (Harrell, 2001).

6.3 CART

CART (Classification and Regression Tree) is a decision tree approach proposed by Breiman et al. (1984). The main difference between CART and C4.5/SEE5.0 is that CART only allows two branches (i.e., two children) to form at each spitting process, while C4.5/SEE5.0 can generate different numbers of branches as required during the induction process. In other words, the decision tree generated by the CART algorithm will always be a binary tree.

As far as the implementation of CART is concerned, when the classes under prediction are continuous in nature, a regression tree may be used to perform the prediction in terms of regression techniques. The information contained in each leaf node will then be expressed as a continuous value. On the other hand, a classification tree can be used to deal with categorical classes. It is noteworthy that the decision tree inducted by the CART algorithm generally tends to overfit. To reduce the overfitting, CART uses test sets to perform the decision tree validation and accordingly to prune the tree (Section 6.6). The measure that CART implements to perform tree induction can be based on either univariate or multivariate criteria as detailed below in this section.

The decision tree induction algorithms discussed in Section 6.1 (ID3 and its derivatives) and Section 6.2 (CHAID) all fall into the category of univariate induction. In this approach, decision boundaries at each tree node are established by the outcomes of a test applied to a single attribute, assessed at every internal node. The data are successively split into two or more subsets based on the test outcomes, continuing until a leaf node is reached, and the corresponding class label is assigned. CART can employ information gain, gain ratio, or the Gini index for univariate decision tree induction.

In univariate decision tree algorithms, each test involves a single feature exclusively. The split through the feature space is constrained to be orthogonal to the axis represented by the chosen feature, as depicted in Figure 6.5, in that the decision boundaries are all perpendicular to the axes defined by attributes, resulting in the feature space being divided into a series of hyper-rectangles. The univariate decision tree is thus similar to the simple parallelepiped classifier in the way it partitions feature space. To overcome this restriction, CART allows the implementation of multivariate criteria to perform decision tree induction as detailed below.

In multivariate decision tree algorithms, node splits are expanded to encompass linear combinations of features (Friedl and Brodley, 1997). At each interior node of a multivariate decision tree, a collection of linear discriminant functions is estimated, and the coefficients for the linear

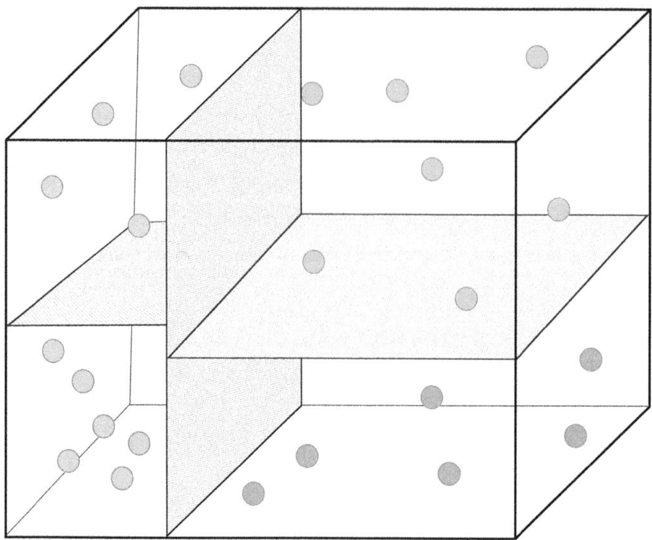

FIGURE 6.5 Example of decision boundaries formed by the univariate decision tree algorithm.

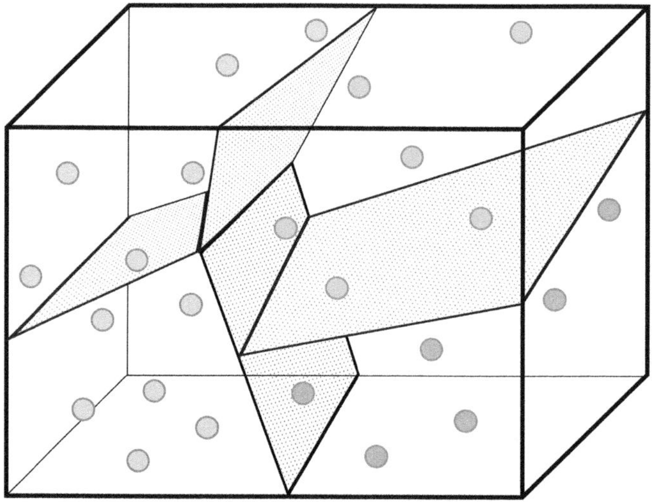

FIGURE 6.6 Decision boundaries formed by the multivariate decision tree algorithm.

discriminant function at each interior node are derived from the training data. The decision boundaries generated by a multivariate decision tree will therefore be more flexible than that formed by a univariate decision tree. Figure 6.6 illustrates an example of decision boundaries formed by a multivariate decision algorithm. In these cases, the multivariate decision boundaries are not necessarily orthogonal to the axes of the feature space, and should provide a more flexible fit to the boundaries of the classes.

When using multivariate tree induction criteria, CART looks for weighted averages of input attributes for splitting. These weighted averages can reveal important database structures and can uncover new critical measures. According to the algorithm proposed by Breiman et al. (1984), the multivariate case search at each node is performed in terms of a linear test (Equation 6.11), which splits the data for that node into two subsets showing the least impurity. Obtaining the coefficients of the splitting function in the CART algorithm is a stepwise procedure. Specifically, during the multivariate decision tree induction process at a given node, the process is performed as follows:

1. **Data Normalization**: The data are normalized by centering all the data samples for the node at their medians and dividing by their inter-quartile ranges.
2. **Univariate Split**: Suppose there are a total of K attributes denoted as X_k, $k=1, 2, ..., K$, and let T denote the splitting threshold. The best univariate split $X_k \leq T$ (for instance, X_k may be attribute *Greenness* and T may be chosen as value 13 as shown in Table 6.2) is found in terms of minimizing the Gini impurity index measure as described earlier.
3. **Perform the Linear Test**: This part of the process involves a number of cycles. A given split

$$\sum_{k=1}^{K} a_k X_k \leq T \tag{6.11}$$

is improved by successively considering splits of the form:

$$\sum_{k=1}^{K} a_k X_k - \delta(X_k + \lambda) \leq T \tag{6.12}$$

where a_k is the weighting coefficient to the corresponding attribute X_k, and δ is the parameter to be optimized by taking λ as the fixed value. The method for updating these coefficients is introduced below. For attribute X_1, one can obtain the best δ^* by calculating

$$\delta^{(i)} = \frac{\sum_{k=1}^{K} a_k x_{ik} - T}{x_{1k} + \lambda}, \quad \text{for} \quad i = 1, 2, ..., N \tag{6.13}$$

where N is the total number of data samples, and x_{ik} denotes the value of the kth attribute for the ith data sample. A possible candidate value for δ^* can be the midpoint of the ordered $\delta^{(i)}$. Accordingly, one can obtain the best split from Equation (6.12). Three values, namely, -0.25, 0.0, and 0.25, are normally proposed as λ values for this procedure. The best $\delta^{(i)}$ among these three values is then taken as the δ^* that is used to update a_k in

$$\sum_{k=1}^{K} a_k X_k \tag{6.14}$$

by the replacement with a_k^{new}, where

$$a_k^{\text{new}} = \begin{cases} a_k & \text{for} \ k \neq 1 \\ a_1 - \delta^* & \text{otherwise} \end{cases} \tag{6.15}$$

The new threshold T^{new} is computed by $T^{\text{new}} = T + \delta^* \times \lambda$. This step is repeated for the other attributes $X_2, X_3, ..., X_K$. At the end of each cycle, one has completed the test

$$\sum_{k=1}^{K} a_k X_k \leq T. \tag{6.16}$$

The next task is to improve the splitting threshold T by finding the best test, fixing a_k, and updating only the threshold T. The process of tree generation is terminated as soon as the decrease in impurity from one cycle to the next is below a predetermined small level.

Readers should note that even though a multivariate decision tree may appear to be more flexible than the univariate tree, several factors affect its performance. These factors include the algorithms implemented to estimate the splitting rule at internal nodes, the various feature selection methods, and the use of different classification algorithms at different nodes (Friedl and Brodley, 1997).

6.4 QUEST

The word QUEST stands for Quick, Unbiased, Efficient, and Statistical Tree algorithm (Loh and Shih, 1997). The decision tree inducted by the QUEST algorithm is a binary tree. When the number of classes is more than two, QUEST will initially group the classes into two clusters to ensure binary splits. This is accomplished through the implementation of a two-means clustering algorithm, designed to minimize the within-cluster sum of squares to the original class means. In general, the two most extreme means serve as the initial cluster centers. Subsequently, QUEST employs quadratic discriminant analysis (QDA) for the splitting process. Unlike most decision tree induction algorithms, QUEST handles split point selection and attribute selection as distinct tasks. In the upcoming discussion, we will break down the QUEST algorithm into two components: split point selection and attribute selection.

6.4.1 Split Point Selection

In the case where the attributes are all numeric variables, let X be the attribute selected to perform a split at node t. If the number of classes in node t is more than two, the two-means clustering algorithm is first applied and groups the classes into superclasses A and B. Let $\bar{x}(A)$ and s_A^2 denote the class mean and variance for class A, and let $\bar{x}(B)$ and S_B^2 denote the corresponding quantities for class B. QDA splits the attribute X into three intervals, namely, $(-\infty, d_1)$, (d_1, d_2), and (d_2, ∞), where d_1 and d_2 are the roots of the following equation:

$$p(A|t)s_A^{-1}\Phi\left[(x-\bar{x}(A))/s_A\right]= p(B|t)s_B^{-1}\Phi\left[(x-\bar{x}(B))/s_B\right], \tag{6.17}$$

where

$$p(j/t)= p(j,t)\Big/ \sum_i p(i,t) \tag{6.18}$$

denotes the estimated posterior probability that a sample belongs to class j in node t, and $\Phi(y)=(2\pi)^{-0.5}\exp(-y^2/2)$ denotes the standard normal density function. The probability $p(j, t)$ is defined as

$$p(j,t)= \pi(j)N_{j,t}/N_j, \tag{6.19}$$

where $\pi(j)$ denotes the prior probability for class j (normally estimated in terms of sample proportion of class j), N_j is the number of samples belonging to class j, and $N_{j,t}$ is the number of samples belonging to class j within node t. QUEST uses only one of the two roots (the one that is closer to the mean of each class) as the split point to ensure a binary split. By taking logs on both sides of Equation (6.17), one can obtain the quadratic equation $ax^2+bx+c=0$, where

$$a = s_A^2 - s_B^2$$

$$b = 2\left(\bar{x}_A \times s_B^2 - \bar{x}_B \times s_A^2\right) \tag{6.20}$$

$$c = \left(\bar{x}_B \times s_A^2\right)^2 -\left(\bar{x}_A \times s_B^2\right)^2 + 2s_A^2 \times s_B^2 \log\left[\left(p(A|t)\times s_B\right)\big/\left(p(B|t)\times s_A\right)\right]$$

If $a=0$ and $\bar{x}_B \neq \bar{x}_B$, there is only one root (i.e., split point d) as given by

$$d = \frac{(\bar{x}_A + \bar{x}_B)}{2} - (\bar{x}_A - \bar{x}_B)^{-1} s_A^2 \log\left[p(A|t)/p(B|t)\right].$$ (6.21)

In the case of $a \neq 0$ and $b^2 - 4ac < 0$, the split point d is defined as $d = (\bar{x}_A + \bar{x}_B)/2$; otherwise, d is defined to be the root of

$$(2a)^{-1}\left(-b \pm \sqrt{b^2 - 4ac}\right).$$ (6.22)

The above derivations assume the attributes being dealt with are numeric. If attributes are categorical variables, QUEST first uses an algorithm called CRIMCOORD to transform categorical variables into numeric ones before starting the process of tree induction. For the detailed derivation of categorical-to-continuous transformation, readers are referred to Loh and Shih (1997).

6.4.2 ATTRIBUTE SELECTION

The previous section explains how a split point is determined to designate a selected attribute. In what follows, how to choose such an attribute is further discussed. The basic idea adopted by QUEST to perform attribute selection is statistical tests. For continuous attributes, a threshold value α is chosen and an analysis-of-variance (ANOVA) F-statistic is computed for every variable. If the largest F-statistic exceeds α, the corresponding attribute is selected to split the node. In the case of categorical attributes, a contingency table and chi-square tests of independence between the class and the categorical attributes are computed. A contingency table is also known as a cross-tabulation, in which the rows denote the information classes, the columns denote the attributes, and each cell in the table then contains the tabulated counts for the corresponding attribute labeled to the corresponding information class. Chi-square tests of independence for a contingency table then test the null hypothesis that the information class and the attributes are independent. Specifically, let $X_1, X_2, \ldots, X_{K1}$ denote continuous attributes, $X_{K1+1}, X_{K1+2}, \ldots, X_K$ are categorical attributes, and let $x_{ik}^{(j)}$ denote the value of the kth attribute for the ith sample in the jth class. The process for attribute selection can be addressed in terms of the following steps:

1. Choose a value $\alpha \in (0, 1)$ to be a prespecified level of significance. Compute an ANOVA F-statistic F_k for each continuous attribute X_k, $k=1$ to K_1. Let $F_{ka} = \max\{F_k : k = 1, 2, \ldots, K_1\}$ and $\alpha_1 = \Pr\{F_{(m_t-1),(N_t-m_t)} > F_{ka}$, where m_t is the number of classes present in node t, N_t is the number of samples in node t, and $F_{u,v}$ denotes the F-distribution with u and v degree of freedom.
2. Calculate the P-value β_k for the chi-square test of independence between class and categorical attributes for categorical attribute X_k, $k=K_1+1$, K_1+2, $\ldots$, K as stored in the contingency table. The degrees of freedom are given by the product $(n_r-1) \times (n_c-1)$, where n_r and n_c are the numbers of rows and columns of the table with nonzero totals. Let $\alpha_2 = \beta_{kb}$, where $\beta_{kb} = \min\{\beta_k : k = K_1+1, K_1+2, \ldots, K\cdot$
3. $k^* = k^a$ if $\alpha_1 \leq \alpha_2$, otherwise define $k^* = k^b$. If $\min\{\alpha_1, \alpha_2\} < \alpha/K$, choose attribute X_{k^*} to split the node.
4. If $\min\{\alpha_1, \alpha_2\} \geq \alpha/K$, compute ANOVA F-statistics F_k^z, $k=1$ to K_1, for the continuous attributes based on the absolute deviations

$$z_{ik}^{(j)} = \left|x_{ik}^{(j)} - \bar{x}_k^{(j)}\right|,$$ (6.23)

where

$$\bar{x}_k^{(j)} = N_{j,t}^{-1} \times \sum_{i=1}^{N_{j,t}} x_{ik}^{(j)}.$$ (6.24)

5. Let

$$F_{k^c}^{(z)} = \max\left\{ F_k^{(z)} : k = 1, 2, \ldots, K_1 \right\}, \qquad (6.25)$$

and

$$\alpha_3 = \Pr\left\{ F_{(m_t-1),(N_t-m_t)} > F_{k^c}^{(z)} \right\} \qquad (6.26)$$

If $\alpha_3 < \alpha/(K_1+K)$, choose attribute X_{k^c} to split the node. Otherwise, choose attribute X_{k^*}.

The QUEST algorithm uses linear combination splits and is essentially a recursive linear discriminant analysis on all the attributes. The process is similar to CART. If the number of classes is greater than two, a prior grouping of the classes into two superclasses is also carried out. In terms of classification accuracy, variability of split points, and tree size, Loh and Shih (1997) show that QUEST has some advantages when univariate splits are used, and that the QUEST trees based on linear combination splits are usually smaller than those generated by other tree induction algorithms.

To present a more comprehensive view, Table 6.3 lists the comparisons of characteristics among the different tree induction algorithms introduced above.

TABLE 6.3

Comparisons among Different Tree Induction Algorithms

Characteristics	QUEST	CART	CHAID	C4.5
Split Types				
Univariate	•	•	•	•
Linear combinations	•	•		
Number of Branches at Each Node				
Always two	•	•	•	•
Two or more				
Missing Value Methods				
Imputation				
Alternate/surrogate splits	•	•	•	•
Missing value branch				
Probability weights				
Tree Size Control				
Stopping rule				
Prepruning	•	•	•	•
Test-sample pruning				
Cross-validation pruning	•	•		
Other Characteristics				
Choice of misclassification costs	•	•	•	
Choice of class prior probabilities	•	•		
Choice of impurity functions	•	•		
Bagging		•		
Error estimation by cross-validation	•	•	•	

6.5   TREE INDUCTION FROM ARTIFICIAL NEURAL NETWORKS

As noted in Chapter 7, artificial neural networks (ANNs) may be more robust than statistical classification methods, but it is hard to find clear roadmaps for explanations of how and why the classification results are achieved. This is due to the fact that the ANN mapping process between the inputs and outputs is performed in a black-box. The reason for this nature lies in the fact that the information within the neural network is retained as real-valued parameters (weights and biases) of the network. This knowledge is encoded in a distributed manner, and the network's learned mapping can be both nonlinear and nonmonotonic (Krishnan et al., 1999). However, it is important that the decisions generated by an ANN are understood, rather than just arbitrarily considering its output. To resolve this issue, one may implement rule extraction algorithms (e.g., Craven, 1996) to uncover the hidden knowledge from ANNs. Extracting rules from neural networks involves the process of acquiring clear explanations that approximate how neural networks make predictions. Rule extraction algorithms serve the dual purpose of interpreting neural networks and uncovering the relationship between input and output variables in data.

Among the numerous symbolic learning algorithms, the methods currently being used to build decision trees from ANNs may be roughly categorized into two types known as decomposition (Duch et al., 2001) and pedagogical (Craven, 1996; Taha and Ghosh, 1999) approaches, respectively. The fundamental concept of the decomposition method is to perform a search process for mapping a well-trained ANN into a decision tree. Such kinds of methods could suffer from a serious computational loading issue because of the increasing dimension of the input vector, increase computational time exponentially, and thus require a pruning procedure to downsize the tree structure. The methods known as SUBSET (Towell and Shavlik, 1994), CRED (Sato and Tsukimoto, 2001), and C-MLP2LN (Duch et al., 2001) belong to this category. Pedagogical approaches, however, treat a trained ANN as a black-box and perform decision tree induction from the ANN by observing the relationship between both the ANN inputs and outputs. Typical methods within this family are referred to as TREPAN (Craven, 1996), ANN-DT (Schmitz et al., 1999), and Binarized Input-Output Rule Extraction (BIO-RE) (Taha and Ghosh, 1999). Although building decision trees from neural networks may be a useful approach and significant progress has been made in this regard, artificial intelligence-based (XAI) tools have recently become more popular for this and are discussed in Chapter 10. This change can be attributed to three major shortcomings in the tree induction process within neural networks: (1) variations in the data samples have the potential to cause the tree induction algorithm to produce a significantly different tree, (2) the size of a decision tree can be extremely large depending on the complexity of the problem, and (3) a single hypothesis is finally presented to the user (Krishnan et al., 1999).

6.6   PRUNING DECISION TREES

A decision tree developed in the manner described in the earlier section will generally be complex. However in some circumstances, the user may use predefined conditions (such as the maximal depth a decision tree can grow, or minimal number of samples a node should hold) to limit the decision tree induction process. In practical applications, the resulting decision tree will most often have long and very uneven paths and so some post-pruning methods will be required. Pruning of the decision tree is performed by using a leaf node to replace a subtree. The general rule to decide whether a subtree should be pruned or not is based on the observation that the expected error rate within the subtree can be reduced by replacing it with a single leaf node. For instance, assume a two-leaf subtree is formed based on three training samples. The left leaf node contains one sample indicating that when the attribute *Wetness*=high, the pixel should be labeled as class *target*. The right leaf node contains two samples showing that when the attribute *Wetness*=low, then the pixel should be labeled as class *non-target*. Later, during the test phase, assume only four test samples are delivered to this subtree domain in which three test samples

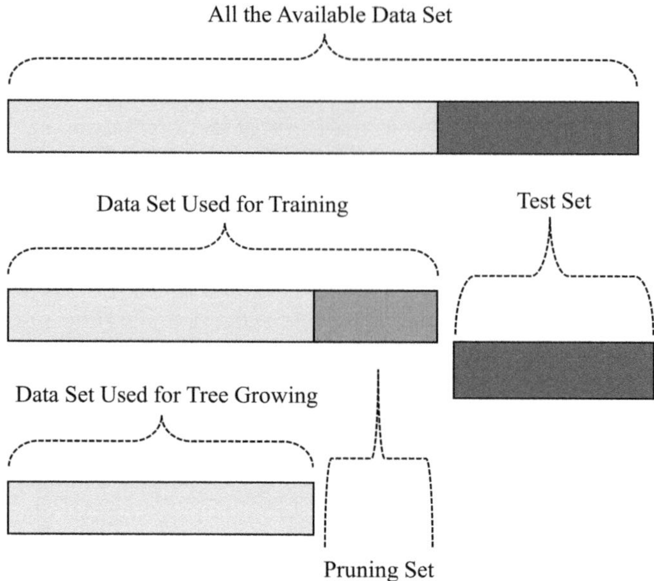

FIGURE 6.7 Typical data partitions for supporting decision tree induction at each stage.

with the attribute *Wetness*=high actually belong to the class *non-target* and one sample with the attribute *Wetness*=low belongs to the class *target*. According to the test result, the misclassification occurring is 4 (all four test samples are misclassified). However, one may consider replacing this subtree with a single leaf node labeled as majority class *non-target*. After this replacement, the classification errors can then be reduced from four to two (since only one training sample and one test sample belong to class *target*).

To perform decision tree pruning, one may require an independent data set called pruning set to support such a pruning process (Esposito et al., 1997). Figure 6.7 shows a typical data partition in supporting the formation of a decision tree at each stage. The data set is first divided into a training and a test set. The training set is further partitioned into two sets, one for decision tree growing during the training phase, and the other (the pruning set) for use in the pruning process.

A variety of pruning methods are available in the literature. For instance, Quinlan (1986, 1993) proposes methods called reduced error pruning (REP), pessimistic error pruning (PEP), and error-based pruning (EBP); Breiman et al. (1984) developed cost complexity pruning (CCP); Cestnik and Bratko (1991) present a Bayesian-based pruning method called minimal error pruning (MEP). These methods are described in the following sections.

6.6.1 REDUCED ERROR PRUNING

Reduced error pruning (REP) produces an optimal pruning of a decision tree. The resulting pruned tree will be the smallest tree among those with a minimal class error with respect to a pruning set. REP is generally performed in two phases. In the first phase, a pruning set is inputted to the decision tree and the accuracies of individual nodes are estimated. Secondly, from a bottom-up direction (i.e., starts from the leaf node), the subtrees, when removed without increasing the error, are pruned away. Specifically, one can express such a pruning process as follows. If S is a subtree rooted at node t, then the gain at t due to pruning subtree S, denoted as $\text{pruning\_gain}_{\text{REP}}(S, t)$, is given by the expression

$$\text{Pruning\_gain}_{\text{REP}}(S,t) = \text{error}(S) - \text{error}(t), \tag{6.27}$$

where error(v) denotes the number of misclassification samples that occurred in v. The node with the largest pruning gain is pruned, and the process is repeated until only nodes with a negative pruning gain remain.

6.6.2 PESSIMISTIC ERROR PRUNING

Pessimistic error pruning (PEP) does not require a pruning set (Figure 6.8). PEP estimates error directly from the training data. The error for leaf node t denoted as $\text{Error}_{\text{PEP}}(t)$ is estimated by

$$\text{Error}_{\text{PEP}}(t) = \frac{N(t) - n_j(t) + 0.5}{N(t)}, \tag{6.28}$$

where $N(t)$ is the number of samples within leaf node t, and $n_j(t)$ is the number of samples belonging to the most likely (i.e., majority) class j within node t. As a subtree S rooted at t is concerned, the error is computed by

$$\text{Error}_{\text{PEP}}(S) = \frac{\sum_{i \in \text{ leaf node of } S} \left(N(i) - n_j(i) + 0.5 \right)}{\sum_{i \in \text{ leaf node of } S} N(i)}. \tag{6.29}$$

Equation (6.29) sums the errors at each leaf node within subtree S. The PEP then compares the error resulting from subtree S and the error that occurred at node t. If $\text{Error}_{\text{PEP}}(t) \leq \text{Error}_{\text{PEP}}(S)$, then pruning is performed. Unlike most pruning methods, PEP runs in a top-down fashion, which tends to be faster since it only has to make one pass and looks at each node only once.

Figure 6.8 shows an example in which node 1 contains a total of 16 training samples with 7 samples belonging to class 1 and 9 samples belonging to class 2 (expressed as 16={7, 9}). Other nodes are described in a similar manner. For node 1, the $\text{Error}_{\text{PEP}}(\text{node 1})$ is

$$\text{Error}_{\text{PEP}}(\text{node 1}) = \frac{7 + 0.5}{16} = \frac{7.5}{16} = 0.47. \tag{6.30}$$

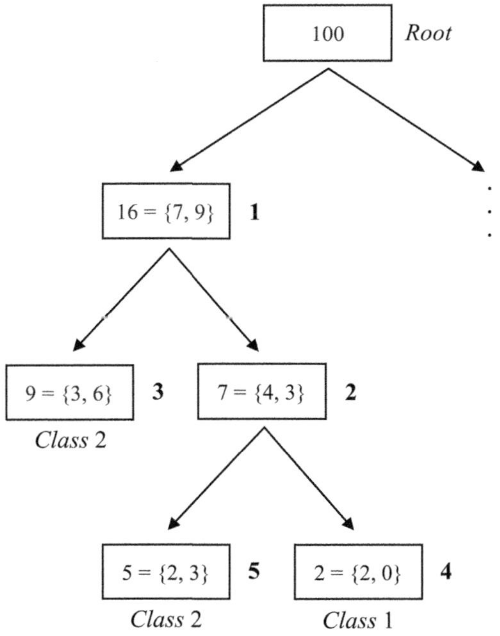

FIGURE 6.8   This example shows a partial decision tree used for pruning experiments.

For the subtree S rooted at node 1, the $\mathrm{Error}_{\mathrm{PEP}}(S)$ is

$$\mathrm{Error}_{\mathrm{PEP}}(S) = \frac{3+0+2+0.5\times 3}{16} = \frac{6.5}{16} = 0.4. \tag{6.31}$$

Since $\mathrm{Error}_{\mathrm{PEP}}$(node 1) is bigger than $\mathrm{Error}_{\mathrm{PEP}}(S)$, the subtree is not pruned.

6.6.3 ERROR-BASED PRUNING

An improved version of PEP is called error-based pruning (EBP). EBP estimates the error by adding the so-called one standard error pruning rule to PEP. For a subtree S rooted at t, pruning is determined by the following equation:

$$\mathrm{Error}_{\mathrm{PEP}}(t) \le \mathrm{Error}_{\mathrm{PEP}}(S) + \mathrm{std\_error}(S) \tag{6.32}$$

where std_error, the standard error, is calculated as

$$\mathrm{Std\_error}(S) = \sqrt{\frac{\mathrm{Error}_{\mathrm{PEP}}(S)\times\left(1-\mathrm{Error}_{\mathrm{PEP}}(S)\right)}{N(t)}}. \tag{6.33}$$

The corresponding standard error for subtree S rooted at node 1 is then calculated as

$$\mathrm{Std\_error}(S) = \sqrt{\frac{0.4(1-0.4)}{16}} = 0.122. \tag{6.34}$$

Since $\mathrm{Error}_{\mathrm{PEP}}(S)=0.4$ (Equation 6.31) and std_error $(S)=0.122$ (Equation 6.34), according to Equation (6.32), the sum of both being 0.522, which is >0.47 ($\mathrm{Error}_{\mathrm{PEP}}$ (node 1) in Equation (6.30)), the result shows that based on EBP (with one standard error rule), the subtree should be pruned.

EBP has a special characteristic called grafting, which is a process whereby an internal node of a decision tree is removed and replaced by one of its subtrees. Figure 6.9 shows the difference between pruning (Figure 6.9a) and grafting (Figure 6.9b) using the decision tree shown in Figure 6.3e. In the case of decision tree pruning, Figure 6.9a shows that the subtree rooted at node 5 is removed and replaced by a leaf node, while in the case of grafting, Figure 6.9b shows that internal node 2 is removed and replaced by the subtree rooted at node 5.

6.6.4 COST COMPLEXITY PRUNING

The cost complexity pruning (CCP) method, which is used in CART, tries to find the best compromise between the errors estimated from a pruning set or cross-validation and the size of the decision tree. In other words, CCP aims to minimize the error and complexity (equal to the number of leaves within a tree) of a decision tree. The total cost of a tree can be represented as

$$\text{Total cost} = \text{Error cost} + \text{complexity cost}. \tag{6.35}$$

Let $R(t)$ denote the error cost at node t, $R(S)$ the error of subtree S, and α the complexity parameter (a measure of how much additional accuracy a split must add to the entire tree to warrant the additional complexity). Following Equation (6.35), for a subtree S, the total cost of S denoted as Total_cost (S) is expressed as

$$\text{Total\_cost}(S) = R(S) + \alpha \times \text{num\_of\_leaves}(S). \qquad (6.36)$$

where num_of_leaves (S) denotes the number of leaves within subtree S. If subtree S is rooted at node t, after S is pruned and replaced by t, the total cost of the leaf node t is expressed as

$$\text{Total\_cost}(t) = R(t) + \alpha. \qquad (6.37)$$

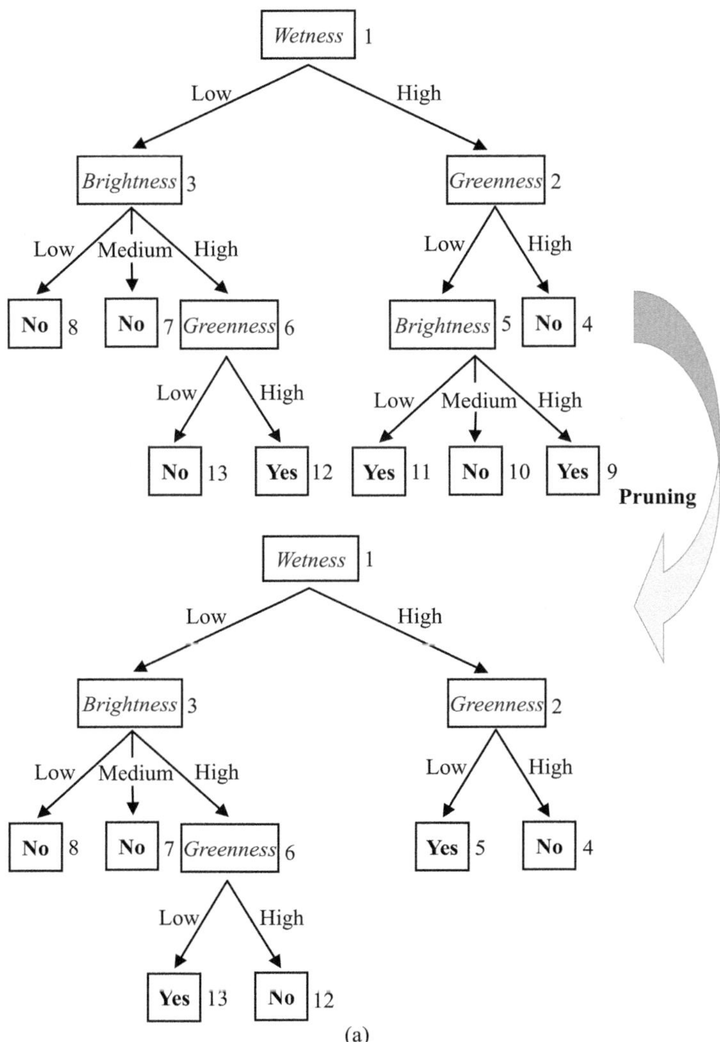

FIGURE 6.9 (a) Decision tree pruning in which the subtree rooted at node 5 is pruned and replaced by a leaf. (b) Decision tree grafting process in which internal node 2 is removed and replaced by the subtree rooted at node 5.

(Continued)

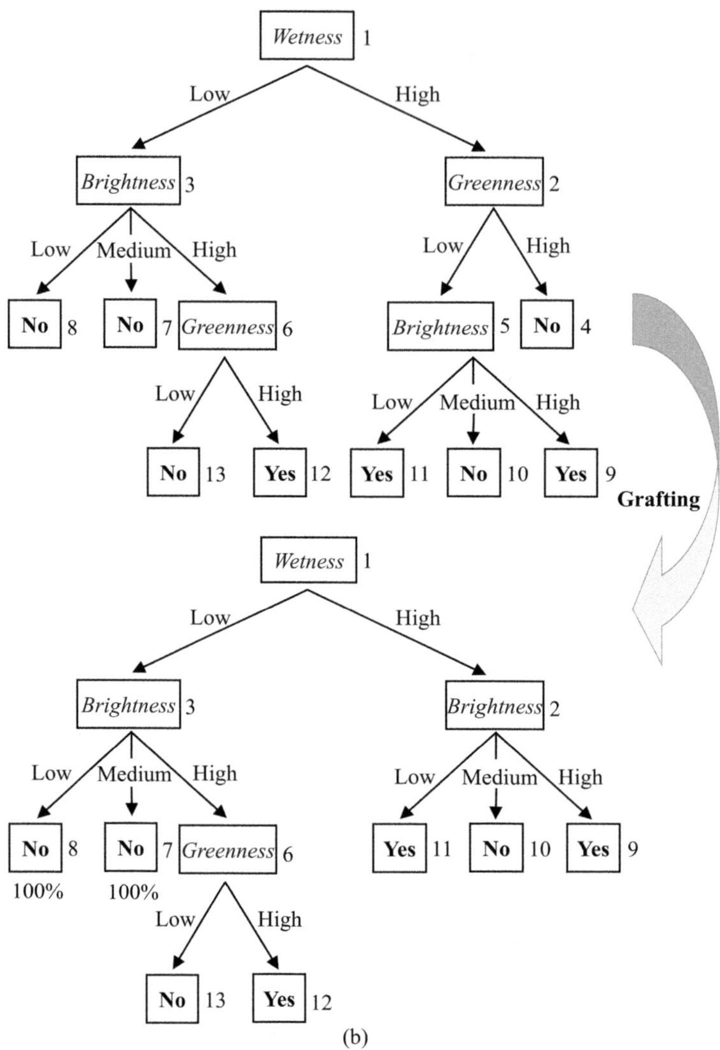

FIGURE 6.9 (Continued)

When the costs of Equations (6.36) and (6.37) are equal, one obtains

$$\alpha = \frac{R(t) - R(S)}{\text{Num\_of\_leaves}(S) - 1},$$

(6.38)

where α can then be regarded as the measure of the increase in apparent error rate per pruned leaf. During the pruning process, α is computed for each node, and the method selects the subtree with the smallest value α. For instance, using the case shown in Figure 6.8, the total number of samples is 100 (shown in the root node) and the error cost for node 1 denoted as $R(node1)$ is 7/100 (i.e., the proportion of error in node 1). The error cost $R(S)$ for the subtree rooted at node 1 is 5/100 (i.e., the proportion of error in node 3 [3 errors], 4 [error 0], and 5 [2 errors]). The value of α is then calculated by

$$\alpha = \frac{7/100 - 5/100}{3 - 1} = 0.01.$$

(6.39)

For both node 2 and the subtree rooted at node 2, the corresponding error costs are 3/100 and 2/100. Parameter α is calculated by

$$\alpha = \frac{3/100 - 2/100}{2-1} = 0.01. \tag{6.40}$$

Although both subtrees (rooted at node 1 and node 2, respectively) take on the same value of α, the best tree pruning policy should prune at node 1, because the reduction in tree size is greater than if the tree is pruned at node 2.

6.6.5 MINIMAL ERROR PRUNING

Minimal error pruning (MEP) also estimates error directly from training data. It uses a bottom-up approach, and after pruning, the estimated classification error can be reduced. The error is estimated using a Bayesian approach, which can be either Laplace or based on m-estimation. Specifically, during the pruning process, the error is first estimated from the subtree, which generates a so-called backed-up error (described in this section). One then assumes that this subtree is replaced by a leaf node and the corresponding expected error for that leaf node is computed. If the expected error is less than the backed-up error, the subtree should be pruned.

For a subtree being replaced by a leaf node i and labeled as class j (note that j is the majority class of samples within leaf node i), the corresponding expected error can be computed as

$$\text{Expected\_error}(i) = \frac{N + c - n_j - 1}{N + c}, \tag{6.41}$$

where N is the number of samples assigned to leaf node i, c is the number of classes, and n_j is the number of samples belonging to the most likely (i.e., majority) class j within node i. Equation (6.41) is called the Laplace error estimate, which is based on the assumption that the probabilities that samples belonging to each class are uniformly distributed. It is apparent from Equation (6.41) that the degree of decision tree pruning is affected by the number of classes.

The expected error for leaf node i can also be calculated in terms of m-estimate, which is expressed by

$$\text{Expected\_error}(i) = \frac{N + m - \text{prior}(j) \times m - n_j}{N + m}, \tag{6.42}$$

where $prior(j)$ is the prior probability of class j determined by the user, and m is a nonnegative parameter generally tuned in terms of experience. However, the general principle for determining value m (generally, m can be as small as 1 and as large as N) is that if data contain some noise, one may choose a lower value of m, which will result in little pruning. On the other hand, if the data contain considerable noise, a higher value of m can be used allowing more pruning. In short, the choice of m depends on the quality of the data. Also note that unlike the Laplace estimate, Equation (6.41) is not sensitive to the number of classes.

Now calculate the error upon node i, which is the father node of the original subtree containing s child nodes. The backed-up error for the subtree can then be calculated by

$$\text{Backed\_up\_error}(i) = \sum_{k=1}^{s} \text{Expected\_error}(k) \times P_k, \tag{6.43}$$

where P_k is the percentage of samples (i.e., the number of samples divided by N) contained in child node k, and Expected_error(k) is calculated according to Equation (6.38).

Again, by using Figure 6.8 as an example, the expected error rate for node 2 (containing $7 = \{4, 3\}$, i.e., 4 samples assigned to class 1 and 3 samples classified as class 2) according to Equation (6.37) is calculated as

$$\text{Expected\_error}(i) = \frac{7+2-4-1}{7+2} = 0.44. \tag{6.44}$$

For nodes 4 and 5, the expected error rates are 0.25 and 0.375, respectively. The backed-up error rate for node 1 according to Equation (6.43) is

$$\text{Backed\_up\_error}(i) = 0.43 \times \frac{5}{7} + 0.25 \times \frac{2}{7} = 0.377. \tag{6.45}$$

Since the backed-up error (0.377) is smaller than the expected error (0.44), the subtree rooted at node 2 is not pruned. For the subtree rooted at node 1, the expected error and backed-up error are, respectively,

$$\text{Expected\_error}(i) = \frac{16+2-9-1}{16+2} = 0.44 \tag{6.46}$$

$$\text{Backed\_up\_error}(i) = 0.377 \times \frac{7}{16} + 0.364 \frac{9}{16} = 0.370. \tag{6.47}$$

Since the backed-up error is greater than the expected error again, the subtree is not pruned.

6.7 ENSEMBLE METHODS

Ensemble learning methods, often referred to as multiclassifier systems or ensembles of classifiers in the literature, are based on constructing a supervised classification model by combining a set of classifier algorithms, each focusing on the same classification problem, and making the final prediction for an unknown sample via votes of the members (Rokach, 2010; Kuncheva, 2014). The main philosophy behind tree-based ensemble models, also called decision forest algorithms, is to construct multiple decision trees using diverse subsets of the initial training data and then forming a final prediction by consolidating their predictions. In recent years, ensemble methods have attracted growing interest within the remote sensing community due to their capability to yield remarkable classification results, and mitigate overfitting concerns, and speed of processing. Several review studies (Safavian and Landgrebe, 1991; Rokach, 2016; Saini and Ghosh, 2017; Zhang et al., 2022; Costa and Pedreira, 2023) have discussed the theory and applications of decision trees for a wide range of remote sensing studies.

6.7.1 BOOSTING

Boosting is an approach to improve the performance of a weak classifier. The basis of the boosting technique is to assign weights to the individual elements of the training data set. The weights of the elements are initially set to be equal. One then compares the classifier's output and the known label of each element of the training data to determine which elements of the training data have been classified incorrectly. These incorrectly classified training data elements are assigned a larger weight, and the classifier is run again. The increased weighting of these training data penalizes any misclassification and should force the classifier to focus on these cases. A method similar to boosting is described by Jackson and Landgrebe (2001). Quinlan (1996) concludes that about ten iterations are the optimum number, and little improvement in accuracy is achieved by performing additional boosting runs. Pal and Mather (2003) find that in the case of dealing with remotely sensed data, 10–15 boosting iterations are sufficient to achieve an improvement in classification accuracy.

Among the boosting algorithms presented in the literature, this section focuses on the widely used AdaBoost algorithm (Freund and Schapire, 1999) for its robustness in solving many of the practical difficulties of the earlier boosting algorithms.

The AdaBoost algorithm takes a training set $(x_1, y_1), (x_2, y_2), \ldots, (x_n, y_n)$ as inputs, where x_i and y_i denote the input features and the corresponding information class, respectively. The classifier is called repeatedly by AdaBoost in a series of iterations denoted as $k=1, 2, \ldots, K$. The core of AdaBoost is to maintain a set of weighting coefficients over the training set. The weight of training sample i on iteration k is denoted by $W_k(i)$. At the start of the training, all training samples are given equal weight, and as the training iterations proceed, the incorrectly classified samples are assigned a larger weighting coefficient to force the classifier to focus on such hard samples. Specifically, in a two-class case, the purpose of AdaBoost is to let the classifier find a so-called weak hypothesis h_k: $x \rightarrow \{+1, -1\}$ appropriate for the weighting set $W(i)$. The goodness of a weak hypothesis is measured in terms of its error ε_k (i.e., the sum of weights of misclassified samples).

$$\varepsilon_k = \sum_{i:H_k(x_i) \neq y_i} W_k(i) \tag{6.48}$$

AdaBoost then chooses a parameter α_k to measure the importance of each weak hypothesis, where α_k is represented by

$$\alpha_k = \frac{1}{2}\ln\left(\frac{1-\varepsilon_k}{\varepsilon_k}\right). \tag{6.49}$$

Note that the larger the value of ε_k, the smaller the value α_k will be. Also note that the process of selecting α_k and $h_k(x)$ can be interpreted as a single optimization step that minimizes the upper bound on the empirical error. Improvement of the bound is guaranteed, provided that $\varepsilon_k < \frac{1}{2}$. The weighting coefficient in the next $k+1$ run is then updated by

$$W_{k+1}(i) = \frac{W_k(i)}{Z_k} \times \begin{cases} e^{-\alpha_k} & \text{if } h_k(x_i) = y_i \\ e^{\alpha_k} & \text{if } h_k(x_i) \neq y_i \end{cases}$$
$$= \frac{W_k(i) \times \exp(-\alpha_i h_k(i) y_i)}{Z_k} \tag{6.50}$$

where Z_K is a normalization factor, usually defined as

$$Z_k = 2\sqrt{\varepsilon_k(1-\varepsilon_k)} \tag{6.51}$$

Each incorrectly classified training sample will be accompanied by an increase in weight. The weight can be regarded as the upper bound on the error of a given training sample. The final hypothesis or decision for an input attribute x, denoted as $H(x)$, is then expressed as

$$H(x) = \text{sign}\left(\sum_{k=1}^{K} \alpha_k h_k(x)\right). \tag{6.52}$$

Figure 6.10 shows the results of using AdaBoost for classifying binary samples. The 150 dot samples within the 2-dimensional space shown in Figure 6.10a are within the range [0.25, 0.75] on each axis, while 150 samples are also drawn from the white area. Results of forming classification boundaries for each iteration k (from 1 to 5) by AdaBoost are displayed. The corresponding errors are shown in Figure 6.10b. It can be seen that AdaBoost converges very quickly within around ten iterations.

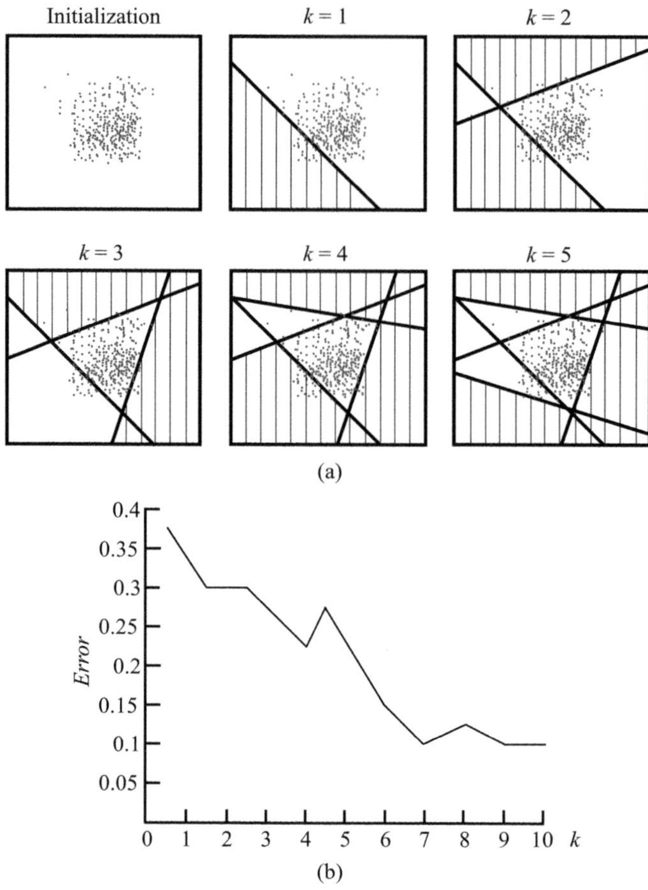

FIGURE 6.10   (a) Progress of classifications using the AdaBoost technique. (b) Classification error at each iteration using AdaBoost.

The case illustrated above is only concerned with binary classification. For the multiclass case, the most straightforward method is to transfer the multiclass issue into a larger binary problem. For instance, for each training sample x, the problem is to determine whether x belongs to class y or class z where class z comprises all classes except y (Schapire and Singer, 1998). Other AdaBoost-relating techniques adopting similar concepts for dealing with the multiclass problem are also provided in the literature. Readers should refer to Freund and Schapire (1999) for details.

6.7.2 Random Forest

The random forest (RF) classifier consists of a combination of tree classifiers where each classifier is generated using a random vector sampled independently from the training set of input vectors, and each tree casts a unit vote for the most popular class in which to place a given input vector (Breiman, 2001). The RF classifier outperforms many image classification techniques because it is nonparametric, capable of handling both continuous and categorical data, simple to configure, robust to overfitting, and unaffected by data noise, making it effective in handling outliers in training data (Kavzoglu, 2017). An important advantage of the method is that feature importance can be easily estimated during training process, which makes it a good choice for

---

Input: Training samples X, Labels y, the number of decision trees: $ntree$, the number of variables to be selected for splitting at each node: m.

Training:

for i to $ntree$

Select a bootstrap sample from training samples with size N for each tree

Grow each tree using 2/3 bootstrapped data (i.e., in-bag samples).

Select m variable at random for splitting at each node and choose the best split

Build each decision tree classifier recursively without pruning

Calculate the accuracy of each tree using the 1/3 bootstrap sample (i.e., out-of-bag samples) and estimate the out-of-bag (OOB) error

end for

Output: decision tree classifier pool $\{D_1, \ldots, D_{ntree}\}$

Classification:

Calculate the probability of an unknown sample x belonging to class c: $p(c/x) = (1/ntree)\sum h_j(c/x)$

Predict class label using majority voting

FIGURE 6.11 Pseudocode for RF algorithm.

embedded feature selection method using high-dimensional data sets (Saeys et al., 2008; Pal and Foody, 2010). The general workflow of pseudocode representing the ensemble model of the RF algorithm is given in Figure 6.11. The technique for generating a RF is generally a combination of the bagging and random subspace methods. Bagging, introduced by Breiman (1996), is a technique to improve classification accuracy and avoid overfitting. Given a training set of size N, bagging generates a number of new training sets each of size n (where $n < N$) by randomly drawing samples with replacement from the original training set. Also suppose the data contain M attributes (e.g., spectral bands). For each node of the tree, m ($m < M$) attributes are randomly chosen to provide the base for calculating the best split at that node. In practical applications, approximately two-thirds of the samples, referred to as in-bag samples, are utilized for training each decision tree, while the remaining one-third, known as out-of-bag samples, are employed in an internal cross-validation approach to assess the accuracy of the ultimate ensemble model. The generalization error is influenced by the accuracy of individual trees, as well as their associations with one another. Once the RF is obtained, each sample is classified into the class taking the majority votes from the tree predictors in the forest. In regression tasks, the mean value of the predictions is considered. The RF algorithm can be expressed as

$$F = \text{Random forest}(X, Y, n\_\text{trees}, m\_\text{features}), \tag{6.53}$$

where F indicates the resulting forest, $n_$trees is the number of trees in the forest, and $m_$features shows the number of randomly selected features for each tree. The general working principles of the RF algorithm (Sahin et al., 2020) are given in Figure 6.12. Configuring the RF algorithm involves user-defined hyperparameters, including the quantity of randomly drawn observations per tree and the choice between drawing with or without replacement. Other considerations involve determining the number of variables selected randomly for each split, specifying the splitting rule, setting the minimum node sample requirement, and establishing the total number of trees. Optimal variable selection hinges on finding a number of variables that minimize correlation while still providing robust predictive power. In fact, identifying the optimal subset size for predictor variables involves exploring a wide range of possibilities, and simple tests can assist in determining the most suitable size (Pal, 2005; Horning, 2010). On the other hand, some typical default values can provide optimal or near-optimal accuracies. For instance, the number of variables drawn randomly for each split is usually set to the $\sqrt{p}$ for classification problems and $p/3$ for regression problems, where p is the number of predictor variables. More discussion and survey on hyperparameter tuning strategies can be found in Oshiro et al. (2012) and Probst et al. (2019).

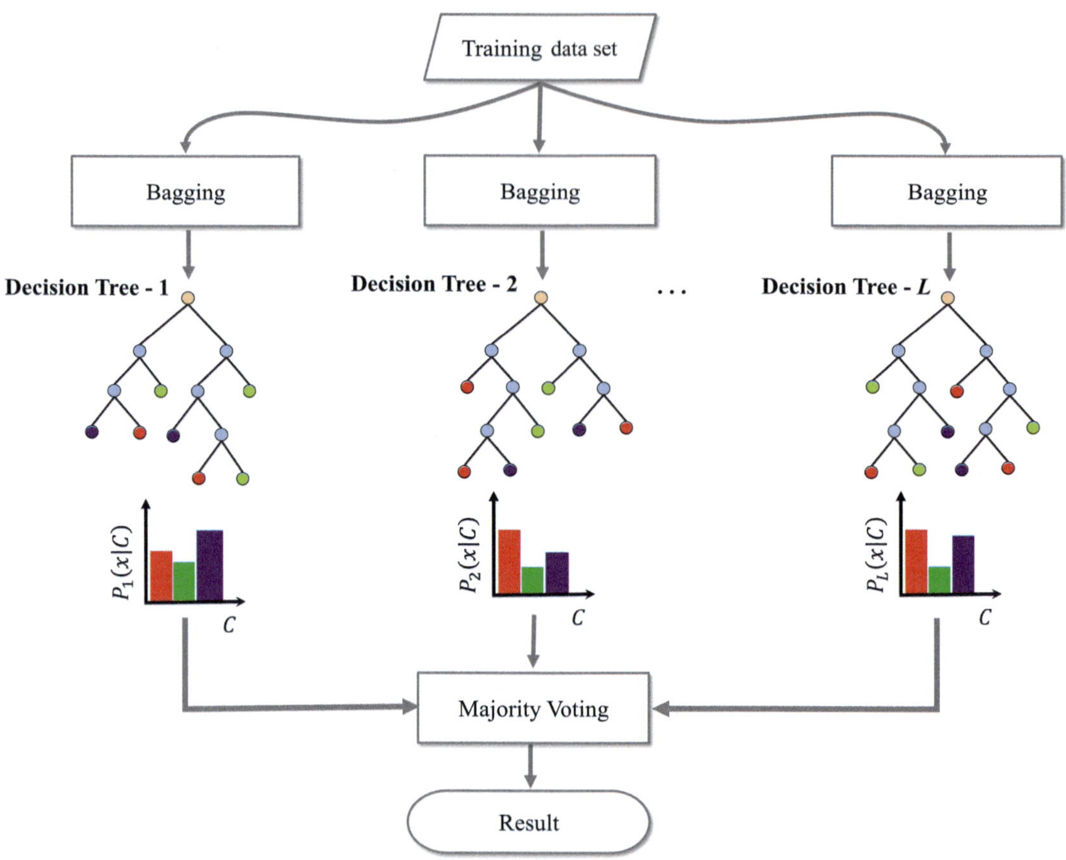

FIGURE 6.12   Working principle of random forest algorithm.

Breiman (2001) outlines numerous advantages of the RF algorithm, including accuracy that rivals AdaBoost and occasionally surpasses it, robustness to outliers and noise, faster execution compared to bagging or boosting, the absence of pruning requirements, and the presentation of informative internal estimates related to error and variable importance. The ensemble of trees created in the classification process can be saved for later application to new data sets. The RF algorithm has received increasing attention and has been successfully applied to many problems due to its accuracy, simplicity, and speed. As a result, it has become one of the benchmark methods in the analysis of remotely sensed data. Given its prevalence, several variants of the RF algorithm, such as rotation forest, canonical correlation forest, and fuzzy RF, have emerged to tackle particular deficiencies and enhance overall performance. The subsequent sections provide an in-depth exploration of the theories behind the rotation forest and canonical correlation forest algorithms. A fuzzy RF is proposed by Bonissone et al. (2010) to combine the robustness of RF with the flexibility of fuzzy logic and fuzzy sets for imperfect data management. A comparative analysis was conducted using multiple data sets to demonstrate the effectiveness of the proposed multiclassifier system, and the results validated the robustness of the proposed multiclassifier system demonstrated. For more reading on RF classifier, the readers should refer to Pal (2005), Belgiu and Drăguţ (2016), Kavzoglu (2017), Sheykhmousa et al. (2020), and Kavzoglu and Teke (2022).

6.7.3 ROTATION FOREST

Rotation forest (RoF), introduced by Rodríguez et al. (2006), is a classifier ensemble technique designed to improve the diversity and performance of its base classifiers, primarily decision trees,

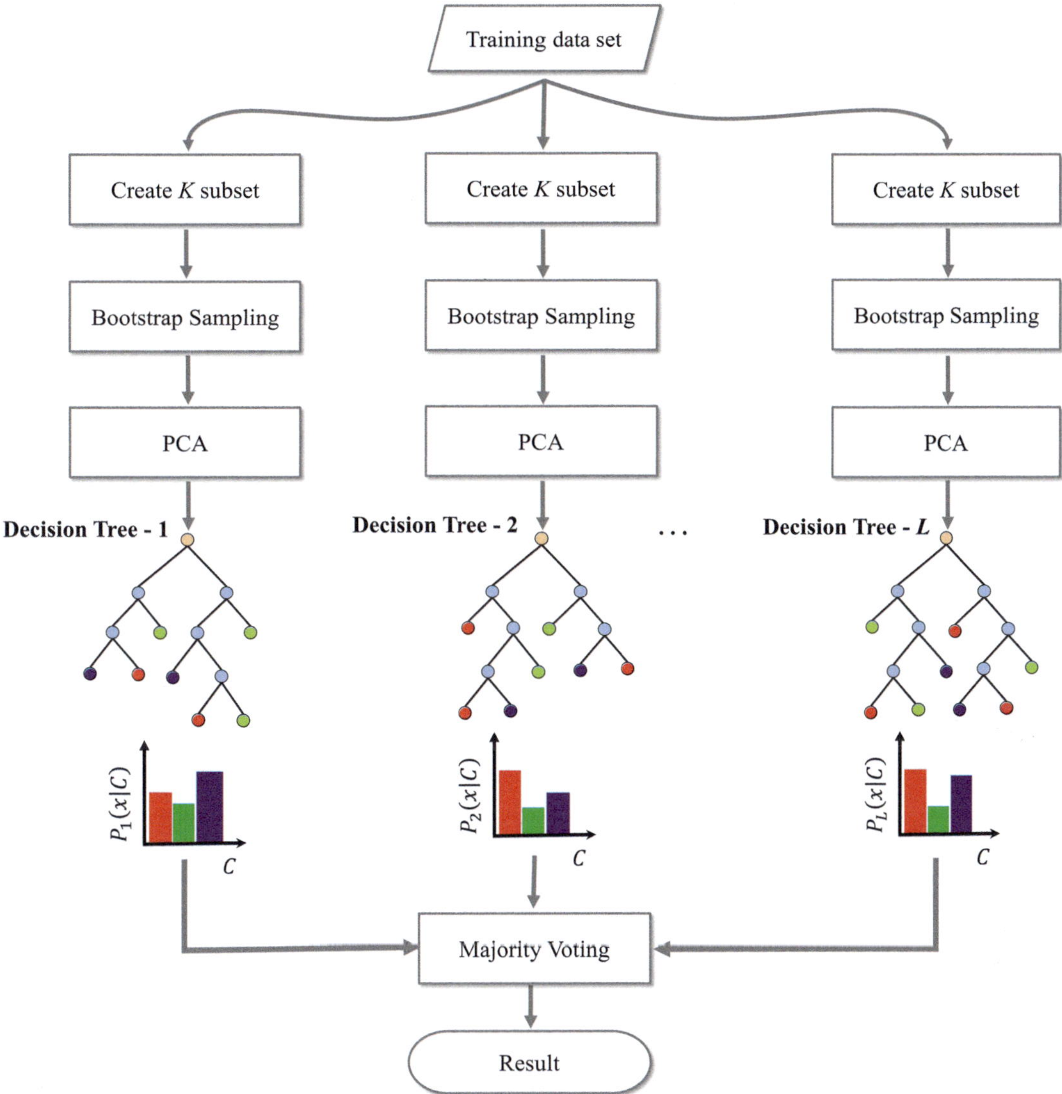

FIGURE 6.13 Operational concept of rotation forest algorithm.

by applying feature transformation through principal component analysis (PCA). The general work-flow representing the ensemble model construction with the RoF algorithm is given in Figure 6.13. The technique is based on a working principle close to RF and shares some similarities in assigning a class label for an unknown pixel. For example, both ensemble methods share the common approach of constructing multiple weak learners and merging their predictions to arrive at a final decision. Similar to RF, the RoF model selects decision trees as the base classifier, leveraging their sensitivity to the rotation of feature axes. Moreover, bootstrapping, a resampling technique where subsets of the training data are created with replacement, is employed by RF and RoF to train individual decision tree construction for enhancing diversity and reducing overfitting. The primary difference in constructing the RoF ensemble model is that the PCA rotates each base classifier's feature axis to ensure the balance between diversity and individual accuracy. For this purpose, the original training data set is randomly split into K subsets, the main parameter of the RoF, and PCA is applied on each subset. This process results in the transformation of the input data set into a new feature space, after which every individual tree within the forest is trained using this modified data set. During the prediction phase, confidence scores for each class are determined through an average combination method (e.g., majority voting), leading to the assignment of the final class label to the class with the highest confidence value.

Let $\mathbf{x} = [x_1, \ldots, x_n]^T$ be a pixel described by n features (i.e., bands), and let $\mathbf{X}$ be the data set containing the training pixels in the form of an $N \times n$ matrix. Let $\mathbf{Y}$ be a vector with class labels for training data, $\mathbf{Y} = [y_1, \ldots, y_N]^T$, y_i takes a value from the set of class labels $\{w_1, \ldots, w_c\}$. Denote by $D_1, \ldots, D_L$ the classifiers in the ensemble and by $\mathbf{F}$, the feature set. To construct the training set for classifier D_i, algorithmic steps are defined as follows by Rodríguez et al. (2006):

1. Split $\mathbf{F}$ randomly into K subsets (K is a parameter of the algorithm), and for simplicity, each feature subset contains $M = n / K$ features.
2. Denote by $F_{i,j}$ the jth subset of features for the training set of classifiers D_i. For each of these subsets, randomly choose a nonempty set of classes and then generate a bootstrap sample of objects, comprising 75 percent of the total data count. Perform PCA exclusively on the M features within $F_{i,j}$ and the chosen subset of $\mathbf{X}$. Store the coefficients of the components, $a_{i,j}^{(1)}, \ldots, a_{i,j}^{(M_i)}$, each of size $M \times 1$. Compile the acquired coefficient vectors into a sparse matrix denoted as the "rotation" matrix R_i.

$$R = \begin{bmatrix} a_{i,1}^{(1)}, \ldots, a_{i,2}^{(M_1)} & [0] & \cdots & [0] \\ [0] & a_{i,2}^{(1)}, \ldots, a_{i,2}^{(M_2)} & \cdots & [0] \\ \vdots & \vdots & \ddots & \vdots \\ [0] & [0] & \cdots & a_{i,K}^{(1)}, \ldots, a_{i,K}^{(M_K)} \end{bmatrix} \tag{6.54}$$

To estimate the training sample set for classifier D_i, the columns R_i are rearranged to R_i^a (size $N \times n$) with respect to the original feature set, and then, the training set for classifier D_i is XR_i^a.

In recent years, there has been a notable expansion in the body of literature focusing on the use of the RoF ensemble learning algorithm in classifying remotely sensed imagery. For example, Xia et al. (2013) investigated the performance of the RoF algorithm for the classification of hyperspectral imagery. Specifically, different feature transformation methods, including maximum noise fraction, independent component analysis, and local Fisher discriminant analysis, were applied as an alternative to PCA. The findings highlighted the effectiveness of the RoF algorithm when applied with PCA transformation in hyperspectral image classification, showcasing its superiority over well-known ensemble models such as bagging, AdaBoost, and RF. Kavzoglu and Colkesen (2013) analyzed the prediction performance of the RoF algorithm in classifying multispectral ASTER data set compared with well-known bagging, AdaBoost, MultiBoosting, DECORATE, random subspace, and RF ensemble learning models. A study by Du et al. (2015) discusses the classification performance of the RoF in the classification of Fully Polarimetric Synthetic Aperture Radar (PolSAR) images. Confirming the superiority of the RoF algorithm, the results underlined that the time required to construct an ensemble model is significantly higher than RF. In addition to the studies examining the pixel-based classification performance of the RoF algorithm, Kavzoglu et al. (2015) analyzed the effectiveness of the algorithm in the object-based classification problem. The results showed that RoF and SVM produced similar predictions on the object-based classification problem, while both outperformed RF and NN classifiers as suggested by McNemar and Wilcoxon's signed rank test statistics. More recently, Ha et al. (2020) conducted a study comparing the performance of maximum likelihood classifier (MLC), RF, RoF, and Canonical correlation forest (CCF) algorithms in mapping seagrass using Sentinel-2 images. Results of the study showed that tree-based techniques outperformed the MLC, and the RoF algorithm was the best-performing one in terms of accuracy metrics and McNemar's statistical test.

6.7.4 CANONICAL CORRELATION FOREST

CCF, proposed by Rainforth and Wood (2015), represents an ensemble learning algorithm that is primarily employed for classification tasks. At its core, this algorithm operates by constructing a collection of canonical correlation trees (CCTs), which are essentially oblique decision trees. Unlike RF, which performs axis-aligned splits, CCF employs canonical correlation analysis (CCA) to identify feature projections that maximize the correlation between features and class labels. The general workflow representing the ensemble model construction with the CCF algorithm is given in Figure 6.14. Each individual CCT within the ensemble model undergoes training through CCA, which aims to identify feature projections that yield the highest correlation between the features and the class labels. Subsequently, the algorithm selects the optimal split within this projected feature space. CCF demonstrates versatility in accommodating both binary and multiclass classification problems. Notably, it differentiates itself from traditional decision tree algorithms by using CCA between features and classes before selecting the split through an exhaustive search in the projected feature space (Xia et al., 2017). In the prediction stage, the classification result is ascertained by combining the predictions of individual CCTs within the ensemble, employing a majority voting rule. While the use of the CCF algorithm for classification purposes has been somewhat

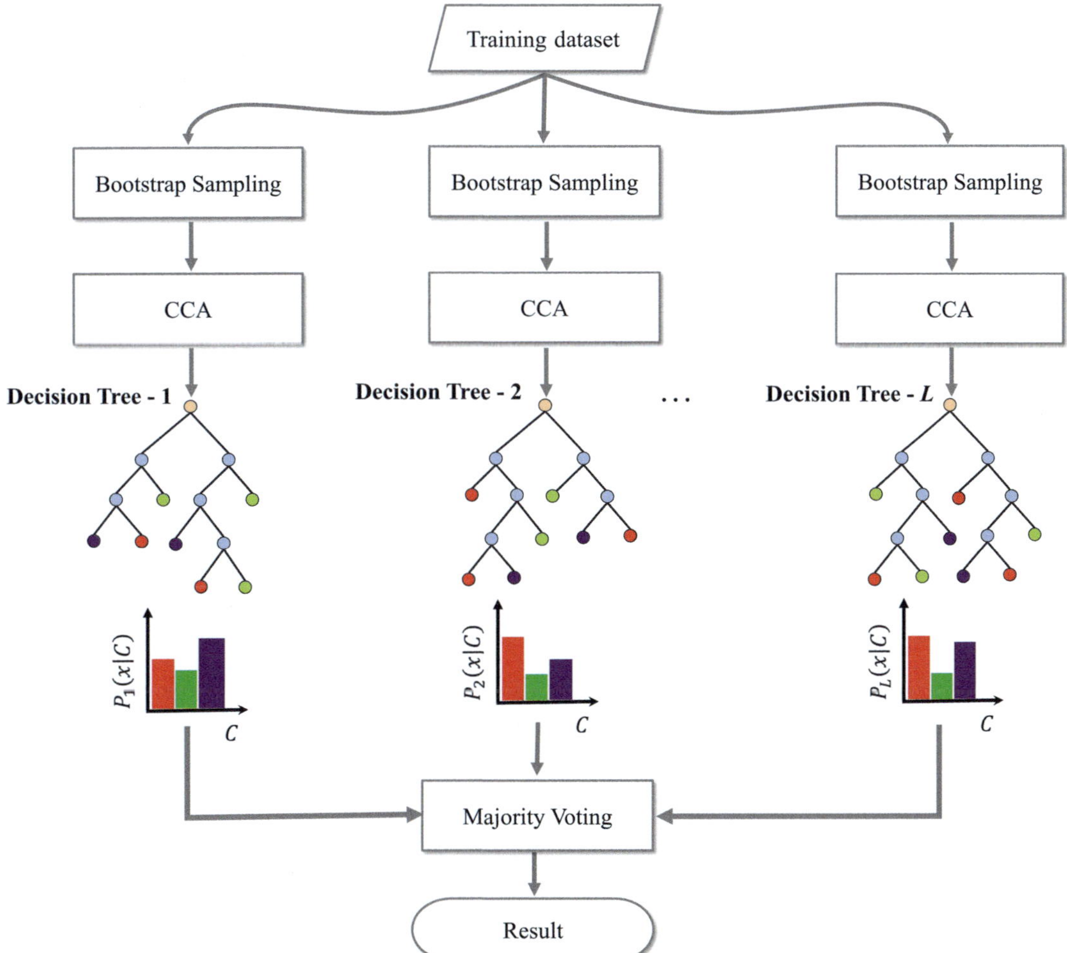

FIGURE 6.14 Workflow of canonical correlation forest algorithm.

limited, its performance has been observed to either surpass or rival that of well-established decision tree-based ensemble methods, as reported by Rainforth and Wood (2015), Xia et al. (2017), Colkesen and Kavzoglu (2019), and Tonbul et al. (2020).

CCF is an ensemble learning algorithm designed for classification tasks, known for its unique approach to feature selection and classification. The main heuristic behind the algorithm is to construct an ensemble of CCTs, which are oblique decision trees. Unlike RF, which performs axis-aligned splits, CCF employs CCA to identify feature projections that maximize the correlation between features and class labels (Figure 6.15). This approach allows CCF to capture complex relationships in the data, making it less sensitive to the number of trees in the ensemble and eliminating the need for parameter tuning. In contrast, RF often requires fine-tuning of hyperparameters to optimize performance. Another key difference is that CCF uses an exhaustive search in the space of projected features to select splits, whereas RF typically employs a random selection process. As a result, CCF can be particularly effective when dealing with high-dimensional data and has demonstrated superior or comparable performance to RF in various classification scenarios, making it a valuable addition to the ensemble learning toolkit.

Till now, several studies have reported the classification performance of the CCF algorithm. For example, Xia et al. (2017) analyzed the performance of CCF for the classification of six benchmark hyperspectral images using different spatial classification strategies (e.g., extended multiattribute profiles). Furthermore, Colkesen and Ertekin (2020) addressed the diversity issue of classification ensemble, as well as the performance of the CCF and RoF algorithms in pixel-based hyperspectral image classification. The study results showed that for the three hyperspectral data sets evaluated, the RoF and CCF algorithms exhibited superior classification performance compared with the RF algorithm. In addition, the diversity between the individual trees in the ensemble models was higher than that in RoF and RF, considering various measures, such as Q-statistic, correlation coefficient, and interrater agreement statistics. In addition to research evaluating the performance of the algorithm in classifying hyperspectral data sets, studies focusing on its performance in classifying multispectral data sets at different spatial resolutions are available in the literature. For instance, Colkesen and Kavzoglu (2017) evaluated the performance of the CCF algorithm for land use/land cover classification using Sentinel-2 and Landsat Operational Land Imager (OLI) imagery, and its performance was compared to RF and RoF algorithms. Furthermore, the object-based classification performance of the CCF was evaluated by Colkesen and Kavzoglu (2019) using very high-resolution WV-2 imagery, suggesting the effectiveness of the CCF for OBIA. More recently, Colkesen and Ozturk (2022) also carried out a comparison of the performance of leading ensemble learning algorithms, including RotF, CCF, CB, XGB, and LightGBM, in the domain of land cover classification utilizing WorldView-2, Sentinel-2, and ROSIS imagery. In summary, the studies mentioned above highlighted the simplicity and effectiveness of the CCF algorithm and suggested the method as a robust algorithm for classifying remotely sensed images.

6.7.5 Extreme Gradient Boosting

Originally introduced by Chen and Guestrin (2016), the use of extreme gradient boosting (XGBoost) has proliferated, potentially positioning it as one of the most frequently employed algorithms in current studies. The primary reason for this widespread adoption can be attributed to its ability to deliver exceptional performance within a scalable framework. The fundamental principle that underlies XGBoost's functionality is the gradual inclusion of weak learners into the model, allowing it to minimize the combination of a training loss and a regularization term, commonly known as the loss function. For binary classification problems, the logistic loss function (i.e., *binary:logistic*) is usually employed to compute the variation between predicted probability and actual label:

$$L\left(y, f\left(x\right)\right) = \log\left(1 + \exp\left(-y * f\left(x\right)\right)\right), \tag{6.55}$$

CCF Algorithm

Input: $\{\mathbf{X}, y\}$: training samples ($N \times D$ matrix), T: the number of decision trees in the ensemble (ensemble size), M: the number of sampling features, D: number of spectral bands, $b \in$ (true, false), whether to projection bootstrap.

Output: CCF $\mathcal{T} = \{t_i\}_{i=1,\dots;T}$

1: Centering $\mathbf{X}$ with zero mean and unit variance

2: **If** $M < D$ **then**

3: $\rightarrow b = true$

4: **else**

5: $\rightarrow b = false$

6: **end if**

7: **for** i=1:T **do**

8:   **If** b **then**

9:   $\{\mathbf{X}', y'\} \leftarrow$ bootstrap sampling from $\{\mathbf{X}, y\}$

10:   **else**

11:   $\{\mathbf{X}', y'\} \leftarrow \{\mathbf{X}, y\}$

12:   **end if**

13: $[., S, L] = \text{GROWTREE}(\mathbf{X}', y', D, M, b)$

14: $t_i = \{S, L\}$

15: **end for**

16: **return** $\mathcal{T} = \{t_i\}_{i=1,\dots;T}$

  Prediction phase

  Input: The ensemble $\mathcal{T} = \{t_i\}_{i=1,\dots;T}$. A new sample x^*

  Output: Class label y^*

Algorithm: GROWTREE

Input: $\{\mathbf{X}^j, y^j\}$: training samples at node j, M: number of sampling features, $b \in$ (true, false),

whether to projection bootstrap, $\mathbb{F}^j$: available features.

Output: sub-tree root node identifier j, sub-tree discriminant nodes Ψ, sub-tree leaf nodes Θ

1: set current node index j to an unique node identifier

2: randomly selecting λ features by taking $\min(M, |\mathbb{F}^j|)$ samples without replacement from $\mathbb{F}^j$

3: **If** b **then**

4: $\{\mathbf{X}', y'\} \leftarrow$ bootstrap sampling from $\{\mathbf{X}^j_{:,\lambda}, y^j\}$

5: **else**

6: $\{\mathbf{X}', Y'\} \leftarrow \{\mathbf{X}^j_{:,\lambda}, y^j\}$

7: **end if**

8: $[\Phi, \cdot] = \text{CCA}(\mathbf{X}', y')$

9: $\mathbf{R}_{1:\lambda,1:\lambda} \leftarrow \Phi, \mathbf{R}_{\lambda:end,\lambda:end} \leftarrow 0$

10: $\mathbf{U} = \mathbf{XR}$

11: $[\xi, z_j, gain] = FINDBESTSPLIT(\mathbf{U})$

12: $\Phi_j = \mathbf{R}_{(:,\xi)}$

13: **If** $\exists\, \mathbf{X}_{(i,:)}\Phi_j \leq z_j$ **then** τ_l **else** τ_r **end if**

14: $\{s_j^1, S_l, L_l\} = \text{GROWTREE}(\mathbf{X}^j_{(\tau_l,:)}, y^j_{(\tau_l,:)}, |\mathbb{F}^j|, M, b)$

15: $\{s_j^2, S_r, L_r\} = \text{GROWTREE}(\mathbf{X}^j_{(\tau_r,:)}, y^j_{(\tau_r,:)}, |\mathbb{F}^j|, M, b)$

16: obtain $\{s_j^1, s_j^2, \Phi_j, z_j\} =$

17: **return** $[j, \{S_l \cup S_r\}, \{L_l \cup L_r\}]$

FIGURE 6.15 Algorithmic steps of the CCF method.

where y is the actual label (either landslide or nonlandslides), and $f(x)$ is the predicted probability. To minimize the logistic loss function, each tree is trained using gradient boosting algorithm. At each iteration, the model computes the negative gradient of the loss function relative to the present predictions, which can be described as

$$-\frac{\partial L\left(y_i, f(x_i)\right)}{\partial f(x_i)} = \frac{y_i}{\left(1 + \exp\left(y_i * f(x_i)\right)\right)}, \tag{6.56}$$

where y_i represents the actual label for the ith data point, and $f(x_i)$ denotes the present estimations. The new decision tree is trained to minimize the following objective function:

$$\text{Obj} = \sum_i \log\left(1 + \exp\left(-y_i\left(f^{t-1}(x_i) + \Phi(x_i, \theta)\right)\right)\right) + \Omega(\theta), \tag{6.57}$$

where $f^{t-1}(x_i)$ is the prediction made by the current ensemble of trees, $\Phi(x_i, \theta)$ is the new tree's prediction, θ is the set of hyperparameters that control the tree's structure and complexity, and $\Omega(\theta)$ represents the regularization term that penalizes complex models.

To stimulate the model to develop a better representation, the logistic loss function applies penalties for inaccurate predictions, thus regulating its performance and mitigating the risk of overfitting to the training data. XGBoost can utilize regularization terms to manage model complexity and mitigate overfitting. L_2, often referred to as ridge regularization, introduces a penalty term that discourages the use of large weights for the features, which can be expressed as

$$\Omega(\theta) = \lambda \sum_j w_j^2 \tag{6.58}$$

The term is calculated by taking the product of a constant λ and the sum of the squared weights $\left(w_j\right)$ of each tree in an ensemble. XGBoost employs a shrinkage scheme to further prevent overfitting, which decreases the effect of each subsequent tree in the ensemble by using a small learning rate (η) for scaling. The learning rate controls the magnitude of the gradient descent algorithm's step at each iteration. With the application of learning rate, the function becomes

$$f^t(x_i) = f^{t-1}(x_i) + \eta h^t(x_i), \tag{6.59}$$

where $f^{t-1}(x_i)$ is the output of the previous $(t-1)$ trees in the ensemble, and $h^t(x_i)$ is the prediction made by the tth tree on the ith sample. The utilization of the logistic function is preferred because it naturally transforms predicted values into the [0, 1] range, allowing for interpretation as probabilities. Finally, when making predictions for a new instance, XGBoost combines the predictions of all trees in the ensemble and applies the logistic function. The final prediction generated by the XGBoost ensemble for the input instance x can be described as

$$p(y = 1 | x) = \frac{1}{\left(1 + \exp(-f(x))\right)}, \tag{6.60}$$

where $p(y = 1 | x)$ is the predicted probability of the positive class, and $f(x)$ is the final prediction of the ensemble.

6.7.6 Light Gradient Boosting Machines

The traditional approach to gradient boosting–based decision trees (GBDT) requires a thorough analysis of all instances in the data set for each feature to determine the optimal split points. Consequently, the computational complexity of such implementations is directly related to both the number of features and instances. As these numbers increase, the computational demand and processing time experience a notable rise, especially when working with high-dimensional data sets. Addressing the critical limitations of GBDT, Ke et al. (2017) proposed LightGBM, which incorporates two specific enhancements: Gradient-based One-Side Sampling (GOSS) and Exclusive Feature Bundling (EFB). The former minimizes the number of samples involved in gradient calculations by prioritizing instances with more substantial gradients. For the second enhancement, EFB clusters distinct features together, thereby accelerating data loading and decreasing cache miss occurrences. These enhancements substantially enhance the scalability and performance of decision trees driven by gradient boosting, thereby establishing LightGBM as a widely favored option for big data applications. In contrast to GBDT, LightGBM integrates GOSS into the computation of variance gain and the process of deciding split points. The mathematical background of the method can be explained as follows (Chen et al., 2019; Wen et al., 2021). Initially, the absolute values of the gradients for the training instances are sorted in a descending manner. The top $\alpha \times 100\%$ of data instances with the highest gradient values are chosen from this sorted list, forming a set denoted as S. A subset, denoted as T, is then randomly sampled from the remaining examples. The size of subset T is b times the size of the complement of S. Finally, the variance $\widetilde{V}_j(d)$ is estimated for the combined set of samples from S and T using the specific formula:

$$\widetilde{(V_j)}(d) = \frac{1}{n} * \left[\frac{\left(\sum_{x_i \in S_l} g_i + \frac{1-\alpha}{b} \sum_{x_i \in T_l} g_i \right)^2}{\left(n_l^j(d) \right)} + \frac{\left(\sum_{x_i \in S_r} g_i + \frac{1-\alpha}{b} \sum_{x_i \in T_r} g_i \right)^2}{\left(n_r^j(d) \right)} \right], \qquad (6.61)$$

where S_l and S_r are the left and right subsets of S, and T_l and T_r are the left and right subsets of T, respectively. $n_l^j(d)$ and $n_r^j(d)$ are the numbers of examples in the left and right subsets, respectively, at node j and depth d. The variance $\widetilde{V}_j(d)$ is then used to split the examples at node j and depth d.

High-dimensional features frequently exhibit sparsity, where many of these sparse features are distinct or unrelated. The EFB method can significantly accelerate the GBDT training process when applied in the LightGBM framework. LightGBM, with the help of EFB, combines these mutually exclusive features into a single feature bundle. This enables the feature scanning algorithm to be adjusted for creating histograms from these bundled features. The consolidation of these mutually exclusive features by LightGBM minimizes the number of distinct features that must be scanned in the training phase, thus accelerating the training process. It should be noted that the feature histograms derived from these bundled features exhibit a close resemblance to those obtained from the original features, thereby preserving the model accuracy. Assuming $F(x)$ represents the final model obtained by combining the predictions of M decision trees, where $h_{i(x)}$ represents the ith decision tree, and w_i is the weight assigned to the prediction of each tree, we can express $F(x)$ as

$$F(x) = \sum_{i=1}^{M} w_i h_{i(x)}. \qquad (6.62)$$

6.7.7 Gradient Boosting Machines

A statistical viewpoint was introduced to boosting by Friedman et al. (2000), who established a connection between the boosting algorithm and loss functions. This iterative approach involves running a classification algorithm on adjusted versions of the training data and then combining the resulting classifiers through a weighted majority vote. Friedman (2001) expanded the application of the boosting algorithm to regression problems by introducing the GBM approach. The GBM operates as a numerical optimization method with the aim of minimizing the loss function by iteratively constructing an additive model. During each step, a new estimator, often referred to as a "weak learner", is systematically incorporated to maximize the reduction in the loss function (Touzani et al., 2018). In other words, GBM builds an ensemble of decision trees that collaborate to reduce the total loss function. The subsequent pseudocode provides a simplified representation of the algorithm:

1. Hyperparameters are set:
   - *max_depth*: the maximum depth of each decision tree, $d \in \mathbb{Z}^+$
   - *num_iterations*: the number of decision trees to create, $K \in \mathbb{Z}^+$
   - *learning_rate*: the step size at which to update the model, $\alpha \in \mathbb{R}^+$
   - *subsample_fraction*: the fraction of data to use for each decision tree, $\eta \in [0,1]$
2. Initialization of the algorithm:
   - Set the residual for the first iteration, $r_0 = y$, where y is the target variable representing landslide susceptibility.
   - Optionally, use the mean value of y as an initial guess for the model.
3. For $k = 1$ to K, do the following:
   - Randomly select a subsample of the data with size $N' = \eta N$, where N is the total number of instances.
   - Train a decision tree T_k of depth d on the subsample, with the target variable equal to the current residue $r_{\{k-1\}}$.
   - Modify the model by incorporating the decision tree with weight α, that is, $F_{k(x)} = F_{\{k-1\}(x)} + \alpha T_{k(x)}$, where x is the input vector of features.
   - Update the residual by subtracting α times the predictions of the decision tree T_k from $r_{\{k-1\}}$, that is, $r_k = r_{\{k-1\}} - \alpha * T_{k(x)}$.
4. End for loop.

6.7.8   Categorical Boosting

Introduced by Yandex in 2017 (Prokhorenkova et al., 2018), the categorical boosting (CatBoost) algorithm has risen to prominence as a gradient boosting decision tree method. CatBoost works by incorporating ordered target statistics and ordered boosting concepts, marking a significant departure from traditional gradient boosting algorithms. These advancements primarily aim to mitigate prediction shift challenges. The concept of target statistics involves dealing with categorical features through the calculation of summary statistics pertaining to the target variable within each category, thereby minimizing information loss (Micci-Barreca, 2001). Converting categorical attributes into numerical representations includes calculating the mean value of the feature for the initial p records in a manner that accounts for the priority of the features and their associated weight coefficient (Huang et al., 2019; Wei et al., 2023).

$$x_{\sigma p,\,k} = \frac{\sum_{j}^{p-1}\left[x_{\sigma j,k} = x_{\sigma p,k}\right]Y_{\sigma j} + a \cdot P}{\sum_{j}^{p-1}\left[x_{\sigma j,k} = x_{\sigma p,k}\right] + a} \tag{6.63}$$

where the value of p is fixed and represents prior information, while a is a positive coefficient that determines the weight of the prior term in the calculation. The notation $[\cdot]$ represents an indicator function that outputs 1 if a certain condition is met and 0 otherwise.

In binary classification scenarios, target encoding assigns each category of a categorical variable a value representing the likelihood that the target variable belongs to a specific class. In contrast to conventional gradient boosting algorithms, CatBoost utilizes an approach in which categorical attributes are encoded during the tree-splitting phase rather than during the preprocessing stage. This feature empowers CatBoost to independently learn how to deal with categorical attributes during the training phase, thus eliminating the requirement for extensive preprocessing steps. The second enhancement in CatBoost, in contrast to typical gradient boosting algorithms, is the introduction of an ordered boosting approach to reduce the influence of target leakage effects. Within CatBoost, the ordered boosting technique involves the preordering of data samples before the creation of individual decision trees. This order helps CatBoost prevent the assessment of a candidate tree with instances that have already been used in building that tree. Consequently, CatBoost systematically incorporates all accessible examples as it advances through the boosting process.

6.7.9 Natural Gradient Boosting

Being a more recent algorithm than its counterparts, the natural gradient boosting (NGBoost), introduced by Duan et al. (2019), is uniquely designed for tasks involving probabilistic predictions and distinguishes itself through its use of a multiparameter boosting strategy and natural gradients. Unlike deterministic models, NGBoost extends its functionality to include the generation of a probability distribution encompassing various possible outcomes. This additional feature provides a deeper insight into the predictions, elucidating both their magnitude and associated uncertainty. The optimization process in NGBoost is facilitated by a natural gradient algorithm that takes into account the geometry of the parameter space and focuses on enhancing a variational lower bound for the log-likelihood. The model begins by initializing the probability distribution using parameter θ based on the marginal distribution of annotations within the data set:

$$\theta(0) = \arg\min_\theta \sum_{i=1}^{n} S(\theta, y_i), \tag{6.64}$$

where S is a proper scoring rule. For each boosting iteration $m = 1, 2, \ldots, M$, the algorithm first computes negative gradients. For each data point i, the negative gradient of the scoring rule S to the current distribution parameter $\theta_i^{(m-1)}$ and the true label y_i is computed:

$$g_i^{(m)} = -\nabla_\theta S\left(\theta_i^{(m-1)}, y_i\right) \tag{6.65}$$

A set of weak learners (in this case, base learners) are then fit to the negative gradients $g_i^{(m)}$, and the output of the weak learner for data point i is denoted by $f^{(m)}(x_i)$. Finally, the algorithm determines the optimal scaling factor $\rho^{(m)}$ and updates the distribution parameter $\theta_i^{(m-1)}$ for each data point i:

$$\rho^{(m)} = \arg\min_\rho \sum_{i=1}^{n} S\left(\theta_i^{(m-1)} - \rho \cdot f^{(m)}(x_i), y_i\right) \tag{6.66}$$

$$\theta_i^{(m)} = \theta_i^{(m-1)} - \eta\left(\rho^{(m)} f^{(m)}(x_i)\right) \tag{6.67}$$

where η is the learning rate. The final output of the algorithm is the set of scaling and base learners $\left\{\rho^{(m)} f^{(m)}\right\}$ for all boosting iterations m.

6.8 CONCLUDING REMARKS

Several decision tree induction algorithms that have been used for remotely sensed image classification are discussed in this chapter. The performance of these algorithms may depend on the choice of criteria to be used or on certain predefined parameters (information gain or Gini index) or statistical tests (chi-square or F-test). Decision boundaries may be either univariate or multivariate. To minimize the risk of overfitting generally seen in the process of decision tree induction, several pruning techniques are discussed in detail. Due to their hierarchical characteristics, decision trees are very useful for data mining and knowledge discovery. This is a great advantage over the currently popular deep learning models, which are known for their black-box nature. Knowledge mining has received considerable attention in recent years, and developments in the use of decision trees in this application area should contribute to the developments in remote sensing image classification. This chapter provides the knowledge necessary for readers to perform image classification using decision trees.

Alongside these fundamental skills, newly introduced advanced decision tree models such as XGBoost, CatBoost, LightGBM, NGBoost, and GBM can be employed to pursue robust learning. It is evident that cutting-edge decision tree models will enhance the already-established reputation of RF, which serves as a benchmark in image classification. To assess the current status of decision trees in classifying remotely sensed images, a literature review was conducted using the Scopus database, focusing on publications from 2015 to October 2023. Since the focus of the investigation was to put forward their use in image classification, the articles published in ten major remote sensing journals, searching their titles, abstracts, and keywords, were derived from the database. The journals selected for this purpose are *GIScience and Remote Sensing, IEEE Geoscience and Remote Sensing Letters, IEEE Journal of Selected Topics in Applied Earth Observations and Remote Sensing, IEEE Transactions on Geoscience and Remote Sensing, International Journal of Remote Sensing, ISPRS Journal of Photogrammetry and Remote Sensing, Photogrammetric Engineering and Remote Sensing, Remote Sensing, Remote Sensing Letters, and Remote Sensing of Environment.* As an initial search in the Scopus database, the "decision tree*" keyword was set to seek the term in the title, abstract, or keywords of the articles. Starting with 103 articles in 2015, the number of studies employing decision trees increased to 313 in 2022, reaching a total of 1,655 articles since 2015. In an in-depth analysis of the use of tree-based methods in remote sensing, separate searches were conducted for eight decision tree models, as shown in Figure 6.16. Some striking results can be drawn when the figure is analyzed. First, the substantial number of publications utilizing the RF (90.70%) indicates its strong preference and reliance among users. Therefore, it can be stated that it has become a benchmark method in image classification. However, the year-wise analysis revealed that the rapid increase in the number of publications has slowed down in recent years, which can be attributed to the growing popularity of deep learning models, which has also been observed in other scientific fields in recent years. Second, the popularity of the XGBoost algorithm (4.44%) can be deduced, although it is a relatively recent technique. Finally, it is evident that the most recent NGBoost algorithm has yet to be applied in classification studies reported in these journals.

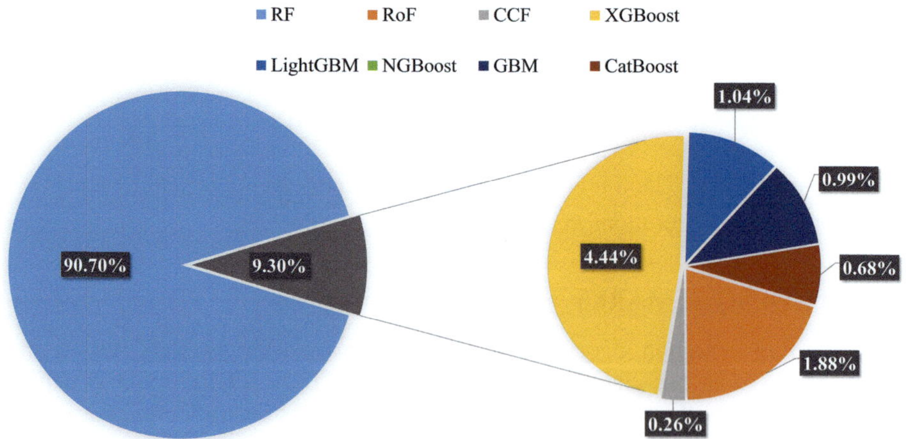

FIGURE 6.16   Literature survey for decision tree-based models.

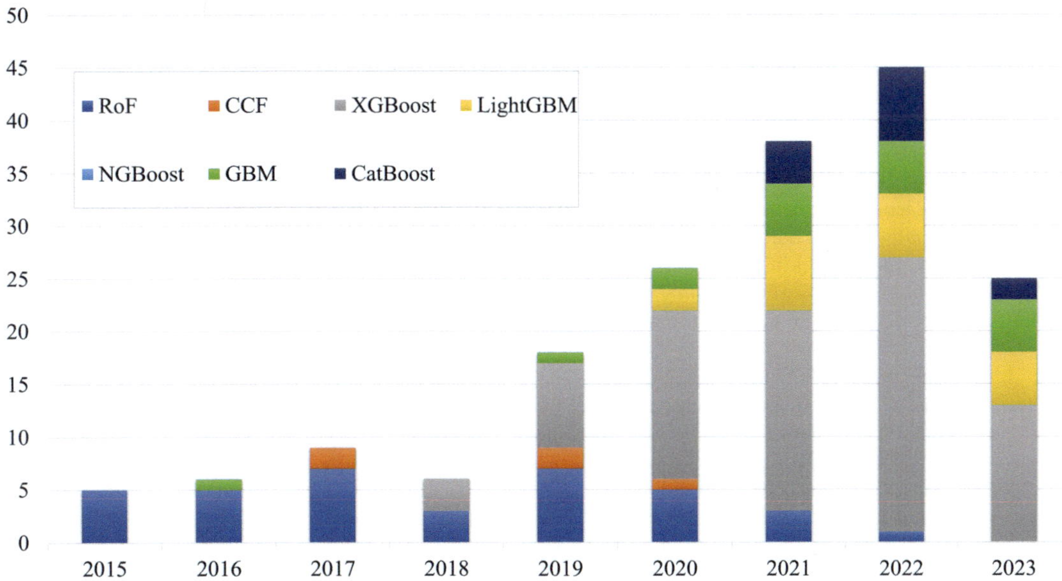

FIGURE 6.17 Number of articles using decision tree-based models.

In addition to the total publication analysis, article statistics for the tree-based methods were examined to reveal their temporal usage trends, starting from 2015 (Figure 6.17). An important observation is that the RoF and CCF methods (improved versions of RF), which have been favored in a small number of publications, have not been used in recent years. Among the latest boosting algorithms, it is evident that XGBoost is among the most popular methods, garnering increased attention from the scientific community. On the other hand, both the GBM and LightGBM methods are favored by users to a similar extent. Given that CatBoost is a relatively newer addition compared with these methods, it has only appeared in several publications since 2021. Considering that the average duration between the introduction of a tree model and its initial publication in remote sensing journals typically spans ~3 years, it is expected that studies employing state-of-the-art tree-based models, particularly NGBoost, for image classification will emerge from this point onward. From the current status of the tree-based algorithms, as discussed above, it can be inferred that the recent ensemble methods are in the promising stage and are more likely to be applied in more research in the remote sensing field. These algorithms present invaluable opportunities to the practitioners because they offer a high level of explainability that helps to understand the model behavior and decision-making process.

REFERENCES

Belgiu, M., and L. Drăguţ. 2016. Random forest in remote sensing: A review of applications and future directions. *ISPRS Journal of Photogrammetry and Remote Sensing* 114:24–31. https://doi.org/10.1016/j.isprsjprs.2016.01.011.

Bonissone, P., J. M. Cadenas, M. C. Garrido, and R. A. Díaz-Valladares. 2010. A fuzzy random forest. *International Journal of Approximate Reasoning* 51:729–747. https://doi.org/10.1016/j.ijar.2010.02.003.

Breiman, L. 1996. Bagging predictors. *Machine Learning* 26:123–140. https://doi.org/10.1007/BF00058655.

Breiman, L. 2001. Random forests. *Machine Learning* 45:5–32. https://doi.org/10.1023/A:1010933404324.

Breiman, L., J. H. Friedman, R. A. Olsen, and C. J. Stone. 1984. *Classification and Regression Trees*. Belmont, CA: Wadsworth. https://doi.org/10.1201/9781315139470.

Cestnik, B., and I. Bratko. 1991. On estimating probabilities in tree pruning. In *Machine Learning: EWSL-91*, (ed.) Y. Kodratoff, pp. 138–150. Berlin, Heidelberg: Springer. https://doi.org/10.1007/BFb0017010.

Chen, C., Q. Zhang, Q. Ma, and B. Yu. 2019. LightGBM-PPI: Predicting protein-protein interactions through LightGBM with multi-information fusion. *Chemometrics and Intelligent Laboratory Systems* 191:54–64. https://doi.org/10.1016/j.chemolab.2019.06.003.

Chen, T., and C. Guestrin. 2016. XGBoost: A scalable tree boosting system. In Proceedings *of the 22nd ACM SIGKDD International Conference on Knowledge Discovery and Data Mining*, August 13–17, San Francisco, CA, pp. 785–794. https://doi.org/10.1145/2939672.2939785.

Colkesen, I., and M. Y. Ozturk. 2022. A comparative evaluation of state-of-the-art ensemble learning algorithms for land cover classification using WorldView-2, Sentinel-2 and ROSIS imagery. *Arabian Journal of Geosciences* 15:942. https://doi.org/10.1007/s12517-022-10243-x.

Colkesen, I., and O. H. Ertekin. 2020. Performance analysis of advanced decision forest algorithms in hyperspectral image classification. *Photogrammetric Engineering & Remote Sensing* 86:571–580.

Colkesen, I., and T. Kavzoglu. 2017. Ensemble-based canonical correlation forest (CCF) for land use and land cover classification using Sentinel-2 and Landsat OLI imagery. *Remote Sensing Letters* 8:1082–1091. https://doi.org/10.1080/2150704X.2017.1354262.

Colkesen, I., and T. Kavzoglu. 2019. Comparative evaluation of decision-forest algorithms in object-based land use and land cover mapping. In *Spatial Modeling in GIS and R for Earth and Environmental Sciences* (eds.) H. R. Pourghasemi, and C. Gokceoglu, pp. 499–517. Amsterdam, Netherlands: Elsevier Academic Press. https://doi.org/10.1016/b978-0-12-815226-3.00023-5.

Costa, V. G., and C. E. Pedreira. 2023. Recent advances in decision trees: An updated survey. *Artificial Intelligence Review* 56:4765–4800. https://doi.org/10.1007/s10462-022-10275-5.

Craven, M. W. 1996. Extracting comprehensible models from trained neural networks. Ph.D. Thesis, University of Wisconsin, Madison, WI.

Du, P., A. Samat, B. Waske, S. Liu, and Z. Li. 2015. Random forest and rotation forest for fully polarized SAR image classification using polarimetric and spatial features. *ISPRS Journal of Photogrammetry and Remote Sensing* 105:38–53. https://doi.org/10.1016/j.isprsjprs.2015.03.002.

Duan, T., A. Avati, D. Y. Ding, S. Basu, K. K. Thai, S. Basu, A., Ng, and A. Schuler. 2019. NGBoost: Natural gradient boosting for probabilistic prediction. In *Proceedings of the 37th International Conference on Machine Learning*, July 13–18, Virtual, pp. 2690–2700.

Duch, W., R. Adamczak, and K. Grabczewski. 2001. A new methodology of extraction, optimization, and application of crisp and fuzzy logical rules. *IEEE Transactions on Neural Networks* 12:277–306. https://doi.org/10.1109/72.914524.

Esposito, F., D. Malerba, and G. Semeraro. 1997. A comparative analysis of methods for pruning decision trees. *IEEE Transactions on Pattern Analysis and Machine Intelligence* 19:476–491. https://doi.org/10.1109/34.589207.

Freund, Y., and R. E. Schapire. 1999. A short introduction to boosting. *Journal of Japanese Society for Artificial Intelligence* 14:771–780.

Friedl, M. A., and C. E. Brodley. 1997. Decision tree classification of land cover from remotely sensed data. *Remote Sensing of Environment* 61:399–409. https://doi.org/10.1016/S0034-4257(97)00049-7.

Friedl, M. A., C. E. Brodley, and A. H. Strahler. 1999. Maximizing land cover classification accuracies produced by decision trees at continental to global scales. *IEEE Transactions on Geoscience and Remote Sensing* 37:969–977. https://doi.org/10.1109/36.752215.

Friedman, J. H. 2001. Greedy function approximation: A gradient boosting machine. *Annals of Statistics* 29:1189–1232. https://doi.org/10.1214/aos/1013203451.

Friedman, J., R. Tibshirani, and T. Hastie. 2000. Additive logistic regression: A statistical view of boosting (with discussion and a rejoinder by the authors). *The Annals of Statistics* 28:337–407. https://doi.org/10.1214/aos/1016120463.

Gislason, P. O., J. A. Benediktsson, and J. R. Sveinsson. 2006. Random forests for land cover classification. *Pattern Recognition Letters* 27:294–300. https://doi.org/10.1016/j.patrec.2005.08.011.

Ha, N. T., M. Manley-Harris, T. D. Pham, and I. Hawes. 2020. A comparative assessment of ensemble-based machine learning and maximum likelihood methods for mapping seagrass using Sentinel-2 imagery in Tauranga Harbor, New Zealand. *Remote Sensing* 12:355. https://doi.org/10.3390/rs12030355.

Harrell, F. E. 2001. *Regression Modeling Strategies: With Applications to Linear Models, Logistic Regression, and Survival Analysis*. New York: Springer.

Horning, N. 2010. Random forests: An algorithm for image classification and generation of continuous fields data sets. In *Proceedings of the International Conference on Geoinformatics for Spatial Infrastructure Development in Earth and Allied Sciences*, December 9–11, Hanoi, Vietnam.

Huang, G., L. Wu, X. Ma, W. Zhang, J. Fan, and X. Yu. 2019. Evaluation of CatBoost method for prediction of reference evapotranspiration in humid regions. *Journal of Hydrology* 574:1029–1041. https://doi.org/10.1016/j.jhydrol.2019.04.085.

Jackson, Q., and D. Landgrebe. 2001. An adaptive classifier design for high dimensional data analysis with a limited training data set. *IEEE Transactions on Geoscience and Remote Sensing* 39:2664–2679. https://doi.org/10.1109/36.975001.

Kass, G. V. 1980. An exploratory technique for investigating large quantities of categorical data. *Applied Statistics* 29:119–127. https://doi.org/10.2307/2986296.

Kavzoglu, T. 2017. Object-oriented random forest for high resolution land cover mapping using quickbird-2 imagery. In *Handbook of Neural Computation*, (eds.) P. Samui, S. Sekhar, and V. E. Balas, pp. 607–619. London, UK: Academic Press. https://doi.org/10.1016/B978-0-12-811318-9.00033-8.

Kavzoglu, T., and A. Teke. 2022. Predictive Performances of ensemble machine learning algorithms in landslide susceptibility mapping using random forest, extreme gradient boosting (XGBoost) and natural gradient boosting (NGBoost). *Arabian Journal for Science and Engineering* 47:7367–7385. https://doi.org/10.1007/s13369-022-06560-8.

Kavzoglu, T., and I. Colkesen. 2013. An assessment of the effectiveness of a rotation forest ensemble for land-use and land-cover mapping. *International Journal of Remote Sensing* 34:4224–4241. https://doi.org/10.1080/01431161.2013.774099.

Kavzoglu, T., I. Colkesen, and T. Yomralioglu. 2015. Object-based classification with rotation forest ensemble learning algorithm using very-high-resolution WorldView-2 image. *Remote Sensing Letters* 6:834–843. https://doi.org/10.1080/2150704X.2015.1084550.

Ke, G., Q. Meng, T. Finley, T. Wang, W. Chen, W. Ma, Q. Ye, and T. Y. Liu. 2017. LightGBM: A highly efficient gradient boosting decision tree. In *Proceedings of the 31st International Conference on Neural Information Processing Systems*, December 4–9, Long Beach, CA, pp. 3147–3155.

Kim, B., and D. A. Landgrebe. 1991. Hierarchical classifier design in high-dimensional, numerous class cases. *IEEE Transactions on Geoscience and Remote Sensing* 29:518–528.

Krishnan, R., G. Sivakumar, and P. Bhattacharya. 1999. Extracting decision trees from trained neural networks. *Pattern Recognition* 32:1999–2009. https://doi.org/10.1016/S0031-3203(98)00181-2.

Kulkarni, A. V., and N. K. Laveen. 1976. An optimisation approach to hierarchical classifier design. In *Proceedings of the Third International Joint Conference on Pattern Recognition*, Piscataway, NJ, pp. 459–466.

Kuncheva, L. I. 2014. *Combining Pattern Classifiers: Methods and Algorithms*, 2nd edition. Hoboken, NJ: John Wiley & Sons. https://doi.org/10.1002/0471660264.

Kuncheva, L. I., and C. J. Whitaker. 2003. Measures of diversity in classifier ensembles and their relationship with the ensemble accuracy. *Machine Learning* 51:181–207. https://doi.org/10.1023/A:1022859003006.

Kurzynski, M. W. 1983. The optimal strategy of a tree classifier. *Pattern Recognition* 16:81–87. https://doi.org/10.1016/0031-3203(83)90011-0.

Lee, T., and J. A. Richards. 1985. A low-cost classifier for multitemporal applications. *International Journal of Remote Sensing* 6:1405–1417. https://doi.org/10.1080/01431168508948286.

Loh, W. Y., and Y. S. Shih. 1997. Split selection methods for classification trees. *Statistica Sinica* 7:815–840.

McIver, D. K., and M. A. Friedl. 2002. Using prior probabilities in decision-tree classification of remotely sensed data. *Remote Sensing of Environment* 81:253–261. https://doi.org/10.1016/S0034-4257(02)00003-2.

Micci-Barreca, D. 2001. A preprocessing scheme for high-cardinality categorical attributes in classification and prediction problems. *SIGKDD Explorations Newsletter* 3:27–32. https://doi.org/10.1145/507533.507538.

Oshiro, T. M., P. S. Perez, and J. A. Baranauskas. 2012. How many trees in a random forest? In *Machine Learning and Data Mining in Pattern Recognition*, (ed.) P. Perner, pp. 154–168. Berlin, Heidelberg: Springer. https://doi.org/10.1007/978-3-642-31537-4_13.

Pal, M. 2005. Random forest classifier for remote sensing classification. *International Journal of Remote Sensing* 26:217–222. https://doi.org/10.1080/01431160412331269698.

Pal, M., and G. M. Foody. 2010. Feature selection for classification of hyperspectral data by SVM. *IEEE Transactions on Geoscience and Remote Sensing* 48:2297–307. https://doi.org/10.1109/TGRS.2009.2039484.

Pal, M., and P. M. Mather. 2003. An assessment of the effectiveness of decision tree methods for land cover classification. *Remote Sensing of Environment* 86:554–565. https://doi.org/10.1016/S0034-4257(03)00132-9.

Probst, P., M. N. Wright, and A.-L. Boulesteix. 2019. Hyperparameters and tuning strategies for random forest. *WIREs Data Mining and Knowledge Discovery* 9:e1301. https://doi.org/10.1002/widm.1301.

Prokhorenkova, L., G. Gusev, A. Vorobev, A. V. Dorogush, and A. Gulin. 2018. CatBoost: Unbiased boosting with categorical features. In *Proceedings of the 32nd International Conference on Neural Information Processing Systems*, December 3–8, Montréal, Canada, pp. 6639–6649.

Quinlan, J. R. 1979. Discovering rules by induction from large collections of examples. In *Expert Systems in the Micro-Electronic Age*, (ed.) D. Michie, pp. 168–201. Edinburgh, Scotland: Edinburgh University Press.

Quinlan, J. R. 1986. Induction of decision trees. *Machine Learning* 1:81–106. https://doi.org/10.1007/BF00116251.

Quinlan, J. R. 1993. *C4.5: Programs for Machine Learning*. San Mateo, CA: Morgan Kaufmann.

Quinlan, J. R. 1996. Bagging, boosting and C4.5. In *Thirteenth National Conference on Artificial Intelligence*, August 4–8, Portland, OR, pp. 725–730.

Rainforth, T., and F. Wood. 2015. Canonical correlation forests. *arXiv preprint*, arXiv:1507.05444. https://doi.org/10.48550/arXiv.1507.05444.

Rodríguez, J. J., L. I. Kuncheva, and C. J. Alonso. 2006. Rotation forest: A new classifier ensemble method. *IEEE Transactions on Pattern Analysis and Machine Intelligence* 28:1619–1630. https://doi.org/10.1109/TPAMI.2006.211.

Rokach, L. 2010. Ensemble-based classifiers. *Artificial Intelligence Review* 33:1–39. https://doi.org/10.1007/s10462-009-9124-7.

Rokach, L. 2016. Decision forest: Twenty years of research. *Information Fusion* 27:111–125. https://doi.org/10.1016/j.inffus.2015.06.005.

Saeys, Y., T. Abeel, and Y. Van der Peer. 2008. Robust feature selection using ensemble feature selection techniques. In *Machine Learning and Knowledge Discovery in Databases* (eds.) W. Daelemans, B. Goethals, and K. Morik, pp. 313–325. Berlin, Heidelberg: Springer. https://doi.org/10.1007/978-3-540-87481-2_21.

Safavian, S. R., and D. Landgrebe. 1991. A survey of decision tree classifier methodology. *IEEE Transactions on Systems, Man and Cybernetics* 21:660–674. https://doi.org/10.1109/21.97458.

Sahin, E. K., I. Colkesen, and T. Kavzoglu. 2020. A comparative assessment of canonical correlation forest, random forest, rotation forest and logistic regression methods for landslide susceptibility mapping. *Geocarto International* 35:341–363. https://doi.org/10.1080/10106049.2018.1516248.

Saini, R., and S. K. Ghosh. 2017. Ensemble classifiers in remote sensing: A review. In 2017 *International Conference on Computing, Communication and Automation (ICCCA)*, May 5–6, Greater Noida, India, pp. 1148–1152. https://doi.org/10.1109/CCAA.2017.8229969.

Sato, M., and H. Tsukimoto, 2001. Rule extraction from neural networks via decision tree induction. *Proceedings of International Joint Conference on Neural Networks* (Cat. No.01CH37222) pp. 1870–1875, https://doi.org/10.1109/IJCNN.2001.938448.

Schapire, R. E., and Y. Singer. 1998. BoosTexter: A system for multiclass multi-label text categorization. *Machine Learning* 39:135–168. https://doi.org/10.1023/A:1007649029923.

Schmitz, G. P., C. Aldrich, and F. S. Gouws. 1999. ANN-DT: An algorithm for extraction of decision trees from artificial neural networks. *IEEE Transactions on Neural Networks* 10:1392–1401. https://doi.org/10.1109/72.809084.

Sheykhmousa, M., M. Mahdianpari, H. Ghanbari, F. Mohammadimanesh, P. Ghamisi, and S. Homayouni. 2020. Support vector machine versus random forest for remote sensing image classification: A meta-analysis and systematic review. *IEEE Journal of Selected Topics in Applied Earth Observations and Remote Sensing* 13:6308–6325. https://doi.org/10.1109/JSTARS.2020.3026724.

Swain, P. H., and H. Hauska. 1977. The decision tree classifier: Design and potential. *IEEE Transactions on Geoscience and Remote Sensing* 15:142–147. https://doi.org/10.1109/TGE.1977.6498972.

Taha, I. A., and J. Ghosh. 1999. Symbolic interpretation of artificial neural networks. *IEEE Transactions on Knowledge and Data Engineering* 11:448–463. https://doi.org/10.1109/69.774103.

Tonbul, H., I. Colkesen, and T. Kavzoglu. 2020. Classification of poplar trees with object-based ensemble learning algorithms using Sentinel-2A imagery. *Journal of Geodetic Science* 10:14–22. https://doi.org/10.1515/jogs-2020-0003.

Touzani, S., J. Granderson, and S. Fernandes. 2018. Gradient boosting machine for modeling the energy consumption of commercial buildings. *Energy and Buildings* 158:1533–1543. https://doi.org/10.1016/j.enbuild.2017.11.039.

Towell, G., and J. Shavlik. 1994. Knowledge-based artificial neural networks. *Artificial Intelligence* 70:119–165. https://doi.org/10.1016/0004-3702(94)90105-8.

Wei, X., C. Rao, X. Xiao, L. Chen, and M. Goh. 2023. Risk assessment of cardiovascular disease based on SOLSSA-CatBoost model. *Expert Systems with Applications* 219:119648. https://doi.org/10.1016/j.eswa.2023.119648.

Wen, X., Y. Xie, L. Wu, and L. Jiang. 2021. Quantifying and comparing the effects of key risk factors on various types of roadway segment crashes with LightGBM and SHAP. *Accident Analysis & Prevention* 159:106261. https://doi.org/10.1016/j.aap.2021.106261.

Xia, J., N. Yokoya, and A. Iwasaki. 2017. Hyperspectral image classification with canonical correlation forests. *IEEE Transactions on Geoscience and Remote Sensing* 55:421–431. https://doi.org/10.1109/TGRS.2016.2607755.

Xia, J., P. Du, X. He, and J. Chanussot. 2013. Hyperspectral remote sensing image classification based on rotation forest. *IEEE Geoscience and Remote Sensing Letters* 11:239–243. https://doi.org/10.1109/LGRS.2013.2254108.

Zhang Y, J. Liu, and W. Shen. 2022. A review of ensemble learning algorithms used in remote sensing applications. *Applied Sciences* 12:8654. https://doi.org/10.3390/app12178654.

7 Deep Learning

The success of human eye-brain interaction in solving pattern recognition issues has inspired researchers to investigate whether computer systems, using a simplified brain model, could surpass the effectiveness of standard statistical classification methods. Ongoing research has led to the creation of artificial neural networks (ANNs), and their application in remote sensing has expanded significantly, covering diverse research problems, including scene classification, object detection, semantic segmentation, and change detection. The significant advantage of neural networks lies in their ability to achieve a high computation rate through massive parallelism. This is made possible by the dense arrangement of interconnected and simple processors (neurons). Due to their nonparametric nature, these methods operate without assuming any statistical distribution of the data. Compared to other machine learning algorithms, they are less dependent on feature engineering. The performance of a neural network is heavily dependent on how well it has been trained to understand the problem and apply this knowledge to new data. During the training process, the neural network learns regularities and inherent relationships in the training data, creating rules based on this distinctive information that can be applied to unseen data. Several developments have made vital contributions to the development of ANNs. First, the introduction of backpropagation, a gradient-based optimization algorithm (Rumelhart et al., 1986), was a crucial breakthrough in neural network training, enabling the training of neural networks with one or two hidden layers and a customizable number of neurons. Second, the use of activation functions that introduce nonlinearity into the network was a critical contribution to continuous development in the field. Last but not least, the introduction of convolutional neural networks and their use as a feature extraction method has opened new avenues in the field. Starting with the fundamental neural network architectures of the multilayer perceptron (MLP) or feedforward neural network, self-organized maps (SOM), Hopfield networks, and adaptive resonance theory (ART), ANNs have recently evolved into deeper forms of networks, including convolutional neural networks (CNNs), recurrent neural networks (RNNs), vision transformers (ViTs), generative adversarial networks (GANs), and deep multilayer perceptron (Deep MLP), which have much more successively connected layers in their structures.

In comparison with shallow neural network models, deep learning (DL), a typical example of representation learning (Bengio et al., 2013), has dominated all scientific disciplines, and its effect on image processing and computer vision has been outstanding. The DL era began in 2012 after the pioneering CNN model AlexNet won the ImageNet competition. The increased number of layers and application of various transformations allow for the exploration of higher-level features since the 2010s and more abstract concepts, revealing intricate and hierarchical relationships within the data. This increased depth enhances the extraction of extensive low-level and high-level semantic information existing in remotely sensed data. The key aspects of a DL model include stacked layers of nonlinear functions that are trained to extract features from input data, allowing predictions based on automatically identified features rather than handcrafted ones. How a model addresses the underlying structure of the data and identifies features is determined by its type and architecture. These models can achieve state-of-the-art performance, especially with high-dimensional and noisy data. Nonetheless, this tendency is not without limits, and the risk of overfitting arises when augmenting the model's depth by adding more layers. In addition, the requirement for more training data and elevated computational costs becomes essential as transformations and complexity levels increase. Despite all these shortcomings, DL has gained popularity, primarily driven by several technical advancements, including robust data processing techniques (e.g., data augmentation, new nonlinear activation functions), high-performance graphics cards, cloud computing, and open data initiatives that provide annotated data for various remote sensing tasks. These advancements

DOI: 10.1201/9781003439172-7

enable the effective computation of nonlinear transformations on the input data, characterized by its capability for end-to-end learning (Kattenborn et al., 2021). Ultimately, the development of deep networks enhances data representations intelligently and automatically.

Standard data analysis approaches in remote sensing typically require feature engineering, involving the heuristic selection of appropriate transformations and the manual derivation of latent variables from the input data before entering the modeling phase. For instance, the use of various spectral indices, application of feature extraction, and selection techniques together with the estimation of other handcrafting features were major agendas in most classification practices. The existence of a large variety of user-dependent choices in the establishment of such variables has been a major challenge when producing classification results and comparing them with others when produced with different choices and settings. This issue has long threatened the reliability of the results because the predictions of most models could not be repeated under the same parameter setting of the machine learning method. On the other hand, DL networks can handle data considerations, including the extraction of numerous low-level and high-level features, and the trained networks carrying all acquired knowledge can be used by any user to conduct the same or similar analysis by benefitting from previous experience. Given these attributes, DL networks have unquestionably leveraged our capabilities to the utmost level for analyzing complex spatial patterns inherent in remotely sensed data. Because of their adaptability and versatility, researchers have developed various designs for handling remote sensing data across spectral, spatial, and temporal dimensions.

Although many DL architectures have been developed over the past decade, CNNs are currently the most popular architecture for pattern recognition. These types of networks have been developed with the goal of learning spatial features, such as edges, corners, textures, or more abstract shapes, which offer the most precise description of the target class. Applying convolutional filters in the image domain, the effectiveness of CNNs lies in their ability to extract abstract features spanning low to high levels from raw images, achieved through the interleaving of convolutional and pooling layers, progressively diminishing the dimensions of the feature maps. The advanced spatial information extraction capabilities of CNNs play a crucial role in achieving remarkably high classification accuracy for remotely sensed data. In recent years, the use of RNNs with their variants (LSTM and GRU) and ViTs has expanded into different applications in remote sensing. In particular, the LSTM model, as a specialized form of RNN, addresses the vanishing gradient issue encountered by traditional RNNs and excels in capturing long-term relationships. Featuring LSTM cells governed by input, forget, and output gates, the model functions similarly to the nodes in a multilayer perceptron. Stacked LSTM cells create layers that enable bidirectional or unidirectional processing in both forward and backward directions. Representing a newer generation of RNNs, the GRU exhibits subtle differences from the LSTM by employing two gates (update and reset) instead of three, with a unique connectivity scheme. Functioning as a streamlined alternative to the LSTM, the GRU is characterized by its lightweight architecture. On the other hand, the recent introduction of ViTs, which incorporates self-attention mechanisms, has been demonstrated to leverage more global information compared with CNN-based models in image classification. This use of increased global information enables ViTs to achieve equivalent or higher performance compared with CNNs across diverse image classification tasks. Recently, another significant contribution to DL has been the emergence of deep MLPs that operate without relying on convolution or attention mechanisms. Despite this, they are capable of delivering performance comparable to CNNs and ViTs. Several recently published papers have reviewed the use of DL models in processing remote sensing data (Zhu et al., 2017; Li et al., 2018; Ma et al., 2019; Song et al., 2019; Yuan et al., 2021).

This chapter delves into the fundamental components, learning methodologies, and key structures of DL models, along with their practical use in analyzing remote sensing images. Detailed descriptions are provided for key architectures that have significantly impacted the field, and diverse applications are illustrated with examples. Furthermore, subsequent sections offer contemporary perspectives on the primary applications of DL in remote sensing, including semantic segmentation, object detection, scene classification, and change detection.

7.1 FUNDAMENTALS

7.1.1 Stochastic Gradient Descent

Batch gradient descent algorithms can become inefficient if there are many samples in the training data set because the entire set is considered for each error function or gradient evaluation in the search for global minimum. Instead of this approach, a stochastic version of the gradient descent, where the error surface is estimated by considering only a single sample at each iteration, is employed. Descending on this stochastic surface results in a substantial enhancement of the ability to navigate flat regions. Consequently, stochastic gradient descent is the most widely used learning algorithm in DL networks (Goodfellow et al., 2016). Assume that $f_i(\mathbf{x})$ is the loss function for the training example of index I, which can be written as follows:

$$f(\mathbf{x}) = \frac{1}{n}\sum_{i=1}^{n} f_i(\mathbf{x}).$$
(7.1)

The gradient of the objective function at $\mathbf{x}$ is computed as

$$\nabla f(\mathbf{x}) = \frac{1}{n}\sum_{i=1}^{n}\nabla f_i(\mathbf{x}).$$
(7.2)

The cost of gradient descent for each iteration with respect to independent variables is $\mathcal{O}(n)$, growing linearly with the size of the training data set. The computational cost is reduced by stochastic gradient descent at each iteration by uniformly sampling an index $i \in \{1,\ldots,n\}$ for data examples at random (Zhang et al., 2023). Then, the gradient $\nabla f_i(\mathbf{x})$ is computed to update $\mathbf{x}$:

$$\mathbf{x} \leftarrow x - \eta\nabla f_i(\mathbf{x}),$$
(7.3)

where η is the learning rate. Thus, the computational cost for each iteration drops from $\mathcal{O}(n)$ of the gradient descent to the constant $\mathcal{O}(1)$. Note that the stochastic gradient $\nabla f_i(\mathbf{x})$ is an unbiased estimate of the full gradient $\nabla f(\mathbf{x})$, which can be expressed as

$$\mathbb{E}_i\nabla f_i(\mathbf{x}) = \frac{1}{n}\sum_{i=1}^{n}\nabla f_i(\mathbf{x}) = \nabla f(\mathbf{x}).$$
(7.4)

The shortcoming of the approach is the estimation of error for each sample, which can take considerable time. To overcome this bottleneck, a minibatch gradient descent algorithm is proposed. Every iteration in minibatch gradient descent computes the error surface for a subset of the entire data set (rather than just one sample). The minibatch size is another hyperparameter, along with the learning rate, for this subset, which is referred to as a minibatch. Minibatches provide a compromise between the local-minima avoidance provided by stochastic gradient descent and the efficiency of batch gradient descent (Buduma and Lacascio, 2017). The mathematical expression of minibatch gradient descent can be given below:

$$\Delta w_{ij} = -\sum_{k\in \min i\, \text{batch}}\int y_i^{(k)}y_j^{(k)}\left(1-y_j^{(k)}\right)\frac{\partial E^{(k)}}{\partial y_j^{(k)}}.$$
(7.5)

7.1.2 Backpropagation

The backpropagation algorithm, introduced by Rumelhart et al. (1986), has been the widely adopted method for updating neuronal activities and interconnection weights in MLPs. As an iterative gradient algorithm, it involves two major steps, called forward and backward propagation, to successfully modify the neural state. Throughout the training period, each sample (such as a feature vector linked to an individual training pixel) is introduced into the input layer, and neuron activities are systematically updated from the input to the output layer through a mapping function. Following the completion of the forward pass, the activities of the output neurons are matched against their expected values. If, for instance, there are five classes, and training pixel i is assumed to belong to class 1, the anticipated output pattern for the five output neurons (one for each class) would be "1 0 0 0 0". Typically, the actual output varies from the expected result, resulting in network errors. These errors are propagated backward through the network, originating from the output layer, and the weights are adjusted. The iterative process of forward and backward passes persists until the network successfully learns the underlying characteristics of all classes, constituting the network training process.

During the process of forward propagation, the activities of neurons are updated in a sequential manner, progressing layer-wise from the input to the output layer, ultimately producing output in the form of activations of the output layer neurons. Let x_j denote the total input for neuron j, which is represented by

$$x_j = \sum_i a_i w_{ji}, \tag{7.6}$$

where a_i is the activity of neuron i, and w_{ji} is the weight of the connection from the ith neuron to the jth neuron. Once the value of x_j is computed from Equation (7.6), it is converted to an output value (for transmission to the next layer if the neuron lies on an intermediate layer) using a mapping function, commonly the sigmoid function. It is a monotonically increasing nonlinear function defined by

$$a_j = f\left(x_j\right) = \frac{1}{1+e^{-\left(x_j/T\right)}}. \tag{7.7}$$

T is a parameter called temperature (normally $T=1$), which renders the sigmoidal curve more abrupt ($T<1$) or gradual ($T>1$). The corresponding pattern is illustrated in Figure 7.1a. After the activity of each neuron within the same layer is computed, a similar process is performed on the next adjacent layer. Note that the input layer is a special case because the neurons in the input layer take the values provided by the training sample. For nodes in the input layer, the activity of neuron j is directly set as the jth component of the input pattern vector.

Sometimes an extra neuron, called the bias, is added to the neural network. It links to all the network layers except the input layer. This bias unit has a constant activity of 1 but affects each neuron j via different weight values (e.g., θ_j in Figure 7.1b). If the bias unit is introduced, Equation (7.6) is modified as follows:

$$x_j = \sum_i a_i w_{ji} - \theta_j \tag{7.8}$$

where θ_j is the bias associated with the neuron j. The effect of the bias term θ is to contribute to a left- or rightward shifting of the sigmoid mapping function, as shown in Figure 7.1b, depending on whether the value θ is negative or positive. The introduction of a bias unit into the network is believed to improve the convergence property of the MLP network.

In backward propagation, the interneuron weights are modified starting from the output layer until the input layer is reached, that is, moving left from the rightmost layer in terms of the layout. Weight updating aims to reduce the identification error of the network. In general, the least mean square error criterion is applied. This is defined as

$$E\left(w\right) = \frac{1}{2} \sum_{j,k} \left(a_{j,k} - o_{j,k}\right)^2, \tag{7.9}$$

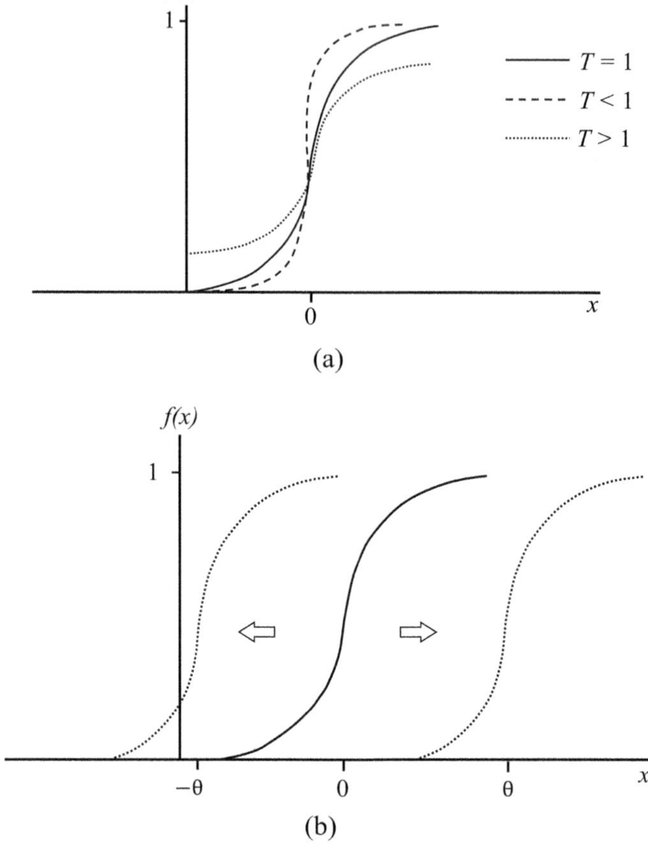

FIGURE 7.1 (a) Shape of the sigmoidal curve varies with parameter T. (b) Location of the sigmoid function has shifted after bias θ is added or subtracted.

where w is a set of weights in a network, $a_{j,k}$ is the jth neuron in the output layer obtained from the kth training sample, and $o_{j,k}$ is the target output at neuron j in the output layer for the kth training sample. Search methods can be applied to find the minimum of the function $E(w)$. One widely used method is that of steepest descent, in which each weight is updated in terms of the expression

$$w_{ji}^{n+1} = w_{ji}^{n} - \eta \frac{\partial E}{\partial w_{ji}^{n}}, \tag{7.10}$$

where w_{ji}^{n} is the weight associated with the connection from neuron i to j at time n, η (the learning rate) is a small positive constant, in the range $0 \leq \eta \leq 1$, controlling the size of the step, and $\partial E / \partial w_{ji}^{n}$ is the gradient at the point w_{ji}^{n}, which can be interpreted as the change in error E that results from a change in weight w_{ji}^{n}. Initial values of the weights, w_{ji}^{O}, are generally small random values. The gradient term in Equation (7.10) can be further expanded to

$$\frac{\partial E}{\partial w_{ji}} = \frac{\partial E}{\partial a_j} \frac{\partial a_j}{\partial x_j} \frac{\partial x_j}{\partial w_{ji}} = \frac{\partial E}{\partial a_j} f'(x_j) a_i = \delta_j a_i \tag{7.11}$$

in terms of the chain rule of calculus, where $f'(x_j)$ is the first derivative of the sigmoid function, a_i is the activity in neurons i, and δ_j is the error value occurring on neurons j. If neurons j are in the output layer, the values of δ_j in Equation (7.11) are simply defined as follows:

$$\delta_j = \left(a_j - o_j\right) f'\left(x_j\right), \tag{7.12}$$

where o_j is the desired output for neurons j. Combining Equations (3.5)–(3.7), the weights w_{ji} for $j=$ output layer can therefore be updated. For the other layers, the δ_j are computed by

$$\delta_j = \left(\sum_k \delta_k w_{kj}\right) f'\left(x_j\right) \tag{7.13}$$

where k denotes the neurons that received output from neurons j. In other words, if neurons j are in the lth layer, then k denotes all neurons in the $(l+1)$th layer. The combination of Equations (7.10), (7.11), and (7.13) determines the adjustment for each weight. This rule for the adjustment of the weights associated with network connections is also known as the generalized δ (delta) rule. Sometimes, a momentum term is added to Equation (7.10) to give

$$w_{ji}^{n+1} = w_{ji}^n - \eta \frac{\partial E}{\partial w_{ji}^n} + \xi\left(w_{ji}^n - w_{ji}^{n-1}\right), \tag{7.14}$$

where $0 \leq \xi \leq 1$. The momentum term is applied to avoid oscillation problems during the search for the minimum value on the error surface and can therefore speed up the convergence procedure.

The effectiveness of an MLP network is influenced by various factors, such as model-specific parameters, network architecture, and the characteristics of the training samples. It is very difficult to choose an optimum combination of those factors to construct an ideal network for a given classification task. The primary challenge associated with the backpropagation learning algorithm is its uncertainty in converging to minimum error, creating the possibility of becoming trapped in local minima on the error surface.

7.1.3 Regularization

The power of a neural network depends on its ability to analyze and make correct predictions on new data after the training process is completed. This is known as the generalization ability of a neural network, which is the main criterion for judging the performance of a network. Producing networks with high generalization capabilities has always been a major agenda in neural network applications. The generalization performance of the model is initially measured using validation and test data sets. Various strategies have been developed specifically to reduce the error level on the test data, often at the expense of higher training errors. The methods used for this purpose are referred to as regularization, which is the most popular technique for preventing overfitting. Currently, there exists a plethora of regularization algorithms to choose from, as substantial efforts in the field have been devoted to improving the efficacy of regularization strategies. When an algorithm performs well on the training data but lacks generalization (i.e., poor performance on test data) indicates overfitting of the model. Improving generalization or avoiding overfitting in neural networks is achieved through regularization, which is a challenging objective for researchers. Numerous regularization strategies are available, ranging from imposing additional constraints on a DL model, such as placing restrictions on parameter values, to adding extra terms in the objective function that can be seen as a soft constraint on parameter values. They may be designed to encapsulate a specific type of prior knowledge or express a general preference for a simpler model class, thereby promoting generalization. Through these penalties and constraints, the underlying problem is comprehended. Another form of regularization, ensemble methods, combines multiple hypotheses explaining the training data (Goodfellow et al., 2016).

Most regularization techniques in DL operate by regulating estimators. Making an estimator more regular involves a trade-off, where additional bias is exchanged for decreased variance. An effective regularization is one that significantly decreases variance without unreasonably elevating bias, demonstrating its success when it results in a profitable transaction. Obtaining the most appropriate model in terms of size and number of parameters is not the only step in determining the complexity of the model. In real-world scenarios, it is often encountered that a large model with appropriate regularization is optimal with minimal generalization error.

Early strategies for detecting overfitting include ensemble techniques that mitigate the risk by combining multiple models, cross-validation methods that identify overfitting by comparing training and validation data in experiments, and penalty methods that penalize complexity models. The ultimate purpose of employing these strategies is to mitigate overfitting during training and enhance the accuracy of test data. Nevertheless, it is essential to be cautious about overfitting when making numerous hyperparameter choices through cross-validation (Ng, 1997). It should be noted that these methods prove particularly advantageous in situations where data are scarce (Moradi et al., 2020). Since it is not known a priori which regularization strategy ensures the highest generalization capability with the lowest possibility of overfitting and underfitting, no single regularization method exists as an optimal solution for the problem at hand. Therefore, some experiments or previous experience on a similar problem is essential to determine the best-performing model with the highest generalization capability.

7.1.3.1 Weight Decay

Adding a penalty function to constrain the model's capacity has been a major regularization approach. Assume that the original objective function J is regularized by a parameter norm penalty $\Omega(\theta)$ to produce a new objective function J', denoted as follows:

$$J'(\theta; X, y) = J(\theta; X, y) + \alpha\Omega(\theta), \tag{7.15}$$

where α is the parameter that controls the degree of contribution of the penalty term Ω. It is possible to apply separate penalty terms to each layer of the network in regularizing neural networks.

The most commonly used weight decay regularization methods are L_1 regularization, also known as *Lasso* regularization, and L_2 regularization, also known as ridge regression or Tikhonov regularization. L_2 regularization adds the sum of the squares of the model parameters to the objective function, whereas the L_1 regularization adds the sum of the absolute values of the model parameters as a penalty term. Regularization using L_1 eliminates irrelevant features, resulting in sparse sets of features, which makes it resilient to outliers because of its intrinsic linear reliance on the model parameters. Thus, in a single optimization step, L_1 regularization integrates effective feature selection and model formation (Demir-Kavuk et al., 2011).

The objective function of a neural network with L_2 is a combination of an unregularized objective and a regularization term

$$L_\lambda(\mathbf{w}) = L(\mathbf{w}) + \lambda\mathbf{w}_2^2. \tag{7.16}$$

The unregularized objective function depends only on the weights through the output of the unit

$$L(\mathbf{w}) = \sum_{i=1}^{N} \ell_i\big(y(X_i; \mathbf{w}, \gamma, \beta)\big) \tag{7.17}$$

where ℓ_i is the loss with respect to the unit's output for a sample i. When using normalization $y(X_i; \alpha\mathbf{w}, \gamma, \beta) = y(X_i; \mathbf{w}, \gamma, \beta)$, the L_2 regularized objective function becomes

$$L(\alpha\mathbf{w}) = \sum_{i=1}^{N} \ell_i\big(y(X_i; \mathbf{w}, \gamma, \beta)\big) + \lambda\mathbf{w}_2^2 = L_{\lambda\alpha^2}(\mathbf{w}). \tag{7.18}$$

This indicates that although the L_2 penalty term drives the weights to decrease, the calculated function does not become simpler because of this regularizing impact (Van Laarhoven, 2017). It aids in decreasing the amplitude of parameters and smoothing out the parameter distribution, making a model less prone to overfitting. Combining L_2 regularization with dropout regularization has shown superior results and is crucial for training deep neural networks (Srivastava et al., 2014). Nevertheless, Van Laarhoven (2017) noted that when batch normalization is incorporated with L_2 regularization, regularization is ineffective. The use of L_1 norms in place of L_2 norms has recently become popular. L_1 regularization has many advantages over L_2 regularization but creates sparse models that are more readily interpreted. For deep neural networks, L_1 regularization and its variations, including group sparsity regularization, are suitable for achieving lower computational costs. The decision to use L_1 or L_2 regularization is contingent on the unique characteristics of the problem, the data involved, and the intended behavior of the model. Practical applications often involve the use of a mixture of L_1 and L_2 regularization, known as elastic net regularization (Zou and Hastie, 2005), to exploit the strengths of both methods and find an optimal compromise between sparsity and weight reduction.

7.1.3.2 Dropout

Investigating the proper size of a neural network for a particular problem has always been a major problem in obtaining high generalization capabilities. In the early days, several dynamic design strategies, including constructive methods (e.g., cascade correlation algorithm by Fahlman and Lebiere, 1989), pruning algorithms (Reed, 1993; Kavzoglu and Mather, 1999), and a combination of both strategies (Hirose et al., 1991), have been developed for this purpose. Pruning is an effective method for removing noncontributing units. With this method, hidden layer nodes or interconnections are removed from the network based on their level of participation in the solution during the training process. A commonly employed method involves utilizing sensitivity analysis to pinpoint nonessential neurons. This process entails setting the value of a specific neuron to zero for all training set inputs and observing the impact on the network output. Consequently, neurons with minimal influence on the network's performance can be identified and eliminated. An alternative approach, introduced by Sietsma and Dow (1991) and termed noncontributing units, is a two-stage method in which the analyst assesses a trained network and decides which neurons should be removed. Mozer and Smolensky (1989a,b) proposed Skeletonization as a method to reduce network size by eliminating nodes in the input and hidden layers using first-order derivatives of the error function. In this approach, the importance of a node is gauged by the alteration in the error function upon removal of the unit. Essentially, the impact of each node in the input and hidden layers on the performance of the network is calculated. Consequently, the less relevant nodes are pruned to form a skeletal version of the network. Castellano et al. (1997) also presented a pruning method to diminish the size of trained feedforward neural networks by progressively eliminating hidden layer neurons and subsequently adjusting the remaining weights to maintain the overall network behavior. This method is formulated using a system of linear equations, and an efficient conjugate gradient algorithm is applied to solve the system in the least squares sense. Sildir et al. (2020) introduced a superstructure-based mixed-integer nonlinear programming approach for optimal structural design, encompassing neuron number selection, pruning, and input selection for neural networks. The method leverages statistical measures, such as the parameter covariance matrix, to enhance test performance while allowing for reduced training performance.

Deep neural networks are highly expressive models that are capable of comprehending intricate interactions between inputs and outputs through multiple hidden layers. Even if the test data are taken from the same distribution, many of these intricate correlations will be missing in the test data because of sampling noise when there are insufficient training data. In turn, this leads to overfitting, which must be avoided by applying regularization techniques that introduce various weight penalties, such as L_1 and L_2 regularization, and stopping the training process as soon as performance on a validation set begins to decrease. An effective regularization method for trimming deep networks is

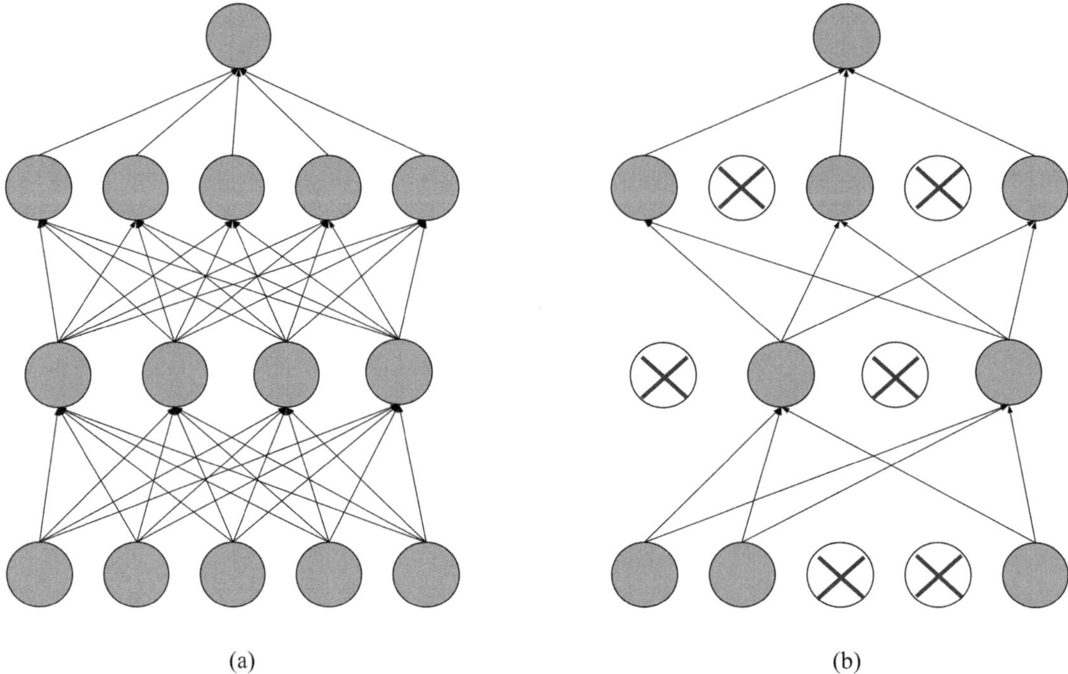

(a)   (b)

FIGURE 7.2 Application of dropout regularization to a neural network model. (a) Standard neural network with two hidden layers, (b) a thinned network produced by applying dropout to the network.

the application of dropout techniques. Dropout prevents overfitting and provides an efficient means of exponentially combining numerous distinct neural network architectures. A unit is randomly dropped out of the network when all connections, including incoming and outgoing connections, are temporarily removed (Figure 7.2). The simplest scenario is to keep each unit independent of the others with a fixed probability p. This may be determined using a validation set, or it can be set at 0.5, which appears to be ideal for most networks. However, the ideal probability of retention for the input units is generally closer to 1 than to 0.5 (Srivastava et al., 2014). Applying dropout to a neural network is identical to taking a sample from a thinned network. All units remaining after dropout constitute the thinning network (Figure 7.2b). You may think of a neural net with n units as a set of 2^n potential neural networks that have been thinned. To maintain an $O(n^2)$ or a lower total number of parameters, all networks share weights. A new thinned network is sampled and trained for each training case. The process of training a neural network with dropout can be viewed as the training of a set of 2^n sparse networks with substantial weight sharing. It is feasible to train several distinct networks in an acceptable amount of time with random dropout. Each presentation of a training case most likely has a distinct network, but all of these networks have the same weights assigned to the hidden units (Hinton et al., 2012).

Dropout regularization consists of an extra hyperparameter, which is the probability of retaining a unit p. This hyperparameter controls the dropout intensity; $p = 1$ suggests that there is no dropout, and low values of p indicate more dropout. Typical values of p for hidden units range from 0.5 to 0.8. When dropout is set at a probability of 0.5, we expect that all models will be equally valued in the eventual combination. For real-valued inputs (e.g., image patches), a value of 0.8 is recommended. For hidden layers, the choice of p is coupled with the choice of the number of hidden units n. Srivastava et al. (2014) also noted that small values of p require large values of n, which can lead to underfitting, whereas large p values may cause insufficient dropout to avoid overfitting.

The use of dropout markedly decreases overfitting while simultaneously enhancing the learning speed of the algorithm, which is achieved by excluding certain training nodes during the training phase. Compared with a regular neural network with the same design, a dropout network usually requires several times longer training time. Each training scenario effectively trains different random architectures. As a result, the calculated gradients do not correspond to the final design employed during testing. It should be noted that the use of dropout regularization is not limited to feedforward neural networks, and it can be applied to graphical models, including Boltzmann machines.

7.1.3.3 Data Augmentation

Deep neural networks have many more layers and interconnections than shallow neural networks in that the total number of parameters is determined by the connections between neurons. Modern deep CNNs can easily encompass millions of parameters. The generalization ability of networks diminishes when they possess a larger number of parameters, causing them to memorize training patterns (Karnin, 1990; Kavzoglu and Mather, 1999). The phenomenon of "tuning to the noise" in classification theory arises when the number of hyperparameters is significantly larger than the training data. In classification problems, inadequate training samples hinder neural networks from capturing class characteristics, whereas an excessive number of training samples may lead to overfitting and prolonged learning times. It is important to emphasize the preference for a larger amount of training data over a smaller amount because insufficient training data impede the network's ability to delineate class characteristics, risking failure in classifying new data (Kavzoglu and Mather, 2003). Conversely, with a large number of training samples, the usual outcome is only a marginal reduction in network performance. According to Staufer and Fischer (1997), the overall classification accuracy, which is used as a metric for generalization performance, tends to increase with the size of the training set, especially in situations where there is overlap between classes. However, the number of training samples is limited in most remote sensing studies. The success of deep neural networks in most applications depends on the presence of large data sets. To increase the performance of the DL model and satisfy the requirements of a large amount of data, data augmentation is an important regularization tool. By applying a sequence of random modifications to the training images, image augmentation creates training instances that are similar yet different, thereby increasing the size of the training set. It is a group of techniques designed to artificially increase the data set by adding transformations or perturbations to the input data. Data augmentation techniques include flipping the images horizontally or vertically, cropping, color jittering, scaling, blur, CutMix, mosaic, and rotations, which are commonly used in image classification applications. Some examples of data augmentation techniques are shown in Figure 7.3.

In the context of object recognition, data set augmentation has proven to be an effective technique. The high-dimensional nature of the images and the multitude of variation factors allow easy simulation. Simple operations, such as translating training images by a few pixels in different directions and rotating and scaling the images, can significantly improve generalization (Goodfellow et al., 2016). Adding noise to the training data is another type of data augmentation method that has long been applied in neural network applications. It should be noted that some unsupervised learning methods, such as the denoising autoencoder discussed in Section 7.2.5.4, use input noise injection at the input level. Since data augmentation strategies can notably contribute to the reduction of the generalization error of a machine learning approach, the same training and test data sets should be employed in the performance comparison of the methods. Otherwise, the comparison will be subjective because of the inclusion of an improved data set, produced with an augmentation strategy. The effectiveness of data augmentation in image classification has been comprehensively discussed by Perez and Wang (2017), Yu et al. (2017), Mikołajczyk and Grochowski (2018), and Oubara et al. (2022).

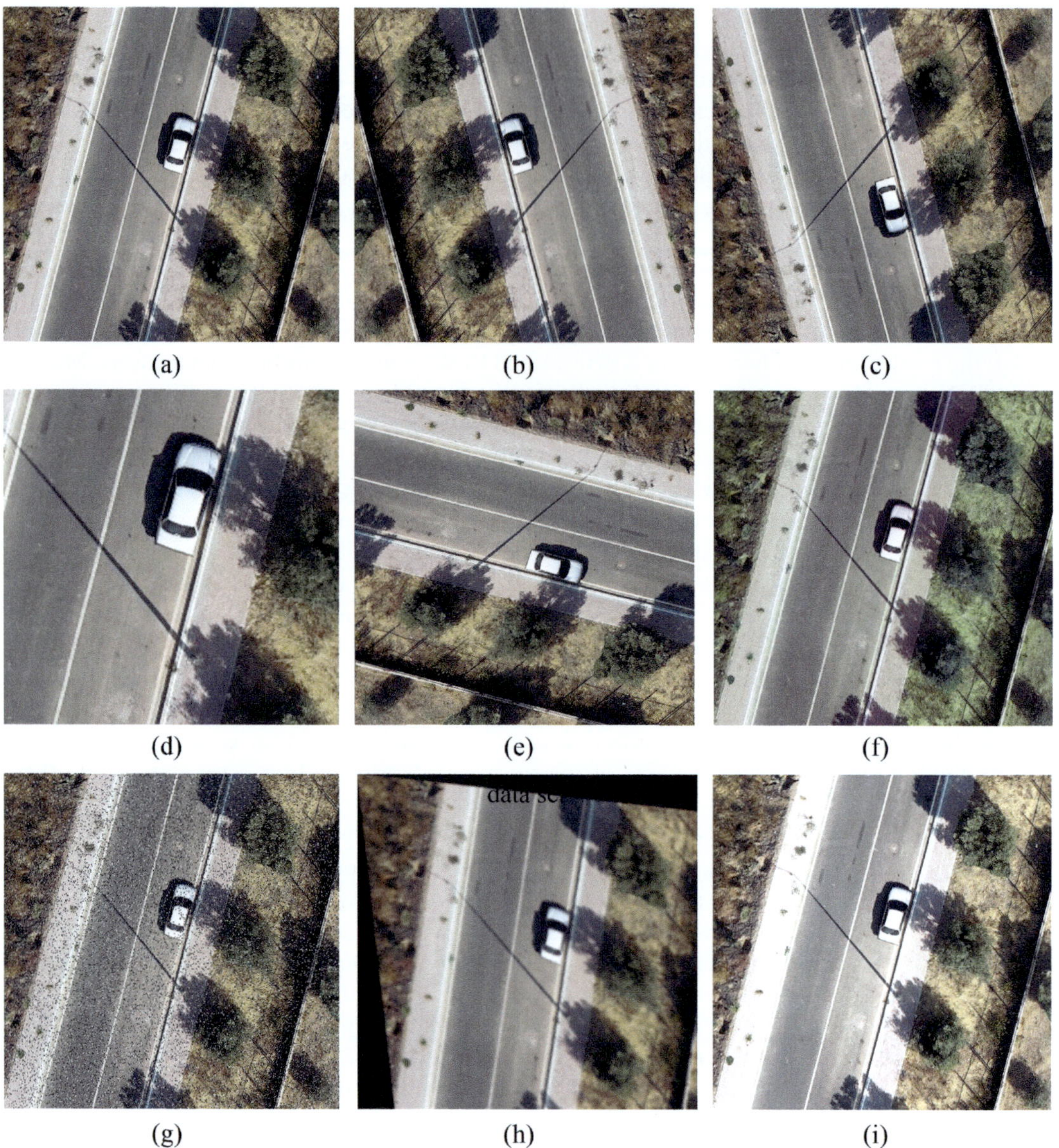

FIGURE 7.3 Examples of data augmentation operations. (a) Original image, (b) horizontal flip, (c) vertical flip, (d) crop, (e) 90° rotate (clockwise), (f) hue, (g) noise, (h) shear, and (i) brightness.

7.1.3.4 Early Stopping

An important issue in the training of neural networks is to define a stopping criterion for the learning process because it is impractical for real-world problems to train a network until the training error decreases to zero. A convergence criterion must be defined to prevent overfitting. When the number of iterations or epochs falls below the optimal value, underfitting occurs (reducing variance but introducing bias). Conversely, if the number of iterations is increased beyond the optimum level (raising variance while decreasing bias), the network tends to overfit (Murugan and Durairaj, 2017). Assessing the optimal point to terminate the learning process for optimal generalization performance involves employing two primary approaches. The prevalent method is to use a validation set to continuously assess the performance of trained networks throughout the learning process. The learning process is typically halted when an increase in the error on the validation set is observed. Early stopping is usually conducted when the mean square error (MSE) is reduced to a certain level or validation loss stops decreasing. According to Ripley (1996), "this is dangerous as it is often encountered examples in which, after an initial drop, the error on the validation set rises slowly for a large number of iterations, then falls dramatically to a small fraction of its previous minimum". Therefore, the ideal solution could be to end the learning process before the network becomes over-specific to the training data and avoid early stopping. Wang et al. (1994) observed that a network demonstrates enhanced generalization performance when learning concludes before reaching the global minimum of the empirical error. Moreover, for a constant number of training samples, a greater ratio of d/n, where d is the number of weights (or nodes) and n is the number of samples, results in a more substantial enhancement in the generalization error if the algorithm is halted before reaching the global minimum. Early stopping is grounded in the underlying philosophy that once learning reaches the global minimum, the network's generalization capabilities diminish as it becomes overly-specific. However, underfitting may occur if early stopping is not employed before reaching a satisfactory learning level or establishing optimal decision boundaries. Some researchers (e.g., Yang et al., 2020b) adopted a strategy in which training is stopped when the training loss is no longer decreased after a certain number of epochs, as suggested by Kavzoglu (2001) and Kavzoglu and Mather (2003). Instead of identifying the early stopping point by considering a deep neural network (DNN) as a whole, Bai et al. (2021) introduced a novel method called progressive early stopping (PES), in which former DNN layers were initially trained by optimizing the network with a relatively large number of epochs. Subsequently, the latter layers are trained with a smaller number of epochs, keeping the preceding layers fixed to counteract the influence of noisy labels. Combining PES with existing approaches for training on noisy labels produced state-of-the-art performance on image classification benchmarks.

7.1.4 Activation Functions

In neural networks, activation functions, as mathematical operations, play a pivotal role in converting the summed weighted input from a node into an output value. This output value is then transmitted to the next hidden layer or serves as an output. Theoretically, any differentiable non-linear function can function as an activation function, serving a crucial role in initiating the training process to estimate hidden layer values. In the context of ANNs, the activation function must exhibit nonlinearity; otherwise, it can only distinguish linearly separable entities, as in the case of the Perceptron, which comprises solely input and output layers. These functions, also known as transfer functions, transform input values into another value. The selection of an activation function is a vital step at the outset of the learning process. Activation functions for different layers may vary because of their distinct roles in the learning procedure. The application of an activation function to the output nodes can yield posterior probabilities. According to Civco and Waug (1994), "an activation function is required to avoid saturation of a processing node, caused by extremely large positive or negative internal summations". Activation functions are applied to reduce the number of iterations, thereby introducing nonlinearity into the network and consequently enhancing generalization performance.

A sigmoid function, also called a logistic function, is the first activation function used for this purpose, which was previously described in more detail in Section 7.1.2. The sigmoid function (Figure 7.4) is expressed as

$$f(x) = \frac{1}{1 + e^{-x}}, \tag{7.19}$$

where x is the sum of the weighted input values to the processing node.

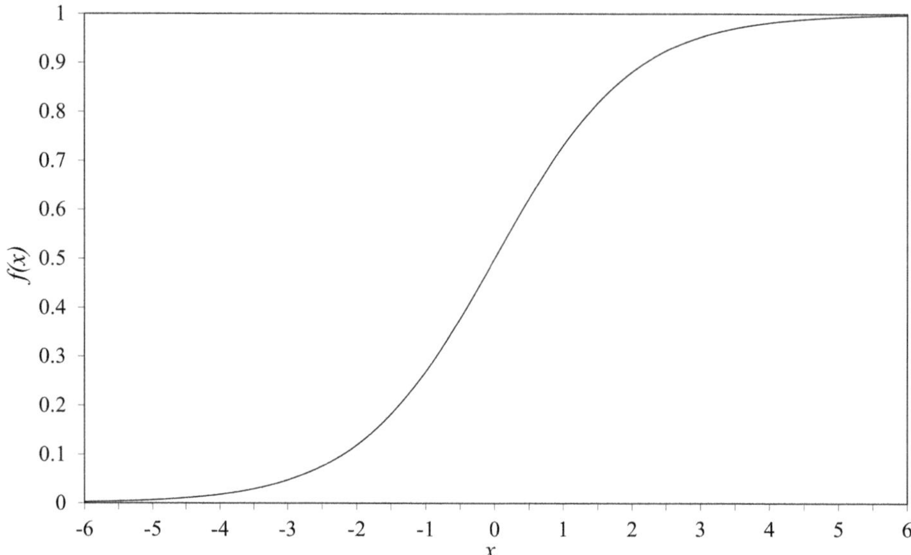

FIGURE 7.4 Sigmoid activation function.

Intuitively, if the logit is extremely small, the output of a logistic neuron is close to 0. Conversely, if the logit is considerably large, the output of the logistic neuron approaches 1. The activation function exhibits an almost linear input-output relationship between its two extreme values. However, as the outputs of a node reach these extremes, the derivative of the activation function decreases. Considering that the change in weights is proportional to the derivative value, only slight modifications occur in the weights. The derivative attains its maximum value when the output is 0.5. Consequently, the weights undergo rapid changes in this scenario, influencing the node to commit to a high or low value. This characteristic likely contributes to the stability of the learning stage (Paola and Schowengerdt, 1995).

Another widely used activation function is the "tanh" function (Figure 7.5). Although the sigmoid and tanh functions share similarities, it is commonly observed that employing the tanh activation function leads to faster convergence of the training algorithms compared with the sigmoid function (Bishop, 1995). The tanh function is in the form given below:

$$f(x) = \tanh(x) = \frac{e^x - e^{-x}}{e^x + e^{-x}}. \tag{7.20}$$

The main difference between these activation functions is that while input data and output classes are coded in the [0, 1] range with the use of sigmoid, they are given in a [−1, 1] range with the tanh function. As can be noticed, the tanh function represents the data in a broader range, which may have a positive effect on the performance of the network. When S-shaped nonlinearities are considered, the tanh neuron is usually the preferred option over the sigmoid neuron, primarily because it is zero-centered.

The recently introduced rectified linear activation function (ReLU) is a piecewise linear function that directly outputs the input if it is positive; otherwise, it outputs zero $f(x) = \max(0, x)$, as shown in Figure 7.6. ReLU has become the default choice for the activation function in many neural networks because the models employing it are easier to train and frequently achieve superior performance. Due to the vanishing gradient problem, the sigmoid and hyperbolic tangent activation functions are not suitable for deployment in networks with many layers. To overcome this problem, the

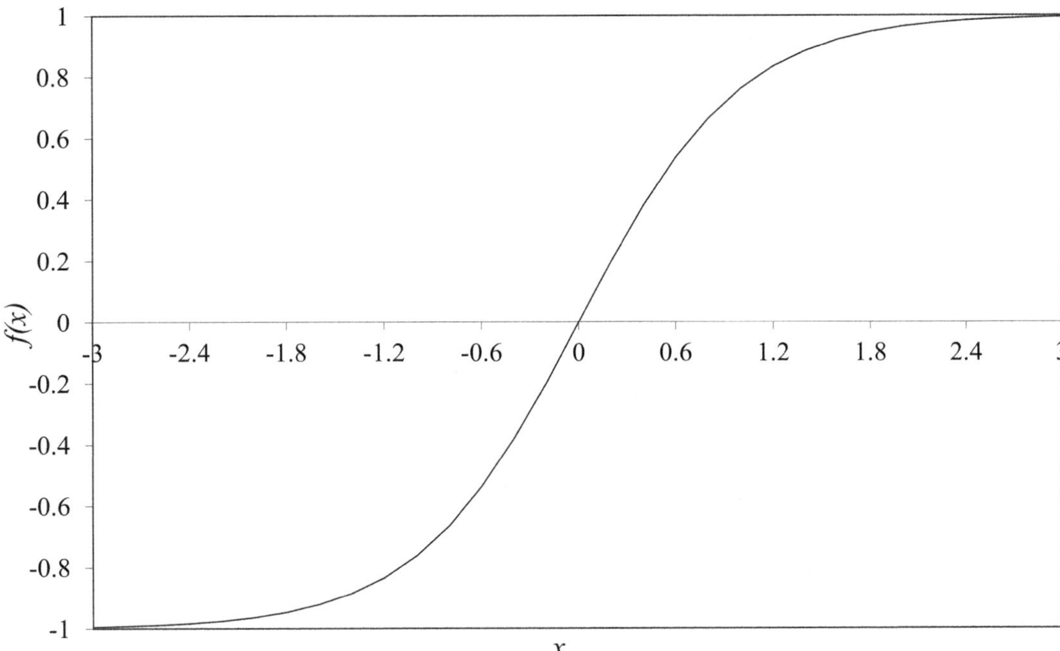

FIGURE 7.5   Tanh activation function.

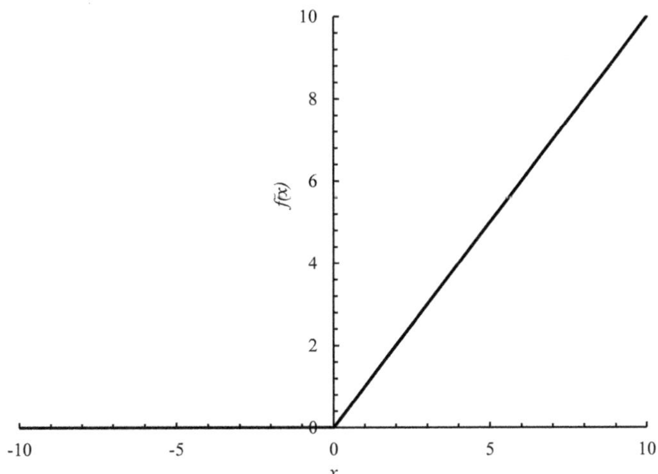

FIGURE 7.6   ReLU activation function.

ReLU activation function is preferred in most applications, particularly when CNNs are employed. The primary issue with ReLU is its transformation of all negative values to zero, resulting in a reduction of the model's ability to fit or train effectively from the data, which is sometimes called a "dying ReLU" or "dead neuron" problem. There are some variants to overcome this drawback. For instance, the leaky ReLU has a small slope to the negative input instead of a flat value of 0 (Figure 7.7). It can be expressed as $f(x) = \max(\text{alpha} * x, x)$, where alpha is a small positive constant. This form of activation function is commonly used in scenarios where sparse gradients may pose challenges, such as during the training of GANs. Xu et al. (2020) argued that the leaky ReLU

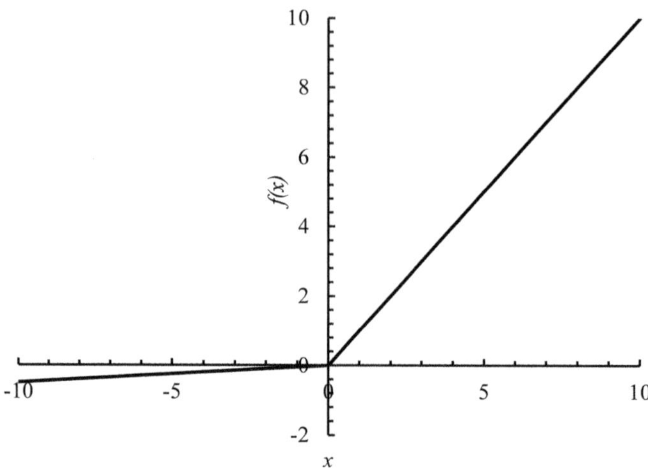

FIGURE 7.7   Leaky ReLU activation function.

activation function performs well, but its effect is not very stable. In addition, compared with ReLU, under what conditions, which one has better performance and effect remains to be discussed.

The Softmax function is another type of activation function used in the output layers of neural networks. It is a combination of multiple sigmoid functions. Similar to the sigmoid function, it produces outputs in the range 0–1, which are treated as probabilities of a particular class. Unlike sigmoid functions, which are used for binary classification, Softmax can be used for multiclass classification problems. A Softmax output layer significantly changes the range of the outputs of a neural network because of its nonlinearity characteristics. The Softmax function can be expressed as

$$f(x_i) = \frac{\exp(x_i)}{\sum_j \exp(x_j)}. \tag{7.21}$$

In constructing a network or model for multiple-class classification, the output layer of the network is designed to contain the same number of neurons as there are classes in the target. This function is commonly used in nearly all output layers of DL architectures.

7.1.5   Loss Functions

The loss function serves as a key element in DL techniques, playing an indispensable role in the creation of machine learning algorithms and the continuous monitoring and refinement of their performance. It is evident that the loss function serves as both a crucial element of the structural risk function and the core component of empirical risk. The selection of an inappropriate loss function for a particular problem undermines the effectiveness of the algorithm. To date, many loss functions have been suggested to optimize the learning process. Wang et al. (2022) reviewed 31 loss functions employed in machine learning practices. Among these loss functions, ten functions are specifically developed for object detection (e.g., focal loss, IoU loss, smooth L_l loss, Dice loss, GIoU loss) and face recognition (e.g., contrastive loss, center loss, triplet loss, A-Softmax loss, AM-Softmax loss, ArcFace loss) using DL models. It is noteworthy that some loss functions are specifically designed for unsupervised learning. The major loss functions serving this purpose are square error, distance error, reconstruction error, and negative variance. Since most DL models for segmenting remote sensing images rely on the current backbone network architecture, they often use cross-entropy as the loss function (Yuan et al., 2021). The sum of the pixel-wise cross-entropy between the real ground truth patches and the predicted label probability is the loss function. For a two-class problem, the cross-entropy can be given as:

$$-y\log(p)-(1-y)\log(1-p),\tag{7.22}$$

where y is a binary indicator of the correct classification of an instance, and p is the probability of an instance for the target class. For multiple classes, cross-entropy is expressed as a weighted summation, as shown below:

$$-\sum_{c=1}^{C} y_c\log(p_c),\tag{7.23}$$

where c denotes the class label. In practice, the outputs from the Softmax function are considered *a posteriori* probability for each class. Due to the typically large size of remote sensing images, especially hyperspectral ones, patches are generated as inputs for deep networks. Consequently, normalized cross-entropy across the entire patch is commonly utilized.

In addition to cross-entropy, the MSE between the input data and the reconstructed output is employed as the learning loss. When dealing with regression tasks that involve continuous values, MSE is usually chosen as a loss function. Henry et al. (2018) adapted CNN by replacing the Softmax with a sigmoid function and employing a class-weighted MSE as the loss function:

$$\frac{1}{N}\sum_{i=1}^{N} w_i\left(y_i - \hat{y}_i\right)^2,\tag{7.24}$$

where y_i denotes the ground truth, $\hat{y}_i$ denotes the sigmoid value of the prediction, and N shows the number of pixels in the image.

Remote sensing images present a higher level of complexity compared with the images utilized in computer vision, primarily due to the multidirectional nature of the same geographical object, various image acquisition geometries, and atmospheric effects. This complexity can result in suboptimal classification accuracy when employing commonly used loss functions initially designed for computer vision challenges. Furthermore, these conventional loss functions excel at distinguishing between distinct classes but struggle to discern individual features within the same class, a challenge frequently encountered in remotely sensed images. In recent years, many attempts have been made to improve loss functions. Yang et al. (2018) adopted multitask loss to train their network. Cross-entropy loss was applied to the classification output and a smooth ℓ_1 loss (least absolute deviations) on the regression output. The summation of both loss functions is preferred to compute prediction with the network. The mathematical expressions of multitask loss can be given as follows:

$$L_{\text{total}} = \text{crosss\_entropy}_{(p,y_{\text{cls}})} + \text{smooth}_{L1}\left(q - y_{\text{reg}}\right)\tag{7.25}$$

$$\text{cross\_entropy}(p,y) = \begin{cases} -\log(p) & \text{if } y=1 \\ -\log(1-p) & \text{otherwise,} \end{cases}\tag{7.26}$$

$$\text{Smooth}_{L1}(x) = \begin{cases} 0.5x^2 & \text{if } |x| < 1 \\ |x|+0.5 & \text{otherwise.} \end{cases}\tag{7.27}$$

Arief et al. (2018) applied a multiclass version of the intersection-over-union (IoU) loss function together with the Softmax function to handle highly imbalanced data sets, which is similar to the function introduced by Rahman and Wang (2016). Srivastava et al. (2017) employed a joint training loss function in networks optimized using two loss functions, corresponding to semantic labeling and nDSM prediction. While the loss corresponds to the average negative log-likelihood for each class in the first case, the network is trained to minimize the average MSE over the minibatch. Gonzalez and Miikkulainen (2020) proposed a loss function discovery and optimization method as a new form of meta-learning. The utilization of genetic loss function optimization (GLO) involves employing genetic programming to construct loss functions in the form of trees, followed by the application of a covariance matrix adaptation evolution strategy to fine-tune their coefficients. The experiments demonstrated notable progress in accuracy, convergence speed, and data requirements. A deeper analysis indicated that these improvements were a consequence of implicit regularization, effectively minimizing overfitting to the data.

7.2 NEURAL NETWORK ARCHITECTURES

Based on their underlying learning strategies and functionalities, DL models can be divided into four major categories: discriminative, generative, representative, and hybrid models (Figure 7.8). Discriminative models learn the relationship between input and output data using regression or classification methods. CNNs, RNNs, and their variants, the gated recurrent unit (GRU) and long- and short-term memory (LSTM) networks, MLPs, and extreme learning machines (ELMs) are examples of discriminative DL models. These architectures can extract discrete features from input signals by nonlinear transformations and then use probabilistic prediction to classify those features into predefined classes. They can be used to model problems involving both feature extraction and classification. Generative models learn the joint distribution and sequential or temporal relations of the data that underlie the model. They are usually applied to augment and enhance training data sets. GANs and variational autoencoders (VAEs) are the two most widely used generative DL models. Improving the generalization performance of generative models often involves adopting a discriminative training approach; however, this approach has the drawback of losing access to unlabeled data. Attempts have been made to merge the advantages of generative and discriminative approaches by proposing heuristic procedures that interpolate between these two extremes using convex combinations of their objective functions (Bishop and Lasserre, 2007). On the other hand, representative DL models are architectures that focus on unsupervised, feature extraction, which may be applied to a variety of tasks, including classification and clustering. Deep belief networks (DBNs), restricted Boltzmann machines (RBMs), and deep autoencoders are the main representative

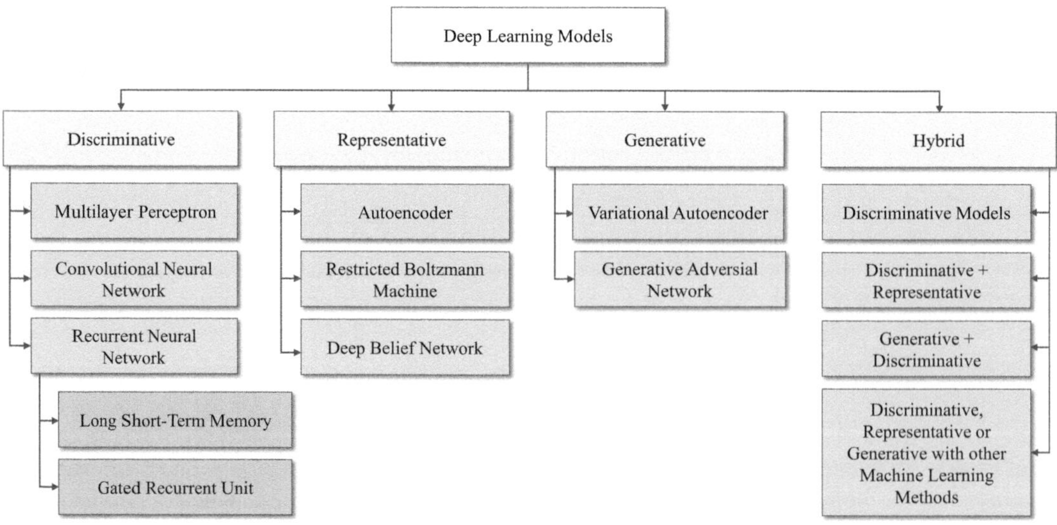

FIGURE 7.8 Taxonomy of deep learning architectures.

DL models. Apart from the aforementioned stand-alone DL models, scientists have also developed hybrid models that combine two or more DL strategies with other machine learning algorithms. Hybrid models incorporate the benefits of discriminative, generative, and discriminative models. A single architecture is combined with two or more DL models in hybrid DL models. Compared with single DL models, superior performances have been reported with the application of hybrid models, as they can overcome the challenges experienced with single DL models. The most widely used DL architectures are discussed in the following subsections. In addition to these architectures, ELMs are models with which high accurate classification results are reported.

An ELM, classified as a discriminative feedforward neural network, serves the purpose of feature extraction and classification. ELM distinguishes itself by its rapid learning process, in which hidden nodes are randomly assigned and require no tuning or updates. The selection of the best randomly initialized neuron parameters in the ELM aids in estimating the required decision boundary. This particular feature provides a substantial advantage over other DL models. However, the accuracy obtained through the combination of random weights is not as high as that achieved by a finely tuned backpropagation neural network. Another positive aspect of ELM is its capability to use a complex activation function without the need for differentiability or training via backpropagation. ELM is frequently used as a rapid classifier that incorporates manually derived features.

7.2.1 Multilayer Perceptron

The multilayer perceptron (MLP) using the backpropagation learning algorithm is the first version of modern feedforward ANNs, consisting of fully connected neurons. A typical three-layer MLP is shown in Figure 7.9a. The leftmost layer of neurons is the input layer, which contains the set of neurons that receive external inputs (in the form of pixel values in the different bands of a multispectral image or other feature values). The input layer performs no computations, unlike the elements of the other layers. The central layer is the hidden layer. There may be more than two hidden layers in complex networks. The rightmost layer of neurons is the output layer that produces the classification results. There are no interconnections between neurons in the same layer; however, all neurons in a given layer are fully connected to neurons in adjacent layers. These interconnections have associated numerical weights w_{jm}, which are adjusted during the learning phase using a gradient descent algorithm. The weights associated with each interconnection are initialized with random values within a certain range. The value held by each neuron is called its activity a_m (Figure 7.9b).

Developed for supervised image classification tasks, MLPs must be "taught" or trained on the key features of the data set under investigation, emphasizing the pivotal role of the learning algorithm in artificial neural network applications. Despite the existence of numerous learning strategies, the backpropagation learning algorithm remains the most widely used, which is comprehensively discussed in Section 7.1.2.

Because of the difficulties in their design and implementation, many researchers (e.g., Foody, 1999; Kavzoglu and Mather, 2003; Mas and Flores, 2008) have noted that the application of ANNs is more complex than that of statistical classifiers. From the design perspective, the number and size of the hidden layer(s) must be specified for the network to be able to learn and become more broadly applicable. The number of hidden layer nodes can change in size since the size of the output layer corresponds to the number of output classes and the size of the input layer typically equals the number of features used for classification. While the general implications of using too small or large network topologies are well recognized, the overall impact of hidden layer sizes on network performance is still uncertain for any type of classification and regressing problems. Determining the ideal number of hidden layer nodes for a given task has been a critical issue, although several strategies exist. Selecting suitable settings for network parameters that significantly impact the learning algorithm's effectiveness is another challenge when using MLPs. Often, finding the parameter values that result in the best classification accuracy requires many trials. Appropriate values for these parameters are frequently determined using a trial-and-error method. This makes

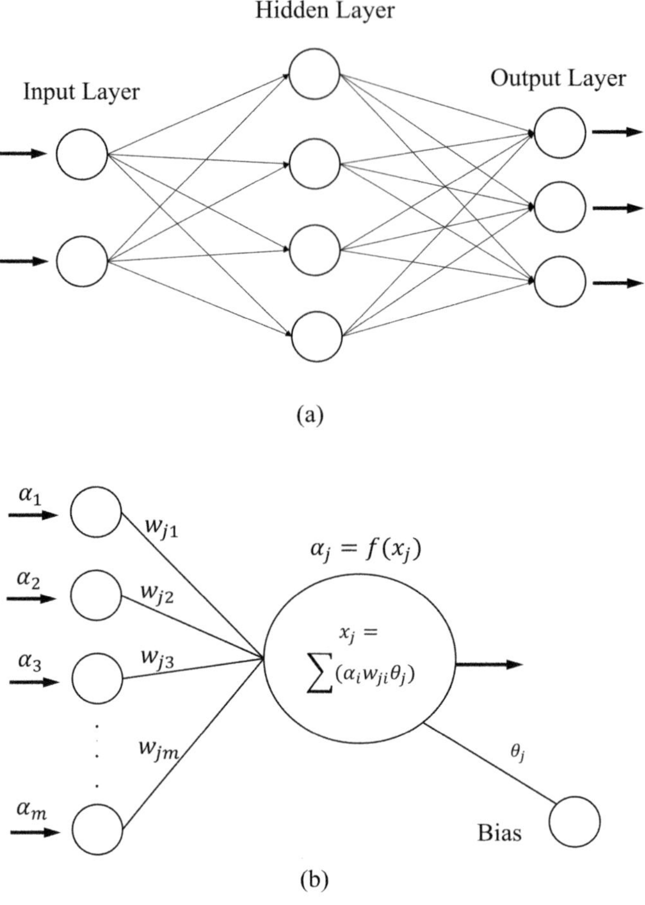

FIGURE 7.9 (a) Typical three-layer multilayer perceptron neural network. (b) Example of the forward propagation procedure on neuron j.

the already-slow network training process even slower. The range of starting weights, learning rate, momentum term value, and number of training iterations are important factors to consider because they all significantly impact the decision of when and how to end the training process. Kavzoglu and Mather (2003) provided some tips for the efficient use of MLPs after investigating the effects of parameter settings and network designs.

7.2.2 Convolutional Neural Networks

Over the past decade, DL algorithms based on CNNs have been widely used to classify remotely sensed data. The use of CNNs has demonstrated significant effectiveness due to their computational efficiency and ability to detect patterns in images. They are widely used DL models, specifically designed to extract local and spatial patterns. The key advantage of CNNs lies in their capacity to automatically identify pertinent features without human intervention, which sets them apart from other machine learning approaches that require handcrafted features for processing. A CNN is a complex model that employs supervised learning. The basic concept is to distribute the weights of feature mappings to different positions in the preceding layer network, thus reducing the number of parameters by using spatial relationships and ultimately improving training performance (Liu et al., 2021). The remote sensing community has extensively reported the use of 1D, 2D, and 3D

CNN-based DL algorithms for semantic segmentation, object detection, and image classification (Paoletti et al., 2019; Roy et al., 2019; Kavzoglu et al., 2021; Kavzoglu and Yilmaz, 2022; Yilmaz et al., 2024).

Since its conception, CNNs have gone through several development stages, including theoretical exploration, experimental advancement, widespread implementation, and extensive research. The notion of receptive fields and neurocognitive machinery in human visual information is one of the most significant stages in the development of an algorithm. Hubel and Wiesel (1962) demonstrated through biological investigation that the transfer of visual information from the retina to the brain occurs through multilayered receptive field stimulation. This is the first proposal of the receptive field concept. Fukushima and Miyake (1982) presented a neurocognitive machine based on receptive field mapping, which is considered to be the first use of CNNs. LeCun et al. (1998) introduced LeNet5, a revolutionary neural network model trained using a gradient-based backpropagation method for supervised learning, which was followed by the experimental development of CNNs. CNNs gained prominence in DL applications with the introduction of the AlexNet network by Krizhevsky et al. (2012), which achieved remarkable success in classifying images from ImageNet. Since then, CNNs have been the main focus of research in computer vision and remote sensing.

A CNN architecture consists of a collection of neural networks organized in a specific order with layers of different dimensions, where each layer performs a different function. The initial layers acquire basic features (low-level), but subsequent layers acquire more complex features (high-level). In CNNs, the three fundamental components are convolutional layers for feature extraction, pooling layers for dimensionality reduction, and fully connected layers for classification. Figure 7.10 serves as an example of the CNN architecture employed for image classification. Feature extraction uses a convolutional operation to identify and categorize the specific properties of an image for analysis. The feature extraction network consists of many pairs of convolutional and/or pooling layers. The fully connected layer uses the output of the convolution process to predict the class of the image based on the features extracted in the previous stages. The purpose of feature extraction is to reduce the dimensionality of a given data set by reducing the number of features. Thus, the process creates new features from existing ones. Finally, a fully connected layer is connected to the feature maps of the pooling layer, usually after flattening them.

7.2.2.1 Convolutional Layers

The convolutional layer is the key element that separates CNNs from their shallow counterparts. The system involves a collection of convolutional filters, which are often referred to as kernels. The input image, represented as N-dimensional data, is convolutionally processed by these filters to produce the output feature map. A kernel is a grid of discrete integers or values. The term "kernel

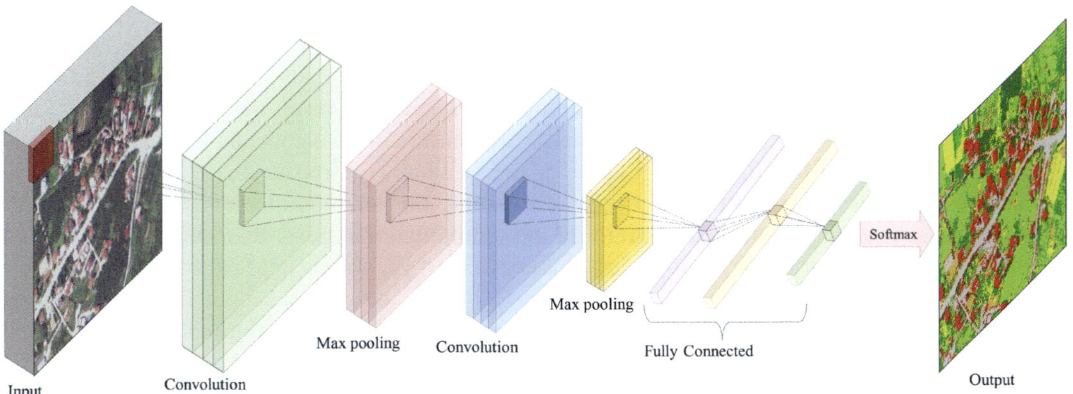

FIGURE 7.10 Architecture of a convolutional neural network.

weight" refers to each value. The goal of each kernel is to identify a particular feature at each input position. As a result, the spatial translation of the input from a feature identification layer to the output is unaffected. Each convolutional layer contains a set of m_1 filters, and the output $Y_i^{(l)}$ represents the result of the lth layer. The layer is composed of $m_1^{(l)}$ feature maps with dimensions $m_2^{(l)} \times m_3^{(l)}$. The ith feature map is calculated as follows:

$$Y_i^{(l)} = B_i^{(l)} + \sum_{j=1}^{m_1^{(l-1)}} K_{ij}^{(l)} * Y_j^{(l-1)}, \tag{7.28}$$

where the matrix of trainable bias is denoted as $B_i^{(l)}$. The filter, $K_{ij}^{(l)}$, has dimensions $\left(2h_1^{(l)+1} \times 2h_2^{(l)+1}\right)$ and connects the jth feature map of the $(l-1)$ layer with the ith feature map of the (l) layer. The symbol $(*)$ represents the 2D discrete convolution function (Naranjo-Torres et al., 2020).

In the initial stage of training, weights are randomly assigned to kernels. Subsequently, the weights are modified during each training epoch, allowing the kernel to acquire important characteristics. The convolutional operation refers to the initial description of the input format in the CNN model. The standard neural network accepts vector format as its input, whereas CNNs take multiband images as its input. For example, a grayscale image is in single-band format, whereas an RGB image is in three-band format. Initially, the kernel traverses the entire image both horizontally and vertically. Then, it computes the dot product between the input image and the kernel, where their respective values are multiplied and aggregated to produce a singular scalar value computed simultaneously. The entire process is repeated until no further shifting occurs. It is important to realize that the computed dot product values correspond to the feature map of the resulting output. In this procedure, the 2D matrix (I) describing the image is convolved with the 2D kernel matrix (K). The resulting mathematical formulation, which includes zero buffering, is given as follows:

$$S_{i,j} = \left(I * K\right)_{i,j} = \sum_{m} \sum_{n} I_{i,j} \cdot K_{i-m,j-n}. \tag{7.29}$$

Figure 7.11 provides a visual representation of the main computations performed at each step. The figure shows the kernel (middle row) and the similarly sized portion of the input image (lower row). The two values are multiplied together, and then, the resulting product values are summed. This sum represents the input result in the output feature map. However, in the previous example, padding is ignored for the input data if a stride of one (indicating the desired step size over all vertical or horizontal positions) is used for the kernel. It is worth noting that an alternative stride value can be used. In addition, by increasing the stride value, it is possible to create a feature map with reduced dimensions. Padding plays an important role in determining the boundary size information associated with the input image. In contrast, the boundary features shift rapidly. Padding is used to increase the size of the input image and the dimensions of the feature map.

Convolutional layers have the advantages of sparse connectivity and weight sharing. In CNNs, sparse connectivity denotes that neurons lack complete connections, resulting in some missing links within the network (Figure 7.12). CNNs have a limited number of weights connecting two successive layers. Therefore, the number of weights or connections required is small, and the memory required to store these weights is similarly small, making this strategy memory efficient. In addition, matrix operations require significantly higher computational costs than point operations in CNNs. In other words, it becomes possible to identify small yet significant features, such as edges, using kernels that contain only a limited number of pixels, despite the input image containing thousands or billions of pixels. This means that we need to store a smaller number of parameters, which results in a reduced memory requirement for the model and improved statistical efficiency. The key characteristic of weight sharing is the reduction of connections between different layers in the network, resulting in fewer network parameters and avoiding overfitting.

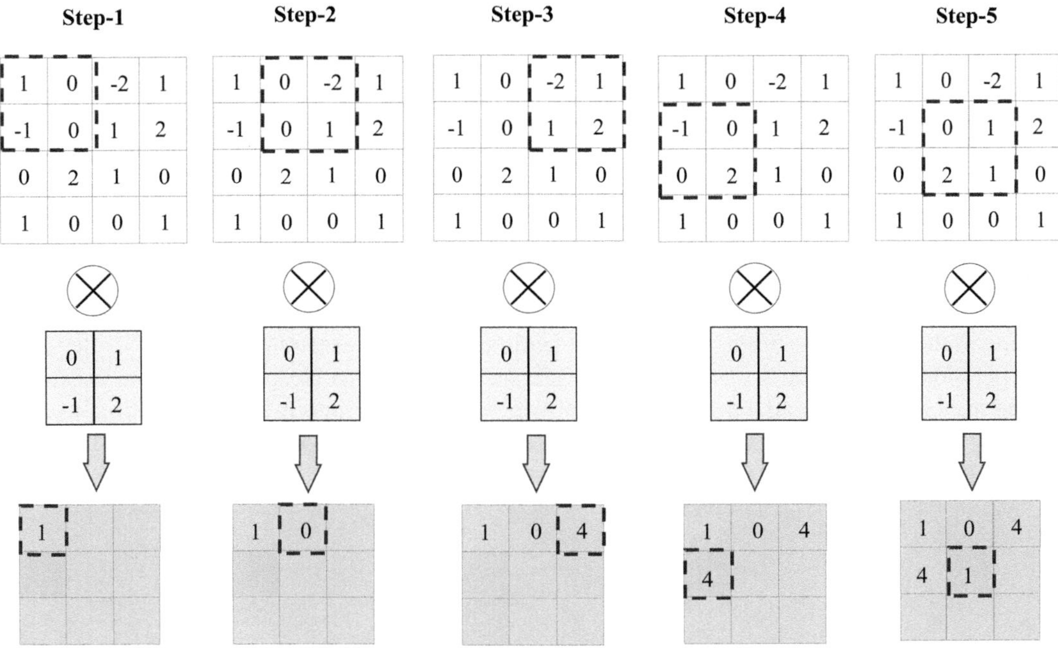

FIGURE 7.11   Instance of the convolutional function with an input image (4×4) and kernel (2×2).

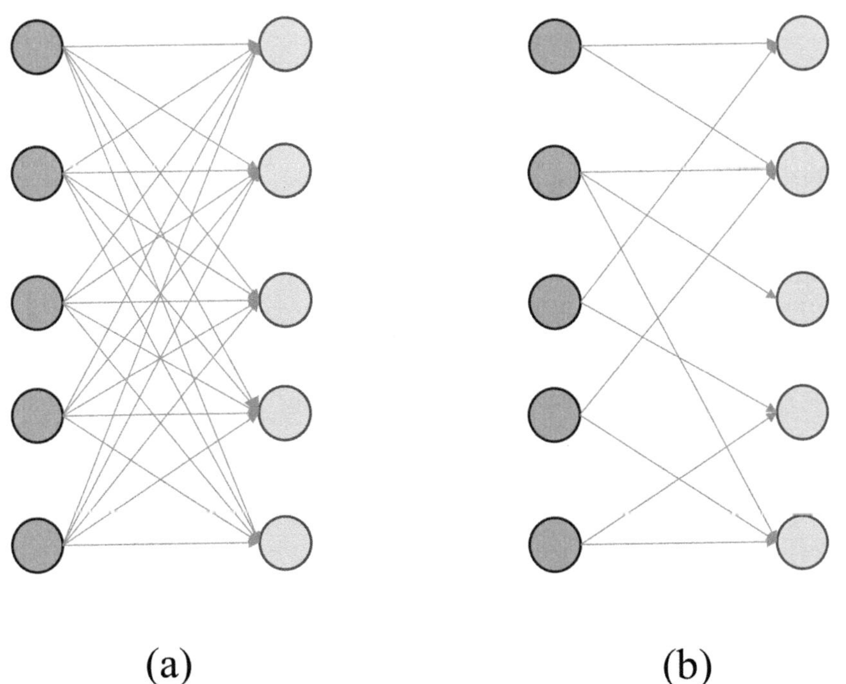

(a)   (b)

FIGURE 7.12   Example of connectivity in CNNs. (a) Dense (full) connectivity, (b) sparse connectivity.

7.2.2.2 Pooling Layers

Once the feature maps have been acquired, it is essential to include a pooling (subsampling) layer after the convolution layer. The pooling layer is an essential element of CNNs, which helps to detect the presence and location of features in the input images. The primary function of the pooling layer is to reduce the spatial dimensions of the convolved feature and the computational cost required to process the data. This also facilitates the extraction of salient features that are independent of both position and rotation, thus maintaining the effectiveness of the model during training (Bhatt et al., 2021). Pooling reduces training duration while preventing overfitting. Nevertheless, it is crucial to acknowledge that the use of this layer may decrease the overall efficiency of CNNs. This limitation arises because the primary focus of the pooling layer is to precisely define the spatial location of features rather than considering their overall significance. As a convolutional process, the pooling operation requires the size of both the stride and the kernel to be specified before execution. Different pooling strategies have been developed, including tree pooling, gated pooling, average pooling, min pooling, max pooling, global average pooling, and global max pooling. The most popular pooling strategies are average and maximum pooling (Figure 7.13).

The max pooling layer retains the highest value from each patch as the filter moves over the feature map. The expression can be expressed as

$$f_{\max}(A) = \max_{n \times m}(A_{n \times m}). \tag{7.30}$$

Typically, the max pooling layer uses 2×2 filters with a stride of 2. The input is downsampled by a factor of two across its dimensions; thus, 75% of the convolutional outputs are discarded. In addition, average pooling calculates the average value for each input patch. The average pooling layer reduces the size of the convolutional activation by dividing the input into pooling regions and calculating the average of each region, which can be expressed as

$$f_{\mathrm{ave}}(A) = \frac{1}{n+m} \sum_{i=1}^{n} \sum_{k=1}^{m}(A_{i,k}). \tag{7.31}$$

7.2.2.3 Fully Connected Layers

The fully connected layer is often positioned at the end of any CNN architecture. Within this network architecture, every neuron is connected to every neuron in the previous layer. The neural network described follows the basic approach of the traditional MLP, which works as a feedforward artificial neural network. The fully connected layer receives its input from the preceding pooling or

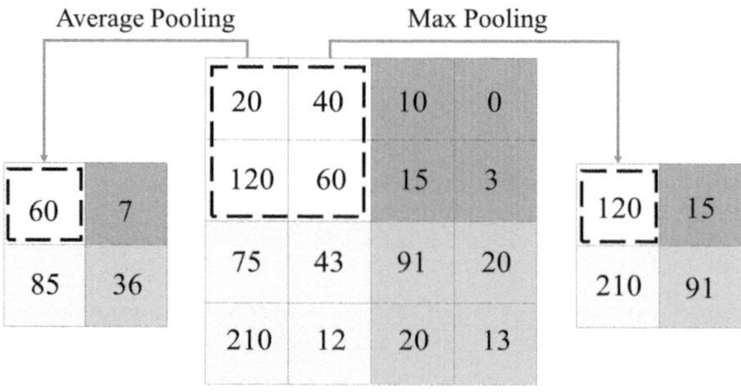

FIGURE 7.13 Operations of average pooling and max pooling.

convolutional layer. The given input is a vector derived from the flattened feature maps. The output of the fully connected layer serves as the final prediction (Figure 7.14).

The input image is compressed into a one-dimensional array and then fed into the fully connected layer. The compressed vector is then further processed in the fully connected layers, where mathematical calculations are performed, and the classification process is conducted. The reason for connecting two layers is that the performance of two fully connected layers exceeds that of a single connected layer.

7.2.2.4 Receptive Field and Feature Map

In neural networks, individual neurons receive input from a certain number of locations within the previous layer. Each neuron in a convolutional layer receives input exclusively from a limited region of the preceding layer, which is called the neuron's receptive field (Figure 7.15). Typically, this

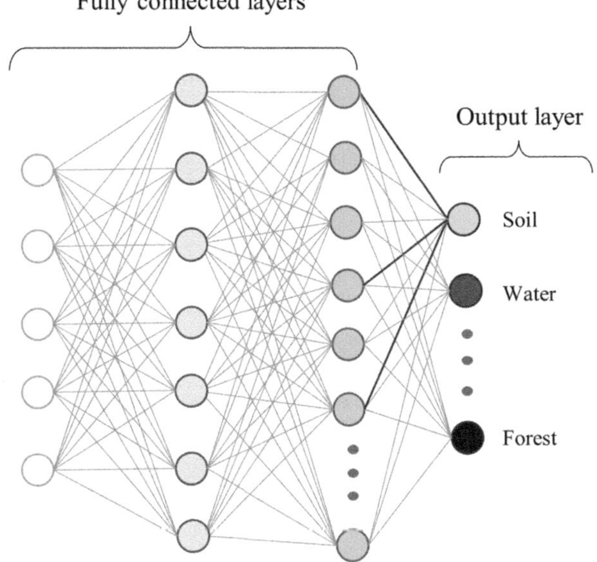

FIGURE 7.14 Representation of a fully connected network.

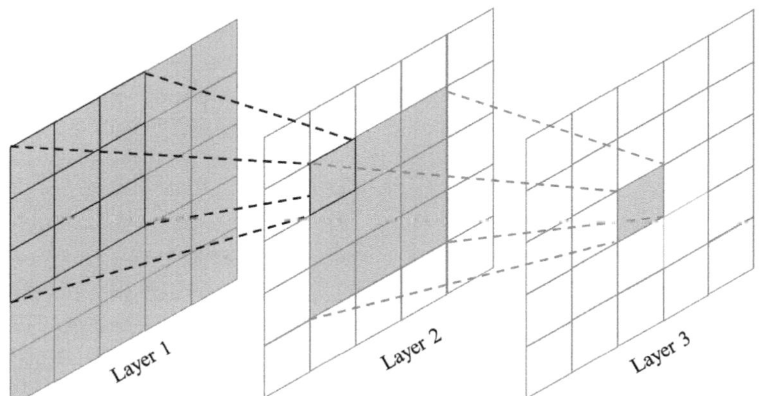

FIGURE 7.15 Receptive field of each convolutional layer determined by a 3×3 kernel. The green region indicates the receptive field of a single pixel in layer 2, whereas the red region indicates the receptive field of a single pixel in layer 3.

region is square (e.g., 3×3 and 5×5). In a fully connected layer, the receptive field encompasses the entire previous layer. Therefore, each neuron in the convolutional layer receives input from a layer than that in the previous layers. This is because of the repeated application of convolution, which considers the value of a pixel with its neighbors. When using extended layers, the total number of pixels within the receptive field remains constant. However, as the size of the field expands because of the cumulative effect of multiple layers, it becomes less densely filled. To control the size of the receptive field, alternatives exist to the conventional convolutional layer. For example, atrous or dilated convolution increases the size of the receptive field by alternating visible and blind areas without increasing the number of parameters. Furthermore, a single dilated convolutional layer can contain filters with different dilation ratios, resulting in a flexible receptive field dimension.

The receptive field denotes the specific area of the input image associated with each pixel on the feature map, indicating the degree to which a pixel perceives the input image. The magnitude of the receptive field directly affects network performance. Typically, a broader receptive field enables the network to gather a greater amount of data, thereby improving its ability to recognize objects. The size of the receptive field is dictated by the architecture of the network, which involves elements such as the dimensions of the convolution filter and a stride. The receptive field of a feature map can be computed iteratively using the following expression:

$$\mathrm{RF}_m = \mathrm{RF}_{m-1} + (k_m - 1) \times \prod_{n=1}^{m-1} s_n. \tag{7.32}$$

The receptive field of the m-layer is determined by adding the receptive field of the $m-1$ layer to the convolution filter of the m-layer. The dimension of the receptive field is computed sequentially from the back to the front layers. Based on the assumption that the receptive field of the input feature is RF_{m-1}, the receptive field following m-layer convolution is RF_m, as shown in Equation (7.32), in which k_m is the size of the convolution filter, and s_n is the nth layer's convolution step.

7.2.2.5 Training CNNs

Training a CNN with multiple layers can be a laborious task due to the nonconvex structure of the loss function. In addition, if the data sets are large, it requires the use of high-performance graphics processing units (GPUs) for computational purposes. The weights of the CNNs are initialized randomly and fine-tuned using a gradient descent algorithm. Transfer learning can be used to achieve effective weight initialization when dealing with a limited training data set. Under such circumstances, it is necessary to solely retrain the classifier or the last layers, thus diminishing the total training duration. By meticulously choosing parameters such as the learning rate, number of iterations, number of training and testing samples, and batch size, a high-quality model can be developed. To enhance the accuracy of the outputs, an ensemble model can be created by training an identical network using many parameter configurations. The network is usually trained using the backpropagation algorithm to update the weights. The training process is hindered by a significant problem known as eliminating gradient flow, which is also referred to as the long time remaining or disappearing gradient problem. This issue makes it more challenging to train the lower layers of a multilayer neural network because it causes the error rate to either decline or exponentially erupt toward the lower layers. This problem can be effectively addressed by carefully choosing a sparse activation function (e.g., sigmoid or tanh). However, the recently introduced rectified linear units (ReLU) function is more effective because of its sparsity and reduced susceptibility to eliminating gradient flow (Goodfellow et al., 2016). In remote sensing applications, including image classification, a commonly employed classifier is the Softmax classifier, which generates probabilities for each class, with the sum of probabilities equal to 1. However, depending on the requirements, alternative classifiers, such as support vector machines or other probabilistic classifiers, may be used.

7.2.2.6 Data Structures in CNNs

In the applications of CNNs, the data typically comprise many spectral bands, where each band represents an observation of a distinct quantity at a specific place in space or time. One benefit of CNNs is their ability to handle inputs of different spatial sizes. It should be noted that traditional matrix multiplication–based neural networks are unable to represent such inputs. CNNs should be utilized even in cases where computing cost and overfitting are not major concerns because they offer a compelling rationale for such operations. For instance, it is ambiguous how to represent a set of images using a weight matrix of a predetermined size, while each image has varying width and height. The resulting output of the convolution process is appropriately scaled. Convolution can be regarded as the process of multiplying matrices. The same convolution kernel produces a distinct circulation matrix of varying size for each input dimension. In certain scenarios, the network permits both the input and output to have varying dimensions. This is particularly useful when assigning a class label to each pixel. In alternative scenarios, such as when there is a desire to assign a singular classification to the entire image, the neural network is required to generate an output of a predetermined size.

To ensure a consistent number of pooled outputs, it is necessary to incorporate extra design measures. One approach is to include a pooling layer that adjusts the size of the pooling regions in accordance with the input size. Convolution can be employed to process inputs with varying sizes, but this is applicable only when the inputs involve different numbers of observations of the same entity type, such as records of varying lengths across time or diverse widths of measurements in space. Convolution is not applicable when the input has a variable size because it can potentially include diverse types of observations without any specific pattern (Goodfellow et al., 2016). For instance, their use of one-dimensional (1D) arrays is infrequent but is steadily increasing in the CNN model. 1D arrays are often similar to pixel-based applications in traditional pixel-based image analysis using spectral information. To clarify, the procedure involves sequentially inputting individual pixels of the images into the input layer of the CNN model and subsequently finding the corresponding output for each pixel (Vali et al., 2020). Conversely, CNNs are commonly employed for the analysis and manipulation of two-dimensional (2D) (e.g., images) and three-dimensional (3D) data (e.g., hyperspectral and volumetric images). Therefore, CNNs are employed in situations with spatial or temporal ordering, regardless of dimensionality (Berman et al., 2019). If the input data are in 2D format, a 2D convolution kernel is applied. Assume that the dimensions of the input image are height×width and have three bands representing the RGB colors. The convolution kernel with dimensions (3, 3) moves across each dimension of the input image (Figure 7.15). To execute a convolution procedure to acquire a value, the value of the image and the value of (height and width) are inputted on each band. The use of a 3D convolutional layer enables the extraction of a more robust volume representation along the three axes and maximizes the benefits of spatial information. The 3D convolution kernel has an additional dimension compared to the 2D convolution kernel, representing the number of 2D slices in hyperspectral images. In the 3D convolutional process, the kernel filter moves the window across the dimensions of height, width, and number of layers on each band (Figure 7.16).

7.2.2.7 Evolving Trends in CNN Design

The architecture of CNNs has witnessed numerous advances from 1989 to the present, falling into distinct categories, including parameter optimization, regularization, structural reformulation, and related techniques. However, the primary focus on improving CNN performance has centered on reconfiguring the processing units and developing novel blocks. Most advances in CNN design have focused on improving depth and spatial exploitation. CNN models can be categorized into eight categories based on architectural modifications: spatial exploitation, multipath, feature map exploitation, width-based multipath, depth-based, exploitation-based, attention mechanisms, and dimension-oriented. CNNs contain various hyperparameters, including learning rate, weights, biases,

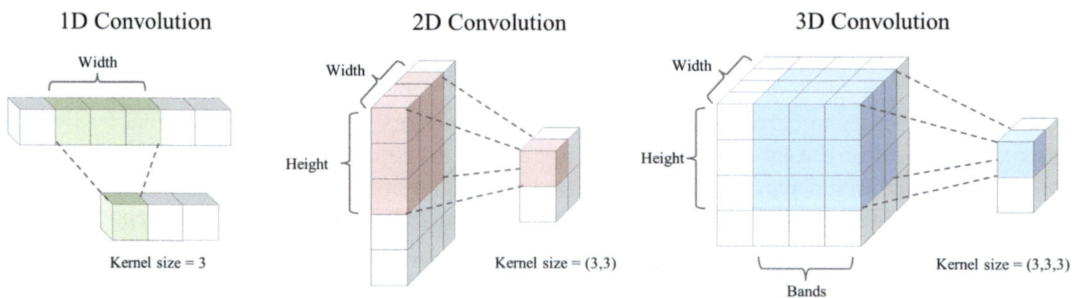

FIGURE 7.16 Data structures for 1D, 2D, and 3D convolutional layers.

number of layers, number of neurons on these layers, activation functions on each layer, stride, and filter size. Different filter sizes allow the study of various levels of correlation, considering the proximity (locality) of the input pixels. A correlation between spatial filters and network learning was recognized when spatial filters were applied in the early 2000s. The analyses also showed that modifying filters could improve CNN performance on coarse- and fine-grained information (Bhatt et al., 2021).

The fundamental concept underlying the deep CNN architecture is the incorporation of additional nonlinear mappings and sophisticated feature structures, allowing the network to effectively mimic the desired function (Bengio et al., 2013). The depth of the network is a crucial parameter for supervised training, with deep networks proving to be more efficient at representing specific functional classes than shallower systems. There is a trade-off between an exponentially large number of neurons and a computationally infeasible result. Delalleau and Bengio (2011) argued that deeper networks can have the same impact as shallower networks, but at a lower cost.

Deep CNNs excel at complex tasks; however, they can face challenges such as declining performance and exploding gradients (vanishing gradients) due to increasing depth rather than overfitting. The vanishing gradient issue results in an increase in both the test and training. The introduction of cross-layer connections or multipath architectures, where skip connections provide direct connections between different layers, tackles these challenges by enabling a tailored flow of information across layers (Hu et al., 2018). The network is divided into parts by cross-layer connections, and the vanishing gradient problem is addressed by propagating gradients to deeper layers. The popularity of CNNs is due to their hierarchical learning and automatic feature extraction capabilities (LeCun et al., 2010). Feature selection (discussed in Chapter 3) has a significant impact on the classification, segmentation, and detection modules. In CNNs, features are selected dynamically through adaptive changes in kernel weights. However, extensive feature sets can introduce noise and lead to network overfitting, highlighting the critical impact of feature map selection in improving network generalization.

From 2012 to 2015, CNN improvements focused on exploiting depth and connection efficiency in network regularization. The discovery by Kawaguchi et al. (2019) emphasized the equal importance of network width (Hanin and Sellke, 2017). Previous research has shown that neural networks using ReLU activation functions require sufficient neurons to maintain a universal approximation property and accommodate an increase in depth (Nguyen et al., 2018). A drawback of DNN topologies is that multiple layers may struggle to acquire meaningful features. While increasing depth allows for diverse feature representations, it does not necessarily increase the learning capacity of the network. Research has shifted from complex designs to expansive and slender architectures to address this problem.

The discriminative ability of neural networks is influenced by different levels of abstraction that prioritize features related to image localization and recognition. This phenomenon, called attention in the human visual system, allows humans to see a scene by combining fragmented views and directing attention to relevant elements. RNN and LSTM networks incorporate attention modules to exclusively focus on relevant features. Researchers employ attention mechanisms in CNNs to improve representation and overcome computational constraints, which are critical for achieving intelligence in object recognition among cluttered backgrounds and complex situations (Bhatt et al., 2021).

7.2.3 Recurrent Neural Networks

Another type of neural network that is frequently employed in sequence analysis is RNNs, which are built to simulate the relationships between different steps in time to extract information from sequences. In comparison with feedforward networks that implement regular functions, recurrent networks, with their internal state loop, share similarities with typical computer programs. RNNs employ sequential data in the network, unlike traditional neural networks. This property, which provides useful information due to the inherent structure in the data sequence, is essential for conducting research on various remote sensing applications. Since neighboring pixels at close spectral bands are strongly dependent on one another and provide contextual and spatial information, image pixels may be viewed as a data sequence. The basic structure of an RNN is shown in Figure 7.17, where the image on the right shows the unrolled version of the model at different time steps. In the figure, x_1, x_2, x_3, x_t represent the input at each time step. RNNs generate output at each layer, and the same output is looped back into the network. An RNN, in contrast to a feedforward network, can deal with dependent, sequential data because it has a recurrent hidden state, the activation of which depends on the previous time at each time step. The network can show dynamic temporal behavior in this manner. RNNs have been widely applied to classify hyperspectral and multitemporal images as spectral sequences of data (Mou et al., 2017; Wu and Prasad, 2017; Lakhal et al., 2018; Liu et al., 2018). Instead of conventional pixel-based classification, RNNs provide an opportunity to consider each pixel with its neighbors (spectral and spatial data) as a vector sequence, thus improving the classification accuracy. The fundamental concept of processing time sequence data is what renders them a popular choice as a DL model for analyzing change detection through temporal remotely sensed data (Lyu et al., 2016; Mou et al., 2018; Sefrin et al., 2021).

Assume that the sequence data $\mathbf{x} = (x_1, x_2, \ldots, x_t)$ are available to represent the information at the tth time step. In image classification using an RNN, x_t corresponds to the spectral value at the tth band. For an RNN, the output of the hidden layer $\mathbf{h}_t$ at time t can be denoted as

$$\mathbf{h}_t = \phi_h\left(\mathbf{W}_{ih}x_t + \mathbf{W}_{hh}\mathbf{h}_{t-1} + \mathbf{b}_h\right) \tag{7.33}$$

where ϕ_h is a nonlinear activation function (e.g., logistic sigmoid or hyperbolic tangent functions), $\mathbf{b}_h$ is the bias vector, $\mathbf{h}_{t-1}$ is the output of the hidden layer at the previous time, and $\mathbf{W}_{ih}$, $\mathbf{W}_{hh}$, and $\mathbf{W}_{ho}$ represent the weight matrices from the current input layer to the hidden layer, the previous hidden layer to the current hidden layer, and the hidden layer to the output layer, respectively. In other words,

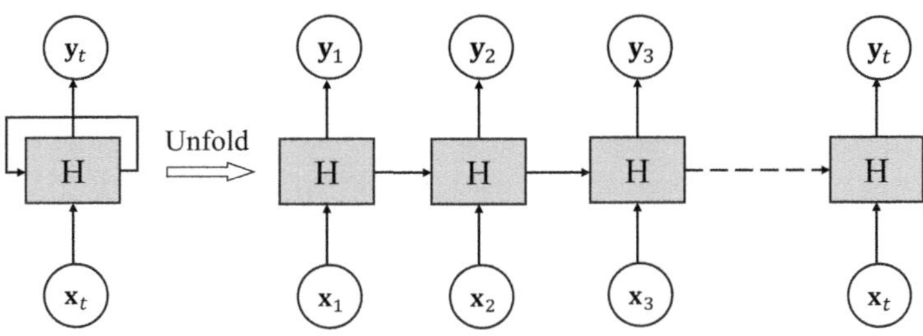

FIGURE 7.17 General structure of a recurrent neural network. Note that $\mathbf{x}_t$ is the input layer, H is the hidden layer, and $\mathbf{y}_t$ is the output layer at time t. The hidden layer H comprises a feedback loop.

the activations from a hidden layer at the previous time step contribute to the input at the current time step, as expressed by the term $\mathbf{W}_{hh}\mathbf{h}_{t-1}$. From Equation (7.33), the contextual relationship can be composed in the time domain. Ideally, the majority of the time information for the sequence data should ideally be captured by $\mathbf{h}_t$. For classification tasks, $\mathbf{h}_t$ is often fed into an output layer, and the probability of the ith class can be derived using the Softmax function (ϕ_o), which can be illustrated as

$$\mathbf{y}_t = \phi_o\left(\mathbf{W}_{ho}\mathbf{h}_t + \mathbf{b}_o\right)$$

$$P\left(\tilde{y} = i\middle|\boldsymbol{\theta},\mathbf{b}\right) = \frac{e^{\theta_i \mathbf{y}_t + b_i}}{\sum_{j=1}^{C} e^{\theta_j \mathbf{y}_t + b_j}} \tag{7.34}$$

where $\mathbf{b}_o$ is a bias vector, $\boldsymbol{\theta}$ and $\mathbf{b}$ are parameters of the Softmax function, and C is the number of classes to discriminate. The weight parameters in Equations (7.33) and (7.34) can be trained using the following loss function:

$$\mathcal{L} = -\frac{1}{N}\sum_{i=1}^{N}\left[y_i \log\left(\tilde{y}_i\right) + \left(1 - y_i\right)\log\left(1 - \tilde{y}_i\right)\right] \tag{7.35}$$

where N is the number of training samples, and y_i and $\tilde{y}_i$ are the true labels and the predicted label of the ith training sample, respectively.

The function in Equation 7.35 can be optimized using a backpropagation through time (BPTT) algorithm (Werbos, 1990), which is used to update the weights of the neuron through a recursive computation of the stochastic gradient descent using the derivative chain rule. The BPTT technique "unfolds" the network in time and propagates error signals backward in time. The algorithm suffers from the vanishing or exploding gradient problem. This problem occurs when the norm of the back-propagated error gradient exponentially decreases with each time step, posing a challenge in learning long-term dependencies within input sequences. More specifically, the repeated occurrence of either large or small derivatives during training can lead to the exponential explosion or vanishing of gradients, as identified by Bengio et al. (1994). This phenomenon ultimately results in weights remaining unchanged in the lower layers, causing the network to lose memorization of initial inputs over time. However, an exploding gradient leads to significant weight updates, introducing instability during training. The ReLU activation function (Section 7.1.4) has been introduced as a solution to the vanishing gradient problem. In ReLU, negative inputs become 0, preventing the backpropagation of errors when encountering negative input values. Leaky ReLU, however, allows for an output between the input and 0.01 when the input is smaller than 0. This ensures that even with a negative activation function, the inputs will not be 0. More details on the training of RNNs and their variants using mathematical foundations can be found in Schuster and Paliwal (1997) and Hang et al. (2019).

7.2.3.1 Long- and Short-Term Memory

LSTM, developed by Hochreiter and Schmidhuber (1997), was proposed to solve the problem of an exploding or vanishing gradient using "memory units" or "gates" that regulate the flow of information. In the network, LSTM includes recurrent connections to memory blocks (Figure 7.18). Every memory block is equipped with multiple memory cells, each having the capacity to retain the sequential states of the network. An LSTM unit consists of two cell states, the memory C_t and the hidden state h_t, and three different gates: input gate (i_t), forget gate (f_t), and output gate (o_t) that controls the flow of information. Throughout the process, the input gate determines which value from the input should be employed to modify the memory, the forget gate identifies the details to be discarded from the block, and the output gate determines how the input and memory of the block combine to produce the output. The three gates combine the current input x_t with the hidden state h_{t-1} coming from the previous time stamp. The two essential functions of the gates include regulating the amount

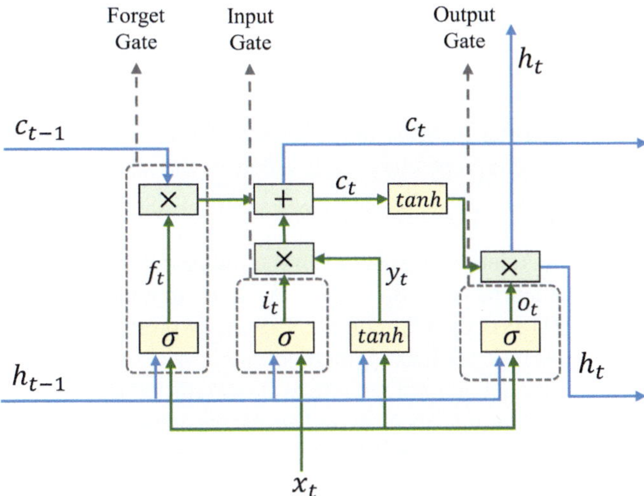

FIGURE 7.18 Basic structure of an LSTM unit. Note that blue lines show the direction in which the information will flow, and green lines indicate internal flows.

of information transmitted or discarded and addressing the issue of vanishing/exploding gradients. These distinctive features enable LSTM to surpass the standard RNN in numerous sequence learning tasks. In a comparative study, Ienco et al. (2017) concluded that while conventional classification methods often overlook temporal correlations, the LSTM model is well suited for extracting long short-term dependencies. However, a drawback of LSTM is its higher number of parameters compared to standard RNNs, making training computationally intensive. To improve their performance further and reduce the redundancy for spatial data, several variants of LSTM that use convolutional layers instead of gate layers have been developed (e.g., Wang et al., 2018; Li et al., 2023a).

7.2.3.2 Gated Recurrent Unit

Another member of the RNN family is the GRU, which was developed by Cho et al. (2014), to reduce the number of gates in the LSTM. GRU, which is characterized by a reduced number of parameters, may be more suitable for the classification of hyperspectral images, especially in scenarios where a limited number of training samples are accessible. The GRU is built on two gating units that play a pivotal role in managing the information flow within the unit. Le et al. (2015) introduced the IRNN, a new type of RNN, showcasing performance on par with LSTM while using an equivalent number of parameters as a standard RNN. The IRNN implements a standard RNN with a reconstructed linear activation function in its recurrent layer. In IRNN, the recurrent weight matrix from hidden to hidden is initialized as the identity matrix, facilitating gradient backpropagation across extended time steps without encountering issues of vanishing or exploding gradients. A notable limitation of IRNN is that the ReLU activation function confines negative input to zero, allowing gradient backpropagation exclusively for positive inputs. If the input is negative, the information associated with the negative input is discarded, and the gradient is prevented from passing through the ReLU. This constraint hinders the network's ability to train effectively. To address the limitations of IRNN, Chandra and Sharma (2017) introduced the Parallel Channel RNN (PC-RNN), utilizing the absolute value of the recurrent layer output. This addition introduces nonlinearity to the proposed RNN while maintaining information flow in both forward and backward propagation steps.

When conducting various image-processing analyses, a common practice is to combine RNN and LSTM networks with other DL models, particularly CNNs. It has been reported that the fusion of CNNs and RNNs could outperform other machine learning methods in land use and land cover classification (Lyu et al., 2016; Song et al., 2018; Mazzia et al., 2019).

7.2.4 Vision Transformers

Initially employed in natural language processing (NLP) with reported success, the vision transformer (ViT) uses the transformer architecture for image classification. Once an image has been divided into patches, a transformer receives the sequence of linear embeddings of these patches as input. In an NLP application, image patches are handled in the same manner as tokens or words. Pretrained on extensive image data sets, the ViT model acquires general image features that can be applied to a wide range of downstream tasks. To determine the final classification, the ViT model extracts information about the relationships between the image patches using a self-attention–based architecture. The ViT model learns to predict labels by showing many images during training. The framework of the standard ViT is depicted in Figure 7.19. After pretraining, a smaller data set is usually fine-tuned for a particular task, such as segmentation or object detection. Promising results have been obtained using the ViT model on multiple image classification benchmarks, and its efficacy has created new avenues for investigating vision-based problems with transformer-based architectures (Dosovitskiy et al., 2021; Pal, 2024). The multihead attention approach is employed by transformers. The memory keys and values in the "encoder-decoder attention" layers originate from the encoder's output, whereas the queries are derived from the previous decoder layer. This enables any location in the input sequence to be addressed to every position in the decoder. Self-attention layers are present in the encoder. The output of the previous layer in the encoder serves as the source of all keys, values, and queries in the self-attention layer. Every point in the encoder has the ability to attend to every position in the encoder's previous layer. The decoder's self-attention layers enable any position to attend to every other position up to and including that position. The decoder inhibits the flow of information to the left to maintain the autoregressive characteristic (Vaswani et al., 2017). Thus, by calculating a weighted total of values from all other portions of the image, self-attention enables the model to capture long-range dependencies between distinct sections of an image. In contrast, multihead attention enables the model to focus on several distinct portions of the input sequence simultaneously and record more intricate connections between various visual elements.

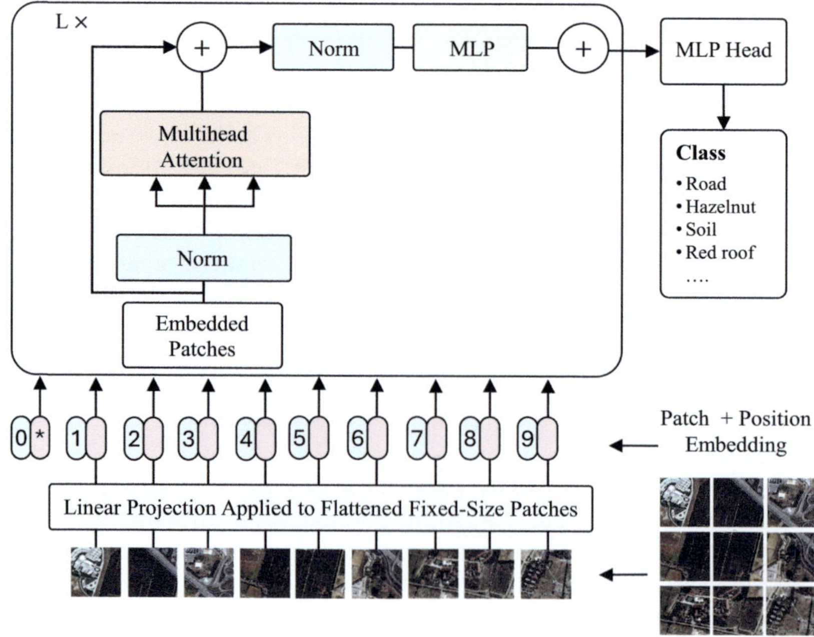

FIGURE 7.19 Vision transformer model.

The ViT model initiates by breaking down the image into a grid of patches or tokens, flattening them, and projecting them linearly to create a sequence of vectors. These vectors are then processed through a succession of transformer blocks, each incorporating a position-wise feedforward network subsequent to a multihead self-attention mechanism. Each multihead attention block in the model comprises numerous parallel self-attention sublayers or heads. These heads individually compute attention weights by projecting the input sequence into three separate vectors: query, key, and value. By taking the dot product of the query matrix and the transpose of the key matrix, a similarity score matrix is formulated. This matrix is then scaled and subjected to a Softmax function to establish the attention weights. The resulting output of the self-attention mechanism for each head is the weighted sum of the value matrix, computed using these weights.

The final output of the multihead attention block is generated by concatenating the outputs of each head and then feeding them through a linear projection layer. Leveraging multiple heads, each equipped with a unique set of learned weights, enables the ViT model to capture more complex and diverse relationships among various elements in the input sequence. The multihead attention method proves valuable in image-processing applications, particularly when high accuracy depends on understanding long-range dependencies between different parts of an image. Through the use of various heads focusing on specific portions, the ViT model effectively captures the intricate and diverse interactions among different patches.

Information loss occurs in ViTs as the input data pass through the model in a downward direction. As a result, the first transformer encoder has more quality information than the subsequent encoders. A modified ViT architecture is created to reduce information loss. The connection scheme involves the first encoder being linked to the last encoder, the second encoder to the one preceding the last, and so forth. Since the final encoder will include both the original and processed data from this process, this model should perform better for small training samples than the ViT model. Scaling up the models becomes challenging as the intricacy of self-attention in ViTs quadratically increases with increasing image size. Bazi et al. (2021) present an extensive evaluation of the ViT model for the classification of remote sensing images. Many attempts have been made to introduce novel ViT models that use multihead attention mechanisms instead of convolutional layers for image classification and object detection. For instance, a hierarchical transformer called the Swin transformer was proposed by Liu et al. (2021). Shifted windows are used to compute the transformer's representation. By restricting self-attention computation to nonoverlapping local windows and permitting cross-window connections, the shifted windowing technique increases efficiency. This hierarchical architecture has a computational cost that is linear with respect to image size and is flexible enough to be modeled at different scales. This approach is suitable for a wide variety of vision tasks, including image classification, semantic segmentation, and object recognition. Chen et al. (2021) proposed a dual-branch transformer called CrossViT to combine image patches (i.e., tokens in a transformer) of different sizes to extract high-level image features. This method handles both large- and small-patch tokens with two distinct branches of varying computational complexity. These tokens are then repeatedly merged with attention to complement one another. In addition, a straightforward but efficient cross-attention token fusion module is used to minimize computation. He et al. (2017a) proposed a model derived from bidirectional encoder representations for the classification of hyperspectral images. This method involves several multihead self-attention layers. Each head encodes a semantic context-aware representation to obtain the discriminative features required for accurate classification.

7.2.5 Deep Multilayer Perceptron

Dominating scientific literature until about 2010, MLP has evolved into a deeper version with end-to-end processing features, namely, CNNs. However, several enhancements have been proposed to MLP to make this legendary architecture a deep MLP model, as an alternative to CNNs and ViTs. Three MLP-based DL architectures, namely, MLP-Mixer (Tolstikhin et al., 2021), S^2-MLP

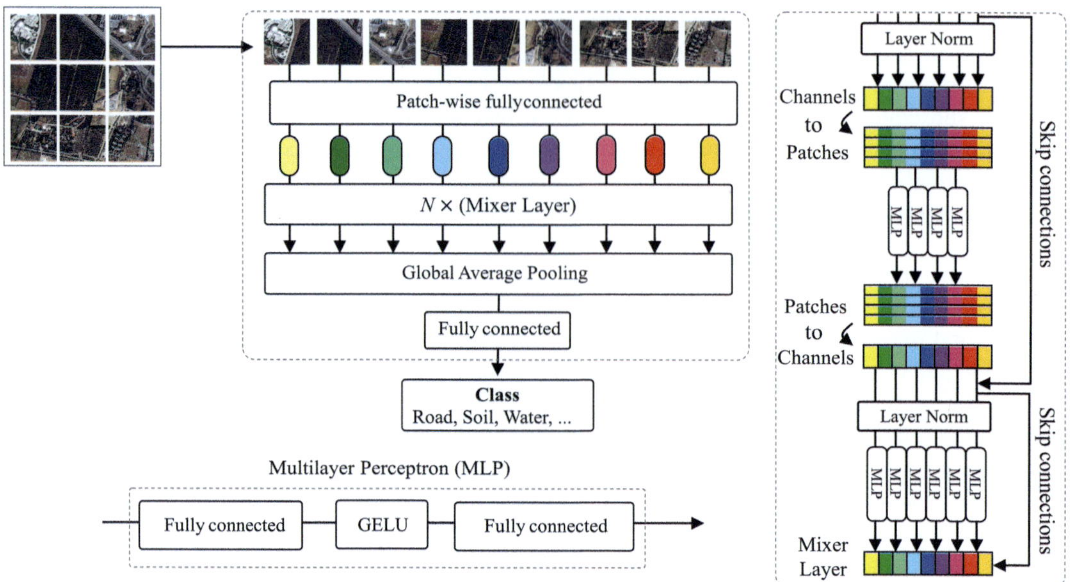

FIGURE 7.20　MLP-Mixer consisting of per-patch linear embeddings, mixer layers, and a classifier head.

(Yu et al., 2022), and HyperMixer (Mai et al., 2022), have been major breakthroughs in the development of deep MLPs. As a simple alternative relying entirely on MLP to reduce inductive biases, the MLP-Mixer introduces two types of layers to replace convolutional layers. The first type entails independently applying MLPs to image patches, thereby "mixing" per-location features with the "token" layer. The second type involves applying MLPs across patches, "mixing" spatial information with the "channel" layer, as depicted in Figure 7.20. Enabling communication between different channels, MLPs for channel mixing independently operate on each token, taking individual rows of the table as inputs (Tolstikhin et al., 2021). On the other hand, MLPs for token mixing facilitate communication between different spatial locations (tokens) by independently operating on each channel and considering individual columns of the table as inputs. The interleaving of these two layers allows interactions between both input dimensions (Pal, 2024). In an extreme scenario, the network architecture can be perceived as a unique CNN employing 1×1 convolutions for channel mixing and single-channel depth-wise convolutions with a complete receptive field. Additionally, it utilizes parameter sharing for token mixing. It is worth noting that convolution introduces added complexity compared to straightforward matrix multiplication in MLPs due to the need for a reduction in matrix multiplication and/or specialized implementation.

Through the channel and token mixing layers, MLPs operate independently on different segments of the data, enabling the network to capture diverse features and their interactions. The preservation of this dimensionality involves taking a series of linearly projected image patches, termed tokens, arranged in a "patches×channels" table, as input. The input data for image classification tasks are commonly organized as a grid of image patches. Each image patch is defined by distinctive features capturing local information about the image content and associated with a specific position or "token" in the input sequence. This iterative process persists until the desired final output is achieved, with the output from each mixer layer being fed into the subsequent layer. One classification label or a collection of probabilities for several labels may be the ultimate result. Research on the application of the MLP-Mixer for image classification tasks indicates that performance degrades when limited training data are available.

Considering the dependency of MLP-Mixer on extremely large-scale data sets, and even producing inferior results than ViT and CNNs for medium-scale data sets (ImageNet-1K and ImageNet-21K),

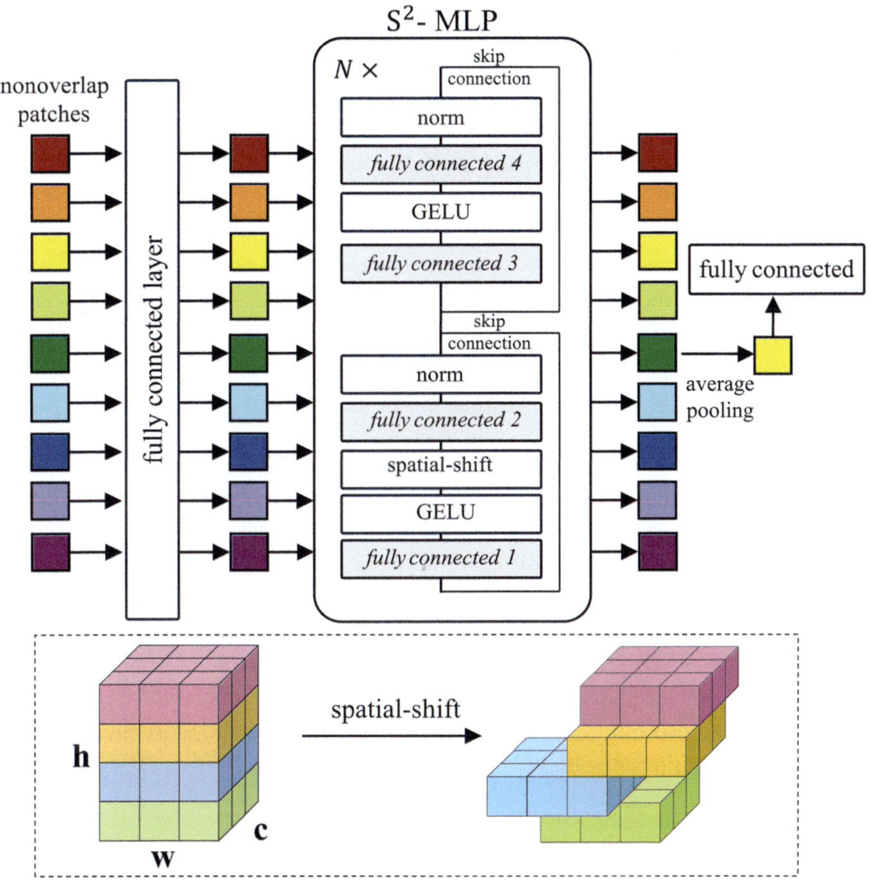

FIGURE 7.21   Architecture of the spatial-shift multilayer perception (S²-MLP) model.

Yu et al. (2022) proposed a novel MLP architecture called spatial-shift MLP (S²-MLP). Since the MLP-Mixer architecture has a spatial-specific property that makes token-mixing MLP prone to over-fitting, S²-MLP only includes channel-mixing MLP in its structure (Figure 7.21). Communication between spatial locations is facilitated by a spatial-shift operation, chosen over a token-mixing MLP for its parameter-free nature and computational efficiency. This spatial-shift operation is spatial-agnostic and maintains a local reception field. In other words, it is parameter-free and efficient for computation. $W \times h$ nonoverlap patches cropped from an image are the input of the model. Passing through a series of S²-MLP blocks, the input is consolidated into a unified feature vector using global average pooling as part of the feature extraction process. Subsequently, this feature vector is directed into a fully connected layer for label prediction. An S²-MLP block contains four fully connected layers, two GELU layers (Hendrycks and Gimpel, 2016), two-layer normalization, two skip connections, and a spatial-shift module. Within each S²-MLP block, the fully connected layer operates similar to the channel-mixing MLP in the MLP-Mixer. The spatial-shift module, an essential part of the proposed method, categorizes c channels into multiple groups and proceeds to shift these distinct channel groups in different directions. When trained on the ImageNet-1K data set, the S²-MLP model achieved higher recognition accuracy than the MLP-Mixer and performed similar to ViT. To improve MLP-Mixer and offer alternatives for deep MLP, some other architectures, including multiscale MLP-Mixer (Zhang et al., 2022), LocoMixer (Yin et al., 2023), and HyperMixer (Mai et al., 2022), have been proposed in the literature.

7.2.6 GENERATIVE ADVERSARIAL NETWORKS

The combination of generative and discriminative perspectives has resulted in the development of a unique machine learning framework called generative adversarial networks (GANs), which was introduced by Goodfellow et al. (2014). GANs, classified as unsupervised DL models, benefit from the capabilities of discriminative models to achieve effective generative models. The GAN model comprises two "adversarial" models engaged in competition: a generative network and a discriminative network (Figure 7.22). GANs, in essence, are built on a game theoretic framework in which the generator network competes against an adversary. Samples $x = g\left(z; \theta^{(g)}\right)$ are directly produced by the generator network. In its adversarial role, the discriminator network differentiates between samples selected from the training data and those created by the generator. The discriminator estimates the probability value given by $d\left(x; \theta^{(d)}\right)$, showing the probability of x being a real training example rather than a fake sample produced by the model (Goodfellow et al., 2016).

In the training process, the generative network is taught to create a mapping from a latent space to a defined data distribution, such as images. Simultaneously, the discriminative network distinguishes between real data and data generated by the generative network. The primary goal in training the generative network is to trick the discriminative network into perceiving the generated examples as realistic representations of the true data distribution. Typically, the discriminative network is structured as a standard convolutional network to generate probabilities. Both networks aim to optimize distinct and opposing loss functions within a zero-sum game, in which a function $v\left(\theta^{(g)}, \theta^{(d)}\right)$ estimates the outcome of the discriminator. The generator receives $-v\left(\theta^{(g)}, \theta^{(d)}\right)$ as its own outcome. Throughout the learning process, each participant attempts to maximize its own outcome, making the convergence of g^* the ultimate objective, as formulated below:

$$g^* = \arg\min_g \max_d v\left(g, d\right) \tag{7.36}$$

The default choice for v is

$$v\left(\theta^{(g)}, \theta^{(d)}\right) = \mathbb{E}_{x \sim p_{\text{data}}} \log d\left(x\right) + \mathbb{E}_{x \sim p_{\text{model}}} \log\left(1 - d\left(x\right)\right). \tag{7.37}$$

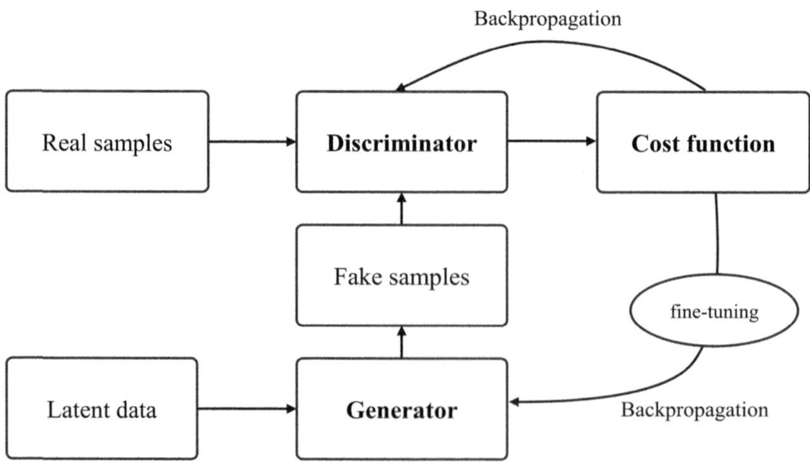

FIGURE 7.22 General framework of standard GANs.

The discriminator is driven to learn the accurate prediction of samples as real or fake, while, concurrently, the generator endeavors to deceive the classifier into perceiving its samples as real. Upon reaching convergence, the generator's samples are indistinguishable from the real data.

Nonconvergence, as identified by Goodfellow et al. (2014), is a potential issue that causes GANs to underfit, particularly due to the limited convergence of the discriminator's cost function. Overall, the simultaneous descent of gradients on the costs of both players is not assured to achieve equilibrium. Hence, careful hyperparameter setting is crucial when applying GANs. It has been shown that the standard cost function derived from the Jensen-Shannon divergence (JS divergence) is not sensitive when handling distributions that are supported by low-dimensional manifolds. The Wasserstein distance, as proposed by Arjovsky et al. (2017), addresses this critical issue in the training of GANs. In addition, Mirza and Osindero (2014) introduced an efficient version of GANs, called conditional GANs. By conditioning the model on additional information, it is possible to direct the data generation process because there is no control over the modes of the data generated in an unconditioned generative model. Such conditioning could be based on class labels or data from different modalities. On the other hand, Mao et al. (2017) proposed a least squares GAN (LSGAN) that replaces the sigmoid cross-entropy with the least squares loss function for the discriminator to mitigate the vanishing gradient problem during training and thus provides a more stable learning process.

Several models based on GANs have been developed, and superior performances have been reported in various remote sensing applications, including image classification, object detection, and data translation. For instance, Springenberg (2015) introduced categorical GAN incorporating unsupervised and semi-supervised learning for hyperspectral image classification. To overcome the difficulty of classifying such images using GANs, resulting from their 1-D spectral structure, Zhan et al. (2018) developed GANs for hyperspectral images (HSGAN) using a semi-supervised framework. When HSGAN was trained using unlabeled hyperspectral data, the generator produced hyperspectral samples that were similar to the real data, whereas the discriminator contained features that could be used to classify hyperspectral data with only a small number of labeled samples. Zhu et al. (2018) proposed a GAN that used a CNN to discriminate the inputs and another CNN to generate the so-called fake inputs. It was reported that this adversarial training approach improves the generalization capability of the discriminative CNN, which is important when the training samples are limited. The proposed methodology was tested on three widely used and freely available hyperspectral data sets, and competitive results were reported. Comprehensive reviews of Dash et al. (2024) and Jozdani et al. (2022) systematically survey the development and applications of GANs in the field of remote sensing.

7.2.7 Deep Autoencoders

An autoencoder, introduced by Rumelhart et al. (1986), is a particular type of neural network that employs an unsupervised algorithm to learn a compressed and informative representation of the input data. Autoencoders, also known as autoassociative neural networks, are trained with the goal of copying their input to their output with the least possible error. These are symmetrical unsupervised neural networks used to learn the internal characteristics of the data set by minimizing the reconstruction error between the input at the encoding layer and its reconstruction at the decoding layer. Employing unsupervised learning, the autoencoder automatically reveals the inner structure of the data by unraveling the sources of variation in the input data. The principal feature of these networks is their capacity to project the original input samples into a new space, creating outputs that are compressed, extended, or even of equal dimensions with minimal distortion (Paoletti et al., 2019). Unlike supervised learning in a neural network, where each training observation x_i has a label or expected value y_i, autoencoders only have unlabeled observations. If we assume that a training data set S_T consists of only unlabeled observations, as shown below:

$$S_T = \left\{ x_i \mid i = 1,\dots,M \right\} \tag{7.38}$$

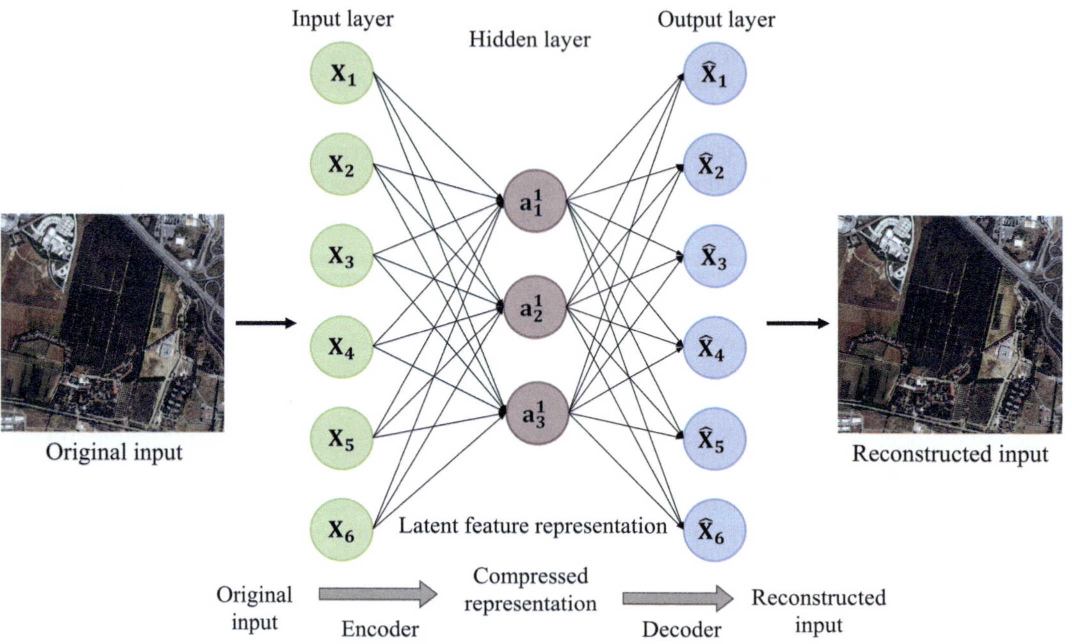

FIGURE 7.23 General architecture of an autoencoder.

where in general $x_i \in \mathbb{R}^n$ with $n \in \mathbb{N}$, an autoencoder learns to reconstruct input observations x_i (Michelucci, 2022). The autoencoder primarily consists of encoder and decoder layers, which are inherently connected by a bottleneck layer, which represents the latent feature representation (Figure 7.23). The encoder is responsible for taking an input (e.g., images, videos, and texts) and mapping it into the latent encoding space, creating a latent feature representation. This encoding represents a compressed form of information, presenting a smaller version of the input data. On the contrary, the decoder takes the output of the encoder and attempts to reconstruct it as accurately as possible. Typically, the encoder and decoder are neural networks with variable network depth and can be trained using custom architectures. Thus, they can be fully connected neural networks or have convolutional and pooling layers. The increased complexity of the architecture allows the autoencoder to learn more complex latent coding.

The encoder function g, which has its own set of learnable parameters, takes an input x_i and creates the latent feature representation $\mathbf{h}_i$ as follows:

$$\mathbf{h}_i = g(x_i) \tag{7.39}$$

where $\mathbf{h}_i \in \mathbb{R}^q$ and $g: \mathbb{R}^n \to \mathbb{R}^q$. Then, the decoder function f then takes the latent feature representation $\mathbf{h}_i$ and creates a reconstructed $\tilde{x}_i$:

$$\tilde{x}_i = f(\mathbf{h}_i) = f(g(x_i)). \tag{7.40}$$

where $\tilde{x}_i \in \mathbb{R}^n$. The problem in training the autoencoder is learning functions $g(\cdot)$ (encoder) and $f(\cdot)$ (decoder) that satisfy

$$\arg\min_{f,g} \left\langle \left[\Delta(x_i, f(g(x_i))) \right] \right\rangle \tag{7.41}$$

where Δ is the reconstruction loss function, which measures the difference between the input and the output of the autoencoder. In the training, the network is penalized by the loss function for producing outputs that differ from the input. On the other hand, $\langle \cdot \rangle$ denotes the average of all observations. It may be possible to find functions f and g based on the autoencoder's design, allowing the autoencoder to achieve perfect output reconstruction. However, this is not particularly useful; thus, autoencoders are designed to be unable to learn to copy perfectly. Therefore, restrictions are applied to autoencoders, allowing them to copy only input data resembling the training data. Due to the necessity of prioritizing the characteristics of the input to copy, the model often learns the beneficial properties of the data (Goodfellow et al., 2016).

7.2.7.1 Undercomplete Autoencoders

Obtaining essential features from the autoencoder is achieved by imposing constraints on its architecture, particularly by limiting the dimension of the latent code, represented as **h**, to be smaller than that of the input data **x**. Such an autoencoder is referred to as undercomplete. The encoder in the undercomplete autoencoder is responsible for projecting the data into a lower-dimensional space, thereby accomplishing dimensionality reduction. The difficulty in performing dimensionality reduction and subsequent reconstruction arises when the input features are independent. Nevertheless, in the presence of an inherent structure in the data, particularly correlations among input features, leveraging and understanding this structure can be achieved by guiding the input through the latent code (Pinaya et al., 2020). Using a "bottleneck" in their architecture, undercomplete autoencoders restrict information flow. The general architecture of the undercomplete autoencoder is shown in Figure 7.24.

The primary limitation of undercomplete autoencoders emerges when employing deep architectures, particularly those with numerous hidden layers, leading to a tendency for the model to overfit the data. Overfitting occurs when the model learns random fluctuations in the training data as primary features, resulting in reduced generalizability as its performance markedly decreases when exposed to new data (Delgado and Oyedele, 2021).

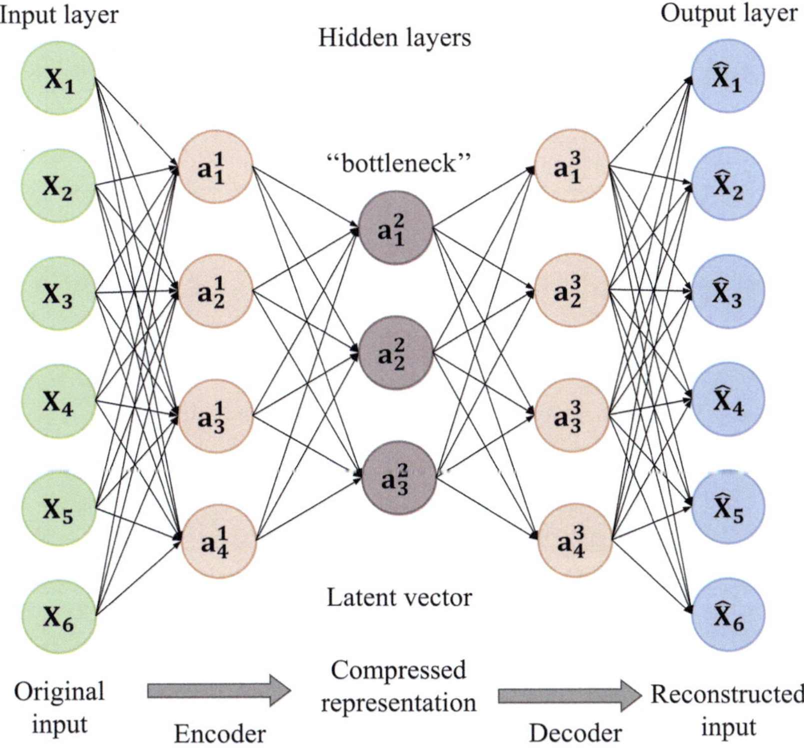

FIGURE 7.24 Architecture of the undercomplete autoencoder.

7.2.7.2   Regularized Autoencoders

The undercomplete autoencoders, characterized by a latent code dimension lower than the input data dimension, have the ability to learn the most crucial features of the data distribution. However, when both the encoder and decoder have excessive capacity, these autoencoders fail to learn any meaningful information. If the latent code is permitted to have a dimension equal to that of the input, and in situations of overcompletion where the latent code dimension exceeds that of the input, a similar issue arises (Goodfellow et al., 2016). Considering this circumstance, regularized autoencoders can be trained successfully by selecting the code dimension and capacities of the encoder and decoder depending on the complexity of the distribution to be modeled. Instead of constraining the model's capacity through shallow encoders and decoders with a small code dimension, regularized autoencoders employ a loss function that promotes additional properties, as well as simply copying input to output. More specifically, representation sparsity, robustness to noise or missing inputs, and small derivatives of the representation are examples of these additional properties. Several options for regularization exist that force the autoencoder to learn a distinct representation of the input data, and these are described below.

7.2.7.3   Sparse Autoencoders

Unlike undercomplete autoencoders, which reduce the number of neurons in the latent code $\mathbf{h}$, sparse autoencoders can also learn meaningful representations with many neurons in the latent code, by using the so-called sparsity constraint. With this constraint, the number of active neurons in the model is limited to impose a bottleneck in the information flow. In this way, the neuron whose output value is close to 1 is activated and the neuron whose output value is close to 0 is inactivated. As illustrated in Figure 7.25, the highlighted nodes are activated in the hidden layers of the sparse autoencoder, and the transparentized nodes will not be used. By applying sparsity to the hidden

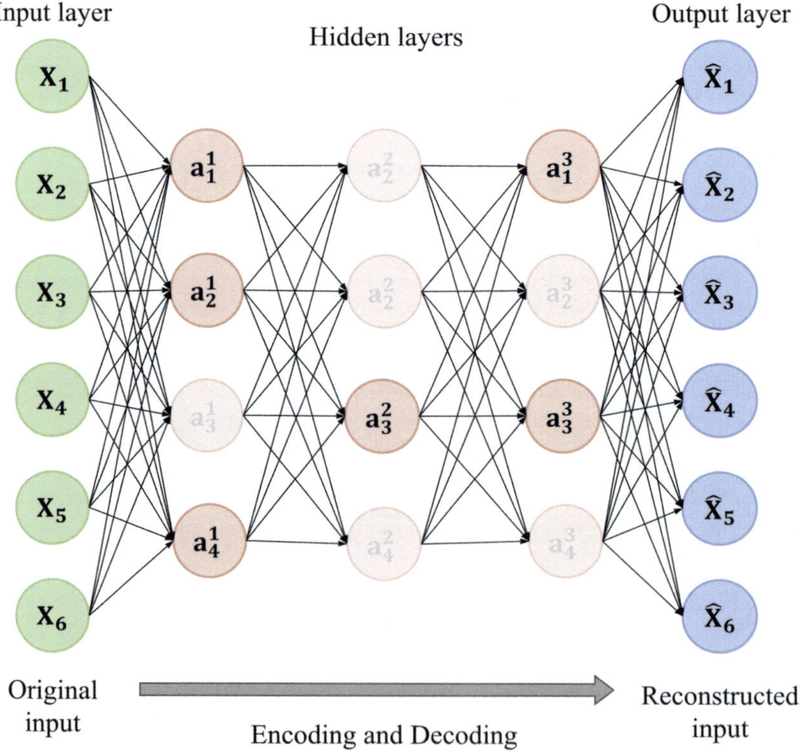

FIGURE 7.25   Structure of the sparse autoencoder.

layers, the autoencoder is forced to represent each input as a combination of a limited number of neurons, even if the latent code **h** has a large dimension. This method prevents the model from learning the noise in the input data and reproducing the input identically, enabling the model to learn the intrinsic features in the input data.

One way to impose sparsity on the latent code is to add a penalty term to the loss function that penalizes the absolute value of activations (L_1-norm). Hence, the optimization objective of such an autoencoder is as follows (Bank et al., 2021):

$$\arg\min_{f,g} \left\langle \left[\Delta\big(x_i, f\big(g(x_i)\big)\big) \right] \right\rangle + \lambda \sum_i |a_i| \tag{7.42}$$

where a_i is the activation at the ith hidden layer and is scaled by a λ parameter.

Another method is to add a Kullback-Leibler (KL) divergence term, which is a measure of the difference between two probability distributions, to the loss function. In such circumstances, the distribution function assumes a value of 1 with a small probability p and a value of 0 with a probability of $1 - p$, similar to that of a Bernoulli distribution. When the activation of neurons deviates significantly from the sparsity rate, as specified by p, there will be a considerable difference between the distributions. This results in a loss function with a high value, which penalizes the solution of this network. In this way, the overall loss function of the model is expressed as follows:

$$\arg\min_{f,g} \left\langle \left[\Delta\big(x_i, f\big(g(x_i)\big)\big) \right] \right\rangle + \sum_j KL\big(p \| \hat{p}_j\big) \tag{7.43}$$

where the aim of the regularization term is matching p to $\hat{p}$. In general, sparse autoencoders are used for various applications, including hyperspectral imagery classification (Tao et al., 2015), hyperspectral unmixing (Ozkan et al., 2018), and scene classification (Othman et al., 2016).

7.2.7.4 Denoising Autoencoders

Denoising autoencoders, as introduced by Vincent et al. (2008), are inspired by the human ability to recognize partially occluded or corrupted images. In these types of autoencoders, the original input data are partially corrupted by adding random noise before feeding it into the model. Specifically, adding noise is a strategy to prevent the autoencoder from simply copying the input to the output. There are several ways to corrupt the original input data by adding noise, including salt-and-pepper noise, masking noise, and additive isotropic Gaussian noise. Another way to add noise is to randomly convert some input values to null values, as depicted in Figure 7.26. For additive isotropic Gaussian noise, the original input x is corrupted to obtain the corrupted version, $\tilde{x}$:

$$\tilde{x} \,|\, x = \mathcal{N}\big(x, \sigma^2 I\big) \tag{7.44}$$

The main objective of training denoising autoencoders is the reconstruction of clean "repaired" input from partially corrupted input. To achieve this, the denoising autoencoder learns to extract useful features from the input data and capture the relationship between the input data. Thus, it reconstructs the input based on the structure of the original input data, ignoring the noise in the input. In this case, denoising autoencoders still aim to minimize the reconstruction loss function, which measures the difference between the input and its reconstruction from the latent code representation. Denoising autoencoders are used in various remote sensing applications, including noise reduction in the spectral domain of hyperspectral images (Zhang et al., 2020), feature extraction and classification of hyperspectral images (Xing et al., 2016), and object-based land cover supervised classification of UAV imagery (Zhang et al., 2017).

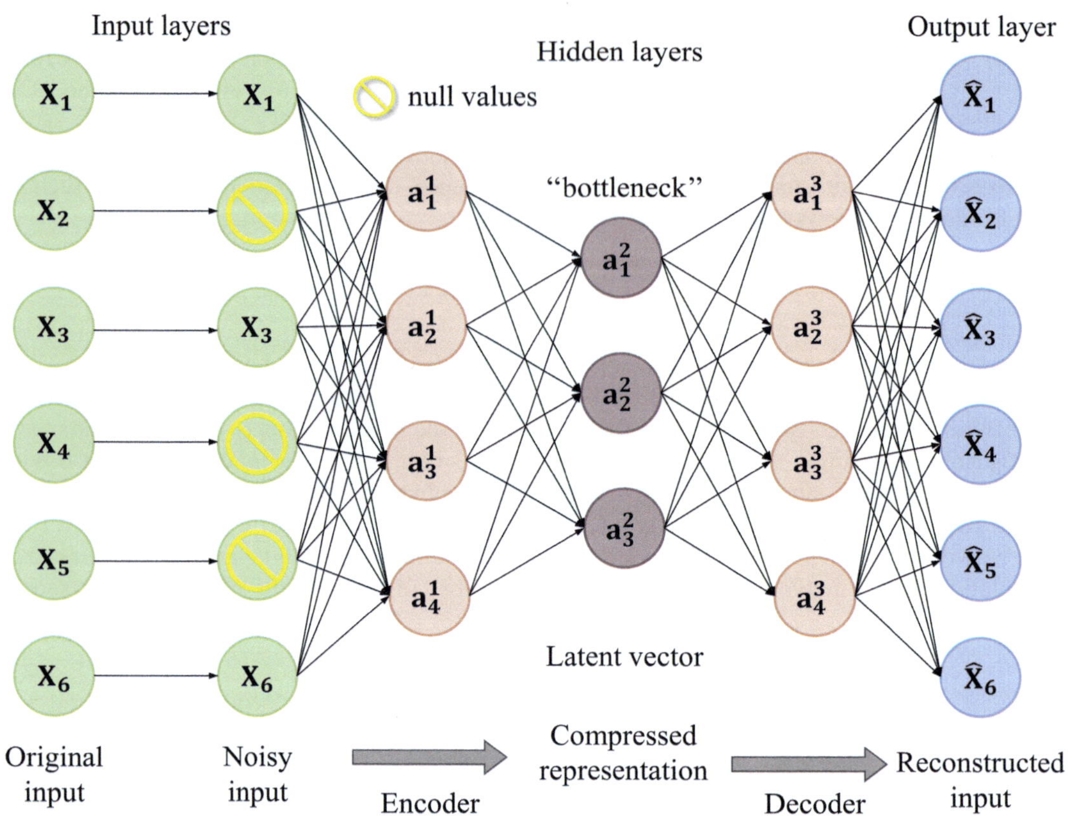

FIGURE 7.26   Structure of the denoising autoencoder.

7.2.7.5 Variational Autoencoders

As generative models, variational autoencoders (VAEs) learn the underlying probability distribution of the data set and solve variational inference (VI) problems by constructing learning representations for high-dimensional distributions (Figure 7.27). Kingma and Welling (2013) and Rezende et al. (2014) created stochastic backpropagation rules that made it possible to backpropagate to both the mean and the covariance parameter using Gaussian distributions with limited variance. Stochastic backpropagation rules with these features are employed in VAE training. VAEs integrate a differentiable generating network with a second neural network, resembling the structure of GANs. In VAEs, the second network acts as a recognition model, performing approximative inference. GANs face challenges in modeling discrete data due to the need for differentiation via visible units, whereas VAEs, which require differentiation through hidden units, cannot accommodate discrete latent variables (Goodfellow et al., 2014). Learning a VAE involves optimizing an objective that guides the latent space to adhere to a predefined prior distribution while balancing the quality of samples autoencoded by stochastic encoders and decoders. VAEs have gained prominence as a favored framework among various generative models. However, VAE models frequently face challenges in reaching a satisfactory balance between the quality of generated samples and the precision of reconstructions (Ghosh et al., 2020).

A probabilistic model $P_\theta(x)$ of sample x in a data space with a latent variable z in a latent space can be written as follows:

$$p_\theta(\mathbf{x}) = \int_z p_\theta(\mathbf{x} \mid z)\, p(z). \tag{7.45}$$

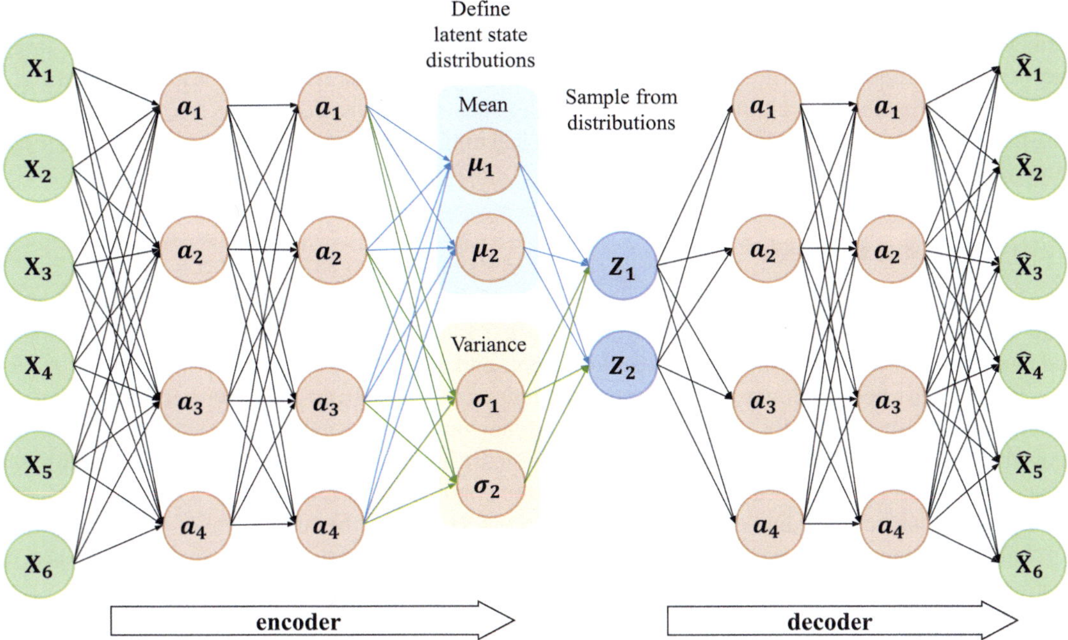

FIGURE 7.27   Structure of the variational autoencoder.

VI can be used to model the evidence lower bound $\log p_\theta(\mathbf{x})$ as follows:

$$\log p_\theta(\mathbf{x}) = \mathbb{E}_{q_\phi(z|\mathbf{x})}\big[\log p_\theta(\mathbf{x}\mid z)\big] - D_{KL}\big(q_\phi(z\mid \mathbf{x})\,\big\|\,p(x)\big) \tag{7.46}$$

where $q_\phi(z\mid\mathbf{x})$ and $p_\theta(\mathbf{x}\mid z)$ are the encoder and decoder models, respectively, and ϕ and θ indicate their parameters.

Several variants of VAEs have been proposed to overcome the major shortcomings of the model. For instance, Burda et al. (2015) introduced the importance-weighted autoencoder (IWAE), which uses a strictly tighter log-likelihood lower bound derived from importance weighting. Within the IWAE model, the recognition network incorporates multiple samples to estimate the posterior, providing greater flexibility for modeling complex posteriors that may not conform to VAE assumptions. They reported that IWAEs develop more sophisticated representations in the latent space than VAEs, contributing to improved test log-likelihood in density estimation benchmarks. Chen et al. (2017) presented a variational lossy autoencoder that offers control over the learning of global latent code by appropriately designing the architecture. The global latent code is directed to eliminate irrelevant details, such as texture information in images, leading to the VAE encoding data in a lossy manner. Other variants of VAEs include vector-quantized variational autoencoder (VQ-VAE) by Van Den Oord and Vinyals (2017), denoising variational autoencoder (DVAE) by Im et al. (2017), generation of small graphs using variational autoencoders (GraphVAE) by Simonovsky and Komodakis (2018), and DAG variational autoencoder (D-VAE) by Zhang et al. (2019a).

7.3   LEARNING PARADIGMS

The scarcity of labeled training data poses a significant challenge that has a significant impact on the performance and generalization capabilities of DL models. In recent decades, innovative learning strategies have been developed to solve the data scarcity problem and improve learning capabilities by presenting solutions to key issues, including the vanishing gradient. Together with the architectural revolution, newly introduced strategies have also greatly contributed to the DL era. Without the use of these robust strategies, it would have been impossible to train such large networks that require a large number of data sets and hyperparameter optimization. To shed light on these well-established approaches, the following subsections are devoted to five primary learning paradigms: transfer learning, semi-supervised learning, reinforcement learning, active learning, and multitask learning.

7.3.1 Transfer Learning

Humans can intelligently apply their knowledge from previous experiences to find solutions to new and unseen problems. Transfer learning, also called knowledge reuse, pretraining, and domain adaptation, is inspired by this fundamental concept, and seeks to enhance the performance of target learners in specific domains by using knowledge extracted from different yet related source domains. This strategy helps to reduce the dependency on a large volume of target domain data for constructing target learners. The process of transfer learning is illustrated in Figure 7.28. Traditional machine learning methods can be effective under the condition that training and testing data adhere to the same feature space and distribution assumptions. However, the fulfillment of this assumption is challenging in many scenarios, prompting the necessity to reconstruct models entirely with newly gathered data sharing a similar distribution. In many practical situations, obtaining sufficient training data is often costly, time-consuming, or even unattainable. This issue is particularly important for CNNs, which usually have many parameters (Windrim et al., 2018). Semi-supervised learning, discussed in the following subsection, provides a partial solution to this issue by alleviating the

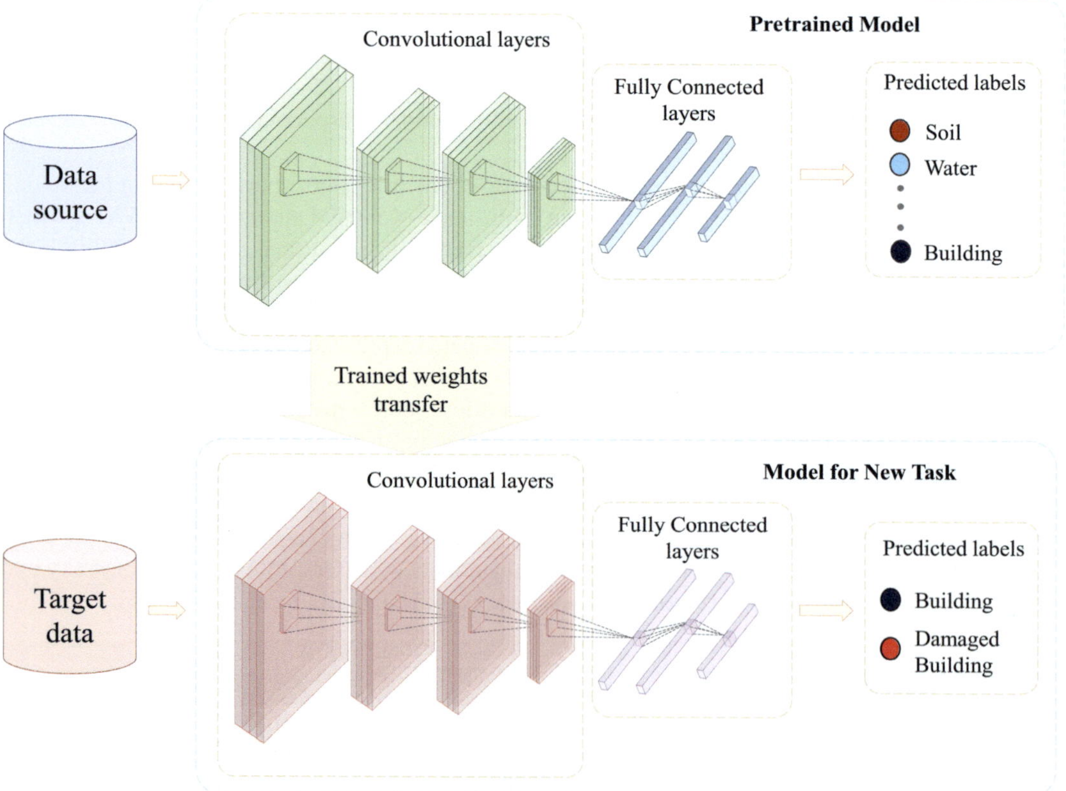

FIGURE 7.28 Schematic diagram of transfer learning.

requirement for an abundance of labeled data. Typically, a semi-supervised approach requires only a small number of labeled samples, leveraging a substantial amount of unlabeled data to enhance learning accuracy. However, gathering unlabeled instances is also challenging in many cases, leading to suboptimal solutions with traditional models. In these situations, the optimal solution involves knowledge transfer or transfer learning between different task domains. Transfer learning also has similarities with multitask learning. While both transfer learning and multitask learning share the common goal of capturing commonalities across tasks, transfer learning concentrates on learning a target task with certain source tasks serving as auxiliary information. In contrast, multitask learning seeks to jointly learn a set of target tasks to enhance the generalization performance of each task without relying on source or auxiliary tasks (Yang et al., 2020a).

Findings from earlier studies have shown that the application of transfer learning approaches is effective in preventing overfitting, speeding up the training process, and reducing the requirement for a large data set (Lyu et al., 2018; Hendrycks et al., 2019; Cui et al., 2020). In transfer learning, knowledge acquired through pretraining on large data sets, such as ImageNet, is leveraged and applied to a target task characterized by limited labeled data. This strategic approach enables models to apply learned representations, effectively transferring knowledge to new tasks; thus, learning performance and generalization are improved significantly. Owing to its adaptable application possibilities, transfer learning has gained prominence as a popular and promising field in machine learning, particularly within the domain of DL. The mathematical foundations and applications of transfer learning have been discussed in several survey studies (e.g., Pan and Yang, 2010; Weiss et al., 2016; Ma et al., 2024).

Among the various strategies designed to address the data scarcity problem in DL, pretraining via transfer learning stands out as a key technique in both research and practical applications. The fundamental issue is to capture important representations from external data and seamlessly transfer the obtained knowledge from the source domain to the target domain, aiming to enhance the efficiency of learning for the target task. Acknowledged as a strategy to use existing knowledge rather than initiating training from scratch, pretraining is well established for reducing the DL model's dependency on the size of the training data set, especially in cases where the target domain data set is limited. It is important to note that for effective transfer learning, the two tasks should have the same type of inputs and share similar characteristics to transfer low-level features or internal representations. One downside of transfer learning is the likelihood of domain mismatch challenges arising between the pretraining and target domains. The widely adopted approach in transfer learning usually includes fine-tuning a pretrained network model on a specific target data set (Windrim et al., 2018). When examining CNNs, it is evident that numerous image-processing tasks necessitate similar foundational low-level features found in the early layers of a DNN. In contrast, the later layers tend to be more task-specific, representing specialized high-level features, and showcasing the adaptability of convolutional networks in the transfer learning context. Thus, pretrained networks are generally integrated into the initial layers, enabling the subsequent layers to be trained to extract more targeted and task-specific features. In their study, Girshick et al. (2014) pretrained a CNN on ImageNet and subsequently adjusted all network parameters through fine-tuning to address data scarcity in the target task, specifically object detection and semantic segmentation. Long et al. (2015) demonstrated that by fine-tuning only the parameters of the last few layers, features tailored to the original task within these layers did not effectively overcome the domain discrepancy. Yang et al. (2017) proposed a CNN characterized by a two-branch architecture that incorporates the spectral and spatial domains. The low and mid-layers of the network were pretrained and transferred from other data sources, and only the top layers were trained using a constrained set of samples obtained from the target scene. The experimental results demonstrated enhanced performance because of the incorporation of transferred features. Research by Wambugu et al. (2021) involved a comprehensive review of studies exploring models and techniques, with a particular emphasis on transfer learning, used to tackle the challenge of insufficient samples in hyperspectral image classification using DL models.

The training of DL models often involves images taken in a specific geographical region, within a short timeframe, or using a particular sensor, all of which define a specific domain for analysis. When applied to classification tasks, spectral shifts between training and test images may occur because of disparities in acquisition conditions, geographical locations, atmospheric variations, and other factors. Therefore, if a model is trained on a particular data set, its accuracy will be high when applied to images within the same domain, considering sensor type and spectral and spatial resolutions. Nevertheless, if this model is used to segment images acquired under different conditions, a notable decrease in performance is expected due to the domain shift between the source and target domains. When faced with such challenges, the incorporation of domain adaptation becomes crucial to tailor models trained in a source domain for a separate yet related target domain. It is noteworthy that domain adaptation can be considered a specific type of transfer learning (Tuia et al., 2016). It is applicable in scenarios where the features of the target space in the domain may differ from those of the source domain. A promising field of investigation for domain adaptation and transfer learning includes delving into GANs. As an example, Benjdira et al. (2019) suggested employing GANs to transform an image from the source domain to the target domain so that the source images emulate the characteristics of the images in the target domain.

In the design of a transfer learning method, there are three critical research questions: when to transfer, what to transfer, and how to transfer. *When to transfer* is related to the cases when the question is whether to transfer between source and target domains, considering the relationship between these domains. It is possible to have a negative transfer effect when the model fails to learn the characteristics of new data due to the irrelevancy of the source domain. Preliminary analysis and prevention of negative transfer are the subjects of the current research. *What to transfer* involves the question of what specific knowledge is transferred between domains. Because some knowledge can be specific to domains, learning algorithms need to be developed to address this crucial issue. *How to transfer*, on the other hand, deals with the way of transferring knowledge between domains. Considering this concept, transfer learning methods can be categorized into four groups: instance-based, feature-based, model-based, and relation-based algorithms (Yang et al., 2020a). Taking into account the differences between domains, transfer learning algorithms can be further classified into two groups: homogeneous and heterogeneous transfer learning. Homogeneous transfer learning is specifically dedicated to enhancing generalization performance within similar domain representations, resulting in a more limited approach. Within these learning approaches, the source and target data share the same feature space. On the other hand, heterogeneous transfer learning, which encompasses knowledge from different domains with distinct feature spaces, has the potential to reveal analogous knowledge in the two domains. Heterogeneous transfer learning empowers the transfer of diverse perspectives or aspects of knowledge from a source to a target domain. It introduces increased flexibility in the selection of a source domain, allowing for choices from either a different feature space or a distinct label space. Considering the characteristics of both types of learning, it can be concluded that heterogeneous transfer learning reflects human intelligence more closely, enabling the easy transfer of knowledge across different types of signals.

7.3.2 SEMI-SUPERVISED LEARNING

Remote sensing classification tasks commonly involve a large number of unlabeled samples and a comparatively modest number of labeled (training) samples. In such cases, traditional supervised classification methods often encounter difficulties, primarily due to overfitting. Semi-supervised learning (SSL) has emerged as a valuable approach to address this overfitting challenge by incorporating learning from both labeled and unlabeled data. This is intimately connected to the fundamental objective of deriving meaningful inferences from the available data. In fact, SSL combines supervised and unsupervised learning techniques, addressing the key challenges of both approaches. The fundamental idea behind SSL is that if both training samples and semi-labeled samples or pseudolabels are used simultaneously to estimate the parameters of the decision rule, better performance with lower bias and variance can be achieved than when using training samples alone (Jackson and Landgrebe, 2001). Starting with an initial training phase on a constrained set of samples, the model is subsequently deployed on unlabeled data to derive pseudolabels. This process is commonly known as semi-label prediction. To collect reliable samples from the unlabeled data, the trained model is applied iteratively. Thus, a new training sample set for classification is generated by combining the most confident predictions with the labeled data. There may be several iterations of the procedure, with each one adding an increasing number of pseudolabels. Until convergence is attained, this process is repeated. Every iteration should improve the performance of the model if the data are appropriate for the procedure. Reducing the amount of labeled data is the main advantage of this learning strategy. However, a major drawback of the method is the inference of inaccurate conclusions due to the selection of irrelevant input features in the training data. Semi-supervised methods are popular when integrated with traditional machine learning techniques. More recently, they have been used for various remote sensing applications within the context of DL. In general, SSL is employed using the five major models. These are (1) generative models, estimating the conditional density for predicting labels of unlabeled samples, (2) self-training schemes, iteratively using the previous classification map to train the classifier, (3) co-training methods, training multiple classifiers with independent subsets of the training set, using unlabeled samples with high reliability to update the training set and train additional classifiers, (4) transductive support vector machines (TSVMs), seeking to maximize the margin for both labeled and unlabeled samples, and (5) graph-based methods that propagate label information from each sample to its neighbors until a globally stable state is reached across all samples (Ma et al., 2016). It should be noted that the SSL methods are mostly based on generative models, which estimate conditional density, adhering to specific statistical distributions, and have been extensively applied to remote sensing image classification.

Many attempts have been made to develop strategies and classifiers based on the SSL philosophy. For example, Blum and Mitchell (1998) introduced a specific type of SSL that employs two classifiers to acquire more objective perspectives on unlabeled samples. The predictions made by each algorithm on the unlabeled examples are used to extend the training set for the other. Jackson and Landgrebe (2001) presented a self-learning and self-improving adaptive classifier designed to regulate the influence of semi-labeled samples. This method assigns full weights to training samples and reduces weights for semi-labeled samples. The results showed that extracting and using additional class label information can improve statistical estimation, leading to enhanced classifier performance and mitigating the Hughes phenomenon. Ma et al. (2016) developed a self-training strategy for unlabeled samples using three specific decisions. Local decisions, influenced by weighted neighborhood information from surrounding samples, are combined with global decisions extracted with DL, determined by the most similar training samples. Subsequently, unlabeled samples with high confidence were selected to augment the training set. The last step involves self-decision, using self-features extracted by DL on the updated training set, to derive spectral-spatial features and create a classification map. More details on the theory and mathematical foundations of SSL can be found in Chapelle et al. (2006), Zhu and Goldberg (2009), and Yang et al. (2023).

Given the limited availability of training samples in image classification tasks, especially in hyperspectral image classification, SSL emerges as a promising approach. The integration of both labeled and unlabeled data during training has the potential to enhance the accuracy and robustness of class predictions. For multispectral image classification, Banerjee and Buddhiraju (2015) suggested a novel self-training–based SSL using an ensemble of classifiers instead of a single classifier with the expectation that the ensemble system will have a lesser generalization error than a single model. In the self-training approach, individual classifiers undergo independent training on separate training subsets, and their predictions on test samples are aggregated using a novel instant run-off voting-inspired method for classifier combination. For hyperspectral image classification, Wu and Prasad (2018) used deep convolutional recurrent neural networks (CRNN) incorporating a SSL framework. The entire training data set, along with their pseudolabels, is employed for the pretraining of a deep CRNN, followed by a fine-tuning phase using the restricted pool of available labeled data. Likewise, Liu et al. (2017a) introduced a semi-supervised CNN for hyperspectral image classification. By incorporating skip connection parameters between the encoder and decoder layers, the network is configured for SSL, with the training process aimed at concurrently minimizing the supervised and unsupervised cost functions. He et al. (2017b) also proposed an SSL model that uses ResNet to extract preliminary image features. Employing ensemble learning, the objective is to construct distinctive image representations by delving into the inherent information in all accessible data. Multiple sets of prototypes are automatically created from labeled and unlabeled data. The prototypes sampled within each set are treated as proxy classes to train the supervised classifiers, resulting in projection functions tailored to the respective classes.

7.3.3 Reinforcement Learning

Reinforcement learning is a dynamic learning paradigm that focuses on interaction to tackle sequential decision-making tasks. Fundamentally, it divides the real world into two main components: an environment and an agent. The agent actively engages with the environment, constantly perceiving its evolving state. Using a trial-and-error methodology, the agent refines its behavior based on the rewards it garners, thereby discerning the repercussions of its actions. The agent is the most crucial component in reinforcement learning, holding the intelligence to make decisions, and suggests the optimal action in any given circumstance (Sewak, 2019). The reward, represented by a singular numerical value, fluctuates with each iteration and plays a pivotal role in formalizing the underlying goal within the reinforcement learning framework. The distinctive feature of reinforcement learning lies in its reliance on reward signals to navigate and guide the learning process. A task, representing a specific instance of the reinforcement learning problem, is defined by the complete specification of an environment. This approach contrasts with supervised learning, where an oracle supervisor provides learning samples. The strength of reinforcement learning lies in interactive settings, where it becomes impractical to acquire examples of desired behavior that are both accurate and representative across all potential situations (Sutton and Barto, 1998). While reinforcement learning offers commendable advantages, it is not without its drawbacks. One notable concern revolves around the potential impact of parameters on the speed of learning. The computational demands and time-consuming nature of reinforcement learning, especially in expansive workspaces, pose significant challenges. Moreover, the issue of generalization in DL models employing reinforcement learning has gained prominence as a critical problem (Packer et al., 2018). In response to these challenges, various algorithms, such as advantage actor-critic, proximal policy optimization, and evolution strategies with policy optimization, have been developed to enhance their effectiveness.

In reinforcement learning, the learner is not provided explicit instructions on the actions to take, which is a distinct characteristic of this learning strategy. Instead, the learner must discern which actions yield the highest reward through experimentation. The challenging scenarios involve actions that influence not only immediate rewards but also subsequent situations, producing trial-and-error search and delayed reward problems. The main challenge is to maintain a balance between exploration and exploitation, which has been thoroughly studied in the literature. While exploitation refers to the maximization of the agent's performance by using existing knowledge, typically assessed through the expected reward, exploration is the act of augmenting existing knowledge through the execution of actions and active interaction with the environment. To maximize the reward, the agent should favor actions tried in the past that proved effective in generating rewards. However, the unknown dynamics of the environment and the need for exploration to accumulate sufficient interaction experiences present challenges (Sutton and Barto, 1998). The dilemma emerges from the fact that neither exclusive exploitation nor exploration guarantees task success. The agent must experiment with various actions and progressively prioritize those deemed most effective (Yang et al., 2020a). In stochastic tasks, each action requires multiple attempts to reliably estimate its expected reward.

A distinctive aspect of reinforcement learning lies in its explicit consideration of the holistic problem involving a goal-driven agent interacting within an uncertain environment. It starts with a complete, interactive agent that is oriented toward achieving specific goals. All agents employing reinforcement learning are explicitly goal-driven, equipped with sensory abilities to perceive environmental aspects, and empowered to make decisions influencing their surroundings. The agent must function even in the presence of considerable uncertainty existing in the environment. In addition to the agent and environment, a reinforcement learning system comprises four essential elements: a policy, a reward function, a value function, and, if needed, a model of the environment. The policy, serving as a guide for the learning agent, determines the learning agent's behavior by mapping perceived environmental states to corresponding actions. The policy serves as the cornerstone of a reinforcement learning agent because it is inherently adequate to determine the agent's behavior. The reward function also plays a pivotal role in defining the objective within a reinforcement learning problem. In fact, this function maps perceived states of the environment to a singular numerical value, called reward, which signifies the intrinsic desirability of the state. The primary goal of the learning agent is to maximize the total received reward. By delineating what events are considered favorable or unfavorable for the agent, the reward function establishes the criteria for assessing the performance of the agent. While reward functions assess immediate desirability, value functions consider long-term benefits, representing the total reward anticipation from a given state. Values guide decision-making, seeking actions leading to states of high value for maximum long-term rewards. Estimating values is crucial, and efficient methods for this estimation form a key component of reinforcement learning algorithms. An extensive search process is usually required to determine the value. Evolutionary algorithms, such as genetic algorithms and simulated annealing (see Chapter 3), have been effectively used for this purpose. However, evolutionary algorithms do not interact with the environment and disregard the crucial structures in reinforcement learning, neglecting the policy's functional nature and actions experienced during the process (Sutton and Barto, 1998). Another crucial component is the model of the environment, which simulates its behavior. This model predicts the next state and reward based on the current state and action.

In reinforcement learning, an agent's decisions hinge on the environment's signal known as its state, which is explained by the Markov property. A reinforcement learning task aligned with the Markov property is denoted as a Markov decision process (MDP) (Bellman, 1957). A Markov process is a discrete stochastic process characterized by the Markov property, simplifying the simulation of the world within a continuous space. When both the state and action spaces are finite, it is termed a finite Markov decision process (finite MDP). While the prevailing theory of reinforcement learning largely centers on finite MDPs, these concepts possess broader applicability. A specific finite MDP is delineated by its state and action sets, along with the one-step dynamics of the environment.

In the framework of DL, reinforcement learning is often used to learn the intrinsic characteristics of the extracted features. Thus, deep reinforcement learning merges the perceptual strengths of DL with the decision-making capabilities of reinforcement learning, showcasing an artificial intelligence method that closely mimics human cognitive processes (Kaelbling et al., 1996). In other words, deep reinforcement learning integrates the feature representation ability of DL with the decision-making ability of reinforcement learning. By incorporating deep learning with reinforcement learning, some successful DL models have been proposed, including the deep Q-network (Mnih et al., 2015), double deep Q-network (DDQN; Van Hasselt et al., 2016), value iteration networks (VIN; Tamar et al., 2016). Comprehensive reviews of deep reinforcement learning and survey of the methods can be found in Arulkumaran et al. (2017), Li (2017), and Wang et al. (2020). Parallel to the substantial amount of research and success reported, deep reinforcement learning has been widely used in image processing and remote sensing applications, including band selection for hyperspectral image classification, image enhancement and denoising, object detection, and image fusion.

7.3.4 ACTIVE LEARNING

Active learning, which is another type of learning in machine learning designed to minimize the labeling workload in supervised learning, can help to obtain superior classification results through the enhancement of training sample quality. This enhancement is accomplished through the iterative selection of informative samples from a subset of unlabeled samples, employing an acquisition or query function. Thus, the use of an active learning method allows for the accelerated training of deep networks with a diminished need for training samples when juxtaposed with conventional SSL methods (Liu et al., 2017b). Active learning is fundamentally driven by the idea that learning efficiency can be enhanced by actively choosing highly relevant data points (MacKay, 1992). This selection is guided by ranking scores derived from model predictions. The chosen candidates are integrated into the training set, and the classifier is subsequently trained using these new samples. The efficiency of training with actively selected samples surpasses that of randomly selected samples, making samples more suitable for training. Incorporating selected samples (i.e., pseudosamples) from model uncertainty areas into the training set forces the model to address regions with low confidence (Tuia et al., 2011). Active learning requires a model with the ability to learn from a small number of samples and convey uncertainty about unseen data, thus imposing constraints on the selection of models viable within the active learning framework. Thus, the emphasis on active learning strategies has predominantly focused on problems with low dimensions, for which kernel- and graph-based methods are usually employed (Gal et al., 2017).

The active learning framework assumes the existence of a budget for the learner to make queries within the relevant domain. However, in many practical applications, this budget may be constrained, leading to the possibility that the labeled data obtained through active learning might not be adequate to develop a highly accurate classifier in the given domain (Yang et al., 2020a). Transfer learning has some similarities with SSL, but there is a fundamental difference between these learning paradigms. Active learning addresses this issue by enhancing the quality of training samples, whereas SSL tackles the problem by augmenting the quantity of training samples. To synthesize both learning strategies, Wang et al. (2017) proposed a novel semi-supervised active learning algorithm in which the representativeness and discriminativeness are discovered using a labeling process based on a supervised clustering technique and classification results. Based on the clustering and classification results, pseudolabels were assigned to the unlabeled data. Meanwhile, unlabeled samples that cannot be confidently assigned pseudolabels in each iteration are recognized as candidates for active learning.

As an effective strategy in machine learning research, active learning is employed to reduce the cost of acquiring a large number of labeled training sets. It is characterized by three main elements: (1) the presence of an initial training set, (2) the existence of a pool set, and (3) the use of a query function. The active learning model typically comprises a learner initially trained with a limited number of labeled samples. The learner iteratively chooses new training examples from the candidate pool, aiming to extract maximum information about the unlabeled data set and eventually enhance the overall model performance. Concurrent updates occur in both the labeled and unlabeled data sets, involving the addition of informative samples to the labeled set and their removal from the unlabeled set. Exclusive labeling of queried samples eliminates redundant and noninformative samples. Due to its critical role in the iterative learning stage, the query function holds significant importance in active learning. Haut et al. (2018), for instance, used the estimation of posterior probabilities of class membership, $p\left(y\middle|\mathbf{x}\right)$, to rank the candidates in the pool set in a Bayesian CNN model. Tuia et al. (2011) surveyed various heuristics and presented a taxonomy for active learning techniques, classifying them into committee-based, large margin–based, and posterior probability-based methods. Similar taxonomic categorization and mathematical theory of transfer learning approaches were presented by Liu et al. (2017b) and Niu et al. (2020). Gal et al.

(2017) reviewed and compared the performances of five Bayesian transfer functions: Shannon's Max entropy, Bayesian active learning by disagreement, variation ratios, mean standard deviation, and random acquisition (i.e., baseline).

The active learning framework has proven to be effective when dealing with large data sets that require an accurate selection of samples. This is particularly advantageous in remote sensing applications, where searching through a substantial number of pixels is necessary and manual definition is both redundant and time-consuming. Active learning algorithms have become popular in remote sensing image processing, and several approaches have been proposed to solve specific remote sensing tasks, including scene classification (Pires de Lima and Marfurt, 2020; Song and Yang, 2022) and land cover classification (Hamrouni et al., 2021; Huang et al., 2021; Hilal et al., 2022), and object detection (Chen et al., 2022b; Li et al., 2022). Current studies and trends in the field show that quantum transfer learning (Otgonbaatar et al., 2023), which provides significant advantages over conventional transfer learning algorithms, will be a hot research topic in the design and application of new types of networks.

7.3.5 Multitask Learning

Most DL models use identical hidden layers across various tasks. Sharing knowledge between tasks is exceptionally beneficial when they are vastly similar. However, when this assumption is not met, there is a considerable drop in performance. Multitask learning (MTL), also known as joint learning, is a powerful paradigm that incorporates an inductive transfer mechanism with the primary aim of enhancing generalization performance, similar to transfer learning. Using relevant information from related learning tasks, MTL jointly learns several target tasks to mitigate data sparsity. From this perspective, transfer learning and MTL share similarities (Yang et al., 2020a). The concept, akin to human learning, emphasizes the advantage of acquiring knowledge in multiple learning tasks simultaneously, where insights gained in one task can contribute to the understanding and performance of other related tasks. Leveraging the domain-specific information inherent in the training data of interconnected tasks is achieved by concurrently training tasks and capitalizing on mutual benefits. The term "multiple tasks" encompasses the learning of multiple output targets from a single input source, the learning of a single output target from multiple input sources, or a blend of both. Multitask networks exhibit the potential for performance improvement when the associated tasks share complementary information or serve as mutual regularizers, as suggested by Vandenhende et al. (2020).

MTL, a type of machine learning method focused on concurrently addressing multiple tasks, proves advantageous in handling the scarcity of data by tapping into valuable insights from related learning tasks (Thung and Wee, 2018). However, simultaneous learning of multiple tasks presents new challenges in terms of design and optimization. Grounded in the assumption that all tasks (or at least a subset) share some level of association, both empirical and theoretical findings support the notion that jointly learning multiple tasks results in superior performance compared with learning them independently. Reports suggest that MTL is beneficial even in scenarios involving tasks with no apparent connection (Romera-Paredes et al., 2012). It incorporates the training signals of related tasks as an inductive bias. The objective of inductive transfer is to use supplementary sources of information to enhance learning performance on the current task. Multitask bias induces the learner to favor hypotheses that provide explanations for multiple tasks. Simultaneously training a network to identify various aspects such as object outlines, shapes, edges, regions, subregions, textures, reflections, highlights, shadows, text, orientation, size, and distance, can lead to improved recognition of complex real-world objects (Caruana, 1997). MTL can be incorporated with supervised learning, unsupervised learning, SSL, active learning, reinforcement learning, multiview learning, and online learning. Similar to MTL, transfer learning transfers knowledge from one task to another. However, transfer learning aims to use one or more tasks to aid a target task, whereas MTL involves the cooperation of multiple tasks to assist each other (Zhang and Yang, 2018).

Crawshaw (2020) classified the current MTL methods into two main groups, employing a well-known dichotomy: hard parameter sharing and soft parameter sharing. Hard parameter sharing involves the sharing of model weights across multiple tasks, ensuring that each weight is trained to simultaneously minimize various loss functions. This strategy effectively reduces the potential for overfitting. Within the framework of soft parameter sharing, individual tasks maintain unique task-specific models with separate weights. Nevertheless, the joint objective function incorporates the distance between the model parameters of different tasks. Although there is no explicit parameter sharing, task-specific models are motivated to exhibit similarity in parameters. Many MTL optimization methods include regularization strategies that penalize parameter distance, and ongoing efforts focus on developing additional regularization techniques. According to Crawshaw (2020), optimization methods fall into six primary categories: loss weighting, regularization, gradient modulation, task scheduling, multiobjective optimization, and knowledge distillation. On the other hand, Zhang and Yang (2018) suggested that a prospective avenue for the evolution of multitask deep models is the development of more flexible architectures capable of handling diverse tasks, including outliers.

In remote sensing classification, a considerable number of studies have been conducted. For instance, Alhichri (2018) employed MTL to classify remote sensing scenes using SqueezeNet, a type of CNN architecture. In the study, each data set (UC Merced, KSA, and AID) represented a learning task. The results revealed the promising capabilities of MTL in sharing information between tasks and improving classification accuracy. A similar study conducted by Liu and Shi (2020) proposed multitask DL for hyperspectral image classification using two deep CNNs for four widely used hyperspectral data sets (Pavia University, Pavia Center, Salinas Valley, and Indian Pines). The incorporation of spectral knowledge was intended to guarantee the similarity of shared features across various domains. The results obtained highlight the efficacy of the proposed method in boosting the performance of deep CNNs, particularly in cases with limited training samples. Zheng et al. (2022) designed a multitask learning network (MTLN) that considered each small-scale data set as a distinct task. This MTLN leveraged complementary information from multiple tasks to enhance generalization. By concurrently learning shared and task-specific features, the MTLN preserved both generalization and discrimination capabilities. This study explored two scenarios: multiple scene classification tasks and joint evaluation of scene classification and semantic segmentation.

7.4 APPLICATION OF DL IN REMOTE SENSING

7.4.1 Semantic Segmentation

Semantic segmentation, which is one of the key applications in image processing and computer vision, is a process that labels each pixel with a class corresponding to its image object. Semantic segmentation annotates segments with semantic information corresponding to the classes of objects within the image. In other words, it extends its utility by incorporating semantic information between the pixels that compose these objects. It aims to produce a segmentation mask that entails the classification of every pixel within a limited number of semantic classifications or labeled classes and the segmentation of the input image based on the associated semantic data. With the emergence of DL methods, semantic segmentation has shown remarkable progress, with numerous studies reporting superior performance over traditional pixel-based image classification. In semantic segmentation, contextual information is important for predicting the label of a pixel. Contextual information for a pixel is contingent upon the size and continuity of the semantic uniform segment to which it belongs, as well as the quantity and density of neighboring segments representing various classes and backgrounds. Consequently, image segmentation is perceived as a multiscale contextual issue, even in cases where predictions are targeted at individual pixels (Hoeser and Kuenzer, 2020). Therefore, the current research focuses more on producing solutions to this issue by exploiting

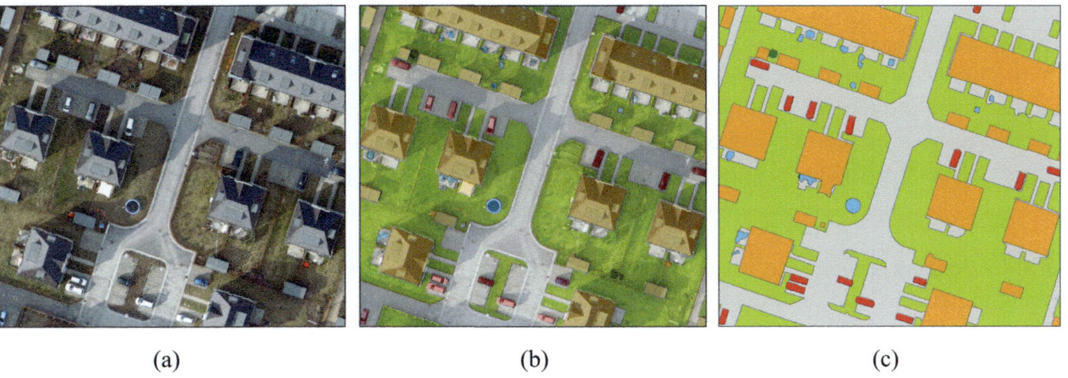

(a)   (b)   (c)

FIGURE 7.29   Example of semantic segmentation. (a) Remotely sensed image, (b) semantic labels for image objects (building – orange color; meadow – green color; vehicles – red color; road – gray color; others – blue color), (c) semantic mask.

features in the DL network. To demonstrate the semantic segmentation concept, the ISPRS 2D semantic label Potsdam/Germany benchmark data set (Figure 7.29) that contains high-resolution airborne images with a spatial resolution of 5 cm can be given as an example. The data set includes red, blue, green, and NIR orthorectified images with corresponding digital surface models (ISPRS, 2024). The semantic mask illustrated in Figure 7.29c is used to detect image objects.

DL-based approaches differ from traditional neural networks by incorporating more hidden layers. These layers play a crucial role in extracting diverse features from the input data, enabling the creation of an accurate representation of the classes. Recent DL-based research in remote sensing has mainly employed CNNs because of the notable achievements reported for semantic segmentation of multidimensional data sets, including very high-resolution (VHR) multispectral and hyperspectral imagery (Kemker et al., 2018; Kavzoglu and Yilmaz, 2022; Sertel et al., 2022). The convolutional layer, known as a filter or kernel, moves across the input image to identify unique features and generate feature maps. A pooling layer is then used to reduce the size of these feature maps. The convolutional and pooling layers together create the hidden layer. In semantic segmentation studies, architectures that include encoder-decoder elements are frequently employed. The encoder module uses pooling and convolutional layers to gradually reduce the spatial dimensions of feature maps to extract extensive semantic information. In the decoder phase, the module employs deconvolution techniques to systematically recover the spatial details that were reduced during encoding. Several deep network architectures have been specially developed for image classification using semantic segmentation. The most popular networks are the fully convolutional network, VGGNet, GoogleNet, SegNet, U-Net, U-Net++, and DeepLab, which are reviewed by Yuan et al. (2021).

Recognizing the importance of VHR remotely sensed imagery, these images have much larger dimensions than those used in computer vision, for which DL models have been developed. The use of the entire image in a DL model is impractical because of the heavy computing load. Therefore, most studies have adopted the use of cropped images (image patches) as input. To address this critical issue in remote sensing, the application of multiscale semantic segmentation has become a practical solution, harnessing the power of global context information at multiple hierarchical levels. For example, after implementing a two-stage multiscale training strategy, Ding et al. (2020) reported enhanced performance in both accuracy and stability compared with training models based on local patches.

Although encouraging results have been reported for DL-based semantic segmentation, several challenges remain for future work. First, defining appropriate features with semantic meaning in high-resolution images is challenging because of complex textures. Second, determining semantic rules becomes challenging due to the scale and hierarchy inherent in VHR images, resulting in the production of objects at different scales (Burnett and Blaschke, 2003). Furthermore, semantic algorithms require a vast amount of training data and tuning of a considerable number of parameters. While the use of pretrained networks successfully prevents overfitting in RGB images with ample labeled data, the remote sensing domain faces a distinct challenge due to the limited availability of labeled data. This scarcity has prompted researchers to adopt unsupervised, semi-supervised feature extraction methods, and also employ transfer learning techniques. When such extraction methods are used to extract deep features for every pixel, pixel-based thematic maps are produced after a classification process. In recent studies, Vision and Swin transformers have been also employed for this particular task. Several reviews exist on semantic image segmentation using DL models (e.g., Liu et al., 2018; Garcia-Garcia et al., 2018; Hoeser and Kuenzer, 2020; Yuan et al., 2021).

7.4.2 Object Detection

Object detection, an important application in remote sensing, has become possible through the use of VHR images acquired by contemporary sensors (particularly mounted on UAVs), complemented by advanced DL algorithms. CNNs and transformers, which constitute the backbone of deep learning–based object detection models, are currently instrumental in shaping the trajectory of this field. These models are essential for providing critical insights that contribute to the semantic understanding of images and videos. Their processing framework essentially involves locating and classifying objects, ultimately drawing a tight bounding box around each object. Object detection is closely tied to other image-processing applications, such as object classification, semantic segmentation, and instance segmentation. It is a significant area of study in computer vision and has widespread applications in many scientific domains, including remote sensing. Detecting objects such as ships, aircraft, and vehicles in high-resolution remote sensing images is challenging due to the intricate dynamics of neighboring relations, complex backgrounds, and varying lighting conditions, potentially causing recognition algorithms to make incorrect inferences for ground and target objects (Yildirim and Kavzoglu, 2022). The effectiveness of object detection in a complex context is contingent on the features obtained from the objects. In contrast to traditional methods that rely on manually crafted features such as shape, size, edges, contours, and outlines, DL models can automatically learn feature representations. This makes DL models highly appropriate for object detection, as they can capture both low-level and high-level features in remote sensing images. These models are proficient in learning hierarchical representations, semantic information, and deeper features from input images or patches. Consequently, they can accurately identify and categorize objects in images and subsequently label them with rectangular bounding boxes along with confidence levels.

DL-driven object detection frameworks can be categorized into two main groups. The first category follows the traditional pipeline, starting with the generation of region proposals, referred to as regions of interest (ROI), and subsequently classifying each proposal into distinct object categories, which represent two-stage object detectors. Methods falling under this category include region-based convolutional neural networks (R-CNN), spatial pyramid pooling (SPP-Net), Fast R-CNN, Faster R-CNN, region-based fully convolutional network (R-FCN), feature pyramid network (FPN), and Mask R-CNN. In the second group of models, the entire process is considered to be either a regression or classification problem, using a unified framework to produce conclusive results in terms of both location and category information. Recognized as region-free detectors, these models typically rely on cell grid approaches to segment the image and predict the class label for each grid cell. The methods in this second group include DetectorNet, OverFeat, you only look once (YOLO), single-shot multibox detector (SSD), EfficientDET, deconvolutional single-shot detector (DSSD), and deeply supervised object detectors (DSOD). Researchers have released several subsequent versions of YOLO, identified as YOLOv2 to YOLOv8, which is the latest model at the time of writing this book. There are also several modified limited versions, such as YOLO-Lite. The initial YOLO architecture comprises 24 convolution layers followed by two fully connected layers. While predicting multiple bounding boxes per grid cell, it employs nonmaxima suppression to select only those bounding boxes with the highest IoU (with respect to a certain threshold) considering the ground truth. R-CNN, a well-known model with numerous adaptations, stands out as another popular model. Its core concept involves converting the object detection challenge into a classification problem by hierarchically grouping similar regions based on color, texture, size, and shape compatibility. The selective search algorithm is employed to segment the image into numerous candidate regions. Subsequently, a CNN is applied to obtain the eigenvectors of these candidate regions, and the final stage involves object detection by a classifier that determines the type of each candidate area. The computational intensity of creating a substantial number of ROIs and assessing them with a CNN in R-CNN has led to the development of alternative versions, such as Fast R-CNN and Faster R-CNN. Conversely, SPP-Net adopts a distinct strategy by incorporating the spatial pyramid pooling layer following the last convolution layer, effectively eliminating repetitive processing, and enabling CNN to handle images of diverse sizes. This modification significantly accelerates the speed of object detection. However, there are faster and simpler methods that follow a one-stage detection approach. SSD is an example of a method that uses a single CNN to predict both object categories and anchor offsets (bounding boxes). During sampling, the object's bounding box is predicted using a priori boxes of different scales and aspect ratios. CNN then extracts features that are directly classified and regressed. SSD uses the feature layer group within the pyramid structure for object detection. The prediction of large targets is accomplished using high-level feature maps with a large receptive field, whereas low-level feature maps with a smaller receptive field are employed for predicting small targets. The model employs a nonmaximum suppression algorithm to generate detection outcomes. Several efforts have been made to refine this computationally efficient technique, including the development of DF-SSD, SE-SSD, and Tiny-SSD. Many review papers (e.g., Jiao et al., 2019; Zhao et al., 2019; Liu et al., 2020; Wu et al., 2020; Kaur and Singh, 2023) have been presented to comprehensively describe and compare the DL-based algorithms used for object detection.

Recently integrated into object detection, transformer-based models have exhibited significant success in various computer vision tasks, attaining performance levels comparable to mainstream CNN-based models. As discussed in Section 7.2.4, the self-attention mechanism enables transformers to overcome the constraint associated with nonparallelizable RNN models, thereby enhancing computational efficiency. In contrast to CNNs, the self-attention mechanism offers a global perceptual field, capturing more comprehensive interactions within the data set. The convolution operation is restricted by its localization to a specific pixel neighborhood, resulting in a lack of consideration for global information from neighboring areas. Conversely, the transformer serves as a fully connected graph modeling approach, enabling the representation of heterogeneous nodes by projecting them into a comparable space for similarity computation. Thus, there exists a robust theoretical basis for employing the transformer in various computer vision tasks, including object detection, due to its versatile modeling capabilities (Li et al., 2023b). Transformer-based models can be classified into two primary categories: transformer neck–based detectors and transformer backbone–based detectors. In transformer neck–based detectors (e.g., DETR, YOLOS, Efficient DETR, and TSP), the determination of class labels and bounding box coordinates involves a set of learnable object queries, without any alteration to the backbone used for feature extraction. Conversely, transformer backbone–based detectors (e.g., PVT and Swin transformer) introduce a universal visual backbone that transforms the image into a sequence, deviating from the typical convolution-based feature extraction approach. To enhance detection accuracy and replace the CNN backbone in traditional detectors, many approaches also integrate multiscale feature fusion.

Novel techniques continuously emerge, requiring rigorous evaluation and comparison with previous results. Various metrics have been introduced to evaluate the effectiveness of detectors based on specific data sets. Typically, the DL models designed for object detection are assessed against established benchmarks such as ImageNet, PASCAL VOC, MS COCO, and Open Images V5. While certain metrics are commonly utilized across all data sets, others are specific to the benchmark data set. Metrics that are commonly employed include detection accuracy, precision, recall, average precision (AP), mean average precision (mAP), IoU, and F1-score. AP estimates the precision averaged over all detected objects, and mean average precision (mAP) represents the overall mean value of AP, indicating the accuracy of object detectors across all classes. In most studies, AP and its derivatives are employed to measure the quality of detection. The calculation involves 10 IoU values ranging from 50% to 95% with 5% increments, commonly expressed as AP@50:5:95. Alternatively, assessment can be performed at specific IoU values, such as the frequently used 50% and 75%, known as AP50 and AP75. Owing to the variations in bounding box annotation formats, researchers commonly present metrics that align with the source code provided by each data set. The requirement for conversions in bounding box annotations across diverse data sets leads researchers to generally avoid evaluating all tested methods using every possible metric, which poses a considerable challenge in current object detection studies. In practice, it would be more significant if techniques trained and tested on one data set could also be evaluated using the metrics used in other data sets.

7.4.3 Scene Classification

With the development of Earth observation technology, high-resolution imagery acquired by aerial and satellite-based sensors has become abundant, making it easier to obtain such data every day. The large amount of data with rich content presents new challenges that require effective comprehension of semantic content and the development of more intelligent methods for the identification and classification of remote sensing imagery. Scene classification, which involves automatically assigning specific semantic labels (such as residential scene or agricultural scene) to each remote sensing image or patch based on its content, has attracted significant attention in the interpretation of remote sensing images. Scene classification for remote sensing images has both theoretical research significance and a wide range of applications, such as land use and land cover (LULC) classification, urban planning, land resource management, and disaster monitoring. The surge in the quantity of extensive image data, representing the dense sampling of varied real-world scenes, coupled with the resurgence of DL techniques capable of learning robust feature representations directly from raw data, has led to significant advancements in scene classification.

The abundance of similar and potentially confusing features and objects in remote sensing images underlines the critical need to extract discriminative features to achieve precise scene classification. Despite extensive efforts in image scene classification, the task remains challenging owing to factors such as complex structures and spatial patterns that influence scene representation and classification. To address this challenge, substantial efforts have been dedicated to the generation of diverse data sets and the exploration of various approaches. Unlike the applications detailed in the earlier sections, scene classification focuses on the automatic allocation of a semantic label to each scene image. Considering the image block as the basic unit in image interpretation is advantageous for harnessing spatial context information and addressing interpretation ambiguity since the scene is a combination of multiple objects, environments, and semantics (Gu et al., 2019). The understanding of remote sensing image scenes extends beyond the recognition of individual objects and includes the perception of the topological distribution of multiple objects within a given scene. In the context of scene classification conducted in feature space, the efficacy of feature representation is a crucial factor influencing the performance of the classification methods.

Current methods for scene classification are generally categorized into three main groups based on the features they utilize: handcrafted feature-based methods that involve low-level features, unsupervised feature learning–based methods, and deep feature-based methods that incorporate high-level features. Early approaches to scene classification primarily employed handcrafted features, holding attributes such as color, texture, shape, and spatial information. These approaches primarily concentrate on using significant engineering skills and domain knowledge to extract various human-engineered features or their combinations that serve as the fundamental attributes of a scene, providing valuable information for scene classification. The main handcrafted features include color histograms, gray-level co-occurrence matrix, scale-invariant feature transform (SIFT), and histogram of oriented gradients (HOG). Dividing low-level features into two categories, some researchers (e.g., Zhang et al., 2019b) have designated another category as mid-level feature approaches that compute a holistic image representation formed by local visual features, such as SIFT. In addition, the bag-of-visual-words (BoVW) model is a well-known mid-level method that has gained widespread adoption for classifying remote sensing image scenes. Initially designed for text analysis, the BoVW method was developed to represent a document based on its word frequency. The model is adapted to represent images at the frequency of "visual words", formed by quantizing local features using a clustering method. It is important to emphasize that the efficacy of BoVW-based methods is closely tied to the extraction of manually designed local features, including local structural points, color histograms, and texture features. To address the drawbacks of manually crafted features, automatic extraction of features from images is viewed as a more viable approach. Unsupervised feature learning from unlabeled input data has emerged as a compelling alternative to handcrafted features and has shown substantial advancements in image scene classification in recent years. The main objective of unsupervised feature learning is to obtain a set of basis functions or filters for feature encoding. These functions take input from either handcrafted features or raw pixel intensity values and yield a set of learned features. Learning features directly from images, rather than relying on manually designed features, allows for the acquisition of more discriminative features that are better suited to the specific problem. Principal component analysis (PCA), sparse coding, and autoencoders are commonly used in unsupervised feature learning. It should be noted that certain models within unsupervised feature learning, such as sparse coding and autoencoders, can be easily combined in a stacked manner to create more intricate unsupervised architectures.

The advent of DL frameworks has motivated researchers to move from handcrafted features to trainable multilayer neural networks. Consequently, several DL models have demonstrated significant capabilities in feature representation, especially in the context of remote sensing image scene classification. The state-of-the-art performance has established the DL approach as the mainstream in scene classification. Unlike low-level and mid-level features, deep learning models have the ability to acquire more robust, abstract, and discriminative features through DL-based models, minimizing the requirement for significant engineering skills and domain expertise. In contrast to unsupervised feature learning methods, DL models, with their multiple processing layers, can capture more complex and discriminative feature representations. Features extracted from the upper layers of deep neural networks possess semantic abstraction properties, making deep features more suitable for semantic-level scene classification. The efficacy of remote sensing image scene classification has been enhanced by the application of CNNs, in which shallow layers capture local low-level features and deep layers extract global high-level semantic features. The use of high-level semantic features plays a crucial role in bridging the semantic gap between different scenes of the same class, resulting in enhanced classification performance. Employing pretrained CNNs, such as ResNet, VGG, and DenseNet, proves to be a successful strategy for remote sensing image scene classification, particularly in situations where there is a scarcity of labeled data sets for training. Recent research suggests that pretrained CNNs can be transferred to aerial scene classification without the need for training modalities such as fine-tuning or training from scratch (Hu et al., 2015; Petrovska et al., 2020; Risojević and Stojnić, 2021). To address the challenge of limited labeled samples, various publicly accessible benchmark data sets are employed for the pretraining of models. These VHR image data sets include UC Merced land use (21 categories with 2,100 images), WHU-RS19 (19 categories with 1,005 images), RSSCN7 (7 categories with 2,800 images), AID (30 types with 10,000 images), NWPU-45 (45 categories with 31,500 images), and OPTIMAL-31 (31 categories with 1,860 images) data sets.

A scene image usually represents a specific image patch manually extracted from large VHR aerial or satellite images, portraying explicit semantic classes, such as residential or commercial areas. Due to the high resolution of these data, various scene images may encompass identical types of objects or exhibit similar spatial arrangements among objects. Without a doubt, the significant variations that could exist in the spatial arrangements and structural patterns of scene images make scene classification a highly challenging task (Zhu et al., 2017). Despite achieving satisfactory performance in remote sensing scene classification through fine-tuning of pretrained CNN models, this approach has limitations. Specifically, the CNN models are trained on data sets of a relatively small and fixed size, which are constrained by the matrix operations in the fully connected layers. Consequently, fine-tuning requires input images to be processed in the same size as those used during pretraining. To tackle the challenges in scene classification, other DL models have also been employed to achieve the highest performance. Bazi et al. (2021) applied transfer learning to enhance classification performance by utilizing a pretrained vision transformer (ViT) model with data augmentation and network pruning. Deng et al. (2022) introduced the CTNet, a hybrid high-performance joint network that combines the strengths of CNNs and ViT to enhance the performance of a single DL model. On the contrary, some researchers have delved into the utilization of GAN models for semi-supervised scene classification, leveraging adversarial training to obtain insights into the underlying data distributions from real training samples. Guo et al. (2021) proposed a new method for semi-supervised scene classification in remote sensing images utilizing GANs. This approach integrated a gating unit, a self-attention gating module, and a pretrained Inception V3 branch into the discriminative network to enhance the feature representation capability, facilitating semi-supervised classification. A comprehensive review of the use of GANs and other DL models for scene classification can be found in Thapa et al. (2023).

7.4.4 Change Detection

Change detection is an important subfield of multitemporal image analysis in remote sensing. It involves a thorough quantitative assessment and identification of changes using remote sensing imagery taken at different dates (time-series images). These changes result from a combination of natural factors, such as droughts, floods, and landslides, alongside human-induced activities such as urbanization, industrialization, and deforestation (Lu et al., 2004; Kavzoglu, 2008; Kavzoglu and Colkesen, 2011). In the conventional image analysis framework, each pixel undergoes assessment, resulting in either a binary label assignment or an analysis of change detection indicating transitions (from-to analysis). Although change detection algorithms have demonstrated advantages across diverse applications, they pose notable challenges. The most significant challenge among these is the substantial variation in data acquisition parameters, as it can influence the identification of relevant changes by introducing irrelevant information into the data set. This undesired change can also occur due to atmospheric features such as fog, clouds, and dust. Sunlight angles introduce another challenge concerning the presence and orientation of shadows in the scene. Additionally, temporal changes in vegetation growth and the surface reflectance of objects can have impacts on the results of change detection analysis. Therefore, an effective change detection method should have the capability to distinguish between significant alterations and inappropriate changes in satellite images, alongside detecting temporal changes.

Motivated by successful applications in various domains, DL models, extracting meaningful insights from data, have been lately employed to address the challenge of change detection, resulting in notable achievements (Khelifi and Mignotte, 2020; Bai et al., 2023). It is worth noting that there are fewer DL models designed for change detection when compared to models developed for semantic segmentation, scene classification, and object detection (Zhu et al., 2019). In general, the existing change detection models employ two primary frameworks. The first employs DL to identify corresponding points in two images, assessing whether changes occur at these points. The second framework considers the change detection problem as a LULC classification problem in that semantic segmentation is used to identify the changed regions, allowing for comparison and classification of these mapped locations. From an experimental viewpoint, semantic segmentation is easier to implement, executes faster, and exhibits higher detection accuracy. DL-based change detection models can be classified into three main categories: fully supervised learning–based methods, fully unsupervised learning–based methods, and transfer learning–based techniques (Khelifi and Mignotte, 2020). Fully supervised models are built on CNNs, which are hierarchical models that transform an image into feature maps ranging from low level to high level. U-Net is the most recognized CNN architecture applied to change detection tasks (Hamdi et al., 2019; Jaturapitpornchai et al., 2019). The general architecture of U-Net is symmetrical, consisting of an encoder for spatial feature extraction and a decoder for generating the segmentation map from the encoded features. The use of the enhanced U-Net architecture (U-Net++) was also implemented for end-to-end change detection in VHR satellite images (Peng et al., 2019). Multiscale feature maps were learned by introducing dense skip connections between different layers in this architecture. In addition, a strategy employing residual blocks was applied to facilitate the gradient convergence of the network. CNN models designed specifically for change detection include fully convolutional neural networks (FCNNs) developed by Daudt et al. (2019), Siamese CNN (S-CNN) by Zhang et al. (2018), combining CNNs with a bidirectional long short-term memory (BiLSTM) network by Liu et al. (2019).

To attain a satisfactory level of accuracy, supervised DL models rely on extensive annotated data. However, many change detection studies face a scarcity of training data, making it challenging to construct such models. The collection of change information for ground objects is not only expensive but also requires considerable time and effort. Therefore, the use of unsupervised, semi-supervised models and models employing transfer learning is not only more feasible but also imperative when compared to supervised models. Unsupervised feature learning methods primarily rely on models that can learn feature representations from patches without requiring supervision. For change detection, DL networks have been developed using unsupervised deep learning models, including autoencoders, DBNs, and RBMs. The last category of DL models relies on transfer learning, which involves transferring knowledge across various tasks or domains. The first approach involves employing the outputs of one or more layers of a CNN network trained on a different task as generic high-dimensional feature detectors and subsequently training a new shallow model based on these features. The alternative method is more complex and requires the fine-tuning of a network pretrained on general images. In this case, not only is the final layer replaced, but the previous layers are also retained. As applies to other DL applications, the future direction in change detection is moving toward the utilization of coupled DL models, leveraging the superior characteristics of different models. Chen et al. (2022a) provided an example using the bitemporal image transformer, employing a ResNet model as the backbone. This model focuses on capturing the long-range context within bitemporal images, which diverges from the conventional strategy of modeling dense pixel relationships. Instead, it represents input images using high-level semantic tokens and addresses context in a compact token-based space-time.

7.5 CONCLUDING REMARKS

Over the past two decades, there has been a significant transformation in scientific methodologies, with a shift from computational science to a more data-driven orientation. This evolutionary shift has resulted in a major transition in remote sensing image analysis. In this chapter, the current state of the art in DL for remote sensing is thoroughly reviewed. Instead of conventional statistical and shallow machine learning algorithms, neural networks with deeper layers and more complex structures have dominated the field, and new models with superior performances are constantly reported. The increased availability of higher-resolution imagery acquired at high temporal frequency, facilitated by advancements in remote sensing imaging technologies, opens up new research opportunities with DL-based methods. Thus, many semantic-level tasks in remote sensing imagery, including object detection, semantic segmentation, scene classification, and building extraction, have witnessed rapid development. In this context, DL serves as an extraordinary toolbox that goes beyond conventional small-scale benchmarking tasks to address large-scale, real-world challenges through implicit models.

Built upon many hierarchical layers that automatically extract significant and distinctive feature representations, DL models have been widely applied for a large variety of remote sensing applications, including preprocessing of remotely sensed imagery, hyperspectral image analysis, semantic segmentation, object detection, multimodal image fusion, scene classification, and change detection. These models can outperform traditional machine learning algorithms (e.g., random forest and support vector machines) using handcrafted features in the analysis of remote sensing images (Teke et al., 2021; Yilmaz et al., 2022; Tonbul et al., 2023).

In recent years, the adoption of DL-based methods in remote sensing has advanced rapidly. However, it can be asserted that the exploration of DL-based image analysis is currently thriving, leaving significant potential for further advancement. As the number of DL methods continues to expand, implementations through frameworks and libraries are also witnessing considerable expansion. The software development ecosystem in this sector is dynamic, featuring a plethora of open-source tools originating from academia, industry, start-ups, and larger open-source communities. Their common objective is to streamline the intricate data analysis process and offer integrated environments based on standard programming languages. It is noteworthy that Python is currently the most popular programming language for research, development, and production. Although there exist a number of DL libraries for developing and deploying models, the most popular ones are TensorFlow, PyTorch, and Keras.

Best practices and new research frontiers in deep learning techniques have long been discussed by researchers, considering the issues ranging from data preparation to hyperparameter optimization of DL models. Although significant improvements have been reported with the introduction of new loss functions, regularization techniques, and hybridization of different learning strategies, issues remain to be solved, particularly for inexperienced users. As of 2024, current challenges and future prospects for DL use in remote sensing can be listed as follows:

- Due to the scarcity of training samples, DL architectures often encounter difficulties in converging, particularly in training all layers of the network. In addition, to effectively compare different DL algorithms, it is necessary to establish benchmark data sets tailored to diverse image-processing tasks and applications. Supervised learning models like CNNs rely on extensive labeled training data, posing challenges in many applications. There is a need for additional publicly available labeled data sets, particularly those obtained by UAVs and hyperspectral sensors, to enhance the training and benchmarking processes. Existing benchmark data sets tend to be specific to limited classes, emphasizing the need for data sets that include not only residential areas but also industrial, agricultural, and semi-urbanized regions. It is also important to acknowledge that DL models, often trained on historical data, encounter challenges in effectively generalizing to new and unseen data sets. In some studies, the creation and training of DL models rely heavily on open data sharing. Researchers can enhance the accuracy and applicability of their models by sharing extensive and diverse data sets. Future initiatives should prioritize the promotion of open data sharing, crowdsourcing (particularly in emergency cases), and collaboration among researchers, industry, and government agencies to advance the field of remote sensing image analysis.
- To tackle the scarcity of labeled data problem, one possible solution is to use multimodal remote sensing data and pretrained multimodal DL models. Another solution could be to integrate domain knowledge into the training and testing stages of DL models to achieve good performance with limited labeled samples. Essentially, transferring knowledge between various domains emerges as a viable solution for accomplishing DL tasks. In such scenarios, two mainstream approaches exist: meta-learning for few-shot classification and transfer learning for cross-domain classification, where the training occurs on a source domain, and testing is conducted on a target domain (Dash et al., 2021, 2022; Lu et al., 2024).

- Although there are many frameworks and libraries available, there is no single library suitable for all problems in remote sensing image analysis, and often, a combination of them is required to implement DL models, leading to potential bugs and failures. Moreover, if a model is developed using an older version of the library, it may become incompatible, preventing the entire process from initiating due to conflicts with the newer version. As noted by El Aoun et al. (2022), 45.85% of the research studies use at least two DL libraries in their workflow. Currently, this is a major challenge for users and developers. As a result, there is a need for developing optimized libraries considering both the requirements of practitioners and the capabilities of current hardware. With numerous DL libraries sharing the same purpose, developers will feel submerged and confused about choosing the most suitable library for their requirements.

- Training and executing DL models, which involve optimizing millions or billions of parameters, demand significant processing capacity, particularly in terms of GPU power, especially when dealing with multidimensional data (e.g., hyperspectral images) due to their computational requirements. To overcome this challenge, a common approach is to train models using a subset of the training data, referred to as image patches, rather than entire images. When there is a scarcity of labeled samples, using pretrained DL models may be preferred as it helps avoid overfitting. Additionally, addressing this challenge involves the development of more efficient DL algorithms and the utilization of cloud computing resources (e.g., Kaggle, Amazon Web Services (AWS), Google Colab) to alleviate the strain on local infrastructure. For future studies, researchers should investigate the development of models designed for training and deployment on mobile or low-power edge computing devices as an alternative solution.

- A major obstacle for students and young professionals is to choose the appropriate DL architecture for a specific task when there are so many models with their variants available. Although CNNs are the most applied DL architecture, hybrid and ensemble models combining the strengths of different models and learning strategies need to be developed to solve complex problems in remote sensing. For a successful DL application, the selected model should be aligned with the specific requirements of the application, computational requirements, and data available for analysis. The selection of the optimal model for a problem also involves the expected accuracy level, which is directly related to the complexity of that model.

- Training and optimizing DL models remain a major agenda for many users. The search for optimal hyperparameters in developing neural networks has always been a significant challenge and has increased to a higher level with the introduction of DL models (see Chapter 9). Typically, a specialist performs numerous hyperparameter tests and conducts fine-tuning to reach an architectural decision, which can sometimes take days. Developing an insight for network tuning is challenging, given the large number of parameters to adjust, considering that the network topology itself is a hyperparameter. Automated Network Architecture Search (NAS) tools and hyperparameter exploration routines have been recently proposed, but they are currently limited to small problems. Therefore, whether network tuning can be simplified for users remains an open question.

- The incorporation of uncertainty estimates in DL models improves the stability and accuracy of the analysis, especially when models are unclear or lack an adequate number of representative samples. These estimates play a critical role in offering a more precise and meaningful comprehension of the produced thematic maps. Uncertainty quantification refers to the process of estimating and understanding the uncertainty associated with the predictions made by a DL model. Deep networks, while powerful, often lack a natural way to express uncertainty in their predictions, which is crucial for many applications, especially in safety-critical domains. A deeper discussion on this issue in the context of the accuracy assessment of classification results is presented in Section 10.5.

- The interpretability of DL models poses a significant challenge, hindering the understanding of their prediction rationale. This lack of transparency, especially in the mathematical and design aspects of DL, creates a black-box effect for many users, including those using high-level tools and packages. Despite the effectiveness of current models and advancements in training strategies, there is a notable requirement for explainability. Considering CNNs as an example, the complex nature of convolutional layers, batch normalization, and activation functions remains elusive, undermining the reliability and trustworthiness of the results. Efforts to visualize and manipulate internal network states have yielded only a partial and limited understanding of how networks operate in high-dimensional feature spaces. Addressing this challenge requires the development of new and robust explainable artificial intelligence (XAI) techniques (see Section 10.3) for interpreting internal model behavior. The visibility and interpretability of DL models can be improved by employing methods such as feature visualization, attention mechanisms, and rule extraction. To mitigate risks associated with prediction generalization, emphasis should be placed on creating more interpretable models and developing visualization methods that enhance the understanding and rationalization of DL model results.

REFERENCES

Alhichri, H. 2018. Multitask classification of remote sensing scenes using deep neural networks. In *Proceedings of IEEE International Geoscience and Remote Sensing Symposium*, Valencia, Spain, July 22–27, pp. 1195–1198. https://doi.org/10.1109/IGARSS.2018.8518874.

Arief, H. A., G.-H. Strand, H. Tveite, and U. G. Indahl. 2018. Land cover segmentation of airborne LiDAR data using stochastic Atrous network. *Remote Sensing* 10:973. https://doi.org/10.3390/rs10060973.

Arjovsky, M., S. Chintala, and L. Bottou. 2017. Wasserstein generative adversarial networks. In *Proceedings of the 34th International Conference on Machine Learning*, Sydney, Australia, vol. 70, pp. 214–223.

Arulkumaran, K., M. P. Deisenroth, M. Brundage, and A. A. Bharath. 2017. Deep reinforcement learning: A brief survey. *IEEE Signal Processing Magazine* 34:26–38. https://doi.org/10.1109/MSP.2017.2743240.

Bai, T., L. Wang, D. Yin, K. Sun, Y. Chen, W. Li, and D. Li. 2023. Deep learning for change detection in remote sensing: A review. *Geo-spatial Information Science* 26:262–288. https://doi.org/10.1080/10095020.2022.2085633.

Bai, Y., E. Yang, B. Han, Y. Yang, J. Li, Y. Mao, Y., G. Niu, and T. Liu. 2021. Understanding and improving early stopping for learning with noisy labels. In *Proceedings of 35th Conference on Neural Information Processing Systems*, vol. 34, pp. 24392–24403.

Banerjee, B., and K. M. Buddhiraju. 2015. A novel semi-supervised land cover classification technique of remotely sensed images. *Journal of the Indian Society of Remote Sensing* 43:719–728. https://doi.org/10.1007/s12524-014-0370-z.

Bank, D., N. Koenigstein, and R. Giryes. 2021. Autoencoders. *arXiv preprint*, arXiv:2003.05991.

Bazi, Y., L. Bashmal, M. M. Al Rahhal, R. Al Dayil, and N. Al Ajlan. 2021. Vision transformers for remote sensing image classification. *Remote Sensing* 13:516. https://doi.org/10.3390/rs13030516.

Bellman, R. 1957. A Markovian decision process. *Journal of Mathematics and Mechanics* 6:679–684. https://www.jstor.org/stable/24900506.

Bengio, Y., A. Courville, and P. Vincent. 2013. Representation learning: A review and new perspectives. *IEEE Transactions on Pattern Analysis and Machine Intelligence* 35:1798–1828. https://doi.org/10.1109/TPAMI.2013.50.

Bengio, Y., P. Simard, and P. Frasconi. 1994. Learning long-term dependencies with gradient descent is difficult. *IEEE Transactions on Neural Networks* 5:157–166.

Benjdira, B., Y. Bazi, A. Koubaa, and K. Ouni. 2019. Unsupervised domain adaptation using generative adversarial networks for semantic segmentation of aerial images. *Remote Sensing* 11:1369. https://doi.org/10.3390/rs11111369.

Berman, D. S., A. L. Buczak, J. S. Chavis, and C. Corbett. 2019. A survey of deep learning methods for cyber security. *Information* 10:122. https://doi.org/10.3390/info10040122.

Bhatt, D., C. Patel, H. Talsania, J. Patel, R. Vaghela, S. Pandya, K. Modi, and H. Ghayvat. 2021. CNN variants for computer vision: History, architecture, application, challenges and future scope. *Electronics* 10:2470. https://doi.org/10.3390/electronics10202470.

Bishop, C. M. 1995. *Neural Networks for Pattern Recognition*. Oxford: Clarendon Press.

Bishop, C. M., and J. Lasserre. 2007. Generative or discriminative? Getting the best of both worlds. *Bayesian Statistics* 8:3–24.

Blum, A., and T. Mitchell. 1998. Combining labeled and unlabeled data with co-training. In *Proceedings of the Eleventh Annual Conference on Computational Learning Theory*, Madison, Wisconsin, pp. 92–100.

Buduma, N. and N. Lacascio. 2017. *Fundamentals of Deep Learning Designing Next-Generation Machine Intelligence Algorithms*. Sebastopol, CA: O'Reilly Media.

Burda, Y., R. Grosse, and R. Salakhutdinov. 2015. Importance weighted autoencoders. arXiv preprint arXiv:1509.00519.

Burnett, C., and T. Blaschke. 2003. A multi-scale segmentation/object relationship modelling methodology for landscape analysis. *Ecological Modelling* 168:233–249. https://doi.org/10.1016/S0304-3800(03)00139-X.

Caruana, R. 1997. Multitask learning. *Machine Learning* 28:41–75.

Castellano, G., A. M. Fanelli, and M. Pelillo. 1997. An iterative pruning algorithm for feedforward neural networks. *IEEE Transactions on Neural Networks* 8:519–531. https://doi.org/10.1109/72.572092.

Chandra, B., and R. K. Sharma. 2017. On improving recurrent neural network for image classification. In *Proceedings of International Joint Conference on Neural Networks*, Anchorage, AK, pp. 1904–1907. https://doi.org/10.1109/IJCNN.2017.7966083.

Chapelle, O., B. Scholkopf, and A. Zien. 2006. *Semi-Supervised Learning*. Cambridge, MA: The MIT Press.

Chen, C. F. R., Q. Fan, and R. Panda. 2021. CrossViT: Cross-attention multi-scale vision transformer for image classification. In *Proceedings of the IEEE/CVF International Conference on Computer Vision*, Seoul, Korea, pp. 357–366.

Chen, H., Z. Qi, and Z. Shi. 2022a. Remote sensing image change detection with transformers. *IEEE Transactions on Geoscience and Remote Sensing* 60:5607514. https://doi.org/10.1109/TGRS.2021.3095166.

Chen, J., J. Sun, Y. Li, and C. Hou. 2022b. Object detection in remote sensing images based on deep transfer learning. *Multimedia Tools and Applications* 81:12093–12109. https://doi.org/10.1007/s11042-021-10833-z.

Chen, X., D. P. Kingma, T. Salimans, Y. Duan, P. Dhariwal, J. Schulman, I. Sutskever, and P. Abbeel. 2017. Variational lossy autoencoder. arXiv preprint arXiv:1611.02731.

Cho, K., B. Van Merriënboer, C. Gulcehre, D. Bahdanau, F. Bougares, H. Schwenk, and Y. Bengio. 2014. Learning phrase representations using RNN encoder-decoder for statistical machine translation. arXiv preprint arXiv:1406.1078.

Civco, D. L. and Y. Waug. 1994. Classification of multispectral, multitemporal, multisource spatial data using artificial neural networks. In *Proceedings of 1994 Annual ASPRS/ACSM Convention*, Reno, NV, pp. 123–133.

Crawshaw, M. 2020. Multi-task learning with deep neural networks: A survey. arXiv preprint arXiv:2009.09796.

Cui, B., X. Chen, and Y. Lu. 2020. Semantic segmentation of remote sensing images using transfer learning and deep convolutional neural network with dense connection. *IEEE Access* 8:116744–116755. https://doi.org/10.1109/ACCESS.2020.3003914.

Dash, A., J. Ye, and G. Wang. 2024. A review of generative adversarial networks (GANs) and its applications in a wide variety of disciplines: From medical to remote sensing. *IEEE Access*. https://doi.org/10.1109/ACCESS.2023.3346273.

Dash, T., S. Chitlangia, A. Ahuja, and A. Srinivasan. 2021. Incorporating domain knowledge into deep neural networks. arXiv preprint arXiv:2103.00180.

Dash, T., S. Chitlangia, A. Ahuja, and A. Srinivasan. 2022. A review of some techniques for inclusion of domain-knowledge into deep neural networks. *Scientific Reports* 12:1040. https://doi.org/10.1038/s41598-021-04590-0.

Daudt, R. C., B. Le Saux, A. Boulch, and Y. Gousseau. 2019. Multitask learning for large-scale semantic change detection. *Computer Vision and Image Understanding* 187:102783. https://doi.org/10.1016/j.cviu.2019.07.003.

Delalleau, O., and Y. Bengio. 2011. Shallow vs. deep sum-product networks. In *Proceedings of Advances in Neural Information Processing Systems* 24:666–674.

Delgado, J. M. D., and L. Oyedele. 2021. Deep learning with small datasets: Using autoencoders to address limited datasets in construction management. *Applied Soft Computing* 112:107836. https://doi.org/10.1016/j.asoc.2021.107836.

Demir-Kavuk, O., M. Kamada, T. Akutsu, and E.-W. Knapp. 2011. Prediction using step-wise L1, L2 regularization and feature selection for small data sets with large number of features. *BMC Bioinformatics* 12:412. https://doi.org/10.1186/1471-2105-12-412.

Deng, P., K. Xu, and H. Huang. 2022. When CNNs meet vision transformer: A joint framework for remote sensing scene classification. *IEEE Geoscience and Remote Sensing Letters* 19:1–5. https://doi.org/https://doi.org/10.1109/LGRS.2021.3109061.

Ding, L., J. Zhang, and L. Bruzzone. 2020. Semantic segmentation of large-size VHR remote sensing images using a two-stage multiscale training architecture. *IEEE Transactions on Geoscience and Remote Sensing* 58:5367–5376. https://doi.org/10.1109/TGRS.2020.2964675.

Dosovitskiy, A., L. Beyer, A. Kolesnikov, D. Weissenborn, X. Zhai, T. Unterthiner, M. Dehghani, et al. 2021. An image is worth 16x16 words: Transformers for image recognition at scale. In *Proceedings of the 9th International Conference on Learning* Representations, May 3–7, Vienna. https://openreview.net/pdf?id=YicbFdNTTy.

El Aoun, M. R., L. N. Tidjon, B. Rombaut, F. Khomh, and A. E. Hassan. 2022. An empirical study of library usage and dependency in deep learning frameworks. arXiv preprint arXiv:2211.15733.

Fahlman, S. E., and C. Lebiere. 1989. The cascade-correlation learning architecture. In *Advances in Neural Information Processing Systems* (NIPS'1989), (ed.) D. S. Touretzky, pp. 524–532. San Mateo, CA: Morgan Kaufmann.

Foody, G. M. 1999. Image classification with a neural network: From completely crisp to fully-fuzzy situations. In *Advances in Remote Sensing and GIS Analysis*, (eds.) P. M. Atkinson, and N. J. Tate, pp. 17–38. New York: John Wiley & Sons.

Fukushima, K., and S. Miyake. 1982. Neocognitron: A self-organizing neural network model for a mechanism of visual pattern recognition. In *Competition and Cooperation in Neural Nets*, (eds.) S. Amari, and M. A. Arbib, pp. 267–285. Berlin: Springer.

Gal, Y., R. Islam, and Z. Ghahramani. 2017. Deep Bayesian active learning with image data. In *Proceedings of the 34th International Conference on Machine Learning,* Sydney, Australia, 70:1183–1192.

Garcia-Garcia, A., S. Orts-Escolano, S. Oprea, V. Villena-Martinez, P. Martinez-Gonzalez, and J. Garcia-Rodriguez. 2018. A survey on deep learning techniques for image and video semantic segmentation. *Applied Soft Computing* 70:41–65. https://doi.org/10.1016/j.asoc.2018.05.018.

Ghosh, P., M. S. Sajjadi, A. Vergari, M. Black, and B. Schölkopf. 2020. From variational to deterministic autoencoders. In *Proceedings of the 8th International Conference on Learning Representations*. https://doi.org/10.48550/arXiv.1903.12436.

Girshick, R., J. Donahue, T. Darrell, and J. Malik. 2014. Rich feature hierarchies for accurate object detection and semantic segmentation. In *Proceedings of IEEE Conference on Computer Vision and Pattern Recognition*, Columbus, OH, pp. 580–587.

Gonzalez, S., and R. Miikkulainen. 2020. Improved training speed, accuracy, and data utilization through loss function optimization. In *Proceedings of 2020 IEEE Congress on Evolutionary Computation*, Glasgow, UK, pp. 1–8. https://doi.org/10.1109/CEC48606.2020.9185777.

Goodfellow, I., J. Pouget-Abadie, M. Mirza, B. Xu, D. Warde-Farley, S. Ozair, A. Courville, and Y. Bengio. 2014. Generative adversarial nets. In *Proceedings of the 27th International Conference on Neural Information Processing Systems*, Montreal, Canada, vol. 2, pp. 2672–2680.

Goodfellow, I., Y. Bengio, and A. Courville. 2016. *Deep Learning*. Cambridge, MA: MIT Press.

Gu, Y., Y. Wang, and Y. Li, 2019. A survey on deep learning-driven remote sensing image scene understanding: Scene classification, scene retrieval and scene-guided object detection. *Applied Sciences* 9:2110. https://doi.org/10.3390/app9102110.

Guo, D., Y. Xia, and X. Luo. 2021. GAN-based semisupervised scene classification of remote sensing image. *IEEE Geoscience and Remote Sensing Letters* 18:2067–2071. https://doi.org/10.1109/LGRS.2020.3014108.

Hamdi, Z. M., M. Brandmeier, and C. Straub. 2019. Forest damage assessment using deep learning on high resolution remote sensing data. *Remote Sensing* 11:1976. https://doi.org/10.3390/rs11171976.

Hamrouni, Y., E. Paillassa, V. Chéret, C. Monteil, and D. Sheeren. 2021. From local to global: A transfer learning-based approach for mapping poplar plantations at national scale using Sentinel-2. *ISPRS Journal of Photogrammetry and Remote Sensing* 171:76–100. https://doi.org/10.1016/j.isprsjprs.2020.10.018.

Hang, R., Q. Liu, D. Hong, and P. Ghamisi. 2019. Cascaded recurrent neural networks for hyperspectral image classification. *IEEE Transactions on Geoscience and Remote Sensing* 57:5384–5394. https://doi.org/10.1109/TGRS.2019.2899129.

Hanin, B., and M. Sellke. 2017. Approximating continuous functions by ReLU nets of minimal width. arXiv preprint. arXiv:1710.11278.

Haut, J. M., M. E. Paoletti, J. Plaza, J. Li, and A. Plaza. 2018. Active learning with convolutional neural networks for hyperspectral image classification using a new Bayesian approach. *IEEE Transactions on Geoscience and Remote Sensing* 56:6440–6461. https://doi.org/10.1109/TGRS.2018.2838665.

He, K., G. Gkioxari, P. Dollár, and R. Girshick. 2017a. Mask R-CNN. In *Proceedings of the IEEE International Conference on Computer Vision*, October 22–29, Venice, Italy, pp. 2961–2969. https://doi.org/ 10.1109/ICCV.2017.322.

He, Z., H. Liu, Y. Wang, and J. Hu. 2017b. Generative adversarial networks-based semi-supervised learning for hyperspectral image classification. *Remote Sensing* 9:1042. https://doi.org/10.3390/rs9101042.

Hendrycks, D., and K. Gimpel. 2016. Gaussian error linear units (GELUS). arXiv preprint arXiv:1606.08415.

Hendrycks, D., K. Lee, and M. Mazeika. 2019. Using pre-training can improve model robustness and uncertainty. In *Proceedings of the 36th International Conference on Machine Learning* 97:2712–2721.

Henry, C., S. M. Azimi, and N. Merkle. 2018. Road segmentation in SAR satellite images with deep fully convolutional neural networks. *IEEE Geoscience and Remote Sensing Letters* 15:1867–1871. https://doi.org/10.1109/LGRS.2018.2864342.

Hilal, A. M., F. N. Al-Wesabi, K. J. Alzahrani, M. Al Duhayyim, M. A. Hamza, M. Rizwanullah, and V. G. Díaz. 2022. Deep transfer learning based fusion model for environmental remote sensing image classification model. *European Journal of Remote Sensing* 55:12–23. https://doi.org/10.1080/22797254.2021.2017799.

Hinton, G. E., N. Srivastava, A. Krizhevsky, I. Sutskever, and R. R. Salakhutdinov. 2012. Improving neural networks by preventing co-adaptation of feature detectors. arXiv preprint arXiv:1207.0580.

Hirose, Y., K. Yamashita, and S. Hijiya, 1991. Back-propagation algorithm which varies the number of hidden units. *Neural Networks* 4:61–66. https://doi.org/10.1016/0893-6080(91)90032-Z.

Hochreiter, S., and J. Schmidhuber. 1997. Long short-term memory. *Neural Computation* 9:1735–1780. https://doi.org/10.1162/neco.1997.9.8.1735.

Hoeser, T., and C. Kuenzer. 2020. Object detection and image segmentation with deep learning on earth observation data: A review-part I: Evolution and recent trends. *Remote Sensing* 12:1667. https://doi.org/10.3390/rs12101667.

Hu, F., G.-S. Xia, J. Hu, and L. Zhang. 2015. Transferring deep convolutional neural networks for the scene classification of high-resolution remote sensing imagery. *Remote Sensing* 7:14680–14707. https://doi.org/10.3390/rs71114680.

Hu, J., L. Shen, and G. Sun. 2018. Squeeze-and-excitation networks. In *Proceedings of the IEEE Conference on Computer Vision and Pattern Recognition*, June 18–23, Salt Lake City, UT, pp. 7132–7141. https://doi.org/10.1109/CVPR.2018.00745.

Huang, Z., C. O. Dumitru, Z. Pan, B. Lei, and M. Datcu. 2021. Classification of large-scale high-resolution SAR images with deep transfer learning. *IEEE Geoscience and Remote Sensing Letters* 18:107–111. https://doi.org/10.1109/LGRS.2020.2965558.

Hubel, D. H., and T. N. Wiesel. 1962. Receptive fields, binocular interaction and functional architecture in the cat's visual cortex. *The Journal of Physiology* 160:106–154. https://doi.org/10.1113/jphysiol.1962.sp006837.

Ienco, D., R. Gaetano, C. Dupaquier, and P. Maurel. 2017. Land cover classification via multitemporal spatial data by deep recurrent neural networks. *IEEE Geoscience and Remote Sensing Letters* 14:1685–1689. https://doi.org/10.1109/LGRS.2017.2728698.

Im, D. J., S. Ahn, R. Memisevic, and Y. Bengio. 2017. Denoising criterion for variational auto-encoding framework. In *Proceedings of the 31th AAAI Conference on Artificial Intelligence* 31:2059–2065. https://doi.org/10.1609/aaai.v31i1.10777.

ISPRS, 2024. ISPRS Urban modelling and semantic labelling benchmark: 2D semantic labeling contest-Potsdam. https://www.isprs.org/education/benchmarks/UrbanSemLab/2d-sem-label-potsdam.aspx (accessed January 2, 2024).

Jackson, Q., and D. A. Landgrebe. 2001. An adaptive classifier design for high-dimensional data analysis with a limited training data set. *IEEE Transactions on Geoscience and Remote Sensing* 39:2664–2679. https://doi.org/10.1109/36.975001.

Jaturapitpornchai, R., M. Matsuoka, N. Kanemoto, S. Kuzuoka, R. Ito, and R. Nakamura. 2019. Newly built construction detection in SAR images using deep learning. *Remote Sensing* 11:1444. https://doi.org/10.3390/rs11121444.

Jiao, L., F. Zhang, F. Liu, S. Yang, L. Li, Z. Feng, and R. Qu. 2019. A survey of deep learning-based object detection. *IEEE Access* 7:128837–128868. https://doi.org/10.1109/ACCESS.2019.2939201.

Jozdani, S., D. Chen, D. Pouliot, and B. A. Johnson. 2022. A review and meta-analysis of generative adversarial networks and their applications in remote sensing. *International Journal of Applied Earth Observation and Geoinformation* 108:102734. https://doi.org/10.1016/j.jag.2022.102734.

Kaelbling, L. P., M. L. Littman, and A. W. Moore, 1996. Reinforcement learning: A survey. *Journal of Artificial Intelligence Research* 4:237–285.

Karnin, E. D. 1990. A simple procedure for pruning back-propagation trained neural networks. *IEEE Transactions on Neural Networks* 1:239–242. https://doi.org/10.1109/72.80236.

Kattenborn, T., J. Leitloff, F. Schiefer, and S. Hinz. 2021. Review on convolutional neural networks (CNN) in vegetation remote sensing. *ISPRS Journal of Photogrammetry and Remote Sensing* 173:24–49. https://doi.org/10.1016/j.isprsjprs.2020.12.010.

Kaur, R., and S. Singh. 2023. A comprehensive review of object detection with deep learning. *Digital Signal Processing* 132:103812. https://doi.org/10.1016/j.dsp.2022.103812.

Kavzoglu, T. 2001. An investigation of the design and use of feed-forward artificial neural networks in the classification of remotely sensed images, Ph.D. thesis, University of Nottingham, UK.

Kavzoglu, T. 2008. Determination of environmental degradation due to urbanization and industrialization in Gebze, Turkey. *Environmental Engineering Science* 25:429–438. https://doi.org/10.1089/ees.2006.0271.

Kavzoglu, T., A. Teke, and E. O. Yilmaz. 2021. Shared blocks-based ensemble deep learning for shallow landslide susceptibility mapping. *Remote Sensing* 13:4776. https://doi.org/10.3390/rs13234776.

Kavzoglu, T., and E. O. Yilmaz. 2022. Analysis of patch and sample size effects for 2D-3D CNN models using multiplatform dataset: Hyperspectral image classification of ROSIS and Jilin-1 GP01 imagery. *Turkish Journal of Electrical Engineering and Computer Sciences* 30:2124–2144. https://doi.org/10.55730/1300-0632.3929.

Kavzoglu, T., and I. Colkesen. 2011. Assessment of environmental change and land degradation using time series of remote sensing images. *Fresenius Environmental Bulletin* 20:274–281.

Kavzoglu, T., and P. M. Mather. 1999. Pruning artificial neural networks: An example using land cover classification of multi-sensor images. *International Journal of Remote Sensing* 20:2787–2803.

Kavzoglu, T., and P. M. Mather. 2003. The use of backpropagating artificial neural networks in land cover classification. *International Journal of Remote Sensing* 24:4907–4938. https://doi.org/10.1080/0143116031000114851.

Kawaguchi, K., J. Huang, and L. P. Kaelbling. 2019. Effect of depth and width on local minima in deep learning. *Neural Computation* 31:1462–1498. https://doi.org/10.1162/neco_a_01195.

Kemker, R., C. Salvaggio, and C. Kanan. 2018. Algorithms for semantic segmentation of multispectral remote sensing imagery using deep learning. *ISPRS Journal of Photogrammetry and Remote Sensing* 145:60–77. https://doi.org/10.1016/j.isprsjprs.2018.04.014.

Khelifi, L., and M. Mignotte. 2020. Deep learning for change detection in remote sensing images: Comprehensive review and meta-analysis. *IEEE Access* 8:126385–126400. https://doi.org/10.1109/ACCESS.2020.3008036.

Kingma, D. P., and M. Welling. 2013. Auto-encoding variational bayes. arXiv preprint arXiv:1312.6114.

Krizhevsky, A., I. Sutskever, and G. E. Hinton. 2012. ImageNet classification with deep convolutional neural networks. In *Proceedings of the Advances in Neural Information Processing Systems*, December 3–6, Lake Tahoe, NV, pp. 1097–1105.

Lakhal, M. I., H. Çevikalp, S. Escalera, and F. Ofli. 2018. Recurrent neural networks for remote sensing image classification. *IET Computer Vision* 12:1040–1045. https://doi.org/10.1049/iet-cvi.2017.0420.

Le, Q. V., N. Jaitly, and G. E. Hinton. 2015. A simple way to initialize recurrent networks of rectified linear units. *arXiv preprint* arXiv:1504.00941.

LeCun, Y., K. Kavukcuoglu, and C. Farabet. 2010. Convolutional networks and applications in vision. In *Proceedings of 2010 IEEE International Symposium on Circuits and Systems*, May 30–June 2, Paris, France, pp. 253–256. https://doi.org/10.1109/ISCAS.2010.5537907.

LeCun, Y., L. Bottou, Y. Bengio, and P. Haffner. 1998. Gradient-based learning applied to document recognition. *Proceedings of the IEEE* 86:2278–2324. https://doi.org/10.1109/5.726791.

Li, B., Q.-W. Wang, J.-H. Liang, E.-Z. Zhu, and R.-Q. Zhou. 2023a. SquconvNet: Deep sequencer convolutional network for hyperspectral image classification. *Remote Sensing* 15:983. https://doi.org/10.3390/rs15040983.

Li, Q., Chen, Y., and Y. Zeng. 2022. Transformer with transfer CNN for remote-sensing-image object detection. *Remote Sensing* 14:984. https://doi.org/10.3390/rs14040984.

Li, Y. 2017. Deep reinforcement learning: An overview. arXiv preprint arXiv:1701.07274.

Li, Y., H. Zhang, X. Xue, Y. Jiang, and Q. Shen. 2018. Deep learning for remote sensing image classification: A survey. *Wiley Interdisciplinary Reviews: Data Mining and Knowledge Discovery* 8:e1264. https://doi.org/10.1002/widm.1264.

Li, Y., N. Miao, L. Ma, F. Shuang, and X. Huang. 2023b. Transformer for object detection: Review and benchmark. *Engineering Applications of Artificial Intelligence* 126:107021. https://doi.org/10.1016/j.engappai.2023.107021.

Liu, B., X. Yu, A. Yu, P. Zhang, and G. Wan. 2018. Spectral-spatial classification of hyperspectral imagery based on recurrent neural networks. *Remote Sensing Letters* 9:1118–1127. https://doi.org/10.1080/2150704X.2018.1511933.

Liu, B., X. Yu, P. Zhang, X. Tan, A. Yu, and Z. Xue. 2017a. A semi-supervised convolutional neural network for hyperspectral image classification. *Remote Sensing Letters* 8:839–848. https://doi.org/10.1080/2150704X.2017.1331053.

Liu, L., W. Ouyang, X. Wang, P. Fieguth, J. Chen, X. Liu, and M. Pietikäinen. 2020. Deep learning for generic object detection: A survey. *International Journal of Computer Vision* 128:261–318. https://doi.org/10.1007/s11263-019-01247-4.

Liu, P., H. Zhang, and K. B. Eom. 2017b. Active deep learning for classification of hyperspectral images. *IEEE Journal of Selected Topics in Applied Earth Observations and Remote Sensing* 10:712–724. https://doi.org/10.1109/JSTARS.2016.2598859.

Liu, R., Z. Cheng, L. Zhang, and J. Li. 2019. Remote sensing image change detection based on information transmission and attention mechanism. *IEEE Access* 7:156349–156359. https://doi.org/10.1109/ACCESS.2019.2947286.

Liu, S., and Q. Shi, 2020. Multitask deep learning with spectral knowledge for hyperspectral image classification. *IEEE Geoscience and Remote Sensing Letters* 17:2110–2114. https://doi.org/10.1109/LGRS.2019.2962768.

Liu, Z., Y. Lin, Y. Cao, H. Hu, Y. Wei, Z. Zhang, S. Lin, and B. Guo. 2021. Swin transformer: Hierarchical vision transformer using shifted windows. In *Proceedings of 2021 IEEE/CVF International Conference on Computer Vision*, pp. 9992–10002. https://doi.org/10.1109/ICCV48922.2021.00986.

Long, M., Y. Cao, J. Wang, and M. I. Jordan. 2015. Learning transferable features with deep adaptation networks. In *Proceedings of the 32nd International Conference on Machine Learning*, Lille, France, pp. 97–105.

Lu, D., P. Mausel, E. Brondizio, and E. Moran. 2004. Change detection techniques. *International Journal of Remote Sensing* 25:2365–2401.

Lu, X., T. Gong, and X. Zheng. 2024. Domain mapping network for remote sensing cross-domain few-shot classification. *IEEE Transactions on Geoscience and Remote Sensing* 62:1–11. https://doi.org/10.1109/TGRS.2024.3352908.

Lyu, H., H. Lu, and L. Mou. 2016. Learning a transferable change rule from a recurrent neural network for land cover change detection. *Remote Sensing* 8:506. https://doi.org/10.3390/rs8060506.

Lyu, H., H. Lu, L. Mou, W. Li, T. Wright, X. Li, X. Li X, et al. 2018. Long-term annual mapping of four cities on different continents by applying a deep information learning method to Landsat data. *Remote Sensing* 10:471. https://doi.org/10.3390/rs10030471.

Ma, L., Y. Liu, X. Zhang, Y. Ye, G. Yin, and B. A. Johnson. 2019. Deep learning in remote sensing applications: A meta-analysis and review. *ISPRS Journal of Photogrammetry and Remote Sensing* 152:166–177. https://doi.org/10.1016/j.isprsjprs.2019.04.015.

Ma, X., H. Wang, and J. Wang. 2016. Semisupervised classification for hyperspectral image based on multi-decision labeling and deep feature learning. *ISPRS Journal of Photogrammetry and Remote Sensing* 120:99–107. https://doi.org/10.1016/j.isprsjprs.2016.09.001.

Ma, Y., S. Chen, S. Ermon, and D. B. Lobell. 2024. Transfer learning in environmental remote sensing. *Remote Sensing of Environment* 301:113924. https://doi.org/10.1016/j.rse.2023.113924.

MacKay, D. J. C. 1992. Information-based objective functions for active data selection. *Neural Computation* 4:590–604.

Mai, F., A. Pannatier, F. Fehr, H. Chen, F. Marelli, F. Fleuret, and J. Henderson. 2022. Hypermixer: An MLP-based low cost alternative to transformers. arXiv preprint arXiv:2203.03691. https://doi.org/10.48550/arXiv.2203.03691.

Mao, X., Q. Li, H. Xie, R. Y. K. Lau, Z. Wang, and S. P. Smolley. 2017. Least squares generative adversarial networks. In *Proceedings of the IEEE International Conference on Computer Vision*, Venice, Italy, pp. 2794–2802.

Mas, J. F., and J. J. Flores. 2008. The application of artificial neural networks to the analysis of remotely sensed data. *International Journal of Remote Sensing* 29:617–663. https://doi.org/10.1080/01431160701352154.

Mazzia, V., A. Khaliq, and M. Chiaberge. 2019. Improvement in land cover and crop classification based on temporal features learning from Sentinel-2 data using recurrent-convolutional neural network (R-CNN). *Applied Sciences* 10:238. https://doi.org/10.3390/app10010238.

Michelucci, U. 2022. An introduction to autoencoders. arXiv preprint, arXiv:2201.03898. https://doi.org/10.48550/arXiv.2201.03898.

Mikołajczyk, A., and M. Grochowski. 2018. Data augmentation for improving deep learning in image classification problem. In *Proceedings of 2018 International Interdisciplinary PhD Workshop*, Świnoujście, Poland, pp. 117–122, https://doi.org/10.1109/IIPHDW.2018.8388338.

Mirza, M., and S. Osindero. 2014. Conditional generative adversarial nets. arXiv preprint arXiv:1411.1784.

Mnih, V., K. Kavukcuoglu, D. Silver, A. A. Rusu, J. Veness, M. G. Bellemare, A. Graves, et al. 2015. Human-level control through deep reinforcement learning. *Nature* 518:529–533. https://doi.org/10.1038/nature14236.

Moradi, R., R. Berangi, and B. Minaei. 2020. A survey of regularization strategies for deep models. *Artificial Intelligence Review* 53:3947–3986. https://doi.org/10.1007/s10462-019-09784-7.

Mou, L., L. Bruzzone, and X. X. Zhu. 2018. Learning spectral-spatial-temporal features via a recurrent convolutional neural network for change detection in multispectral imagery. *IEEE Transactions on Geoscience and Remote Sensing* 57:924–935. https://doi.org/10.1109/TGRS.2018.2863224.

Mou, L., P. Ghamisi, and X. Zhu. 2017. Deep recurrent neural networks for hyperspectral image classification. *IEEE Transactions on Geoscience and Remote Sensing* 55:3639–3655. https://doi.org/10.1109/TGRS.2016.2636241.

Mozer, M. C., and P. Smolensky. 1989a. Skeletonization: A technique for trimming the fat from a network via relevance assessment. In *Advances in Neural Information Processing Systems*, (ed.) D. S. Touretzky, pp. 107–115. San Mateo, CA: Morgan Kaufmann.

Mozer, M. C., and P. Smolensky. 1989b. Using relevance to reduce network size automatically. *Connection Science* 1:3–16.

Murugan, P., and S. Durairaj. 2017. Regularization and optimization strategies in deep convolutional neural network. arXiv preprint arXiv:1712.04711. https://doi.org/10.48550/arXiv.1712.04711.

Naranjo-Torres, J., M. Mora, R. Hernández-García, R. J. Barrientos, C. Fredes, and A. Valenzuela. 2020. A review of convolutional neural network applied to fruit image processing. *Applied Sciences* 10:3443. https://doi.org/10.3390/app10103443.

Ng, A. Y. 1997. Preventing "overfitting" of cross-validation data. In *Proceeding of the 14th International Conference on Machine Learning*, Nashville, Tennessee, pp. 245–253.

Nguyen, Q., M. C. Mukkamala, and M. Hein. 2018. Neural networks should be wide enough to learn disconnected decision regions. In *Proceedings of International Conference on Machine Learning*, July 3, Stockholm, Sweden, pp. 3740–3749.

Niu, S., Y. Liu, J. Wang, and H. Song. 2020. A decade survey of transfer learning (2010-2020). *IEEE Transactions on Artificial Intelligence* 1:151–166. https://doi.org/10.1109/TAI.2021.3054609.

Otgonbaatar, S., G. Schwarz, M. Datcu, and D. Kranzlmüller. 2023. Quantum transfer learning for real-world, small, and high-dimensional remotely sensed datasets. *IEEE Journal of Selected Topics in Applied Earth Observations and Remote Sensing* 16:9223–9230. https://doi.org/10.1109/JSTARS.2023.3316306.

Othman, E., Y. Bazi, N. Alajlan, H. Alhichri, and F. Melgani. 2016. Using convolutional features and a sparse autoencoder for land-use scene classification. *International Journal of Remote Sensing* 37:2149–2167. https://doi.org/10.1080/01431161.2016.1171928.

Oubara, A., F. Wu, A. Amamra, and G. Yang. 2022. Survey on remote sensing data augmentation: Advances, challenges, and future perspectives. In *Advances in Computing Systems and Applications*, (eds.) M. R. Senouci, S. Y., Boulahia, and M. A. Benatia. Cham: Springer. https://doi.org/10.1007/978-3-031-12097-8_9.

Ozkan, S., B. Kaya, and G. B. Akar. 2018. EndNet: Sparse autoencoder network for endmember extraction and hyperspectral unmixing. *IEEE Transactions on Geoscience and Remote Sensing* 57:482–496. https://doi.org/10.1109/TGRS.2018.2856929.

Packer, C., K. Gao, J. Kos, P. Krähenbühl, V. Koltun, and D. Song. 2018. Assessing generalization in deep reinforcement learning. arXiv preprint arXiv:1810.12282.

Pal, M. 2024. Deep learning algorithms for hyperspectral remote sensing classifications: An applied review. *International Journal of Remote Sensing* 45:451–491. https://doi.org/10.1080/01431161.2023.2297178.

Pan, S. J., and Q. Yang. 2010. A survey on transfer learning. *IEEE Transactions on Knowledge and Data Engineering* 22:1345–1359. https://doi.org/10.1109/TKDE.2009.191.

Paola, J. D., and R. A. Schowengerdt. 1995. A review and analysis of backpropagation neural networks for classification of remotely-sensed multi-spectral imagery. *International Journal of Remote Sensing* 16:3033–3058. https://doi.org/10.1080/01431169508954607.

Paoletti, M. E., J. M. Haut, J. Plaza, and A. Plaza. 2019. Deep learning classifiers for hyperspectral imaging: A review. *ISPRS Journal of Photogrammetry and Remote Sensing* 158:279–317. https://doi.org/10.1016/j.isprsjprs.2019.09.006.

Peng, D., Y. Zhang, and H. Guan. 2019. End-to-end change detection for high resolution satellite images using improved UNet++. *Remote Sensing* 11:1382. https://doi.org/10.3390/rs11111382.

Perez, L., and J. Wang. 2017. The effectiveness of data augmentation in image classification using deep learning. arXiv preprint arXiv:1712.04621.

Petrovska, B., E. Zdravevski, P. Lameski, R. Corizzo, I. Štajduhar, and J. Lerga. 2020. Deep learning for feature extraction in remote sensing: A case-study of aerial scene classification. *Sensors* 20:3906. https://doi.org/10.3390/s20143906.

Pinaya, W. H. L., S. Vieira, R. Garcia-Dias, and A. Mechelli. 2020. Autoencoders. In Machine Learning, (eds.) A. Mechelli, and S. Vieira, pp. 193–208. Cambridge, MA: Academic Press. https://doi.org/10.1016/B978-0-12-815739-8.00011-0.

Pires de Lima, R., and K. Marfurt. 2020. Convolutional neural network for remote-sensing scene classification: Transfer learning analysis. *Remote Sensing* 12:86. https://doi.org/10.3390/rs12010086.

Rahman, M. A., and Y. Wang. 2016. Optimizing intersection-over-union in deep neural networks for image segmentation. In *International Symposium on Visual Computing*, Las Vegas, NV, pp. 234–244. Berlin, Heidelberg: Springer.

Reed, R. 1993. Pruning algorithms: A survey. *IEEE Transactions on Neural Networks* 4:740–747.

Rezende, D. J., S. Mohamed, and D. Wierstra. 2014. Stochastic backpropagation and approximate inference in deep generative models. *Proceedings of the 31st International Conference on Machine Learning*, 32:1278–1286.

Ripley, B. D. 1996. *Pattern Recognition and Neural Networks*. Cambridge: Cambridge University Press.

Risojević, V., and V. Stojnić. 2021. Do we still need ImageNet pre-training in remote sensing scene classification? arXiv preprint arXiv:2111.03690.

Romera-Paredes, B., A. Argyriou, N. Bianchi-Berthouze, and M. Pontil. 2012. Exploiting unrelated tasks in multi-task learning. In *Proceedings of the 15th International Conference on Artificial Intelligence and Statistics*, April 21–23, La Palma, Canary Islands, vol. 22, pp. 951–959.

Roy, S. K., G. Krishna, S. R. Dubey, and B. B. Chaudhuri. 2019. HybridSN: Exploring 3-D-2-D CNN feature hierarchy for hyperspectral image classification. *IEEE Geoscience and Remote Sensing Letters* 17:277–281. https://doi.org/10.1109/LGRS.2019.2918719.

Rumelhart, D. E., G. E. Hinton, and R. J. Williams. 1986. Learning internal representations by error propagation. In *Parallel Distributed Processing*: *Explorations in the Microstructure of Cognition*, (eds.) D. E. Rumelhart, J. L. Mcclelland, and PDP Research Group, pp. 318–362. Cambridge, MA: MIT Press.

Schuster, M., and K. K. Paliwal. 1997. Bidirectional recurrent neural networks. *IEEE Transactions on Signal Processing* 45:2673–2681. https://doi.org/10.1109/78.650093.

Scfrin, O., F. M. Riese, S. Keller. 2021. Deep learning for land cover change detection. *Remote Sensing* 13:78. https://doi.org/10.3390/rs13010078.

Sertel, E., B. Ekim, P. E. Osgouei, and M. E. Kabadayi. 2022. Land use and land cover mapping using deep learning based segmentation approaches and VHR Worldview-3 images. *Remote Sensing* 14:4558. https://doi.org/10.3390/rs14184558.

Sewak, M. 2019. *Deep Reinforcement Learning: Frontiers of Artificial Intelligence*. Singapore: Springer.

Sietsma, J., and R. J. F. Dow. 1991. Creating artificial neural networks that generalize. *Neural Networks* 4:67–79. https://doi.org/10.1016/0893-6080(91)90033-2.

Sildir, H., E. Aydin, and T. Kavzoglu. 2020. Design of feedforward neural networks in the classification of hyperspectral imagery using superstructural optimization. *Remote Sensing* 12:956. https://doi.org/10.3390/rs12060956.

Simonovsky, M., and N. Komodakis. 2018. GraphVAE: Towards generation of small graphs using variational autoencoders. In *Proceedings of the 27th International Conference on Artificial Neural Networks*, Rhodes, Greece, October 4–7, pp. 412–422.

Song, A., J. Choi, Y. Han, and Y. Kim. 2018. Change detection in hyperspectral images using recurrent 3D fully convolutional networks. *Remote Sensing* 10:1827. https://doi.org/10.3390/rs10111827.

Song, H., and W. Yang. 2022. GSCCTL: A general semi-supervised scene classification method for remote sensing images based on clustering and transfer learning. *International Journal of Remote Sensing* 43:5976–6000. https://doi.org/10.1080/01431161.2021.2019851.

Song, J., S. Gao, Y. Zhu, and C. Ma. 2019. A survey of remote sensing image classification based on CNNs. *Big Earth Data* 3:232–254. https://doi.org/10.1080/20964471.2019.1657720.

Springenberg, J. T. 2015. Unsupervised and semi-supervised learning with categorical generative adversarial networks. arXiv preprint, arXiv:1511.06390.

Srivastava, N., G. Hinton, A. Krizhevsky, I. Sutskever, and R. Salakhutdinov. 2014. Dropout: A simple way to prevent neural networks from overfitting. *The Journal of Machine Learning Research* 15:1929–1958.

Srivastava, S., M. Volpi, and D. Tuia. 2017. Joint height estimation and semantic labeling of monocular aerial images with CNNs. In *Proceedings of IEEE International Geoscience and Remote Sensing Symposium*, July 23–28, Fort Worth, TX, pp. 5173–5176.

Staufer, P., and M. M. Fischer. 1997. Spectral pattern recognition by a two-layer perceptron: Effects of training set size. In *Neurocomputation in Remote Sensing Data Analysis*, (eds.) I. Kanellopoulos, G. G. Wilkinson, F. Roli, and J. Austin, pp. 105–116. London: Springer.

Sutton, R. S., and A. G. Barto. 1998. *Reinforcement Learning: An Introduction*. Cambridge, MA: MIT Press.

Tamar, A., Y. Wu, G. Thomas, S. Levine, and P. Abbeel. 2016. Value iteration networks. In *Proceedings of Advances in Neural Information Processing Systems* (NIPS'2016), Barcelona, Spain, p. 29.

Tao, C., H. Pan, Y. Li, and Z. Zou. 2015. Unsupervised spectral-spatial feature learning with stacked sparse autoencoder for hyperspectral imagery classification. *IEEE Geoscience and Remote Sensing Letters* 12:2438–2442. https://doi.org/10.1109/LGRS.2015.2482520.

Teke, A., E. O. Yilmaz, and T. Kavzoglu. 2021. Comparative assessment of deep learning and machine learning models in shallow landslide susceptibility. In *Proceedings of International Symposium on Applied Geoinformatics*, December 2–3, Riga, Latvia.

Thapa, A., T. Horanont, B. Neupane, and J. Aryal. 2023. Deep learning for remote sensing image scene classification: A review and meta-analysis. *Remote Sensing* 15:4804. https://doi.org/10.3390/rs15194804.

Thung, K.-H., and C.-Y. Wee. 2018. A brief review on multi-task learning. *Multimedia Tools and Applications* 77:29705–29725. https://doi.org/10.1007/s11042-018-6463-x.

Tolstikhin, I. O., N. Houlsby, A. Kolesnikov, L. Beyer, X. Zhai, T. Unterthiner, J. Yung, et al. 2021. MLP-mixer: An all-MLP architecture for vision. arXiv:2105.01601.

Tonbul, H., E. O. Yilmaz, and T. Kavzoglu. 2023. Comparative analysis of deep learning and machine learning models for burned area estimation using Sentinel-2 Image: A case study in Mugla-Bodrum, Turkey. In *Proceedings of 10th International Conference on Recent Advances in Air and Space Technologies*, June 7–9, Istanbul, Turkey.

Tuia, D., C. Persello, and L. Bruzzone. 2016. Domain adaptation for the classification of remote sensing data: An overview of recent advances. *IEEE Geoscience and Remote Sensing Magazine* 4:41–57. https://doi.org/10.1109/MGRS.2016.2548504.

Tuia, D., M. Volpi, L. Copa, M. Kanevski, and J. Munoz-Mari. 2011. A survey of active learning algorithms for supervised remote sensing image classification. *IEEE Journal of Selected Topics in Signal Processing* 5:606–617. https://doi.org/10.1109/JSTSP.2011.2139193.

Vali, A., S. Comai, and M. Matteucci. 2020. Deep learning for land use and land cover classification based on hyperspectral and multispectral earth observation data: A review. *Remote Sensing* 12:2495. https://doi.org/10.3390/rs12152495.

Van Den Oord, A., and O. Vinyals. 2017. Neural discrete representation learning. In *Proceedings of 31st Conference on Neural Information Processing Systems*, December 4–9, Long Beach, CA.

Van Hasselt, H., A. Guez, and D. Silver, D. 2016. Deep reinforcement learning with double Q-learning. In *Proceedings of the 30th AAAI Conference on Artificial Intelligence*, p. 30. https://doi.org/10.1609/aaai.v30i1.10295.

Van Laarhoven, T. 2017. L2 regularization versus batch and weight normalization. arXiv preprint arXiv:1706.05350.

Vandenhende, S., S. Georgoulis, M. Proesmans, D. Dai and L. Van Gool. 2020. Revisiting multi-task learning in the deep learning era. arXiv preprint arXiv:2004.13379.

Vaswani, A., N. Shazeer, N. Parmar, J. Uszkoreit, L., Jones, A. N., Gomez, L. Kaiser, and I. Polosukhin. 2017. Attention is all you need. arXiv:1706.03762.

Vincent, P., H. Larochelle, Y. Bengio, and P. A. Manzagol. 2008. Extracting and composing robust features with denoising autoencoders. In *Proceedings of the 25th International Conference on Machine Learning*, July 5–9, Helsinki, Finland, pp. 1096–1103. https://doi.org/10.1145/1390156.1390294.

Wambugu, N., Y. Chen, Z. Xiao, K. Tan, M. Wei, X. Liu, and J. Li. 2021. Hyperspectral image classification on insufficient-sample and feature learning using deep neural networks: A review. *International Journal of Applied Earth Observation and Geoinformation* 105:102603. https://doi.org/10.1016/j.jag.2021.102603.

Wang, H. N., N. Liu, Y. Y. Zhang, D. W. Feng, F. Huang, D. S. Li, and Y. M. Zhang. 2020. Deep reinforcement learning: A survey. *Frontiers of Information Technology & Electronic Engineering* 21:1726–1744.

Wang, L., X. Xu, H. Dong, R. Gui, R. Yang, and F. Pu. 2018. Exploring convolutional LSTM for PolSAR image classification. In *Proceedings of IEEE International Geoscience and Remote Sensing Symposium*, July 2–27, Valencia, Spain, pp. 8452–8455. https://doi.org/10.1109/IGARSS.2018.8518517.

Wang, Q., Y. Ma, K. Zhao, and Y. Tian. 2022. A comprehensive survey of loss functions in machine learning. *Annals of Data Science* 9:187–212. https://doi.org/10.1007/s40745-020-00253-5.

Wang, Z., B. Du, L. Zhang, L. Zhang, and X. Jia. 2017. A novel semisupervised active-learning algorithm for hyperspectral image classification. *IEEE Transactions on Geoscience and Remote Sensing* 55:3071–3083. https://doi.org/10.1109/TGRS.2017.2650938.

Wang, Z., C. Di Massimo, M. T. Tham, and A. J. Morris. 1994. A procedure for determining the topology of multilayer feedforward neural networks. *Neural Networks* 7:291–300.

Weiss, K., T. M. Khoshgoftaar, and D. Wang. 2016. A survey of transfer learning. *Journal of Big Data* 3:1–40. https://doi.org/10.1186/s40537-016-0043-6.

Werbos, P. J. 1990. Backpropagation through time: What it does and how to do it. *Proceedings of the IEEE* 78:1550–1560. https://doi.org/10.1109/5.58337.

Windrim, L., A. Melkumyan, R. J. Murphy, A. Chlingaryan, and R. Ramakrishnan. 2018. Pretraining for hyperspectral convolutional neural network classification. *IEEE Transactions on Geoscience and Remote Sensing* 56:2798–2810.

Wu, H., and S. Prasad. 2017. Convolutional recurrent neural networks for hyperspectral data classification. *Remote Sensing* 9:298. https://doi.org/10.3390/rs9030298.

Wu, H., and S. Prasad. 2018. Semi-supervised deep learning using pseudo-labels for hyperspectral image classification. *IEEE Transactions on Image Processing* 27:1259–1270. https://doi.org/10.1109/TIP.2017.2772836

Wu, X., D. Sahoo, and S. C. Hoi. 2020. Recent advances in deep learning for object detection. *Neurocomputing* 396:39–64. https://doi.org/10.1016/j.neucom.2020.01.085.

Xing, C., L. Ma, and X. Yang. 2016. Stacked denoise autoencoder based feature extraction and classification for hyperspectral images. *Journal of Sensors* 2016:3632943. https://doi.org/10.1155/2016/3632943.

Xu, J., Z. Li, B. Du, M. Zhang, and J. Liu. 2020. Reluplex made more practical: Leaky ReLU. In *Proceedings of IEEE Symposium on Computers and Communications*, Rennes, France, pp. 1–7. https://doi.org/10.1109/ISCC50000.2020.9219587.

Yang, B., W. Luo, and R. Urtasun. 2018. PIXOR: Real-time 3D object detection from point clouds. In *Proceedings of the IEEE conference on Computer Vision and Pattern Recognition*, Salt Lake City, UT, pp. 7652–7660.

Yang, J., Y.-Q. Zhao, and J. C.-W. Chan. 2017. Learning and transferring deep joint spectral-spatial features for hyperspectral classification. *IEEE Transactions on Geoscience and Remote Sensing* 55:4729–4742. https://doi.org/10.1109/TGRS.2017.2698503.

Yang, Q., Y. Zhang, W. Dai, and S. J. Pan. 2020a. *Transfer Learning*. Cambridge, UK: Cambridge University Press.

Yang, X., Z. Song, I. King, and Z. Xu. 2023. A survey on deep semi-supervised learning. *IEEE Transactions on Knowledge and Data Engineering* 35:8934–8954. https://doi.org/10.1109/TKDE.2022.3220219.

Yang, Y., H. Gu, Y. Han, and H. Li. 2020b. An end-to-end deep learning change detection framework for remote sensing images, In *Proceedings of 2020 IEEE International Geoscience and Remote Sensing Symposium*, Waikoloa, HI, pp. 652–655, https://doi.org/10.1109/IGARSS39084.2020.9324076.

Yildirim, E., and T. Kavzoglu. 2022. Ship detection in optical remote sensing images using YOLOv4 and Tiny YOLOv4. In (eds) M.B. Ahmed, A.A. Boudhir, İ.R. Karaş, V. Jain and S. Mellouli, *Innovations in Smart Cities Applications: Lecture Notes in Networks and Systems*, vol. 393, pp. 913–924. Switzerland: Springer Nature.

Yilmaz, E. O., A. Teke, and T. Kavzoglu. 2022. Performance evaluation of depthwise separable CNN and random forest algorithms for landslide susceptibility prediction. In *Proceedings of IEEE International Geoscience and Remote Sensing Symposium (IGARSS'2022)*, July 17–22, Kuala Lumpur, Malaysia, pp. 5477–5480.

Yilmaz, E. O., H. Tonbul, and T. Kavzoglu. 2024. Marine mucilage mapping with explained deep learning model using water-related spectral indices: A case study of Dardanelles Strait, Turkey. *Stochastic Environment Research Risk Assessment* 38:51–68. https://doi.org/10.1007/s00477-023-02560-8.

Yin, M., Z. Chang, and Y. Wang. 2023. LocoMixer: A local context MLP-like architecture for image classification, In *Proceedings of the 35th International Conference on Tools with Artificial Intelligence*, Atlanta, GA, pp. 762–769, https://doi.org/10.1109/ICTAI59109.2023.00117.

Yu, T., X. Li, Y. Cai, M. Sun, and P. Li 2022. S2-MLP: Spatial-shift MLP architecture for vision. In *Proceedings of the IEEE/CVF Winter Conference on Applications of Computer Vision*, Waikoloa, HI, pp. 297–306.

Yu, X., X. Wu, C. Luo, and P. Ren. 2017. Deep learning in remote sensing scene classification: A data augmentation enhanced convolutional neural network framework. *GIScience & Remote Sensing* 54:741–758.

Yuan, X., J. Shi, and L. Gu. 2021. A review of deep learning methods for semantic segmentation of remote sensing imagery. *Expert Systems with Applications* 169:114417. https://doi.org/10.1016/j.eswa.2020.114417.

Zhan, Y., D. Hu, Y. Wang, and X. Yu. 2018. Semisupervised hyperspectral image classification based on generative adversarial networks. *IEEE Geoscience and Remote Sensing Letters* 15:212–216. https://doi.org/10.1109/LGRS.2017.2780890.

Zhang, A., Z. C. Lipton, M. Li, and A. J. Smola. 2023. *Dive into Deep Learning*. Cambridge, UK: Cambridge University Press.

Zhang, C., L. Zhou, Y. Zhao, S. Zhu, F. Liu, and Y. He. 2020. Noise reduction in the spectral domain of hyperspectral images using denoising autoencoder methods. *Chemometrics and Intelligent Laboratory Systems* 203:104063. https://doi.org/10.1016/j.chemolab.2020.104063.

Zhang, H., Z. Dong, B. Li, and S. He. 2022. Multi-scale MLP-Mixer for image classification. *Knowledge-Based Systems* 258:109792. https://doi.org/10.1016/j.knosys.2022.109792.

Zhang, M., S. Jiang, Z. Cui, R. Garnett, and Y. Chen. 2019a. D-VAE: A variational autoencoder for directed acyclic graphs. In *Proceedings of 33rd Conference on Neural Information Processing Systems*, Vancouver, Canada.

Zhang, W., P. Tang, and L. Zhao. 2019b. Remote sensing image scene classification using CNN-CapsNet. *Remote Sensing* 11:494. https://doi.org/10.3390/rs11050494.

Zhang, X., G. Chen, W. Wang, Q. Wang, and F. Dai. 2017. Object-based land-cover supervised classification for very-high-resolution UAV images using stacked denoising autoencoders. *IEEE Journal of Selected Topics in Applied Earth Observations and Remote Sensing* 10:3373–3385. https://doi.org/10.1109/JSTARS.2017.2672736.

Zhang, Y., and Q. Yang. 2018. An overview of multi-task learning. *National Science Review* 5:30–43.

Zhang, Z., G. Vosselman, M. Gerke, D. Tuia, and M. Y. Yang. 2018. Change detection between multimodal remote sensing data using Siamese CNN. arXiv preprint arXiv:1807.09562.

Zhao, Z. Q., P. Zheng, S. T. Xu, and X. Wu. 2019. Object detection with deep learning: A review. *IEEE Transactions on Neural Networks and Learning Systems* 30:3212–3232. https://doi.org/10.1109/TNNLS.2018.2876865.

Zheng, X., T. Gong, X. Li, and X. Lu, 2022. Generalized scene classification from small-scale datasets with multitask learning. *IEEE Transactions on Geoscience and Remote Sensing* 60:5609311. https://doi.org/10.1109/TGRS.2021.3116147.

Zhu, L., Y. Chen, P. Ghamisi, and J. A. Benediktsson. 2018. Generative adversarial networks for hyperspectral image classification. *IEEE Transactions on Geoscience and Remote Sensing* 56:5046–5063. https://doi.org/10.1109/TGRS.2018.2805286.

Zhu, M., Y. He, and Q. He. 2019. A review of researches on deep learning in remote sensing application. *International Journal of Geosciences* 10:1–11. https://doi.org/10.4236/ijg.2019.101001.

Zhu, X. X., D. Tuia, L. Mou, G. S. Xia, L. Zhang, F. Xu, and F. Fraundorfer. 2017. Deep learning in remote sensing: A comprehensive review and list of resources. *IEEE Geoscience and Remote Sensing Magazine* 5:8–36. https://doi.org/10.1109/MGRS.2017.2762307.

Zhu, X., and A. B. Goldberg. 2009. *Introduction to Semi-Supervised Learning*. New York: Springer. https://doi.org/10.1007/978-3-031-01548-9.

Zou, H., and T. Hastie. 2005. Regularization and variable selection via the elastic net. *Journal of the Royal Statistical Society Series B: Statistical Methodology* 67:301–320. https://doi.org/10.1111/j.1467-9868.2005.00503.x.

8 Object-Based Image Analysis

The spatial and temporal dynamics captured by a series of satellite images, depicting the complexity of the landscape, involve a combination of natural and artificial elements, along with potential distortions like shadows or occlusions. Pixels in these images commonly represent a mixture of landscape features, including both man-made and natural objects, influenced by the spatial resolution of the sensor and the specific structure of the studied area (Kavzoglu et al., 2017). To identify objects in an image, it is crucial to have a pixel size considerably smaller than the size of the object. Given that urban objects are significantly smaller than natural features, a considerably small pixel size is necessary for urban applications. Recent advancements in remote sensing data acquisition systems coupled with a growing demand for semantic information in GIS applications have resulted in substantial improvements in the spatial resolution of satellite and UAV images. This enhancement has led to the production of very high-resolution (VHR) images (<2.0 m) with modern sensors (e.g., IKONOS, QuickBird, WorldView series, GaoFen, GeoEye-1, KOMPSAT, SPOT-6 and SPOT-7, Göktürk-1, Pléiades). To be more specific, with the first generation of satellite images, such as 30-m resolution images from Landsat TM and ETM+ instruments covering $900\,m^2$ on the ground, a single pixel could encompass several buildings; conversely, the images acquired by the current VHR sensors can cover a single building with tens and hundreds of pixels. The acquisition of VHR images from many sources, particularly by the cameras mounted on UAVs, has greatly facilitated the creation and updating of GIS map databases, including polygonal data.

The increased demand for accurate and timely information requires prompt and accurate classifications, which calls for a reliable and unbiased analysis of certain phenomena. Image processing techniques developed since the 1970s, which rely solely on pixel-based analysis, have become insufficient to fulfill these expectations. In pixel-based classification, spatial relations, shape, and context information are overlooked. In addition, the traditional classification approach is not well adapted, especially when studying areas covered with natural and semi-natural vegetation, where there is a continuous variation in ground surface cover. From a technical standpoint, a pixel can be comprehended on the basis of its spectral characteristics, the combination of spectral endmembers, or its neighboring environment. Considering that neighboring pixels are highly likely to belong to the same land cover class in VHR images, an individual pixel cannot be assigned to a valid earth object. Instead, a group of homogeneous neighboring pixels, referred to as image objects or segments, can more accurately reflect the reality on the ground. Avoiding pixel-based layouts and providing meaningful image objects open up a new dimension in rule-based automatic image analysis. In this approach, image objects can be directly labeled using a set of features, including spatial features, or they can be used to model complex classes based on their spatial relationships (Lang, 2008). To address these challenges, object-based image analysis (OBIA), also known as geographical object-based image analysis (GEOBIA), has evolved as a new paradigm. It is noteworthy that some researchers (e.g., Liu et al., 2006; Yan et al., 2006; Baatz et al., 2008) have suggested the term "object-oriented" instead of "object-based", considering that objects are not regarded as primitive information but are modeled with expert knowledge to identify classes. Although "object-based" has been the more prevalent term in the literature, primarily to distinguish it from object-based programming in computer science, these terms are often used interchangeably.

The concept of OBIA has been applied in various contexts involving Earth observation data. It has been employed in research that emphasizes the integration of ground observation images and the extraction of associated geographical information. OBIA can be viewed as an emulation of human perception for the analysis of the environment to produce tangible information from imagery, considering the spectral, textural, geometrical, and contextual features of objects.

DOI: 10.1201/9781003439172-8

The traditional pixel-based approach has been replaced by the OBIA approach, which has emerged as a vibrant field focusing on advancing land cover classification using high spatial resolution remote sensing imagery. In contrast to the pixel-based approach, the object-based approach involves segmenting images into homogeneous regions, considering spectral, spatial, and textural features. This methodology enhances the foundation for image analysis and introduces a vital connection between remote sensing and GIS. Although the object-based data representation and classification paradigm were initially proposed in the 1970s, its development faced delays primarily because of the computational power constraints of computers. During its early development, the remote sensing community dedicated considerable efforts to promote the incorporation of object-based technology into land cover mapping (Blaschke and Strobl, 2001; Dorren et al., 2003; Blaschke et al., 2004; Yu et al., 2006; Walker and Blaschke, 2008) and develop software to conduct the first OBIA applications (e.g., Buck et al., 1999; De Kok et al., 1999; Niemeyer et al., 1999). However, since the release of the eCognition software in 2000, the number of OBIA-based studies has surged, contributing to the solidification of its position and establishing it as a standard tool in the classification of remotely sensed data. As evident from recent international conferences and special issues in peer-reviewed journals focused on OBIA techniques and a wide range of applications, there has been a noticeable increase in remote sensing research.

OBIA also plays an important role in image interpretation. The entire process of image analysis can be characterized by the transformation of knowledge (Lang and Blaschke, 2006). It is also crucial for extracting significant LULC features from aerial and satellite images, thus helping to distinguish LULC classes. It also provides a conceptual foundation and theoretical framework for employing multiscale approaches in the analysis of land cover features, which, in most cases, cannot be modeled with a single scale. The main objective of OBIA is to provide efficient and automated methods and tools for replicating and even surpassing the human interpretation of remote sensing images in an automated or semi-automated manner. This aims to enhance repeatability and productivity while minimizing subjectivity, labor, and time costs. The methodological framework of OBIA facilitates the machine-based interpretation of complex classes, encompassing spectral, spatial, and structural properties, as well as hierarchical characteristics (Lang, 2008). The OBIA process involves the grouping of pixels, which makes it possible to analyze homogeneous clusters of pixels. This analysis includes considerations of size, shape, and texture in addition to spectral features, and the ability to query objects based on their specific location and neighborhood. Owing to these robust characteristics, OBIA has the potential to deal with complex image analysis tasks that consider internal spatial structures and produce more realistic thematic maps, minimizing the salt-and-paper effect (Kavzoglu, 2017; Tonbul, 2021). While the effectiveness of OBIA is broadly acknowledged, the unresolved challenge lies in selecting a robust classification method, dealing with the diversity of data sources, choice of segmentation parameters, size of the training set, and selection of relevant object features. Although the incorporation of various features of image objects is one of the most important aspects of the OBIA approach, the availability of hundreds of features raises the issue of the curse of dimensionality. Using a large number of features in the OBIA process not only increases the required computational time but also requires more samples for training, which is unavailable in most cases. Therefore, the feature selection techniques, discussed in Section 3.2, can serve as robust analytical tools for identifying the most relevant features while excluding those that are irrelevant or make only a marginal contribution. Many studies have applied feature selection methods, such as filter, wrapper, embedded, or evolutionary algorithms, and reported improvements in classification accuracy (e.g., Laliberte et al., 2012; Colkesen and Kavzoglu, 2018; Kavzoglu et al., 2018).

The process of OBIA can be divided into three stages: segmentation, classification, and accuracy assessment. The theoretical background of the major classification techniques and object-based accuracy assessment are discussed in Chapters 2, 5–7, and Section 10.1.7. Image segmentation is the first and crucial step in which meaningful, nonoverlapping, and homogeneous image objects are created based on a huge set of features that can include spectral, contextual, geometric, and textural information. Image objects have additional spectral features, such as variance, minimum, mean, and maximum values of each band, band ratios, and other spatial features, which help to increase the delineation of LULC classes. In fact, segmentation establishes the foundation for an OBIA classification by creating objects (their number being much fewer than the pixels in the image) that serve as the fundamental processing units in the classification process. Instead of finding the most appropriate label for a pixel, OBIA deals with the labeling of objects created using a segmentation approach. The key to successful image classification lies in obtaining image objects that are representative of real objects. In essence, the accuracy achieved in image classification depends mainly on the quality of segmentation (the rate of overlap between image objects and real objects), which depends on the choice of segmentation parameters (Radoux and Defourny, 2008; Kim et al., 2009; Gao et al., 2011; Kavzoglu and Tonbul, 2018a,b; Ming et al., 2018). Employing an effective segmentation method and setting its parameters with optimal values are difficult tasks because of the texture, size, and complex structure of earth objects. It should also be noted that segmentation quality is related to image quality, the number of image bands, the spatial resolution of the imagery, and the complexity of the scene (Belgiu and Drăguţ, 2014). Also, the concept of image segmentation itself is elusive, with no established rules on how homogeneity in an object can be consistently defined across the entire image. Currently, no universally accepted method or algorithm is available for determining the optimal parameters for a specific problem. However, it is essential to perform a cognitive evaluation of the created segments to assess the quality of segmentation. As noted earlier, the effect of the choice of parameter values on the accuracy achieved by classification is undeniable. Therefore, attempts have been made to develop strategies for determining optimal parameter values using local and global measures, focusing on intra-segment homogeneity and inter-segment heterogeneity. Several comprehensive reviews exist on the determination of segmentation parameters for OBIA applications, including Im et al. (2014), El-naggar (2018), and Ez-zahouani et al. (2023b).

By diminishing the image complexity and level of detail, image segmentation contributes to a clearer interpretation of the image content. Image segmentation aims to divide an image into meaningful regions that have a strong correlation with objects or areas of the real world (Kavzoglu and Yildiz, 2014). In other words, the purpose of segmentation is to create image objects by maximizing intra-segment homogeneity and inter-segment heterogeneity. In the ideal case, the constructed segment boundaries should perfectly overlap with the real objects on the ground. However, while theoretically plausible, creating the perfect image objects corresponding exactly to real land cover objects is an unattainable and utopic target due to the complex structure of remotely sensed data and the continuum of LULC classes in nature. When segmentation methods are applied to images, a two-faced problem arises depending on the creation of fewer or more than the optimal number of segments due to inappropriate selection of segmentation parameters. Over-segmentation occurs when more image objects than the optimal number of segments are created for real objects. In other words, the image objects are smaller than the reference objects. Under-segmentation is the reverse case where there are insufficient numbers of segments; that is, image objects are larger than reference objects. The ideal case can thus be redefined such that over-segmentation and under-segmentation should be at a minimum level, and image objects correspond to ground reference objects. It is more advantageous to have a slight over-segmentation and no under-segmentation because merging segments later is a more intricate process than splitting them (Castilla and Hay, 2008). To date, many segmentation algorithms have been developed for image segmentation, which has been reviewed by Pal and Pal (1993), Blaschke (2010), Ma et al. (2017), Hossain and Chen (2019), Kotaridis and Lazaridou (2021), and Ez-zahouani et al. (2023a). Based on the strategies for partitioning an image into objects, Schiewe (2002) categorized the segmentation methods into four main categories: (1) point-based, (2) edge-based, (3) region-based, and (4) a combination of the former. Segmentation methods can also be categorized based on different criteria, basically considering the information and the methodology used in image partitioning (Figure 8.1). While supervised methods require a manually produced segmentation map called a reference map, unsupervised methods, which are more frequently used, estimate quality scores based on the segmented image. Apart from

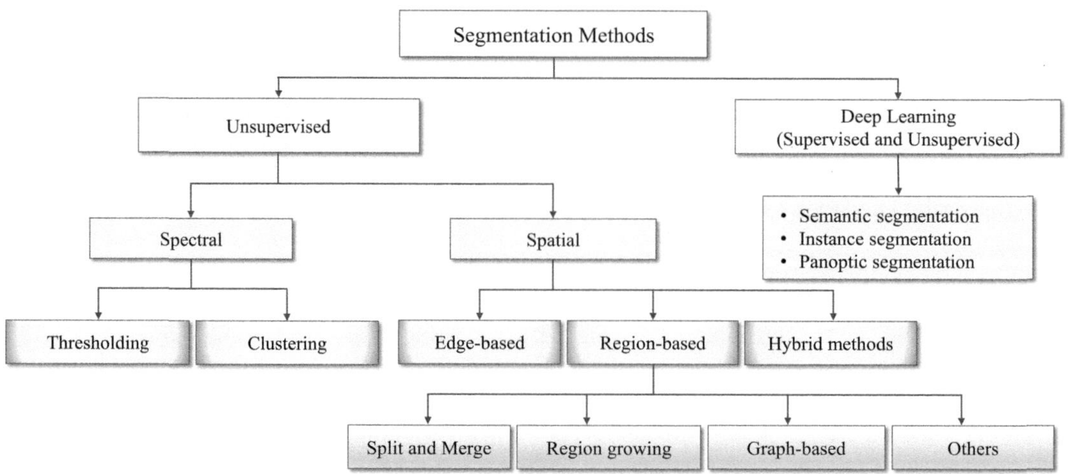

FIGURE 8.1 Taxonomy of the methods used in image segmentation.

semantic segmentation with deep learning, the segmentation methods can be divided into three broad categories depending on whether they focus on spectral information, spatial information, or a combination of both.

This chapter presents the fundamental theory of the OBIA approach with special emphasis on segmentation methods, which are discussed considering the current taxonomy. With the popularity of deep learning models in the analysis of remotely sensed data, their application to object-based image analysis is a popular research topic. The chapter also delves into the discussion of semantic segmentation providing powerful feature extraction capabilities. The metrics and algorithms used to assess segmentation quality are also reviewed under the headings of supervised and unsupervised evaluation.

8.1 CLUSTERING-BASED SEGMENTATION

Clustering-based segmentation relies on the fundamental idea that pixels corresponding to a real-world object are closer to each other in the spectral feature space, constituting a cluster. Clustering techniques have long been employed in pattern recognition and remote sensing, particularly for unsupervised classification. Therefore, their extension to image segmentation can be considered a natural process because many techniques have been developed to perform clustering in pixel-based image analysis. Clustering of the image feature space involves a statistical summary of the data through the histogram of pixel values in the imagery. Analyzing the histogram establishes the homogeneity criterion based on the number of pixels, thereby obtaining the corresponding segmentation. The clustering process, in its general form, starts with portioning the feature space and then grouping the pixels with similar locations into clusters without considering their spatial relationship in the image. Differentiating from the conventional clustering approach used in classification processes, a second step is conducted to merge the neighboring pixels in the same cluster into the final clusters, theoretically corresponding to real objects in the image. At the end of the process, a label is assigned to every pixel, and adjacent pixels that share the same label constitute a homogeneous region. Determining the appropriate number of clusters (known as the cluster validity problem) and disregarding the spatial information in the first stage are the major shortcomings of clustering-based segmentation approaches (Cheng et al., 2001). Although these methods are fast and easy to implement, they are likely to produce over-segmented results. Additional post-processing steps are required to refine the regions and produce smother and spatially homogeneous image objects. The mathematical theories of two popular clustering algorithms, namely, *k*-means and fuzzy C-means, are thoroughly discussed previously in Sections 2.3.1.1 and 2.3.1.2, respectively. However, the fundamentals of the mean-shift and superpixel segmentation algorithms are discussed in the following sections.

8.1.1 Mean-Shift Algorithm

The mean-shift algorithm (Fukunaga and Hostetler, 1975) relies on nonparametric density estimation, assuming that local maxima of the density correspond to clusters. Due to its ability to achieve fast convergence, the algorithm has found widespread use in image segmentation. The algorithm does not require a priori knowledge about the number of clusters in the data and can identify the positions of local maxima through a series of iterations. These iterations can be interpreted as shifting the points toward the modes where convergence is achieved when they reach a certain mode. The shift method uses a kernel with a scale parameter that determines the extent of local smoothing performed during intensity estimation. The mean-shift method applied to image segmentation uses distinct kernels for the spectral and spatial domains. The scale parameter for the spectral domain can be estimated by maximizing the average likelihood of the retained data. On the contrary, the scale for the spatial domain can be chosen according to the amount of compactness or over-segmentation desired in the image or can be determined using geographical statistics (e.g., semi-variogram–based estimates).

Given n data points $x_i = 1,\ldots,n$ in a d-dimensional space, the multivariate kernel density estimate was obtained with kernel $K(x)$, which is taken to be a radially symmetric, nonnegative function centered at zero and integrating to one, and bandwidth (window radius) h is (Comaniciu et al., 2001)

$$\hat{f}(x) = \frac{1}{nh^d} \sum_{i=1}^{n} K\left(\frac{x - x_i}{h}\right) \tag{8.1}$$

Based on Equation (8.1), the bandwidth h can be varied in two ways. First, by selecting a different bandwidth $h = h(x)$ for each estimation point x, one can define the balloon density estimator as

$$\hat{f_1}(x) = \frac{1}{nh(x)^d} \sum_{i=1}^{n} K\left(\frac{x - x_i}{h(x)}\right) \tag{8.2}$$

The estimate of f at x is the average of identically scaled kernels centered at each data point. By selecting a different bandwidth $h = h(x_i)$ for each data point x_i, then the sample point density estimator can be obtained:

$$\hat{f_2}(x) = \frac{1}{n} \sum_{i=1}^{n} \frac{1}{h(x_i)^d} K\left(\frac{x - x_i}{h(x_i)}\right) \tag{8.3}$$

where the estimate of f at x is the average of differently scaled kernels centered at each data point. The sample point estimators, on the other hand, are themselves densities, being nonnegative and integrating into one. Their most attractive property is that a particular choice of $h(x_i)$ reduces the bias considerably. Indeed, when $h(x_i)$ is taken to be reciprocal to the square root of $f(x_i)$,

$$h(x_i) = h_0 \left[\frac{\lambda}{f(x_i)}\right]^{1/2}, \tag{8.4}$$

the bias becomes proportional to h^4, while the variance remains unchanged, proportional to $n^{-1}h^{-d}$. Since $f(x_i)$ is unknown, it has to be estimated from the data.

The final estimate (Equation 8.3) is influenced by the choice of the proportionality constant λ, which divides the range of density values into low and high densities. When the local density is low, that is, $\tilde{f}(x_i) < \lambda$, $h(x_i)$ increases relative to h_0 implying more smoothing for point x_i. For data points that verify $\tilde{f}(x_i) > \lambda$, the bandwidth becomes narrower. A good initial choice is to take λ as the geometric mean of $\left\{\tilde{f}(x_i)\right\}_{i=1\ldots n}$. For superior results, a certain degree of tuning is required for λ (Comaniciu et al., 2001). However, in most instances, the sample point estimator demonstrated significantly superior performance compared with the fixed bandwidth estimator.

After these definitions, we can also define the variable bandwidth mean shift using an adaptive estimator. To simplify notations, we denote the profile of a kernel K as a function $k : [0,\infty] \to R$ such that $K(x) = k\left(\|x\|^2\right)$. We also denote $h_i \equiv h(x_i)$ for all $i = 1 \ldots n$. Then, the sample point estimator (Equation [8.3]) can be written as

$$\hat{f}_K(x) = \frac{1}{n} \sum_{i=1}^{n} \frac{1}{h_i^d} k\left(\frac{x - x_i}{h_i}^2\right).$$
(8.5)

A natural estimator of the gradient of f is the gradient of $\hat{f}_K(x)$

$$\hat{\nabla} f_K(x) \equiv \nabla \hat{f}_K(x) = \frac{2}{n} \sum_{i=1}^{n} x - x_i h_i^{d+2} k'\left(\|x - x_i h_i\|^2\right)$$

$$= \frac{2}{n} \sum_{i=1}^{n} \frac{x_i - x}{h_i^{d+2}} g\left(\left\|\frac{x - x_i}{h_i}\right\|^2\right)$$
(8.6)

$$= \frac{2}{n} \left[\sum_{i=1}^{n} \frac{1}{h_i^{d+2}} g\left(\left\|\frac{x - x_i}{h_i}\right\|^2\right)\right] \times \left[\frac{\sum_{i=1}^{n} \frac{X_i}{h_i^{d+2}} g\left(\left\|\frac{x - x_i}{h_i}\right\|^2\right)}{\sum_{i=1}^{n} \frac{1}{h_i^{d+2}} g\left(\left\|\frac{x - x_i}{h_i}\right\|^2\right)} - x\right],$$

where we describe $g(x)$ as

$$g(x) = -k'(x),$$
(8.7)

The last bracket in Equation (8.6) represents the variable bandwidth mean-shift vector

$$M_v(x) \equiv \frac{\sum_{i=1}^{n} \frac{x_i}{h_i^{d+2}} g\left(\left\|\frac{x - x_i}{h_i}\right\|^2\right)}{\sum_{i=1}^{n} \frac{1}{h_i^{d+2}} g\left(\left\|\frac{x - x_i}{h_i}\right\|\right)} - x.$$
(8.8)

To see the significance of Equation (8.8), we define first the kernel G as

$$G(x) = Cg\left(\|x\|^2\right),$$
(8.9)

where C is a normalization constant that forces G to integrate into one. By employing Equation (8.4), the term that multiplies the mean-shift vector in Equation (8.6) can be written as

$$\frac{2}{n}\left[\sum_{i=1}^{n}\frac{1}{h_i^{d+2}}g\left(\left\|\frac{\mathbf{x}-\mathbf{x}_i}{h_i}\right\|^2\right)\right] = \frac{2}{C}\left[\frac{\sum_{i=1}^{n}\tilde{f}(\mathbf{x}_i)}{n\lambda h_0^2}\right]\hat{f}_G(\mathbf{x}),\tag{8.10}$$

where

$$\hat{f}_G(\mathbf{x}) \equiv C\frac{\sum_{i=1}^{n}\tilde{f}(\mathbf{x}_i)\frac{1}{h_i^d}\cdot g\left(\left\|\frac{\mathbf{x}-\mathbf{x}}{h_i}\right\|^2\right)}{\sum_{i=1}^{n}\tilde{f}(\mathbf{x}_i)},\tag{8.11}$$

where $\hat{f}_G(\mathbf{x})$ is nonnegative and integrates into one, representing an estimate of the density of the data points weighted by the pilot density values $\tilde{f}(\mathbf{x}_i)$.

Finally, using Equations (8.6), (8.8), and (8.10), a generalized equation for the fixed bandwidth mean shift can be obtained.

$$M_v(\mathbf{x}) = \frac{\lambda}{n^{-1}\sum_{i=1}^{n}\overline{f}(\mathbf{x}_i)}\frac{h_0^2}{2/C}\frac{\hat{\nabla}f_K(\mathbf{x})}{\hat{f}_G(\mathbf{x})}.\tag{8.12}$$

For more details on the mathematical foundations and properties of the adaptive mean-shift algorithm, the readers should refer to Comaniciu et al. (2001).

8.1.2 SUPERPIXEL SEGMENTATION

Superpixel segmentation is recognized as one of the leading segmentation methods that employ the perceptual grouping of image pixels into "superpixels". Superpixels encapsulate more information than individual pixels and conform more closely to image edges than rectangular image patches (Neubert and Protzel, 2012). They represent statistically significant, homogeneous image regions forming small, local, and consistent clusters based on specific criteria like color and texture. As a preliminary processing stage, an over-segmented image is generated for more advanced analysis. Various algorithms have been suggested for performing superpixel segmentations. A popular member of superpixel segmentation is the Simple Linear Iterative Clustering (SLIC) method, proposed by Achanta et al. (2012), generally producing higher quality segments while requiring less processing time and memory compared to other state-of-the-art methods. The objective of the SLIC algorithm, a spatially localized variant of the k-means algorithm, is to cluster pixels that share similar spatiospectral characteristics, thereby producing an effective representation of the image. By balancing accuracy with efficiency, SLIC leverages the advantages of both superpixel and clustering techniques. It has been widely used in the segmentation of remotely sensed data (Ren and Malik, 2003; Kavzoglu and Tonbul, 2017a, 2018a). The algorithm consists of a single parameter (k), which is the desired equal-numbered superpixel size, and has a simple structure. In the SLIC algorithm, pixels are clustered based on their color similarity and proximity in the image plane using the five-dimensional [labxy] space, where [lab] is the pixel color vector in CIELAB color space (also known as CIE L*a*b*), as defined by the International Commission on Illumination (CIE). It describes color as three numerical unique values in which L* specifies the lightness of a color, and a* and b* represent the hue and saturation of red/green and blue/yellow components, respectively.

The SLIC clustering process begins with an initialization step in which k initial cluster centers $C_i = [l_i a_i b_i x_i y_i]^T$ are sampled on a regular grid spaced S pixels apart. To produce almost equally sized superpixels, the grid interval is $S = \sqrt{N/k}$. The centers are relocated to seed locations that align with the lowest gradient position in a 3×3 neighborhood. This approach is adopted to prevent the centering of a superpixel on an edge and to decrease the probability of initiating a superpixel with a pixel that may be affected by noise. To speed up the clustering process, a maximum distance measure D is introduced, which determines the nearest cluster center for each pixel. Because the expected spatial extent of a superpixel is a region of approximate size $S \times S$, the search for similar pixels is conducted in a region $2S \times 2S$ around the superpixel center (Achanta et al., 2012). In the conventional k-means algorithm, distances are computed from each cluster center to every pixel in the image. However, SLIC only computes the distances from each cluster center to pixels within a $2S \times 2S$ region (Figure 8.2). This methodology not only diminishes the requirement for distance computations but also guarantees that the complexity of SLIC remains unaffected by the number of superpixels.

The SLIC method starts by assigning each pixel to the nearest cluster center using a distance measure D (Equation 8.13) that combines the color proximity distance (Equation 8.14) and the spatial proximity distance (Equation 8.15) by their respective maximum distances within a cluster, N_c and N_S:

$$D = \sqrt{\left(\frac{d_c}{N_c}\right)^2 + \left(\frac{d_s}{N_S}\right)^2},\tag{8.13}$$

where

$$d_c = \sqrt{(l_j - l_i)^2 + (a_j - a_i)^2 + (b_j - b_i)^2}\tag{8.14}$$

$$d_S = \sqrt{(x_j - x_i)^2 + (y_j - y_i)^2}\tag{8.15}$$

The assignment and update steps can be iteratively repeated until the error converges. Subsequently, a post-processing stage is implemented to ensure connectivity, in which disjoint pixels are reassigned to neighboring superpixels. Superpixel homogeneity is governed by color distance, whereas spatial distance is used to promote superpixel compactness (Toro et al., 2015; Tonbul and Kavzoğlu, 2018). Despite its simplicity, the algorithm effectively adheres to boundaries, delivering performance

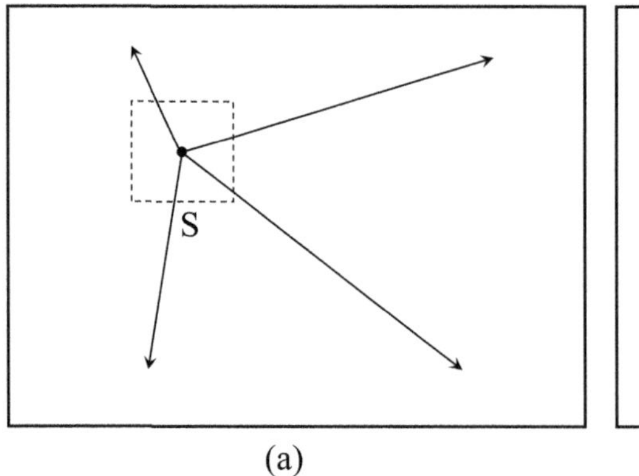

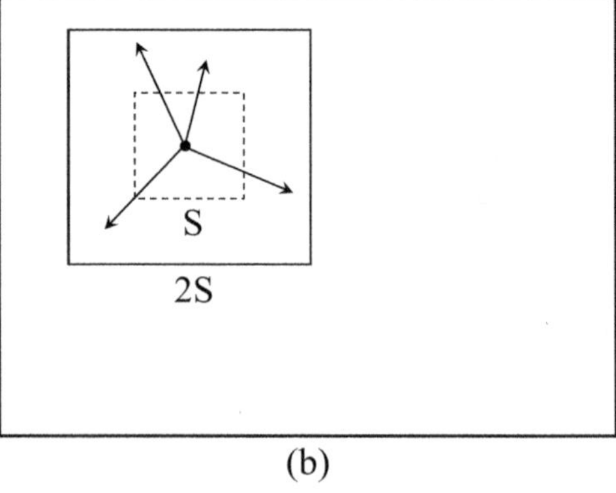

(a)   (b)

FIGURE 8.2   Search strategy of (a) standard k-means clustering and (b) SLIC superpixel algorithm.

comparable to or exceeding that of alternative methods. It smoothly adjusts to a multispectral domain, ensuring that the resultant over-segmentation is uniformly distributed across the entire image (Achanta et al., 2012).

8.2 THRESHOLDING-BASED SEGMENTATION

Thresholding is a widely recognized and frequently applied method in the domain of image segmentation. The underlying assumption of these algorithms is that pixels associated with specific objects possess a unique feature, marked by values significantly different from those representing the background. The categorization of thresholding methods involves two major groups: global and local thresholding. It is important to note that within research studies, these techniques are also denoted by different terms, such as single thresholding, multiple thresholding, adaptive thresholding, and optimal thresholding. Global thresholding involves selecting a single threshold based on the overall information about the image. On the contrary, local thresholding determines the threshold based on the specific properties of pixels within the local regions of the image. Single thresholding is the most straightforward method of image segmentation. When using a multimodal histogram, it becomes possible to establish several thresholds to delineate distinct regions corresponding to various real-world objects. This strategy is referred to as multiple thresholding. Adaptive thresholding, on the other hand, involves automatically adjusting and selecting the threshold value based on the image pixels and layout, rather than relying on manual specification (Kaur and Singh, 2013). Another type of algorithm, the multithreshold segmentation (MTS) algorithm, available in eCognition Developer, combines histogram-based methods with homogeneity measurement from multiresolution segmentation. This combination calculates a threshold to divide the chosen pixel set into two subsets to maximize heterogeneity. Users have the flexibility to specify the pixel value threshold or enable a self-adaptive mode when using the automatic threshold algorithm.

A crucial aspect of the thresholding process is the determination of an appropriate threshold value. The selection of the threshold value is frequently performed in an ad hoc manner, especially when only one attribute is considered. However, optimal values can be determined using exhaustive or stochastic search processes. These methods seek to identify values that optimize specific criteria related to the shape or statistical characteristics of the object's histogram. The criteria commonly employed in this context usually encompass the objective of decreasing within-class variance while simultaneously increasing between-class variation. Stochastic search methods are essential when the identification of several thresholds is required, as exhaustive searches become impractical because of exponential escalation in potential values, making them computationally infeasible. However, despite the careful selection of thresholds based on optimization criteria linked to the statistical distribution of pixel attributes, the actual outcomes of image segmentation are not always guaranteed. This limitation arises due to the lack of spatial information integration as judgments are made independently for each pixel. Consequently, thresholding is commonly employed as a preprocessing tool, followed by the application of other post-processing techniques, especially morphological operations, to enhance the results produced by pixel-based thresholding algorithms. Overall, clustering-based segmentation methods are not extensively used for segmenting remote sensing images, particularly in urbanized regions, which can be attributed to the substantial variation in histograms and the existence of hidden clustering challenges (Beveridge et al., 1989). However, new approaches have been proposed that use nonlinear multidimensional thresholding based on fuzzy rules and evolutionary algorithms.

8.3 EDGE-BASED SEGMENTATION

The fundamental concept behind edge-based segmentation (Haralick, 1981; Kundu and Pal, 1986) is to detect edges and subsequently connect them through contouring algorithms. This approach operates under the assumption that there is a sudden change in pixel properties at edges. From this perspective, edges are regarded as boundaries between objects and where changes occur. Therefore, segmentation methods based on edges aim to identify spatially adjacent pixels with spectral similarity by using edge detection operators to delineate boundaries between neighboring segments. These edges represent locations in the image where there are discontinuities in properties such as gray level, color, and texture. Nonetheless, the outcome of edge detection does not directly serve as a segmentation result. To form edge chains aligned with the image borders, it is necessary to incorporate additional processing steps connecting the edges. The primary objective is to attain, at a minimum, a partial segmentation wherein local edges are grouped into an image, preserving only those edge chains corresponding to existing objects or image segments (Wang, 2008).

Various algorithms are available for delineating object edges, capturing both the geometric and physical attributes of image objects. Moreover, a variety of edge detectors have been employed for edge detection within images. The edge detection process can be divided into three steps: filtering, enhancement, and detection. Different filtering methods have been proposed to produce minimum blurring and displacement of edges. Commonly used edge detectors compute the gradient of a gray-level image, with edge points usually indicating the locations of local maxima in the gradient. Each edge is characterized by both magnitude and direction. Due to the presence of numerous edges in images compared with object boundaries, a subsequent edge selection process is required following the edge detection (Brejl, 1999). Various operators, including the Roberts, Sobel, and Prewitt operators, are known as the first difference operators, whereas the Laplacian operator is categorized as the second difference operator (Jensen, 2005). The primary distinction among these operators arises from the weights assigned to each element in the convolution kernel mask. These operators demand substantial alteration in the gray level between neighboring points, enabling them to detect sudden changes in edges between regions. However, they are incapable of discerning ill-defined edges resulting from a gradual shift in the gray level along the edge. Various approaches exist for connecting edges to an image boundary, with factors such as the size of the edge and the continuity of its direction typically taken into consideration. One approach to connecting edges to object boundaries is edge relaxation.

In addition to traditional edge-based segmentation approaches, diverse soft computing methods, including those based on fuzzy logic, genetic algorithms, and neural networks, have been applied. It is noteworthy that all operators generate fragmented edges and may omit certain crucial edges. The effectiveness of each operator is measured by considering parameters such as false edges, missing edges, edge angles, distance from the true edge, and distortion (Lucchese and Mitra, 2001). Following the identification of the edges, the subsequent phase involves converting these edges into closed boundaries. This typically entails eliminating edges generated by noise, bridging gaps where no edge is detected, and making decisions about connecting edge segments that constitute a single object. Several approaches for linking edges have been proposed to address situations in which edges are not fully connected. To address this issue, the Hough transform is used to identify the optimal edges that most accurately match the partial edges. A neighborhood search is also applied to identify suitable candidates for linking edge pixels. In summary, although many algorithms have been put forward to detect edges and establish connections for object generation, the difficulty lies in identifying ideal edges to construct image objects.

8.4 WATERSHED SEGMENTATION

Among the commonly employed segmentation methods, the watershed segmentation method integrates features from both edge-based and region-based techniques. Using mathematical morphology for edge detection, the watershed transformation offers robust segmentation tools, proving more efficient than traditional edge detection algorithms. Primarily rooted in the hydrological basin concept, basin transformation involves filling basins with water, starting from a local minimum. Dams are erected at points where water from diverse basins accumulates, and the process ends when the water level reaches its highest point. Similarly, basins separated by the dam are segmented into horizontal zones or so-called basin lines. The original watershed transformation applied to segmentation problems resulted in over-segmentation of the image. The morphological gradient of an image can be defined as

$$g(f) = (f \oplus B) - (f \ominus B) \tag{8.16}$$

where $(f \oplus B)$ and $(f \ominus B)$ are the elementary dilation and erosion of f by the smallest regular structuring element B defined on the digitization grid, respectively. Geodesic reconstruction is often combined with the watershed approach to simplify the gradient images and prevent over-segmentation. Merging over-segmented regions through region coherency is essential for ensuring that segmentation is as accurate and meaningful as possible.

To overcome this shortcoming, several improvements have been suggested. For instance, Meyer and Beucher (1990) proposed an improved version called marker-controlled segmentation. The foundation of this approach lies in the recognition that machine vision systems often roughly "know" the location of the objects to be segmented. The execution of this method follows the following steps. First, the properties used to designate objects are defined as object markers. The same procedure is applied to the background, identifying the image objects when there is no pixel associated with any object. These markers form background markers. The following steps are simple and universally applicable: the gradient image undergoes alterations to retain only the most notable contours within the defined regions of interest established by the markers. Adjustment of the gradient involves altering the homotopy of the function. Subsequently, the ultimate search for contours is conducted on the modified gradient image using watershed transformation. The final segmentation requires no supervision, parameters, or heuristics. The parameterization that directs segmentation is concentrated in the marker construction step, providing a more straightforward and confirmable control mechanism. Ji and Park (1998) proposed a two-step segmentation in which watershed segmentation is implemented in the first step and artificial neural networks are used in the second step to reduce the over-segmentation. Wang and Li (2014) proposed edge-constrained watershed segmentation, which includes a two-stage segmentation process. Using edge-constrained watershed segmentation and edge allocation, initial small segments, or subobject primitives (sub-OPs), are obtained. These segments undergo a progressive merging process into larger segments until the edge-controlled thresholds are met, giving rise to the initial OPs. The second step involves nonconstrained merging of the OPs, resulting in the final segmentation. Additionally, a repetitive pairwise segment merging approach is implemented. The mathematical theories of watershed and marker-controlled segmentation can be found in Meyer and Beucher (1990).

8.5 REGION-BASED SEGMENTATION

Region-based segmentation is based on the fundamental concept of creating homogeneous regions that satisfy a certain homogeneity criterion. This criterion can be based on image features such as grayscale, color, texture, and shape. These approaches can be categorized into two main groups: full and partial image segmentation. In the former, the goal is to produce regions that are as large as possible and that can be directly associated with real-world objects. In the latter, the goal is to produce consistent regions; however, they can provide flexibility based on variations in the region (Wang, 2008). The initiation of region-based segmentation algorithms involves starting from an object seed and expanding outward until the object boundaries are encompassed. In principle, edge-based and region-based representations serve as alternate representations of the same object. However, region-based techniques can yield results that differ markedly from those obtained through edge-based approaches (Kavzoglu and Tonbul, 2017b). The foundational operations within region-based segmentation approaches include merging and splitting. The fundamental procedure for region-based image segmentation is as follows: (1) obtaining an initial segmentation of the image, (2) merging or splitting adjacent segments based on their similarity or dissimilarity, and (3) repeating the previous step until no segments remain that require merging or splitting (Bins et al., 1996). Improved segmentation performance is achieved by partitioning the image into smaller tiles, potentially owing to a decrease in image heterogeneity. The reduction in tile size for segmentation effectively diminishes the inherent pattern complexity of the image, resulting in a more accurate delineation of target objects (Drăguţ et al., 2019).

8.5.1 Region Splitting and Merging

The region splitting (top-down) and merging (bottom-up) approach includes a hierarchical process for region-based image segmentation. The algorithms based on this methodology start by considering the entire image by splitting it into subregions until all the created segments satisfy the homogeneity criteria. Combining both a bottom-up and a top-down approach, these methods form objects through the merging of pixels in the bottom-up phase. During the top-down phase, the entire image is split into regions of a predefined shape (usually square) based on heterogeneity criteria. They were developed to overcome the imbalanced performance problems of global measures in which incorrect segmentations are produced at the edges of the split regions. In cases where the seed is not homogeneous, the splitting method divides the seed into four square subregions. These subregions then act as seeds at the succeeding level, persisting until homogeneity is achieved across all subregions. The general criterion for stopping the segmentation process is that the features of a newly segmented pair should not differ from the threshold value of the original region. The main problem with these algorithms lies in determining the optimal location for implementing the division. In most cases, segmentation is used as the first stage of a split-and-merge algorithm. To improve the split-and-merge methodology, some robust strategies have been proposed. For instance, Kelkar and Gupta (2008) introduced an improved quadtree method for representing split images, and then, a neighbor naming method is applied to the neighbor list of every quadtree node. Manousakas et al. (1998) suggested modifications to the 2D split-and-merge method, employing simulated annealing and controlled boundary elimination methods. It was reported that the proposed approach reduced the number of regions significantly. Alshehhi and Marpu (2017) proposed hierarchical graph-based image segmentation in which features are extracted using Gabor and morphological filtering and graph-based segmentation considering color, and shape features for road extraction from urban areas.

8.5.2 Region Growing

Region growing has been the most popular approach for region-based segmentation practices. Creating an image segment that grows by region typically starts with a single pixel as a seed and adds adjacent pixels in succession until no adjacent pixels meet the similarity criterion. A new seed pixel is then selected from the unlabeled pixels, and the process is repeated until all pixels are labeled. In other words, the growing process stops when adjacent pixels do not meet the similarity criterion. The homogeneity criterion plays an important role in deciding whether a pixel should be part of the growing region. The seed selection process is recognized for its impact on increasing computational costs and time requirements. To mitigate these challenges, Wang et al. (2010) employed the k-means clustering algorithm to generate seeds in a region-based image segmentation algorithm, whereas Verma et al. (2011) proposed a single-seeded region growing technique using an adaptive thresholding strategy. Byun et al. (2011) presented an approach employing a block-based seed selection method, combining modified seeded region growing and region merging approaches. Zhang et al. (2014), on the other hand, introduced a hybrid region merging method for segmenting high-resolution remote sensing images, in which the advantages of global-oriented and local-oriented region merging strategies are combined.

To implement the region growing approach, two criteria must be specified by the analyst: an algorithm for generating seeds and a similarity criterion to be used to merge pixels. In theory, different seed positions lead to different sections. The common approach is to take the first seed as the top left pixel of the image, with subsequent seeds selected as pixels adjacent to the already labeled regions. Alternatively, one can select regions that represent the pixel characteristics used. For example, pixels with gray values corresponding to peaks in the gray-level histogram can be used as seeds.

In the region growing approach, the optimal merging strategy works by examining the adjacent pixels formed around each seed pixel. Adams and Bischof (1994) proposed an efficient region growing algorithm that, given n seeds (groups of connected pixels) as input, assigns each pixel in the image to one of the regions corresponding to the seeds. This is accomplished using an ordered list of all unassigned pixels adjacent to regions ranked by similarity criteria calculated for each pixel relative to the adjacent region. The following steps are then repeated until all pixels are assigned to a region: the pixel with the highest similarity measure is added to its adjacent region, the average gray value of the region is updated, and the pixel list is updated to include unassigned pixels adjacent to the newly expanded region. It is also possible to assign pixels between two regions to a boundary pixel class. It is important to note that objects smaller than the spatial resolution of the image cannot be identified during segmentation. Conversely, when the size of objects exceeds the spatial resolution, they are fragmented into pixels. Many region growing algorithms, including multiresolution segmentation, mean-shift algorithm, hierarchical stepwise optimization, and recursive hierarchical segmentation, are all discussed in the following sections. However, literature surveys have shown that the most popular method has been the MRS algorithm in segmentation studies.

8.5.3 Multiresolution Segmentation

The multiresolution image segmentation (MRS) introduced by Baatz and Schäpe (2000) is a bottom-up region merging algorithm, which is the most popular image segmentation method. The algorithm is available in the eCognition software package. Starting with each pixel forming an independent image object, the algorithm progressively merges pairs of image objects into larger ones. The decision to merge is determined by local homogeneity criteria, which assess the similarity of adjacent image objects. Two fundamental characteristics of the MRS algorithm are given as follows: decision heuristics guiding the selection of image objects to merge in each step and the specification of homogeneity for image objects and determining the level of compatibility between a pair of image objects. Several heuristics, such as local mutual best fitting, global mutual best fitting, and distributed treatment order, are also defined by Baatz and Schäpe (2000) for the region merging process. A neighboring segment can be randomly selected and merged if the homogeneity criterion is satisfied. Alternatively, the neighboring segment that best satisfies the homogeneity criteria can be selected. For mutual best fitting, both the neighboring segment and the segment of interest are evaluated separately with respect to their neighbors that best satisfy the homogeneity criterion of a possible merger. If the neighboring segment and the segment of interest describe each other as the best fit, the merger is performed (Fourie, 2011). Baatz and Schäpe (2000) define a d-dimensional feature space for the degree of fitting (h) as follows:

$$h = \sqrt{\sum_d \left(f_{1d} - f_{2d}\right)^2}, \tag{8.17}$$

where f_{1d} and f_{2d} represent the values of the two adjacent objects. The distances can be standardized by the standard deviation (σ) over all segments of the feature in each dimension:

$$h = \sqrt{\sum_d \left(\frac{f_{1d} - f_{2d}}{\sigma_{fd}}\right)^2}. \tag{8.18}$$

When two segments are evaluated for a possible merge, the change in the degree of connectivity (change of heterogeneity) is measured as follows:

$$h_{\text{diff}} = h_m - \frac{h_1 + h_2}{2}, \tag{8.19}$$

where h_1 and h_2 are the degree of connectivity before virtual merging, and m is the connectivity after virtual merging. The variation of the degree of connectivity measure is extended by including the object size, defined by n, as follows:

$$h_{\text{diff}} = h_m - \frac{h_1 n_1 + h_2 n_2}{n_1 + n_2} \tag{8.20}$$

The object size should be used as a weighting factor in measuring heterogeneity and should yield results in the following manner:

$$h_{\text{diff}} = (n_1 + n_2) h_m - (n_1 h_1 + n_2 h_2) \tag{8.21}$$

This definition can be extended to a random number of c channels, each with a weight of w_c:

$$h_{\text{diff}} = \sum_c w_c \left(n_1 \left(h_{mc} - h_{1c} \right) + n_2 \left(h_{mc} - h_{2c} \right) \right) \tag{8.22}$$

Baatz and Schäpe (2000) proposed extensions for the homogeneity criterion by suggesting the integration of two additional heterogeneity measures based on segment shape. The cohesion criterion is defined as the relationship between the segment's boundary length and the square root of the number of pixels (n) within the segment and is expressed as follows:

$$h_{\text{compactness}} = \frac{1}{\sqrt{n}} \tag{8.23}$$

The smoothness criterion is defined as the relationship between the segment's boundary length and the perimeter of the bounding box (b) of the object:

$$h_{\text{smoothness}} = \frac{1}{b} \tag{8.24}$$

The heterogeneity criteria contribute with varying weights during the segmentation process, depending on the parameters set by the user. The algorithm has three user-defined parameters that need to be adjusted by the user:

- The scale parameter is a unitless measure that governs the relative segment sizes defined by h_{diff} (Equation 8.22).
- The shape or color parameter describes the weights (in the range 0–1) that a segment should have when determining segment merging.
- The compactness parameter describes the weights (in the range 0–1) that the different shape measures should have to guide the segment merging.

The MRS segmentation algorithm is generally used as a multiscale analysis tool. Therefore, it is essential that the image objects conform to the desired scale level. As the scale parameter increases, the object size also increases, leading to a proportional decrease in the total number of objects. In general, it is more effective to select the scale parameter so that the object boundaries can

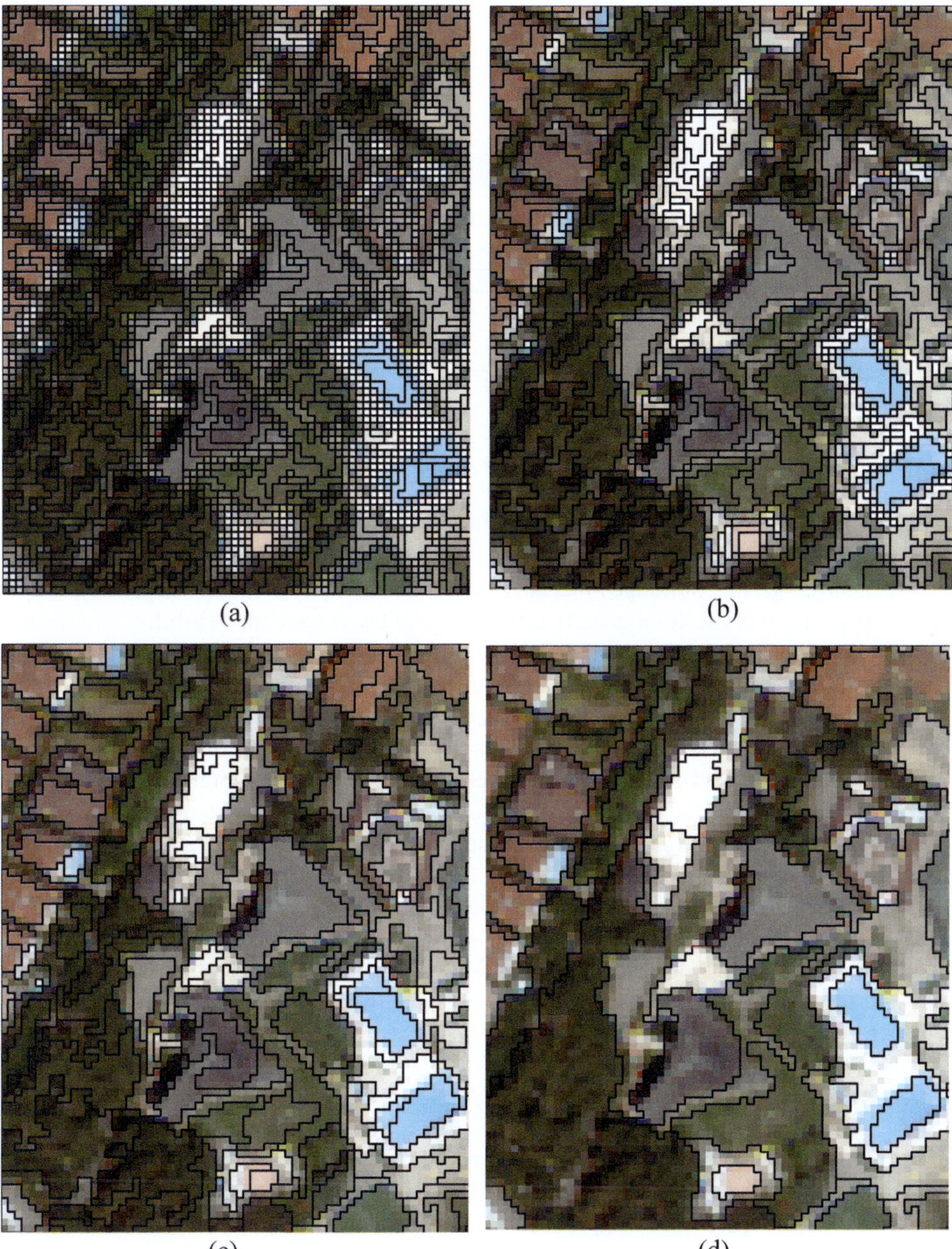

FIGURE 8.3   Segmentation with scale values of (a) 5, (b) 10, (c) 20, and (d) 50.

be distinguished. Figure 8.3 depicts the effect of the scale parameter on the object sizes using a high-resolution Worldview-2 satellite image.

Starting with one-pixel objects, the segmentation process evolves through the stepwise merging of similar neighboring objects until a heterogeneity threshold, dictated by a scale parameter (SP), is attained (Benz et al., 2004). SP, a key parameter in the multiresolution segmentation algorithm, defines the maximum allowable variation in heterogeneity during segmentation, directly

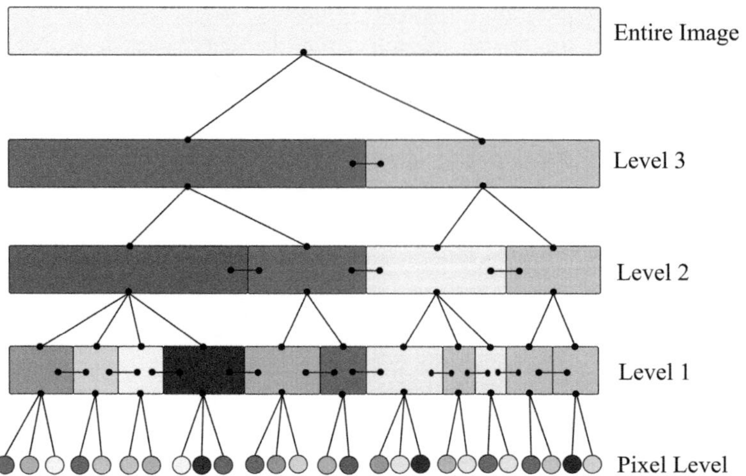

FIGURE 8.4 Multiresolution image segmentation object hierarchy.

influencing the average size of image objects. The SP, in relation to object homogeneity, is specified within the composition of the homogeneity criterion field. Users have the flexibility to customize the homogeneity criterion, adjusting the weighting shape and compactness criteria. The SP is defined by the composition of the homogeneity criterion, incorporating two essential aspects: spectral homogeneity and shape homogeneity. To be more precise, it includes assigning weights to both spectral heterogeneity and shape heterogeneity, where shape heterogeneity is further defined by weights of smoothness and compactness. Depending on the characteristics of the region under study, classification can be performed using the hierarchy of objects obtained with scales defined at different levels. The segmentation hierarchy structure and levels according to the multiresolution segmentation concept are shown in Figure 8.4.

As shown in the figure, image objects are connected to the network at the next level, starting at the pixel level. Each level is created directly connected to its subobjects; that is, subobjects are combined with larger image objects at the next level. Thus, each image object is connected to its neighbor and to its child and parent object. It is possible to identify relationships between objects and consider local context information.

8.6 HYBRID SEGMENTATION

Hybrid segmentation employing edge-based and region-based segmentations has recently attracted significant attention in the scientific community. These methods aim to overcome the inherent constraints of edge- and region-based approaches by integrating strategies that combine information from these diverse approaches with the expectation of improved segmentation results. As discussed above, edge-based methods are good at extracting the edges of objects, but they encounter challenges in creating closed segments. On the other hand, approaches that prioritize specific regions are proficient at restricting geographical areas; however, this efficiency is attained at the cost of precisely defining segment borders. In this context, the current trend in the field of image segmentation involves the adoption of hybrid models that incorporate both boundary and spatial information. The hybrid techniques effectively utilize edge-based processes to outline segments and subsequently integrate them through the application of region-based techniques. The commonly adopted procedure is to initiate segment creation through edge-based methods and then merge these segments through region-based methodologies. Hybrid models adeptly employ border pixels to delineate initial segments and interior pixels to merge segments with similar characteristics. Various approaches propose different possibilities for combining edge- and region-based techniques, such as computing regions and edges separately and integrating the information in subsequent steps, or directly calculating edges or regions using complementary information. For instance, Kermad and Chehdi (2002) proposed a hybrid image segmentation approach that was tested on agricultural landscapes and forestry vegetation, integrating the information resulting from edge- and region-based segmentation. The method, which is almost automatic and unsupervised, checks the coherence of the results by comparing the two segmentations. The process starts with the over-segmentation results, and both approaches iterate, loosening certain constraints, until they converge to stable and cohesive outcomes. The coherence in the results is attained through the minimization of dissimilarity measures between the edges and boundaries of the regions. The goal is to obtain the optimal solution, emphasizing compatibility among the segmentation results. Mueller et al. (2004), on the other hand, combined edge- and region-based techniques by preserving and extracting straight region boundaries while suppressing superfluous image details. The proposed method considering the key information of shape knowledge has two parts to be implemented: extraction of model edges and edge-guided, region-based segmentation. Li et al. (2010) introduced a segmentation algorithm that gives priority to texture and is grounded in graph models incorporating region adjacency and nearest neighbor graphs. They devised a composite distance metric, factoring in texture, spectral, and shape features, to measure the similarity between nodes, ensuring consistent semantic descriptions for texture objects. Employing the combined distance on the graph models, the iterative process of fast merging yields the ultimate segmentation result.

Most region merging methods employ a single global parameter to control the iterative merging process of segments, providing users with control over both under- and over-segmentation. Some researchers, in response, have explored alternative strategies (e.g., Johnson and Xie, 2011; Chen et al., 2014). The use of local measures to identify segments experiences under- and over-segmentation at the selected optimal SP. These segments are then refined through suitable splitting and merging procedures. The implementation of this local refinement strategy is effective in advancing segmentation quality by resolving under- and over-segmentation concerns. However, a challenge arises in the execution of local refinement in an operational context during the further steps of splitting and merging (Yang et al., 2016). In addition, these approaches consider heterogeneity between adjacent segments as the merging criteria. Therefore, it is essential to give equal consideration to both homogeneities within the segments and heterogeneities between them. To address this issue, Wang et al. (2018) introduced a novel hybrid segmentation approach that incorporates objective heterogeneity and relative homogeneity as criteria for region merging. Within the merging process, two primary concerns arise: the criteria for merging and the order in which merging occurs. Existing literature has utilized various merging criteria such as variance, area-weighted variance, Moran's I, spectral angle, F-measure, and spectral and geometric properties. To determine adjacent relationships, region adjacency and nearest neighbor graphs have been commonly employed in many studies (Hossain and Chen, 2019). Although hybrid models demonstrate considerable promise, their practical application poses significant challenges. Hence, recent research has concentrated on enhancing hybrid methods to make them more feasible for practitioners.

8.7 EVALUATION OF SEGMENTATION QUALITY

Image objects, created through a segmentation process, need to have intrinsic size, shape, and geographical relationships with the scene they represent. Their boundaries should be well defined, ensuring coherence and a good representation of the actual objects in the scene. However, both theoretical analysis and practical application indicate that the segmentation quality of remote sensing images is contingent upon the specific land cover types and segmentation algorithms employed, underlining the importance of incorporating multiscale and multiregion segmentation strategies (Gao et al., 2011; Johnson and Xie, 2011; Kavzoglu et al., 2016, 2017). A widely accepted standard for good segmentation has been defined with four criteria: (1) regions of an image segmentation should be uniform and homogeneous concerning some characteristics such as gray tone or texture (intra-region uniformity); (2) adjacent regions should have significant differences with respect to the characteristics in which they are uniform (inter-region disparity); (3) region interiors should be simple and without holes; and (4) boundaries of each segment should be simple, not ragged, and spatially accurate (Haralick and Shapiro, 1985). The first two criteria are defined as Characteristic Criteria, and the last two are defined as Semantic Criteria. These criteria are widely accepted as the *de facto* standard for assessing unsupervised image segmentation performance (Zhang et al., 2008). While some metrics are based on each of these criteria, others combine the four criteria in a special manner, such as the weighted sum of inter- and intra-region metrics.

Evaluation metrics for image segmentation measure the quality of a segmented image using supervised or unsupervised criteria. Evaluation of segmentation is crucial for enhancing the effectiveness of current segmentation algorithms and creating new, robust segmentation algorithms. As a preprocessing step before classification, segmentation directly influences the accuracy of the classification process through the quality of object formation. While each segmentation method requires different parameters to be set by the analyst, inaccurate setting of segmentation parameters can lead to the formation of image objects that deviate from the desired outcome. Therefore, qualitative and quantitative assessments of segmentation results are crucial for selecting an appropriate segmentation method and its optimal parameters. Segmentation quality metrics enable the evaluation of segmentation results, benchmarking algorithms for specific applications, and automatic calibration of algorithm parameters. There is no standard for determining the segmentation method and its parameters for a particular problem. Similarly, there is no universally accepted standard for assessing segmentation quality. It is frequently observed that the parameters are determined through a trial-and-error approach. Kotaridis and Lazaridou (2021) conducted an extensive literature review that revealed that 46% of OBIA studies employed qualitative evaluation methods and trial-and-error strategies. These approaches are notably unsuitable for remote sensing applications, especially when the goal is to generate reliable and accurate thematic maps that can be employed in GIS databases for decision-making processes.

Repeatability and interpretability of the methods are the most important issues in recent studies (see Chapter 10 for detailed discussion). Moreover, the evaluation of segmentation quality is highly valuable for making comparisons between algorithms or estimating parameters for a specific algorithm. However, when considering the complexity of high-resolution data, the varying dimensions, shapes, and spatial distributions of geographical features, it becomes challenging to establish a global- or local-SP model for adjusting segmentation parameters. The techniques used to evaluate the quality of segmentation can be divided into two main categories: supervised and unsupervised. Supervised methods involve comparing segmentation results with predefined ground truth or reference objects. The foundation of these approaches lies in modeling geometric and/or arithmetic inconsistencies between reference objects and their corresponding segments (Clinton et al., 2010). On the other hand, unsupervised methods operate without the necessity for reference objects in assessing the quality of segmented images, relying instead on human intuition or visual comparisons. These methods are suitable for real-time segmentation evaluations and automatically determining parameter options. While the literature suggests various metrics, the most common ones are presented here.

8.7.1 Supervised Approach

The difference between a segmented image and the reference objects is the underlying philosophy of supervised approaches, which evaluate the segmentation result through numerical comparison of created image objects and reference objects. Ideally, an exact match between the image segments created with reference objects is expected. Existing supervised evaluation techniques are based on modeling the geometric and/or arithmetic relationship between the considered image and reference objects. The difference between reference objects and segmented images is characterized using certain criteria. The general approach in the quality assessment of segmentation is based on estimating over-segmentation (i.e., image objects are smaller than the reference objects) and under-segmentation (i.e., image objects are much larger than the reference objects). To date, some quality metrics have been suggested and applied by researchers to determine segmentation goodness (Zhang, 1996, 2001; Costa et al., 2018; Ye et al., 2018; Jozdani and Chen, 2020). Grouped into analytical, empirical goodness, and empirical discrepancy categories, these methods highlight the superior suitability and utility of empirical approaches for evaluating segmentation algorithm performance. Among the empirical methods, the discrepancy methods were found to be superior in comparative evaluations. However, it should be mentioned that the performance comparisons made in the domain of computer vision may not be valid for the segmentation of remotely sensed data.

The assessment of the difference between the obtained segmentation and the ground truth segmentation can be performed using any measure of agreement or similarity between the two partitions. Beauchemin and Thomson (1997) categorized discrepancy approaches into four groups, considering the pixel, area, point-pair, and boundary. The widely adopted pixel-based discrepancy approach includes counting the misclassified pixels in the segmentation output in relation to a reference partition. Area-based methods specifically consider the overlapping area between corresponding segments, while boundary-based methods specifically contrast the perimeters of the segments. The point-pair discrepancy method quantifies the agreement between two segmented images without directly addressing the correspondence problem between the regions. It is of crucial importance that the performance of segmentation algorithms is influenced by many factors; therefore, only one evaluation method would be not enough to judge all properties of an algorithm, and different methods should be cooperated. Hence, for a complete evaluation and comparison of segmentation techniques, the use of a set of performance measures rather than a single measure is suggested. Finding the most appropriate set of analysis tools for segmentation quality remains an open question and requires further investigation. Because of these considerations, no universally accepted metric or method is available for the evaluation of segmentation quality. The most widely used criteria in segmentation evaluations are as follows:

- Number of misclassified pixels or area-based analysis
- Physical location of the misclassified pixels
- Number of segments in the image
- Characteristics of segmented objects.

Numerous supervised methods have been introduced to evaluate the quality of image segmentation. The most widely used metrics are shown in Table 8.1, where $A_{r(i)}$ denotes the total area of the reference objects, and $A_{s(j)}$ shows the total area of the corresponding image objects. In addition, m indicates the number of reference polygons, and v shows the number of corresponding segments. OS and US denote over-segmentation and under-segmentation, respectively.

Over-segmentation (OS) is essentially the case in which the resulting segments are smaller than they should be. This is used as a metric obtained by dividing the intersection of the reference objects and the corresponding segments by the reference object area. The OS metric takes values between zero and one, and the ideal segmentation result is achieved for a value of zero. Under-segmentation (US) is essentially the case in which the resulting segments are larger than they should be. It is estimated by dividing the intersection of the reference objects and the corresponding segments by the corresponding object area. The US metric takes values between zero and one, and the ideal segmentation result is achieved for a value of zero. OS and US represent the fundamental problems in segmentation; hence, they are used in the estimation of other robust metrics. As an example, D metric takes the root mean square of the over- and under-segmentation metrics. It takes a value between 0 and 1, where a value of 0 indicates ideal segmentation. As one of the most popular segmentation evaluation techniques, AFI calculates the degree of overlap between the reference objects and the corresponding image objects (i.e., considering both under-segmentation and over-segmentation). AFI value of zero indicates that the reference object and the segment overlap exactly. In other words, when the AFI value is close to zero, it can be interpreted that segmentation is of high quality. Qr is

TABLE 8.1

Quality Metrics Used to Evaluate Segmentation Goodness

Metric	Formula	Range	Authors
Over-segmentation	$OS_{ij} = 1 - \dfrac{A_{r(i)} \cap A_{s(j)}}{A_{r(i)}}$	$[0,1]$	Clinton et al. (2010)
Under-segmentation	$US_{ij} = 1 - \dfrac{A_{r(i)} \cap A_{s(j)}}{A_{s(j)}}$	$[0,1]$	Clinton et al. (2010)
Index D	$D_{ij} = \sqrt{\dfrac{OS_{ij}^2 + US_{ij}^2}{2}}$	$[0,1]$	Levine and Nazif (1985)
Area fit index	$AFI_{ij} = \dfrac{A_{r(i)} - A_{s(j)}}{A_{r(i)}}$	$[-\infty,+\infty]$	Lucieer and Stein (2002)
Quality rate	$Qr_{ij} = \dfrac{A_{r(i)} \cap A_{s(j)}}{A_{r(i)} \cup A_{s(j)}}$	$[0,1]$	Weidner (2008)
Potential segmentation error	$PSE = \dfrac{\sum \left| A_{s(j)} - A_{r(i)} \right|}{\sum A_{r(i)}}$	$[0,\infty]$	Liu et al. (2012)
Number of segments ratio	$NSR = \dfrac{\left| m - v \right|}{m}$	$[0,\infty]$	Liu et al. (2012)
Euclidean distance 2	$ED2 = \sqrt{\left(PSE^2 + NSR^2 \right)},$	$[0,\infty]$	Liu et al. (2012)
Euclidean distance 3	$ED3 = \sqrt{\dfrac{\left(OS2_{ij} \right)^2 + \left(US2_{ij} \right)^2}{2}}$	$[0,1]$	Yang et al. (2014)
Over-segmentation 2	$OS2_{ij} = \sum_i \sum_j \left(1 - \dfrac{A_{r(i)} \cap A_{s(j)}}{A_{r(i)}} \right),$	$[0,1]$	Yang et al. (2014)
Under-segmentation 2	$US2_{ij} = \sum_i \sum_j \left(1 - \dfrac{A_{r(i)} \cap A_{s(j)}}{A_{s(i)}} \right)$	$[0,1]$	Yang et al. (2014)
Fitness function	$F_{ij} = \dfrac{\left(A_{r(i)} + A_{s(j)} \right) - 2 \times \left(A_{r(i)} \cap A_{s(j)} \right)}{A_{r(i)}}$	$[0,\infty]$	Costa et al. (2008)
Edge location	$ED_{ij} = 1 - \dfrac{\mathrm{perim}\left(s_{(i)} \right) \cap \mathrm{perim}\left(r_{(j)} \right)}{\mathrm{perim}\left(s_{(i)} \right)}$	$[0,1]$	Persello and Bruzzone (2010)

used to estimate the intersection of reference objects and their corresponding image objects divided by the union of reference objects and corresponding image objects. The metric takes values between zero and one, and ideal segmentation is achieved for a Qr value of one. As another measure of segmentation quality, a PSE metric value of zero indicates ideal segmentation, while an increase in this value indicates an increase in under-segmentation and a decrease in segmentation quality. The NSR metric is calculated as the absolute difference between the number of reference objects and the number of corresponding segments divided by the number of reference objects. A value of zero for this metric indicates an ideal relationship between the reference objects and the corresponding image segments. The ED2 metric is a measure of the Euclidean distance that quantifies both geometric and arithmetic differences. Theoretically, a zero value for ED2 implies an ideal spatial match between reference objects and the generated segments. Yang et al. (2014) proposed another version of the Euclidean distance estimation (ED3), based on the idea that values of ED2 are most reliable when PSE and NSR have a similar order of magnitude, which does not occur when many objects are over-segmented. To address this issue, they proposed three new indices consisting of OS2, US2, and ED3 based on the average values of local over-segmentation and under-segmentation metrics. On the other hand, Costa et al. (2008) suggested a measure for the fitness function that compares the reference segments $r_{(i)}$ with a parameter vector $s_{(j)}$. It should be noted that $F_{ij} = 0$ indicates a perfect match between the reference and output segments. Edge index (ED_{ij}) measures the precision of the object edges recognized in the classification map concerning those of the actual object. A perfect match in the borders of the two compared regions leads to an index value equal to 0, whereas a large mismatch among the region edges results in error values that are close to 1.

8.7.2 Unsupervised Approach

The reliability of supervised methods in segmentation quality assessment depends on the plausibility and representativeness of the reference objects used in comparison with the constructed image objects. Not only the number of polygons but also their correctness is of critical importance. A discouraging issue is that different quality estimates are computed with different selected reference objects. Therefore, supervised methods lack universality, are less adaptable to continuously varying LULC classes, and require high costs for field study and/or office work. Perhaps, the most important problem in the application of supervised methods is that the reference objects, usually digitized from a map, are user choices and are produced only for a regularly shaped LULC class (e.g., buildings). Therefore, the assessment solely shows the quality of segmentation for the target class, assuming that reference polygons are collected as representative of the entire study site with accurate digitization. To avoid the above issues, unsupervised methods rely on statistical and mathematical measures to determine the quality of segmentation. Because these methods do not require reference polygons, they are more objective and repeatable and are also advantageous in terms of time and cost. These methods evaluate different segmentation algorithms by computing quality metrics based on the segmented image without prior knowledge of the correct segmentation. Another advantage of these methods is that they can be applied to determine the optimal SP for segmentation. Two fundamental concepts adopted in the development of unsupervised methods are the assessment of local variance (intra-segment homogeneity) and its change with varying scale values and the use of a global objective function such as Moran's I (inter-segment heterogeneity), which is used to estimate spatial autocorrelation. Other considerations for unsupervised metrics include using a semi-variogram (Ming et al., 2012), foreground similarity measurement (Li et al., 2014), energy function (Yang et al., 2015), and Geary's index (Cánovas-García and Alonso-Sarría, 2015).

8.7.2.1 Estimation of the Scale Parameter

As discussed earlier, the effectiveness of segmentation is directly associated with the optimal selection of segmentation parameters, requiring proper parameterization by the user. These parameters govern the size, shape, and boundaries of the segmented entities. The parameters vary from one segmentation method to another. In the case of the multiresolution segmentation, these parameters are scale, shape, and compactness; however, scale, which relates to the range of study, resolution ratio, heterogeneity, cartographic scale, and semantic granularity, is the most important of these parameters. Therefore, the scale issue has emerged as a major problem, particularly for OBIA studies conducted using multiscale segmentation methods. In many OBIA applications, significant efforts are devoted to selecting the appropriate scale value for the segmentation algorithm to obtain optimized segmentation results. Having two developed versions, estimation of scale parameter (ESP), proposed by Drăguț et al. (2010, 2014) as a plugin for eCognition software, is based on the fundamental philosophy that the local variance (LV) provides a measure of object heterogeneity within a scene, offering guidance on the most suitable scale level. ESP tools present a statistical method to determine the scale parameter values for segmentation across various layers. These tools have been the most preferred method for determining the optimized segmentation scale for multiresolution segmentation. Both ESP tools are programmed in the Cognition Network Language (CNL) environment of the eCognition Developer software and are used for multi-resolution image segmentation. While the first version of the ESP tool (Drăguţ et al., 2010) can only process single-band images, the improved version (ESP-2), introduced as a fully automated methodology for the selection of suitable SPs, can simultaneously process up to 30-band images (Drăguţ et al., 2014). Considering the availability of an abundance of remotely sensed data, particularly hyperspectral images, and the incorporation of spectral, spatial, and textural features, the 30-band limitation is an important obstacle, which requires special consideration to reduce the size of the data set. It should be noted that only intra-segment homogeneity is considered in the ESP tool, which appears to be a drawback of the method. As underlined by Ming et al. (2018), intra-segment homogeneity plays a more important role in GEOBIA, which theoretically proves the rationality of the ESP tool.

In ESP estimations, automatic image segmentation is performed according to the increment amount specified by the user (1, 10, and 100), and LVs are calculated as the average of the standard deviations during the extraction of each object. The results are exported as text files and used to produce plots for the estimation of optimal scale values. LV plots are used to evaluate the appropriate SPs according to the data characteristics of the image. The threshold values in the rates of change of LV indicate the scale parameter at which the significant objects will be segmented. An LV-RoC plot is obtained by determining the rate of change of the LV value calculated for each scale parameter using Equation (8.25).

$$\text{RoC} = \left[\frac{\text{LV}_{(L)} - \text{LV}_{(L-1)}}{\text{LV}_{(L)}} \right] \cdot 100, \tag{8.25}$$

where $\text{LV}_{(L)}$ denotes the LV of the target object, and $\text{LV}_{(L-1)}$ shows the LV of the next subobject level.

The LV-RoC graph is shaped with sudden oscillations between peaks and plunges, on descendant trends, whereas LV graphs are far smoother. Theoretically, one of the first peaks in the graph indicates the optimum SP that can be selected. At these peaks, the segments should match the types of objects characterized by relatively equal degrees of homogeneity. It must be underlined that the method does not indicate a certain scale value as the optimal solution but suggests candidate optimization scales. It is also noteworthy that shape and compactness parameters can also have a significant impact on the segmentation results, particularly in classifications facing the challenge of spectrally similar objects. The ESP tool can be run iteratively to determine the optimal scale values using various shape and compactness combinations. The main shortcoming of this approach is that the user has to decide which peaks in the curve suggest an optimal SP and how many candidate scales to test in the accuracy assessment. Moreover, approaches seeking to find the optimal parameter combination for all parameters (e.g., genetic algorithm, Taguchi optimization) should be preferred (Costa et al., 2008; Tonbul and Kavzoglu, 2020b). On the other hand, it was reported by Drăguţ et al. (2019) that segmentation quality could be improved using small tiles instead of the segmentation of whole images in which optimal scale estimates are performed for each tile. Instead of tiling a scene without considering the landscape characteristics, Kavzoglu et al. (2017) suggested a two-stage approach to extract distinct regions of a scene that are separated by natural landscape boundaries using a broad-scale approach and then estimated the optimal SP for each region to conduct segmentation.

An example LV-RoC plot produced for a sample data set is shown in Figure 8.5. The graph depicts the changes in LV (red) and rate of change (RoC) (blue) with increasing SP. The vertical lines indicate the two potential optimal scale parameters selected for the sample image. As can be seen from the figure, the value of 20 can be determined as the finest scale value as the first peak point where the sudden change and the first jump are seen. As an alternative value, the second peak (29) can also be used, as suggested by Drăguţ et al. (2010, 2014).

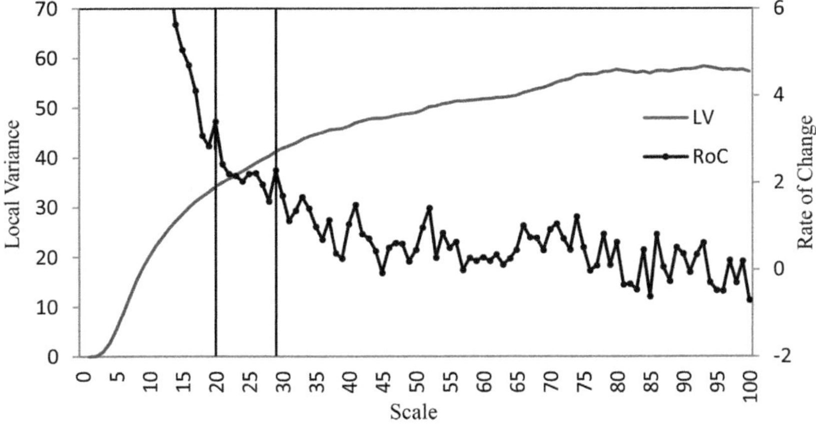

FIGURE 8.5 ESP tool output for a sample data set.

8.7.2.2 Global Score

Among the methods for assessing the quality of segmentation, approaches that include the evaluation of intra-segment and inter-segment heterogeneity to identify optimal parameters are prominently exemplified by methods incorporating Moran's I (MI) and variance calculations. The methodology proposed by Espindola et al. (2006) involves utilizing an empirical goodness method, integrating metrics for both intra-segment homogeneity (variance) and inter-segment heterogeneity (spatial autocorrelation). This approach is specifically designed for region merging segmentation algorithms, serving as a global evaluation step after segmentation. To be exact, the approach considers heterogeneity among objects and homogeneity within objects. The metric for segmentation quality within each image partition is the variance weighted by each partition area, which is calculated as follows:

$$V = \frac{\sum_{i=1}^{n} a_i \cdot v_i}{\sum_{i=1}^{n} a_i},$$

(8.26)

where v_i is the variance of the segment, and a_i is the area of the segment. The global MI technique, ranging between $[-1, +1]$, is used to calculate the quality measure between image segments, which is calculated by the following formula:

$$\mathrm{MI} = \frac{n \sum_{i=1}^{n} \sum_{j=1}^{n} w_{ij} (y_i - \bar{y})(y_j - \bar{y})}{\sum_{i-1}^{n} (y_i - \bar{y})^2 \left(\sum_{i \neq j} \sum w_{ij} \right)},$$

(8.27)

where n is the total number of segments, w_{ij} shows the spatial proximity measure, y_i is the mean spectral value of region R_i, and $\bar{y}$ is the mean spectral value of the image. Each weight w_{ij} represents the measure of the spatial adjacency of regions R_i and R_j. If these regions are adjacent, $w_{ij} = 1$. Intra-segment and inter-segment values can be in different ranges. To use both values together, it is necessary to reduce them to the range $[0, +1]$ and perform normalization. Normalization is performed as below:

$$F(V, \mathrm{MI}) = \frac{x - x_{\min}}{x_{\max} - x_{\min}}$$

(8.28)

where x_{min} and x_{max} denote the maximum and minimum values of MI or variance, respectively. To assign an overall "Global Score" (GS) to each image segmentation, the normalized weighted variance and MI values are calculated as follows (Espindola et al., 2006; Johnson and Xie, 2011):

$$GS = V_{norm} + MI_{norm}, \tag{8.29}$$

where V_{norm} and MI_{norm} denote the normalized variance and MI values, respectively. The optimum scale value is calculated according to the minimum GS value. The use of the average GS value ensured an equal contribution from all bands in the segmentation assessment. The optimal segmentation was determined as the one with the smallest average GS, because it signifies the least combined weighted variance and spatial autocorrelation across the three spectral bands (Johnson and Xie, 2011). With similar considerations in formulation, Zhang et al. (2021) suggested a local optimal segmentation index (LOSI) metric that uses normalized forms of the standard deviation rate change and the local Moran's index of each segment. A low LOSI value indicates both intra-segment homogeneity and inter-segment heterogeneity. It was reported that the approach could outperform and generate less error than the global optimal segmentation approaches (ESP and GS). The LOSI index demonstrates sensitivity to land cover discrepancies and provides good performance on cross-scale segmentation. On the other hand, Cánovas-García and Alonso-Sarría (2015) proposed replacing Moran's coefficient with the Geary coefficient, which allows the variability of the objects to be considered. Instead of a global approach using the entire image, the use of a local approach considering uniform spatial units is suggested.

8.7.2.3 Overall Goodness F-Measure

Various methods can be used to combine under-segmentation and over-segmentation goodness measures to calculate the "overall goodness" of each segmentation level. Initiated from the finding that the F-measure has a greater sensitivity to over- and under-segmentation (Zhang et al., 2015), combining goodness measures of Moran's spatial autocorrelation and area-weighted variance using the F-measure is suggested by Johnson et al. (2015). Overall goodness (OG_f) can thus be calculated from:

$$OG_f = \left(1 + a^2\right) \frac{MI_{norm} \cdot V_{norm}}{a^2 \cdot MI_{norm} + V_{norm}}, \tag{8.30}$$

where a is the value that controls the weighting of MI_{norm} and V_{norm}. OG_f is calculated for each scale to determine the best single scaling and three-level hierarchical segmentation parameters. OG_f values range from 0 to 1, with higher values indicating better segmentation quality. Unlike the other unsupervised evaluation metrics, weight adjustment can be conducted by weighting the under-segmentation and over-segmentation goodness metrics to control the weight of the within- and between-object homogeneity metrics. While values $a > 1$ may be appropriate for selecting the parameters for finer segmentation levels (assigning greater weight to the within-object measure), values of $a < 1$ may be more appropriate for selecting parameters for coarser segmentation levels (assigning greater weight to the between-object measure) (Johnson et al., 2015). By incorporating weights, the methodology allows for the regularization of over- and under-segmentation. Analysts, however, must test different weight values to determine the ideal combination for their objectives. The trial-and-error process may compromise the objectivity of segmentation and increase processing time. By setting the weight parameter to 3, 1, and 0.33 for levels 1–3, Grybas et al. (2017) concluded that the OG_f metric was more effective than RoC-LV in their investigation, and the adjustable weight facilitated the determination of multiple scale parameters for segmentation. The overall goodness F-measure has been widely employed as an unsupervised segmentation quality measure (e.g., Tonbul and Kavzoglu, 2019, 2020a; Wang et al., 2019).

8.8 CONCLUDING REMARKS

As a remote sensing image analysis paradigm, OBIA has dominated the remote sensing field with its superior advantages over pixel-based image classification since the early 2000s. As the first and most important step in the creation of image objects before the image classification stage, image segmentation can directly affect the performance of the image classification. Therefore, it is of great importance to identify effective segmentation approaches and define optimum segmentation parameters for OBIA studies. In other words, not only selecting a robust segmentation method but also setting the optimal values for the parameters is a critical issue in practical applications. Despite the use of many approaches and techniques in the literature, advanced approaches and methods are required to qualitatively and quantitatively evaluate the segmentation results of high-resolution images, especially those containing dense image pixels with heterogeneous structures and similar spectral characteristics.

Segmentation approaches mentioned in this chapter make crisp decisions about the definition of regions in imagery. Nonetheless, the definition of regions within an image is not always distinctly outlined, introducing potential uncertainty at each phase of image analysis and pattern recognition. Fuzzy set theory provides a framework for dealing with uncertainty and ambiguity, and its relevance extends to image segmentation, where fuzzy operators and inference rules are widely utilized. Alternatively, a graph-based segmentation approach has also been employed for classifying remotely sensed data. In this approach, image segmentation is viewed as a problem of graph partitioning, with nodes commonly representing individual pixels or regions, and edges establishing connections between spatially adjacent nodes. Graph-based segmentation algorithms increase the volume of data to be processed, and higher computational complexity has limited their implementation in practical applications. In recent years, the implementation of semantic segmentation in image analysis applications has been a hot topic in the remote sensing community. Complementary properties of OBIA and CNNs have been employed in different forms to extract hierarchical and semantic features of image objects (Kucharczyk et al., 2020). The application of machine learning, especially deep learning (discussed in Chapter 7), holds great potential for the analysis of earth observation data as it can identify patterns within extensive data sets. However, these models are regarded as black-boxes, and understanding their underlying decision-making process is unclear although there are recent studies to explain or interpret the decision-making mechanism of the model, thoroughly discussed in Chapter 10. Moreover, the absence of labeled samples for specific applications hinders their effectiveness, as the quality of the input data employed in training influences the superiority of the results.

Acknowledging that a comprehensive understanding of OBIA and the basics of image segmentation is crucial for their effective application, image segmentation continues to be a subject of substantial interest within the remote sensing community. However, future studies and more efforts for model development and evaluation of segmentation quality are needed to make further steps for the standardization of OBIA applications, particularly for novice practitioners.

REFERENCES

Achanta, R., A. Shaji, K. Smith, A. Lucchi, P. Fua, and S. Süsstrunk. 2012. SLIC superpixels compared to state-of-the-art superpixel methods. IEEE *Transactions on Pattern Analysis and Machine Intelligence* 34:2274–2282. https://doi.org/10.1109/TPAMI.2012.120.

Adams, R., and L. Bischof. 1994. Seeded region growing. *IEEE Transactions on Pattern Analysis and Machine Intelligence* 16:641–647. https://doi.org/10.1109/34.295913.

Alshehhi, R., and P. R. Marpu. 2017. Hierarchical graph-based segmentation for extracting road networks from high-resolution satellite images. *ISPRS Journal of Photogrammetry and Remote Sensing* 126:245–260. https://doi.org/10.1016/j.isprsjprs.2017.02.008.

Baatz, M., and A. Schäpe. 2000. Multiresolution segmentation: An optimization approach for high quality multi-scale image segmentation. In *Angewandte Geographische Informations-Verarbeitung* XII, (eds.) J. Strobl, T. Blaschke, and G. Griesbner, pp. 12–23. Karlsruhe, Germany: Wichmann Verlag.

Baatz, M., C. Hoffmann, and G. Willhauck. 2008. Progressing from object-based to object-oriented image analysis. In *Object-Based Image Analysis: Spatial Concepts for Knowledge-Driven Remote Sensing Applications*, (eds.) T. Blaschke, S. Lang, and G. J. Hay, pp. 29–42. Berlin, Heidelberg: Springer.

Beauchemin, M., and K. P. B. Thomson. 1997. The evaluation of segmentation results and the overlapping area matrix. *International Journal of Remote Sensing* 18:3895–3899. https://doi.org/10.1080/014311697216720.

Belgiu, M., and L. Drăguţ. 2014. Comparing supervised and unsupervised multiresolution segmentation approaches for extracting buildings from very high-resolution imagery. *ISPRS Journal of Photogrammetry and Remote Sensing* 96:67–75. https://doi.org/10.1016/j.isprsjprs.2014.07.002.

Benz, U. C., P. Hofmann, G. Willhauck, I. Lingenfelder, and M. Heynen. 2004. Multi-resolution, object-oriented fuzzy analysis of remote sensing data for GIS-ready information. *ISPRS Journal of Photogrammetry and Remote Sensing* 58:239–258. https://doi.org/10.1016/j.isprsjprs.2003.10.002.

Beveridge, J. R., J. Griffith, R. R. Kohler, A. R. Hanson, and E. M. Riseman. 1989. Segmenting images using localized histograms and region merging. *International Journal of Computer Vision* 2:311–347. https://doi.org/10.1007/BF00158168.

Bins, L. S. A., L. G. Fonseca, G. J. Erthal, and F. M. Ii. 1996. Satellite imagery segmentation: A region growing approach. In *Proceedings of VIII Simpósio Brasileiro de Sensoriamento Remoto,* April 14–19, Salvador, Brasil, pp. 677–680.

Blaschke, T. 2010. Object-based image analysis for remote sensing. *ISPRS Journal of Photogrammetry and Remote Sensing* 65:2–16. https://doi.org/10.1016/j.isprsjprs.2009.06.004.

Blaschke, T., and J. Strobl. 2001. What's wrong with pixels? Some recent developments interfacing remote sensing and GIS. *Zeitschrift für Geoinformationssysteme* 14:12–17.

Blaschke, T., C. Burnett, and A. Pekkarinen. 2004. Image segmentation methods for object-based analysis and classification. In *Remote Sensing Image Analysis: Including the Spatial Domain*, (eds.) S. M. De Jong, and F. D. Van der Meer, pp. 211–236. Dordrecht: Kluver Academic Publishers.

Brejl, M. 1999. Automated initialization and automated design of border detection criteria in edge-based image segmentation. Ph.D. Thesis, The University of Iowa, Iowa City, IA.

Buck, A., R. De Kok, T. Schneider, and U. Ammer. 1999. Improvement of a forest GIS by integration of remote sensing data for the observation and inventory of protective forests in the Bavarian Alps. In *Proceedings of IUFRO Conference on Remote Sensing and Forest Monitoring*, June 1–3, Rogow, Poland.

Byun, Y., D. Kim, J. Lee, and Y. Kim. 2011. A framework for the segmentation of high-resolution satellite imagery using modified seeded-region growing and region merging. *International Journal of Remote Sensing* 32:4589–4609. https://doi.org/10.1080/01431161.2010.489066.

Cánovas-García, F., and F. Alonso-Sarría. 2015. A local approach to optimize the scale parameter in multiresolution segmentation for multispectral imagery. *Geocarto International* 30:937–961. https://doi.org/10.1080/10106049.2015.1004131.

Castilla, G., and G. J. Hay. 2008. Image objects and geographic objects. In *Object-Based Image Analysis: Spatial Concepts for Knowledge-Driven Remote Sensing Applications*, (eds.) T. Blaschke, S. Lang, and G. J. Hay, pp. 91–110. Berlin, Heidelberg: Springer. https://doi.org/10.1007/978-3-540-77058-9_5.

Chen, J., M. Deng, X. Mei, T. Chen, Q. Shao, and L. Hong. 2014. Optimal segmentation of a high-resolution remote-sensing image guided by area and boundary. *International Journal of Remote Sensing* 35:6914–6939. https://doi.org/10.1080/01431161.2014.960617.

Cheng, H. D., X. H. Jiang, Y. Sun, and J. Wang. 2001. Color image segmentation: Advances and prospects. *Pattern Recognition* 34:2259–2281. https://doi.org/10.1016/S0031-3203(00)00149-7.

Clinton, N., A. Holt, J. Scarborough, L. Yan, and P. Gong. 2010. Accuracy assessment measures for object-based image segmentation goodness. *Photogrammetric Engineering & Remote Sensing* 76:289–299. https://doi.org/10.14358/PERS.76.3.289.

Colkesen, I., and T. Kavzoglu. 2018. Selection of optimal object features in object-based image analysis using filter-based algorithms. *Journal of the Indian Society of Remote Sensing* 46:1233–1242. https://doi.org/10.1007/s12524-018-0807-x.

Comaniciu, D., V. Ramesh, and P. Meer. 2001. The variable bandwidth mean shift and data-driven scale selection. In Proceedings *of Eighth IEEE International Conference on Computer Vision*, July 7–14, Vancouver, Canada, pp. 438–445. https://doi.org/10.1109/ICCV.2001.937550.

Costa, G. A. O. P., R. Q. Feitosa, T. B. Cazes, and B. Feijó. 2008. Genetic adaptation of segmentation parameters. In *Object-Based Image Analysis: Spatial Concepts for Knowledge-Driven Remote Sensing Applications*, (eds.) T. Blaschke, S. Lang, and G. J. Hay, pp. 679–695. Berlin, Heidelberg: Springer. https://doi.org/10.1007/978-3-540-77058-9_37.

Costa, H., G. M. Foody, and D. S. Boyd. 2018. Supervised methods of image segmentation accuracy assessment in land cover mapping. *Remote Sensing of Environment* 205:338–351. https://doi.org/10.1016/j.rse.2017.11.024.

De Kok, R., T. Schneider, and U. Ammer. 1999. Object-based classification and applications in the alpine forest environment. *International Archives of Photogrammetry and Remote Sensing,* Hannover, September 27–30, 1–6.

Dorren, L. K., B. Maier, and A. C. Seijmonsbergen. 2003. Improved Landsat-based forest mapping in steep mountainous terrain using object-based classification. *Forest Ecology and Management* 183:31–46. https://doi.org/10.1016/S0378-1127(03)00113-0.

Drăguţ, L., D. Tiede, and S. R. Levick. 2010. ESP: A tool to estimate scale parameter for multiresolution image segmentation of remotely sensed data. *International Journal of Geographical Information Science* 24:859–871. https://doi.org/10.1080/13658810903174803.

Drăguţ, L., M. Belgiu, G. Popescu, and P. Bandura. 2019. Sensitivity of multiresolution segmentation to spatial extent. *International Journal of Applied Earth Observation and Geoinformation* 81:146–153. https://doi.org/10.1016/j.jag.2019.05.002.

Drăguţ, L., O. Csillik, C. Eisank, and D. Tiede. 2014. Automated parameterisation for multi-scale image segmentation on multiple layers. *ISPRS Journal of Photogrammetry and Remote Sensing* 88:119–127. https://doi.org/10.1016/j.isprsjprs.2013.11.018.

El-naggar, A. M. 2018. Determination of optimum segmentation parameter values for extracting building from remote sensing images. *Alexandria Engineering Journal* 57:3089–3097. https://doi.org/10.1016/j.aej.2018.10.001.

Espindola, G. M., G. Camara, I. A. Reis, L. S. Bins, and A. M. Monteiro. 2006. Parameter selection for region-growing image segmentation algorithms using spatial autocorrelation. *International Journal of Remote Sensing* 27:3035–3040. https://doi.org/10.1080/01431160600617194.

Ez-zahouani, B., A. Teodoro, O. El Kharki, L. Jianhua, I. Kotaridis, X. Yuan, and L. Ma. 2023a. Remote sensing imagery segmentation in object-based analysis: A review of methods, optimization, and quality evaluation over the past 20 years. *Remote Sensing Applications: Society and Environment* 32:101031. https://doi.org/10.1016/j.rsase.2023.101031.

Ez-zahouani, B., O. El Kharki, S. Kanga Idé, and M. Zouiten. 2023b. Determination of segmentation parameters for object-based remote sensing image analysis from conventional to recent approaches: A review. *International Journal of Geoinformatics* 19:23–42. https://doi.org/10.52939/ijg.v19i1.2497.

Fourie, C. 2011. A one class object-based system for sparse geographic feature identification. M.Sc. Thesis, Stellenbosch University, South Africa.

Fukunaga, K., and L. D. Hostetler. 1975. The estimation of the gradient of a density function, with applications in pattern recognition. IEEE *Transactions on Information Theory* 21:32–40. https://doi.org/10.1109/TIT.1975.1055330.

Gao, Y., J. F. Mas, N. Kerle, and J. A. N. Pacheco. 2011. Optimal region growing segmentation and its effect on classification accuracy. *International Journal of Remote Sensing* 32:3747–3763. https://doi.org/10.1080/01431161003777189.

Grybas, H., L. Melendy, and G. R. Congalton. 2017. A comparison of unsupervised segmentation parameter optimization approaches using moderate- and high-resolution imagery. *GIScience & Remote Sensing* 54:515–533. https://doi.org/10.1080/15481603.2017.1287238.

Haralick, R. M. 1981. Edge and region analysis for digital image data. In *Image Modeling,* (ed.) A. Rosenfeld, pp. 171–184. New York: Academic Press.

Haralick, R. M., and L. G. Shapiro. 1985. Image segmentation techniques. *Computer Vision, Graphics, and Image Processing* 29:100–132. https://doi.org/10.1016/S0734-189X(85)90153-7.

Hossain, M. D., and D. Chen. 2019. Segmentation for object-based image analysis (OBIA): A review of algorithms and challenges from remote sensing perspective. *ISPRS Journal of Photogrammetry and Remote Sensing* 150:115–134. https://doi.org/10.1016/j.isprsjprs.2019.02.009.

Im, J., L. J. Quackenbush, M. Li, and F. Fang. 2014. Optimum scale in object-based image analysis. In *Scale Issues in Remote Sensing*, (ed.) Q. Wang, pp. 197–214. Hoboken, NJ: John Wiley & Sons. https://doi.org/10.1002/9781118801628.ch10.

Jensen, J. R. 2005. *Introductory Digital Image Processing: A Remote Sensing Perspective*, 3rd edition. Upper Saddle River, NJ: Prentice-Hall.

Ji, A., and H. W. Park. 1998. Image segmentation of color image based on region coherency. In Proceedings *of IEEE International Conference on Image Processing (ICIP'98)*, October 7, Chicago, IL, pp. 80–83. https://doi.org/10.1109/ICIP.1998.723425.

Johnson, B., and Z. Xie. 2011. Unsupervised image segmentation evaluation and refinement using a multi-scale approach. *ISPRS Journal of Photogrammetry and Remote Sensing* 66:473–483. https://doi.org/10.1016/j.isprsjprs.2011.02.006.

Johnson, B. A., M. Bragais, I. Endo, D. B. Magcale-Macandog, and P. B. M. Macandog. 2015. Image segmentation parameter optimization considering within- and between-segment heterogeneity at multiple scale levels: Test case for mapping residential areas using Landsat imagery. *ISPRS International Journal of Geo-Information* 4:2292–2305. https://doi.org/10.3390/ijgi4042292.

Jozdani, S., and D. Chen. 2020. On the versatility of popular and recently proposed supervised evaluation metrics for segmentation quality of remotely sensed images: An experimental case study of building extraction. *ISPRS Journal of Photogrammetry and Remote Sensing* 160:275–290. https://doi.org/10.1016/j.isprsjprs.2020.01.002.

Kaur, G., and K. A. Singh. 2013. Comparative study of image segmentation using thresholding techniques. In *Proceedings of National Conference on Advanced Computing Technologies*, March 30, Rohtak, India.

Kavzoglu, T. 2017. Object-oriented random forest for high resolution land cover mapping using Quickbird-2 imagery. In *Handbook of Neural Computation*, (eds.) P. Samui, S. S. Roy, and V. E. Balas, pp. 607–619. San Diego, CA: Academic Press. https://doi.org/10.1016/B978-0-12-811318-9.00033-8.

Kavzoglu, T., and H. Tonbul. 2017a. Selecting optimal SLIC superpixels parameters by using discrepancy measures. In *Proceedings of 38th Asian Conference on Remote Sensing*, October 23–27, New Delhi, India, pp. 212–218.

Kavzoglu, T., and H. Tonbul. 2017b. A comparative study of segmentation quality for multi-resolution segmentation and watershed transform. In *Proceedings of 8th International Conference on Recent Advances in Space Technologies (RAST)*, June 19–22, Istanbul, Turkey, pp. 113–117. https://doi.org/10.1109/RAST.2017.8002984.

Kavzoglu, T., and H. Tonbul. 2018a. An experimental comparison of multi-resolution segmentation, SLIC and K-means clustering for object-based classification of VHR imagery. *International Journal of Remote Sensing* 39:6020–6036. https://doi.org/10.1080/01431161.2018.1506592.

Kavzoglu, T., and H. Tonbul. 2018b. Segmentation quality assessment for varying spatial resolutions of very high resolution satellite imagery. *Geography and Natural Resources* 2:109–114.

Kavzoglu, T., and M. Yildiz. 2014. Parameter-based performance analysis of object-based image analysis using aerial and Quikbird-2 images. *ISPRS Annals of Photogrammetry, Remote Sensing and Spatial Information Sciences* 31–37. https://doi.org/10.5194/isprsannals-II-7-31-2014.

Kavzoglu, T., H. Tonbul, M. Y. Erdemir, and I. Colkesen. 2018. Dimensionality reduction and classification of hyperspectral images using object-based image analysis. *Journal of the Indian Society of Remote Sensing* 46:1297–1306. https://doi.org/10.1007/s12524-018-0803-1.

Kavzoglu, T., M. Y. Erdemir, and H. Tonbul. 2016. A region-based multi-scale approach for object-based image analysis. In *Proceedings of XXIII International Society for Photogrammetry and Remote Sensing Congress*, July 12–19, Prague, Czech Republic, pp. 763–769. https://doi.org/10.5194/isprs-archives-XLI-B7-241-2016.

Kavzoglu, T., M. Y. Erdemir, and H. Tonbul. 2017. Classification of semiurban landscapes from very high-resolution satellite images using a regionalized multiscale segmentation approach. *Journal of Applied Remote Sensing* 11:035016. https://doi.org/10.1117/1.JRS.11.035016.

Kelkar, D., and S. Gupta. 2008. Improved quadtree method for split merge image segmentation. In *Proceedings of 2008 First International Conference on Emerging Trends in Engineering and Technology*, July 16–18, Nagpur, India, pp. 44–47. https://doi.org/10.1109/ICETET.2008.145.

Kermad, C. D., and K. Chehdi. 2002. Automatic image segmentation system through iterative edge-region co-operation. *Image and Vision Computing* 20:541–555. https://doi.org/10.1016/S0262-8856(02)00043-4.

Kim, M., M. Madden, and T. A. Warner. 2009. Forest type mapping using object-specific texture measures from multispectral IKONOS imagery: Segmentation quality and image classification issues. *Photogrammetric Engineering & Remote Sensing* 75:819–829. https://doi.org/10.14358/PERS.75.7.819.

Kotaridis, I., and M. Lazaridou. 2021. Remote sensing image segmentation advances: A meta-analysis. *ISPRS Journal of Photogrammetry and Remote Sensing* 173:309–322. https://doi.org/10.1016/j.isprsjprs.2021.01.020.

Kucharczyk, M., G. J. Hay, S. Ghaffarian, and C. H. Hugenholtz. 2020. Geographic object-based image analysis: A primer and future directions. *Remote Sensing* 12:2012. https://doi.org/10.3390/rs12122012.

Kundu, M. K., and S. K. Pal. 1986. Thresholding for edge detection using human psychovisual phenomena. *Pattern Recognition Letters* 4:433–441. https://doi.org/10.1016/0167-8655(86)90041-3.

Laliberte, A. S., D. M. Browning, and A. Rango. 2012. A comparison of three feature selection methods for object-based classification of sub-decimeter resolution UltraCam-L imagery. *International Journal of Applied Earth Observation and Geoinformation* 15:70–78. https://doi.org/10.1016/j.jag.2011.05.011.

Lang, S. 2008. Object-based image analysis for remote sensing applications: modeling reality: Dealing with complexity. In *Object-Based Image Analysis: Spatial Concepts for Knowledge-Driven Remote Sensing Applications*, (eds.) T. Blaschke, S. Lang, and G. J. Hay, pp. 3–27. Berlin, Heidelberg: Springer. https://doi.org/10.1007/978-3-540-77058-9_1.

Lang, S., and T. Blaschke. 2006. Bridging remote sensing and GIS-What are the main supportive pillars? In *International Archives of Photogrammetry, Remote Sensing and Spatial Information Sciences*, (eds.) S. Lang, T. Blaschke, and E. Schöpfer, vol. XXXVIII-4/C42, July 4–5, 2006, Salzburg University, Austria, pp. 1–6.

Levine, M. D., and A. Nazif. 1985. Rule-based image segmentation: A dynamic control strategy approach. *Computer Vision, Graphics, and Image Processing* 32:104–126. https://doi.org/10.1016/0734-1 89X(85)90004-0.

Li, H., F. Meng, B. Luo, and S. Zhu. 2014. Repairing bad co-segmentation using its quality evaluation and segment propagation. IEEE *Transactions on Image Processing* 23:3545–3559. https://doi.org/10.1109/TIP.2014.2330759.

Li, N., H. Huo, and T. Fang. 2010. A novel texture-preceded segmentation algorithm for high-resolution imagery. IEEE *Transactions on Geoscience and Remote Sensing* 48:2818–2828. https://doi.org/10.1109/TGRS.2010.2041462.

Liu, Y., L. Bian, Y. Meng, H. Wang, S. Zhang, Y. Yang, X. Shao, and B. Wang. 2012. Discrepancy measures for selecting optimal combination of parameter values in object-based image analysis. *ISPRS Journal of Photogrammetry and Remote Sensing* 68:144–156. https://doi.org/10.1016/j.isprsjprs.2012.01.007.

Liu, Y., M. Li, L. Mao, F. Xu, and S. Huang. 2006. Review of remotely sensed imagery classification patterns based on object-oriented image analysis. *Chinese Geographical Science* 16:282–288. https://doi.org/10.1007/s11769-006-0282-0.

Lucchese, L., and S. K. Mitra. 2001. Color image segmentation: A state-of-the-art survey. *Proceedings-Indian National Science Academy - Part A*, 67:207–222.

Lucieer, A., and A. Stein. 2002. Existential uncertainty of spatial objects segmented from satellite sensor imagery. IEEE *Transactions on Geoscience and Remote Sensing* 40:2518–2521. https://doi.org/10.1109/TGRS.2002.805072.

Ma, L., M. Li, X. Ma, L. Cheng, P. Du, and Y. Liu. 2017. A review of supervised object-based land-cover image classification. *ISPRS Journal of Photogrammetry and Remote Sensing* 130:277–293. https://doi.org/10.1016/j.isprsjprs.2017.06.001.

Manousakas, I. N., P. E. Undrill, G. G. Cameron, and T. W. Redpath. 1998. Split-and-merge segmentation of magnetic resonance medical images: Performance evaluation and extension to three dimensions. *Computers and Biomedical Research* 31:393–412. https://doi.org/10.1006/CBMR.1998.1489.

Meyer, F., and S. Beucher. 1990. Morphological segmentation. *Journal of Visual Communication and Image Representation* 1:21–46. https://doi.org/10.1016/1047-3203(90)90014-M.

Ming, D., T. Ci, H. Cai, L. Li, C. Qiao, and J. Du. 2012. Semivariogram-based spatial bandwidth selection for remote sensing image segmentation with mean-shift algorithm. *IEEE Geoscience and Remote Sensing Letters* 9:813–817. https://doi.org/10.1109/LGRS.2011.2182604.

Ming, D., W. Zhou, L. Xu, M. Wang, and Y. Ma. 2018. Coupling relationship among scale parameter, segmentation accuracy, and classification accuracy in GeOBIA. *Photogrammetric Engineering & Remote Sensing* 84:681–693. https://doi.org/10.14358/PERS.84.11.681.

Mueller, M., K. Segl, and H. Kaufmann. 2004. Edge- and region-based segmentation technique for the extraction of large, man-made objects in high-resolution satellite imagery. *Pattern Recognition* 37:1619–1628. https://doi.org/10.1016/j.patcog.2004.03.001.

Neubert, P., and P. Protzel. 2012. Superpixel benchmark and comparison. In *Forum Bildverarbeitung*, (eds.) F. P. Leaon, and M. Heizmann, pp. 1–12. Karlsruhe, Germany: KIT Scientific Publishing.

Niemeyer, I., M. J. Canty, and M. Baatz. 1999. Fractal-hierarchical pattern recognition for safeguard purposes. In Proceedings *of 2nd International Symposium on Operationalization of Remote Sensing*, April 19–23, Enschede, The Netherlands.

Pal, N. R., and S. K. Pal. 1993. A review on image segmentation techniques. *Pattern Recognition* 26:1277–1294. https://doi.org/10.1016/0031-3203(93)90135-J.

Persello, C., and L. Bruzzone. 2010. A novel protocol for accuracy assessment in classification of very high resolution images. IEEE *Transactions on Geoscience and Remote Sensing* 48:1232–1244. https://doi.org/10.1109/TGRS.2009.2029570.

Radoux, J., and P. Defourny. 2008. Quality assessment of segmentation results devoted to object-based classification. *Object-Based Image Analysis: Spatial Concepts for Knowledge-Driven Remote Sensing Applications*, (eds.) T. Blaschke, S. Lang, and G. J. Hay, pp. 257–271. Berlin, Heidelberg: Springer. https://doi.org/10.1007/978-3-540-77058-9_14.

Ren, X., and J. Malik. 2003. Learning a classification model for segmentation. In *Proceedings of Ninth IEEE International Conference on Computer Vision*, October 13–16, Nice, France, pp. 10–17. https://doi.org/10.1109/ICCV.2003.1238308.

Schiewe, J. 2002. Segmentation of high-resolution remotely sensed data: Concepts, applications and problems. In Proceedings *of Joint ISPRS Commission IV Symposium: Geospatial Theory, Processing and Applications*, July 9–12, Ottawa, Canada.

Tonbul, H. 2021. Optimization of region-based segmentation algorithm based on natural land cover boundaries in object-based classification. Ph.D. Thesis, Gebze Technical University, Turkey.

Tonbul, H., and T. Kavzoğlu. 2018. Image segmentation approaches in object-based image analysis and analysis of segmentation quality. *Map Journal* 84:12–23.

Tonbul, H., and T. Kavzoglu. 2019. The effect of spectral bands on segmentation quality analysis with optimum parameter determination. In Proceedings *of X. Turkish National Photogrammetry and Remote Sensing Union (TUFUAB' 2019)*, April 25–27, Aksaray, Türkiye.

Tonbul, H., and T. Kavzoglu. 2020a. A spectral band based comparison of unsupervised segmentation evaluation methods for image segmentation parameter optimization. *International Journal of Environment and Geoinformatics* 7:132–139. https://doi.org/10.30897/ijegeo.641216.

Tonbul, H., and T. Kavzoglu. 2020b. Semi-automatic building extraction from Worldview-2 imagery using Taguchi optimization. *Photogrammetric Engineering & Remote Sensing* 86:547–555. https://doi.org/10.14358/PERS.86.9.547.

Toro, C. A. O., C. G. Martín, Á. G. Pedrero, and E. M. Ruiz. 2015. Superpixel-based roughness measure for multispectral satellite image segmentation. *Remote Sensing* 7:14620–14645. https://doi.org/10.3390/rs71114620.

Verma, O. P., M. Hanmandlu, S. Susan, M. Kulkarni, and P. K. Jain. 2011. A simple single seeded region growing algorithm for color image segmentation using adaptive thresholding. In *Proceedings of 2011 International Conference on Communication Systems and Network Technologies*, June 3–5, Katra, India, pp. 500–503. https://doi.org/10.1109/CSNT.2011.107.

Walker, J. S., and T. Blaschke. 2008. Object-based landcover classification for the Phoenix metropolitan area: Optimization vs. transportability. *International Journal of Remote Sensing* 29:2021–2040. https://doi.org/10.1080/01431160701408337.

Wang, M., and R. Li. 2014. Segmentation of high spatial resolution remote sensing imagery based on hard-boundary constraint and two-stage merging. IEEE *Transactions on Geoscience and Remote Sensing* 52:5712–5725. https://doi.org/10.1109/TGRS.2013.2292053.

Wang, Y., Q. Meng, Q. Qi, J. Yang, and Y. Liu. 2018. Region merging considering within- and between-segment heterogeneity: An improved hybrid remote-sensing image segmentation method. *Remote Sensing* 10:781. https://doi.org/10.3390/rs10050781.

Wang, Y., Q. Qi, Y. Liu, L. Jiang, and J. Wang. 2019. Unsupervised segmentation parameter selection using the local spatial statistics for remote sensing image segmentation. *International Journal of Applied Earth Observation and Geoinformation* 81:98–109. https://doi.org/10.1016/j.jag.2019.05.004.

Wang, Z. 2008. An automatic region-based image segmentation system for remote sensing applications. Ph.D. Thesis, University of South Carolina, Columbia, SC.

Wang, Z., J. R. Jensen, and J. Im. 2010. An automatic region-based image segmentation algorithm for remote sensing applications. *Environmental Modelling & Software* 25:1149–1165. https://doi.org/10.1016/j.envsoft.2010.03.019.

Weidner, U. 2008. Contribution to the assessment of segmentation quality for remote sensing applications. In *Proceedings of 21st Congress for the International Society for Photogrammetry and Remote Sensing*, July 3–11, Beijing, China.

Yan, G., J. -F. Mas, B. H. P. Maathuis, Z. Xiangmin, and P. M. Van Dijk. 2006. Comparison of pixel-based and object-oriented image classification approaches-a case study in a coal fire area, Wuda, Inner Mongolia, China. *International Journal of Remote Sensing* 27:4039–4055. https://doi.org/10.1080/01431160600702632.

Yang, J., P. Li, and Y. He. 2014. A multi-band approach to unsupervised scale parameter selection for multi-scale image segmentation. *ISPRS Journal of Photogrammetry and Remote Sensing* 94:13–24. https://doi.org/10.1016/j.isprsjprs.2014.04.008.

Yang, J., Y. H. He, J. Caspersen, and T. Jones. 2015. A discrepancy measure for segmentation evaluation from the perspective of object recognition. *ISPRS Journal of Photogrammetry and Remote Sensing* 101: 186–192. https://doi.org/10.1016/j.isprsjprs.2014.12.015.

Yang, J., Y. He, and J. Caspersen. 2016. A self-adapted threshold-based region merging method for remote sensing image segmentation. In *Proceedings of 2016 IEEE International Geoscience and Remote Sensing Symposium (IGARSS)*, July 10–15, Beijing, China, pp. 6320–6323. https://doi.org/10.1109/IGARSS.2016.7730652.

Ye, S., R. G. P. Pontius Jr., and R. Rakshit. 2018. A review of accuracy assessment for object-based image analysis: From per-pixel to per-polygon approaches. *ISPRS Journal of Photogrammetry and Remote Sensing* 141:137–147. https://doi.org/10.1016/j.isprsjprs.2018.04.002.

Yu, Q., P. Gong, N. Chinton, G. Biging, M. Kelly, and D. Schirokauer. 2006. Object-based detailed vegetation classification with airborne high spatial resolution remote sensing imagery. *Photogrammetric Engineering & Remote Sensing* 72:799–811. https://doi.org/10.14358/PERS.72.7.799.

Zhang, H., J. E. Fritts, and S. A. Goldman. 2008. Image segmentation evaluation: A survey of unsupervised methods. *Computer Vision Image Understanding* 110:260–280. https://doi.org/10.1016/j.cviu.2007.08.003.

Zhang, L., H. Liu, X. Li, and X. Qian. 2021. Optimizing the segmentation of a high-resolution image by using a local scale parameter. *Photogrammetric Engineering & Remote Sensing* 87:503–511. https://doi.org/10.14358/PERS.87.7.503.

Zhang, X., P. Xiao, X. Feng, J. Wang, and Z. Wang. 2014. Hybrid region merging method for segmentation of high-resolution remote sensing images. *ISPRS Journal of Photogrammetry and Remote Sensing* 98:19–28. https://doi.org/10.1016/j.isprsjprs.2014.09.011.

Zhang, X., X. Feng, P. Xiao, G. He, and L. Zhu. 2015. Segmentation quality evaluation using region-based precision and recall measures for remote sensing images. *ISPRS Journal of Photogrammetry and Remote Sensing* 102:73–84. https://doi.org/10.1016/j.isprsjprs.2015.01.009.

Zhang, Y. J. 1996. A survey on evaluation methods for image segmentation. *Pattern Recognition* 29:1335–1346. https://doi.org/10.1016/0031-3203(95)00169-7.

Zhang, Y. J. 2001. A review of recent evaluation methods for image segmentation. In *Proceedings of Sixth International Symposium on Signal Processing and its Applications*, August 13–16, Kuala Lumpur, Malaysia, pp. 148–151. https://doi.org/10.1109/ISSPA.2001.949797.

9 Hyperparameter Optimization

Until now, many machine learning algorithms have been introduced, each tailored for distinct problem domains or data sets, as mentioned in Chapters 5–7. In recent years, these algorithms have undergone substantial evolution, characterized by increasingly intricate internal architectural designs aimed at more adeptly elucidating hidden patterns within multidimensional real-world data. This can be partly attributed to the advancements in computational power with the development of new-generation processors, graphics processing units (GPUs), and other hardware components. As computational resources have become more powerful and accessible, researchers and practitioners have been able to develop and experiment with larger and more complex models. In this context, the structural sophistication of modern machine learning models extends beyond simple linear constructs. Instead, they involve complex architectures, sometimes governing hierarchical model designs (e.g., decision trees) or determining the number of base learners participating in ensemble schemes (e.g., random forest). Recent deep neural networks are significantly influenced by an extensive array of hyperparameter decisions related to the architecture, regularization, and optimization of neural networks.

Almost all artificially intelligent algorithms comprise two fundamental elements: (1) parameters and (2) hyperparameters (i.e., meta-parameters). The former group includes the internal variables for which a machine learning model learns from the training data, such as coefficients in linear regression, weights in neural networks, or split points in decision trees. The objective of linear regression is to discover the optimal-fitting line that minimizes the difference between predicted and observed values. The coefficients of this line are the internal variables that the model learns from the training data. The connections between the nodes of the neural network have weights that the model adjusts during training. Likewise, decision trees make decisions by recursively splitting the data on the basis of certain features. The split points are the internal variables that the model learns from the training data, which helps to partition the data into subsets for better predictions. These values are adjusted during the training process, allowing the model to adapt and optimize its performance on the basis of the patterns and relationships present in the provided data set. In other words, parameters are the internal settings that the algorithm fine-tunes to best fit the observed data. The latter group, on the other hand, is generally higher-level structural elements in nature that guide the learning process and influence the behavior, speed, complexity, and other aspects of the models (Bischl et al., 2023). These are essentially external configuration settings that are not learned from the data but are set before the training process. Hence, it is crucial to configure them based on prior experience before initiating the training process (Yu and Zhu, 2020). The learning rate in gradient boosting, the number of hidden layers in a neural network, and the depth of a decision tree are all examples of hyperparameters. The learning rate is a crucial hyperparameter that influences the convergence speed and model performance. Setting it too high can cause the model to overshoot the minimum while setting it too low can lead to slow convergence. It needs to be configured on the basis of data characteristics and prior experience to ensure optimal model performance. The number of hidden layers is a structural decision that influences the model's generalization capacity to learn complex patterns. It is set based on the complexity of the task and the available data. Too few hidden layers may result in underfitting, whereas too many can lead to overfitting. Similarly, the depth of a decision tree is a hyperparameter that needs to be set before training. It controls the trade-off between model complexity and simplicity. Excessive depth in a tree may result in the inclusion of noise from the training data, hindering its capacity to generalize effectively to new and unseen data.

In the past decade, machine learning has undergone significant advancements; however, the efficacy of numerous machine learning methods continues to hinge heavily on manually

DOI: 10.1201/9781003439172-9

engineered features or finely tuned hyperparameter settings. This reliance on human intervention remains a critical determinant, often demarcating the divide between ordinary and state-of-the-art performance in machine learning applications (Hutter et al., 2015). Therefore, constructing an optimal machine learning model is a nontrivial task that necessitates a thorough exploration of potential configurations. This endeavor, that is, designing an ideal model architecture, is commonly called hyperparameter optimization (HPO) or hyperparameter tuning. This intricate process involves systematically traversing the multifaceted landscape of hyperparameter values to discern the combination that elicits peak performance from the model. However, this is not always a routine procedure. The process of hyperparameter tuning exhibits variation across distinct machine learning algorithms because of the diverse nature of the hyperparameters employed. These hyperparameters can be categorized into different types, including categorical, discrete, and continuous variants. In response to these challenges, there has been a growing trend in research dedicated to the development of new and automated techniques for optimizing hyperparameters. This research domain seeks to mitigate the limitations of manual tuning by leveraging algorithmic approaches that can systematically explore the hyperparameter space, thereby enhancing the efficiency and efficacy in the pursuit of optimal model configurations. The core philosophy behind this evolution is to streamline the hyperparameter tuning process, making it more scalable and adaptable to the complexities inherent in contemporary machine learning algorithms. There are several compelling reasons to apply HPO techniques to machine learning models (Feurer and Hutter, 2019):

- **Reduction of Human Effort**: HPO alleviates the substantial human effort traditionally invested in hyperparameter tuning. This is particularly valid for large data sets or complex machine learning algorithms replete with an extensive array of hyperparameters. This automation allows developers to focus more on the conceptual aspects of model design and problem-solving, as opposed to laborious parameter tweaking.
- **Enhanced Model Performance**: The diverse landscape of machine learning hyperparameters often entails different optimal configurations for achieving peak performance across various data sets or problem domains. HPO systematically explores this multidimensional space, leading to improved model performance by identifying configurations that optimally align with the specific nuances of a given data set or problem.
- **Enhanced Reproducibility**: Implementing consistent hyperparameter tuning processes through HPO enhances the reproducibility of the models and research outcomes. This standardization is pivotal for fair comparisons among different machine learning algorithms. Applying the same HPO methodology across diverse algorithms facilitates the identification of the most suitable model for a particular problem, fostering transparency and comparability in research endeavors, which are critical issues for the interpretability of black-box methods of machine learning.

There exists a plethora of optimization algorithms designed for different types of problems. However, the selection of a suitable technique for problem consideration is challenging. For instance, traditional optimization methods may prove unsuitable for HPO problems, given the nonconvex or nondifferentiable nature of many HPO problems, often resulting in local rather than global optima. Gradient descent–based approaches, which are one of the most popular traditional optimization algorithms, are adept at tuning continuous hyperparameters by computing their gradients. In contrast, numerous optimization techniques better suit the complexities of HPO problems than traditional methods such as gradient descent. These include Bayesian optimization, multifidelity optimization, decision-theoretic approaches, and metaheuristic algorithms. Beyond addressing continuous hyperparameters, many of these algorithms exhibit efficacy in identifying discrete, categorical, and conditional hyperparameters.

The intricacies of remote sensing data, including varying spatial resolutions, spectral characteristics, and temporal dynamics, necessitate a fine-tuned model that can effectively capture the hidden patterns inherent in satellite or aerial imagery. Given their significant impact on the prediction process, HPO also plays a key role in remote sensing applications. One of the primary implications of neglecting HPO in remote sensing applications is the risk of suboptimal model performance. Remote sensing data, characterized by its multifaceted nature and inherent variability, demand careful consideration of hyperparameter configurations to ensure that models can effectively adapt to the nuances present in different environmental conditions, sensor characteristics, and temporal variations. In terms of interpretability, careful selection of hyperparameters can influence how well the model's decision-making process can be elucidated. In remote sensing applications, where stakeholders often require insights into the rationale behind predictions, the absence of HPO may impede the transparent communication of model outputs and hinder informed decision-making processes. HPO, often facilitated by techniques such as random search, grid search, or more advanced methods, is imperative to strike a balance between underfitting and overfitting. This ultimately improves the overall performance and reliability of remote sensing applications. Consequently, HPO methods are applied to various remote sensing tasks, such as vehicle identification (Xie et al., 2024), ecosystem productivity prediction (Huang et al., 2024), mineral prospectivity mapping (Lin et al., 2021), semantic segmentation (Igonin et al., 2020), and natural disaster mapping (Liao et al., 2024), among others.

This section presents the fundamental principles of HPO along with an overview of the general HPO process. Following this foundation, an exploration of the pivotal hyperparameters inherent in common machine learning models that warrant tuning is presented. Subsequently, a comprehensive examination is undertaken to delve into the diverse array of state-of-the-art optimization approaches specifically designed to address HPO challenges. This contemplative exploration contributes to the ongoing evolution of HPO techniques, ensuring their relevance and efficacy in the dynamic landscape of machine learning. In conclusion, this section concludes with key remarks that encapsulate the core findings and implications derived from the exploration of mathematical optimization, and contemporary approaches shaping the optimization landscape.

9.1   WHAT IS HYPERPARAMETER OPTIMIZATION?

Optimization plays a central role in addressing decision-making challenges, whether they pertain to engineering or everyday situations. The decision-making process involves selecting from various, sometimes numerous, alternatives. Optimization is the process of seeking a solution to a problem in which the objective is to maximize or minimize one or multiple objective functions within a domain that encompasses acceptable values for variables while adhering to specific constraints. At the core of machine learning, mathematical optimization models include objective functions and constraints formulated as a system of equations or inequalities. The search domain may encompass numerous sets of variables, each attempting to minimize or maximize the objective function(s) while meeting the specified restrictions. These sets are denoted as feasible solutions, and the most superior among them is recognized as the optimal solution to the problem.

In the context of machine learning, the fundamental objective is to create an optimization model for the refinement of learning parameters and model hyperparameters using available data. The selection of the best hyperparameter configuration directly influences the resulting classification performance. Numerous papers have been published, and more continue to be released, comparing the performances of optimal configurations with those of other settings. With the immense increase in both data volume in remote sensing and model complexity, the optimization of machine learning involves intricate challenges for many tasks. Unfortunately, the relationship between a particular hyperparameter configuration and related performance is uncertain and not known in advance. Therefore, the fine-tuning process, which involves continuously adjusting hyperparameters and training models with the adapted values, can be viewed as a somewhat opaque skill ("black art"), requiring expert experience, rules of thumb, or sometimes exhaustive search techniques. There is substantial interest in automated approaches capable of optimizing the performance of any given learning algorithm for a specific problem (Snoek et al., 2012). The success of certain optimization methods in machine learning has inspired researchers to address even more intricate machine learning challenges and to create new and enhanced versions of existing ones.

In the iterative design process of machine learning models, the exploration of hyperparameter space, also called search space, through optimization techniques plays a pivotal role in identifying the optimal configuration for these models. This systematic search allows for the fine-tuning of hyperparameters, leading to a more refined and performance-optimized machine learning model. The efficacy of this process lies in its ability to navigate in the complex and multidimensional hyperparameter space, ultimately guiding the model toward configurations that yield superior results in terms of accuracy and generalization. A regular HPO process involves four key components: an estimator, which can be either a regressor or a classifier, with its associated objective function; a search space, often referred to as the configuration space, defining the range of potential hyperparameter values; a search or optimization method employed to identify optimal hyperparameter combinations within the specified space; and an evaluation function designed to systematically compare and assess the performance of different hyperparameter configurations.

To illustrate the process, assume a machine learning algorithm M and data set $\mathcal{D}$. The goal is to maximize or minimize a given metric (mostly accuracy assessment metrics such as overall accuracy and mean square error [MSE]), depending on whether the objective is to maximize accuracy, minimize error, or achieve some other desired metric. The objective can be represented as an objective function $f(\theta)$ where θ denotes the vector of hyperparameters for a given machine learning model M. The aim can be basically defined as

$$
\begin{aligned}
\text{Maximize/minimize} \quad & f(\theta) \\
\text{Subject to} \quad & \theta \in \mathbb{R}
\end{aligned}
\tag{9.1}
$$

In Equation (9.1), the goal is to find the optimal hyperparameter set θ^* that minimizes or maximizes $f(\theta)$. The components within the search space θ can exhibit various types, encompassing continuous, integer-valued, categorical, and occasionally conditional hyperparameters. Each type reflects the nature of the parameter it represents, and these diverse types add complexity and nuance to the HPO process. For instance, continuous hyperparameters can take any real value within a specified range, such as the learning rate in gradient descent–based optimization algorithms that control the number of steps taken during the optimization process. In contrast, integer-valued hyperparameters are constrained to discrete, whole-number values, such as the number of estimators in a random forest algorithm. Categorical hyperparameters represent a set of distinct categories or options, with the hyperparameter assuming one of these predefined values, such as the activation function in a neural network (e.g., ReLU, sigmoid, or tanh). Moreover, certain hyperparameters may be conditional, meaning that their applicability depends on the values of other hyperparameters. For instance, in deep learning architectures, the hyperparameters associated with a particular layer may only come into play if the network indeed has that specific layer. Consider another scenario in which you have a machine learning framework with multiple algorithms, each with a set of subsidiary hyperparameters. The activation of these subsidiary hyperparameters is contingent on the selection of the corresponding algorithm. For example, if a decision tree algorithm is chosen, only then the associated hyperparameters (e.g., maximum depth of the tree) become relevant or "active".

In straightforward scenarios, all hyperparameters can assume unrestrained real values, and the feasible set of hyperparameters can be represented as a real-valued n-dimensional vector space. However, in most cases, hyperparameters in machine learning models often possess specific domains and are subject to various constraints, transforming their optimization problems into intricate, constrained optimization challenges. For example, consider the regularization strength (λ) in a linear regression model. In this case, λ should be nonnegative, and its feasible range is constrained by the characteristics of the data set and the desired model complexity. Another example is the learning rate (α) in gradient descent optimization. For stable convergence, α typically should be within a certain range, for example, between 0 and 1. Categorical hyperparameters further illustrate this complexity. For instance, the choice of the kernel function in a SVM could be limited to options, such as "linear", "poly", or "rbf", constraining the feasible values of the hyperparameter to a predefined set.

Given all conditions, the primary objective of HPO is to obtain the optimal or near-optimal performance of a machine learning model by tuning hyperparameters within predefined resource constraints or budgets. The mathematical expression of the objective function, denoted as $f(\theta)$, varies depending on the specific objective function of the chosen machine learning algorithm and the associated performance metric. Various metrics can be employed to evaluate the performance of the machine learning model. The choice of metric is contingent on the nature of the problem and the desired outcomes. In practical applications, time budgets constitute a crucial constraint in the optimization of HPO models and require careful consideration. Optimizing the objective function of a machine learning model with a reasonable number of hyperparameter configurations can be a time-intensive process. Each evaluation of a hyperparameter value involves retraining the entire machine learning model, followed by processing the validation set to generate a score reflective of the model's performance. The main generic process of HPO typically involves the following steps:

- **Defining the Search Space**: Identify the crucial hyperparameters for the performance of the machine learning model. Specify the feasible range or values for each hyperparameter within the search space, also called the configuration space.
- **Choosing an Optimization Algorithm**: Select an optimization algorithm to effectively and intelligently explore the hyperparameter space and then find optimal configurations. Common algorithms include Bayesian optimization, random search, genetic algorithms, and particle swarm optimization, all of which are discussed in the following sections.
- **Defining the Objective Function**: Describe the objective function $f(\theta)$ which is typically the performance metric of the machine learning model based on the validation set.
- **Initializing the Optimization Process**: Start the optimization algorithm with an initial set of hyperparameter configurations. Evaluate the objective function for each configuration in the initial set.
- **Iterative Optimization**: Iteratively refine the hyperparameter configurations by generating new candidate configurations based on the optimization algorithm's strategy. Evaluate the objective function for each candidate configuration. Update the model of the objective function based on the evaluations.
- **Convergence and Stopping Criteria**: Determine a stopping criterion, such as a maximum number of iterations or a threshold for improvement. Conclude the optimization process when the algorithm converges or when the stopping criterion is met.
- **Retrieve Optimal Hyperparameters**: Extract the hyperparameter configuration that corresponds to the optimal or near-optimal value of the objective function.
- **Evaluation on Test Set (Optional)**: Validate the final model with the selected hyperparameters on a separate test set to assess its generalization performance.
- **Fine-tuning and Refinement (Optional)**: Perform further fine-tuning or refinement if additional resources or time allows for a more exhaustive search.
- **Documenting and Interpreting Results**: Document the optimal hyperparameter configuration and associated performance metrics. Interpret the results to understand the impact of hyperparameter choices on model performance.

This iterative and systematic process aims to efficiently navigate the hyperparameter space and discover configurations that optimize the machine learning model's performance within specified resource constraints.

9.2 HYPERPARAMETER OPTIMIZATION TECHNIQUES

HPO techniques, diverse in nature, explore the hyperparameter space, seeking configurations that optimize the performance of machine learning models. The diversity in these techniques stems from the recognition that no single method is universally superior; different algorithms may perform better under different circumstances or for specific types of problems. Each algorithm adheres to a unique methodology for delving into the hyperparameter space and making decisions on the subsequent configurations to scrutinize. These techniques can handle myriad types of hyperparameters, ranging from continuous to discrete, categorical, and conditional. Their strength lies in their adaptability to the nuances of the hyperparameter space, considering constraints, dependencies, and the diverse values that each hyperparameter can take. The modalities by which these techniques navigate the hyperparameter space diverge significantly. Some methods employ grid search, methodically exploring combinations within predefined ranges, whereas others, exemplified by random search or Bayesian optimization, embrace more stochastic and probabilistic frameworks. Purposefully designed to align with the complexities of optimization problems, HPO techniques exhibit variability in their effectiveness due to certain factors. These encompass the dimensionality of the hyperparameter space, the nature of objective functions, and the computational resources at one's disposal. Further diversity is evident in the efficacy of parallelization techniques and their ability to efficiently handle expansive optimization challenges. Certain methods lend themselves to parallel execution, facilitating the concurrent evaluation of multiple configurations. The capability to grapple with constraints emanating from conditional hyperparameters or restrictions on specific values diverges across HPO techniques. Some approaches seamlessly incorporate constraints into their framework, whereas others necessitate supplementary considerations. In their discerning approach, techniques may turn to surrogate models to approximate the underlying objective function, thereby assisting in determining which hyperparameter configurations warrant further evaluation. The selection of surrogate models includes Gaussian processes and tree-structured Parzen estimators.

9.2.1 MODEL-FREE ALGORITHMS

9.2.1.1 Trial-and-Error (Manual Testing)

Trial-and-error, often regarded as manual testing or empirical searching, is a straightforward approach for identifying the hyperparameters of the models. This method involves the iterative adjustment of model hyperparameters based on empirical and experiential observations of the model's performance made by users. The simplicity of trial-and-error facilitates rapid insights into the behavior of a model as hyperparameter values change. However, the lack of clarity regarding the number of diverse hyperparameter variations made during this process poses a significant challenge. In scenarios with numerous hyperparameters and complex models, manual exploration becomes more challenging as the hyperparameter count increases, resulting in exponential growth in the search space. This issue can further pose a challenge in determining the optimal configuration from the search space, particularly in scenarios with several hyperparameters and complex models, since manual exploration becomes exponentially more burdensome as the number of hyperparameters increases. Within this iterative process, empirical observations serve as the guiding force, with practitioners leveraging experiential knowledge to fine-tune model parameters. This introduces the potential issue of subjectivity and bias, as the effectiveness of hyperparameter choices may be influenced by the proficiency and expertise of the users. Furthermore, the reproducibility dimension of the obtained results is another issue of the trial-and-error process when taking into consideration the stochastic nature of many algorithms. Moreover, a lack of generalizability could undermine the applicability of the insights gained to a broad-scale problem through this approach, and experiential knowledge gained in one context may not easily transfer to different data sets or models.

9.2.1.2 Grid Search

Grid search is a systematic model-free optimization technique that explores predefined hyperparameter combinations across a grid. Initially, beginning with a comprehensive search space and step size exploration, the algorithm fine-tunes these parameters based on the success of prior hyperparameter configurations. Within the defined search space, this method rigorously tests all combinations of hyperparameters, utilizing an exhaustive exploration strategy that evaluates the Cartesian product of a finite set of user-specified values. Thus, grid search can offer better reproducibility because the search space is predefined and the process is automated compared to the manual testing approach. The algorithmic steps of the grid search method are shown in Figure 9.1. On the other hand, regardless of their impact on the training phase of machine learning, all hyperparameters are considered to have equal weights (Alibrahim and Ludwig, 2021). Consequently, a manual procedure is necessary to pinpoint global optimums. In reality, certain hyperparameters may have a more significant influence, and giving them equal weight might result in suboptimal performance. However, its major drawback lies in computational inefficiency, especially when dealing with a high-dimensional parameter space, because it requires the model to be evaluated at all grid points (Belete and Huchaiah, 2022).

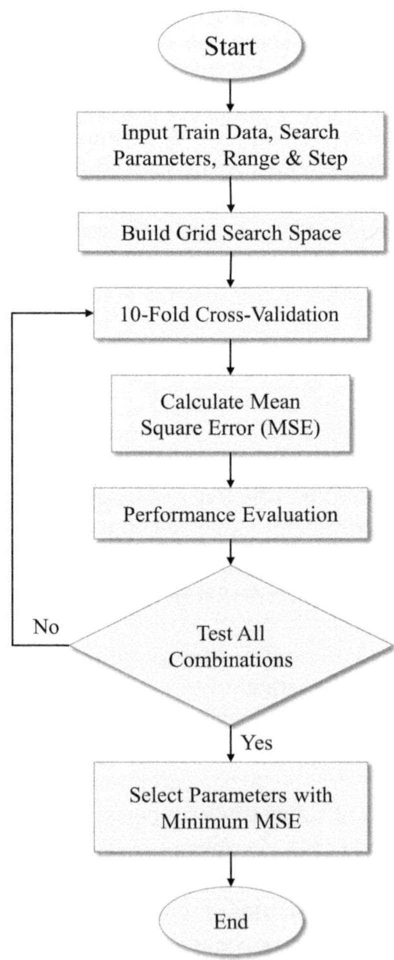

FIGURE 9.1 Flowchart for algorithmic steps of grid search.

In a d-dimensional hyperparameter configuration pool with n_i grid points for the ith hyperparameter, the total number of combinations to evaluate is given by the product of the grid sizes

$$N = \prod_{i=1}^{d} n_i \tag{9.2}$$

The grid search algorithm evaluates the objective function f_θ at each grid point θ_j, where $j = 1, 2, 3, \ldots, N$. In this way, the optimization problem is to find

$$\theta^* = \arg\min_\theta f(\theta) \tag{9.3}$$

The computational complexity of the grid search is $O(N)$, where N is the total number of evaluations. This makes it computationally expensive, especially in high-dimensional spaces. As such, grid search is hindered by the curse of dimensionality since the exponential growth in joint values with hyperparameter increase makes it unsuitable for optimizing numerous hyperparameters (Andonie, 2019).

9.2.1.3 Random Search

Random search is a model-free and decision-theoretic method developed to overcome the constraints linked to the grid search strategy. In contrast to grid search, random search randomly selects instances within specified minimum and maximum limits as potential hyperparameter values from a predefined configuration space (Figure 9.2). Subsequently, these selected values are trained until a predetermined threshold is reached. Empirical and theoretical evidence suggests that randomly chosen trials are more efficient for HPO than trials conducted on a grid. The fundamental principle guiding random search is that by exploring a sufficiently extensive search space, global optima, or at least an approximate range, can be identified (Yang and Shami, 2020). This method enables time efficiency and ease of parallelization, primarily because each trial is drawn and evaluated independently from the others. Nevertheless, the primary drawback of grid search lies in its inefficiency when dealing with hyperparameter configuration spaces of high dimensionality. This inefficiency stems from the exponential increase in the number of evaluations as the number of hyperparameters increases. Using an equivalent computational budget, random search typically outperforms grid search or more intricate HPO techniques (Bergstra et al., 2011). The computational complexity of random search depends on the number of iterations and the time required to evaluate the objective

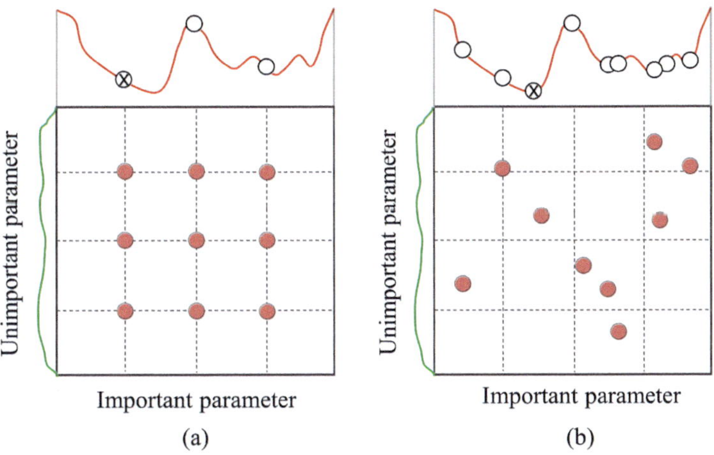

FIGURE 9.2 Comparison of (a) grid search and (b) random search for nine trials to minimize a function containing one important and one unimportant parameter.

function at each iteration. In particular, in higher-dimensional spaces, random search methods require significantly fewer computational resources than grid search (Lemley et al., 2016). After extensive experiments, Bergstra and Bengio (2012) also concluded that a comparison with neural networks configured by a strict grid search revealed that random search within the same domain could uncover models that were just as good or superior, all within a fraction of the computational time. A notable drawback of the algorithm is the absence of consideration for prior knowledge regarding a well-performing search space, in contrast to Bayesian optimization. Therefore, some researchers suggested extended versions of the random search to compensate for these limitations. For example, Florea and Andonie (2019) developed a weighted random search (WRS) algorithm to address the deficiency of prior knowledge consideration in a regular random search. WRS generates new hyperparameter values by assigning a probability of change (p), incorporating the best values found so far with a probability of $1 - p$ based on the relative importance of the hyperparameters in optimizing the objective function, thereby enhancing the exploration of promising parameter combinations. Similarly, Bohan (2020) proposed an improved random search method that incorporates hyperparameter space separation, resulting in performance comparable to that of traditional random search with fewer samplings. By dividing the hyperparameter space and sampling a point for each subarea, the algorithm enhances the probability of discovering optimal configurations, making it a more efficient optimization approach.

9.2.2 Gradient-Based Optimization

Originally conceived as a mathematical optimization technique, gradient descent iteratively updates its solution using the first derivative of the function until it converges to an optimal solution. The gradient of a function at a given point provides insight into the direction of the most significant increase in the function, intending to decrease the objective function estimation. Similar to this idea, the hyperparameter gradients can be computed, and then gradient descent can be utilized to optimize these hyperparameters. With this working principle, the algorithm tends to converge more rapidly toward the point in the parameter space where the objective function achieves a minimum value within a specific neighborhood. Despite their efficiency, gradient-based optimization methods are applicable only to the optimization of continuous hyperparameters (Yang and Shami, 2020). In addition, the efficiency of gradient-based optimization methods is more assured in the case of convex functions, as they may face challenges in consistently finding the global optimum for non-convex functions (Zöller and Huber, 2021). Therefore, there is no guarantee that machine learning algorithms will consistently identify global optima. Instead, they might converge on local optima.

In gradient descent, the update rule for the hyperparameters at each iteration can be computed from

$$\theta_{k+1} = \theta_k - \alpha \nabla f(\theta_k), \tag{9.4}$$

where $\nabla f(\theta_k)$ is the gradient of the objective function with respect to the hyperparameters at iteration k, and α is the learning rate. To use gradient-based optimization, the gradients of the objective function with respect to the hyperparameters are needed:

$$\nabla f(\theta_k) = \frac{\partial f}{\partial \theta_1}, \frac{\partial f}{\partial \theta_2}, \dots, \frac{\partial f}{\partial \theta_d} \tag{9.5}$$

The computational complexity of gradient-based optimization depends on the number of iterations, the time required to compute gradients at each iteration, and the choice of hyperparameters.

More details on the mathematical foundations of the gradient descent approach can be found in Section 5.3.4, where its use in the search for optimal parameters of SVMs is explained in detail. In addition, readers should refer to Section 7.1.1 for the mathematical theory of stochastic gradient descent, which is widely used in the training of deep learning models.

Input: priors, $P(f)$, $P(\theta)$, criterion $C(.)$, target $f(.)$
where θ represents the hyper-parameters (the neurons and the learning rate)
f represents the objective function $= [f(x_1)f(x_2)\dots f(x_n)]$
Output: X^*
Consider the training data T includes both the features (x) and the response (y)

    1. for $i = 1, 2, \dots \boldsymbol{do}$
    2. fit probabilistic model for $f(x)$ on data T_{t-1}
    3. select new $\boldsymbol{x}_{t+1}$ by maximizing the acquisition function (expected improvement)
    4. $x_{t+1} = \underset{x \in T_i}{\text{argmax}}\, u(x|P(f|T))$
    5. Evaluate the noisy and expensive function $\boldsymbol{y}_{t+1}$
    6. $\boldsymbol{y}_{t+1} = f(x_{t+1}) + N(0, \sigma^2)$
    7. Data augmentation with $\boldsymbol{X}_{t+1}$ and $\boldsymbol{y}_{t+1}$:$\boldsymbol{T}_{t+1} = \{T_t, (X_{t+1}, y_{t+1})\}$,
    8. Perform Bayesian update to obtain $P(f)$ and $P(\theta)$
End

FIGURE 9.3   Pseudocode for Bayesian optimization technique.

9.2.3 BAYESIAN OPTIMIZATION

Bayesian HPO algorithms are generally used to optimize functions that are computationally expensive; the derivatives are difficult to evaluate and lack a closed-form expression (Tran et al., 2020; Yang and Shami, 2020). By constructing a probabilistic model, these algorithms can effectively choose the next data point to evaluate. To this end, the algorithms consider the information gained from previous evaluations (Shahriari et al., 2016). Hence, the need for iterative calculations or simulations might be decreased. The focus is to simulate an objective function (e.g., typically the performance metric of the artificially intelligent model) as a probabilistic surrogate. Later, the immediate pool of hyperparameters is iteratively chosen to assess predicated on the surrogate model taken into consideration. The pseudocode for the Bayesian optimization is given in Figure 9.3 (Snoek et al., 2012). Similar to a typical hyperparameter tuning problem, let $f(\theta)$ be an objective function to be optimized in which x corresponds to the hyperparameters of the model under investigation. Subject to the nature of the problem, the main aim could be to minimize or maximize the $f(\theta)$. The underlying theory includes merging of the prior distribution (f) of the function $f(\theta)$ with the obtained sample information to derive the post-distribution of the function $f(\theta)$. The posterior distribution is given by Bayes' theorem:

$$P(f \mid D) = \frac{P(D \mid f) \cdot P(f)}{P(D)} \tag{9.6}$$

where $P(f \mid D)$ is the posterior distribution, $P(D \mid f)$ is the likelihood of the observed data given the function values, and $P(D)$ is the marginal likelihood. Subsequently, this posterior information is employed to identify the points where the function attains maximum or minimum values based on a specified criterion (Wu et al., 2019). The determination of the subsequent assessment point is made by maximizing the acquisition function:

$$x_{\text{next}} = \arg\max_x \alpha(x) \tag{9.7}$$

The surrogate models taken into consideration serve as a proxy for the true objective function, thereby guiding the search path for optimal configuration. Rather than directly optimizing the expensive and time-consuming true objective function, using a surrogate model can allow for more efficient exploration of the hyperparameter search space. For instance, when the Gaussian process (GP) is considered as the surrogate model, the GP provides a distribution over functions and allows for the incorporation of prior knowledge and uncertainty. The prior distribution over functions is typically assumed to be a Gaussian process:

$$f \sim \text{GP}\big(m(x), k(x, x')\big), \tag{9.8}$$

where $m(x)$ is the mean function, and $k(x, x')$ is the covariance (kernel) function. The posterior distribution $P(f \mid D)$ is updated based on observed data:

$$P(f \mid D) \sim \text{GP}\big(m_{\text{post}}, k_{\text{post}}\big), \tag{9.9}$$

where m_{post} is the updated mean function, and k_{post} is the updated covariance function.

The conventional strategy in Bayesian optimization relies on Gaussian processes for modeling the target function, leveraging their expressive features, smooth uncertainty estimates, and the computational ease of closed-form predictions. However, it is important to recognize some limitations. One of these limitations is their cubic scaling concerning the number of data points, which limits their suitability for scenarios with an abundance of function evaluations. This cubic scaling limitation can be overcome by scalable Gaussian process approximations, such as sparse Gaussian processes (Feurer and Hutter, 2019). This approach offers an approximation of the full Gaussian process by using a subset of the original data set as inducing points for constructing the kernel k. Another challenge associated with Gaussian processes using standard kernels is their limited scalability in high-dimensional settings. As a solution, several extensions have been proposed to efficiently handle the inherent properties of configuration spaces featuring many hyperparameters. These solutions include approaches such as the integration of random embeddings (Wang et al., 2016), cylindrical kernels (Oh et al., 2018), and additive kernels (Gardner et al., 2017). Owing to the superior scalability and flexibility of certain machine learning models compared with Gaussian processes, substantial research has been dedicated to tailoring these models for Bayesian optimization. For example, Snoek et al. (2015) introduced the Deep Networks for Global Optimization (DNGO) approach. This novel method uses neural networks to model distributions over functions, taking advantage of the remarkable flexibility and scalability of neural networks as opposed to Gaussian processes. In this framework, neural networks serve as feature extractors for preprocessing inputs, and the outputs of the final hidden layer are used as basis functions for Bayesian linear regression. Alternatively, researchers have attempted to employ random forest, known for its capability to manage extensive, categorical, conditional configuration spaces, and SVMs for HPO (Wenzel et al., 2017; Feurer and Hutter, 2019).

The techniques mentioned above generally support the application of optimization algorithms based on Gaussian processes to various real-world scenarios. However, each technique often has its own limitations, including approximations that result in a decrease in prediction performance. Concerning alternative strategies, there have been proposals for optimization algorithms not based on Gaussian processes, with the tree-structured Parzen estimator (TPE) standing out as one of the most notables (Bergstra et al., 2011). TPE is well suited for computationally intensive problems because it employs an efficient search strategy with a surrogate model and is easily parallelizable. Its surrogate, built on Parzen estimators (Parzen, 1962), can naturally handle complex search spaces, scale to tens of variables, and accommodate at least one thousand observations. These attributes make TPE more effective than Gaussian process–based optimization algorithms in certain problem domains (Ozaki et al., 2022). TPE employs a kernel density estimator to model probability density functions for both effective and ineffective hyperparameters. The algorithm samples hyperparameters on the basis of these estimated probability densities and evaluates the model's performance with the selected hyperparameters. Through the calculation of the expected improvement of the objective function, TPE iteratively updates the probability densities.

In the TPE algorithm, if presented with a hyperparameter vector θ and an objective function $f(\theta)$, the main focus is to maximize the expected improvement (EI) of the objective function f at the hyperparameter vector θ (Omotehinwa et al., 2023). The definition of the expected improvement is as follows:

$$\mathrm{EI}(\theta) = \int_{-\infty}^{f(\theta_{\min})} \big(f(\theta_{\min}) - f(\theta)\big) P\big(f(\theta) \,|\, D\big) df(\theta), \tag{9.10}$$

where $\theta_{\min}$ denotes the vector of hyperparameters maximizing the observed objective function f in the preceding iterations, D is the set of previous observations of the objective function, and $P\big(f(\theta) \,|\, D\big)$ shows the probability density function for the previous observations D.

The algorithm models the probability density functions $P(\theta \,|\, x)$ and $P(x)$, where x is the value of the objective function $f(\theta)$ for a given hyperparameter vector θ. The probability density functions $P(\theta \,|\, x)$ and $P(x)$ are modeled as

$$\begin{cases} l(\theta), & \text{if } x \geq x^* \\ g(\theta), & \text{if } x < x^* \end{cases} \tag{9.11}$$

$$P(x) = \int P(x \,|\, \theta) P(\theta) d\theta, \tag{9.12}$$

where x^* is the maximum observed value of the objective function, $l(\theta)$ and $g(\theta)$ are the probability density functions of θ given the previous observations D, and $P(x \,|\, \theta)$ is the likelihood of observing the value x for the given hyperparameter vector θ.

Most real-world problems involve simultaneous optimization of multiple objectives within intricate search spaces. These objectives frequently contradict each other, require computationally expensive evaluations, and possess a black-box quality that lacks available analytical forms. To address these issues, the TPE algorithm has been extended to advanced variants, including multifidelity TPE (Falkner et al., 2018), multiobjective TPE (Ozaki et al., 2022), and meta-learning TPE (Watanabe et al., 2023).

9.2.4 MULTIFIDELITY OPTIMIZATION

Multifidelity optimization incorporates two crucial evaluation strategies: low fidelity and high fidelity. The low fidelity involves evaluating a restricted subset at a lower cost but with limited generalization ability, whereas the latter involves assessing a relatively large subset with higher generalization ability at a greater computational expense (Kavzoglu and Teke, 2022). In the upcoming sections, we delve into the core principles of Successive Halving and Hyperband algorithms, which are two widely used methods in multifidelity optimization.

9.2.4.1 Successive Halving

Successive Halving is a resource-aware multifidelity optimization strategy that balances the trade-off between two fundamental concepts: exploration and exploitation. The former concept means discovering and trying out new possibilities, while the latter focuses on known, favorable solutions or options in the search space. The goal is to identify the best configuration within a small initial budget of resources, such as computational time or number of evaluations. To this end, the algorithm also applies early stopping criteria for the worst half configurations by progressively eliminating suboptimal ones, subsequently doubles the budget, and iterates through the identical procedure until it attains the maximum available resources or only a few candidates remain (Lee et al., 2023). The approach employed by the Successive Halving algorithm to rapidly evaluate a

---

Input : Budget B,

       The number of arms n

Output: Singleton element of $S_{\lceil \log_2(n) \rceil}$

Define: $\ell_{i,k}$ denotes the kth loss from the ith arm

  1. $S_0 = [n]$

  2. **for** $k \leftarrow 0$ **to** $\lceil \log_2(n) \rceil - 1$ **by** 1 **do**

    3.   Pull each arm in S_k for $r_k = \left\lfloor \dfrac{B}{|S_k|\lceil \log_2(n) \rceil} \right\rfloor$ additional times and set

           $R_k = \sum_{j=0}^{k} r_j.$

    4.   Let σ_k be a bijection on S_k such that

           $\ell_{\sigma_k(1),R_k} \leq \ell_{\sigma_k(2),R_k} \leq \cdots \leq \ell_{\sigma_k(|S_k|),R_k}.$

    5. $S_{k+1} = \{i \in S_k \mid \ell_{\sigma_k(1),R_k} \leq \ell_{\sigma_k(\lfloor |S_k|/2 \rfloor),R_k}\}.$

  6. **end**

  7. **return** $S_{\lceil \log_2(n) \rceil}$

FIGURE 9.4   Pseudocode for Successive Halving optimization method.

group of potential models resembles the format of a single-elimination tournament (Soper, 2023). The algorithmic steps of the technique in pseudocode form are given in Figure 9.4. The number of configurations and rounds n_s is determined by a user-defined constant η

$$n_s = \frac{n_0}{\eta^{s-1}}, \tag{9.13}$$

where s is the round number, and n_0 is the initial number of configurations. In each round, configurations receive exponentially increasing resources, and the total budget is distributed across rounds to strike a balance between thorough exploration and focused exploitation. The total resources allocated to a configuration in the round can be defined as follows:

$$s : R_s = \eta^{s-1} \tag{9.14}$$

After each round, the algorithm eliminates the worst-performing half of configurations based on their performance on the allocated resources. The algorithm systematically adapts to the available computational budget, making it a powerful and resource-efficient approach for HPO.

9.2.4.2   Hyperband

Within artificial intelligence–oriented remote sensing applications, Bayesian optimization algorithms have recently become the preferred method for selecting hyperparameter configurations that yield optimal performance for a given machine learning model. Despite excelling at identifying superior configurations more efficiently than standard baselines, these algorithms face the formidable challenge of concurrently fitting and optimizing a high-dimensional, nonconvex function characterized by uncertain smoothness and potentially noisy evaluations (Li et al., 2018). Successive Halving HPO is effective when there are a large number of hyperparameters, limited computational resources, and a need for efficient resource allocation. The pseudocode for the algorithm is given in Figure 9.5. However, there is no guarantee of converging to the optimal configuration, even with infinite resources. This is because it tends to eliminate poorly performing configurations at lower budgets, possibly excluding the best configuration achievable at higher budgets (Awad et al., 2021). To solve these problems, Hyperband, developed by Li et al. (2018), is introduced as a variant of Successive Halving, which efficiently allocates resources through random sampling of configurations. By formulating HPO as a pure-exploration adaptive resource allocation problem, the approach embraces a bracketed structure to systematically discard underperforming candidates,

---

Input: R, η default $\eta = 3$

Initalization: $s_{max} = \lfloor \log_n(R) \rfloor$, $B = (s_{max} + 1)R$

1. **for** $s \in \{s_{max}, s_{max} - 1, ..., 0\}$ **do**

2. $\qquad n = \left\lceil \dfrac{B}{R} \dfrac{\eta^s}{(s+1)} \right\rceil, \qquad r = R\eta^{-s}$

$\qquad$ // begin SuccessiveHalving with (n, r) inner loop

3. $\qquad T = get\_hyperparameter\_configuration(n)$

4. $\qquad$ **for** $i \in \{0, ..., s\}$ **do**

5. $\qquad\qquad n_i = \lfloor n\eta^{-i} \rfloor$

6. $\qquad\qquad r_i = r\eta^i$

7. $\qquad\qquad L = \{run\_then\_return\_val\_loss(t, r_i) : t \in T\}$

8. $\qquad\qquad T = top\_k(T, L, \lfloor n_i/\eta \rfloor)$

9. $\qquad$ **end**

10. **end**

11. **return** *Configuration with the smallest intermediate loss seen so far.*

FIGURE 9.5 Pseudocode for Hyperband optimization method.

thereby enhancing the efficiency of the optimization process. It addresses the challenge of balancing the number of configurations (n) tried against the resources allocated to each configuration (B/n). Unlike Bayesian optimization algorithms, it makes minimal assumptions compared to previous configuration evaluation approaches. The algorithm performs a grid search over feasible values of n, with each value of n associated with a minimum resource (r) allocated to all configurations before some are discarded. The algorithm consists of two components: an inner loop invoking Successive Halving for fixed values of n and r, and an outer loop iterating over different values of n and r. Hyperband approximately uses B total resources for each bracket, representing different trade-offs between n and (B/n). The algorithm requires two inputs: the maximum resource allocated to a single configuration (R) and a parameter (η) controlling the proportion of configurations discarded in each round of Successive Halving. Hyperband starts with the most aggressive bracket, setting n to maximize exploration while ensuring at least one configuration is allocated R resources. Subsequent brackets reduce n by a factor of approximately η until the final bracket, where every configuration is allocated R resources (classical random search). This geometric search in the average budget per configuration eliminates the need to select n for a fixed budget, but it requires approximately s max $+ 1$ times more work than running Successive Halving for a single value of n. Many studies have reported the rapid process capabilities and effectiveness of the Hyperband optimization technique in terms of classification accuracy (e.g., Kavzoglu and Teke, 2022; Azedou et al., 2023). In contrast, Falkner et al. (2017) emphasized that its dependence on randomly drawn configurations may hinder the rapid identification of optimal configurations. To overcome this shortcoming, they proposed a combination of Hyperband with Bayesian optimization by maintaining a probabilistic model that captures the density of good configurations in the input space instead of using a random search strategy.

9.2.5 METAHEURISTIC ALGORITHMS

9.2.5.1 Genetic Algorithm

The genetic algorithm, a population-based search technique discussed in the context of feature selection (see Section 3.2.5.1), mimics the hypothetical mechanism of natural selection, using reproduction, crossover, and mutation operators to evolve a population of candidate solutions (Kavzoglu and Mather, 2002). The concept of this robust and popular algorithm was first presented by Holland (1992). During the reproduction process, genetic information is copied from the current generation's individuals (chromosomes) to create the next generation. The genetic algorithm evaluates the performance of each set of hyperparameters (individual) based on a specified fitness (or objective)

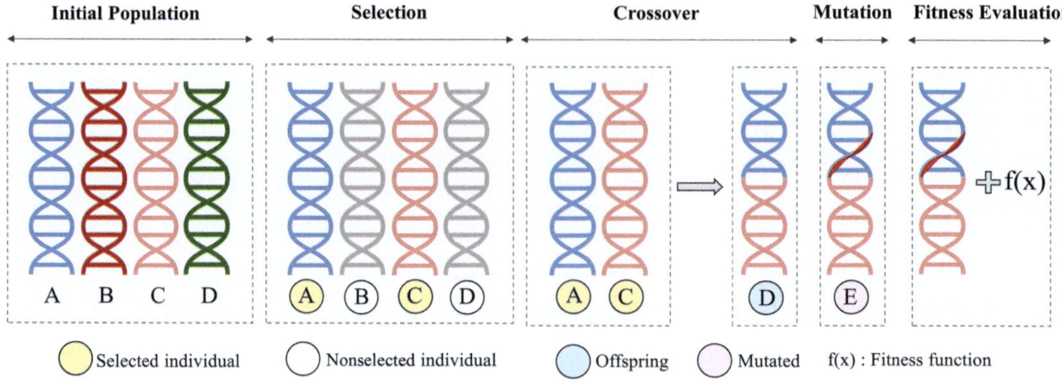

FIGURE 9.6 Working principle of genetic algorithm.

function, often derived from the model's performance on a validation data set (Figure 9.6). The fitness function is a nonlinear function used to estimate the quality (fitness level) of each individual in the population, usually considering the estimated MSE. Crossover, sometimes called recombination, introduces diversity by combining genetic information from two parent hyperparameter sets to generate new configurations. In the context of HPO, this operation explores the solution space by blending characteristics of successful hyperparameter combinations. By mimicking the concept of genetic recombination, crossover helps the algorithm discover novel combinations that might lead to improved model performance. Mutation introduces random changes to individual hyperparameters within a set. While reproduction and crossover focus on preserving and combining successful traits, mutation introduces an element of exploration by allowing for random variations. This randomness helps prevent the algorithm from converging prematurely and enables the discovery of potentially beneficial hyperparameter configurations that might not be immediately apparent through the other operators (Kavzoglu and Teke, 2022). It contributes to the overall adaptability of the algorithm and facilitates a more thorough exploration of the hyperparameters. A flowchart of the entire process of the genetic algorithm is shown in Figure 3.11 (Chapter 3). In the genetic algorithm concept, the problem can be expressed as

$$\min_{y} p(y)$$

$$\text{Subject to } g(y) = 0$$

$$h(y) \leq 0 \tag{9.15}$$

$$y^l \leq y_i \leq y^u : i = 1,\ldots,n$$

$$y = \left[y_1, y_2, \ldots, y_n \right]^T$$

where y is the set of n integer variables with lower and upper limits of y^l and y^u, $p(y)$ denotes the fitting function, and $g(y)$ and $h(y)$ indicate the equality and inequality constraints, respectively.

The genetic algorithm essentially draws inspiration from evolutionary theory positing that individuals, represented as chromosomes, bearing attributes conducive to adept environmental adaptation in a given generation are more inclined to perpetuate their traits to subsequent generations. This search methodology finds widespread application in the pursuit of optimal solutions, or, at the very least, their approximations. The genetic algorithm procedure begins with the generation of a randomized initial population comprising multiple solutions. The primary objective at this stage lies in finding solutions throughout the search space, fostering population diversity, and maximizing the prospect of engendering individuals well suited to the task at hand (Mirjalili, 2019). Subsequently, a fitness or objective function is employed to evaluate the effectiveness of individuals within the population for a specific problem. Given that an inaccurate definition of the fitness function could lead to the search process and impact the final outcome, it is imperative to accurately specify this function (Kavzoglu et al., 2015). To maintain a consistent population size, individuals with the highest fitness scores are more likely to be passed on to the next generation. During this phase, parents produce offspring carrying their most advantageous traits through a crossover operation, which involves the recombination and exchange of a certain percentage of genes across different chromosomes.

Genetic algorithms offer easy implementation without the need for precise initializations because their selection, crossover, and mutation operations mitigate the risk of overlooking the global optimum. This proves beneficial when a data analyst lacks substantial experience in defining a suitable initial search space for hyperparameters. However, a limitation lies in the algorithm introducing additional hyperparameters for configuration, encompassing the fitness function type, population size, crossover rate, and mutation rate. Additionally, the technique operates in a sequential manner, posing challenges for parallelization. Lobo et al. (2000) underlined that genetic algorithm with perfect mixing had time complexities of $O(m)$ and $O(m^2)$, respectively, for the two extreme cases of building block scaling, uniform and exponential. As a result, in some cases, GA may be inefficient because of its low convergence speed.

9.2.5.2   Particle Swarm Optimization

Being another evolutionary population-based search algorithm, particle swarm optimization (PSO) (Eberhart and Kennedy, 1995) commences with a set of randomly generated solutions (i.e., particles). The theory of the PSO algorithm and its application to feature selection are also discussed in Section 3.2.5.2. The algorithm mimics the collective behavior seen in biological organisms or specifically animal swarms (e.g., insects, herds, birds, and fishes), fostering cooperation for efficient food finding. Each swarm member adapts its search pattern based on its own learning experiences and those of fellow members. It draws inspiration from two key research areas: evolutionary algorithms and artificial life (Wang et al., 2018a). The former employs a swarm approach for simultaneous exploration of a vast solution space, whereas the latter focuses on the study of artificial systems exhibiting life-like characteristics. Initially designed to simulate birds seeking food in a "cornfield vector", the algorithm involved social cooperation within a bird's neighborhood, and it was later expanded to a multidimensional search, using topological neighborhoods instead of Euclidean ones (Shi, 2004). The initial PSO algorithm is delineated using the following expressions:

$$v_{t+1} = v_i + c_1\beta_1 (p_i - x_i) + c_2\beta_2 (p_g - x_i) \tag{9.16}$$

$$x_{t+1} = x_i + v_{t+1} \tag{9.17}$$

In these equations, c_1 and c_2 are positive constants that adjust the relative impact of individual and group knowledge on the overall behavior of the system. β_1 and β_2 are the random functions producing values in the range [0, 1], p_i is the particle's own best position, and p_g is the global best position on the particle's velocity update. The x_i is the current position of ith particle in the dimension considered. The sign in the brackets drives particle acceleration, striking a balance between local and global optimization analogous to the exploration-exploitation trade-off in genetic algorithms, where control involves intricate parameters such as population size and mutation rate (Banks et al., 2007). PSOs commonly fall into either the global or local variants. In the global version, particles adjust velocity based on personal and global best performance, potentially converging quickly but with a risk of local minima. In the local version, particles consider personal and neighborhood best, providing slower but more diverse solutions (Shi, 2004). The algorithm is a widely used optimization technique, especially for multidimensional search spaces, Banks et al. (2007) reported that it faces challenges in swarm stagnation and dynamic environments. The acknowledged weakness of stagnation requires strategies to keep particles moving, including velocity restriction and hybridization with other computing paradigms.

9.3 CHALLENGES IN HYPERPARAMETER OPTIMIZATION

While the utilization of the HPO algorithms unquestionably enhances the performance of machine learning models, it is imperative to acknowledge that certain facets, notably their computational complexity, remain ripe for refinement. Advances in model performance through HPO must be tempered by recognizing the ongoing challenges related to the computational demands of these optimization processes. Furthermore, the diverse landscape of HPO models introduces a layer of complexity and underlying assumptions because each model possesses unique advantages tailored to specific problems. A comprehensive overview of these HPO models is essential for judiciously selecting optimization algorithms that align with the diverse array of machine learning models and problem domains. Such an overview ensures a nuanced approach that considers the distinctive characteristics and requirements of different machine learning models, thereby facilitating the optimal selection of HPO methodologies for specific use cases. In essence, while HPO algorithms offer notable improvements in model performance, ongoing efforts to address computational complexities and a nuanced understanding of the diverse HPO landscape are essential for refining and tailoring these optimization techniques to the ever-evolving challenges within the realm of machine learning.

Major challenges that should be considered by the analysts to select the most suitable algorithm for their problem can be given as follows:

- **High-Dimensional Search Space**: The hyperparameter space is frequently both high-dimensional and complex, including a mixture of continuous, categorical, and conditional hyperparameters. This complexity increases the computational expense and the challenges inherent in the optimization problem. In addition, determining which hyperparameters of an algorithm should undergo optimization and specifying suitable ranges for optimization is not always straightforward. The limited availability of training data stands out as a major challenge to accomplishing effective HPO, particularly in terms of generalization estimations. When working with a relatively small validation set, the uncertainty in choosing the best model through cross-validation may be more significant than the uncertainty associated with evaluating the performance of any individual model on the test set. Within the decision-theoretic paradigm, methods are centered around defining a hyperparameter search space and then identifying the best-performing combination within that space. Grid search, as an integral part of this approach, exhaustively explores the optimal configuration within a predefined hyperparameter domain. Bayesian optimization models, unlike grid search, leverage past results of tested values to determine the next hyperparameter value, minimizing the need for unnecessary evaluations. This efficiency allows Bayesian optimization to identify the optimal hyperparameter combination with fewer iterations.
- **Nonconvexity**: Conventional optimization approaches may not be suitable for HPO problems because of the nonconvex or nondifferentiable nature of many problems, leading to the emergence of local rather than global optima (Luo, 2016). In cases where the objective function $f(\theta)$ exhibits nonconvexity, encountering multiple local minima/maxima is inevitable. Traditional optimization methods, such as gradient descent, may become trapped in suboptimal solutions (local minima) instead of reaching the global optimum. Hence, robust HPO techniques, including multifidelity optimization methods and metaheuristic algorithms, are preferable. These algorithms not only address continuous hyperparameters but also demonstrate proficiency in effectively handling discrete, categorical, and conditional hyperparameters (Yang and Shami, 2020).

- **Resource Constraints**: HPO frequently demands substantial computational resources and time, presenting challenges for real-time applications and extensive data sets, especially when considering complex machine learning models. Consequently, HPO techniques commonly employ data sampling to obtain approximate values of the objective function. Therefore, effective optimization methods for HPO problems should be adept at using these approximations. Many black-box optimization models tend to neglect the function evaluation time, making them unsuitable for optimizing problems with strict time and budget constraints. Multifidelity optimization algorithms, particularly bandit-based ones, have been developed to address challenges arising from limited resources. Hyperband, a widely recognized bandit-based optimization technique, is considered to be an improved iteration of random search. It creates smaller data sets and allocates an equal budget to each hyperparameter combination. During each iteration, Hyperband eliminates poorly performing hyperparameter configurations to conserve time and resources.
- **Noisy Objective Functions**: The presence of noise in the objective function, arising from randomness or limited data, introduces challenges in differentiating real trends from random fluctuations. It is essential to recognize that grid search, random search, and Hyperband face a significant constraint – they treat each hyperparameter independently and neglect hyperparameter correlations (Wang et al., 2018b). Consequently, these methods exhibit inefficiency when applied to machine learning algorithms with conditional hyperparameters, such as SVM, DBSCAN, and logistic regression (Yang and Shami, 2020). HPO algorithms demonstrate the capacity to navigate through local minima and display robustness when confronted with noisy evaluations of the objective function. These attributes share similarities with extremum-seeking algorithms. Bayesian optimization, with its solid mathematical foundation, performs well in handling noisy objective functions compared with other HPO methods (Snoek et al., 2012).

9.4 CONCLUDING REMARKS

Every machine learning model comes with its hyperparameters, and the primary aim of machine learning practices is the automatic determination of the best configuration for optimal performance. The success of an algorithm depends on the rapid and precise identification of key dimensions, underscoring the importance of a systematic HPO in machine learning applications. The effectiveness of a machine learning method, whether it is state-of-the-art or exhibits weaker performance, is closely tied to the discovery of the optimal hyperparameter configuration. Using default hyperparameter values from a software package is a common practice that often results in suboptimal solutions. High performance is not guaranteed across all problem types with such a setting strategy. HPO offers notable advantages, including a reduction in human effort for applying machine learning methods, improved prediction performance, and increased process reliability due to enhanced reproducibility. Techniques such as Bayesian optimization are implemented to autonomously explore optimal hyperparameter settings, framing the task as a black-box optimization problem. Despite the capabilities of contemporary machine learning methods, particularly deep neural networks, their extensive training times and substantial computational demands pose limitations. The impracticality becomes evident when individual function evaluations take days or weeks, making the application of black-box optimizers computationally unfeasible. Therefore, selecting the HPO method is a critical issue in machine learning applications.

Numerous algorithms for HPO have been devised to tackle a range of machine learning objectives. The theoretical frameworks and algorithmic steps of the key HPO techniques are detailed in the preceding sections of this chapter. Implementing optimization with grid search is straightforward and requires minimal code infrastructure. Similarly, random search is just as easy to execute, utilizing the same tools and seamlessly integrating into the existing workflow. In contrast, adaptive search algorithms have a higher level of code complexity. The use of sequential model-based optimization methods, such as Bayesian optimization, for hyperparameter tuning in machine learning algorithms is hindered by significant computational bottlenecks, limiting their widespread adoption. Nevertheless, modern multifidelity methods, such as Successive Halving and Hyperband methods, deploy less complex objective functions to provide approximate evaluations of hyperparameter settings. Recent studies have illustrated their state-of-the-art performance, which results in faster processing. Given the diverse advantages and drawbacks of different HPO methods arising from their algorithmic complexity, a thorough understanding of these characteristics is crucial. In addition to these, there are many alternative algorithms in the literature, including Taguchi optimization (Tonbul and Kavzoglu, 2020) and whale optimization (Ali et al., 2023).

Deep neural networks extensively rely on a diverse set of hyperparameter choices, covering aspects such as architecture, regularization, and optimization. The formulation of neural architecture is predominantly guided by the researcher's existing knowledge and experience. Inexperienced users may find it challenging to adjust the neural architecture to suit their particular needs. Furthermore, existing knowledge and fixed thinking paradigms may act as constraints, potentially limiting the exploration of new neural architectures. To address this issue, neural architecture search (NAS) methods have been developed to autonomously design a neural architecture that achieves the best possible performance within computing constraints, minimizing human involvement (Ren et al., 2021; Baymurzina et al., 2022). Unlike the classical HPO, NAS is associated with higher computational expenses, necessitating additional efforts to maintain practical feasibility. Notably, the use of reinforcement learning strategy in the design of neural network architectures has resulted in state-of-the-art classification performances, as evidenced by the work of Zoph and Le (2016).

REFERENCES

Ali, Y. A., E. M. Awwad, M. Al-Razgan, and A. Maarouf. 2023. Hyperparameter search for machine learning algorithms for optimizing the computational complexity. *Processes* 11:349. https://doi.org/10.3390/pr11020349.

Alibrahim, H., and S. A. Ludwig. 2021. Hyperparameter optimization: Comparing genetic algorithm against grid search and Bayesian optimization. In *Proceedings of IEEE Congress on Evolutionary Computation*, 28 June–01 July, Kraków, Poland, pp. 1551–1559. https://doi.org/10.1109/CEC45853.2021.9504761.

Andonie, R. 2019. Hyperparameter optimization in learning systems. *Journal of Membrane Computing* 1:279–291. https://doi.org/10.1007/s41965-019-00023-0.

Awad, N. H., N. Mallik, and F. Hutter. 2021. DEHB: Evolutionary hyperband for scalable, robust and efficient hyperparameter optimization. arXiv:2105.09821.

Azedou, A., A. Amine, I. Kisekka, S. Lahssini, Y. Bouziani, and S. Moukrim. 2023. Enhancing land cover/land use (LCLU) classification through a comparative analysis of hyperparameters optimization approaches for deep neural network (DNN). *Ecological Informatics* 78:102333. https://doi.org/10.1016/j.ecoinf.2023.102333.

Banks, A., J. Vincent, and C. Anyakoha. 2007. A review of particle swarm optimization. Part I: Background and development. *Natural Computing* 6:467–484. https://doi.org/10.1007/s11047-007-9049-5.

Baymurzina, D., E. Golikov, and M. Burtsev. 2022. A review of neural architecture search. *Neurocomputing* 474:82–93.

Belete, D. M., and M. D. Huchaiah. 2022. Grid search in hyperparameter optimization of machine learning models for prediction of HIV/AIDS test results. *International Journal of Computers and Applications* 44:875–886. https://doi.org/10.1080/1206212X.2021.1974663.

Bergstra, J., and Y. Bengio, Y. 2012. Random search for hyper-parameter optimization. *Journal of Machine Learning Research* 13:281–305.

Bergstra, J., R. Bardenet, Y. Bengio, and B. Kégl. 2011. Algorithms for hyper-parameter optimization. In *Proceedings of Advances in Neural Information Processing Systems (NIPS'2011)*, December 12–17, Granada, Spain.

Bischl, B., M. Binder, M. Lang, T. Pielok, J. Richter, S. Coors, J. Thomas, et al. 2023. Hyperparameter optimization: Foundations, algorithms, best practices, and open challenges. *WIREs Data Mining and Knowledge Discovery* 13:e1484. https://doi.org/https://doi.org/10.1002/widm.1484.

Bohan, L. 2020. Random search plus: A more effective random search for machine learning hyperparameters optimization. Master's Thesis, University of Tennessee.

Eberhart, R., and J. Kennedy. 1995. A new optimizer using particle swarm theory. In *Proceedings of the Sixth International Symposium on Micro Machine and Human Science*, October 4–6, Nagoya, Japan, pp. 39–43.

Falkner, S., A. Klein, and F. Hutter. 2017. Combining hyperband and Bayesian optimization. In *Proceedings of 31st Conference on Neural Information Processing Systems (NIPS 2017)*, Long Beach, CA.

Falkner, S., A. Klein, and F. Hutter. 2018. BOHB: Robust and efficient hyperparameter optimization at scale. In *Proceedings of International Conference on Machine Learning*, July 10–15, Stockholm, Sweden.

Feurer, M., and F. Hutter. 2019. Hyperparameter optimization. In *Automated Machine Learning: Methods, Systems, Challenges*, (eds.) F. Hutter, L. Kotthoff, and J. Vanschoren, pp. 3–33. Cham: Springer International Publishing. https://doi.org/10.1007/978-3-030-05318-5_1.

Florea, A. -C., and R. Andonie. 2019. Weighted random search for hyperparameter optimization. *International Journal of Computers, Communication and Control* 14:154–169. https://doi.org/10.15837/ijccc.2019.2.3514.

Gardner, J., C. Guo, K. Weinberger, R. Garnett, and R. Grosse. 2017. Discovering and exploiting additive structure for Bayesian optimization. *Proceedings of the Seventeenth International Conference on Artificial Intelligence and Statistics* 54:1311–1319.

Holland, J. H. 1992. *Adaptation in Natural and Artificial Systems: An Introductory Analysis with Applications to Biology, Control, and Artificial Intelligence*. Cambridge, MA: MIT Press.

Huang, C., B. Chen, C. Sun, Y. Wang, J. Zhang, H. Yang, S. Wu, et al. 2024. Synergistic application of multiple machine learning algorithms and hyperparameter optimization strategies for net ecosystem productivity prediction in Southeast Asia. *Remote Sensing* 16:17. https://doi.org/10.3390/rs16010017.

Hutter, F., J. Lücke, and L. Schmidt-Thieme. 2015. Beyond manual tuning of hyperparameters. *KI - Künstliche Intelligenz* 29:329–337. https://doi.org/10.1007/s13218-015-0381-0.

Igonin, D. M., P. A. Kolganov, and Y. V. Tiumentsev. 2020. Choosing hyperparameter values of the convolution neural network when solving the problem of semantic segmentation of images obtained by remote sensing of the Earth's surface. *Optical Memory and Neural Networks* 29:317–329. https://doi.org/10.3103/S1060992X20040086.

Kavzoglu, T., and A. Teke. 2022. Advanced hyperparameter optimization for improved spatial prediction of shallow landslides using extreme gradient boosting (XGBoost). *Bulletin of Engineering Geology and the Environment* 81:201. https://doi.org/10.1007/s10064-022-02708-w.

Kavzoglu, T., and P. M. Mather. 2002. The role of feature selection in artificial neural network applications. *International Journal of Remote Sensing* 23:2919–2937. https://doi.org/10.1080/01431160110107743.

Kavzoglu, T., E. K. Sahin, and I. Colkesen. 2015. Selecting optimal conditioning factors in shallow translational landslide susceptibility mapping using genetic algorithm. *Engineering Geology* 192:101–112. https://doi.org/10.1016/j.enggeo.2015.04.004.

Lee, H., G. Lee, J. Kim, S. Cho, D. Kim, and D. Yoo. 2023. Improving multi-fidelity optimization with a recurring learning rate for hyperparameter tuning. In *Proceedings of the IEEE/CVF Winter Conference on Applications of Computer Vision*, January 2–7, Waikoloa, HI, pp. 2309–2318.

Lemley, J., F. Jagodzinski, and R. Andonie. 2016. Big holes in big data: A Monte Carlo algorithm for detecting large hyper-rectangles in high dimensional data. In *Proceedings of IEEE 40th Annual Computer Software and Applications Conference*, June 10–14, Atlanta, GA, pp. 563–571. https://doi.org/10.1109/COMPSAC.2016.73.

Li, L., K. Jamieson, G. DeSalvo, A. Rostamizadeh, and A. Talwalkar. 2018. Hyperband: A novel bandit-based approach to hyperparameter optimization. *Journal of Machine Learning Research* 18:1–52.

Liao, M., H. Wen, L. Yang, G. Wang, X. Xiang, and X. Liang. 2024. Improving the model robustness of flood hazard mapping based on hyperparameter optimization of random forest. *Expert Systems with Applications* 241:122682. https://doi.org/10.1016/j.eswa.2023.122682.

Lin, N., Y. Chen, H. Liu, and H. Liu. 2021. A comparative study of machine learning models with hyperparameter optimization algorithm for mapping mineral prospectivity. *Minerals* 11:159. https://doi.org/10.3390/min11020159.

Lobo, F. G., D. E. Goldberg, and M. Pelikan. 2000. Time complexity of genetic algorithms on exponentially scaled problems. In *Proceedings of the 2nd Annual Conference on Genetic and Evolutionary Computation*, July 10–12, Las Vegas, NV, pp. 151–158.

Luo, G. 2016. A review of automatic selection methods for machine learning algorithms and hyper-parameter values. *Network Modeling Analysis in Health Informatics and Bioinformatics* 5:1–16. https://doi.org/10.1007/s13721-016-0125-6.

Mirjalili, S. 2019. Genetic algorithm. In *Evolutionary Algorithms and Neural Networks: Theory and Applications*, (ed.) S. Mirjalili, pp. 43–55. Cham: Springer International Publishing. https://doi.org/10.1007/978-3-319-93025-1_4.

Oh, C., E. Gavves, and M. Welling. 2018. BOCK: Bayesian optimization with cylindrical kernels. In *Proceedings of the 35th International Conference on Machine Learning*, Stockholmsmässan, Stockholm Sweden, vol. 80, pp. 3868–3877.

Omotehinwa, T. O., D. O. Oyewola, and E. G. Dada. 2023. A light gradient-boosting machine algorithm with tree-structured parzen estimator for breast cancer diagnosis. *Healthcare Analytics* 100218. https://doi.org/10.1016/j.health.2023.100218.

Ozaki, Y., Y. Tanigaki, S. Watanabe, M. Nomura, and M. Onishi. 2022. Multiobjective tree-structured Parzen estimator. *Journal of Artificial Intelligence Research* 73:1209–1250. https://doi.org/10.1613/jair.1.13188.

Parzen, E. 1962. On estimation of a probability density function and mode. *The Annals of Mathematical Statistics* 33:1065–1076.

Ren, P., Y. Xiao, X. Chang, P. Y. Huang, Z. Li, X. Chen, and X. Wang. 2021. A comprehensive survey of neural architecture search: Challenges and solutions. *ACM Computing Surveys* 54:1–34. https://doi.org/10.1145/3447582.

Shahriari, B., K. Swersky, Z. Wang, R. P. Adams, and N. de Freitas. 2016. Taking the human out of the loop: A review of Bayesian optimization. *Proceedings of the IEEE* 104:148–175. https://doi.org/10.1109/JPROC.2015.2494218.

Shi, Y. 2004. Particle swarm optimization. *IEEE Connections* 2:8–13.

Snoek, J., H. Larochelle, and R. P. Adams. 2012. Practical Bayesian optimization of machine learning algorithms. In *Proceedings of Advances in Neural Information Processing Systems 25* (NIPS'2012), December 3–6, Lake Tahoe, Nevada.

Snoek, J., O. Rippel, K. Swersky, R. Kiros, N. Satish, N. Sundaram, M. Patwary, et al. 2015. Scalable Bayesian optimization using deep neural networks. In *Proceedings of the 32nd International Conference on Machine Learning*, Lille, France, vol. 37, pp. 2171–2180.

Soper, D. S. 2023. Hyperparameter optimization using successive halving with greedy cross validation. *Algorithms* 16:17. https://doi.org/10.3390/a16010017.

Tonbul, H., and T. Kavzoglu. 2020. Semi-automatic building extraction from Worldview-2 imagery using Taguchi optimization. *Photogrammetric Engineering & Remote Sensing* 86:547–555. https://doi.org/10.14358/PERS.86.9.547.

Tran, N., J.-G. Schneider, I. Weber, and A. K. Qin. 2020. Hyper-parameter optimization in classification: To-do or not-to-do. *Pattern Recognition* 103:107245. https://doi.org/https://doi.org/10.1016/j.patcog.2020.107245.

Wang, D., Tan, D., Liu, L. 2018a. Particle swarm optimization algorithm: An overview. *Soft Computing* 22:387–408. https://doi.org/10.1007/s00500-016-2474-6.

Wang, J., J. Xu, and X. Wang. 2018b. Combination of hyperband and Bayesian optimization for hyperparameter optimization in deep learning. arXiv preprint arXiv:1801.01596.

Wang, Z., F. Hutter, M. Zoghi, D. Matheson, and N. de Feitas. 2016. Bayesian optimization in a billion dimensions via random embeddings. *Journal of Artificial Intelligence Research* 55:361–387. https://doi.org/10.1613/jair.4806.

Watanabe, S., N. Awad, M. Onishi, and F. Hutter. 2023. Speeding up multi-objective hyperparameter optimization by task similarity-based meta-learning for the tree-structured Parzen estimator. arXiv:2212.06751.

Wenzel, F., T. Galy-Fajou, M. Deutsch, and M. Kloft. 2017. Bayesian nonlinear support vector machines for big data. In *Machine Learning and Knowledge Discovery in Databases*: *Lecture Notes in Computer Science*, (eds.) M. Ceci, J. Hollmén, L. Todorovski, C. Vens, and S. Džeroski, p. 10534. Cham: Springer. https://doi.org/10.1007/978-3-319-71249-9_19.

Wu, J., X.-Y. Chen, H. Zhang, L.-D. Xiong, H. Lei, and S.-H. Deng. 2019. Hyperparameter optimization for machine learning models based on Bayesian optimization. *Journal of Electronic Science and Technology* 17:26–40. https://doi.org/https://doi.org/10.11989/JEST.1674-862X.80904120.

Xie, H., Y. Zhang, Y. He, K. You, P. Dai, B. Fan, D. Yu, et al. 2024. On-road high-emitting vehicle identification by an automatic hyperparameter optimization model based on a remote sensing system. *Measurement* 225:113938. https://doi.org/https://doi.org/10.1016/j.measurement.2023.113938.

Yang, L., and A. Shami. 2020. On hyperparameter optimization of machine learning algorithms: Theory and practice. *Neurocomputing* 415:295–316. https://doi.org/10.1016/j.neucom.2020.07.061.

Yu, T., and H. Zhu. 2020. Hyper-parameter optimization: A review of algorithms and applications. arXiv Preprint arXiv2003.05689.

Zöller, M.-A., and M. F. Huber. 2021. Benchmark and survey of automated machine learning frameworks. *Journal of Artificial Intelligence Research* 70:409–472. https://doi.org/10.1613/jair.1.11854.

Zoph, B., and Q. V. Le. 2016. Neural architecture search with reinforcement learning. arXiv preprint arXiv:1611.01578.

10 Accuracy Assessment and Model Explainability

Accuracy assessment is the final and crucially important step in the process of generating thematic maps from remotely sensed data before these end-products are used in scientific research and decision-making processes. A rigorous accuracy assessment is essential in any mapping practice to evaluate the uncertainty and accuracy associated with the produced maps. This evaluation is vital to establish the level of confidence in the results and verify whether the analysis objectives have been fulfilled. It is also critical to assess the generalization capabilities of the trained models, which can provide a direct measurement about the transferability of the models. Since the beginning of the first practices with the introduction of satellite images in the 1970s, theory, methods, and applications of accuracy assessment and associated other issues have been at the center of research in remote sensing. Discussions have mainly focused on the assessment of map-level and class-level accuracies, conducting effective sampling schemes and sampling units in the establishment of ground reference data, and lately, interest has shifted toward developing new assessment strategies for advanced classifiers (described in the earlier chapters). Although there is an abundance of high spatial, spectral, and temporal remotely sensed data, coupled with a wide variety of techniques, the issues related to accuracy assessment continue to be a major challenge in classification applications. Issues related to good and bad practices in accuracy assessment have been thoroughly discussed in recent studies (Stehman and Czaplewski, 1998; Olofsson et al., 2014; Ye et al., 2018; Morales-Barquero et al., 2019; Stehman and Foody, 2019; Maxwell et al., 2021a,b). The importance of the subject has been emphasized in numerous studies over the years, and some guidelines for best practices have been put forward. The study by Morales-Barquero et al. (2019) reported that only 32% of the 282 selected papers published between 1998 and 2017 on image classification included reproducible accuracy assessment results. The most striking result is that no accuracy measure was given in 11.2% of the reviewed studies. Similar results were also reported by Trodd (1995). The insightful and stimulating findings of these studies indicate that there remains a need for further efforts to standardize accuracy assessment protocols. Given the current state of image classification applications, it can be affirmed that the issues related to rigorous accuracy assessment will become even more significant in future applications using cutting-edge deep learning models in classification practices where different terminology and accuracy measures are employed.

Accuracy is quantified empirically, involving the selection of samples, ideally random and independent samples, from the thematic map. The selected pixels are then compared to reference data preferably obtained through a field study. Ground reference data set, often referred to as ground truth, is divided into different parts to be used in the training and validation stages of the classification. The accuracy assessment involves using testing pixels, which are selected samples, to estimate the classifier's correct identification percentage for pixels in each class within the image. It also helps determine the proportions of pixels from each class that are mistakenly categorized into other classes (misclassification). The accuracy achieved by a supervised classification is largely dependent upon several factors, including the representatives of the samples employed in both training and testing stages, the complexity of the classification problem, and the robustness of the employed algorithm. The accuracy assessment of thematic maps is closely connected to the representativeness or fidelity of data sets utilized in the classification process, with the size and quality of these data sets playing a key role in the evaluation. Along with the sample size, an appropriate sampling scheme, which represents the spatial variation of the ground reference data (i.e., sampling locations), is also a critical issue for the success and reliability of the classification results. If the training and test samples are not drawn from the same distribution, the whole process may fail, which will be noticed after the accuracy assessment. Therefore, when an expected level of accuracy is not achieved, the typical approach involves a comprehensive examination of the entire process, starting from the establishment of ground reference data to determine the major causes of confusion among the classes. Being aware of the classes that exhibit confusion with each other can guide the analyst in devising remedies, including the integration of additional features (such as images from different sources and dates, and ancillary data) and enhancing the size and quality of the samples through an improved sampling design. It should be noted that the data set considered in a classification is not the only reason for the failure of a classification; improper design and hyperparameter setting of the classifier may be also responsible for it. It is important to note that the characteristics of training data are more important in machine learning and deep learning methods as they consider all individual samples. In contrast, the statistical classifiers rely solely on descriptive statistics (e.g., mean or variance of the pixel values at each band).

In the last two decades, the classification of remotely sensed data has evolved from traditional statistical classifiers to machine learning and deep learning methods due to their adaptable and robust nature in dealing with complex data and capturing inherent patterns and nonlinear relationships among variables. This shift brings substantial benefits, such as enhanced prediction accuracy, automation, and versatility. However, most of these algorithms tend to be opaque, which restricts the comprehensive understanding of the models they create. While reproducibility and transparency constitute the twin pillars of effective accuracy assessment practices, as highlighted by Stehman and Foody (2019), the crucial factor influencing the credibility of the classification model lies in its interpretability or explainability. The lack of interpretability can lead to concerns about the reliability and trustworthiness of the generated thematic maps. When there is no clear understanding of how the model reaches its conclusions, it becomes difficult to validate its accuracy, assess potential biases or limitations, and substantiate the reliability of its outputs. This can undermine confidence in image classification endeavors and obstruct effective decision-making for planning and scientific studies. White-box predictive models, such as rule-based models, logistic regression, and decision trees, offer explicit insights into the decision-making process by revealing their weights or decision nodes. However, black-box or opaque models, which often require the optimization of a large number of parameters using training data, do not offer such transparency. Their internal mechanisms are often kept hidden and lack transparency, which obstructs the understanding and interpretation of the decision-making process and creates challenges in comprehending the reasoning behind their conclusions (Kavzoglu, 2009). Moreover, it can be challenging to articulate how attributes interact with each other to arrive at a final prediction, which usually requires a post hoc analysis (Arrieta et al., 2020). Since the applications of machine learning and deep learning models have become popular in many critical areas of research, the issues of transparency and explainability are increasingly being recognized as critically important (Angelov et al., 2021). To alleviate the significant issues imposed by black-box models, explainable artificial intelligence (XAI) has lately received much scientific interest from diverse disciplines, including geosciences (Mamalakis et al., 2022; Dahal and Lombardo, 2023), agriculture (Ryo, 2022), and susceptibility mapping of natural hazards (Kavzoglu and Teke, 2022; Yilmaz et al., 2024). The primary objective of XAI methods is to convert the internal processes of a black-box model into a more interpretable state, considering a trade-off between explainability and accuracy. Thus, it can provide insights into model design, decision transparency, and accuracy-interpretability trade-off.

This chapter provides a comprehensive exploration of all facets of accuracy assessment, encompassing topics like sampling design, sample size, scale, the spatial distribution of data, and the representativeness of ground reference information. It introduces the mathematical principles underpinning accuracy metrics used in pixel-based, polygon-based classifications, and in deep learning applications. The chapter also delves into the statistical tests used for comparing thematic maps. Additionally, the theory and mathematical foundations of recent approaches developed for interpreting trained models are provided. To elaborate the applications of accuracy assessment and the benefits of applying explainability models, discussed in this chapter, a case study employing the XGBoost algorithm for pixel-based image classification is also presented.

10.1 ACCURACY ASSESSMENT

No classification is complete until its accuracy has been assessed. In this context, the term accuracy means the level of agreement between labels assigned by the classifier and class allocations based on ground data collected by the user, known as test data. Ground reference data do not necessarily represent reality, due to observation and recording errors, mislocation of test data sites, differences caused by changes in land cover between the time of observation and the date of imaging, and so forth. Reference data are split into two subsets: training and testing data in supervised image classification. While the training data are introduced to the classification model to learn the underlying characteristics of the problem (training or learning phase), the test data set is utilized to estimate how much the model learned about the problem (validation phase). Independence between training and testing data is a fundamental principle to objectively assess the predictive power of the model. In addition to these data sets, machine learning and deep learning methods require another independent data set, called validation data, which is sometimes misrepresented as the test data set in remote sensing literature. The purpose of using validation data in the model development stage is to control the learning process to tune the model's hyperparameters and avoid two bottlenecks, namely, underfitting and overfitting, which are the main causes of poor performance. It is critical to stop the learning process when the algorithm becomes too specific to the training data (overfitting) and loses its generalization capabilities. Generalization may be defined as the ability of an algorithm to interpolate and extrapolate to data that it has not seen before (Atkinson and Tatnall, 1997). To achieve high generalization capability, in an ideal case, a learning process should be terminated when the error rate starts to increase for the validation data set (Figure 10.1). Therefore, in classification practices, the error rate is continuously monitored throughout the learning process. As can be seen from the figure, the optimal epoch to stop the learning process (known as earl-stopping, discussed in Section 7.1.3.4) should be defined to avoid underfitting and overfitting, both of which are undesirable in machine learning applications. Underfitting occurs when the model is unable to learn the underlying characteristics of the input data and capture the relationship between input and output data. In contrast to underfitting, overfitting happens when a model fits to training data too closely; that is, it becomes too specific to the training data as it memorizes specific features. Although high training performance is achieved

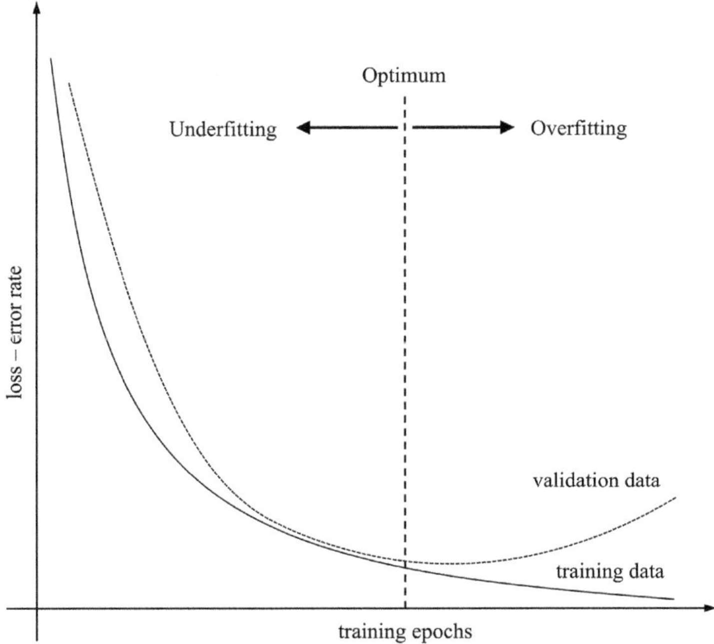

FIGURE 10.1 Relationship between error rate and training epochs for training and validation data sets.

with overfitting, poor generalization is evident on new and unseen data. To obtain high generalization capability for machine learning methods, especially deep learning models, a variety of regularization techniques, discussed in Chapter 7, are employed.

Until recently, assessing the classification accuracy of remotely sensed data was traditionally viewed as a straightforward task, encompassing the calculation of several conventional accuracy metrics. With the emergence of new classification approaches, accuracy analyses have become tailored to specific classification methods, such as per-pixel, per-polygon, or fuzzy classification. Using appropriate measures of classification accuracy can offer map users a reliable assessment of classification performance. In the classification of remotely sensed data, there is no universally recognized standard for accuracy level, but the common goal is to attain an overall accuracy of about 85% (Anderson et al., 1976; Scepan, 1999; Wulder et al., 2006; Foody, 2008; Congalton and Green, 2019), or up to 90% (Lins and Kleckner, 1996; Bourgeau-Chavez et al., 2015), and an individual class accuracy of no <70% (Thomlinson et al., 1999). However, some researchers (Trodd, 1995; Rogan et al., 2003; Ge et al., 2007) reported that the accuracies estimated for land cover classification and change detection rarely meet this commonly recommended target. In addition, reviewing 84 classifications given in 25 articles, Trodd (1995) reported that the mean value of the producer's accuracy of the classifications was 59%. As underlined by Stehman and Foody (2019), the accuracy metrics and their targets should be determined before the analysis and these targets need to align with the specific requirements of the application. Therefore, the 85% accuracy threshold does not hold a universally accepted status, even though it is occasionally treated as such. A detailed discussion on this issue is provided by Foody (2008).

As underlined by Stehman and Czaplewski (1998), accuracy assessment includes three basic components that should be followed as consecutive steps. The first step is the sampling scheme (design) used to select the reference sample. The second step is called the response design used to obtain the reference land cover classification for each sampling unit through field visits, or image analysis. The last step includes the accuracy analysis through various metrics and procedures. A sampling scheme describes how sample pixels are selected from the image to characterize the thematic classes of interest. Both the sampling scheme and sample size are also of importance in assessing the accuracy of the thematic map derived from remotely sensed data. Samples may be derived from field observations, from farm records (in the case of agricultural crops), or from maps or air photographs. It is also possible to produce endmembers using a spectral library consisting of the spectral signatures of the land use and land cover objects (Kavzoglu et al., 2024). There are certain restrictions on sampling, including cost, availability of source information such as maps and air photographs, and accessibility. Therefore, the current trend is toward interpreting higher spatial resolution imagery to assemble reference data sets instead of field-based in-situ data collection (Morales-Barquero et al., 2019). If the area of interest is large, then it may not be possible to conduct a thorough and statistically valid sampling procedure close to the time of the satellite overpass. When clouds are a problem, the logistics of sampling are made more difficult by the need to be on call for a number of overpass times. Attention must be paid to the questions of sampling scheme and sample size, particularly if statistical methods of pattern recognition are employed. Furthermore, if the target classes are of a temporally changing nature, care must also be taken to ensure that the sample data adequately represent the temporal state of the phenomena being observed.

As noted earlier, the whole ground reference data set is divided into three parts in the applications of state-of-the-art algorithms. Training data are used in supervised methods of pattern recognition to "teach" a classifier the main characteristics of each class. Campbell (1987) points out that:

- The number of sample observations has a direct relationship with the confidence interval on the estimate of the accuracy of a classification, and on the estimates of statistical parameters used in the particular, chosen, classifier. For example, estimates of the mean vector and variance-covariance matrices of the individual classes are required by the maximum likelihood classifier.
- There should be a minimum number of samples that are assigned to every class, to ensure that each class is properly represented (avoiding unbalanced data problem).

The evaluation of classification accuracy involves the utilization of a separate test data set, which is set aside until the production of the thematic map. Subsequently, the samples in the data set are classified using the same procedure employed in generating the thematic map. A comparison between the assigned label by the classifier and the label provided by a human observer enables an assessment of the classification accuracy. This topic is discussed in Section 10.1.1. Where it is not possible to acquire test data, then the use of cross-validation is a possible solution.

Some authors describe their test data as ground truth. This may be a misleading and inaccurate description. It is human to err, and one presumes that this aspect of human behavior extends to the collection of test and training data in remote sensing. Some of the errors in a thematic map may not be due to the fallibility of the classifier but may be due to the result of mislocation of some elements of the test data, or due to faulty identification of the attributes of the test data. Consequently, the term "ground reference data" is more appropriate when describing the collected sample data for image classification.

10.1.1 SAMPLING SCHEME AND SPATIAL AUTOCORRELATION

The sampling scheme aims to establish a statistically sound sampling framework that is cost-effective and accommodates the diverse goals arising from various users. As noted by Stehman (2004a), employing a scientifically rigorous sampling design provides a secure foundation for accuracy assessment practices. Several sampling schemes or designs are described in the literature (e.g., Berry and Baker, 1968; Borak and Strahler, 1999; Ginevan, 1979; Fitzpatrick-Lins, 1981; Stehman, 1992; Wang et al., 2005; Congalton and Green, 2019). Congalton (1988) suggests that both random sampling without replacement and stratified unaligned random sampling generally provide satisfactory results. In the estimation of individual class accuracies, Chen and Wei (2009) observed that stratified sampling achieves better precision than simple random sampling and systematic sampling with smaller sample sizes, especially for estimating a small class. Stehman (1992) suggests, however, that the usual formulae for deriving the value of the Kappa coefficient may perform poorly when stratified sampling methods are used. Congalton and Plourde (2000) hold a more lenient perspective but make the point that the placement of samples is as important as the sampling scheme. According to Hedger et al. (2001), for a given sample size, the systematic scheme exhibited fewer errors in comparison with the random scheme, and when targeting a particular level of error, the systematic approach required smaller sample sizes than the random approach. Chen and Stow (2002) compared the classification performances in relation to three sampling designs: single pixel, seed, and polygon, and concluded that the sample size, image resolution, and degree of autocorrelation inherent within each class influence the performance of sampling strategies. Obtaining representative training data through the use of polygons may be challenging since the collected data may show autocorrelation and may not adequately represent informational classes (Muchoney and Strahler, 2002). Stehman (2009) evaluated the sampling schemes in terms of seven desirable sampling design criteria and concluded with the practical reality that when dealing with limited resources, the essential approach is to prioritize objectives when designing a project. The key is to select an appropriate design, rather than a perfect design, by acknowledging the strengths and weaknesses of various design options and comprehending the trade-offs among objectives and desirable design criteria.

The collection of single, noncontiguous pixels may introduce more variance compared to polygonal sample collection and avoid the associated autocorrelation effects (May 2006). Zhen et al. (2013) investigated the effects of four sampling schemes and samples on classification accuracy in supervised object-based classification. The study findings indicated that the accuracy estimates for overall, user's, and producer's accuracies with test data were about 10% higher than the actual accuracy for the study region. This overestimation was slightly more pronounced when the test sample points were permitted to be within the same polygons as the training data points. Another important issue is the sampling design when object-based image analysis is to be applied. As underlined in the study of Waldner et al. (2017), in cases where the sampling units are objects and the reference data are pixel-based, it is advisable to modify training sample proportions through a sampling strategy, such as the synthetic minority oversampling technique (SMOTE), to ensure alignment between pixel proportions and object proportions.

Atkinson (1991, 1996) notes that the standard statistical rules of sampling, as outlined in conventional statistical texts such as Cochran (1977), do not hold for spatial data because locations in space are in fixed positions and their attributes are therefore autocorrelated. He proposes the use of geostatistical methods, which take into account the spatial relationships between the pixels and use a systematic scheme. Van der Meer et al. (1998) present the results of an investigation of mapping accuracy using geostatistical methods. They show that the optimum sampling distance varies with the nature of the target, being of the order of 100 m for vegetated areas. For soil, however, the optimum sampling distance is the highest achievable spatial resolution of the instrument. Guidance on the use of geostatistical methods in remote sensing is provided by Curran (1988) and Woodcock et al. (1988a, b).

In a random sampling scheme, the sampled pixels are randomly chosen from the image. The term without replacement means that no pixel is selected twice. Although this approach is quite easy to apply, it might not be always suitable in practice because some small information classes may be omitted. Moreover, if the number of samples selected is not sufficiently large, some information classes may be undersampled (or oversampled). Hence, stratified unaligned sampling may be preferred. One may choose several unaligned subareas for each class from the image, and perform random sampling within each subarea. However, Congalton and Green (2019) note that the assumptions on which the computation of the Kappa coefficient is based are satisfied only by random sampling.

A third sampling scheme, cluster sampling, can also be applied. In this sampling, groups of patch-like pixels, representing, for example, an agricultural field, are selected. Adopting this approach enables the efficient collection of a large number of samples. Nonetheless, the utilization of extensive cluster samples is discouraged, as pixels within a cluster lack mutual independence. Therefore, a sample of 40 contiguous pixels, for instance, does not constitute 40 independent samples, owing to the autocorrelation effect highlighted earlier in this section. For instance, Congalton (1988) suggested that no cluster should be larger than ten pixels.

The concept of local spatial autocorrelation signifies the tendency for values in adjacent pixels to be more analogous than those separated by larger spatial distances. In addition to the spatial resolution of the sensor, spatial autocorrelation in an image is related to object size, spacing, and shape (Jupp et al., 1988). The existence of spatial autocorrelation suggests a departure from the assumption of independently distributed data, which is crucial for many statistical techniques. Consequently, most conventional statistical methods may not be suitable for analyzing such data (Spiker and Warner, 2007). On the other hand, spatial autocorrelation can offer an opportunity to improve the classification performance by integrating spatial or contextual attributes into the model (Dobbertin and Gregory, 1996; Hyppänen, 1996; Chen et al., 2012; Fan and Myint, 2014; Karasiak et al., 2022), also to determine the optimum segmentation scale in object-based image analysis (Espindola et al., 2006; Kim et al., 2008). However, the issue of spatial autocorrelation is frequently ignored in the creation of training, validation, and test data sets, and thus overestimated generalization capabilities have been reported in the literature (Friedl et al., 2000; Roberts et al., 2017; Meyer et al., 2019; Schratz et al., 2019; Karasiak et al., 2022). When training and testing samples are collected from the same fields or randomly selected from a set of samples by ignoring their spatial distribution in the scene (issue also called "spatial balance" by Stevens and Olsen (2004) and Fowler et al. (2020)), they can be high likely to be correlated; that is, they are not independent because of spatial autocorrelation. In addition to geostatistical approaches, such as semi-variograms, global Moran's I (MI) (Moran, 1950) can also be used to estimate the autocorrelation that exists in reference data. The estimated MI index reveals valuable insights into the magnitude of spatial autocorrelation and the significance of the observed patterns. By considering these statistical indicators, the analyst can determine the spatial characteristics of the data and assess their potential implications for spatial data leakage. Global Moran's I (MI) can be calculated using the following formula (Johnson and Xie, 2011):

$$\text{MI} = \frac{n \sum_{i=1}^{n} \sum_{j=1}^{n} w_{ij} \left(y_i - \bar{y} \right)\left(y_j - \bar{y} \right)}{\sum_{i=1}^{n} \left(y_i - \bar{y} \right)^2 \left(\sum_{i \neq j} \sum w_{ij} \right)} \tag{10.1}$$

where n shows the total number of regions, w_{ij} indicates a measure of the spatial proximity, y_i is the mean spectral value of R_i region, and $\bar{y}$ shows the mean spectral value of the image. Each weight w_{ij} is a measure of the spatial adjacency of regions R_i and R_j.

The index is limited to a finite range, characterized by a specific benchmark value and variations between negative and positive spatial autocorrelation. A value of -1 signifies perfect negative spatial autocorrelation, while 1 represents perfect positive spatial autocorrelation. Values approaching zero are typical when samples are chosen randomly. In theory, the traditional Moran's I index should have a consistent range of $[-1,+1]$, but in practice, this is observed in only a limited number of cases. The variability of the interval seems to be contingent on the population under consideration. Tillé et al. (2018) recommend the adoption of a normalized version of the index (MI_N) to maintain a steady and defined variation range of $[-1,+1]$.

Let $\mathbf{y}$ be the vector of y_i, 1 be a vector of ones, $\bar{\mathbf{y}} = \bar{y}\mathbf{1} = \mathbf{11}^T\mathbf{y}/n$, $\mathbf{W} = w_{ij}$, and $\mathbf{P} = \mathbf{I} - \mathbf{11}^T\mathbf{y}/n$. Equation (10.1) can therefore be written as

$$\mathrm{MI}_N = \frac{(\mathbf{y}-\bar{\mathbf{y}})^T\,\mathbf{W}(\mathbf{y}-\bar{\mathbf{y}})}{(\mathbf{y}-\bar{\mathbf{y}})^T(\mathbf{y}-\bar{\mathbf{y}})\mathbf{1}^T\mathbf{W}\mathbf{1}} = \frac{\mathbf{y}^T\mathbf{PWPy}}{\mathbf{y}^T\mathbf{Py1}^T\mathbf{W}\mathbf{1}}. \tag{10.2}$$

Fowler et al. (2020) used the MI_N index to determine which route to survey in in-situ data collection and studied its relationship with classification accuracy considering 270 training sets, derived from 36 strata. It was observed that the overall accuracy demonstrated a substantial negative relationship with the MI_N index, indicating that classification accuracy improved with lower values of MI_N. The presence of a greater number of training samples did not assure improved accuracy, especially when their spatial autocorrelation was poor $(\mathrm{MI}_N > 0)$. In accuracy assessment practices, MI_N index can be used to evaluate the quality of reference data and thus revise the sampling scheme in the study.

10.1.2 SAMPLE SIZE, SCALE, AND SPATIAL VARIABILITY

The performance of any classifier is significantly affected by the number of training samples incorporated in the learning phase. This issue is perhaps more important for machine learning and deep learning models than conventional statistical classifiers because their performance relies entirely on the attributes of the provided training data. Huang et al. (2002), in a comparative investigation featuring maximum likelihood, support vector machines, decision trees, and neural network classifiers, concluded that the size of the training data had a more significant influence on classification accuracy than the algorithms. It is commonly agreed that sample size needs to be specified considering the characteristics of the project (Foody, 2009a; Zhu et al., 2016). In addition to the size of the training data, the characteristics and distribution of the data, as well as the sampling approach employed, are other crucial factors. Class proportions within the reference data should receive special attention because classifiers can demonstrate greater sensitivity to these proportions than to sample size (Waldner et al., 2017). This brings the issue of the representativeness of the samples collected for a study area. The higher the representativeness of the samples introduced into the classification process, the greater the potential for obtaining accurate and dependable classification results. Therefore, a prosperous machine learning application requires a delicate balance between the quality and quantity of training samples. An inadequate number of training samples hinder classifiers from capturing class characteristics, while excessively large samples may lead to overfitting and prolonged learning times. Therefore, choosing a larger volume of training data is consistently recommended over a smaller set. The key rationale behind this recommendation is that if a classifier fails to accurately learn the characteristics of classes within the training data, it is bound to face challenges when classifying new data. In scenarios with large training samples, the consequence is often a marginal decrease in performance. As noted by Hush and Horne (1993), "the more training data we have, the more incorrect functions we are able to reject, and the more likely we are to find the correct function".

As stated earlier, the incorporation of additional features, such as spectral bands, calls for a greater number of training samples. This correlation can be conceptualized as linear. The question, in most cases, is how many training samples are required to produce optimal (or near-optimal) classification results. Providing a conclusive answer to this question is challenging as it varies based on several factors, including the complexity of the problem, the characteristics of the training data, and the sophistication of the classification model implemented. On the other hand, different accuracy measures exhibit significant variations in the minimum required sample size to achieve similar precision. Chen and Wei (2009) observed that the overall accuracy requires fewer samples, whereas the Kappa coefficient requires the most samples to achieve a similar level of precision. Since the early days of remote sensing, sample size has remained a focal point of debate and discussion in the realm of classification problems (e.g., Hord and Brooner, 1976; van Genderen and Lock, 1977; Hay, 1979; Fitzpatrick-Lins, 1981; Rosenfield et al., 1982; Mather and Koch, 2011). Hay (1979) proposed the use of stratified sampling with a minimum of 50 samples per class. It is noteworthy that this guideline is still applied in many cases. Congalton (2004) suggests that a minimum of 50 samples should be employed for each class, and large study sites' sample size should be increased to 75–100 samples. Mather and Koch (2011) suggest as a rule of thumb that the number of training data pixels per class should be at least $30 \times N_f$ (the number of features). This rule is based on the notion that in univariate studies, a sample size of 30 or more is considered "large". This rule could be modified to state that the sample size per class should be at least 30 times the number of parameters to be estimated. Better-founded advice employs a theoretical model of the data distribution to predict the sample size needed to attain a specified level of accuracy. (Fitzpatrick-Lins, 1981; Congalton, 1991). After an extensive number of experiments, Kavzoglu and Mather (2003) suggest that the number of training samples should be at least ten times the number of weights ($10 \times N_w$) in the network or $30 \times N_f \times (N_f + 1)$ for multilayer perceptron neural networks.

Congalton and Green (2019) present a method for estimating sample size that is based on the multinomial distribution. The sample size n is derived from the relationship $n = B\Pi_i(1 - \Pi_i)/b_i^2$, where b_i is the required precision (expressed as a proportion so that 0.05 is equivalent to 5% precision), B is the upper $(\alpha/k) \times 100$th percentile of the chi-square distribution with 1 degree of freedom, k is the number of classes, Π_i is the proportion of the area covered by class i, and α is the required confidence level. Vieira (2000) gives the following example: at a confidence level of 95% and a desired precision of 0.05, find the sample size necessary for the derivation of a valid confusion matrix. There are $k=7$ classes, and class *wheat* occupies ~46% of the map area ($\Pi_{\text{wheat}}=0.46$). The value of chi-square for the probability level (0.05/7)=0.00714 with 1 degree of freedom is 7.348. The minimum training data sample size is therefore 7.348(0.46)(1−0.46)/(0.05)2=730. Vieira (2000) notes that this value is quite close to the value of $30 \times 3 = 90$ samples per class (or $7 \times 90 = 630$ total samples) suggested by Mather and Koch (2011).

The size of the sample required for statistically valid measures of classification accuracy to be computed is not the only criterion that should be considered. Statistical classifiers require that certain parameters be estimated accurately. In the case of the maximum likelihood classifier, these parameters are the mean vector and the variance-covariance matrix for each class. The number of features in the feature space has an impact on the sample size. As mentioned before, when the dimensionality of the data increases while keeping the sample size constant, the accuracy of parameter estimates diminishes, leading to a reduction in classifier efficiency, which is known as the *Hughes phenomenon*. For low-dimensional data, the suggestion that there should be a minimum size of $30 \times$ number of wavebands will give satisfactory results in most cases. However, as the dimensionality of the data increases (e.g., when hyperspectral data sets are used), then the required sample size will be unfeasibly large, and some method of dimensionality reduction, such as an orthogonal transform (Sections 3.1–3.3) or the use of feature selection methods (Section 3.2), will be required. These remarks are directed toward the use of statistical classifiers, which (in the case of the maximum likelihood classifier) operate by defining a model of the data distribution, such as the multivariate normal distribution, and then estimating the parameters of the model from the training data.

Other classifiers, such as decision trees, artificial neural networks, or support vector machines, are nonparametric and require that training data sets are large enough to represent the characteristics of each class. These methods are called nonparametric because they do not involve the estimation of statistical parameters. Kavzoglu and Mather (2003) observed that the performance of multilayer perceptron significantly increased with the addition of training samples from two study sites (up to 6%), but the increase in classification accuracy was not in a linear form. Evans (1998) notes that decision tree classifiers are susceptible to large changes in accuracy when only small changes are made in the composition of the training samples, indicating that both the size and the nature of the training samples assume considerable importance when such classifiers are used. There is some evidence that the neural networks, decision trees (random forest, rotation forest, canonical forest), and support vector machines perform better than statistical classifiers even with small training data sets (Foody et al., 1995; Foody and Mathur, 2004, 2006; Kavzoglu and Colkesen, 2012; Sahin et al., 2020), while the selection of training data sets from interclass boundary areas (i.e., mixed pixels) helps to improve classifier performance (Foody, 1999; Foody and Mathur, 2006; Kavzoglu and Reis, 2008). Foody et al. (2006) adopt four principles, namely, intelligent selection of the most informative training samples, selective class exclusion, acceptance of imprecise descriptions for spectrally distinct classes, and the adoption of a one-class classifier to reduce the training sample size. By employing the four-stage strategy, the necessary training set size was substantially lowered, going well below the levels suggested by conventional widely used heuristics, with minimal impact on accuracy.

The scarcity of data sets poses the primary obstacle in the application of deep learning models. Although traditional machine learning methods require training from scratch, various learning approaches have been developed benefiting from the advanced capabilities of deep learning. The training data utilized to train deep learning models are extensive, comprising millions of accurately labeled samples from various remotely sensed imagery. Acquiring comprehensive and adequately labeled data sets is vital for enabling deep learning models to obtain valuable representations. To tackle this challenge, recent advancements include the development of transfer and self-supervised learning techniques. Transfer learning is an effective strategy when dealing with a target data set of considerable size that is nonetheless inadequate for training an entirely new network from scratch. This method involves using the parameters of a pretrained model from an earlier task (referred to as the upstream task) as a starting point for learning a different task of interest (known as the downstream task). The initial layers focus on capturing broader characteristics such as edges, corners, circles, and solid color blocks, while deeper layers delve into finer-grained features. In transfer learning, the activations of the first layers are not updated during the training process (they are "frozen" layers), whereas the last layers carrying problem-specific information are fine-tuned with small sample data sets. In the field of remote sensing, this technique is extensively utilized, where features extracted from pretrained and publicly accessible data sets are transferred to new deep learning networks. Transfer learning not only reduces the effort required to train a high-performing model from the ground up but also eliminates the necessity for substantial computational resources in the case of a large network with numerous samples, thereby considerably reducing training time. Nonetheless, supervised pretraining approaches in transfer learning continue to face the challenge of requiring substantial annotated data sets for the earlier task. On the other hand, self-supervised learning, an emerging approach for learning from unlabeled data, has the potential to overcome this obstacle by extracting representations directly from the data using a pretraining objective. This optimization process produces transferable representations without relying on manual labeling. More information on transfer and self-supervised learning is given in Chapter 7. The use of transfer and self-supervised learning is discussed in Pires de Lima and Marfurt (2020), Alem and Kumar (2022), Gadiraju and Vatsavai (2023), and Muhtar et al. (2023). Lastly, it should be mentioned that an alternative technique for enlarging the labeled data set with deep learning is to apply data augmentation, a process that entails creating additional samples by modifying the initial data. These modifications can encompass shifts, flips, rotations, translations, shearing, scaling, and alterations in brightness.

The sample size is also related to the scale of observation (which determines the objects in the landscape that are deemed significant), and it also has an influence on observed variability both within and between classes. If a sample is to be representative, then its size should be related to the variance of the class. The scale of observation is generally proportional to variability; for example, consider the difference between two images with spatial resolutions of 1 km and 1 m, respectively. The latter shows more detail, in the sense that what appear to be homogeneous regions at 1-km resolution become fragmented and more variable as resolution increases (Woodcock and Strahler, 1987). Detailed discussion on the issue of scale in remote sensing image classification is given in Section 1.1.4.

There is also an obvious link between the concept of spatial variability, mentioned in the preceding paragraph, and image texture. One can, in fact, consider the dependence between spatial scale/resolution and spatial variance from several aspects. For example, the multifractal dimension is based on the concept of sets each with its own fractal dimension (Pecknold et al., 1997). The discrete wavelet transform is a second example of a hierarchical or multiscale procedure. Unlike the Fourier transform, which decomposes an image into its frequency components, the wavelet transform takes into account both the spatial and the frequency characteristics of an image. In other words, the wavelet transform can deal with nonstationary series. It is described as multiscale because it operates by decomposing the image into scale components in a recursive fashion. Zhu and Yang (1998) use wavelets to characterize image texture and demonstrate the utility of such texture measures in classification. Moreover, Tso and Olsen (2005) use wavelets to detect multiscale edge information to improve classification accuracy. The applications of wavelets in remote sensing are described in Section 4.1.1.5 of Chapter 4.

10.1.3 Adequacy of Training and Testing Data

Before conducting a classification process, the training and test data sets must undergo a careful examination to ensure their representativeness. This examination should consider the result of the applied selection strategy by incorporating the spectral variability of each class to include all spectral subclasses, resulting from different planting dates, seed properties, soil conditions, shadowing, and other effects. In this sense, the analyst must ensure that the data sets accurately reflect the typical characteristics of each class. One approach to assess the adequacy of the data set is to analyze the number of pixels left unclassified. If there are a considerable number of unclassified or misclassified pixels, it implies that the data set lacks adequate representation of class characteristics. In such cases, it is advisable to check the land cover types depicting the bulk of the study site and introduce more samples into the training data set. To test the spectral resemblance of the classes and ensure that all samples are correctly labeled, separability indices (Chapter 3) can be employed as a preliminary step before starting the training process. Kavzoglu (2009) suggested a two-stage refinement process aimed at acquiring more representative samples. This involves spectral histogram analysis and boundary analysis through dimensionality reduction. Significant accuracy improvement was reported for a neural network classifier by improving the class separability. Many studies (e.g., Huang et al., 2002; Pal and Mather, 2003; Foody, 2009b) point out that the adequacy of training and test samples in characterizing the spectral variability of the classes is the foremost important factor in image classification.

The adequacy of training data is paramount as it serves to educate a supervised classifier. It is crucial to verify that the training data do not include erroneous or unrepresentative samples. The presence of such outliers will have a lesser impact on supervised classifiers that are based on mean values alone, as one outlier will not have a particularly large influence on the value of the mean of a large sample (large means >30 measurements on the given feature or variable). However, the variances and covariances are likely to be more badly affected because their computation includes the square of the difference between a given observation and the mean of its class. On the other hand, the direct training of deep learning models on sample data makes them prone to the effects of unrepresentative (atypical) samples, unless the network architecture incorporates specific considerations. The presence of outliers can be accommodated by the use of robust statistical estimators, which are not as severely affected by outliers as are conventional estimators. Mather and Koch (2011) provide details of a method of weighting the observations used in the calculation of means and variance-covariance matrices. The weights are proportional to the Mahalanobis distance of the observation from its class mean. Since the position of the class mean vector in feature space will move as the weights change, the procedure iterates until the weight estimates converge. The use of the Mahalanobis distance implies that this weighting algorithm is appropriately applied to training data sets that are being used in statistical classifiers such as maximum likelihood. An alternative approach could use cross-validation procedures, in which the training data set is subdivided into a number of mutually exclusive groups, for example, into ten groups each containing 10% of the training data. The classifier is trained on 90% of the available training data (i.e., nine subgroups combined) and is applied to the remaining 10%. If the label given to an observation by the classifier does not correspond to the class allocated by the analyst, then that observation is eliminated. The procedure is repeated until all ten subsets have been labeled. An added sophistication is to use several classifiers and to eliminate only those training data observations that are mislabeled by all classifiers, or by a majority of the classifiers (Brodley and Friedl, 1996, 1999).

In most remote sensing projects, it is not always possible to collect representative and balanced training samples due to the cost, time, and access restrictions. Mislabeled samples in the ground reference data set also create a further challenge in the learning of the problem. Either a refinement process or an algorithm that is less sensitive to such issues should be applied to obtain a representative data set, which is studied by some researchers. For instance, Foody (2010) investigated the consequences of different imperfections in ground data on the estimation of change accuracy, including producer's and user's accuracy, as well as change extent. The outcomes from the analysis of 12 simulated data sets demonstrated that even minimal errors in ground reference data could result in significant inaccuracies in change detection analyses, emphasizing the need to be wary of considering data as a gold standard reference. By intentionally mislabeling 10% of the training samples to introduce noise into the data, Rogan et al. (2008) observed a reduction of 6% in artificial neural network (ARTMAP) classification accuracy and a corresponding 7% decrease in the classification accuracy of decision trees. Pal and Mather (2003) investigated the effect of training sample size on the performance of decision trees and reported that accuracy improved from 78.3% to 84.2% for a univariate decision tree, and from 78.15% to 83.9% for a multivariate decision tree when training sample size was increased from 700 to 2,700. Similarly, Millard and Richardson (2015) assessed how a random forest classifier performed with respect to classification accuracy, taking into account training data characteristics like sample size, class proportions, and spatial autocorrelation. They observed that using a larger training sample size produced lower out-of-bag accuracy and independent assessment error rates. In the cases where equal numbers of samples cannot be collected for training data, the lower and upper bounds of the samples for each class can be defined to maintain a balance for larger and smaller (rare) classes (Zhu et al., 2016).

Another significant issue associated with the adequacy of reference data is the handling of imbalanced data, which is a serious challenge in the field of remote sensing. In cases where the proportions of classes are distributed unevenly or are imbalanced compared to the actual land cover proportions, all classification algorithms, whether statistical or machine learning–based, may exhibit bias. It is noteworthy that the existence of imbalance significantly undermines the learning process, as machine learning methods tend to prioritize the dominant class and disregard the rare classes (Japkowicz and Stephen, 2002; Das et al., 2018). Under such circumstances, the classification process might exhibit a bias toward the class that represents the largest proportion in the sample set, leading to subjective interpretations of accuracy estimates and degradation in classifier performances (Foody and Mathur, 2004; He and Garcia, 2009; Maxwell et al., 2018). Classes that are disproportionately abundant in the training data can have a strong influence on the final classification result, while classes that are inadequately represented in the training data (rare classes) may similarly lack representation in the classification. In these cases, the extent of this bias is a function of the training data class imbalance (Millard and Richardson, 2015). In the use of simple random sampling, the likelihood of selecting a class is proportional to the size of the class area in the whole study site. This may result in the collection of fewer samples from rare classes in the training data. For instance, the Indian Pines data set, which is a widely used benchmark hyperspectral data for remote sensing studies, includes only 20 samples of oats and 28 samples of grass-pasture-wood classes, whereas it comprises 2,455 pixels for the soybean-mintill class. An equalized stratified random sampling can be applied to avoid this problem (Stehman and Foody, 2009). As mentioned earlier, this may not be possible for various economic and logistical reasons. When working with imbalanced data sets in which a class forms a minor portion of the training data, several sampling strategies such as oversampling, undersampling, and cluster-based sampling can be applied to generate data sets with enhanced class balance. For instance, Waldner et al. (2017) applied the SMOTE technique, developed by Chawla et al. (2002), to generate new samples for rare classes synthetically and reported an increase in the accuracy of up to 30%. More information on this issue and possible solutions for imbalanced data can be found in Fernández et al. (2017), Buda et al. (2018), Das et al. (2018), and Paoletti et al. (2023).

10.1.4　Conventional Accuracy Analysis

Accuracy assessment is typically expressed in the form of a square matrix, called an error matrix, confusion matrix, or contingency table in which the assignment of samples to each class is shown. The matrix includes both correctly and incorrectly classified pixels, from which various accuracy metrics can be estimated. It is with these accuracy measures that the produced thematic maps gain meaning and reliability. The correct interpretation of accuracy measures derived from confusion matrices is of great importance to determine the efficiencies and deficiencies of a classification being carried out. In particular, examining class-level accuracies derived from a confusion matrix provides a valuable means to further refine training and test samples, potentially leading to improved classification accuracy. A critical assumption in the use of error matrices is that the error distribution within the contingency table accurately represents the types of misclassifications occurring throughout the entire study area. Without this assumption, it is not possible to make any inferences about the probabilities of correctly or incorrectly assigning pixels to land use and land cover categories (Prisley and Smith, 1987). This assumption is perhaps the most vulnerable element in the use of confusion matrices.

Map-level accuracy measures are typically single global estimates providing a general measure for the quality of classification over the study site under investigation. Classification accuracy is often expressed as a single percentage value, computed by comparing the areas covered by each category in the classification map and the ground reference data. The error matrix and accuracy metrics derived from it provide no information about the spatial distribution of the error in a study site (Steele et al., 1998; Foody, 2005; Comber et al., 2012). Therefore, estimated overall accuracy measures from the error matrix might not be well suited for subregions, as the local error rates could vary considerably from the static global estimates, mainly resulting from the spatial variations in landscape features, deficiencies of image acquisition systems, and representativeness problems of ground reference data result in variations in local accuracy levels. Without considering any other factors, one might expect that erroneously classified pixels should be randomly distributed over the study area. Observations of the actual distribution of such erroneously classified pixels may show that this is not the case. For instance, in a per-pixel classification, it might be seen that erroneously classified pixels are distributed around field boundaries, or along spatial features such as roads and railways, which are not included in the classification. Therefore, the use of non–site-specific accuracy assessment might be misleading when interpreting classification results. To address this problem, several enhancements have been proposed. For instance, Persello and Bruzzone (2010) proposed two separate accuracy estimates to objectively assess the overall thematic accuracy: one from homogeneous areas and the other from the borders of the objects. The procedure involves the computation of two separate confusion matrices. Another effective solution proposed by Foody (2005) is to conduct an accuracy assessment in specific subregions of the study area. Enabling confusion matrices to be derived locally and globally provides the ability to estimate accuracy and characterize the nature of misclassification locally and globally. This approach can only be applied when a large ground reference data regularly cover the entire study site. By dividing the study site into 48 regions and estimating local accuracies in those regions, Foody (2005) produced maps of spatial variation in classification accuracies using geographically weighted regression. He drew attention to extreme variations in map-level (53.33%–100%) and class-level (0%–100%) accuracy measures. Inspired by this study, Comber et al. (2012) employ geographically weighted approaches to explain the spatial variation in the accuracy of Boolean and fuzzy classifications of remotely sensed data. They also propose a "portmanteau accuracy" measure to describe Boolean land cover accuracy and fuzzy difference measures to describe the accuracy of fuzzy land cover. On the other hand, Vieira and Mather (1999) present some methods of visualizing the spatial pattern of classification error, and of analyzing these patterns using simple methods of spatial statistics. The same authors show how cartographic methods can be used to depict the reliability of classifications derived from remotely sensed data. The spatial distribution of errors in agricultural crop classifications was found to be closely related to the positions of the field boundaries. The use of a buffering operation within a geographical information system (GIS) improved classification accuracy without compromising the integrity of the results and removed autocorrelated errors. To incorporate the spatial accuracy information into the resulting thematic map, Kavzoglu and Mather (2002) suggest the use of different color tones for each class, depending on class membership grade to create a more informative classification map. The study also suggests the production of an additional thematic map showing the probability level of prediction for each pixel. Thus, deficiencies in the classification protocol can be easily observed at the study site scale.

The reliability of classification as a major source of land use and land cover data is primarily determined by its accuracy. A thorough quality analysis of ground reference data is necessary to detect outliers and mixed pixels, which can compromise result reliability and introduce inaccuracies in class boundary definition, leading to a decrease in classification accuracy. This can be a major issue for the classifiers assuming that the pixels are pure representations of land cover types (Fisher and Pathirana, 1990). The occurrence of mixed pixels is a consequence of the interaction between the spatial resolution of the sensor and the complex composition of the landscape. In natural and semi-natural landscapes, mixed pixels are mostly located at the boundaries of two or more discrete classes, and along a continuum from gradual to abrupt (Wood and Foody, 1989; Foody and Boyd, 1999). In the traditional approach, samples are collected from near the centers of polygons or fields to avoid the boundary pixels, which are high likely to be a mixture of two or more classes. Thus, possible localization errors in geometric rectification and resampling outcomes can be prevented.

Since the 1980s, the confusion (or error) matrix has been the most common, in fact, a standard descriptive tool used for classification accuracy assessment. A confusion matrix used to estimate a number of descriptive metrics is a square array of dimension $n \times n$, where n is the number of classes. The matrix shows the relationship between two samples of measurements taken from the area that has been classified. The first set represents test data that have been collected via field observation, inspection of agricultural records, air photo interpretation, or other similar means. The second sample is composed of the labels of the pixels, allocated by the classifier, that correspond to the test data points. The columns in a confusion matrix represent test data, while rows represent the labels assigned by the classifier. Let x_{ij} denote the number of samples classified into class i $(i = 1, 2, \ldots, r)$ in the thematic map, and class j $(j = 1, 2, \ldots, r)$ in the reference data set (Figure 10.2).

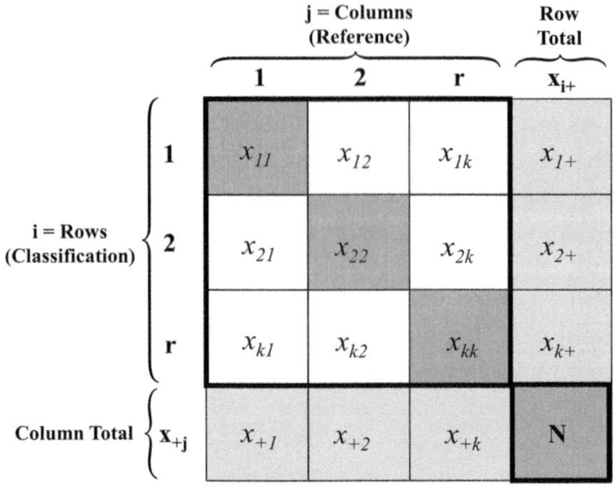

FIGURE 10.2   Graphical representation of a confusion (error) matrix.

The overall accuracy can be computed by dividing the total number of correctly classified pixels (x_{ii}) (i.e., the sum of the diagonal elements of the confusion matrix) into the total number of pixels (N) in the training data set, as follows:

$$\text{Overall Accuracy}\,(\text{OA}) = \frac{\sum_{i=1}^{r} x_{ii}}{N}. \tag{10.3}$$

Individual class accuracies can be calculated by dividing the total number of correctly classified pixels for each class by the corresponding column or row totals (marginals). As a result, two accuracy measures, producer's accuracy and user's accuracy, can be calculated. For each information class i in a confusion matrix, the producer's accuracy, a measure of omission error, is calculated by dividing the entry (i, i) by the sum of column i (Equation 10.4), while the user's accuracy is obtained by dividing the entry (i, i) by the sum of row i (Equation 10.5). Thus, the producer's accuracy tells us the proportion of pixels in the test data set that are correctly recognized by the classifier. The user's accuracy measures the proportion of pixels identified by the classifier as belonging to class i that agree with the test data.

$$\text{Producer's Accuracy}\,(\text{PA}) = \frac{x_{jj}}{x_{+j}} \tag{10.4}$$

$$\text{User's Accuracy}\,(\text{UA}) = \frac{x_{ii}}{x_{i+}} \tag{10.5}$$

The accuracy metrics of overall accuracy, producer's accuracy, and user's accuracy, though quite simple to use, are based on either the principal diagonal, columns, or rows of the confusion matrix only, which does not use the information from the whole confusion matrix. A multivariate index called the Kappa coefficient (Cohen, 1960) has found favor. The Kappa coefficient uses all of the information in the confusion matrix in order for the chance allocation of labels to be taken into consideration (though Foody (2000) suggests that chance agreement is overestimated, and accuracy is underestimated. He presents an alternative formulation in Foody (1992)).

The Kappa coefficient is defined by

$$\hat{k} = \frac{N \sum_{i=1}^{r} x_{ii} - \sum_{i=1}^{r} (x_{i+} \cdot x_{+i})}{N^2 - \sum_{i=1}^{r} (x_{i+} \cdot x_{+i})}. \tag{10.6}$$

In this equation, $\hat{k}$ is the Kappa coefficient, r is the number of columns (and rows) in a confusion matrix, x_{ii} is the entry (i, i) of the confusion matrix, x_{i+} and x_{+i} are the marginal totals of row i and column j, respectively, and N is the total number of observations (Congalton and Green, 2019). For computational purposes, the following form is often used:

$$\hat{k} = \frac{\theta_1 - \theta_2}{1 - \theta_2},$$
(10.7)

where

$$\theta_1 = \frac{\sum_{i=1}^{r} x_{ii}}{N},$$
(10.8)

and

$$\theta_2 = \frac{\sum_{i=1}^{r} x_{i+}x_{+i}}{N^2}.$$
(10.9)

The large sample variance of Kappa is (Congalton and Green, 2019)

$$\mathrm{Var}\left(\hat{k}\right) = \frac{N(x_{i+} - x_{ii})}{\left[x_{i+}(N - x_{+i})\right]^3}\left[(x_{i+} - x_{ii})(x_{i+}x_{+i} - Nx_{ii}) + Nx_{ii}(N - x_{i+} - x_{+i} + x_{ii})\right].$$
(10.10)

The distribution of the ratio $\dfrac{\hat{k}}{\mathrm{var}\left(\hat{k}\right)}$ is approximately Gaussian, and so it can be used as a test statistic for the null hypothesis that the observed accuracy differs from zero only as a result of sampling from a large population. Individual class values of Kappa (conditional Kappa values) can be derived as follows:

$$\hat{k}_{\mathrm{cond}} = \frac{Nx_{ii} - x_{i+}x_{+i}}{Nx_{i+} - x_{i+}x_{+i}}.$$
(10.11)

The Kappa coefficient $\hat{k}$ takes not just the principal diagonal entries but also the off-diagonal entries into consideration. The higher the value of Kappa, the better the classification performance. If all information classes are correctly identified, Kappa takes the value 1.

It should be noted that the interpretation of the Kappa statistic is based on the assumption of a multinormal sampling model. If the test data are not chosen properly, the above assessments become less reliable. Another consideration relates to sample size and sampling scheme; the consensus view appears to be that simple random sampling is required for the use of the Kappa coefficient, and that a minimum sample size is needed in order to ensure a specific, predefined level of accuracy.

Although the Kappa coefficient continues to be a standard metric for map-level accuracy assessment, as being employed in 50.4% of the studies reviewed by Morales-Barquero et al. (2019) and 40% of the OBIA studies reviewed by Ye et al. (2018), it has been subjected to the harshest criticism in the literature. The central argument behind these criticisms stems from the notion that Kappa provides redundant accuracy information with well-established overall accuracy (Liu et al., 2007; Olofsson et al., 2014; Stehman and Foody, 2019). Another important issue is the overestimation of the degree of chance agreement, which results in an underestimation of classification accuracy (Foody, 1992, 2008). Stehman (1997) underlines that the significant dependency of Kappa on the marginal proportions within the error matrix casts uncertainty on its appropriateness for making comparisons. Foody (2020) argues that chance agreement holds no relevance in an accuracy assessment as it is unsuitably modeled in the computation of a Kappa coefficient for remote sensing applications. Several improvements for Kappa have been suggested to overcome the deficiencies (Foody, 1992; Pontius Jr, 2000, 2002; van Vliet et al., 2011; Pontius Jr and Millones, 2008). However, in their seminal work titled "Death to Kappa", Pontius Jr and Millones (2011) investigated the reliability and validity of five different Kappa indices and concluded that the assessed Kappa indices were useless, misleading and/or flawed for the practical applications in remote sensing. They, along with Foody (2020), strongly discourage the use of Kappa indices for accuracy assessment and map comparison. Instead, they suggest employing two simpler summary parameters: quantity disagreement and allocation disagreement. Providing a more positive perspective, Congalton and Green (2019) recommend that Kappa's strength lies in its capability to assess whether one error matrix exhibits statistically significant differences from another, rather than merely presenting an accuracy value as an alternative accuracy measure. Nonetheless, Foody (2020) raises a contrary view that the comparison of Kappa coefficients becomes notably challenging when classes vary in abundance. This complexity arises because the magnitude of a Kappa coefficient captures both agreement in labeling and the properties of the populations being studied.

10.1.5 Accuracy Analysis for Machine Learning

In the last decade, we have witnessed the increasing trend of using machine learning and deep learning (DL) methods in remote sensing practices. In particular, DL offers a powerful framework for incorporating textural and contextual information, as well as providing feature extraction capabilities. DL-based studies employing techniques for the processing of remotely sensed data have been published in journals in both remote sensing and artificial intelligence fields. As a repercussion of this tendency, terminology and metrics used in the artificial intelligence field have been adopted and used by the remote sensing community (Maxwell et al., 2021a,b). This transition influenced the accuracy assessment concept in conventional remote sensing literature. The terminology typically used in the evaluation of classification accuracy in DL studies is conducted through the binary confusion matrix, where the class of interest is termed the positive instance, otherwise the negative instance (Table 10.1).

The inclusion of recall (i.e., the correct prediction of event responses) and specificity (i.e., the correct prediction of no event responses) in the analysis is crucial as they offer valuable insights into the classifier's deficiencies in terms of missed targets and false-positive results. Recall refers to the ratio of true positives to the acceptance class (Equation 10.12), while specificity represents the ratio of genuine negatives to the rejection class (Equation 10.13). In other words, misclassification may occur as a result of diminished sensitivity or specificity. Precision quantifies the number of positive class predictions that actually belong to the positive class (Equation 10.14). A significant distinction lies in the fact that the F-measure (F-score, F1-score, or Hellden's Index), which is the weighted harmonic mean of the precision and recall, does not consider the number of negative examples classified or even the presence of negative examples in the data set. Conversely, the balanced accuracy metric assigns equal importance to correctly identifying positive and negative instances. In cases

TABLE 10.1

Conceptual Framework of a Binary Confusion Matrix

	Reference Data	
	Positive	**Negative**
Positive	True positive (TP)	False positive (FP)
Negative	False negative (FN)	True negative (TN)

where imbalance and uneven class distributions exist in data sets, F-measure should be thus preferred. Information about other map-level and class-level accuracy measures that have less use in applications can be found in review studies of Liu et al. (2007) and Maxwell et al. (2021a).

$$\text{Recall (sensitivity)} = \frac{\text{TP}}{\text{TP} + \text{FN}} \tag{10.12}$$

$$\text{Specificity} = \frac{\text{TN}}{\text{TN} + \text{FP}} \tag{10.13}$$

$$\text{Precision} = \frac{\text{TP}}{\text{TP} + \text{FP}} \tag{10.14}$$

$$\text{Balanced accuracy} = \frac{1}{2}\left(\text{recall} + \text{specificity}\right) \tag{10.15}$$

$$F\text{-measure} = \frac{2 \times \text{Precision} \times \text{recall}}{\text{Precision} + \text{recall}} \tag{10.16}$$

While one can assess the model performance using individual accuracy metrics, it is always worth examining how they relate to each other to gain a broader understanding. Two types of curves have been widely used in machine learning studies to show the prediction capability of a classification model. The first measure that has been recently employed in remote sensing is the AUC (area under the curve) ROC (receiver operating characteristics) curve, called the AUC-ROC or ROC curve in the literature. It is a graphical plot of the true-positive (TP) rate (y-axis) (i.e., recall or sensitivity) against the false-positive (FP) rate (x-axis) (i.e., 1–specificity) and helps to select optimal models when two or more models are tested. It measures the overall performance of the binary classification model. In the multiclass case, multiple ROC curves and associated AUC-ROC values are estimated sequentially using the one-versus-all methodology. If the estimated AUC value is close to 1.0, the algorithm has a high capacity to differentiate between classes. If it is close to 0.5, the algorithm has a low capacity for discriminating capacity (Kavzoglu and Teke, 2022). ROC curve analysis (see Figure 10.3) is an effective way for the comparative evaluation of classification results. It can also be employed to analyze the behavior of algorithms and optimize model parameters. It is specifically developed for two-class problems, but it can be extended to multiclass problems using the formulations proposed by Hand and Till (2001).

Another curve used in the assessment of prediction performances of DL models is the area under precision-recall curve (P-R curve). The P-R curves help determine how precision varies as recall levels change. In other words, rather than relying on a single value, it takes into account different probability thresholds to convey the trade-off between the TP rate and positive predictive value in evaluating a predictive model. In classification problems, a threshold value is usually set to determine the class label considering the estimated probability value. If the estimated probability is higher than the threshold, it is labeled; otherwise, it is unlabeled or labeled as "unknown". In the P-R curve estimations, the threshold value is changed to estimate the power of the classifier. Given the inverse correlation between precision and recall, the curve typically exhibits a nonlinear pattern, signifying that elevating one metric leads to a reduction in the other, yet the reduction may not be proportional. In the P-R curve graph, a baseline model or classifier is shown as a horizontal line representing very low precision (Figure 10.4). The main difference between the ROC curve and the P-R curve is that the former considers the FP rates, whereas the latter is based on precision. It is important to note that ROC curve analysis is more appropriate when the observations are balanced between each class, whereas P-R curves should be employed for imbalanced data sets. Similar to the

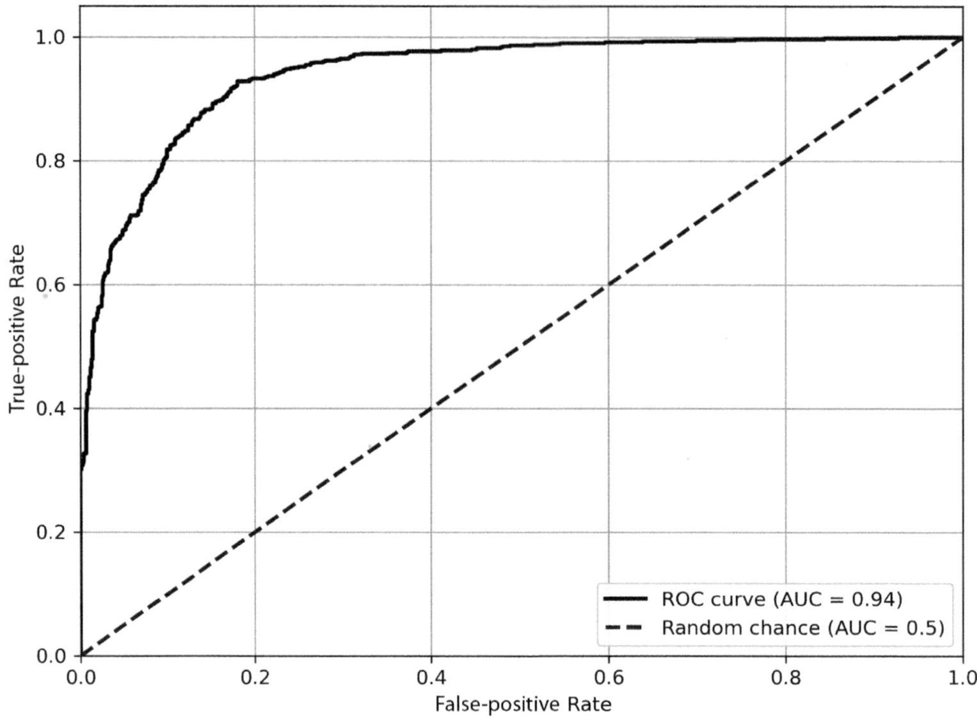

FIGURE 10.3   ROC curve for a classification with an AUC of 0.94.

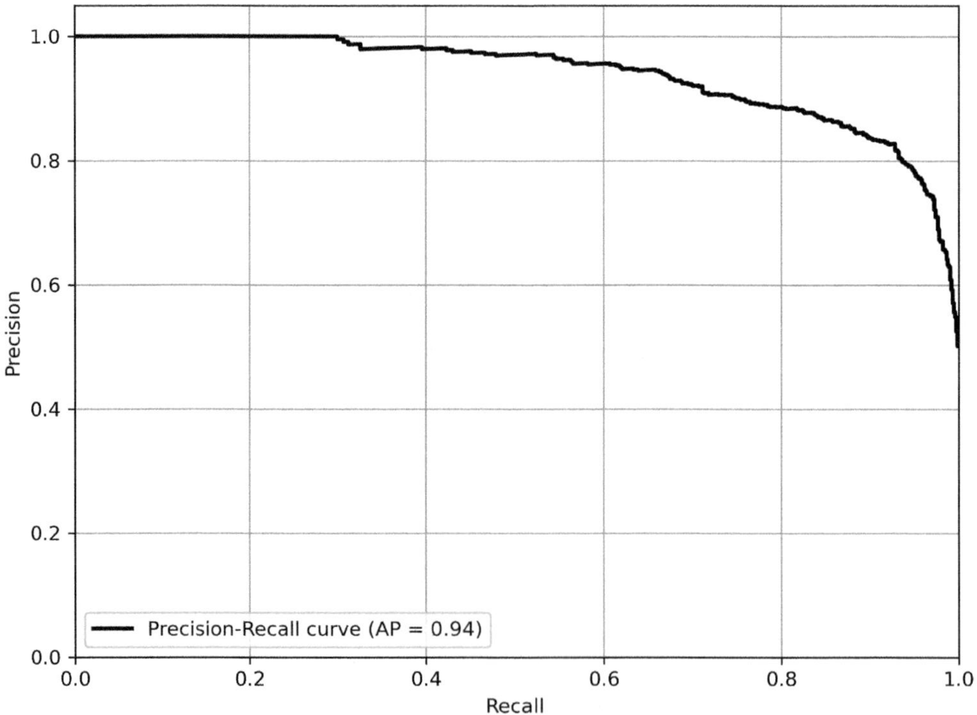

FIGURE 10.4   Precision-recall curve for a classification with an average precision of 0.94.

AUC-ROC curve, the area under the P-R curve (AUC-PR) can be estimated and used as a separate accuracy measure, which is referred to as average precision (AP) or, in some studies, mean average precision (mAP), which is the main measure among COCO evaluation metrics. Note that COCO (Common Objects in Context) is a large-scale object detection, segmentation, and captioning data set, developed with the goal of advancing the state-of-the-art in object recognition.

10.1.6 Fuzzy Accuracy Assessment

When the goal of classification is to unambiguously establish the pixel memberships, specifically in scenarios where there is uncertainty regarding the varying levels of class presence (mixed pixels), "hard" classification methods become irrelevant. This led to the introduction of the concept of "soft" or "fuzzy" classification. Soft classification diverges from the conventional practice of directly assigning pixels to specific classes by estimating the probability of a pixel's membership (membership grade) for each class. Besides employing fuzzy classifiers, it is also feasible to transform the results of traditional hard classifiers into a fuzzy representation of land cover (Foody, 1996). This approach yields more intricate membership details, enabling a deeper comprehension of the characteristics of each pixel. Compared to hard classification, the soft classification approach is better suited for addressing the classification challenges associated with land cover features that typically exhibit a continuous form in nature. The most common conventional method for fuzzy classification is the fuzzy C-means algorithm, which can be used either in an unsupervised or in a supervised classification fashion (see Section 2.3.1.2). In the current applications of machine learning and DL methods, produced results are in the form of fuzzy representations that require different accuracy assessments than that of hard classifications. Gopal and Woodcock (1994) suggested a fuzzy reference labeling protocol using the linguistic scale in that the sample unit is assigned a label: 1 = absolutely wrong, 2 = understandable but wrong, 3 = reasonable or acceptable answer, 4 = good answer, and 5 = absolutely right. Based on this labeling protocol, two fuzzy accuracy measures, namely, MAX and RIGHT operators, were suggested. To assess the precision of fuzzy classifications within the context of a probabilistic understanding of class membership values, Conese and Maselli (1993) and Finn (1993) employed conditional entropy and mutual information, while Foody (1996) introduced a symmetric measure of information similarity. Ricotta (2005) introduced several measures that are based on the notion of nonspecificity for quantifying the pixel-level categorical uncertainty associated with nonprobabilistic fuzzy classifications of remotely sensed images.

Considering that the conventional error matrix and the derived accuracy measures are only appropriate for hard classification, Binaghi et al. (1999) proposed an accuracy assessment approach that uses the fuzzy set theory to extend the applicability of the traditional error matrix method to the evaluation of soft classifiers (Figure 10.5). This approach is structured to handle situations where classification and/or reference data are presented in multimembership form, with membership grades indicating varying degrees of approximation to intrinsically vague classes. The methodology relies on the assumption that membership values within classes are known for the set of reference data. In the classification of remotely sensed data, the membership grades within a specific land cover category are directly related to the coverage percentages within individual pixels. The membership values for the reference data represent subpixel land cover estimates derived from ground truth labeling procedures, which can be generated using images of diverse resolutions, or through a comprehensive field study.

Let R_n and C_m be the sets of reference and classification data assigned to class n and m, respectively, where r is the number of classes, $1 \leq n \leq r$, $1 \leq m \leq r$. R_n and C_m form two hard partitions of the sample data set X. The process by which individuals from a given sample data set X are determined to be either members or nonmembers of the classes n and m is defined by the characteristic or discrimination function of the sets R_n and C_m. The element of the error matrix M in row m and column n represents the cardinality of the intersection set $C_m \cap R_n$:

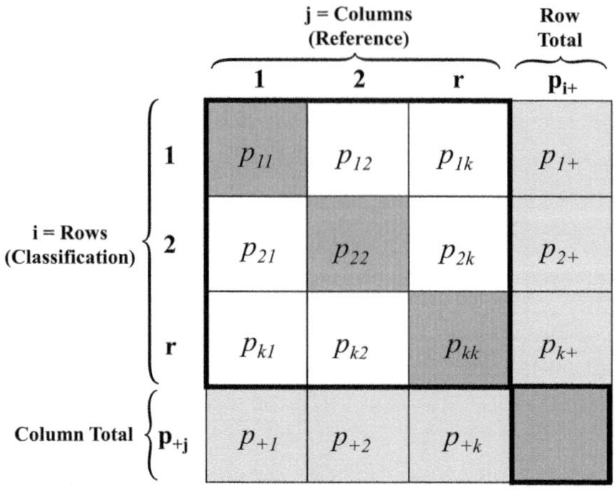

FIGURE 10.5 Fuzzy error matrix.

$$M(m,n) = C_m \cap R_n = \sum x \in X\mu_{C_m \cap R_n}(x), \tag{10.17}$$

with the character function:

$$\mu_{C_m \cap R_n}(x) = 1 \;\; \text{iff} \;\; x \in C_m \wedge x \in R_n, \;\; 0 \;\; \text{otherwise.} \tag{10.18}$$

The generic error matrix is shown in Figure 10.5, where $p_{m,m}$ represents the cardinality of the intersection set $C_m \cap R_n$ computed according to Equation (10.17); p_{i+} and p_{+i} are the total assignment to the ith class for classification and reference data, respectively.

Stehman et al. (2007) introduced the formulations of accuracy measures similar to overall, user's, and producer's accuracies obtained from a fuzzy error matrix for stratified random sampling. A mathematical representation of the fuzzy-weighted Kappa coefficient was given by Huang and Lees (2007). In addition, Lewis and Brown (2001) introduced a generalized area-based confusion matrix with the assumption that subpixel classification can be interpreted as area estimation.

10.1.7 Object-Based Accuracy Assessment

Traditional pixel-based accuracy assessment suffers from two fundamental underlying problems. Firstly, pixels are not meaningful representations of Earth objects as they are arbitrary delineations of rectangular areas defined by the IFOV of the recording sensor. Secondly, there exist some varying degrees of positional errors for pixels in an image because of the geometric correction process employing a limited number of known coordinates. As we perceive our surroundings as groups of objects, object-based image analysis is usually a more logical way of analyzing remotely sensed data. With the introduction of images at higher spatial resolution, object-based image analysis has evolved as a new image processing paradigm. It seems that the latest developments are oriented toward integration with convolutional neural networks (CNNs) (geographical object-based convolutional neural networks). As described in more detail in Chapter 8, OBIA has two main processing steps: creating image objects (called segments) through some intelligent strategies and classifying these objects using a classification algorithm. Created image objects are considered as the spatial unit in the classification process. In OBIA, it is crucial to assess the degree of correspondence between digital image objects extracted from high-resolution images and reference objects existing in the study site (Blaschke, 2010; Lizarazo, 2013). It has been reported that the segmentation quality measured with some metrics (discussed in Section 8.7) has a direct influence on the resulting classification accuracy (Kim et al., 2009; Marpu et al., 2010; Kavzoglu et al., 2017).

Accuracy assessment in the object or polygon domain requires some additional considerations compared to pixel-based accuracy assessment. In many studies, researchers have employed pixel-based accuracy metrics by ignoring the objects, but the pixels in those objects. In a recent study, Ye et al. (2018) observed that 107 of the 209 selected OBIA articles applied-per-pixel approaches, 93 studies used polygon-based accuracy analysis, and the rest employed both approaches. Pixel-based accuracy assessment, favored for its simplicity and ongoing experience, has been criticized and thus slowly replaced by polygon-based accuracy analysis (Ye et al., 2018). Creation of reference polygons for accuracy assessment is an ill-posed problem, and no universal approach exists for practitioners. While some researchers collect reference polygon data through a field study, in which spatial resolution of the image should be considered to avoid border or mixed pixels, others create reference polygons from the created segments that are not representative of the natural boundaries of the objects. When a polygon is associated with multiple categories in accordance with the reference information, researchers must figure out how to handle its affiliation with multiple categories. On the other hand, some researchers are in favor of employing pixel-based accuracy assessment to evaluate OBIA classifications. For instance, Marpu (2009) opposes the utilization of image objects as fundamental entities in accuracy assessment, based on two key arguments. Firstly, object-based methods are typically compared to pixel-based classification methods, where pixels are the basic entities, and their data values represent the actual recorded measurements of remote sensing sensors, in contrast to objects formed through segmentation methods. Secondly, accuracy assessment relies on statistical estimations that demand a sufficient number of samples for reliable testing. In OBIA applications, there is often a scarcity of samples, such as a limited number of buildings as regularly shaped objects in a study site, making accurate testing challenging.

A method to construct a confusion matrix for objects that have partial membership to multiple categories was suggested by Pontius Jr and Connors (2009). Also, since each object in the image has a varying size, instead of a traditional error matrix, an area-based error matrix can be applied. Because the accuracy evaluation of thematic maps created through an OBIA approach should take into account the reference unit's spatial extent, the individual cells reflect the total area of the reference units that fall into that cell (MacLean and Congalton, 2012). The concept of an area-based error matrix is shown in Table 10.2.

Considering the area-based error matrix, overall accuracy can be estimated as

$$\text{OA}_{\text{OBIA}} = \frac{\sum_{i=1}^{k} S_{ii}}{S} \tag{10.19}$$

TABLE 10.2
Area-Based Error Matrix

		Reference Data (j)				
		1	2	$\cdots$	k	Total Area
Classified data (i)	1	S_{11}	S_{12}	$\cdots$	S_{1k}	S_{1+}
	2	S_{21}	S_{22}	$\cdots$	S_{2k}	S_{2+}
	$\vdots$	$\vdots$	$\vdots$	$\cdots$	$\vdots$	$\vdots$
	k	S_{k1}	S_{k2}	$\cdots$	S_{kk}	S_{k+}
	Total Area	S_{+1}	S_{+2}	$\cdots$	S_{+k}	S

Note that S_{ij} shows the total area of the reference data in the thematic map for class i and reference data class j, and S is the total area of the reference data.

where S_{ii} represents the major diagonal cells, and S is the total area. Using the same methodology as the traditional error matrix calculations, the producer's and user's accuracies can be also calculated, as described by Radoux et al. (2010).

Lizarazo (2014) proposed a novel similarity index, called STEP, to estimate both geometric and thematic accuracies of OBIA classification. In this novel approach, four similarity metrics, namely, shape similarity, theme similarity, edge similarity, and position similarity, are used to form a STEP similarity matrix, which is a rectangular array structured in rows and columns that expresses the similarity between classified objects and their corresponding reference objects. A value for similarity in the range of [0, 1] is assigned to each classified object that matches (intersects) a given reference object. If a classified object does not intersect with a particular reference object, there is no similarity to calculate, leading to an empty cell value. Rows and columns of the STEP matrix are aggregated. An area-weighted error matrix that includes the weighted area of reference units corresponding to cells is then created.

A popular index used to compare image objects, also prevalent to analyze DL results, is the intersection-over-union (IoU), also called the Jaccard index. It is a ratio of the intersection of the reference and classified samples with the union of the two groups (Equation 10.20). Since the IoU index has a similar formulation to that of the F-measure (Equation 10.16), a nonlinear correlation exists between the two metrics. IoU can be computed either by considering pixels or by using bounding boxes that enclose individual objects within the image. IoU is frequently employed in tasks related to object recognition and instance segmentation. It is also applied to pixels on the classified image in the applications of semantic segmentation.

$$\text{Intersection Over Union} \left(\text{IoU} \right) = \frac{\text{TP}}{\text{TP} + \text{FP} + \text{FN}} \tag{10.20}$$

Object-based accuracy assessment still has some fundamental and unresolved conceptual challenges. There is a pressing need for the profession to further develop methodologies concerning sampling design, reference design, and accuracy assessment for OBIA, as underlined by several researchers (e.g., Stehman and Wickham, 2011; Ye et al., 2018; Kucharczyk et al., 2020; Maxwell et al., 2021b). This need is evident because the existing literature lacks a straightforward method for evaluating the accuracy of OBIA maps, considering the fact that these maps comprise objects of diverse shapes and sizes. Current status and future directions for OBIA incorporation with CNNs are discussed in a review paper by Kucharczyk et al. (2020).

10.2 COMPARISON OF THEMATIC MAPS

Investigating the performances of the methods has been a major agenda in remote sensing studies. Therefore, assessing the quality and superiority of the produced thematic maps is essential because the analyst and the users are interested in the reliability and rigor of the end product, which is related to many issues discussed throughout this chapter. Major scenarios that the researchers have been involved for thematic map comparison may be given as

- Comparing the performance of a newly proposed classifier, an improved version of a classifier, or a series of classifiers against that of a conventional algorithm. This is particularly important when they are applied to benchmark data sets.
- Comparing the performance of an algorithm for different data considerations, including studies on the contribution of ancillary data, investigating the effects of feature extraction and selection in the case of high-dimensional data sets, or analyzing the effects of sampling design and sample size.
- Analyzing the performance characteristics of an algorithm in relation to hyperparameter setting.

The comparison of the thematic maps produced for the above scenarios is conducted either qualitatively (visual assessment) or quantitatively. Quantitative analyses have primarily concentrated on evaluating the magnitude of difference in the estimated accuracies. However, determining the superiority of a method based on a higher accuracy estimate is not conclusive, as accuracy metrics can be influenced by sampling variability (De Leeuw et al., 2006; Stehman and Foody, 2019). A straightforward comparison of the estimated classification accuracies can be subjective and misleading; although it is widely practiced, a rigorous assessment through hypothesis testing methods should be employed. These methods estimate the statistical significance of the difference (inequality) or similarity in the estimated accuracies. Although researchers have discussed many statistical tests for thematic map comparison, the theoretical bases of five popular hypothesis tests are discussed in the following sections with their underlying assumptions. In comparative studies, the characteristics of the chosen hypothesis test should be carefully considered. Assumption of normal distribution (parametric or nonparametric), minimum sample size requirement, and independent sample constraints must all be taken into account for reliable and rigorous comparison. Results will be unreliable when the underlying assumptions are violated. For instance, the fundamental assumption of the z-test (significance test of the difference between two independent Kappa coefficients) is that the data sets used in both Kappa estimations (i.e., construction corresponding error matrices) must be independent. Since this is often not the case in remote sensing applications, independent data sets should be considered in the estimation of the z-test, or other statistical tests (e.g., McNemar's test) should be preferred (Foody, 2004, 2009a). On the other hand, for some hypothesis testing methods, a binomial test can be a better option to analyze the case of a limited number of test samples. Another important issue related to hypothesis testing is that these tests can provide a binary output whether a null hypothesis is rejected or not. However, the testing does not provide any insights into the degree of difference in accuracy. Foody (2009a) suggests that the confidence interval for the difference in classification accuracy can be utilized as an alternative or supplementary approach to conventional hypothesis testing, offering a more comprehensive understanding of the distinctions in classification accuracy values.

An alternative to hypothesis testing is to quantify the amount of mutual or shared information that the compared thematic maps include. To employ this methodology evolved from the information theory, Finn (1993) proposed Average Mutual Information (AMI) to evaluate a different aspect of the problem than does either overall accuracy or Kappa coefficient, measuring "consistency" rather than "correctness". AMI can be seen as a metric to assess the extent to which the classes on the map reveal information about the real classes present in the landscape, and the degree of information shared in the agreement between the mapped and actual classes. In the context of information theory, the information content of a map is equivalent to its uncertainty, reflecting the uncertainty about potential values (classes) at each point on the map. This quantification is represented by the entropy of the map. If a map $\mathbf{A}$ has n classes, and the proportion of area in the ith class is p_i, then the entropy of the map is:

$$H(\mathbf{A}) = -K \sum_{i=1}^{n} p_i \log(p_i), \tag{10.21}$$

where K is a constant used to modify the units of $H(\mathbf{A})$. $H(\mathbf{A})$ is maximal when all n classes are equally probable; that is, p_i is the same for all I (the greatest uncertainty), and is at a minimum when one class occupies the entire map (the least uncertainty). For a pair of maps of the same location, $\mathbf{A}$ and $\mathbf{B}$, let there be n classes in Map $\mathbf{A}$ and m classes in Map $\mathbf{B}$. Further, let $p(a_j)$ be the proportion (probabilities) of Map $\mathbf{A}$ in class j, and let $p(b_i)$ be the proportion of Map $\mathbf{B}$ in class i. At this point, the entropy of each map can be calculated as $H(\mathbf{A})$ and $H(\mathbf{B})$. By overlapping Map $\mathbf{A}$ on to Map $\mathbf{B}$, an $m \times n$ contingency table can be constructed with n columns representing each class in Map $\mathbf{A}$ and m rows representing each class in Map $\mathbf{B}$, and with elements of the matrix representing the total area of the map that is class j on Map **A** and class i on Map **B**. The joint probability of being in class j on Map **A** and classification Map **B**, $p(a_j, b_i)$, is given by dividing the (i, j) element of the contingency table, c_{ij}, by the total of all elements in the table (i.e., the total area of the map).

The amount of information that Map **B** contains about Map **A** is reflected in the reduction in uncertainty about Map **A** when Map **B** is known. That is, if a certain point in Map **B** is known to be class i, then this may increase the probability that the point is class j in Map **A**. This can be expressed as a conditional probability, $p(a_j | b_i)$, the probability of a point being class j in Map **A** given that in Map **B**, the point is class i. From probability theory:

$$p(a_j | b_i) = p(a_j, b_i) / p(b_i) .$$
(10.22)

The AMI metric for thematic map comparison can then be estimated as

$$\mathrm{AMI} = K \sum_j \sum_i p(b_i, a_j) \log\left[p(b_i | a_j) / p(b_i) \right].$$
(10.23)

Describing AMI as robust to combining classes, Salk et al. (2018) list several advantages of the technique in map comparison. Firstly, it typically rises when a map's legend is disaggregated, highlighting the value of the data provided by additional categories. Secondly, this tendency becomes more prominent as the categories become more evenly distributed, given that common classes convey a greater amount of information compared to rare classes. Besides these two characteristics, AMI generally represents the ratio of correctly classified points, except in cases of heavily flawed maps. AMI can be also used to identify mislabeled classes. Despite the listed advantages, it was stated that AMI still cannot be thought of as the "perfect" map accuracy metric.

10.2.1 McNemar's Test

McNemar's test serves as a valuable tool for assessing the statistical significance of differences in modeling performances between multivariate and bivariate methods. It has been a popular hypothesis testing method when compared to the other methods, particularly against the z-test due to its less sensitivity to sample dependency (Foody, 2004, 2009a; De Leeuw et al., 2006). McNemar's chi-square statistic is a nonparametric test applied to a 2×2 contingency table (Table 10.3). Each cell in the table indicates the number of observed outcomes in each category. For instance, n_{ii} represents the number of samples with the same results by Classifier I and Classifier II. The null hypothesis assumes that both classifiers have the same performance and the same error rates $(n_{ij} = n_{ji})$.

The following statistic (Equation 10.24) has a χ^2 distribution with 1 degree of freedom. A term called "continuity correction" (−1 in the formulation) is used to account for the fact that the statistic is discrete, while the χ^2 distribution is continuous (Dietterich, 1998).

$$\chi^2 = \frac{\left(|n_{ij} - n_{ji}| - 1 \right)^2}{n_{ij} + n_{ji}},$$
(10.24)

TABLE 10.3

Contingency Table for McNemar's Test

	Classification (Prediction) I	
Classification (Prediction) II	**Correct (i)**	**Incorrect (j)**
Correct (i)	n_{ii}	n_{ij}
Incorrect (j)	n_{ji}	n_{jj}

---

where n_{ij} shows the number of samples that are misclassified by method i, but correctly classified by method j. n_{ji} represents the number of samples that are misclassified by method j, but correctly classified by method i.

If the estimated statistical value is larger than the critical value defined by the desired level of significance, the null hypothesis can be rejected. In other words, it can be inferred that the two predictions are statistically different from each other, implying that the best-performing method is statistically superior. Note that the critical table value is 3.841459 for a 95% confidence level or 0.05 significance level $\left(\chi^2_{1,0.05}\right)$. It is important to state that if n_{ij}, n_{ji}, or both are small $\left(n_{ij} + n_{ji} < 20\right)$, then McNemar's statistic is not approximated by the χ^2 distribution, then should be approximated by the use of a binomial distribution (Japkowicz and Shah, 2011).

10.2.2 z-Test

z-Test, proposed by Cohen (1960), is a two-tailed statistical test used to compare estimated error matrices based on derived Kappa coefficients and corresponding estimates of variance. It is a parametric test, assuming that samples are derived from normally distributed populations.

Let $\hat{k}_1$ and $\hat{k}_2$ denote the estimates of the Kappa coefficient for two error matrices. Let also $\mathrm{var}\left(\hat{k}_1\right)$ and $\mathrm{var}\left(\hat{k}_2\right)$ be the corresponding estimates of the variance as computed from the appropriate equations. The test statistic for testing whether two independent error matrices are significantly different is given by Congalton and Green (2019):

$$z = \frac{\left|\hat{k}_1 - \hat{k}_2\right|}{\sqrt{\mathrm{var}\left(\hat{k}_1\right) + \mathrm{var}\left(\hat{k}_2\right)}}. \tag{10.25}$$

Given the null hypothesis $H_0 : \left(\hat{k}_1 - \hat{k}_1\right) = 0$ and the alternative hypothesis $H_1 : \left(\hat{k}_1 - \hat{k}_1\right) \neq 0$. The null hypothesis $\left(H_0\right)$ is rejected if $z \geq z_{\alpha/2}$. The z-test has a critical value of 1.96 for a 95% confidence level. As thoroughly discussed in Section 10.1.4, there are many studies against the use of the Kappa coefficient in accuracy assessment. Since the z-test is based on a comparison of two estimated Kappa coefficients and their variances, the test should be employed with care considering the underlying assumption that the data used to construct the two error matrices must be independent samples. Since training and validation samples are usually selected from ground reference data in most cases, the same samples are employed in the comparison of classification methods; hence, samples are not independent of each other.

10.2.3 Wilcoxon Signed-Ranks Test

Wilcoxon signed-ranks test, proposed by Wilcoxon (1946), is another nonparametric test used to assess whether sample mean ranks differ. The test can be considered as a nonparametric alternative to the matched-pair t-test. The test ranks the differences in performances of two classifiers, ignoring the signs, and compares the ranks for positive $\left(T_+\right)$ and negative differences $\left(T_-\right)$. If calculated test statistic z_{wilcox} (Equation 10.26) is smaller than the critical table value for the desired level of significance, the null hypothesis is rejected.

$$z_{\mathrm{wilcox}} = \frac{\min\left(T_+, T_-\right) - \dfrac{n(n+1)}{4}}{\sqrt{\dfrac{n(n+1)(2n+1)}{24}}}. \tag{10.26}$$

The Wilcoxon signed-ranks test operates under the assumption that performance measures can be qualitatively compared, giving more importance to larger differences, without considering the magnitude of these differences (Kavzoglu et al., 2015).

10.2.4 5×2-Cross-Validation t-Test

The main idea behind the 5×2 cross-validation t-test lies in the use of five replications of two-fold cross-validation instead of ten-fold cross-validation. In each replication, the available data are randomly partitioned into two equal-sized sets, S_1 and S_2. Each learning algorithm (A or B) is trained on each set and tested on the other set. Thus, the process results in four error estimates: $p_A^{(1)}$ and $p_B^{(1)}$ (trained on S_1 and tested on S_2) and $p_A^{(2)}$ and $p_B^{(2)}$ (trained on S_2 and tested on S_1). Subtracting corresponding error estimates gives us two estimated differences: $p^{(1)} = p_A^{(1)} - p_B^{(1)}$ and $p^{(2)} = p_A^{(2)} - p_B^{(2)}$. Let s_i^2 be the variance estimated for the ith replication, and $p_1^{(1)}$ be the $p^{(1)}$ from the first of the five replications. Then, the following estimate called the 5×2 cv $\tilde{t}$ statistic can be estimated (Dietterich, 1998):

$$\tilde{t} = \frac{p_1^{(1)}}{\sqrt{\dfrac{1}{5}\sum_{i=1}^{5} s_i^2}}. \tag{10.27}$$

Under the null hypothesis, $\tilde{t}$ has approximately a t distribution with 5 degrees of freedom. The null hypothesis is rejected when the estimated statistic is >2.571 for 95% confidence. In a comparative study, Dietterich (1998) reviewed and compared five statistical tests used for comparison of classifier performances and concluded that the 5×2 cross-validation t-test was slightly more powerful than McNemar's test. After reporting a deficiency for the test, which is dependent on the $p_1^{(1)}$ chosen for the test, Alpaydın (1999) proposed an improved version called 5×2 cv F that is based on a random choice and combines the results of the ten possible statistic promises. The estimated statistic (f) is approximately F-distributed with 10 and 5 degrees of freedom. The null hypothesis is rejected when the statistic f is >4.74.

$$f = \frac{\sum_{i=1}^{5}\sum_{j=1}^{2}\left(p_i^{(j)}\right)^2}{2\sum_{i=1}^{5} s_i^2} \tag{10.28}$$

10.2.5 Friedman Test

The Friedman test, proposed by Friedman (1937), may be considered as the nonparametric counterpart of the one-way ANOVA test. The test is used to assess whether there are any statistically significant differences between the performances of three or more classifiers. The test should be employed when the assumption of normal distribution for the one-way repeated-measures ANOVA test is not met or when the dependent variable is ordinal in nature. The process of the Friedman test is as follows (Japkowicz and Shah, 2011):

- Consider n data sets and k classifiers to be compared.
- Each algorithm is ranked for each data set separately, from the best-performing classifier to the worst-performing classifier.
- Let R_{ij} be the rank of classifier f_j on data set S_i.
- Compute the following estimates:
- The mean rank of classifiers f_j on all data sets:

$$\bar{R}_{.j} = \frac{1}{n} \sum_{i=1}^{n} R_{ij} \tag{10.29}$$

The overall mean rank:

$$\bar{R} = \frac{1}{nk} \sum_{i=1}^{n} \sum_{j=1}^{k} R_{ij} \tag{10.30}$$

The "sum of squares total" denoting the variation in the ranks:

$$SS_{\text{Total}} = n \sum_{j=1}^{k} \left(\bar{R}_{+j} - \bar{R} \right)^2 \tag{10.31}$$

The "sum of squares error" denoting the error variation:

$$SS_{\text{Error}} = \frac{1}{n(k-1)} \sum_{i=1}^{n} \sum_{j=1}^{k} \left(\bar{R}_{ij} - \bar{R} \right)^2 \tag{10.32}$$

Friedman statistic:

$$\chi_F^2 = \frac{SS_{\text{Total}}}{SS_{\text{Error}}} \tag{10.33}$$

The null hypothesis assumes that all the classifiers are equivalent in their performance, and hence, their average ranks $\bar{R}_{+j}$ should be equal. The Friedman statistic χ_F^2 follows a χ^2 distribution with $k-1$ degrees of freedom for large n (usually > 15) and k (usually > 5).

10.3   EXPLAINABILITY METHODS

The internal operational mechanisms of contemporary AI systems typically exhibit a black-box nature, a phenomenon acknowledged in many scholarly works. These AI models excel at crafting intricate designs underpinned by thousands of learned parameters, involving complex mathematical representations. This complexity contributes to their impressive performance, yet it comes at the expense of transparency and explainability. The inner workings and logical processes within these models remain obscured, presenting a notable drawback as it obstructs users, whether experts or novices, from the ability to validate, comprehend, and make sense of the decision-making process of the model. While the lack of transparency may not always be problematic in scenarios where performance is the foremost concern, it can become a critical issue particularly when decisions with potential consequences for human lives are at stake. Additionally, this opacity can be particularly problematic in applications involving sensitive personal data, such as healthcare or finance, where the fairness and ethical implications of AI decisions are paramount. In the context of remote sensing applications, the lack of transparency can pose significant challenges, particularly when AI algorithms are used in decision-making processes related to environmental monitoring, disaster management, or infrastructure planning, as these decisions can have far-reaching impacts on communities and ecosystems. The proliferation of such complex AI algorithms has raised concerns about potential algorithmic biases and the challenge of explaining or interpreting model outputs (Adadi and Berrada, 2018; Fel et al., 2022). These issues pose significant risks to the reliability and transparency of AI systems in various domains. In remote sensing applications, the opaqueness inherent in intricate AI models poses significant impediments, particularly when striving to achieve accurate and reliable results. For instance, AI-driven remote sensing applications may offer predictions on crop health, yet the rationale behind identifying specific areas as diseased or stressed lacks clarity, impeding the adaptability of interventions. Consequently, the opacity intrinsic to AI systems' crafting intricate designs in remote sensing applications necessitates concerted efforts to enhance interpretability, ensuring that these technologies can be effectively harnessed for vital environmental and agricultural management tasks.

In response to the aforementioned problems, explainable AI (XAI) has emerged as a field dedicated to transforming opaque AI systems into more transparent constructs by offering a range of tools. Whether XAI represents a resurgence of a traditional concept, or an entirely novel field remains a subject of debate. Nevertheless, it has recently become a crucial area of research spanning multiple disciplines, driven by the imperative to enhance trust and accountability in algorithmic decision-making. As underlined by Silva et al. (2018), a good explanation should maximize the following properties (known as the three Cs of interpretability): completeness, correctness, and compactness.

In the domain of land cover classification, XAI methodologies can elucidate how AI algorithms discern between different land cover types. They can pinpoint the specific spectral features, textural patterns, or contextual cues that drive the algorithm's decisions, shedding light on the rationale behind land classification outcomes. Furthermore, XAI can be instrumental in the identification of potential sources of bias or errors in remote sensing models, enhancing their robustness and reliability. Additionally, in precision agriculture, XAI can clarify the parameters and variables influencing crop yield predictions derived from remote sensing data. Farmers and agricultural experts can gain a comprehensive understanding of why AI systems recommend specific cultivation practices or irrigation strategies based on satellite-derived information. This transparency aids in optimizing resource allocation, maximizing crop yields, and minimizing environmental impact. Recognizing the significance of interpretability in machine learning, it becomes evident that there is a compelling necessity to enhance the emphasis on research in this domain, with the goal of fostering progress and unifying scientific insights.

XAI represents a concerted effort to unveil the intricate internal mechanisms and behavioral patterns of black-box AI systems, furnishing a diverse array of tools catering to both local and global interpretability contexts. Local interpretability elucidates the micro-level decision-making process within AI models. This facet delves into granular inquiries such as the identification of influential attributes within the data set, elucidating the specific data features that played a pivotal role in shaping the decisions of the learning algorithm. Moreover, it strives to unravel the rationale behind the algorithm's specific actions taken in response to individual input instances. In essence, local explanations provide a detailed, instance-specific breakdown of AI decision logic, facilitating a deep understanding of model behavior at a fine-grained level. Conversely, global interpretability provides a macroscopic perspective on AI model functionality, offering a high-level overview of how the model operates and arrives at decisions across the data set as a whole. Rather than scrutinizing individual predictions, global explanations aim to uncover overarching patterns, relationships, and decision pathways that characterize the model's behavior on a broader scale. This holistic viewpoint is indispensable for grasping the systemic principles governing AI model performance and behavior. To sum up, both local and global interpretability facets are indispensable components of the XAI framework. Local interpretability delves into the minutiae of model decisions, dissecting them at the instance level, while global interpretability offers a panoramic view of model functioning, revealing the systemic mechanisms underpinning its decision-making processes. These dual perspectives are essential for achieving a comprehensive and nuanced understanding of the inner workings of AI models in remote sensing applications and other domains. Many global and local interpretation methods, which are either model-specific or model-agnostic, have been suggested to explain machine learning and DL methods. Explainable models that utilize DL can be divided into two main methods: Activation-Based Methods (ABMs) and Backpropagation-Based Methods (BBMs). Major ABMs include Class Activation Maps and Gradient-Class Activation Maps (You et al., 2023), while BBMs encompass approaches like Layer-wise Relevance Propagation,

DeepLIFT, and SmoothGrad saliency maps. A good discussion and taxonomy of the models used in explaining models are presented by Carvalho et al. (2019) and Tjoa and Guan (2021). As examples of global and local XAI, five popular model-agnostic techniques have been discussed in the following sections.

10.3.1 SHapley Additive exPlanations

SHapley Additive Explanations (SHAP), based on Shapley values from game theory (Shapley, 1953), provides a robust framework for evaluating the magnitude and influence of individual explanatory variables on the outcome of any machine learning model. The SHAP analysis, introduced by Lundberg and Lee (2017), leverages an additive feature attribution technique, in which the model's output is construed as a linear sum of its input parameters. For a given prediction, SHAP computes the importance of each feature. Shapley values are subsequently computed based on the mean incremental contribution across all conceivable permutations of the features. These SHAP values encompass two primary components: "game" and "players", where the former pertains to the model's outcome and the latter corresponds to the model's features. Consequently, Shapley values provide a global interpretability of the constructed model by quantifying the cumulative effect of each feature on the model's output. Moreover, they facilitate local interpretability by offering this insight for each individual observation.

SHAP analysis furnishes explanations for opaque machine learning models by computing SHAP values for each feature within the classification process. Shapley regression values represent feature importance in linear models when multicollinearity is present. To obtain these values, the model needs to be retrained using different feature subsets $S \subseteq F$, where F is the set of all features. An importance value is assigned to each feature, indicating the influence of adding that feature on the model's prediction. To estimate this effect, a model $f_{S \cup \{i\}}$ is trained by including the feature in the model, and another model f_S is trained with the feature withheld. Then, predictions from the two models are compared based on the current input $f_{S \cup \{i\}}\left(x_{S \cup \{i\}}\right) - f_S(x_S)$, where x_S represents the values of the input features in the set S. Given that the consequences of omitting a feature are interrelated with the presence of other features in the model, the differences are calculated for every conceivable combination of feature subsets $S \subseteq F/\{i\}$. The Shapley values are then calculated and employed as feature attributions. They are a weighted average of all possible differences (Lundberg and Lee, 2017):

$$\phi_i = \sum_{S \subseteq F\setminus\{i\}} \frac{|S|!\left(|F|-|S|-1\right)!}{|F|!}\left[f_{S \cup \{i\}}\left(x_{S \cup \{i\}}\right) - f_S(x_S)\right]. \tag{10.34}$$

where ϕ_i represents the contribution of feature i, S denotes any feature subset of F excluding feature i. When an input feature has a negative SHAP value, it implies a decrease in the model output, and conversely, a positive SHAP value indicates an increase. The magnitude of the SHAP value is directly proportional to the impact of the input feature on the predicted value.

Although global interpretation techniques for tree-based machine learning models (e.g., random forest, gradient boosted trees), which provide an overview of how input features affect the entire model, have a well-established history, there has been relatively little focus on local explanations that clarify how input features influence individual predictions (such as those for a single sample). For estimating local feature interactions of tree-based models, Lundberg et al. (2020) proposed an improved SHAP analysis with a new set of tools and showed the effectiveness of global and local explanations using different data sets. They concluded that local explanations would become an essential tool for many machine learning tasks in the future. In the remote sensing domain, SHAP analysis has been recently applied to analyze the machine learning and DL models in the analysis of remotely sensed data, including classification (Abdollahi and Pradhan, 2021; Temenos et al., 2023), water quality estimation (Zhu et al., 2022), and object detection (Kawauchi and Fuse, 2022).

10.3.2 Partial Dependence Plot

Partial dependence plot (PDP) serves as a paramount instrument within the realm of machine learning. It primarily elucidates the correlation between a singular input attribute, denoted as x, and the resultant prediction outcome y. Importantly, it accomplishes this task while meticulously holding the values of all other input attributes constant. This depiction of partial dependence is conventionally symbolized as $P(y|x)$. The PDP is a globally applicable statistical artifact that effectively delineates the influence of specific attributes or pairs of attributes on the output generated by a machine learning algorithm. Notably, these PDPs are versatile, transcending the confines of a particular model, hence earning the title of "model-agnostic". To derive such model-agnostic PDPs, a systematic approach is employed. This involves the meticulous variation of the attribute of interest, while steadfastly maintaining the remaining input attributes at a fixed, unchanging value. In essence, this process facilitates the identification of pivotal factors within a model's functionality. This identification is made possible by scrutinizing the shape and magnitude of the PDP corresponding to each attribute. Moreover, it furnishes the capability to illustrate the manner in which a model's predictions evolve as the value of a specific attribute fluctuates, while all other attributes remain in a state of equipoise.

The methodology entails the sustenance of all attribute values in x at their mean values, with the exception of the attribute i, which is set to its mean value, denoted as $x_j = \text{avg}(x_j)$ for all j not equal to i. Subsequently, a range of values for attribute i is selected, such as $x_i = [x_{\min}, x_{\max}]$, and a grid of values, x_{grid} is meticulously crafted within the chosen range. For each value encapsulated within x_{grid}, the model's output, denoted as $f(x_{\text{grid}})$, is predicted. Here, $x_{\text{grid}}[i]$ assumes the value of x_i, and $x_{\text{grid}}[j]$ is assigned the mean value of $\text{avg}(x_j)$. Effectively, for each value contained within the selected attribute, the model's output is predicted by substituting that specific attribute value, while preserving all other attributes at their mean values. To culminate, y_{grid} is visualized as a function of x_{grid}. A simplified representation of a PDP can be mathematically expressed as follows:

$$y_{\text{grid}} = f\left(x_{\text{grid}}[i] = x_i, \ x_{\text{grid}}[j] = \text{avg}(x_j) \ \text{for all} \ j! = i\right) \ \text{for} \ x_i \ \text{in} \ [x_{\min}, x_{\max}] \tag{10.35}$$

It is noteworthy that PDPs are frequently presented in conjunction with confidence intervals, which functions as a testament to the inherent uncertainty embedded within the predictions. These confidence intervals serve the critical purpose of delineating the range within which the genuine average response of the model is anticipated to fall, for each specific value of x. Additionally, PDPs can be instrumental in unveiling interaction effects between attributes. For instance, a PDP may unveil that the impact of attribute x on the ultimate result differs across various levels of another attribute, say z. This revelation encapsulates pivotal insights into the operational dynamics of the model and can furnish invaluable guidance for feature engineering or the enhancement of overall model performance.

In certain scenarios, a solitary attribute's relationship with the outcome may not suffice. This necessitates the conception of higher-order PDPs, encompassing two-dimensional or even three-dimensional PDPs, to comprehend the interactions between attribute pairs or groups of attributes. Pertaining to the interpretation of a PDP, it is of paramount importance to grasp the notion of effect size. The effect size inherently embodies the impact wrought by variations in the attribute x upon the predicted outcome. The magnitude and direction of the effect size can be ascertained by discerning the gradient and orientation of the PDP curve.

10.3.3 Pairwise Interaction Importance

Pairwise interaction importance plays an important role in assessing the significance of interactions between predictors in a machine learning model. This concept is quantified through the feature importance rank measure approach, as described by Greenwell et al. (2018). It serves as a means to identify and evaluate interactions that have a substantial impact on a model's predictive performance. When two predictors exhibit a robust interaction, it implies that the relationship between one of these predictors and the target variable significantly depends on the values of the other predictor, as elucidated by Ryo (2022). In simpler terms, the strength of the association between one predictor and the response variable experiences noteworthy alterations contingent on the values assumed by the other predictor.

The mathematical formulation for pairwise interaction importance can be represented through a straightforward calculation: it involves comparing the model's predictions when considering the pair of features together versus when considering each feature individually. This quantification is expressed as follows:

$$\text{Interaction}\_\text{importance}(i,j) = P(i,j) - P(i) - P(j) \tag{10.36}$$

where i and j are indices that represent the two features under consideration. $P(i, j)$ denotes the model's prediction when both features i and j are taken into account, $P(i)$ signifies the prediction when only feature i is considered, and $P(j)$ represents the prediction when only feature j is considered. The outcome of this calculation serves as a measure of the significance of the interaction between the two features.

In practical terms, this quantification allows machine learning practitioners and data scientists to discern which pairs of predictors or features exert a notable combined influence on the predictions. It aids in identifying interactions that might otherwise be overlooked when analyzing features in isolation, thereby contributing to a more comprehensive understanding of the model behavior and the factors that drive its predictive performance.

10.3.4 Permutation-Based Feature Importance

Variable importance metrics delineate the degree to which a prediction model's predictive power is contingent upon the information encapsulated within each covariate. Introduced by Breiman (2001), the permutation-based feature importance approach involves the random shuffling of values for an individual variable and subsequently evaluating the model's performance on a separate holdout data set. The change in the model performance before and after shuffling the values is utilized to ascertain the significance of each feature. A feature's importance is assessed by measuring the increase in the model's prediction error that occurs when the feature's values are randomly permuted. If shuffling a feature's values leads to a higher model error, the feature is considered "important" because the model heavily relied on it for making predictions. Conversely, if the model's error remains unchanged after shuffling a feature's values, the feature is considered "unimportant" because the model disregards the feature in making predictions. Hence, the permutation-based feature importance is instrumental in identifying features that possess minimal or negligible predictive significance, enabling a more tangible interpretation of the model. Its value is especially pronounced in the context of nonlinear models, which encompass a considerable number of features.

The concept of permutation feature importance was originally introduced by Breiman (2001) for random forests. Building on this concept, Fisher et al. (2019) proposed a model-agnostic version of feature importance, which they termed "model reliance". They also introduced more advanced ideas regarding feature importance, including a model-specific version that accounts for the fact that various prediction models can perform well on the data.

10.3.5 Local Interpretable Model-Agnostic Explanations (LIME)

Local interpretable model-agnostic explanations (LIME), introduced by Ribeiro et al. (2016), represent a pragmatic instantiation of local surrogate modeling. Rather than formulating a comprehensive model to forecast the behavior of an entire data set, LIME focuses on the development of localized models aimed at elucidating the decision-making processes behind specific data points. Perturbations within the locality of the explained instance are employed to generate explanations, following a concept akin to locally weighted regression. It defines explanations as an optimization task and seeks to discover a trade-off between the local fidelity of explanation and its interpretability. By virtue of its model-agnostic architecture, LIME is adaptable to any machine learning algorithm. The essence of LIME involves a transformation of the original data set into a synthetically generated counterpart, accomplished through the localized perturbation of instances. Subsequently, the newly generated data set is subjected to a fitting process utilizing a straightforward, interpretable algorithm, such as Lasso or linear regression. During this fitting phase, the sampling methodology takes into consideration the proximity between the perturbed features and the data point of interest (Ward et al., 2021). In image classification, LIME returns a list of the most influential areas in the image for particular prediction (Robnik-Šikonja and Bohanec, 2018). The operational framework of LIME's explanation mechanism can be concisely expressed as follows (Kuzlu et al., 2020):

$$\text{Explanation}(x) = \underset{g \in G}{\arg\min} \, L\left(f, \, g, \pi_x\right) + \Omega(g). \tag{10.37}$$

The model g is structured as a linear combination of the features present in x, with coefficients derived through the calibration of the model against the predictions of f within the local vicinity of x. The overarching objective is the simplification of the output generated by the intricate model f, transforming it into a more interpretable representation. This transformation is achieved by minimizing the measure $L(f, g, \pi_x)$, which gauges the closeness of the explanation model g to the predictions produced by the model f.

10.4 A CASE STUDY FOR ACCURACY ASSESSMENT AND XAI

To clarify the theory of accuracy assessment and model explainability, presented in this chapter, a pixel-wise classification was performed using high-resolution satellite imagery of WorldView-3 for a study area of Akyazı district of Sakarya province, Türkiye, through the application of the XGBoost algorithm. The imagery, a Level 2A product, underwent rigorous radiometric and geometric corrections and was georeferenced to the WGS-84 datum and UTM projection system. The sensor featured one panchromatic band (0.31 m), eight bands in the visible and near-infrared spectrum (1.24 m), and eight bands in the shortwave infrared spectrum (3.7 m). Before the classification process, the spectral bands were converted to top-of-atmosphere radiance and the Gram-Schmidt sharpening algorithm was applied to enhance the spatial resolution of the 3.7-m bands to 1.24 m.

By considering the spectral attributes specific to the study area, a total of 11 land use and cover (LULC) classes, covering the bulk study area, were identified and delineated. Before the classification process, three data sets comprising training, validation, and test subsets were generated to facilitate the training and evaluation of the classifier. After the training stage, the XGBoost algorithm was applied to produce a thematic map of the study area (Figure 10.6). The trained model was applied to the test data set, and thus, accuracy assessment was conducted to produce an error matrix (Table 10.4). It was estimated that overall accuracy and Kappa coefficient were 90.60% and 89.66%, respectively. In addition, user's and producer's accuracies were estimated for each class. Omission and commission errors can be easily observed from the error matrix and estimated class-level accuracies. The lowest user's accuracy was estimated for the red roof class, which is mainly confused with the bare soil and road classes. The same confusion can be noticed from the classification of bare soil pixels in which 301 pixels were classified as red roof.

To provide a deeper understanding of the XGBoost performance, an additional class-level accuracy assessment was conducted, yielding valuable insights into how well the model distinguishes between different classes. The results, presented in Table 10.5, illustrate the precision, recall, and F1-score for each class. The table demonstrates varying levels of performance across different classes. For instance, in classifying the red roof, the model exhibits high precision at 86.18%, indicating that its

FIGURE 10.6 Thematic map produced by pixel-based classification using XGBoost classifier.

TABLE 10.4
Map-Level Accuracy Assessment Results

	Young Poplar	Hazelnut	Corn	Pasture	Poplar	Red Roof	Shadow	Road	Bare Soil	White Roof	Water
Young Poplar	**3,933**	68	39	27	13	0	0	0	0	0	0
Hazelnut	81	**3,557**	179	55	181	16	9	0	2	0	0
Corn	22	429	**3,365**	76	151	25	12	0	0	0	0
Pasture	63	17	16	**3,841**	4	24	12	27	65	4	7
Poplar	10	98	102	1	**3,815**	5	49	0	0	0	0
Red Roof	0	59	43	108	1	**3,192**	109	200	327	36	5
Shadow	0	77	27	1	89	20	**3,859**	1	1	0	5
Road	0	6.	2	63	0	52	2	**3,708**	152	49	46
Bare Soil	0	15	9	122	1	301	0	51	**3,555**	25	1
White Roof	0	2	4	68	0	6	0	95	8	**3,897**	0
Water	0	13	4	1	1	63	24	15	19	0	**3,940**
User's Acc.	96.40	87.18	82.48	94.14	93.50	78.24	94.58	90.88	87.13	95.51	96.57
Producer's Acc.	95.72	81.94	88.79	88.03	89.64	86.18	94.68	90.51	86.10	97.16	98.40
Overall Acc.					90.60						
Kappa Coef.					89.66						

Bold values show the number of correctly classified pixels

positive predictions are often accurate, though the recall of 78.24% implies some missed instances. Similarly, the model's precision for corn class is relatively high at 88.79%, but the lower recall of 82.48% suggests that it struggles to identify a significant portion of actual corn instances. In addition, young poplar, road, bare soil, shadow, and water show well-balanced performance. It can be concluded that the class-level metrics provide valuable insights into the model's strengths and areas where it may require further refinement to achieve improved performance for specific classes.

TABLE 10.5

Class-Level Accuracy Assessment Results

Class	Precision (%)	Recall (%)	F1-Score (%)
Young poplar	95.72	96.40	96.06
Hazelnut	81.94	87.18	84.48
Corn	88.79	82.48	85.51
Pasture	88.04	94.14	90.99
Poplar	89.64	93.50	91.53
Red roof	86.18	78.24	82.01
Shadow	94.68	94.58	94.63
Road	90.51	90.88	90.69
Bare soil	86.10	87.13	86.61
White roof	97.16	95.51	96.33
Water	98.40	96.57	97.48

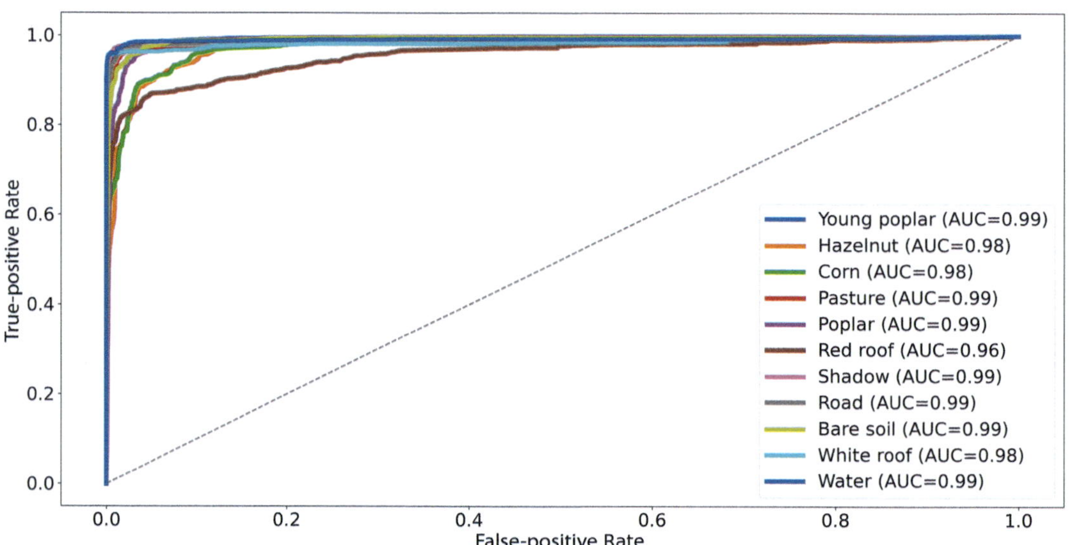

FIGURE 10.7 ROC curves and AUC scores of each class.

The results of the classification performance were also analyzed using ROC curves with their respective AUC scores (Figure 10.7), and P-R curves with average precision scores (Figure 10.8). Both curves provide significant information about the effectiveness of the model in distinguishing various LULC classes within the study area. The estimated highest scores suggest that the model performs exceptionally well for the majority of classes, achieving high AUC and average precision scores, which are indicative of their ability to discriminate between classes accurately. Classes such as young poplar, pasture, poplar, shadow, road, bare soil, and water demonstrate particularly robust discrimination with AUC scores of 0.99, indicating a high TP rate and a low FP rate. This implies that the model excels in correctly classifying these land cover types and maintaining a high level of precision.

The P-R curves verify this performance, with high average precision scores reflecting the classifier's ability to deliver accurate and reliable results for these classes. However, it is noteworthy that certain classes, such as hazelnut, red roof, and corn, exhibit slightly lower AUC and average precision scores compared to the others. This could be attributed to various factors, such as the spectral similarity of the hazelnut class to corn or bare soil may lead to some misclassifications.

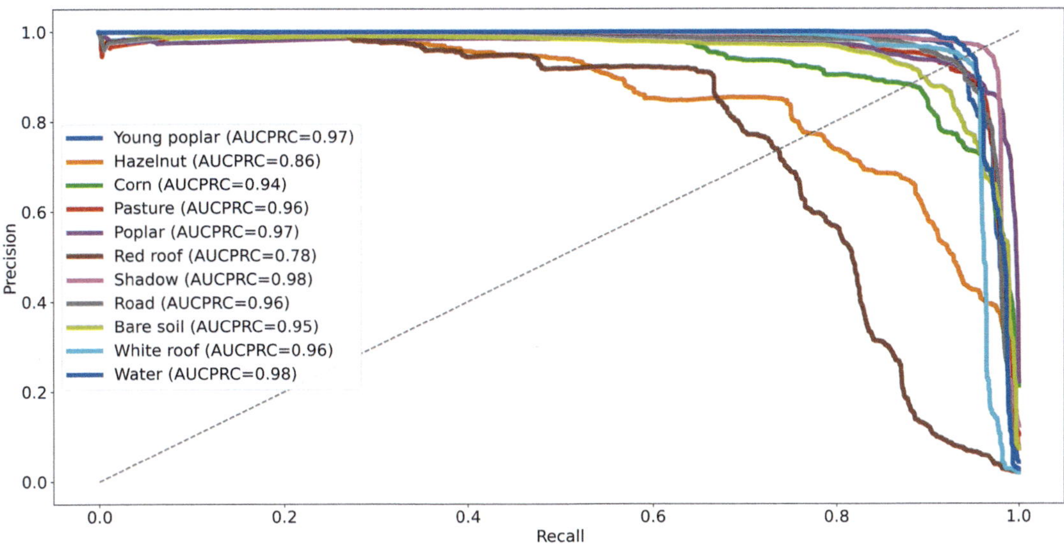

FIGURE 10.8 Precision-recall curves for each class.

After rigorous accuracy assessment analysis of the LULC map, a total of five XAI analyses, namely, permutation-based feature importance, pairwise interaction importance, partial dependence, LIME, and SHAP, were applied to shed light on the decision-making process of the XGBoost model. The results obtained through permutation-based feature importance analysis using the *eli5* library in Python programming language revealed the relative importance of spectral bands for classifying the poplar class in the data set (Figure 10.9). Among the features considered, the red band emerged as the most influential band, with a feature weight of 0.2957±0.0007, indicating its pivotal role in the accurate discrimination of poplar trees. This prominence can be attributed to the red band's sensitivity to chlorophyll content, a key indicator of plant health. Similarly, the red-edge 2 band, with a feature weight of 0.1335±0.0023, showcased its significance in poplar classification, as it is highly responsive to changes in chlorophyll content and biochemical properties. The green band, with a weight of 0.0912±0.0009, also played a substantial role, given its relevance to the reflectance properties of healthy vegetation. Other bands, such as coastal blue, SWIR-1, and SWIR-8, while not as dominant, still contributed to the classification process, possibly reflecting information related to moisture and soil characteristics. In contrast, some bands, including five SWIR bands, exhibited lower feature weights, suggesting a lesser impact on the classification of poplar species.

In addition to the permutation-based feature importance tool, pairwise interaction importance was employed to delve deeper into the potential hidden relationships between spectral bands. This analysis was conducted using the *vip* package in the R programming language. The resulting pairwise band interaction importance plot, depicted in Figure 10.10, showcases the top ten most influential band interactions on the XGBoost model. Upon reviewing the pairwise interaction plots, it becomes evident that the XGBoost model exhibited the highest sensitivity to the interactions between the SWIR-1 and SWIR-8 bands. These interactions had a significant impact on the model's performance. Notably, the strength of interaction between SWIR-1 and SWIR-8, reaching a value of 0.100, surpassed that of other pairwise relationships. This discovery is intriguing as it reveals that SWIR-1 and SWIR-8 frequently interact with other factors and play a pivotal role in the decision-making mechanism of the XGBoost model. Remarkably, they managed to exert this influence even though they were not ranked among the top four factors in the singular permutation-based feature importance analysis. On the contrary, the red-edge band, which was ranked as the second most important band in the permutation-based feature importance analysis, only interacted twice with the SWIR-1 and green bands. This finding suggests that the red-edge band's role in

Feature	Weight
Red	0.2957 ±0.0007
Red edge	0.1335 ±0.0023
Green	0.0912 ±0.0009
Coastal Blue	0.0886 ±0.0009
SWIR-1	0.0755 ±0.0014
SWIR-8	0.0487 ±0.0011
Blue	0.0434 ±0.0014
NIR-2	0.0373 ±0.0008
SWIR-4	0.0192 ±0.0007
NIR-1	0.0164 ±0.0005
Yellow	0.0106 ±0.0003
SWIR-2	0.0022 ±0.0002
SWIR-6	0.0019 ±0.0002
SWIR-5	0.0010 ±0.0003
SWIR-3	0.0004 ±0.0002
SWIR-7	0.0001 ±0.0001

FIGURE 10.9 Permutation-based feature importance of spectral bands for poplar class.

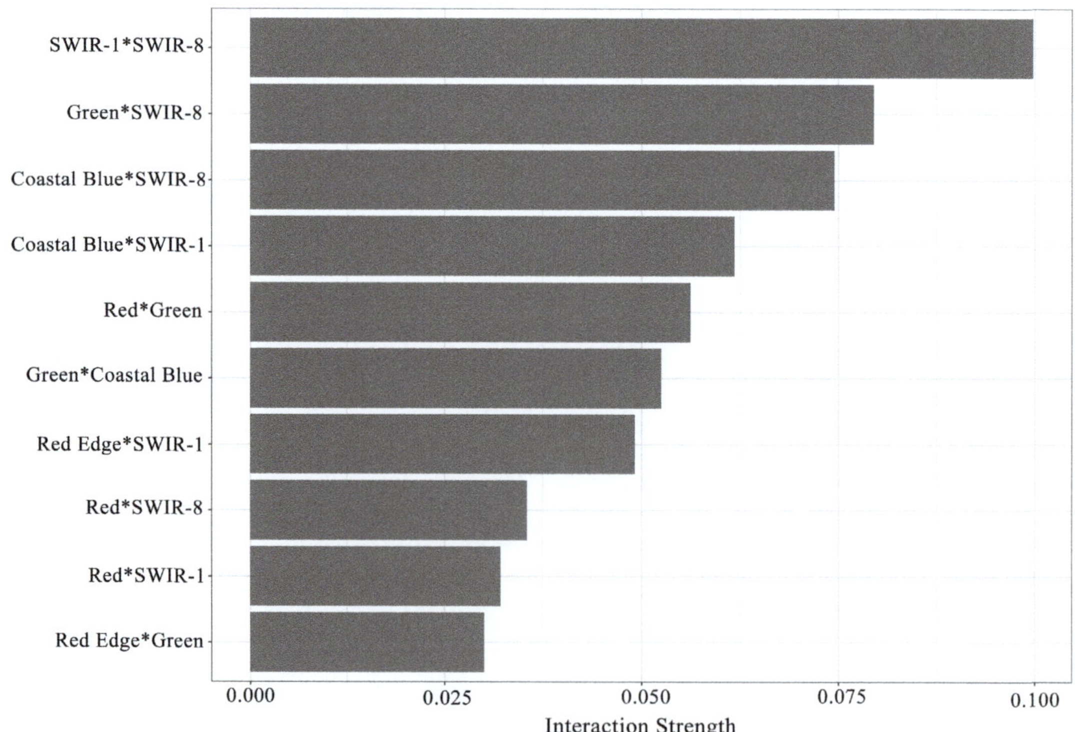

FIGURE 10.10 Pairwise band interaction importance plot.

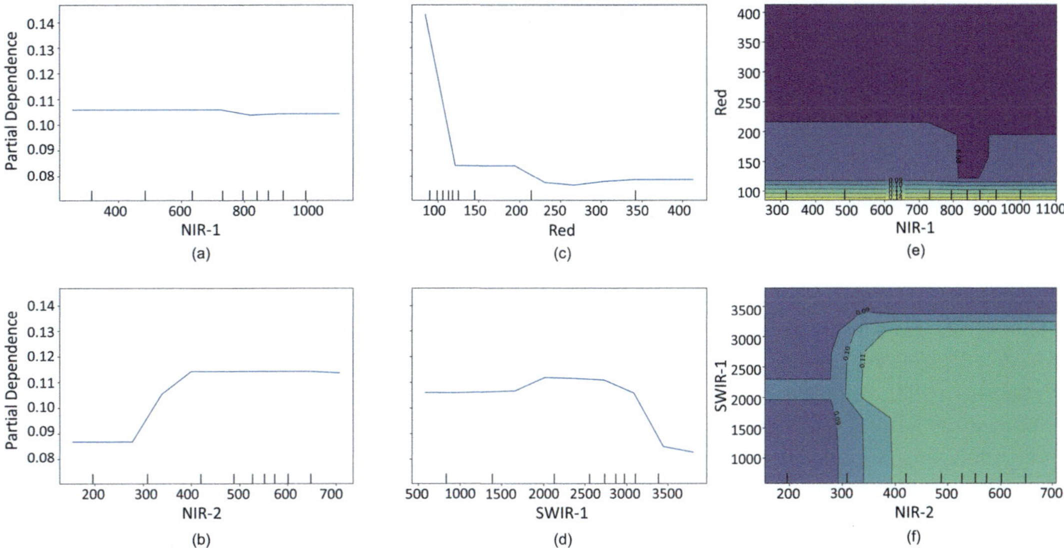

FIGURE 10.11   One-way partial dependence plots: (a) NIR-1, (b) NIR-2, (c) red, (d) SWIR-1, and 2D partial dependence plots for pairwise comparison of spectral bands: (e) red and NIR-1 and (f) SWIR-1 and NIR-2.

the decision-making process is more specific, focusing on select interactions with other bands, as opposed to broad interactions across various bands.

One-way PDPs by using the *alibi* library in Python were also applied to conduct a thorough analysis of the specific band values that exerted a significant influence on poplar classification (Figure 10.11). A notable pattern emerged from these plots, with red, NIR-2, and SWIR-1 showing distinct upward and downward trends at particular values, while NIR-1 exhibited a nearly linear relationship with poplar classification values. When focusing on the red band, a significant decrease in partial dependence scores was evident after pixel values surpassed 125, followed by a relatively stagnant phase up to around 190, and, subsequently, a gradual decline in scores until they stabilized around the pixel value of 220. This pattern suggests that the influence of the red band on poplar classification is influenced by specific factors. One plausible explanation for these observations is that the spectral characteristics of the red band may initially lead to a drop in partial dependence scores at higher pixel values. This phenomenon may be related to the absorption of healthy vegetation in the red waveband. Examining the NIR-2 band, it became apparent that the partial dependence scores initially followed a linear trend up to approximately pixel values of 280. However, it suddenly increased to a partial dependence score of about 0.12 and subsequently resumed an almost linear trend up to the maximum pixel value. It may be linked to a specific characteristic in the NIR-2 band's spectral information that gives higher reflectance for healthy vegetation. In the case of the SWIR-1 band, the PDP indicated an initial near-constant partial dependence score up to the pixel value of 1,650. A slight increase up to approximately pixel values of 2,000 was observed, following a consistent pattern up to the pixel values of around 2,800. However, it subsequently showed a gradual decrease up to the maximum pixel value. This behavior might be associated with changing the moisture properties of the leaves in the SWIR-1 band.

A 2D partial dependence analysis was also conducted to provide a more comprehensive insight into the interplay between spectral bands, specifically examining the relationship between red and NIR-1, as well as SWIR-1 and NIR-2. The two-way partial dependence plot reveals that the connection between red and NIR-1 is more pronounced under certain conditions. This relationship becomes notably stronger when the pixel values of the red band exceed ~210, and simultaneously, the pixel values of NIR-1 fall within the range of 820–910. This observation suggests that specific combinations of pixel values in these two spectral bands are indicative of a more significant influence on the poplar classification. When exploring the relationship between SWIR-1 and NIR-2, it was noted that higher pixel values in both bands have a more substantial impact on the classification of poplar trees, while their lower values are less influential. This suggests that certain combinations of higher pixel values in the SWIR-1 and NIR-2 bands are indicative of a stronger influence on the decision of the model regarding poplar classification. In short, these findings point out the importance of specific high-value ranges within these two spectral bands in accurately classifying poplar samples.

To illuminate the decision-making process of the XGBoost model in poplar classification using the LIME approach with the *lime* library in Python, a randomly selected poplar sample from the study area was used as an illustrative example (Figure 10.12). LIME provides a comprehensive breakdown of the explanation results in the form of a plot, which comprises three primary components: most relevant features, prediction probabilities, and feature values. At the right part of the figure, a table is presented, listing the features for interpreting the model's decision as prediction probabilities. These probabilities are generated by the XGBoost algorithm and serve to depict the relationships among all features in the training model. They offer insights into how different features, including the ones listed in the table, contribute to the final classification decision. The subfigure in the left section also includes all the bands used in the analysis, along with their specific values. This information helps in comprehending the significance of each spectral band and its impact on the model's prediction for the specific poplar sample under consideration.

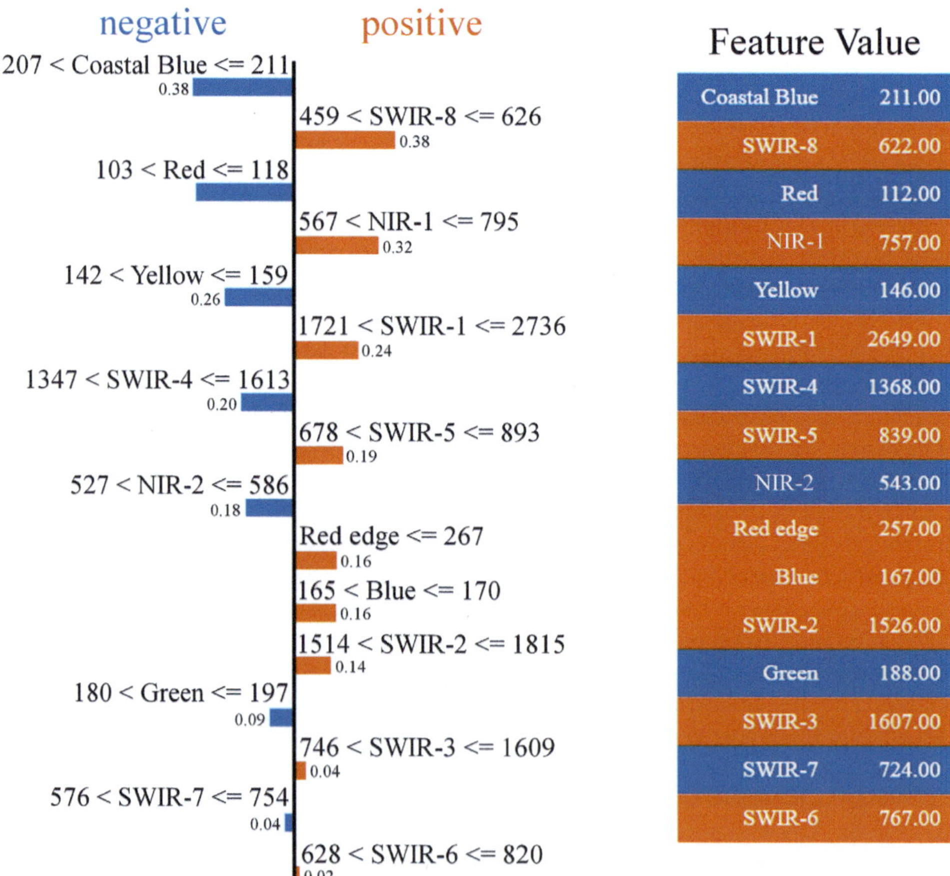

FIGURE 10.12   Local explanation with LIME for a poplar instance.

A randomly selected poplar sample from the data set was used to provide a local explanation of the influential spectral bands on the XGBoost model's prediction. In this particular case, nine bands were found to positively support the model's prediction, contributing significantly to its decision. These bands include SWIR-8, NIR-1, SWIR-1, SWIR-5, red-edge, Blue, SWIR-2, SWIR-3, and SWIR-6. Their spectral characteristics played a vital role in the local decision of the model for this specific sample. Conversely, seven spectral bands were identified as contradicting the model's prediction for the same poplar sample. These bands, namely, coastal blue, red, yellow, SWIR-4, NIR-2, green, and SWIR-7, appeared to be at odds with the model's assessment. The local explanations further revealed specific insights. For example, they indicated that pixel values of the coastal blue band between 207 and 211 would, on average, decrease the predicted probability of the poplar class by 0.38. In contrast, pixel values within the range of 459–626 in the SWIR-8 band would increase the predicted probability by 0.38. This suggests that the coastal blue band at lower values had a negative impact on the model's prediction for the poplar class, while the SWIR-8 band within the specified range had a positive impact.

The SHAP plot, derived through the summation of localized interpretations acquired from the SHAP tree annotation function, ranks the significance of spectral bands. Essentially, the SHAP methodology assesses the anticipated incremental contribution of each feature. As depicted in Figure 10.13, the y-axis represents the spectral bands within the WV-3 data set, the x-axis denotes

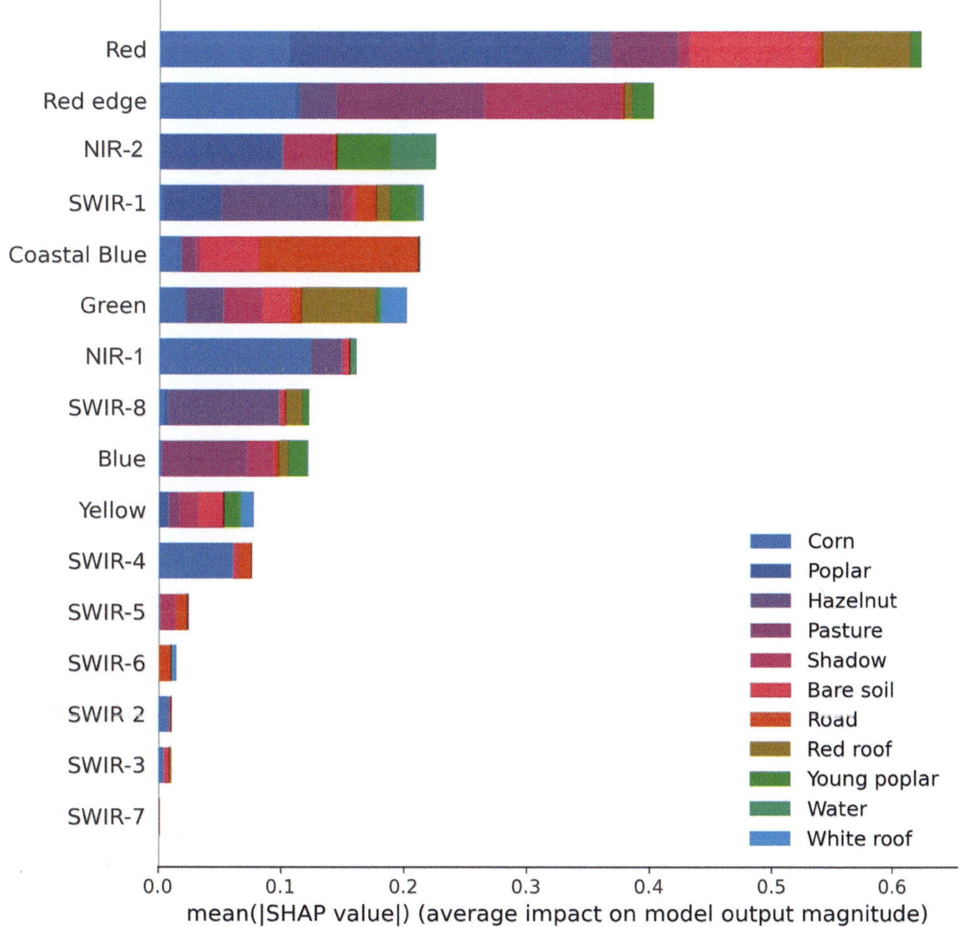

FIGURE 10.13 Global SHAP analysis for the model explanations.

the estimated Shapley value, and the coloration (blue, red, purple, pink, orange, and green) signifies the extent of influence exerted by the spectral bands on the LULC classes. As can be seen, the influence of the red band in the WV-3 image exhibited a notably greater effect on the output of the XGBoost classifier compared to the remaining spectral bands, as pointed out by permutation-based feature importance. This observation emphasizes that changes in this particular band value can lead to considerable impacts on the model prediction. Following the red band, the subsequent two bands of notable significance are the red-edge and NIR-2 bands, respectively. The rationale behind the elevated Shapley values associated with these spectral bands can be attributed to the intense coverage of vegetation prevalent in the study area. On the other hand, upon scrutinizing the graph, it becomes evident that the spectral bands with relatively lower significance are SWIR-5, SWIR-6, SWIR-2, and SWIR-3, which is similar to the findings in premutation-based feature importance. In summary, the results reveal that the predominant land cover classes in the study area are primarily vegetation, and certain SWIR bands exert a comparatively lesser influence on the classifier's capacity for detection.

The XGBoost classifier underwent a training process, followed by the generation of SHAP beeswarm plots to assess the significance of individual features in classifying the poplar class (Figure 10.14). These plots incorporate three key attributes: (1) the vertical position along the y-axis

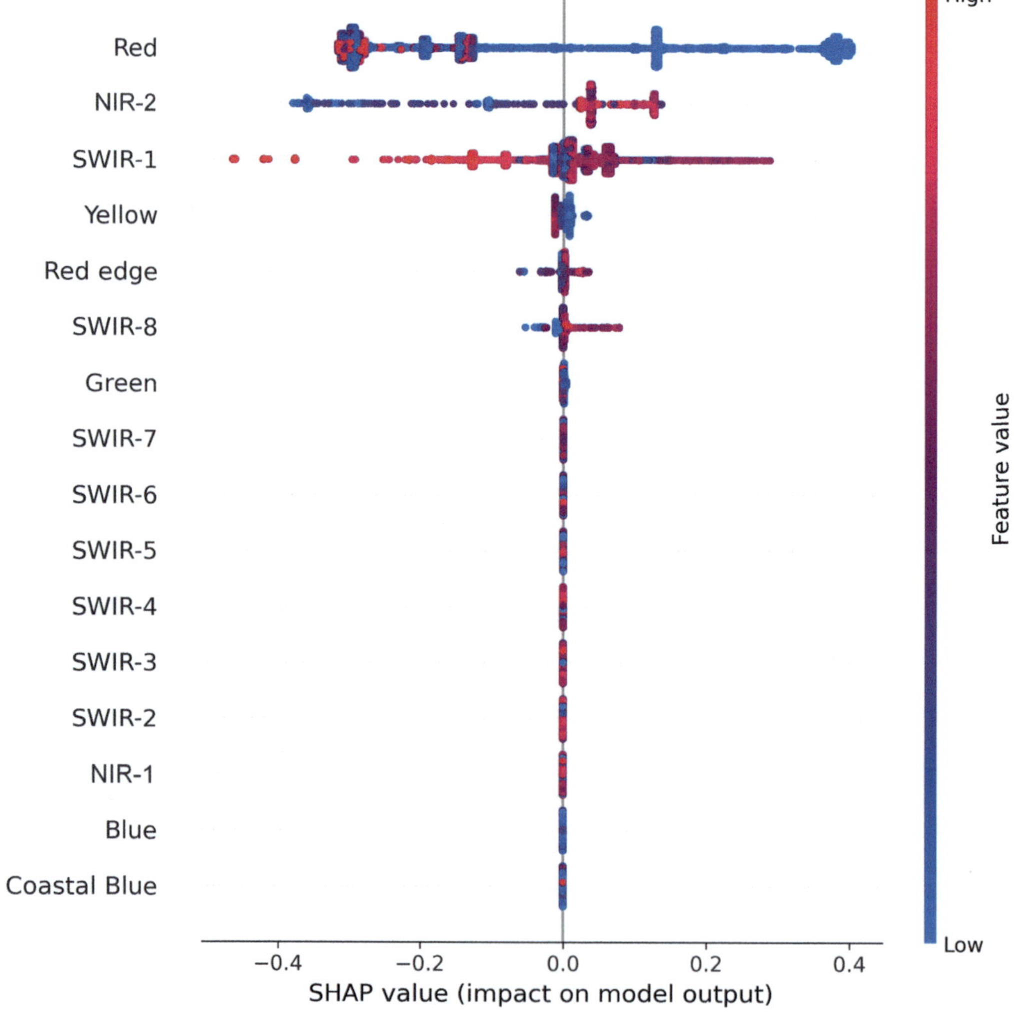

FIGURE 10.14 Beeswarm SHAP graph for poplar class.

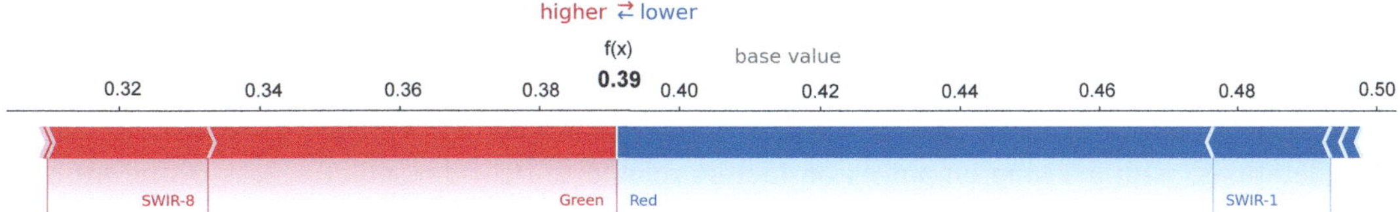

FIGURE 10.15   Force plot SHAP graph of a poplar observation.

reflects the factor's level of importance, (2) coloration distinguishes whether the factor exerts a high (red) or low (blue) influence on a given data row, and (3) the horizontal axis denotes the impact of the Shapley value, indicating whether it leads to a contribution or irrelevancy in the prediction. It is noteworthy that each point on the plot corresponds to a specific row within the training data set. Given the elevated Shapley values, it is evident that the red, NIR-2, and SWIR-1 bands exhibit a notable disparity in their efficacy for discriminating poplar planted areas. Nevertheless, it is important to note that the red band demonstrates limited predictive capability for discerning poplar plantations, as evidenced by the concentration of data points with high values predominantly situated on the left side of the graph. The NIR-2 band exhibits a positive influence on the poplar class prediction, signifying the efficacy of this index in discriminating the poplar class from other pixel categories. It was observed that elevated values of the SWIR-1 band, as indicated by the red dots, demonstrated both positive and negative impacts on the accurate identification of poplar areas. Ultimately, it can be deduced that the remaining spectral bands exert minimal influence on the detection of poplar areas, which can also be seen from the figure.

The SHAP force plot provides an alternative means of assessing the impact of each feature on the prediction for a specific observation. In this representation, positive SHAP values are depicted on the left side (in red), while negative values appear on the right side (in blue). The emphasized value corresponds to the prediction made for that particular observation (Figure 10.15). Within the graphical representation, factors highlighted in red (e.g., Green and SWIR-8) exert positive effects on the predictive performance of the model, whereas factors presented in blue (e.g., Red and SWIR-1) have a detrimental influence on the likelihood of poplar class prediction. Moreover, the magnitude of the color scale associated with the factors correlates with the significance of the respective feature.

10.5   CONCLUSIONS AND GUIDELINES FOR BEST PRACTICE

The estimation of thematic map accuracy and validation of the applied methodology are essential components of any classification exercise. No classification is complete unless it includes validation and accuracy estimation stages. Without a doubt, the realm of accuracy assessment has made significant methodological advancements in recent years. Nonetheless, new issues arise with the use of state-of-the-art machine learning approaches, particularly DL models employed in image classification, object detection, instance segmentation, and panoptic segmentation problems. There is a growing consensus for the improvement of accuracy reporting in the creation of thematic and change detection maps produced through remote sensing analysis. This change is meant to align more effectively with new classification techniques and enhance the transparency of reporting (Olofsson et al., 2014; Castilla, 2016; Morales-Barquero et al., 2019; Stehman and Foody, 2019). Therefore, accuracy assessment should remain in a state of continuous improvement to fulfill the requirements of scientific credibility. This might entail, among other actions, the elimination of flawed practices and the establishment of more rigorous criteria for providing more reliable accuracy estimates considering the uncertainties and error propagation in the entire processing chain. Stehman and Czaplewski (1998) identified three key components (sampling design, response design, and accuracy analysis) that serve as the foundation for accuracy analysis in this chain. Therefore, achieving best practices entails not only focusing on these principles individually but also taking into account the interactions between these components. Sampling design is vital for subsequent statistical analysis to give objective and valid statistical inferences. The quality of a thematic map depends on two major factors: the data incorporated into the training and validation phases and the degree of robustness of the classification method employed. It should be underlined that the accuracy that can be achieved in a classification project is significantly influenced by a variety of preclass allocation considerations, with a strong emphasis on factors related to the representativeness of the training data. Undoubtedly, the quality of the training data can have a greater effect on classification accuracy than the classification method per se.

To achieve high accurate results, the characteristics of the samples and the design and parameterization of the classification method should be studied with care. A critical issue that undermines the reliability or trustworthiness of the thematic maps is that significantly different results may be obtained when different data sets and classifiers are employed in the classification of land cover features in the same site. Thematic maps become meaningful, and gain validity and credibility when their accuracy is sufficient and transparent to users. This requires providing clear definitions for all the steps involved in a classification process. This problem calls out three vital principles for good practices in accuracy assessment: reproducibility, reliability, and transparency. Achieving reproducibility in classification tasks requires a comprehensive definition of sampling parameters (e.g., sampling unit, sampling methodology, and class-specific sample sizes), as well as comprehensive explanations of the considerations involved in the selection and parameterization of the chosen classifier. Reliability in accuracy assessment is related to the stability of the results or the variation in accuracy estimates. A reliable evaluation would be demonstrated by low standard errors in accuracy and area estimates, coupled with limited variability among predictors when assigning reference class labels. The inclusion of standard errors or confidence intervals in the report serves as a significant measure of reliability. The reporting of standard errors or confidence intervals is a crucial component of assessing reliability, as it quantifies the extent of consistency in the estimates (Stehman and Foody, 2019). The lack of transparency, coupled with deficiencies in rigorous evaluation, is also a cause for major concern. This concern is particularly relevant as remote sensing analysis becomes increasingly sophisticated and automated in delivering spatial data products and information services across various disciplines and applications (Morales-Barquero et al., 2019).

DL algorithms have demonstrated effectiveness in various remote sensing applications, while it is crucial to acknowledge that their predictions may encompass uncertainty due to several factors, such as inadequate representations of spatial data and its underlying distribution, data noise, model complexity, and distribution shifts, among others, ultimately resulting in less reliable outcomes. In particular, when comparing two thematic maps produced by DL algorithms, the need to address and quantify the uncertainties becomes paramount. For example, consider a DL algorithm trained to classify land cover types from satellite imagery. During the training and validation process, the algorithm performs exceptionally well, achieving high accuracy. When this model is applied to testing data, its performance may degrade, leading to uncertainties in the accuracy of the thematic map for areas not well represented in the training set. Differences between the training and testing data distributions can lead to uncertainty in the model's predictions. Another example can be given that the training data set may not adequately represent the diversity of land cover types in a particular region. For instance, if a DL model is trained primarily on urban areas, its performance in mapping rural or forested regions may be less reliable. The model's predictions for underrepresented classes or regions may be less accurate, introducing uncertainty when comparing thematic maps across diverse landscapes. As a result, limited or biased representation of certain classes in the training data can result in uncertainties when classifying underrepresented classes in real-world scenarios. To address and quantify uncertainties in DL models, various uncertainty quantification algorithms can be employed. In the literature, a variety of uncertainty quantification algorithms has been proposed for different DL-based tasks, including Monte Carlo dropout (Milanés-Hermosilla et al., 2021; Choubineh et al., 2023; Zhao et al., 2024) and Bayesian DL or Bayesian neural networks (Izmailov et al., 2019; Olivier et al., 2021; Fernández et al., 2022). These approaches provide measures of confidence or uncertainty associated with model predictions.

In addition to these criteria, the explainability or interpretability of the models is of critical importance to understanding the decision-making process of the models. Without a clear understanding of how the model arrives at its conclusions, it becomes problematic to validate its accuracy, assess potential biases or limitations, and substantiate the reliability of its outputs. Opaque or black-box models provide a limited degree of transparency. Their internal mechanisms are often hidden and lack clarity, obstructing the comprehension and interpretation of the decision-making process and creating difficulties in understanding the rationale behind their conclusions.

Keeping all the preceding critical observations and associated suggestions in mind, particularly from the recently published seminal studies of Morales-Barquero et al. (2019), Stehman and Foody (2019), and Maxwell et al. (2021a,b), we thus choose to present some guidelines for best practice and reporting standards for accuracy assessment in remote sensing field, which can be also adopted by other fields.

- Detailed information about the sampling scheme, sample unit (e.g., pixels, image patches, clusters), sample sizes for each class in training, validation and test data sets, and class definitions should be provided. Employing independent data collected through a probability sampling design, avoiding spatial autocorrelation, is recommended.
- Ground reference data can be presented as a separate thematic map showing the distributions of the reference data. If certain categories are omitted despite their presence in the study area, it is important to provide explanatory information and reasons for this decision.
- The recommendation is to utilize balanced data in both the training and validation phases. Unbalanced data present a skewed distribution of observations across targeted population subgroups. Their use in model training and statistical tests may cause serious problems that the analyst must be aware of. Using a significantly larger number of samples for classes with distinct spectral features, in contrast to other classes (e.g., the water class in land cover classification) may lead to overestimated map-level accuracies, as these classes will carry more weight in accuracy estimations.
- When reporting the results of accuracy assessment, map-level and class-level accuracy metrics together with a complete (raw) error matrix (except for the cases where too large error matrices are produced, e.g., scene classification) should be provided. It should be borne in mind that map-level accuracy measures can be misleading for subregions of the image, where significantly higher or lower error rates may exist. In addition, normalization by dividing the values by the total numbers in the reference category (normalized error matrix) should be avoided due to its potential to introduce bias and imprecision into accuracy estimates (Stehman, 2004b; Stehman and Foody, 2019).
- Although it is one of the most widely used map-level accuracy measures, many studies (e.g., Pontius Jr and Millones, 2011; Olofsson et al., 2014; Stehman and Foody, 2019; Foody, 2020) urge to abandon the use of Kappa coefficient in accuracy assessment. Therefore, the use of the Kappa coefficient should be avoided due to the well-grounded criticism that it can be misleading and flawed for practical applications in remote sensing. It should be also underlined that overall accuracy and the Kappa coefficient are highly correlated accuracy metrics.
- In the analysis of per-class accuracies, omission and commission errors are particularly important when the class balance cannot be achieved. Reporting one-sided accuracy measures by ignoring either omission or commission errors (e.g., providing only the user's accuracy not the producer's accuracy, or presenting recall not precision) can result in a partial understanding of the resulting prediction. In this perspective, only average values of user's and producer's accuracies should not be reported.
- With the availability of more reference data for classification and application of cross-validation approaches, accuracy estimates are performed on different subsets. Hence, the practitioners need to report variance and confidence intervals together with the accuracy measures.

- Studies including per-polygon assessments or OBIA experiments should present information about the segmentation quality (Section 8.7) along with the polygon-based accuracy metrics.
- In DL-based studies, loss and accuracy learning curves for training and validation data sets should be also provided to analyze the learning process whether underfitting or overfitting occurred.
- The use of ROC and P-R curves is suggested to analyze the performance of the models in their predictions.
- DL-based models can exhibit superior performances in classification problems. In many studies, map-based and class-based accuracies of over 95% have been reported when different DL models or DL models with various hyperparameter settings and architectural considerations are compared. However, discrepancies in classification results may arise for the areas outside ground reference data, highlighting the importance of the representativeness of ground reference data and the generalization capabilities of DL models. To ensure the highest level of reliability in the obtained results, it is crucial to evaluate uncertainty quantification, as discussed earlier, in scenarios involving DL-based classification.
- Given the advancements and widespread adoption of black-box machine learning and DL techniques in remote sensing, it should be a common practice to include global and local explainability models when reporting the outcomes of image analysis. These models help to understand the overall behavior of the trained models and the contribution of each feature to the decisions of the model. On the other hand, local explainability may be required in cases where a single instance (e.g., a segment or a polygon) or a particular section of the study area is a major concern.
- If a specific code is developed, or a data set is created for conducting a study, it is suggested that both codes and data be shared on a public platform to meet the requirements of reproducibility, reliability, and transparency. Thus, the impact of the paper will increase and both code and data will be objectively tested by other researchers.
- Finally, any limitations related to the data and issues related to any stage of the process should be reported.

REFERENCES

Abdollahi, A., and B. Pradhan. 2021. Urban vegetation mapping from aerial imagery using explainable AI (XAI). *Sensors* 21:4738. https://doi.org/10.3390/s21144738.

Adadi, A., and M. Berrada. 2018. Peeking inside the black-box: A survey on explainable artificial intelligence (XAI). *IEEE Access* 6:52138–52160. https://doi.org/10.1109/ACCESS.2018.2870052.

Alem, A., and S. Kumar. 2022. Transfer learning models for land cover and land use classification in remote sensing image. *Applied Artificial Intelligence* 36:2014192. https://doi.org/10.1080/08839514.2021.2014192.

Alpaydın, E. 1999. Combined 5×2 cv F test for comparing supervised classification learning algorithms. *Neural Computation* 11:1885–1892. https://doi.org/10.1162/089976699300016007.

Anderson, J. R., E. E. Hardy, J. T. Roach, and R. E. Witmer. 1976. A land-use and land-cover classification system for use with remote sensor data. In *US Geological Survey Professional Paper 964*, Washington, DC: US Government Printing Office. https://doi.org/10.3133/pp964.

Angelov, P. P., E. A. Soares, R. Jiang, N. I. Arnold, and P. M. Atkinson. 2021. Explainable artificial intelligence: An analytical review. *Wiley Interdisciplinary Reviews: Data Mining and Knowledge Discovery* 11:e1424. https://doi.org/10.1002/widm.1424.

Arrieta, A. B., N. Díaz-Rodríguez, J. Del Ser, A. Bennetot, S. Tabik, A. Barbado, S. Garcia, et al. 2020. Explainable artificial intelligence (XAI): Concepts, taxonomies, opportunities and challenges toward responsible AI. *Information Fusion* 58:82–115. https://doi.org/https://doi.org/10.1016/j.inffus.2019.12.012.

Atkinson, P. M. 1991. Optimal ground-based sampling for remote sensing investigations: Estimating the regional mean. *International Journal of Remote Sensing* 12:559–567. https://doi.org/10.1080/01431169108929672.

Atkinson, P. M. 1996. Optimal sampling strategies for raster-based geographical information systems. *Global Ecology and Biogeographical Letters* 5:271–280. https://doi.org/10.2307/2997795.

Atkinson, P. M., and A. R. L. Tatnall. 1997. Introduction to neural networks in remote sensing. *International Journal of Remote Sensing* 18:699–709. https://doi.org/10.1080/014311697218700.

Berry, B. J. L., and A. M. Baker. 1968. Geographical sampling. In *Spatial Analysis*, (eds.) B. J. L. Berry, and D. F. Marble, pp. 91–100. Englewood Cliffs, NJ: Prentice-Hall.

Binaghi, E., P. A. Brivio, P. Ghezzi, and A. Rampini. 1999. A fuzzy set-based accuracy assessment of soft classification. *Pattern Recognition Letters* 20:935–948. https://doi.org/10.1016/S0167-8655(99)00061-6.

Blaschke, T. 2010. Object based image analysis for remote sensing. *ISPRS Journal of Photogrammetry and Remote Sensing* 65:2–16. https://doi.org/10.1016/j.isprsjprs.2009.06.004.

Borak, J. S., and A. H. Strahler. 1999. Feature selection and land cover classification of a MODIS-like data set for semi-arid environment. *International Journal of Remote Sensing* 20:919–938. https://doi.org/10.1080/014311699212993.

Bourgeau-Chavez, L., S. Endres, M. Battaglia, M. E. Miller, E. Banda, Z. Laubach, P. Higman, et al. 2015. Development of a bi-national great lakes coastal wetland and land use map using three-season PALSAR and Landsat imagery. *Remote Sensing* 7:8655–8682. https://doi.org/10.3390/rs70708655.

Breiman, L. 2001. Random forests. *Machine Learning* 45:5–32. https://doi.org/10.1023/A:1010933404324.

Brodley, C., and M. A. Friedl. 1996. Improving automated land cover mapping by identifying and eliminating mislabelled observations from training data. In *Proceedings of the International Geoscience and Remote Sensing Symposium (IGARSS'96)*, May 27–31, Lincoln, NE, pp. 1382–1384. https://doi.org/10.1109/IGARSS.1996.516669.

Brodley, C., and M. A. Friedl. 1999. Identifying mislabelled training data. *Journal of Artificial Intelligence Research* 11:131–167. https://doi.org/10.1613/jair.606.

Buda, M., A. Maki, and M. A. Mazurowski. 2018. A systematic study of the class imbalance problem in convolutional neural networks. *Neural Networks* 106:249–259. https://doi.org/10.1016/j.neunet.2018.07.011.

Campbell, J. B. 1987. *Introduction to Remote Sensing*. London: The Guilford Press.

Carvalho, D. V., E. M. Pereira, and J. S. Cardoso. 2019. Machine learning interpretability: A survey on methods and metrics. *Electronics* 8:832. https://doi.org/10.3390/electronics8080832.

Castilla G. 2016. We must all pay more attention to rigor in accuracy assessment: Additional comment to "The improvement of land cover classification by thermal remote sensing". *Remote Sensing* 8:288. https://doi.org/10.3390/rs8040288.

Chawla, N. V., K. W. Bowyer, L. O. Hall, and W. P. Kegelmeyer. 2002. SMOTE: Synthetic minority over-sampling technique. *Journal of Artificial Intelligence Research* 16:321–357. https://doi.org/10.1613/jair.953.

Chen, D., and D. Stow. 2002. The effect of training strategies on supervised classification at different spatial resolutions. *Photogrammetric Engineering and Remote Sensing* 68:1155–1162.

Chen, D., and H. Wei. 2009. The effect of spatial autocorrelation and class proportion on the accuracy measures from different sampling designs. *ISPRS Journal of Photogrammetry and Remote Sensing* 64:140–150. https://doi.org/10.1016/j.isprsjprs.2008.07.004.

Chen, Y. X., K. Qin, Y. Liu, S. Z. Gan, and Y. Zhan. 2012. Feature modelling of high resolution remote sensing images considering spatial autocorrelation. *The International Archives of the Photogrammetry, Remote Sensing and Spatial Information Sciences* 39:467–472. https://doi.org/10.5194/isprsarchives-XXXIX-B3-467-2012.

Choubineh, A., J. Chen, F. Coenen, and F. Ma. 2023. Applying Monte Carlo dropout to quantify the uncertainty of skip connection-based convolutional neural networks optimized by big data. *Electronics* 12:1453. https://doi.org/10.3390/electronics12061453.

Cochran, W. G. 1977. Sampling *Techniques*. New York: John Wiley & Sons.

Cohen, J. 1960. A coefficient of agreement for nominal scales. *Educational and Psychological Measurement* 20:37–46. https://doi.org/10.1177/001316446002000104.

Comber, A., P. Fisher, C. Brunsdon, and A. Khmag. 2012. Spatial analysis of remote sensing image classification accuracy. *Remote Sensing of Environment* 127:237–246. https://doi.org/10.1016/j.rse.2012.09.005.

Conese, C., and F. Maselli. 1993. Selection of optimum bands from TM scenes through mutual information analysis. *ISPRS Journal of Photogrammetry and Remote Sensing* 48:2–11. https://doi.org/10.1016/0924-2716(93)90059-V.

Congalton, R. G. 1988. A comparison of sampling schemes used in generating error matrices for assessing the accuracy of maps generated from remotely sensed data. *Photogrammetric Engineering and Remote Sensing* 54:593–600.

Congalton, R. G. 1991. A review of assessing the accuracy of classifications of remotely sensed data. *Remote Sensing of Environment* 37:35–46. https://doi.org/10.1016/0034-4257(91)90048-B.

Congalton, R. G. 2004. Putting the map back in map accuracy assessment. In *Remote Sensing and GIS Accuracy Assessment*, (eds.) R. S. Lunetta, and J. G. Lyon, pp. 1–12. Boca Raton, FL: CRC Press. https://doi.org/10.1201/9780203497586.ch1.

Congalton, R. G., and K. Green. 2019. *Assessing the Accuracy of Remotely Sensed Data: Principles and Practices*, 3rd edition. Boca Raton, FL: CRC Press. https://doi.org/10.1201/9780429052729.

Congalton, R. G., and L. C. Plourde. 2000. Sampling methodology, sampling placement and other important factors in assessing the accuracy of remotely sensed forest maps. In *Accuracy 2000*, (eds.) G. B. M. Heuvelink, and M. J. P. M. Lemmens, pp. 117–124. Amsterdam: Delft University Press.

Curran, P. J. 1988. The semi-variogram in remote sensing: An introduction. *Remote Sensing of Environment* 24:493–507. https://doi.org/10.1016/0034-4257(88)90021-1.

Dahal, A., and L. Lombardo. 2023. Explainable artificial intelligence in geoscience: A glimpse into the future of landslide susceptibility modeling. *Computers & Geosciences* 176:105364. https://doi.org/10.1016/j.cageo.2023.105364.

Das, S., S. Datta, and B. B. Chaudhuri. 2018. Handling data irregularities in classification: Foundations, trends, and future challenges. *Pattern Recognition* 81:674–693. https://doi.org/10.1016/j.patcog.2018.03.008.

De Leeuw, J., H. Jia, L. Yang, X. Liu, K. Schmidt, and A. K. Skidmore. 2006. Comparing accuracy assessments to infer superiority of image classification methods. *International Journal of Remote Sensing* 27:223–232. https://doi.org/10.1080/01431160500275762.

Dietterich, T. G. 1998. Approximate statistical tests for comparing supervised classification learning algorithms. *Neural Computation* 10:1895–1923. https://doi.org/10.1162/089976698300017197.

Dobbertin, M., and S. B. Gregory. 1996. A simulation study of the effect of scene autocorrelation, training sample size and sampling method on classification accuracy. *Canadian Journal of Remote Sensing* 22:360–367. https://doi.org/10.1080/07038992.1996.10874660.

Espindola, G. M., G. Câmara, I. A. Reis, L. S. Bins, and A. M. Monteiro. 2006. Parameter selection for region-growing image segmentation algorithms using spatial autocorrelation. *International Journal of Remote Sensing* 27:3035–3040. https://doi.org/10.1080/01431160600617194.

Evans, F. 1998. An investigation into the use of maximum likelihood classifiers, decision trees, neural networks and conditional probabilistic networks for mapping and predicting salinity. M.Sc. Thesis, Department of Computer Science, Curtin University of Technology, Australia.

Fan, C., and S. Myint. 2014. A comparison of spatial autocorrelation indices and landscape metrics in measuring urban landscape fragmentation. *Landscape and Urban Planning* 121:117–128. https://doi.org/10.1016/j.landurbplan.2013.10.002.

Fel, T., D. Vigouroux, R. Cadène, and T. Serre. 2022. How good is your explanation? Algorithmic stability measures to assess the quality of explanations for deep neural networks. In *Proceedings of the 2022 IEEE/CVF Winter Conference on Applications of Computer Vision (WACV)*, January 3–8, Waikoloa, HI, pp. 720–730. https://doi.org/10.1109/WACV51458.2022.00163.

Fernández, A., S. del Río, N. V. Chawla, and F. Herrera. 2017. An insight into imbalanced big data classification: Outcomes and challenges. *Complex & Intelligent Systems* 3:105–120. https://doi.org/10.1007/s40747-017-0037-9.

Fernández, J., M. Chiachío, J. Chiachío, R. Muñoz, and F. Herrera. 2022. Uncertainty quantification in neural networks by approximate Bayesian computation: Application to fatigue in composite materials. *Engineering Applications of Artificial Intelligence* 107:104511. https://doi.org/10.1016/j.engappai.2021.104511.

Finn, J. T. 1993. Use of the average mutual information index in evaluating classification error and consistency. *International Journal of Geographical Information Systems* 7:349–366. https://doi.org/10.1080/02693799308901966.

Fisher, A., C. Rudin, and F. Dominici. 2019. All models are wrong, but many are useful: Learning a variable's importance by studying an entire class of prediction models simultaneously. *Journal of Machine Learning Research* 20:1–81.

Fisher, P. F., and S. Pathirana. 1990. The evaluation of fuzzy membership of land cover classes in the suburban zone. *Remote Sensing of Environment* 34:121–132. https://doi.org/10.1016/0034-4257(90)90103-S.

Fitzpatrick-Lins, K. 1981. Comparison of sampling procedures and data analysis for a land-use and land-cover map. *Photogrammetric Engineering and Remote Sensing* 47:343–351.

Foody, G. M. 1992. On the compensation for chance agreement in image classification accuracy assessment. *Photogrammetric Engineering and Remote Sensing* 58:1459–1460.

Foody, G. M. 1996. Approaches for the production and evaluation of fuzzy land cover classifications from remotely-sensed data. *International Journal of Remote Sensing* 17:1317–1340. https://doi.org/10.1080/01431169608948706.

Foody, G. M. 1999. The significance of border training patterns in classification by a feedforward neural network using back propagation learning. *International Journal of Remote Sensing* 20:3549–3562. https://doi.org/10.1080/014311699211192.

Foody, G. M. 2000. Accuracy of thematic maps derived by remote sensing. In *Accuracy 2000*, (eds.) G. B. M. Heuvelink and M. J. P. M. Lemmens, pp. 217–224. Amsterdam: Delft University Press.

Foody, G. M. 2004. Thematic map comparison: Evaluating the statistical significance of differences in classification accuracy. *Photogrammetric Engineering & Remote Sensing* 70:627–633. https://doi.org/10.14358/PERS.70.5.627.

Foody, G. M. 2005. Local characterization of thematic classification accuracy through spatially constrained confusion matrices. *International Journal of Remote Sensing* 26:1217–1228. https://doi.org/10.1080/01431160512331326521.

Foody, G. M. 2008. Harshness in image classification accuracy assessment. *International Journal of Remote Sensing* 29:3137–3158. https://doi.org/10.1080/01431160701442120.

Foody, G. M. 2009a. Classification accuracy comparison: Hypothesis tests and the use of confidence intervals in evaluations of difference, equivalence and non-inferiority. *Remote Sensing of Environment* 113:1658–1663. https://doi.org/10.1016/j.rse.2009.03.014.

Foody, G. M. 2009b. Sample size determination for image classification accuracy assessment and comparison. *International Journal of Remote Sensing* 30:5273–5291. https://doi.org/10.1080/01431160903130937.

Foody, G. M. 2010. Assessing the accuracy of land cover change with imperfect ground reference data. *Remote Sensing of Environment* 114:2271–2285. https://doi.org/10.1016/j.rse.2010.05.003.

Foody, G. M. 2020. Explaining the unsuitability of the Kappa coefficient in the assessment and comparison of the accuracy of thematic maps obtained by image classification. *Remote Sensing of Environment* 239:111630. https://doi.org/10.1016/j.rse.2019.111630.

Foody, G. M., and A. Mathur. 2004. Toward intelligent training of supervised image classifications: Directing training data acquisition for SVM classification. *Remote Sensing of Environment* 93:107–117. https://doi.org/10.1016/j.rse.2004.06.017.

Foody, G. M., and A. Mathur. 2006. The use of small training sets containing mixed pixels for accurate hard image classification: Training on mixed spectral responses for classification by a SVM. *Remote Sensing of Environment* 103:179–189. https://doi.org/10.1016/j.rse.2006.04.001.

Foody, G. M., A. Mathur, C. Sanchez-Hernandez, and D. S. Boyd. 2006. Training set size requirements for the classification of a specific class. *Remote Sensing of Environment* 104:1–14. https://doi.org/10.1016/j.rse.2006.03.004.

Foody, G. M., and D. S. Boyd. 1999. Fuzzy mapping of tropical land cover along an environmental gradient from remotely sensed data with an artificial neural network. *Journal of Geographical Systems* 1:23–35. https://doi.org/10.1007/s101090050003.

Foody, G. M., M. B. McCullagh, and W. B. Yates. 1995. The effect of training set size and composition on artificial neural net classification. *International Journal of Remote Sensing* 16:1707–1723. https://doi.org/10.1080/01431169508954507.

Fowler, J., F. Waldner, and Z. Hochman. 2020. All pixels are useful, but some are more useful: Efficient in situ data collection for crop-type mapping using sequential exploration methods. *International Journal of Applied Earth Observation and Geoinformation* 91:102114. https://doi.org/10.1016/j.jag.2020.102114.

Friedl, M. A., C. Woodcock, S. Gopal, D. Muchoney, A. H. Strahler, and C. Barker-Schaaf, C. 2000. A note on procedures used for accuracy assessment in land cover maps derived from AVHRR data. *International Journal of Remote Sensing* 21:1073–1077. https://doi.org/10.1080/014311600210434.

Friedman, M. 1937. The use of ranks to avoid the assumption of normality implicit in the analysis of variance. *Journal of the American Statistical Association* 32:676–701. https://doi.org/10.1080/01621459.1937.10503522.

Gadiraju, K. K., and R. R. Vatsavai, 2023. Remote sensing based crop type classification via deep transfer learning. *IEEE Journal of Selected Topics in Applied Earth Observations and Remote Sensing* 16:4699–4712. https://doi.org/10.1109/JSTARS.2023.3270141.

Ge, J., J. Qi, B. M. Lofgren, N. Moore, N. Torbick, and J. M. Olson. 2007. Impacts of land use/cover classification accuracy on regional climate simulations. *Journal of Geophysical Research: Atmospheres* 112:1–12. https://doi.org/10.1029/2006JD007404.

Ginevan, M. E. 1979. Testing land-use map accuracy: Another look. *Photogrammetric Engineering and Remote Sensing* 45:1371–1377.

Gopal, S., and C. E. Woodcock. 1994. Theory and methods for accuracy assessment of thematic maps using fuzzy sets. *Photogrammetric Engineering and Remote Sensing* 60:181–188.

Greenwell, B. M., B. C. Boehmke, and A. J. McCarthy. 2018. A simple and effective model-based variable importance measure. arXiv preprint, arXiv:1805.04755.

Hand, D. J., and R. J. Till. 2001. A simple generalisation of the area under the ROC curve for multiple class classification problems. *Machine Learning* 45:171–186. https://doi.org/10.1023/A:1010920819831.

Hay, A. M. 1979. Sampling designs to test land-use map accuracy. *Photogrammetric Engineering and Remote Sensing* 45:529–533.

He, H., and E. A. Garcia. 2009. Learning from imbalanced data. *IEEE Transactions on Knowledge and Data Engineering* 21:1263–1284. https://doi.org/10.1109/TKDE.2008.239.

Hedger, R. D., P. M. Atkinson, and T. J. Malthus. 2001. Optimizing sampling strategies for estimating mean water quality in lakes using geostatistical techniques with remote sensing. *Lakes & Reservoirs: Research & Management* 6:279–288. https://doi.org/10.1046/j.1440-1770.2001.00159.x.

Hord, R. M., and W. Brooner. 1976. Land use map accuracy criteria. *Photogrammetric Engineering and Remote Sensing* 42:671–677.

Huang, C., L. S. Davis, and J. R. G. Townshend. 2002. An assessment of support vector machines for land cover classification. *International Journal of Remote Sensing* 23:725–749. https://doi.org/10.1080/01431160110040323.

Huang, Z., and B. G. Lees. 2007. Assessing a single classification accuracy measure to deal with the imprecision of location and class: Fuzzy weighted Kappa versus Kappa. *Journal of Spatial Science* 52:1–12. https://doi.org/10.1080/14498596.2007.9635096.

Hush, D. R., and B. G. Horne. 1993. Progress in supervised neural networks. *IEEE Signal Processing Magazine* 10:8–39. https://doi.org/10.1109/79.180705.

Hyppänen, H. 1996. Spatial autocorrelation and optimal spatial resolution of optical remote sensing data in boreal forest environment. *International Journal of Remote Sensing* 17:3441–3452. https://doi.org/10.1080/01431169608949161.

Izmailov, P., W. J. Maddox, P. Kirichenko, T. Garipov, D. Vetrov, and A. G. Wilson. 2019. Subspace inference for Bayesian deep learning. arXiv preprint, arXiv:1907.07504.

Japkowicz, N., and M. Shah. 2011. *Evaluating Learning Algorithms: A Classification Perspective*. New York: Cambridge University Press. https://doi.org/10.1017/CBO9780511921803.

Japkowicz, N., and S. Stephen. 2002. The class imbalance problem: A systematic study. *Intelligent Data Analysis* 6:429–449. https://doi.org/10.3233/IDA-2002-6504.

Johnson, B., and Z. Xie. 2011. Unsupervised image segmentation evaluation and refinement using a multi-scale approach. *ISPRS Journal of Photogrammetry and Remote Sensing* 66:473–483. https://doi.org/10.1016/j.isprsjprs.2011.02.006.

Jupp, D. L. B., A. H. Strahler, and C. E. Woodcock. 1988. Autocorrelation and regularization in digital images. I. Basic theory. IEEE *Transactions on Geoscience and Remote Sensing* 26:463–473. https://doi.org/10.1109/36.3050.

Karasiak, N., J. F. Dejoux, C. Monteil, and D. Sheeren. 2022. Spatial dependence between training and test sets: Another pitfall of classification accuracy assessment in remote sensing. *Machine Learning* 111:2715–2740. https://doi.org/10.1007/s10994-021-05972-1.

Kavzoglu, T. 2009. Increasing the accuracy of neural network classification using refined training data. *Environmental Modelling and Software* 24:850–858. https://doi.org/https://doi.org/10.1016/j.envsoft.2008.11.012.

Kavzoglu, T., and A. Teke. 2022. Predictive performances of ensemble machine learning algorithms in landslide susceptibility mapping using random forest, extreme gradient boosting (XGBoost) and natural gradient boosting (NGBoost). *Arabian Journal for Science and Engineering* 47:7367–7385. https://doi.org/10.1007/s13369-022-06560-8.

Kavzoglu, T., and I. Colkesen. 2012. The effects of training set size for the performance of support vector machines and decision trees. In *Proceedings of the 10th International Symposium on Spatial Accuracy Assessment in Natural Resources and Environmental Sciences*, July10–13, Florianópolis, Brazil, pp. 127–132.

Kavzoglu, T., and P. M. Mather. 2002. The role of feature selection in artificial neural network applications. *International Journal of Remote Sensing* 23:2919–2937. https://doi.org/10.1080/01431160110107743.

Kavzoglu, T., and P. M. Mather. 2003. The use of backpropagation artificial neural networks in land cover classification. *International Journal of Remote Sensing* 24:4907–4938. https://doi.org/10.1080/0143116031000114851.

Kavzoglu, T., and S. Reis. 2008. Performance analysis of maximum likelihood and artificial neural network classifiers for training sets with mixed pixels. *GIScience & Remote Sensing* 45:330–342. https://doi.org/10.2747/1548-1603.45.3.330.

Kavzoglu, T., I. Colkesen, A. Atesoglu, H. Tonbul, E. O. Yilmaz, S. Ozlusoylu, and M. Y. Ozturk. 2024. Construction and implementation of in-situ measurement based poplar spectral library considering phenological stages for land cover classification using high-resolution satellite images. *International Journal of Remote Sensing* (Accepted), 45, pp. 2049–2072. https://doi.org/10.1080/01431161.2024.2326800

Kavzoglu, T., I. Colkesen, and T. Yomralioglu. 2015. Object-based classification with rotation forest ensemble learning algorithm using very-high-resolution WorldView-2 image. *Remote Sensing Letters* 6:834–843. https://doi.org/10.1080/2150704X.2015.1084550.

Kavzoglu, T., M. Y. Erdemir, and H. Tonbul. 2017. Classification of semiurban landscapes from very high-resolution satellite images using a regionalized multiscale segmentation approach. *Journal of Applied Remote Sensing* 11:035016. https://doi.org/10.1117/1.JRS.11.035016.

Kawauchi, H., and T. Fuse. 2022. SHAP-based interpretable object detection method for satellite imagery. *Remote Sensing* 14:1970. https://doi.org/10.3390/rs14091970.

Kim, M., M. Madden, and T. Warner. 2008. Estimation of optimal image object size for the segmentation of forest stands with multispectral IKONOS imagery. In *Object-Based Image Analysis*, (eds.) T. Blaschke, S. Lang, and G. J. Hay, pp. 291–307. Berlin, Heidelberg: Springer. https://doi.org/10.1007/978-3-540-7 7058-9_16.

Kim, M., M. Madden, and T. Warner. 2009. Forest type mapping using object-specific texture measures from multispectral IKONOS imagery: Segmentation quality and image classification issues. *Photogrammetric Engineering and Remote Sensing* 75:819–830. https://doi.org/10.14358/PERS.75.7.819.

Kucharczyk, M., G. J. Hay, S. Ghaffarian, and C. H. Hugenholtz. 2020. Geographic object-based image analysis: A primer and future directions. *Remote Sensing* 12:2012. https://doi.org/10.3390/rs12122012.

Kuzlu, M., U. Cali, V. Sharma, and Ö. Güler. 2020. Gaining insight into solar photovoltaic power generation forecasting utilizing explainable artificial intelligence tools. *IEEE Access* 8:187814–187823. https://doi.org/10.1109/ACCESS.2020.3031477.

Lewis, H. G., and M. Brown. 2001. A generalized confusion matrix for assessing area estimates from remotely sensed data. *International Journal of Remote Sensing* 22:3223–3235. https://doi.org/10.1080/014311160152558332.

Lins, K. S., and R. L. Kleckner. 1996. Land cover mapping: An overview and history of the concepts. In *Gap Analysis: A Landscape Approach to Biodiversity Planning*, (eds.) J. M. Scott, T. H. Tear, and F. Davis, pp. 57–65. Bethesda, MD: American Society for Photogrammetry and Remote Sensing.

Liu, C., P. Frazier, and L. Kumar. 2007. Comparative assessment of the measures of thematic classification accuracy. *Remote Sensing of Environment* 107:606–616. https://doi.org/10.1016/j.rse.2006.10.010.

Lizarazo, I. 2013. Meaningful image objects for object-oriented image analysis. *Remote Sensing Letters* 4:419–426. https://doi.org/10.1080/2150704X.2012.744485.

Lizarazo, I. 2014. Accuracy assessment of object-based image classification: Another STEP. *International Journal of Remote Sensing* 35:6135–6156. https://doi.org/10.1080/01431161.2014.943328.

Lundberg, S. M., and S. I. Lee. 2017. A unified approach to interpreting model predictions. In *Proceedings of the 31st International Conference on Neural Information Processing Systems (NIPS'17)*, December 4–9, Long Beach, CA, pp. 4765–4774.

Lundberg, S. M., G. Erion, H. Chen, A. DeGrave, J. M. Prutkin, B. Nair, R. Katz, et al. 2020 From local explanations to global understanding with explainable AI for trees. *Nature Machine Intelligence* 2:56–67. https://doi.org/10.1038/s42256-019-0138-9.

MacLean, M. G., and R. G. Congalton. 2012. Map accuracy assessment issues when using an object-oriented approach. In *Proceedings of the American Society for Photogrammetry and Remote Sensing 2012 Annual Conference*, March 19–23, Sacramento, CA, pp. 19–23.

Mamalakis, A., E. A. Barnes, and I. Ebert-Uphoff. 2022. Investigating the fidelity of explainable artificial intelligence methods for applications of convolutional neural networks in geoscience. *Artificial Intelligence for the Earth Systems* 1:1–21. https://doi.org/10.1175/aies-d-22-0012.1.

Marpu, P. R. 2009. Geographic object-based image analysis. PhD Thesis, University of Freiberg. https://doi.org/10.23689/fidgeo-249.

Marpu, P. R., M. Neubert, H. Herold, and I. Niemeyer. 2010. Enhanced evaluation of image segmentation results. *Journal of Spatial Science* 55:55–68. https://doi.org/10.1080/14498596.2010.487850.

Mather, P. M., and M. Koch. 2011. *Computer Processing of Remotely-Sensed Images: An Introduction*, 4th edition. Chichester: John Wiley & Sons.

Maxwell, A. E., A. W. Timothy, and F. Fang. 2018. Implementation of machine-learning classification in remote sensing: An applied review. *International Journal of Remote Sensing* 39:2784–2817. https://doi.org/10.1080/01431161.2018.1433343.

Maxwell, A. E., T. A. Warner, and L. A. Guillén. 2021a. Accuracy assessment in convolutional neural network-based deep learning remote sensing studies-Part 1: Literature review. *Remote Sensing* 13:2450. https://doi.org/10.3390/rs13132450.

Maxwell, A. E., T. A. Warner, and L. A. Guillén. 2021b. Accuracy assessment in convolutional neural network-based deep learning remote sensing studies-Part 2: Recommendations and best practices. *Remote Sensing* 13:2591. https://doi.org/10.3390/rs13132591.

May, D. Z. 2006. A multispectral remote sensing investigation of leaf area index at black rock forest, NY. M.Sc. Thesis, College of Arts and Science, Ohio University, Athens, OH.

Meyer, H., C. Reudenbach, S. Wöllauer, and T. Nauss. 2019. Importance of spatial predictor variable selection in machine learning applications: Moving from data reproduction to spatial prediction. *Ecological Modelling* 411:108815. https://doi.org/10.1016/j.ecolmodel.2019.108815.

Milanés-Hermosilla, D., R. T. Codorniú, R. López-Baracaldo, R. Sagaró-Zamora, D. Delisle-Rodriguez, J. Villarejo-Mayor, and J. R. Núñez-Álvarez. 2021. Monte Carlo Dropout for uncertainty estimation and motor imagery classification. *Sensors* 21:7241. https://doi.org/10.3390/s21217241.

Millard, K., and M. Richardson. 2015. On the importance of training data sample selection in random forest image classification: A case study in peatland ecosystem mapping. *Remote Sensing* 7:8489–8515. https://doi.org/10.3390/rs70708489.

Morales-Barquero, L., M. B. Lyons, S. R. Phinn, and C. M. Roelfsema. 2019. Trends in remote sensing accuracy assessment approaches in the context of natural resources. *Remote Sensing* 11:2305. https://doi.org/10.3390/rs11192305.

Moran, P. A. 1950. Notes on continuous stochastic phenomena. *Biometrika* 37:17–23. https://doi.org/10.2307/2332142.

Muchoney, D. M., and A. H. Strahler. 2002. Pixel-and site-based calibration and validation methods for evaluating supervised classification of remotely sensed data. *Remote Sensing of Environment* 81:290–299. https://doi.org/10.1016/S0034-4257(02)00006-8.

Muhtar, D., X. Zhang, P. Xiao, Z. Li, and F. Gu. 2023. CMID: A unified self-supervised learning framework for remote sensing image understanding. *IEEE Transactions on Geoscience and Remote Sensing* 61:5607817. https://doi.org/10.1109/TGRS.2023.3268232.

Olivier, A., M. D. Shields, and L. Graham-Brady. 2021. Bayesian neural networks for uncertainty quantification in data-driven materials modeling. *Computer Methods in Applied Mechanics and Engineering* 386:114079. https://doi.org/10.1016/j.cma.2021.114079.

Olofsson, P., G. M. Foody, M. Herold, S. V. Stehman, C. E. Woodcock, and M. A. Wulder. 2014. Good practices for estimating area and assessing accuracy of land change. *Remote Sensing of Environment* 148:42–57. https://doi.org/10.1016/j.rse.2014.02.015.

Pal, M., and P. M. Mather. 2003. An assessment of the effectiveness of decision tree methods for land cover classification. *Remote Sensing of Environment* 86:554–565. https://doi.org/10.1016/S0034-4257(03)00132-9.

Paoletti, M. E., O. Mogollon, S. Moreno, J. C. Sancho, and J. M. Haut. 2023. A comprehensive survey of imbalance correction techniques for hyperspectral data classification. *IEEE Journal of Selected Topics in Applied Earth Observations and Remote Sensing* 16:5297–2314. https://doi.org/10.1109/JSTARS.2023.3279506.

Pecknold, S., S. Lovejoy, D. Scherter, and C. Hooge. 1997. Multifractals and resolution dependence of remotely sensed data: GSI to GIS. In *Scale in Remote Sensing and GIS*, (eds.) D. A. Quattrochi, and M. F. Goodchild, pp. 361–394. New York: Lewis Publishers.

Persello, C., and L. Bruzzone. 2010. A novel protocol for accuracy assessment in classification of very high resolution images. IEEE *Transactions on Geoscience and Remote Sensing* 48:1232–1244. https://doi.org/10.1109/TGRS.2009.2029570.

Pires de Lima, R., and K. Marfurt. 2020. Convolutional neural network for remote-sensing scene classification: Transfer learning analysis. *Remote Sensing* 12:86. https://doi.org/10.3390/rs12010086.

Pontius Jr, R. G. 2000. Quantification error versus location error in comparison of categorical maps. *Photogrammetric Engineering and Remote Sensing* 66:1011–1016.

Pontius Jr, R. G. 2002. Statistical methods to partition effects of quantity and location during comparison of categorical maps at multiple resolutions. *Photogrammetric Engineering and Remote Sensing* 68:1041–1049.

Pontius Jr, R. G., and J. Connors. 2009. Range of categorical associations for comparison of maps with mixed pixels. *Photogrammetric Engineering & Remote Sensing* 75:963–969.

Pontius Jr, R. G., and M. Millones. 2008. Problems and solutions for kappa-based indices of agreement. In Proceedings *of International Conference on Studying, Modeling and Sense Making of Planet Earth*, June 1–6, Lesvos, Greece.

Pontius Jr, R. G., and M. Millones. 2011. Death to Kappa: Birth of quantity disagreement and allocation disagreement for accuracy assessment. *International Journal of Remote Sensing* 32:4407–4429. https://doi.org/10.1080/01431161.2011.552923.

Prisley, S. P., and J. L. Smith. 1987. Using classification error matrices to improve the accuracy of weighted land-cover models. *Photogrammetric Engineering and Remote Sensing* 53:1259–1263.

Radoux, J., R. Bogaert, D. Fasbender, and P. Defourny. 2010. Thematic accuracy assessment of geographic object-based image classification. *International Journal of Geographical Information Science* 25:895–911.

Ribeiro, M. T., S. Singh, and C. Guestrin. 2016. Why should I trust you? Explaining the predictions of any classifier. In *Proceedings of the 22nd ACM SIGKDD International Conference on Knowledge Discovery and Data Mining (KDD'16)*, August 13–17, San Francisco, CA, pp. 1135–1144. https://doi.org/10.1145/2939672.2939778.

Ricotta, C. 2005. On possible measures for evaluating the degree of uncertainty of fuzzy thematic maps. *International Journal of Remote Sensing* 26:5573–5583. https://doi.org/10.1080/01431160500285175.

Roberts, D. R., V. Bahn, S. Ciuti, M. S. Boyce, J. Elith, G. Guillera-Arroita, S. Hauenstein, et al. 2017. Cross-validation strategies for data with temporal, spatial, hierarchical, or phylogenetic structure. *Ecography* 40:913–929. https://doi.org/10.1111/ecog.02881.

Robnik-Šikonja, M., and M. Bohanec. 2018. Perturbation-based explanations of prediction models. In *Human and Machine Learning*, (eds.) J. Zhou, and F. Chen, pp. 159–175. Berlin, Heidelberg: Springer. https://doi.org/10.1007/978-3-319-90403-0_9.

Rogan, J., J. Franklin, D. Stow, J. Miller, C. Woodcock, and D. Roberts. 2008. Mapping land-cover modifications over large areas: A comparison of machine learning algorithms. *Remote Sensing of Environment* 112:2272–2283. https://doi.org/10.1016/j.rse.2007.10.004.

Rogan, J., J. Miller, D. Stow, J. Franklin, L. Levien, and C. Fischer. 2003. Land-cover change monitoring with classification trees using Landsat TM and ancillary data. *Photogrammetric Engineering & Remote Sensing* 69:793–804. https://doi.org/10.14358/PERS.69.7.793.

Rosenfield, G. H., K. Fitzpatrick-Lins, and H. S. Ling. 1982. Sampling for thematic map accuracy testing. *Photogrammetric Engineering and Remote Sensing* 48:131–137.

Ryo, M. 2022. Explainable artificial intelligence and interpretable machine learning for agricultural data analysis. *Artificial Intelligence in Agriculture* 6:257–265. https://doi.org/https://doi.org/10.1016/j.aiia.2022.11.003.

Sahin, E. K., I. Colkesen, and T. Kavzoglu. 2020. A comparative assessment of canonical correlation forest, random forest, rotation forest and logistic regression methods for landslide susceptibility mapping. *Geocarto International* 35:341–363. https://doi.org/10.1080/10106049.2018.1516248.

Salk, C., S. Fritz, L. See, C. Dresel, and I. McCallum. 2018. An exploration of some pitfalls of thematic map assessment using the new map tools resource. *Remote Sensing* 10:376. https://doi.org/10.3390/rs10030376.

Scepan, J. 1999. Thematic validation of high-resolution global land-cover data sets. *Photogrammetric Engineering and Remote Sensing* 65:1051–1060.

Schratz, P., J. Muenchow, E. Iturritxa, J. Richter, and A. Brenning. 2019. Hyperparameter tuning and performance assessment of statistical and machine-learning algorithms using spatial data. *Ecological Modelling* 406:109–120. https://doi.org/10.1016/j.ecolmodel.2019.06.002.

Shapley, L. S. 1953. Stochastic games. *Proceedings of the National Academy of Sciences* 39:1095–1100. https://doi.org/10.1073/pnas.39.10.1095.

Silva, W., K. Fernandes, M. J. Cardoso, and J. S. Cardoso. 2018. Towards complementary explanations using deep neural networks. In *Understanding and Interpreting Machine Learning in Medical Image Computing Applications*, (eds.) D. Stoyanov, et al., pp. 133–140. Berlin, Heidelberg: Springer. https://doi.org/10.1007/978-3-030-02628-8_15.

Spiker, J. S., and T. A. Warner. 2007. Scale and spatial autocorrelation from a remote sensing perspective. In: *Geo-Spatial Technologies in Urban Environments*, (eds.) R. R. Jensen, J. D. Gatrell, and D. McLean, pp. 197–213. Berlin, Heidelberg: Springer. https://doi.org/10.1007/978-3-540-69417-5_10.

Steele, B. M., J. C. Winne, and R. L. Redmond. 1998. Estimation and mapping of misclassification probabilities for thematic land cover maps. *Remote Sensing of Environment* 66:192–202. https://doi.org/10.1016/S0034-4257(98)00061-3.

Stehman, S. V. 1992. Comparison of systematic and random sampling for estimating the accuracy of maps generated from remotely sensed data. *Photogrammetric Engineering and Remote Sensing* 58:1343–1350.

Stehman, S. V. 1997. Selecting and interpreting measures of thematic classification accuracy. *Remote Sensing of Environment* 62:77–89. https://doi.org/10.1016/S0034-4257(97)00083-7.

Stehman, S. V. 2004a. Sampling design for accuracy assessment of large-area, land-cover maps: Challenges and future directions. In *Remote Sensing and GIS Accuracy Assessment*, (eds.) R. S. Lunetta, and J. G. Lyon, pp. 13–29. Boca Raton, FL: CRC Press.

Stehman, S. V. 2004b. A critical evaluation of the normalized error matrix in map accuracy assessment. *Photogrammetric Engineering & Remote Sensing* 70:743–751. https://doi.org/10.14358/PERS.70.6.743.

Stehman, S. V. 2009. Sampling designs for accuracy assessment of land cover. *International Journal of Remote Sensing* 30:5243–5272. https://doi.org/10.1080/01431160903131000.

Stehman, S. V., and G. M. Foody. 2009. Accuracy assessment. In *The SAGE Handbook of Remote Sensing*, (eds.) T. A. Warner, M. D. Nellis, and G. M. Foody, pp. 297–309. London, UK: SAGE.

Stehman, S. V., and G. M. Foody. 2019. Key issues in rigorous accuracy assessment of land cover products. *Remote Sensing of Environment* 231:111199. https://doi.org/10.1016/j.rse.2019.05.018.

Stehman, S. V., and J. D. Wickham. 2011. Pixels, blocks of pixels, and polygons: Choosing a spatial unit for thematic accuracy assessment. *Remote Sensing of Environment* 115:3044–3055. https://doi.org/10.1016/j.rse.2011.06.007.

Stehman, S. V., and R. L. Czaplewski. 1998. Design and analysis for thematic map accuracy assessment: Fundamental principles. *Remote Sensing of Environment* 64:331–344. https://doi.org/10.1016/S0034-4257(98)00010-8.

Stehman, S. V., M. K. Arora, T. Kasetkasem, and P. K. Varshney. 2007. Estimation of fuzzy error matrix accuracy measures under stratified random sampling. *Photogrammetric Engineering & Remote Sensing* 73:165–173. https://doi.org/10.14358/PERS.73.2.165.

Stevens Jr, D. L., and A. R. Olsen. 2004. Spatially balanced sampling of natural resources. *Journal of the American Statistical Association* 99:262–278. https://doi.org/10.1198/016214504000000250.

Temenos, A., N. Temenos, M. Kaselimi, A. Doulamis, and N. Doulamis. 2023. Interpretable deep learning framework for land use and land cover classification in remote sensing using SHAP. *IEEE Geoscience and Remote Sensing Letters* 20:1–5. https://doi.org/10.1109/LGRS.2023.3251652.

Thomlinson, J. R., P. V. Bolstad, and W. B. Cohen. 1999. Coordinating methodologies for scaling landcover classifications from site-specific to global: Steps toward validating global map products. *Remote Sensing of Environment* 70:16–28. https://doi.org/10.1016/S0034-4257(99)00055-3.

Tillé, Y., M. M. Dickson, G. Espa, and D. Giuliani. 2018. Measuring the spatial balance of a sample: A new measure based on Moran's I index. *Spatial Statistics* 23:182–192. https://doi.org/10.1016/j.spasta.2018.02.001.

Tjoa, E., and C. Guan. 2021. A Survey on explainable artificial intelligence (XAI): Toward medical XAI. *IEEE Transactions on Neural Networks and Learning Systems* 32:4793–4813. https://doi.org/10.1109/TNNLS.2020.3027314.

Trodd, N. M. 1995. Uncertainty in land cover mapping for modelling land cover change. In *Proceedings of RSS95 Remote Sensing in Action*, (eds.) P. J. Curran, and Y. C. Robertson, pp. 1138–1145. Nottingham: Remote Sensing Society.

Tso, B., and R. C. Olsen. 2005. A contextual classification scheme based on MRF model with improved parameter estimation and multiscale fuzzy line process. *Remote Sensing of Environment* 97:127–136. https://doi.org/10.1016/j.rse.2005.04.021.

Van der Meer, F., K. Scholte, S. De Jong, and M. Dorrestijn. 1998. Scaling to MERIS resolution: Mapping accuracy and spatial variability. In *EARSeL Workshop on Imaging Spectroscopy*, October 6–8, Zurich, Switzerland, pp. 147–153.

Van Genderen, J. L., and B. F. Lock. 1977. Testing land use map accuracy. *Photogrammetric Engineering and Remote Sensing* 43:1135–1137.

Van Vliet, J., A. K. Bregt, and A. Hagen-Zanker. 2011. Revisiting Kappa to account for change in the accuracy assessment of land-use change models. *Ecological Modelling* 222:1367–1375. https://doi.org/10.1016/j.ecolmodel.2011.01.017.

Vieira, C. A. O. 2000. Accuracy of remote sensing classification of agricultural crops: A comparative study. Ph.D. Thesis, School of Geography, The University of Nottingham, Nottingham, UK.

Vieira, C. A. O., and P. M. Mather. 1999. Assessing the accuracy of classifications using remotely sensed data. In Proceedings *of the 4th International Airborne Remote Sensing Conference/21st Canadian Symposium on Remote Sensing*, June 21–24, Ottawa, Canada, pp. 823–830.

Waldner, F., D. C. Jacques, and F. Löw. 2017. The impact of training class proportions on binary cropland classification. *Remote Sensing Letters* 8:1122–1131. https://doi.org/10.1080/2150704X.2017.1362124.

Wang, G., G. Gertner, and A. B. Anderson. 2005. Sampling design and uncertainty based on spatial variability of spectral variables for mapping vegetation cover. *International Journal of Remote Sensing* 26:3255–3274. https://doi.org/10.1080/01431160500114748.

Ward, I. R., L. Wang, J. Lu, M. Bennamoun, G. Dwivedi, and F. M. Sanfilippo. 2021. Explainable artificial intelligence for pharmacovigilance: What features are important when predicting adverse outcomes? *Computer Methods and Programs in Biomedicine* 212:106415. https://doi.org/https://doi.org/10.1016/j.cmpb.2021.106415.

Wilcoxon, F. 1946. Individual comparisons of grouped data by ranking methods. *Journal of Economic Entomology* 39:269–270. https://doi.org/10.1093/jee/39.2.269.

Wood, T. F., and G. M. Foody. 1989. Analysis and representation of vegetation continua from Landsat Thematic Mapper data for lowland heaths. *International Journal of Remote Sensing* 10:181–191. https://doi.org/10.1080/01431168908903855.

Woodcock, C. E., A. H. Strahler, and D. B. Jupp. 1988a. The use of variogram in remote sensing: I. Scene models and simulated images. *Remote Sensing of Environment* 25:323–348. https://doi.org/10.1016/0034-4257(88)90108-3.

Woodcock, C. E., A. H. Strahler, and D. B. Jupp. 1988b. The use of variogram in remote sensing: II. Real digital images. *Remote Sensing of Environment* 25:349–379. https://doi.org/10.1016/0034-4257(88)90109-5.

Woodcock, C. E., and A. H. Strahler. 1987. The factor of scale in remote sensing. *Remote Sensing of Environment* 21:311–332. https://doi.org/10.1016/0034-4257(87)90015-0.

Wulder, M. A., S. E. Franklin, J. C. White, J. Linke, and S. Magnussen. 2006. An accuracy assessment framework for large-area land cover classification products derived from medium-resolution satellite data. *International Journal of Remote Sensing* 27:663–683. https://doi.org/10.1080/01431160500185284.

Ye, S., R. G. Pontius Jr, and R. Rakshit. 2018. A review of accuracy assessment for object-based image analysis: From per-pixel to per-polygon approaches. *ISPRS Journal of Photogrammetry and Remote Sensing* 141:137–147. https://doi.org/10.1016/j.isprsjprs.2018.04.002.

Yilmaz, E. O., H. Tonbul, and T. Kavzoglu. 2024. Marine mucilage mapping with explained deep learning model using water-related spectral indices: A case study of Dardanelles Strait, Turkey. *Stochastic Environmental Research and Risk Assessment* 38:51–68. https://doi.org/10.1007/s00477-023-02560-8.

You, J., J. Zhao, X. Huang, G. Zhang, A. Chen, M. Hou, and J. Cao. 2023. Explainable convolutional neural networks driven knowledge mining for seismic facies classification. IEEE *Transactions on Geoscience and Remote Sensing* 61:5911118. https://doi.org/10.1109/TGRS.2023.3280364.

Zhao, R., K. Wang, Y. Xiao, F. Gao, and Z. Gao. 2024. Leveraging Monte Carlo Dropout for uncertainty quantification in real-time object detection of autonomous vehicles. *IEEE Access* https://doi.org/10.1109/ACCESS.2024.3355199.

Zhen, Z., L. J. Quackenbush, S. V. Stehman, and L. Zhang. 2013. Impact of training and validation sample selection on classification accuracy and accuracy assessment when using reference polygons in object-based classification. *International Journal of Remote Sensing* 34:6914–6930. https://doi.org/10.1080/01431161.2013.810822.

Zhu, C., and X. Yang. 1998. Study of remote sensing image texture analysis and classification using wavelets. *International Journal of Remote Sensing* 13:3167–3187. https://doi.org/10.1080/014311698214262.

Zhu, X., H. Guo, J. J. Huang, S. Tian, W. Xu, and Y. Mai. 2022. An ensemble machine learning model for water quality estimation in coastal area based on remote sensing imagery. *Journal of Environmental Management* 323:116187. https://doi.org/10.1016/j.jenvman.2022.116187.

Zhu, Z., A. L. Gallant, C. E. Woodcock, B. Pengra, P. Olofsson, T. R. Loveland, S. Jin, S. Dahal, L. Yang, and R. F. Auch. 2016. Optimizing selection of training and auxiliary data for operational land cover classification for the LCMAP initiative. *ISPRS Journal of Photogrammetry and Remote Sensing* 122:206–221. https://doi.org/10.1016/j.isprsjprs.2016.11.004.

Index